Faber and Kell's Heating and Air-conditioning of Buildings

The authors

P. L. Martin joined Oscar Faber & Partners in 1947 as an assistant to the late J. R. Kell; he became a Partner in 1960 and was a Consultant to the practice from 1983 to 1985. His formal association with this book began in 1971 when he was invited by Mr Kell to become a co-author for the fifth edition. He was Chairman of the Heating & Ventilating Research Association (later BSRIA) for 1967–8, President of the Institution of Heating & Ventilating Engineers (later CIBSE) for 1971–2 and Chairman of the Association of Consulting Engineers for 1983–4. He served as a member of the Engineering Council for 1982–6 and, for 11 years from 1974, was a Visiting Professor at the University of Strathclyde.

Mr Martin was made a CBE in 1980 and awarded the honorary degree of DSc in 1985; he holds the gold, silver and bronze medals awarded by IHVE for technical papers and other services to the industry.

D. R. Oughton joined Oscar Faber & Partners in 1967 as an assistant in their Belfast office. He was made an Associate in 1975; became a Partner in 1981 and, following changes in the structure of the firm, a Director of Oscar Faber Consulting Engineers in 1983 and of Facet Ltd in 1984. He was awarded the degree of MSc by the University of Bristol in 1979 after presentation of a research paper dealing with energy consumption and energy targets for air-conditioned office buildings. He served on the Council of BSRIA for 1975–87, was a member of the CIBSE Technical Publications Committee for 1975–85 and has served on the CIBSE Professional Practice Committee since 1992. He is currently the Chairman of the European Intelligent Building Group and has been the author of a number of technical papers presented at conferences, etc.

Mr Oughton is a Chartered Engineer, a Fellow of CIBSE and a Member of both the Institute of Energy and the Institute of Refrigeration. He is also a Member of the Association of Consulting Engineers.

Faber and Kell's Heating and Air-conditioning of Buildings

With some notes on combined heat and power

Eighth edition revised by

P. L. Martin CBE DSc CEng MConsE

D. R. Oughton MSc CEng MConsE

Architectural Press

Architectural Press
An imprint of Butterworth-Heinemann Ltd
225 Wildwood Avenue, Woburn, MA 01801-2041
Linacre House, Jordan Hill, Oxford OX2 8DP
A division of Reed Educational and Professional Publishing Ltd

 A member of the Reed Elsevier plc group

OXFORD AUCKLAND BOSTON
JOHANNESBURG MELBOURNE NEW DELHI

First published 1936
Second edition 1943
Reprinted 1945, 1948, 1951
Third edition 1957
Reprinted 1958, 1961
Fourth edition 1966
Fifth edition 1971
Reprinted 1974
Sixth edition 1979
Reprinted (with amendments) 1988
Seventh edition 1989
Eighth edition 1995
Reprinted 1995
Paperback edition 1997
Reprinted 1999, 2000

British Library Cataloguing in Publication Data
Martin, P. L.
Faber and Kell's Heating and
Air-conditioning of Buildings. – 8Rev.ed
I. Title II. Oughton, D. R.
697

ISBN 0 7506 3778 1

Library of Congress Cataloguing in Publication Data
Martin, P. L. (Peter Lewis)
Faber and Kell's heating and air-conditioning of buildings – Eighth ed./revised by
P. L. Martin, D. R. Oughton
1. Heating. 2. Air conditioning. 3. Ventilation. I. Oughton,
D. R. II. Faber, Oscar. Heating and air-conditioning of buildings.
TH7222.M35 94–16759

Composition by Genesis Typesetting, Rochester, Kent
Printed in Great Britain by St Edmundsbury Press Limited, Bury St Edmunds, Suffolk

Contents

Preface xi
Preface to first edition xiii

1 **Fundamentals** **1**
Units and quantities 2
Criteria which affect human comfort – definitions 4
Thermal indices 9
Methods of measurement 12
Application 14

2 **The building in winter** **18**
Conservation of energy 19
Heat losses 19
Legislation 27
Atypical construction features 31
Condensation 37
Air infiltration 41
Temperature difference 47
Miscellaneous allowances 52
Continuous versus intermittent operation 56
Steady state and dynamic response 60

3 **The building in summer** **65**
Conservation of energy 66
Solar heat gains 68
Air infiltration and ventilation 85
Miscellaneous heat gains 86
Temperature difference 89
Intermittent operation 91

4 **Survey of heating methods** **94**
Condensation – a warning 96
Electrical off-peak storage systems 96
Direct systems – solid fuel 96
Direct systems – liquid fuel (primarily convective) 97
Direct systems – gaseous fuels (primarily radiant) 98

Direct systems – gaseous fuels (primarily convective) 101
Direct systems – electrical (primarily radiant) 104
Direct systems – electrical (primarily convective) 107
Control of direct systems 109
Indirect systems – liquid media 109
Indirect systems – vapour media 110
Indirect systems – gaseous media 111
Factors affecting choice 111

5 Electrical storage heating 116
Methods of storage 116
Capacity of the heat store 117
Capacity of room stores 119
Capacity of central stores 122
Equipment for room stores 124
Equipment for central stores 130
Electrical supply to storage equipment and installations 138

6 Indirect heating systems 139
Water systems – characteristics 139
Water systems – piping arrangements 146
Steam systems – characteristics 153
Steam systems – piping arrangements 157
Air systems – characteristics 159
Pipework heat emission 162
External distribution systems 163

7 Heat emitting equipment 172
Principal criteria 172
Radiant heating 174
Combined radiant and convective heating 183
Convective heating 189

8 Pumps and other auxiliary equipment 199
Pumps 199
System pressurisation 207
Non-storage calorifiers 212
Condensate handling equipment 214
Steam pressure reduction 218
Heat meters 220
Air venting, etc. 223
Corrosion 224

9 Piping design for indirect heating systems 227
Water systems – principles 227
Water systems – applications 231
Water systems – footnotes 241
Water systems – gravity circulation 243
Steam systems – principles 246
Provision for thermal expansion 251

10 Boilers and firing equipment 255
Basic considerations 255
Criteria for boiler selection 257
Boiler houses – size and location 262
Types of boiler 263
Boiler fittings and mountings 270
Boiler firing – solid fuel 272
Boiler firing – oil fuel 276
Boiler firing – gaseous fuel 281
Miscellaneous burner equipment 287
Instrumentation 289

11 Fuels, storage and handling 290
Solid fuel 290
Liquid fuel 293
North Sea natural gas 302
Liquified petroleum gas 303
Electricity 304
Miscellaneous fuels 306

12 Combustion and chimneys 312
Combustion processes 312
Sample calculations 316
Chimneys 318
Chimney construction 326
Special methods for flue gas disposal (Natural gas) 329
Air supply to boiler houses 332

13 Ventilation 334
Air supply for human emissions 335
Air supply for other reasons 337
Air change rates 339
Legislation and other rules 339
Criteria for air supply to occupied rooms 340
Methods of ventilation 343
Kitchens 354
Special applications 360
Recirculation units 363

14 Air-conditioning 365
General principles 366
Traditional systems 368
High velocity systems 373
All-air systems 374
Air–water systems 381
Other systems 387
Alternative methods of cooling 396
Summary of systems and application 398

15 Air distribution 400
General principles 400
Distribution for air-conditioning 407

16 Ductwork design 423
Ductwork 424
Ductwork components and auxiliaries 429
Pressure distribution in ducts 434
Thermal insulation of ducts 444
Sound control 445
Noise dispersal 452
Commissioning and measurement 457

17 Fans and air treatment equipment 462
Fan types and performance 462
Position of outside air intake 471
Air filtration 472
Air humidification 481
Air heating and cooling coils 485
Packaged air handling plant 489
Air-to-air heat exchangers 490

18 Calculations for air-conditioning design 501
Heat gains 501
Psychrometry 502
Application 505
Design calculations for other systems 514
System diagrams and automatic controls 515

19 Refrigeration: water chillers and heat pumps 516
Mechanical refrigeration 516
Refrigerating media 519
Types of refrigeration plant 521
Choice of refrigeration plant 527
Refrigeration plant components 531
Heat recovery 539
Heat pumps 542

20 Hot water supply systems 547
Local systems 548
Central systems 554
Cylinders, indirect cylinders and calorifiers 559
Requirements for storage capacity and boiler power 564
Feed cisterns 570
Unvented hot water systems 571
Water treatment 573
Materials, etc. 574
Solar collectors 576

21 Piping design for central hot water supply systems 578
Primary circulations 578
Secondary outflow and return pipework 581
System arrangements 594

22 Automatic controls and building management systems 601
Sensing devices 605
Control devices 608
Controller modes of operation 616
Systems controls 619
Heating system controls 621
Air-conditioning system controls 624
Building management systems 636

23 Running costs 644
Space heating 646
Period of use 648
Mechanically ventilated buildings 655
Air-conditioned buildings 656
Hot water supply 659
Energy targets 660
Maintenance 663
Investment appraisals 664

24 Combined heat and power (CHP) 671
Basic considerations 671
CHP systems – criteria for selection 680
Conclusion 682

Appendix I Temperature levels 683

Appendix II SI unit symbols 684

Appendix III Conversion factors 685

Index 687

Preface

Each time a new edition of this book appears, it is proper that the authors should acknowledge the permission of the Chartered Institution of Building Services Engineers to use data from the current edition of the *CIBSE Guide*. To this invaluable work, extensive reference is made, both to present facts and to encourage the reader who may wish to pursue matters further. In some cases the presentation of those data may differ slightly from that of the source, simplifications being used in instances where these produce results adequate to match the inexactitudes of building construction on site.

It is remarkable that, in a time of deep recession in the construction industry, so many changes in design techniques and consequent developments in manufactured products are taking place in the building services sector. Some of these have resulted from the belated interest by Government in energy conservation, but the majority have come about in consequence of response by system designers to the lead given by the engineering profession in calling for action to preserve the global environment.

In several chapters of this present edition reference is made to proposed revisions of Part L of Building Regulations which, in a draft dated 26 January 1993, were circulated for comment. It is now understood that the less contentious sections of these are likely to be published for implementation in 1995.

Still outstanding, however, is the firm intention of Government to provide, in the Regulations, means whereby the installation of mechanical ventilation or air conditioning may be restricted to situations where it is 'reasonably necessary'. The profession is uniquely placed here to contribute expertise to the establishment of realistic criteria, against which a judgement may be made, and thus to avoidance of conflicts in implementation which would result from the use of such platitudes as may be found in the BREEAM documents.

In conclusion, our thanks are due to many colleagues and friends in the industry who have provided willing assistance; to patient secretaries who have produced repeated drafts; to artists whose care and skill have prepared finished drawings; and, in particular, to Harold Smith, our Librarian colleague, who has demonstrated to us from time to time that, in Cowper's words, 'The truth lies somewhere, if we knew but where.'

St Albans
1995

P. L. Martin
D. R. Oughton

Preface

[illegible]

Preface to first edition

During 1935 we contributed a series of articles on this subject in *The Architects' Journal*. In view of the considerable interest which these aroused, it has been thought desirable to reproduce them in book form. The material has been amplified, and many new sections added.

We are indebted to Mr G. Nelson Haden and to Mr R. E. W. Butt for many helpful suggestions, also to Mr F. G. Russell for reading through the text. Our thanks are also due to Mr J. R. Harrison, for much work in the preparation of the book, and to Mr H. J. Sharpe for reading the proofs.

Oscar Faber
J. R. Kell
1936

'It's snowing still,' said Eeyore gloomily.

'So it is.'

'*And* freezing.'

'Is it?'

'Yes,' said Eeyore. 'However,' he said, brightening up a little, 'we haven't had an earthquake lately.'

'What's the matter, Eeyore?'

'Nothing, Christopher Robin. Nothing important. I suppose you haven't seen a house or whatnot anywhere about?'

'What sort of a house?'

'Just a house.'

'Who lives there?'

'I do. At least I thought I did. But I suppose I don't. After all, we can't all have houses.'

'But, Eeyore, I didn't know—I always thought—'

'I don't know how it is, Christopher Robin, but what with all this snow and one thing and another, not to mention icicles and such-like, it isn't so Hot in my field about three o'clock in the morning as some people think it is. It isn't Close, if you know what I mean—not so as to be uncomfortable. It isn't stuffy. In fact, Christopher Robin,' he went on in a loud whisper, 'quite-between-ourselves-and-don't-tell-anybody, it's Cold.'*

From *The House at Pooh Corner,* with grateful acknowledgements to
A. A. Milne and Methuen Children's Books

* This quotation, a favourite of the original authors, appeared on the flyleaf to the first edition of this book. It is repeated here since a number of readers of later editions have regretted its absence.

Chapter 1

Fundamentals

Fulfilment of the need for satisfactory environmental conditions within a building, whether these be required for human comfort or in support of some manufacturing process, is a task which has faced mankind throughout history. With the passage of time, a variety of forms of protection against the elements has been provided by structures suited to individual circumstances, the techniques being related to the severity of the local climatic conditions, to the materials which were available and to the skills of the builders. In this sense, the characteristics of those structures provided a form of inherent yet coarse control over the internal environment: finer control had to wait upon the progressive development, over the last two centuries, of systems able to moderate the impact of the external climate still further.

The human body produces heat, the quantity depending upon the level of physical activity, and for survival this must be in balance with a corresponding heat loss. When the rate of heat generation is greater than the rate of loss, then the body temperature will rise. Similarly, when the rate of heat loss exceeds that of production, then the body temperature will fall. If the level of imbalance is severe, *heat stress* at one extreme and *hypothermia* at the other will result and either may prove fatal.

The processes of heat loss from the body are:

- Conduction to contact surfaces and to clothing.
- Convection from exposed skin and clothing surfaces.
- Radiation from exposed surfaces to the surroundings.
- Exhalation of breath.
- Evaporation by sweating.

Involuntary control of these processes is by constriction or dilation of blood vessels, variation in the rate of breathing and variation in the level of sweating, voluntary and involuntary. The individual may assist by removing or adding insulating layers in the form of clothing.

Keeping warm or keeping cool are primitive instincts which have been progressively refined as more sophisticated means have become available to satisfy them. For example, once facilities for the exclusion of the extremes of climate became commonplace for buildings in temperate zones, fashion introduced lighter clothing; this led to greater thermal sensitivity and, in consequence, to less tolerance of temperature variation. By coincidence, however, in the same time span, architectural styles changed also and the substantial buildings of the past, which could moderate the effects of solar heat and winter chill, were succeeded by wholly glazed structures having little or no such thermal capacity.

Before any part of this subject is pursued in further detail, however, it would seem appropriate to discuss some of the quantities and units that are relevant, with particular reference to the meaning and use applied to them in the present context.

Units and quantities

The notes included here are not intended to be a comprehensive glossary of terms included in the *Système International d'Unités* but, rather, an *aide mémoire* covering those which are specific to the subject matter of this book but are not necessarily in everyday use. The four basic units are the kilogram (kg) for mass, the metre (m) for length, the second (s) for time and the kelvin (K) for thermodynamic temperature, all with their multiples and sub-multiples. From these, the following secondary units are derived:

Force. The unit here is the newton (N), which is the force necessary to accelerate a mass of one kilogram to a velocity of one metre per second in one second, that is, 1 N = 1 kg m/s^2. When a mass of one kg is subjected to acceleration due to gravity, the force then exerted is 9.81 N.
Heat. The unit of energy, including heat energy, is the joule (J), which is equal to a force of one newton acting through one metre.
Heat Flow. The rate of heat flow is represented by the watt (W), which is equal to one joule produced or expended in one second, that is, 1 W = 1 J/s = 1 Nm/s.
Pressure. The standard unit is the newton per square metre (N/m^2), also known, more conveniently, as the pascal (Pa). The bar continues to be used in some circumstances and 1 bar = 100 kPa.
Specific Heat Capacity. This is the quantity of heat required to raise the temperature of one kilogram of a substance through one kelvin, the units being kJ/kg K. Where heat flow in unit time is involved, the unit becomes kJ/s kg K = kW/kg K.
Specific Mass (Density). The unit for this quantity is the kilogram per cubic metre (kg/m^3).
Volume. The cubic metre is the preferred unit but the litre (1 dm^3) is in general use since it is much more convenient in terms of a comprehensible size. To avoid printing confusion between the figure 1 and the lower case letter l, this book will spell out the word *litre* in full.

Properties of materials

Table 1.1 lists a variety of materials and provides details of some relevant physical properties. These serve as a source of reference against the following notes:

Latent heat. When the temperature of water, at atmospheric pressure, is raised from freezing to boiling point, i.e. through 100°C, the heat added is 420 kJ/kg (4.2 × 100). To convert this hot water to steam, however, still at atmospheric pressure and still at 100°C, will require the addition of a significantly greater quantity of heat, i.e. 2257 kJ/kg.* This value is the *latent heat of evaporation* of water and its magnitude, 5.4 times that needed to raise water through 100°C, shows its importance. A similar phenomenon occurs when water at 0°C becomes ice at the same temperature; this change of state releasing 330 kJ/kg, the *latent heat of fusion* of water.

* If water is evaporated at 20°C, the latent heat of evaporation is 2450 kJ/kg.

Table 1.1 Properties of materials

Material	*Specific mass (density)* (kg/m^3)	*Specific heat capacity* (kJ/kg K)	*Thernal conductivity (k)* (W/m K)	*Coeff. of linear expansion per kelvin* ($\times 10^6$)
Metals				
aluminium (sheet)	2700	0.89	240	25.6
brass (cast)	8500	0.37	100	18.7
copper (sheet)	8900	0.39	390	17.5
iron (cast)	7900	0.45	75	10.1
lead (cast)	11300	0.13	35	29.0
mild steel (ingot)	7800	0.48	35	11.3
tin (sheet)	7300	0.23	65	21.4
zinc (sheet)	7100	0.39	110	26.1
Building materials				
asbestos cement (sheet)	700	0.84	0.35	9.9
asbestos cement (decking)	1500	0.84	0.36	9.9
asphalt	1600	1.7	0.43	–
brickwork (protected)	1700	0.8	0.62	2.2
brickwork (exposed)	1700	0.8	0.84	2.2
concrete (no fines)	2000	1.0	1.13	9.9
concrete (pre-cast, heavy)	2100	0.84	1.4	9.9
concrete (pre-cast, light)	1200	1.0	0.38	9.9
concrete block (heavy)	2300	1.0	1.63	9.9
concrete block (light)	600	1.0	0.19	9.9
external render	1300	1.0	0.5	9.9
glass (sheet)	2500	0.84	1.05	8.4
granite	2650	0.92	2.93	7.9
limestone	2000	0.89	1.53	6.3
plaster (dense)	1300	1.0	0.5	–
plaster (lightweight)	600	1.0	0.16	–
plasterboard	950	0.84	0.16	–
roofing felt layers	1000	1.0	0.5	–
screed (sand–cement)	1200	1.0	0.41	–
slate	2700	0.75	1.9	6.0
stone chippings	1800	1.0	0.96	–
tiles (clay)	1900	0.80	0.84	–
Timber				
deal	610	1.21	0.13	5.0
pitch pine	660	1.21	0.14	6.7
plywood	530	1.21	0.14	7.0
studding	650	1.2	0.14	7.0
Miscellaneous				
water at normal pressure				
at 4°C	999.9	4.206	0.576	–
at 20°C	998.2	4.183	0.603	–
at 100°C	958.3	4.219	0.681	–
ice	918	2.04	2.24	–
air at normal pressure				
dry and at 20°C	1.205	1.012	0.026	–
dry and at 100°C	0.88	1.017	0.03	–

Note: For insulating materials see Table 2.3 (p.27).

Specific heat capacity (gases). Whereas, for practical purposes, solids and liquids each have a single specific heat capacity which does not vary significantly, gases have one value for a condition of constant pressure and another for a condition of constant volume. Both values will vary with temperature but, for a given gas, the ratio between them will remain the same. For example, with dry air at 20°C, the value at constant pressure is 1.012 kJ/kg K and the value at constant volume is 0.722 kJ/kg K. At 100°C, the value at constant pressure is 1.017 kJ/kg K.

Thermal expansion (solids and liquids). With very few exceptions, materials expand when heated to an extent which varies directly with their dimensions and with the temperature difference. The examples in Table 1.1, which are averages that are only valid between about 15 and 100°C, show that metals expand more than most building materials, with the result that, for example, care must be taken to accommodate the differential movement between a long straight pipe and a supporting building structure. For a temperature change from 10 to 80°C (70 K), a 10 m length of steel pipe will increase in length by 7.9 mm ($70 \times 10 \times 0.0113$).

Coefficients for superficial (area) and cubic (volume) expansion are ιaken, respectively, as twice and three times those listed for linear movement.

Thermal expansion (gases). A perfect gas will conform to Boyle's and Charles's laws which state that the pressure (P), the volume (V), and the thermodynamic temperature (T) are related such that PV/T is a constant. The coefficient of cubic expansion is therefore temperature dependent and does not have a single value. Most gases conform very closely to the properties of a perfect gas when at a temperature remote from that at which they liquify.

Vapour pressure. Dalton's law of the partial pressures states that if a mixture of gases occupies a given volume at a given temperature, then the total pressure exerted by the mixture will be the sum of the pressures exerted by the components.

Criteria which affect human comfort – definitions

The ancients taught that humans had seven senses,* but it is no more than coincidence that the principal influences which affect human comfort are also seven in number:

- Temperature.
- Conduction, convection and radiation.
- Air volume and movement.
- Activity and clothing.
- Air purity.
- Humidity.
- Ionisation.

Temperature

The direction of heat flow from one substance to another is determined by the temperature of the first relative to that of the second. Thus, in that sense, temperature is akin to a pressure potential and is a relative term: the temperature of boiling water is higher than that of water drawn straight from a cold tap and the temperature of the latter is higher than

* Animation: feeling: hearing: seeing: smelling: speaking: tasting.

that of ice. Ice, however, may be said to be hot, or at a high temperature, relative to liquid air at −190°C.

Following the adoption of the *Système International*, temperature is measured in units of °C or K. The *Celsius* scale has a convenient false zero (0°C), which corresponds to the temperature at which water freezes, and has equal intervals above this to the temperature at which water boils under atmospheric pressure (100°C). The thermodynamic or *absolute* scale, established from the study of pressure effects upon gases, uses the same intervals as the Celsius scale but with a true zero corresponding to the minimum possible temperature obtainable. The intervals here are *kelvin* (K) and, with this scale, water freezes at 273 K.

In order to avoid confusion in terminology, it was the accepted convention in Imperial units that temperature *level* (or potential) should be expressed in terms of *°F*, whereas temperature *difference* (or interval) was in terms of *deg. F.* Similarly, under strict SI rules, temperature level is expressed in °C and difference in K and, although this usage is not obligatory, it is an aid to clarity and will be adopted throughout this book. For the benefit of those readers who, like the present authors, were educated prior to the adoption of metric units, a conversion table, Celsius to Fahrenheit, will be found in Appendix I (p. 683). Figure 1.1 shows the relationship between the various scales.

Reference will be made later in this chapter to a variety of different temperature notations: *dry-bulb*; *wet-bulb*; *globe* and *radiant*. These relate to methods of measurement for particular purposes and all use the Celsius or absolute scales noted above.

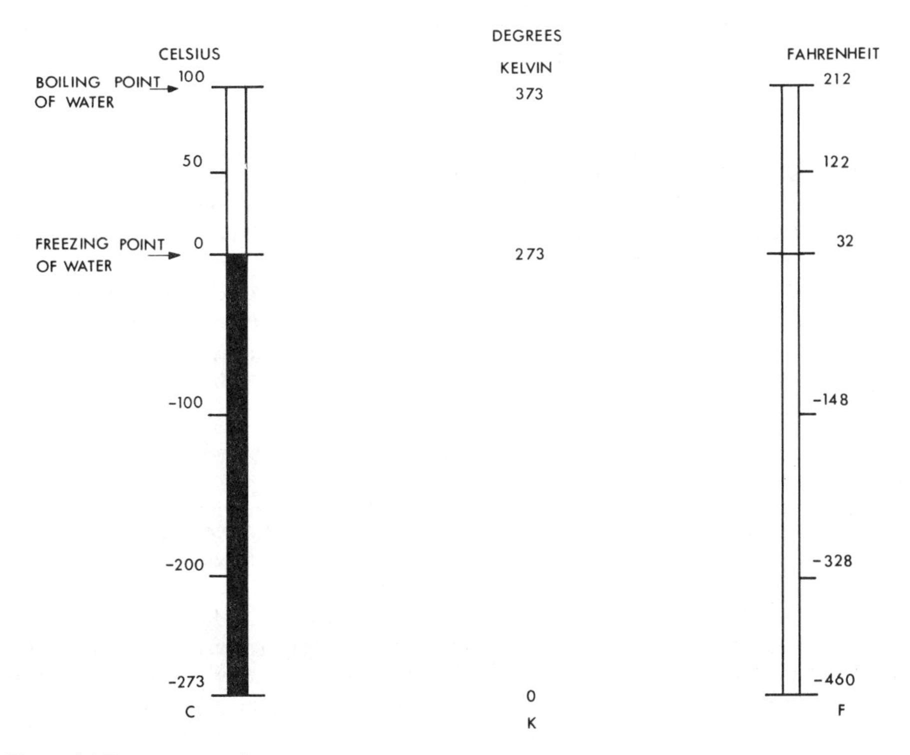

Figure 1.1 Temperature scales

Conduction, convection and radiation

These three terms have already been mentioned as being associated with heat transfer and it is necessary, for the discussions which follow, that there should be a clear perception of the difference between them.

Conduction

This may be described as heat transfer from one particle to another by contact. For example, if two blocks of metal, one hot and one cold, were to be placed in contact, then heat would be conducted from the one to the other until both reached an intermediate temperature. If both blocks were of the same metal, then this temperature could be calculated by the simple process of relating the mass and temperature of one to those of the other: but if the materials were not the same, it would be necessary to take account of the different *specific heat capacities* as noted earlier.

Conductivity is a measure of the quantity of heat that will be transferred through unit area and thickness in unit time for a unit temperature difference ($\mathrm{J\,m/s\,m^2\,K} = \mathrm{W/m\,K}$). Table 1.1 lists values of this property for various materials and it will be noted that metals have a high conductivity, whereas, at the other end of the scale, materials known as *insulators* have a low conductivity. The conductivity of many materials varies widely with temperature and thus only those values that fall within the range to which they apply should be used. Thermal conductivity, which is the property discussed here, should not be confused with electrical conductivity which is a quite separate quantity.

As far as building materials are concerned, those having higher densities are usually hard and are not particularly good insulators. Porous materials are bad conductors when dry and good conductors when wet, a fact which is sometimes overlooked when a newly constructed building is occupied before the structure has been able to dry out properly, which may take some months in the winter or spring.

Convection

Convection is a process in which heat transfer involves the movement of a fluid medium to convey energy, the particles in the fluid having acquired heat by conduction from a hot surface. An illustration commonly used is that of an ordinary (so-called) radiator which warms the air immediately in contact with it: this expands as it is heated, becomes lighter than the rest of the air in the room and rises to form an upward current from the radiator. A second example is water heated by contact with the hot surfaces around the furnace of a boiler, leading then to expansion and movement upwards as in the preceding instance with air. A medium capable of movement is thus a prerequisite for conduction, which cannot, in consequence, occur in a vacuum where no such medium exists.

Radiation

This is a phenomenon perhaps more familiar in terms of light and was, in Newton's time, explained as being the result of bombardment by infinitesimal particles released from the source of heat. At a later date, radiation, whether of light or heat, was supposed to be a wave action in a *subtile* medium (invented by mathematicians for the purpose) known as the *ether*. In the present context, it is enough to state that radiation is the transfer of energy by an electromagnetic process at wavelengths which correspond to, but extend marginally beyond, the infra-red range (10^{-6} to 10^{-4} m). Radiation is completely independent of any intermediate medium and will occur just as readily across a vacuum as across an air space: intensity varies with the square of the distance between the point of origin and the receiving surface.

The amount of radiation emitted by surfaces depends upon their texture and colour, matt black surfaces having an *emissivity* rated as unity in an arbitrary scale. Two values only need to be considered in the present context: 0.95 which represents most dull metals or the materials used in building construction and 0.05 which applies to highly polished materials such as aluminium foil. Surfaces which radiate heat well are also found to be good adsorbers; thus, a black felted or black asphalt roof is often seen to be covered with hoar frost on a cold night, due to radiation to space when nearby surfaces having other finishes are unaffected.

Air volume and movement

Although these two subjects interact, their characteristics may be considered separately.

Volume

It is necessary, when discussing the matter of air volume, to make a clear distinction between the total quantity in circulation and the proportion of it which is admitted from outside a building. It was traditional practice to refer to the latter component as *fresh air* but, since pollution in one form or another is a feature of the urban atmosphere, the term *outside air* is now preferred. In many instances the outside air volume may have entered a building by infiltration, in which case it is usually referred to in terms of *air changes* or *room volumes* per hour but, when handled by some form of mechanical equipment, this is rated in either m^3/s or, if realism prevails and decimal places are avoided, in litre/s.

Movement

It is not always well understood that air movement within a room is a positive rather than a negative effect. The source will be the position and velocity of admission since the location of an opening for removal has virtually no effect upon distribution. Air movement is measured in terms of air velocity (m/s) and must be selected within the limits of draughts at one extreme, and of stagnation at the other.

Activity and clothing

The interaction between human comfort, activity and clothing has been mentioned earlier. In most instances, the variables derive from the nature of the enclosure and the purpose for which it is to be occupied and thus allow for a group classification with like situations.

In the abstract, human activity is graded according to the level of physical exertion which is entailed and to the body area, male or female, ranging from a heat output of about 65–70 W when sleeping, through sitting (80–100 W), standing (100–160 W) to whole-body movement at varying rates of effort (160–230 W). As to clothing, this is graded according to insulation value, the unit adopted being the *clo*. Unity on this scale represents a male European business suit, with a value of $0.155\,m^2\,K/W$: zero in the scale is a minimum swim suit and light summer wear has a value of about 0.5 clo.

Air purity

Pollution can derive from sources outside a building or as a result of contaminants generated within it. In the former case, if air enters the building by simple infiltration, then dusts and fume particles, probably mainly carbonaceous in urban areas, will enter with it. Mechanical plant, on the other hand, is usually provided with air filtration equipment but

Table 1.2 Man-made indoor contaminants

Basement car parks	*Furnishings*	*Office machines*[a]
carbon monoxide	artificial fibres	ammonia
Cleaning agents	formaldehyde	formaldehyde
bleaches	carpet dusts	methanol
deodorants	*Occupants*	nitropyrene
disinfectants	carbon dioxide	ozone
solvents	clothing fibres	paper dust
Cooling towers	footwear dirt	trichloroethane
L. pneumophila	tobacco smoke	trinitrofluorenone

[a] Photo-printing, copying, duplicating and correcting fluids, etc.

this, as a result of indifferent or absent maintenance effort, may well be ineffective. Pollution generated within a commercial building may be from any combination of the sources listed in Table 1.2. Airborne bacteria use dust particles as a form of transport and, if housekeeping is neglected, may present a health hazard. Contaminants are identified by size, the micrometre (μm) being the common unit used for air filter rating.

Humidity

The humidity of air is a measure of the water vapour which it contains and *absolute humidity* is expressed in terms of the mass of water (or water vapour) per unit mass of dry air (kg/kg) and not per unit mass of the mixture. The greatest mass of moisture which, at atmospheric pressure, can exist in a given quantity of air is dependent upon temperature as may be seen from Table 1.3, the effect of pressure variation, for two extreme values, being given in a footnote. This condition is known as a state of *saturation* and if air, virtually saturated at an elevated temperature, be cooled, then a temperature is soon reached where the excess moisture becomes visible in the form of a mist, or as *dew* or rain.

In most practical situations, both externally and within a building, the air will not be saturated and a water vapour content which may be measured relative to that situation will

Table 1.3 Moisture content of saturated air at various temperatures (kg/kg of dry air)

Temperature (°C)	*Moisture* (kg/kg)	*Temperature* (°C)	*Moisture* (kg/kg)	*Temperature* (°C)	*Moisture* (kg/kg)
0	0.0038	18	0.0129	36	0.0389
2	0.0044	20	0.0148	38	0.0437
4	0.0050	22	0.0167	40	0.0491
6	0.0058	24	0.0190	42	0.0551
8	0.0067	26	0.0214	44	0.0617
10	0.0077	28	0.0242	46	0.0692
12	0.0088	30	0.0273	48	0.0775
14	0.0100	32	0.0308	50	0.0868
16	0.0114	34	0.0346	52	0.0972

Note: at 20°C, the moisture contents at saturation for altitudes of −1000 m (113.9 kPa) and + 2000 m (79.5 kPa) are 0.0131 kg/kg and 0.0189 kg/kg respectively.

exist. Two terms are used to quantify the moisture content held, the familiar *relative humidity* and the less well-known *percentage saturation*. In fact, the precise definition of the former is far removed from popular usage and the latter, which is a simple ratio between the moisture content at a given condition and that at saturation, is now more generally accepted for use in calculations.

Ionisation

Ionisation of gases in the air creates groups of atoms or molecules which have lost or gained electrons and have acquired a positive or negative charge in consequence. In clean, unpolluted air, ions exist in the proportion of 1200 positive to 1000 negative per cm^3 of air but, in a city centre, the quantities and proportion change to 500 positive to 300 negative per cm^3. In a building ventilated through sheet metal ductwork, and furnished with synthetic carpets and plastics furniture, the quantities and proportion will reduce still further to perhaps 150 positive to 50 negative ions per cm^3.

It has been suggested that low concentrations of negative ions (or high concentrations of positive ions) lead to malaise and lethergy. The *Guide Section A1* states that present evidence in support of this theory is inconclusive but the advice of the Health and Safety Commission, that adoption of 'an agnostic attitude would be appropriate', praises with even fainter damns!

Thermal indices

Numerous attempts have been made to devise a scale against which comfort may be measured and the following deserve attention.

Equivalent temperature

This scale combines the effects of air temperature, radiation and air movement, all as measured by a laboratory instrument named the *Eupatheoscope*. The late Mr A. F. Dufton, who developed this instrument and the scale at the Building Research Establishment, defined equivalent temperature as 'that temperature of a uniform enclosure in which, in still air, a sizeable black body would lose heat as in the environment, the surface of the body being one-third of the way between the temperature of the enclosure and 100°F' (i.e. approximately body temperature). The scale may be represented mathematically but, as it takes no account of variations in humidity, it fell out of use some 20 years ago and is mentioned here only to record the work on hand at Garston during the 1920s.

Effective temperature

Devised and developed in the USA for particular application to air-conditioning design, this scale combines the effects of air temperature, humidity and air movement but has no point of reference to radiation. It was produced following a long series of subjective tests carried out on a number of people in a wide variety of differing environmental situations and has been revised from time to time. The results are presented in an envelope of curves superimposed on a nomogram, each curve relating to an air velocity. The *Effective Temperature Lines* may be adapted to provide a basis for a chart, in the form shown in Figure 1.2, where comfort zones appropriate to the British Isles are outlined for winter and summer. This scale has limited application for circumstances where no more than a simple heating system is envisaged since, within the temperature range 15–20°C, variations in relative humidity between 40 and 70% have small effect upon comfort.

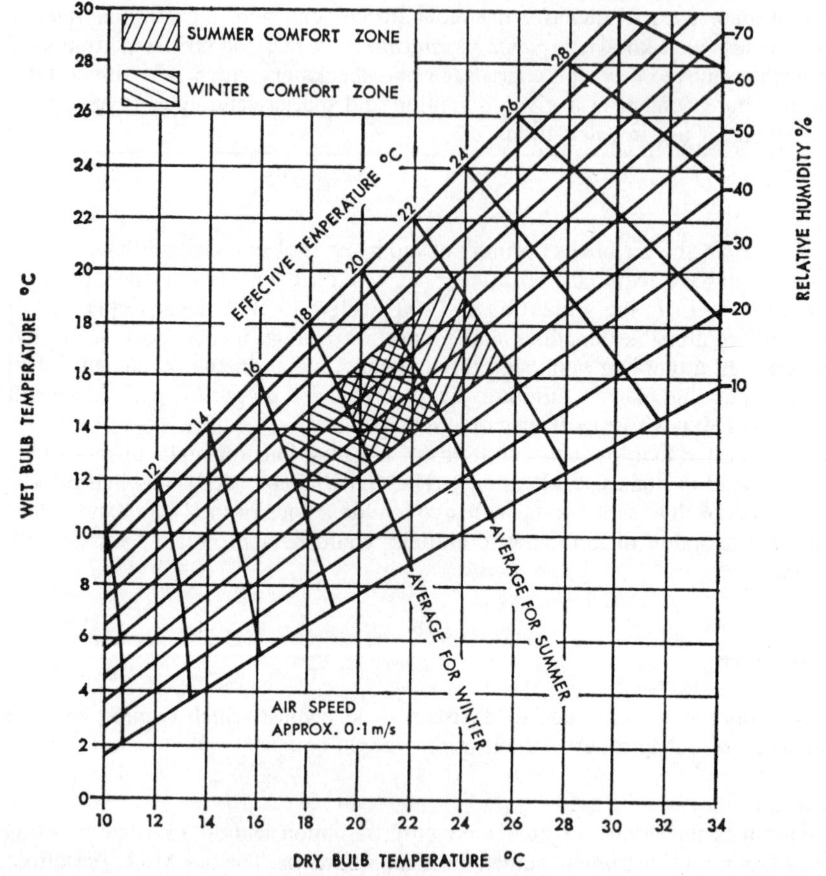

Figure 1.2 Effective temperature chart. This chart is postulated from a variety of sources as being applicable to the British Isles only. It provides a range of conditions for winter and summer within which most people would feel comfortable, provided that they were not transient

Corrected effective temperature

This is a later version of the scale noted above, modified to include the effect of radiation. It is, as before, presented in the form of a nomogram but its reliability is restricted to the central part of the scale between about 15 and 19°C corrected effective temperature.

Fanger's comfort criteria

More recent work in Denmark has produced a series of comfort charts and tables based again on analysis of the results of subjective tests which, in this case, took account of two further variables, namely the metabolic rate for various activities and the clothing worn. The volume of data provided is extremely large and enables predictions to be made with a high level of accuracy: for everyday use, however, the calculations which are required seem disproportionate, having regard to the application of the end result.

Subjective temperature

This is an approach not dissimilar to that proposed by Fanger but simpler and more realistic to apply since account is taken of fewer variables.

Dry resultant temperature
This is the comfort index now included in the *Guide Section A1*, to take account of the effects of temperature, radiation and air movement. The full expression for evaluation of this temperature is:

$$t_{res} = [t_{ri} + t_{ai}(10v)^{0.5}]/[1 + (10v)^{0.5}]$$

where

t_{res} = dry resultant temperature (°C)
t_{ri} = mean radiant temperature (see explanation on p. 13) (°C)
t_{ai} = room air temperature (°C)
v = air velocity (m/s)

It will be noted that, when the air velocity is 0.1 m/s, this expression may be simplified greatly for use in the general run of calculations as $t_{res} = (0.5\,t_{ri} + 0.5\,t_{ai})$.

In addition to the items noted, *wet bulb globe temperature, wet resultant temperature* and the *equitorial comfort index* could be added. These, however, have narrow applications to specific circumstances and are not in general use. Also excluded from the list are two scales which are not comfort indices but which relate to the calculation of heat losses and gains. The more important of these is *environmental temperature*, a concept shown by the Building Research Establishment to provide a valuable simplification of the relationship between air and mean radiant temperatures within a space. The second is *sol–air temperatures*, a scale which increments outside air temperature to take account of solar radiation.

Wind-chill indices
Although the expression *wind-chill* will be familiar to most readers, as a result of recent use by TV weather forecasters, it is not generally appreciated that the concept of wind-enhanced cooling pre-dates the First World War. A paper by Dixon and Prior* provides a full history, including a digest of both the empirical and theoretical evaluations and suggests that analyses of wind-chill indices by wind direction may be useful in deciding upon orientation and layout of new buildings, shelter belts, etc. Wind-chill equivalent temperatures for wind speeds of 2–20 m/s (5–45 mph) and air temperatures between –10°C and +10°C are listed in Table 1.4.

Table 1.4 Wind-chill equivalent temperatures (°C)

	Air temperature (°C)							
Wind speed (m/s)	–10.0	–5.0	0	2.0	4.0	6.0	8.0	10.0
2	–11.0	–6.0	–1.0	1.0	3.0	5.0	7.5	9.5
4	–14.5	–9.0	–3.5	–1.5	1.0	3.0	5.0	7.5
6	–17.5	–11.5	–6.0	–3.5	–1.0	1.0	3.5	5.5
8	–20.0	–14.0	–8.0	–5.5	–3.0	–0.5	2.0	3.0
10	–22.0	–16.0	–9.5	–7.0	–5.0	–2.0	–0.5	3.0
12	–24.0	–17.5	–11.0	–8.5	–6.0	–3.5	–1.0	2.0
15	–26.5	–19.5	–13.0	–10.0	–7.5	–5.0	–2.0	0.5
20	–29.5	–22.5	–15.5	–12.5	–10.0	–7.0	–4.0	–1.5

* Dixon, J. C. and Prior, M. J., 'Wind-chill indices – a review'. *The Meteorological Magazine*, 1987, **116**, 1.

Methods of measurement

The descriptions which follow relate to very basic instruments which may in many respects have been superseded by electronic or other devices. They do, nevertheless, serve to illustrate requirements and to warn of pitfalls.

Temperature

The readings taken from an ordinary mercury-in-glass thermometer provide, in general, temperature values relating to the gas, liquid or solid in which the instrument is immersed. When applied to the air volume within a building, however, the situation is much more complex. Here, the scale reading will portray a situation of equilibrium, taking account of not only air temperature but also of a variety of heat exchanges between surrounding surfaces. These latter may be partly by conduction, partly by convection and partly by radiation.

If such a thermometer were to be immersed in a hot liquid, conduction and convection would account for practically the whole effect, but if the same instrument were used to provide the temperature of room air, then the effect of radiation might predominate. Thus, in summer, the sun temperature might be 37°C, while the shade temperature reached only 27°C: the *air* temperature in each case might well be the same but in the former case the thermometer would be exposed to the sun, while in the latter case it would not.

In practice, the nature of the source of radiant heat affects questions of measurement. The so-called *diathermic* property of glass, in this case the bulb of the thermometer, permits the inward passage of high temperature radiation but is impervious to an outward passage at lower temperature. A radiation shield, which may be no more than a piece of aluminium cooking foil shading the thermometer bulb, will negate this influence and permit air temperature alone to be read.

Measurement of radiant effects thus depends upon a whole variety of circumstances. A *solar* thermometer, for measurement of high temperature radiation, consists of a glass

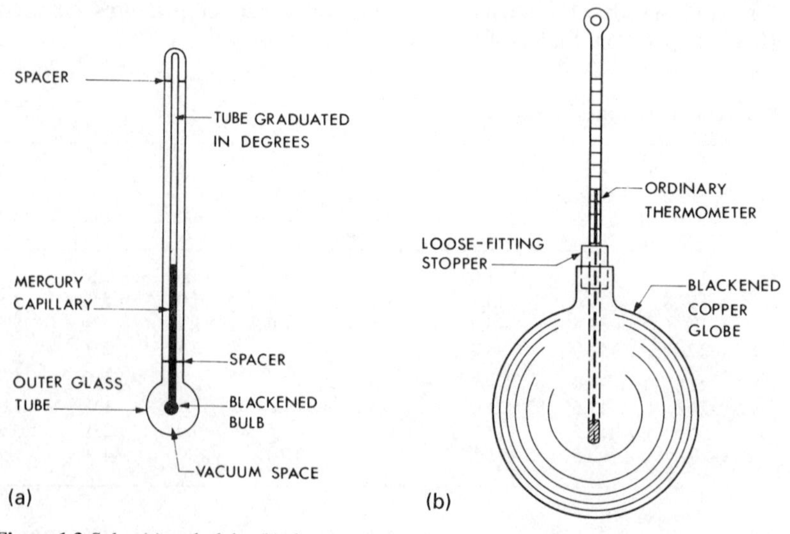

Figure 1.3 Solar (a) and globe (b) thermometers

sphere and tube within which a vacuum has been drawn, containing a simple mercury-in-glass instrument having a blackened bulb, as Figure 1.3(a). Similarly, a *globe* thermometer, used for measurement of low temperature radiation, consists of a hollow copper sphere or cylinder about 150 mm in diameter, blackened on the outside and having a simple mercury-in-glass instrument projecting into it such that the bulb is in the centre, as Figure 1.3(b). The mean radiant temperature (t_{ri}) at a single point in a room may, with difficulty, be measured with a globe thermometer or may be calculated from the arithmetic mean of the surrounding areas, each multipled by the relevant surface temperature.

Humidity

The instrument most commonly used to measure the moisture content of air in a room is a *psychrometer* which makes use of what is called *wet bulb temperature*. This is measured by means of a simple mercury-in-glass thermometer which has its bulb kept wet by means of a water-soaked wick surrounding it. As the water evaporates, it will draw heat from the mercury with the result that a lower temperature will be shown. The rate of evaporation from the wetted bulb depends upon the humidity of the air, i.e. very dry surrounding air will cause a more rapid evaporation – and a lower temperature in consequence – than air which is more moist, although the temperature shown by an ordinary dry bulb thermometer would be the same in each case. The difference between dry and wet bulb temperatures may thus be used as a measure of humidity, individual values being known as the *wet bulb depression.*

The rate of evaporation, and hence the extent of the depression, depends also upon the manner in which the wetted bulb is exposed to the air. Records of external dry and wet bulb temperatures are kept by meteorological authorities world-wide and for this purpose measurement is made using instruments placed in the open air within a louvred box called a *Stevenson screen.* The louvres are arranged so as to allow a natural circulation of air around the thermometers but, of course, they cannot exclude the effect of radiant heat completely, with the result that what are known as *screen* wet bulb temperatures are always about 0.5 K higher than those read from the alternative instruments used by engineers.

The more common of these is the *sling psychrometer* which consists of two thermometers mounted side by side in a frame which is fitted with a handle such that it

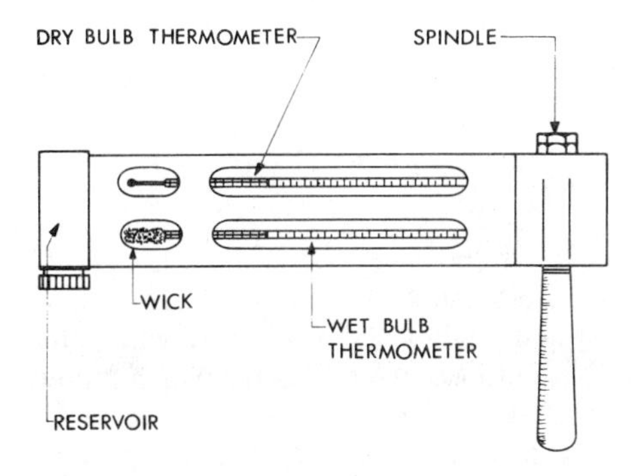

Figure 1.4 A sling psychrometer

may be whirled by hand, as shown in Figure 1.4. One thermometer is fitted with a wick which is fed with water from a small reservoir in the frame and both the water, preferably distilled, and the wick must be clean. Readings are taken after the instrument has been whirled vigorously to simulate an air speed above about 3 m/s. An alternative instrument, the *Assmann* type, consists of a miniature air duct within which the two thermometers are mounted, air being drawn over them by a small fan, clockwork or electrically driven, at a known velocity.

Air velocity

Requirements for reading air velocity fall into two very distinctly different categories. One relates to measurements made for the purpose of determining air volume in a duct or at delivery to, or extract from, a space, and the other to the identification of movement of air within a room. In the former case, velocities are not likely to be less than 4 m/s and (unless things have gone badly astray) the direction of flow should be easily identifiable. For this purpose the traditional field instruments have been *vane anemometers* of one type or another as will be noted in more detail in a later chapter (p. 459).

Within a room, however, circumstances are different and it is generally accepted that a tolerable level of air velocity at the back of the neck is not much in excess of 0.1 m/s when the dry resultant temperature is 20°C, although, in extreme summer conditions, without cooling, a transient increase to 0.3 m/s may be acceptable. For measurement at these low levels, use must be made of either an instrument known as the *Kata thermometer* or of one of the more sophisticated anemometers such as a *hot-wire* type.

The Kata instrument is made in four patterns, and that most used for measurement of low air velocities has a very large silvered bulb filled with coloured alcohol. In order to take a reading, the bulb is first warmed in hot water, causing the alcohol level to rise, and the instrument is then located in the position at which an observation is required. The time taken for the level of the spirit to fall between two marks on the scale, as a result of cooling due to the air movement, is noted and this, together with a reading from a dry bulb thermometer, is applied to a nomogram calibrated for the particular instrument to provide a record of velocity.

As may be appreciated, the Kata thermometer is best suited to laboratory research work and is tedious to use elsewhere. The hot-wire instrument is much more convenient to use than the Kata but neither provides any absolute indication of the *direction* in which the air is moving. It is thus necessary to rely upon a cold smoke, produced chemically, to supply a visual appreciation but this is not wholly satisfactory either, since the smoke tends to diffuse rather too rapidly.

Application

From what has been said in preceding paragraphs it will be clear that, while it is possible to define the various factors which relate to comfort and to list the methods used to measure them, application of this knowledge to system design cannot proceed without the more detailed examination of the various aspects, as outlined in the chapters which follow. This brief résumé of the fundamentals may thus be best summarised by reference to Tables 1.5 and 1.6 which provide, respectively, recommended values of dry resultant temperature (t_{res}) and volume of outside air for most common applications.

Table 1.5 Proposed values for dry resultant temperature $t_{(res)}$

Type of enclosure	t_{res} (°C)
Swimming pool halls	28
Bathrooms; sports changing rooms; hotel bedrooms	22
Bed sitting rooms; banqueting rooms; domestic living rooms; dressing rooms; hotel public rooms; operating theatres[a]	21
Art galleries; banking halls; canteens and cafeterias; dining rooms; laboratories; law courts; libraries; office accommodation generally; reading rooms	20
Factories (sedentary work)	19
Auditoria; churches; domestic bedrooms; exhibition halls; hospital wards and day rooms; lobbies, foyers and corridors; police cells; public bars; school classrooms; shops and stores	18
Cloakrooms; factories (light work); gymnasia; lobbies in domestic buildings; sports halls; working spaces in warehouses and stores	16
Factories (heavy work); storage areas in warehouses	13

[a] Variable between 18 and 21°C.

Table 1.6 Proposed rates for supply of outside air to avoid contamination and odour

	Outside air supply (litre/s)	
Type of space	*Per person*	*Per m² floor area*
Minimum, as appropriate to large lofty enclosures sparsely occupied for short periods (e.g. exhibition halls)	5	0.2
Hotel bedrooms; museums; shopping malls		1.0
Banking halls		1.5
Supermarkets		3.0
Auditoria: theatres	8	6.5
Open plan offices		2.0
Department stores		3.0
Hotel public rooms		4.0
Art galleries (if used for receptions)	10	5.0
Cellular offices		2.0
Individual shops		3.0
Conference meeting rooms		4.5
Cafeteria canteens		5.0
Dance halls	13	6.5
Executive offices		2.0
Banqueting rooms		5.0
Restaurants		8.0
Board rooms; executive conference rooms		9.0
Public bars; sports changing rooms		10.0
Discothèques	17	13.5

Notes:
1. All tabulated values relate to **non-smoking** areas. Where any level of smoking is permitted, increase by 50%; where smoking is likely to be more intense, increase by 100%.
2. Volume rates in the final column relate to the gross floor area.
3. Data base is an updated version of that used for the original printing of Table A9.24 in the 1970 edition of the *Guide*. Each of the values takes account of density of occupation, likely level of activity, possible extraneous odours and other relevant factors.
4. For all accommodation in hospitals, refer to DHSS Building Notes.

The use of resultant temperature as a design criterion takes account of radiation in the general sense but not of particular local effects which may arise from *asymmetrical exposure* of the body to:

- Cold radiation in winter to a single-glazed window.
- Excessive insolation from unshaded glazing.
- Exposure to some internal source of high radiant intensity.

A simplification of the data given in the *Guide Section A1* suggests that, in the first of these situations for a typical window height of 2 m, an occupant seated within a distance represented by the square root of the window width (in m) will experience discomfort when the outside temperature is at about freezing level. In the case of the discomfort arising from the second situation, there are so many variables that a similar simplification is not possible; in an extreme case, however, a rise in radiant temperature (t_{ri}) of 10–15 K might be anticipated and the provision of some form of shading to the glazing, preferably external, is the correct solution.

As to humidity, too high a level will reduce the ability of the body to lose heat by evaporation, with resultant lassitude; too low a level will produce a sensation of coolness on exposed flesh, a parched throat and dry eyes. Where temperatures are high, these extremes are of great importance but in the maritime climate of the British Isles, variations between 40 and 60% saturation are usually acceptable, although 50% saturation is the usual design target for human comfort. Industrial processes often require a better standard of humidity control to within as little as ±5%.

There are two quite extraneous aspects which relate to humidity in any building, the first being the likely incidence of condensation on single glazed windows when their surface temperature falls, in very cold weather, to below the dewpoint temperature of the building air content. The second is the problem of a build up in static electricity, leading to electrostatic shock when occupants touch earthed building components. This is a function of humidities below about 40% coupled with the material and backing of the floor coverings.

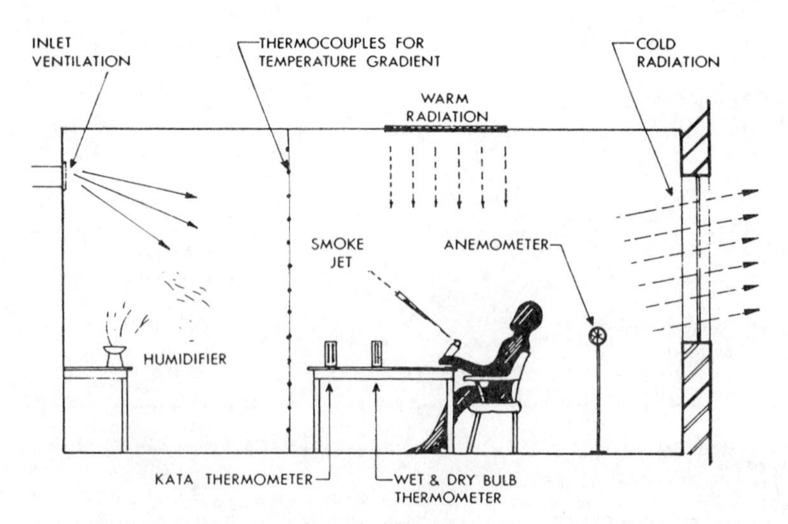

Figure 1.5 Determination of a state of comfort

Conclusion

As will be appreciated, the manner in which air temperature, radiation, humidity and air movement interrelate with one another is very complex and nothing short of an exhaustive investigation of each, as illustrated in Figure 1.5, will suffice to establish the precise situation existing within an enclosure. Even so, when the investigation is complete and all results have been demonstrated to be within a hair's breadth of best practice, the individualities of human sensation are such that it is unlikely that many more than 90% of the occupants of a building will be entirely satisfied with their environment!

Chapter 2

The building in winter

As a general principle when approaching the question of space heating, it is desirable that the building and the heating system should be considered as a single entity. The form and construction of the building will have an important effect not only upon the method to be adopted to provide heating service, but also upon subsequent recurrent energy costs. The amount of heat required to maintain a given internal temperature may be greatly reduced by thermal insulation and by any steps taken to reduce an unwanted intake of outside air. Large areas of glass impose very considerable loads upon any heating system and run counter both to the provision of comfort conditions and to any prospect of energy efficient operation.

The mass of the building structure, light or heavy, has a direct bearing upon the choice of the most appropriate form of system since, in the former case, changes in external temperature will be reflected very quickly within the building and a system having a response rate to match will be the one that is most suitable. On the other hand, a building of traditional heavy construction may well be best served by a system which produces a slow steady output. Tall, multi-storey blocks of offices and dwellings introduce problems related to exposure to wind and solar radiation as well as those related to *chimney or stack effects* within the building itself. The form and design of the heating system must take these aspects into account.

Extraneous influences

Within the space to be heated, the energy which is consumed by electrical lighting and by a variety of other items of current-consuming equipment will be released, as will heat from the occupants. The total of these internal emissions will contribute in some measure towards maintenance of the desired space temperature. In addition, even in winter, heat from the sun may sometimes be enough to cause problems of excessive temperature in those rooms exposed to radiation, while at the same time others, in shade, are not so favoured. These fortuitous effects cannot be overlooked, although it is not possible to rely upon them to make a consistent contribution.

It is a matter of importance to consider how far these heat gains should be taken into account: if they are ignored, then the heating system may be so oversized that it will be unwieldy and will rarely, if ever, run at full capacity. On the other hand, if certain reasonable assumptions are made as to the proportion of the total gains which will coincide and, due possibly to a change in building use, these do not apply, then the heating system may well be undersized. In instances where, despite the last comment, some allowance is made, particular thought must be given to the needs of an intermittently heated building where a *pre-occupancy boost* may be required during the time when no internal or solar gains are available.

Past practice in design, for other than off-peak electrical heating systems, has been to ignore the effect of such gains entirely in so far as calculations for the heat necessary to maintain a given internal temperature are concerned. In calculations for energy consumption and running costs per annum, however, the importance of these gains has been brought out by studies of actual fuel use in buildings. The case for adequate thermostatic controls is self evident.

Conservation of energy

The past situation, when a seemingly limitless supply of fuels of one sort or another was available has come to an end. In the new age which dawned following the 'energy crisis' of the mid-1970s, economic forces have led to a startling rise in the cost of supplies, with the result that many of the old standards of comparison no longer apply. The more recent stabilisation of fuel prices, largely determined by the price of oil in the international market, may be considered as no more than a temporary pause in the inevitable upward spiral of energy costs.

As a result of this situation, energy conservation in the sense of fuel saving is now a doctrine of political importance as well as one of realism. In the context of this book, it has been estimated that between 40 and 50% of the national annual consumption of primary energy is used in services to buildings. By the introduction of sensible economy measures, without real detriment to environmental standards or the quality of life, it is possible to make significant savings in the context of heating and air-conditioning systems. But the most dramatic attack upon energy consumption must come about as a result of reconsideration of the building structure, readjustment of capital cost allocations and improved maintenance of buildings and equipment.

It follows that the first step in embarking upon the assessment of heat requirements for a building should be to ensure that they have been reduced to an economic minimum. This will involve collaboration between the architect and the building services engineer in consideration of the building orientation, selection of materials, addition of thermal barriers and reduction in window areas. Ideally, this collaboration should start at an even earlier stage when the basic plan form of the building is being considered, bearing in mind that the major component of the total thermal load is through the perimeter surfaces.* In this sense, the most economical shape for maximum volume with minimum surface area is a sphere and although this is hardly a practicable shape for a building, the nearer to it the better. A tall shallow slab building is obviously one of the worst in this respect.

Heat losses

The conventional basis for design of any heating system is the estimation of heat loss and, for the purpose of calculation, it is assumed that a *steady state* exists between inside and outside temperatures although, in fact, such a condition rarely obtains. In the past, air temperature difference has been the sole criterion although the mean radiant temperature within the enclosure, if considered, may well call for a higher or a lower air temperature

* Page, J. K., *Energy Requirements for Buildings.* Public Works Congress 1972.
Jones, W. P., *Designing Air-Conditioned Buildings to Minimise Energy Use.* RIBA/IHVE Conference 1974.

for equal comfort. The method of calculation recommended in the current edition of the *Guide Sections A5 and A9* is in terms of dry resultant temperature within the space to be heated and this, as was explained in the preceding chapter (p. 11), takes account of the mean radiant temperature. This refinement will be discussed subsequently but, in this preliminary introduction, the air temperature difference will be used wherever appropriate.

Each room of a building is taken in turn and an estimate is made of the amount of heat necessary to maintain a given steady temperature within the space, assuming a steady lower air temperature outside. The calculation falls into two parts: one relating to conduction through the various surrounding structural surfaces, walls, floor and ceiling; and the other to the heat necessary to warm to room temperature that outside air which, by accident or design, has infiltrated into the space.

Adjacent rooms maintained at the same temperature will have no heat transfer through the partitions or other surfaces between them and these may thus be ignored. Furthermore, if certain surfaces, such as the ceiling or floor, are used for heat output, then these will also not be taken into account so far as heat loss from the room is concerned; they will, however, have inherent losses upward or downward to unheated areas and those will have to be allowed for separately.

The conduction element is calculable from known properties of the building materials, but the infiltration element presents problems, in that what is called the *air change rate* or, alternatively in energy terms, the *ventilation allowance*, is not easy to assess other than by experience. This air change rate is no more than a natural ventilation effect, arising from a number of extraneous circumstances but without which a space would quickly cease to be habitable, and although this element must be dealt with empirically, the ground rules are reasonably well established.

It might be thought that, with so many assumptions and 'rule of thumb' estimates, heat loss calculations are very little better than guesswork. In practice, however, they have proved to be a reliable basis for an overall assessment, partly due to the fact that all areas in the building are treated in a like manner and are thus consistent in response. In addition, the building structure is itself a moderator, as a result of its *thermal inertia*, which remains as a significant factor, even in a lightly constructed building which still has floor slabs, partitions, furniture, etc., to absorb and emit heat and thus smooth out any violent fluctuations.

Conduction losses

The conduction of heat through any material depends upon the conductivity of the material itself and upon the temperature difference between the two surfaces. Ignoring for the moment any heat transfer by radiation within a space, it is the air-to-air transfer of heat through building materials which is relevant. On either side of a slab of building material, it may be supposed that there is a film or relatively dead layer of air which retards the flow of heat. This is illustrated, in Figure 2.1, for a thin material such as a single sheet of glass. The air within the room, being relatively still, offers a higher resistance to heat flow than that outside where wind effects, etc., have to be considered.

The effect of these boundary layers is defined in terms of what is described as *surface resistance*, and representative values are as set out in Table 2.1. The notation here is that given in the *Guide Section A3* and, as noted later, the value of the outside surface resistance (R_{so}) varies with the degree of exposure such that a sheltered surface has a higher resistance to heat flow than one which is exposed to severe wind and other effects. In the extreme case of a very tall building, it might well be supposed that wind

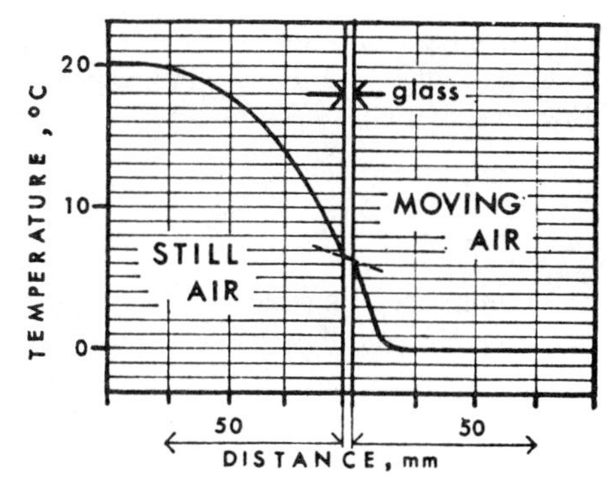

Figure 2.1 Heat transmission gradient through glass

forces were such that the boundary layer is totally dispersed and that the value of the surface resistance is zero, i.e. the temperature of the external surface is, effectively, that of the outside air. The figures listed for internal surface resistances (R_{si}) vary, it will be noted, only with respect to the disposition of the area concerned and the direction of heat flow.

The values noted in Table 2.1 for external and internal surface resistances relate to normal building materials, most of which have high emissivity values. An exception may arise in the case of the resistance offered by an air gap in a cavity construction (R_a) where

Table 2.1 Surface resistances ('normal' exposure)

	Surface resistance (m^{2}K/W)	
Building surface	*Emissivity 0.95*	*Emissivity 0.05*
Outside (R_{so})		
walls	0.06	–
roofs	0.04	–
floors	0.04	–
Inside (R_{si})		
walls (horizontal)	0.12	–
roofs (to above)	0.10	–
ceilings (to above)	0.10	–
floors (to below)	0.14	–
Unventilated air gap (R_a)		
horizontal	0.18	0.35
above	0.17	0.35
below	0.22	1.06
Ventilated air gap (R_a)		
cavity wall (horizontal)	0.18	0.35
behind tiles on hung tile wall	0.12	0.30
loft space over flat ceiling	0.14	0.40
void under unsealed pitched roof	0.16	0.40
void under sealed pitched roof	0.18	0.35
void within flat roof	0.14	0.40

two sets of values are given, one for the situation where the facing surfaces within the air gap have high emissivity and the other to meet the case where a bright metallic surface, such as aluminium foil, is inserted to provide low emissivity.

Conductivities

Values for the thermal properties of a selection of common building materials were given in Table 1.1, and others are listed in the *Guide Section A3*. The following definitions may be helpful:

Conductivity
The standardised value, in watt for one metre thickness per square metre and kelvin:

$$k = \text{W m/m}^2\,\text{K} = \text{W/m K}$$

Conductance
The value for a stated thickness (k/L):

$$C = \text{W/m}^2\,\text{K}$$

Resistivity
The reciprocal of the conductivity ($1/k$):

$$r = \text{m K/W}$$

Resistance
The reciprocal of the conductance (L/k):

$$R = \text{m}^2\,\text{K/W}$$

For example, the thermal conductivity (k) of expanded polystyrene is 0.035 W/m K and the thermal conductance (C) of a slab of this material, 25 mm thick, is thus 0.035/0.025 = 1.4 W/m^2 K. Similarly, the thermal resistivity (r) of the material is 1/0.035 = 28.6 m K/W and the thermal resistance (R) of the slab is 1/1.4 (or 28.6 × 0.025) = 0.71 m^2 K/W.

Moisture in masonry materials

A review at the Building Research Establishment* has shown that the conductivities of all masonry materials bear the same general relationship to their dry densities and follow the same proportional pattern with increasing moisture content. Corrected practical values for moisture content by volume were selected for the *Guide Section C3* in consequence, as follows:

- Brickwork, protected from rain 1%.
- Concrete, protected from rain 3%.
- Both materials, exposed to rain 5%.

* Loudon, A. G., 'U values for the 1970 Guide', *JIHVE*, 1968, **36**, 167.

For the inner skin material of a cavity wall and for internal partitions, etc., the protected values are used. In circumstances where materials are exposed to either driving rain or condensation, the 5% content noted above is no longer valid and it would seem that conductivities increase by about 4% for each 1% increase in moisture content.

U values

The rate of heat transmission, in watt per square metre and kelvin ($W/m^2 K$) is, for the purpose of heat loss calculation, termed the thermal transmittance coefficient (U). This is the reciprocal of the sum of all individual resistances, thus:

$$U = 1/(R_{si} + R_{so} + r_1 L_1 + r_2 L_2 + r_3 L_3 + R_a)$$

where

R_{si} = inside surface resistance ($m^2 K/W$)
R_{so} = outside surface resistance ($m^2 K/W$)
R_a = air space resistance ($m^2 K/W$)
r_1, r_2, etc. = resistivities (m K/W)
L_1, L_2, etc. = thicknesses (m)

Pre-calculated values for transmittance coefficients, applicable to a range of typical forms of construction, are provided in Table 2.2 which appears at the end of this chapter (p. 61). The values listed are for what is called *normal* exposure but the much more comprehensive list of examples given in the *Guide Section A3* includes alternative data for other conditions, defined as follows:

Sheltered Up to 3rd floor of buildings in city centres.
Normal 4th to 8th floors of buildings in city centres; up to 5th floor of suburban and rural buildings.
Severe 9th floor and above in city centres; 6th floor and above in suburban and rural districts; most buildings on coastal or hill sites.

In general terms, the values listed in Table 2.2 will be increased by up to 20% for *severe* exposure but, in the particular case of single glazing, the increase is much greater at about 45%.*

Calculation of a *U* value

The wide range of constructions currently encountered, built up from the many composite elements available, often requires that transmittance coefficients (U values) are calculated from first principles. For example, consider a curtain wall construction comprising:

Outside 40 mm thick pre-cast concrete panel
Air gap 25 mm wide
Insulation 50 mm glass fibre slab
Inside 75 mm lightweight concrete block†
Plaster 13 mm lightweight

* These increased values are of particular relevance only with respect to heat losses in UK winters. Winds in summer may be assumed to have less influence and values for normal exposure are thus appropriate.
† 0.19 (from Table 1.1) + 15% for sand/cement joints = 0.22 W/m K.

The resistances, from the physical data provided in Tables 1.1, 2.1 and 2.3, may be summated as follows:

Outside surface, R_{so}		$= 0.06$
Concrete panel	$(1/1.4) \times (40/1000)$	$= 0.029$
Air space, R_a		$= 0.18$
Insulation	$(1/0.035) \times (50/1000)$	$= 1.43$
Concrete block	$(1/0.22) \times (75/1000)$	$= 0.34$
Plaster	$(1/0.16) \times (20/1000)$	$= 0.125$
Inside surface, R_{si}		$= 0.12$
	ΣR	$= 2.284\ \mathrm{m^2\,K/W}$

and thus

$$U = 1/2.28 = 0.44\ \mathrm{W/m^2\,K}$$

Surface temperatures

It is often necessary, for a variety of reasons, to establish either a surface or an interface temperature for some form of composite construction. The temperature gradient across the structure may be plotted, as shown in Figure 2.2, using the resistances of the various elements, with the simple case of an unplastered and uninsulated 210 brick wall being illustrated in part (a). In this case, the overall resistance would be made up as follows:

Outside surface, R_{so}		$= 0.06$
Brickwork	$(1/0.84) \times (210/1000)$	$= 0.25$
Inside surface, R_{si}		$= 0.12$
	ΣR	$= 0.43\ \mathrm{m^2\,K/W}$

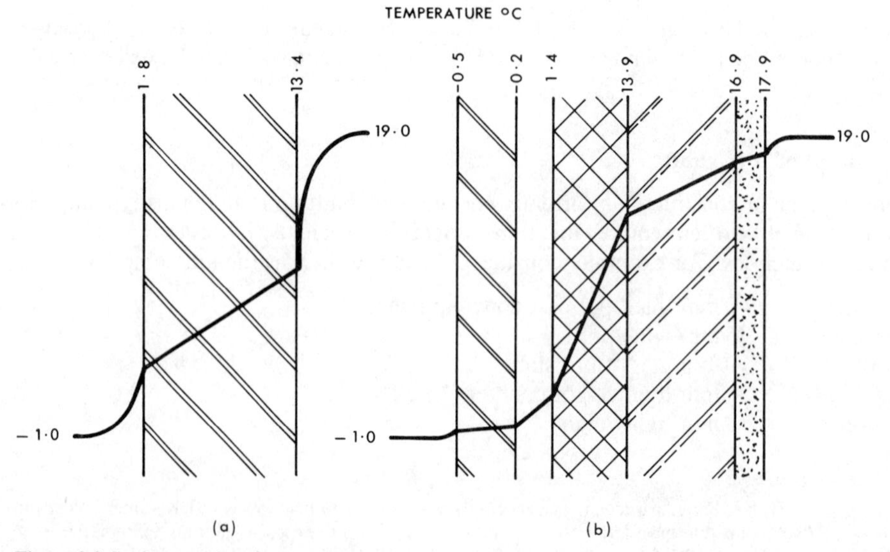

Figure 2.2 Surface and interface temperatures

The temperature gradient, including the components at the surfaces, will be *pro rata* to the resistances and, for inside and outside air temperatures of 19°C and –1°C, respectively, would be:

Outside air to outside surface	=	20 × (0.06/0.43)	=	2.8°C
Across the brickwork	=	20 × (0.25/0.43)	=	11.6°C
Inside surface to inside air	=	20 × (0.12/0.43)	=	5.6°C

The surface temperature of the brickwork inside the enclosure would thus be 19 – 5.6 = 13.4°C.

The more complex case of the curtain wall construction in the previous example is dealt with in the same way, the surface and the interface temperatures being as shown in part (b) of Figure 2.2.

Application

Having established the transmittance coefficients for the various structural elements enclosing a space, it is then possible to evaluate the conduction component of the heat requirement. This process is best illustrated by an example.

A simple building having a roof which is part exposed to the outside air and part to a heated room above it, is shown in Figure 2.3. It is to be heated to provide an internal air temperature of 19°C when the outside air temperature is –1°C. The construction is as follows:

Walls	210 mm brickwork, unplastered.
Roof	19 mm asphalt on 75 mm screed and 150 mm concrete, unplastered.
Floor	Solid *in situ* concrete on earth.
Windows	Single glazing in metal frames without a thermal break.

Surface						*Area (A)* (m^2)			*U* ($W/m^2\,K$)		*AU* (W/K)
Floor	=	14.5	×	8.5	=	123					
		6.0	×	8.5	=	51	174	×	0.58	=	101
Roof	=	6.0	×	8.5	=		51	×	2.16	=	110
Windows	=	12.0	×	2.0	=		24	×	5.7	=	137
Walls	=	48.5	×	4.0	=	194					
		less windows			=	24	170	×	2.32	=	394
							419				742

Thus, the conduction loss, *air to air*, is 742 × 20 = 14.84 kW and $\Sigma(AU)/\Sigma A$ = 742/419 = 1.8 $W/m^2\,K$.

Building insulation

Consideration of thermal insulation has now come to be regarded as an essential routine during the design process for a new building and a matter for earnest consideration in

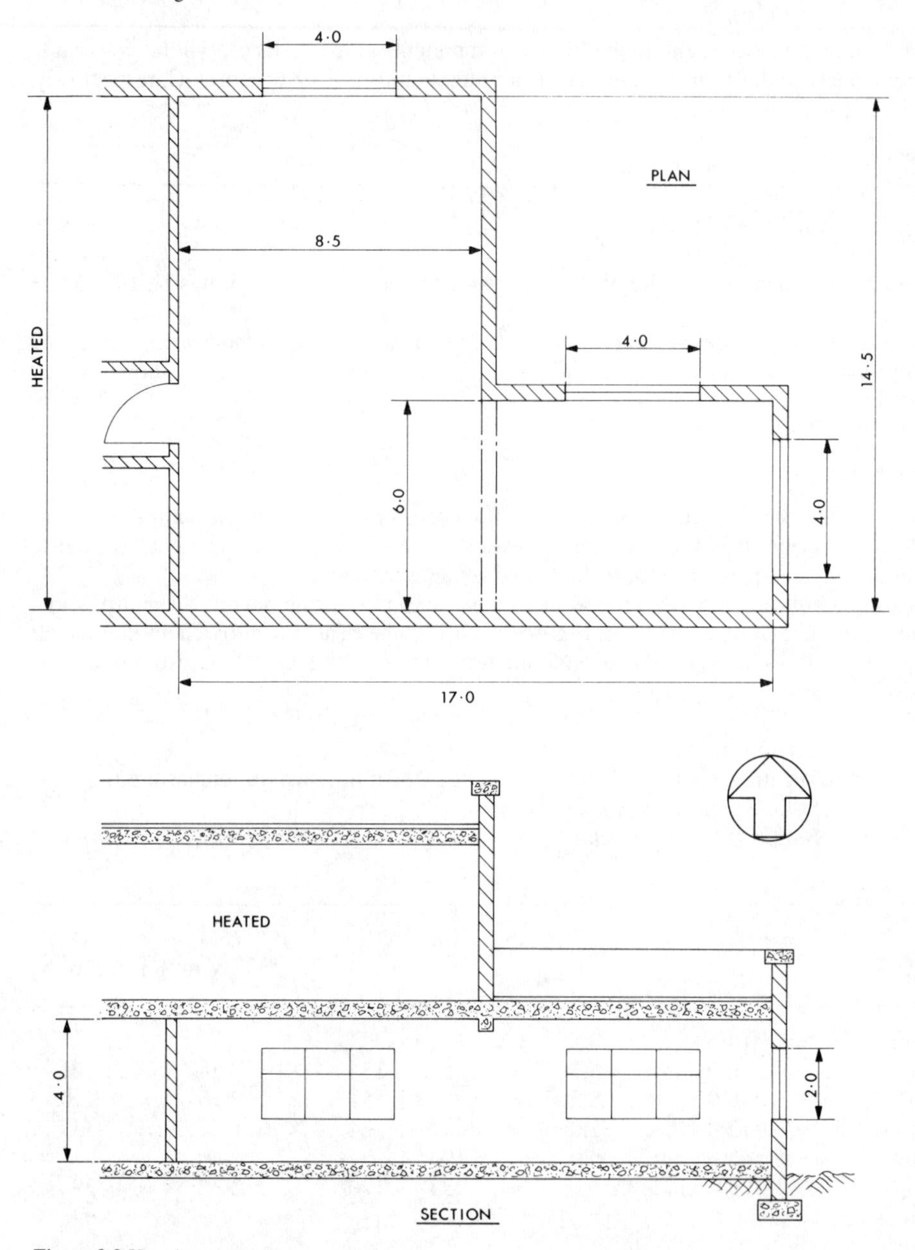

Figure 2.3 Heat loss example

rehabilitation or change of use of any building constructed more than a decade ago. Insulating materials tend to be porous by nature and hence are structurally weak: they are most commonly used as infill material or as an inner skin protected from the weather. When used with flat roofs, insulants are prone to absorb condensation and hence a *vapour seal* or some means of venting is required: recent building methods use a form of construction known as the inverted or 'upside down' roof in which the insulating

Table 2.3 Thermal insulating materials for buildings

Material	*Specific mass (density)* (kg/m³)	*Specific heat capacity* (kJ/kg K)	*Thermal conductivity (k)* (W/m K)
Concrete block, aerated[a]	700	1.0	0.27
Concrete block, aerated[a]	500	1.0	0.18
Concrete block, lightweight	600	1.0	0.19
Fibreboard	300	0.7	0.06
Glass fibre (quilt or mat)	12	0.84	0.04
Glass fibre (slab)	25	1.0	0.035
Mineral fibre (blown)	12	0.84	0.04
Mineral fibre (mat)	50	1.0	0.037
Mineral fibre (slab)	30	1.0	0.033
Phenolic foam (board)	30	1.38	0.018
Polystyrene (expanded, slab)	15	1.38	0.035
Polystyrene (extruded, slab)	15	1.4	0.025
Polyurethane, CFC blown (board)	30	1.4	0.019
Polyurethane, PHA blown (board)	30	1.4	0.021
Urea formaldehyde (foam)[b]	10	1.4	0.031
Vermiculite (granules)	100	0.83	0.065

Notes:
The values quoted here do not allow for any heat bridging. See text for approximate practical adjustments.
[a] Protected, moisture content 3% by volume.
[b] For use only where there is a continuous barrier which will minimise the passage of fumes into the occupied space.

membrane is placed on top of the structure below the water-proofing layer. Nevertheless, it is frequently possible, by the selection of the right materials and techniques, to achieve a high degree of insulation for little or no overall cost.

Table 2.3 provides details of a range of common types of insulating material.

Legislation

Viewed in the light of today (and tomorrow) it may seem strange that it was less than twenty years ago that the influence of *Building Regulations* began to have a direct and substantial bearing upon those aspects of construction which affect the thermal behaviour of a building. The Health and Safety at Work Act (1974), Part III, provided power whereby the Secretary of State may make Regulations with respect to the design and construction of (all) buildings for the purpose of '*furthering the conservation of fuel and power*'. This was a considerable advance in that previous powers existed only under the Thermal Insulation (Industrial Buildings) Act (1957), which provided for a mediocre standard of insulation to factory roofs, and under the Public Health Acts which were unrelated to energy conservation and limited to the construction of dwellings.

The Building Regulations of 1976 (Second Amendment) applied also to dwellings alone and it was not until July 1979 that the Regulations of 1978 (First Amendment) extended that influence to embrace all other categories of building. A later issue of 1985, makes specific reference to a requirement from Part L of Schedule 1 to the effect that '*Reasonable provision shall be made for the conservation of fuel and power in dwellings and other buildings whose floor area exceeds 30 m²*.' The detailed requirements were yet again revised in 1991/2.

Building Regulations (1991) Part L1 (amended 1992)

This issue followed the established pattern of introducing revised criteria to limit heat losses through the building fabric. It imposes maximum U values for the opaque structural elements and places limits upon the proportionate area of single, double or treble glazed windows and roof lights *vis à vis* either the floor, wall or roof as appropriate. Tables 2.4 and 2.5 summarise these general requirements and Figure 2.4 illustrates the definition of a 'semi-exposed' building element.

For dwellings only, the Regulations permit double glazing to be used as an alternative to the levels of insulation required to produce the U values listed in Table 2.5. Thus, if half the windows were to be double glazed then the exposed walls may have a U value of $0.6\,W/m^2\,K$. Further, if all the windows were to be double glazed then the exposed walls may have a U value of $0.6\,W/m^2\,K$, the roof may have a U value of $0.35\,W/m^2\,K$ and the floor may remain uninsulated. This concession (*trade-off*) cannot be used in circumstances where the larger window areas, resulting from the use of double or coated glazing as allowed by footnotes (1) and (2) to Table 2.4, are proposed.

There are two further procedures permitted for buildings in general, including domestic, as alternatives to a simple application of the provisions given in Tables 2.4 and 2.5. The first allows variation, within certain limits, in the requirements for restriction of glazing areas, etc., and for levels of insulation, provided that the rate of heat loss for the building proposed is shown, by calculation, to be not greater than that of a 'notional building', having the same size and shape and which does comply with those requirements. The second method depends upon a comparison of the annual energy use of the building proposed, under normal conditions of occupation and taking account of any heat gains, with that of a similar building designed precisely in accordance with the limiting

Table 2.4 Maximum permissible areas of single glazed windows and roof-lights (Regulations 1992)

Type of building	*Windows*	*Roof-lights*
Dwellings	Windows and roof-lights together 15% of total floor area	
Other residential (including hotels and institutional)	25% of exposed wall area	20% of roof area
Places of assembly, offices and shops	35% of exposed wall area	20% of roof area
Industrial and storage	15% of exposed wall area	20% of roof area

Notes:
1. In any building, the maximum glazed area may be doubled where double glazing is used and trebled where treble glazing is used.
2. Double glazing provided with a low emissivity coating may be considered equivalent to treble glazing. (Emissivity $\leqslant 0.2$).
3. Display windows in shops do not count towards the maximum single glazed area.

Table 2.5 Maximum permissible U values ($W/m^2\,K$) with single glazing (Regulations 1992)

Type of building	*Exposed walls, exposed floors, ground floors*	*Roofs*	*Semi-exposed walls and floors*
Dwellings	0.45	0.25	0.6
All other types	0.45	0.45	0.6

Notes:
1. Any part of a roof having a pitch of 70° or more may have the same U value as a wall.
2. For loft conversions in existing buildings it would be reasonable to have a roof U value of $0.35\,W/m^2\,K$.

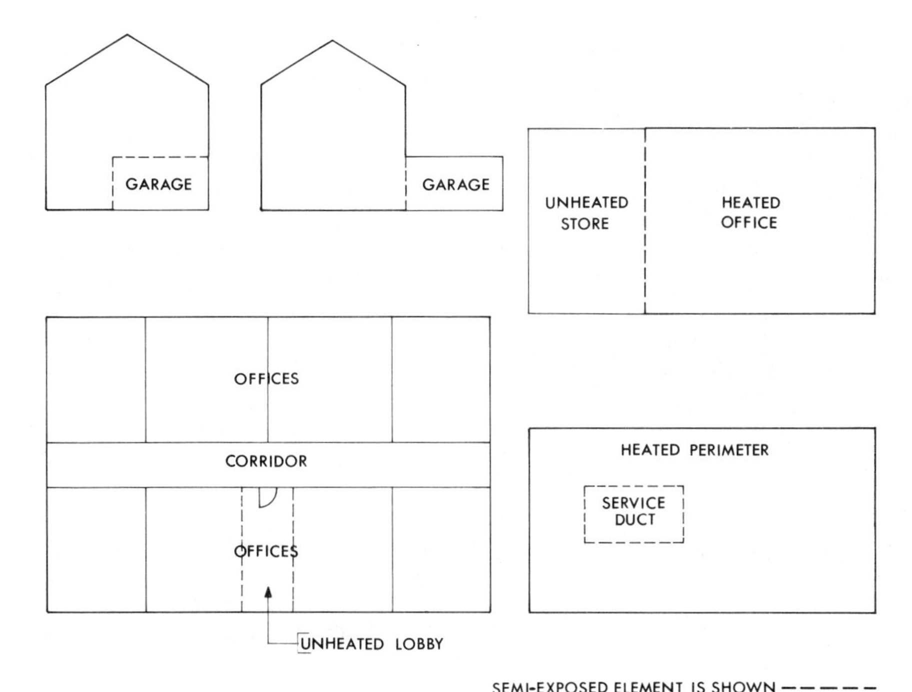

Figure 2.4 Semi-exposed surfaces

conditions tabulated. In this second case, the formal basis for comparison in the case of dwellings is that developed by the Building Research Establishment.* For other buildings, use is made of the *CIBSE Energy Code*.

The Regulations allow that compliance with requirements for a proposed building may be through simple demonstration, by an 'approved person', that the calculations or physical details noted above have been taken into account.

Building Regulations (Draft dated 26 January 1993)

This draft is in circulation for comment and may well be published in its present form, or as revised, before this book goes to press. It would therefore be inappropriate and indeed dangerous to attempt to produce a working *précis* of the requirements of this document save to say that it is by far the most interventionist of the family that has preceded it.

The principal provisions of the draft, in the present context, are:

- It is proposed that double glazing, in the form of a metal framed window with a thermal break, will supersede single glazing as the norm. Compensating provision will be made for the use of windows with treble glazing, and for those having low emissivity coatings or argon filling of the cavity between the panes. Doors are to be included in the same category as windows and roof-lights and be subject to the same requirements for insulation.

* BREDEM Worksheet, *Conservation of Fuel and Power – The 'Energy Target' Method of Compliance for Dwellings.* BRE Report BR150: 1989.

- As a result of adoption of double glazing as the norm, the option of using this as trade off against reduced standards of wall, floor and roof insulation will no longer be available.
- Although criteria for maximum *U* values have not been materially altered and remain much as noted in Table 2.4, an exception is the proposal that roofs in residential buildings (e.g. children's homes, boarding schools, hospitals, hotels, nursing homes and the like) will be subject to an improved requirement if they contain a loft space.
- The scope of the requirements is increased to take account of heat bridges at mortar joints in insulating materials and at lintels, cills and door jambs. Similarly, proposals are made to cover the use of weather stripping of doors and windows and for sealing a variety of gaps in the building fabric, drawing attention in particular to the necessity for sealing dry linings at any and all intersections with floors, ceilings and other surfaces of the building structure.
- As a new requirement, it is proposed that when a material alteration is made to an existing building and/or there is a material change of use, then some of the provisions of the draft Regulations will be applied.
- At some time in the future, energy ratings will be provided for new and converted dwellings. These ratings will originate from a Standard Assessment Procedure (SAP) proposed by Government (100 = good, 0 = poor), purporting to indicate the annual cost to occupiers of providing heating and hot water. It is proposed that, if the rating were to be below 60, the thermal performance of the building fabric should be improved by insulation. The basis for the rating scale is considered by many to be suspect.

The draft proposes three methods for use in demonstrating that a proposed building meets with the Regulations: by calculations based on requirements set out in the text of Part L (the so-called elemental approach); an annual energy use approach as introduced in the 1992 issue and described here under that heading, and lastly a new concept related to a 'target *U* value'. The last of these requires that the weighted average *U* value of the proposed building should not exceed a target *U* value calculated from the expression:

$$U_t = Y(W/S) + Z$$

where

U_t = Target U value (W/m^2 K)
W = Exposed wall area inclusive of windows and small doors (m^2)
S = Total exposed area of fabric, walls, windows, roof and ground floor (m^2)
Y = A factor (e.g. 0.86 for office premises etc.)
Z = A factor (e.g. 0.74 for office premises etc.)

Provision is made that this target *U* value may be adjusted further in the case of domestic buildings to take account of solar heat gains and also, where equipment of this type is provided, of the use of a high-efficiency boiler.

It seems to be intended that compliance with the Regulations is to be dealt with rather more formally than before, a certificate by an 'approved person' being required by the building control authority. In this context, it seems unfortunate that the new draft advocates, for demonstration of the elemental approach, the continuing use of a long-winded and obscure method of calculation, tentatively introduced for a part of the 1992 issue. This employs a number of tables which list factors representing a function of difference in conductivity: these are restrictive in application, an invitation to inaccuracy and neither easier nor quicker to use than the simple sums based upon the established data which they purport to replace.

Atypical construction features

The thermal characteristics of the majority of materials and the various composite construction elements built up from them are not very complex and their thermal transmittance, for steady state energy flow, may be calculated by application of simple arithmetic as has been shown earlier on p. 24. Brief notes on some of the exceptions follow here.

Non-homogenous constructions

These present unique problems arising from discontinuities at corners, at junctions and at heat bridges generally. A rule-of-thumb approach to their solution, which errs on the side of safety, suggests that where insulating blocks are jointed with sand/cement mortar, and where insulation in a timber framed wall is bridged by studs and noggins, the conductivity of the block in the first case and that of the insulation fill in the second, both as quoted in Table 2.3, should be increased notionally by about 15%.

As for hollow blocks, the *Guide Section A3* provides an analytical method of calculation to take account of the resistance of the enclosed air gap (or gaps) but this does not make allowance for mortar infill due to careless workmanship on site. Observation suggests that it is wise to add 50% to any quoted conductivity to allow for the cumulative effect of jointing and of mortar dropped into those cavities.

Ground floors, solid and suspended

In cases where a building covers other than a small area, the heat loss from the floor will occur largely around the perimeter since the temperature of the earth at the centre will, over a period of time, approach that within the room. This situation applies whether the floor is a slab on the ground, a suspended concrete construction or suspended joists and boarding. Whereas previous practice was to select a U value according to the floor shape

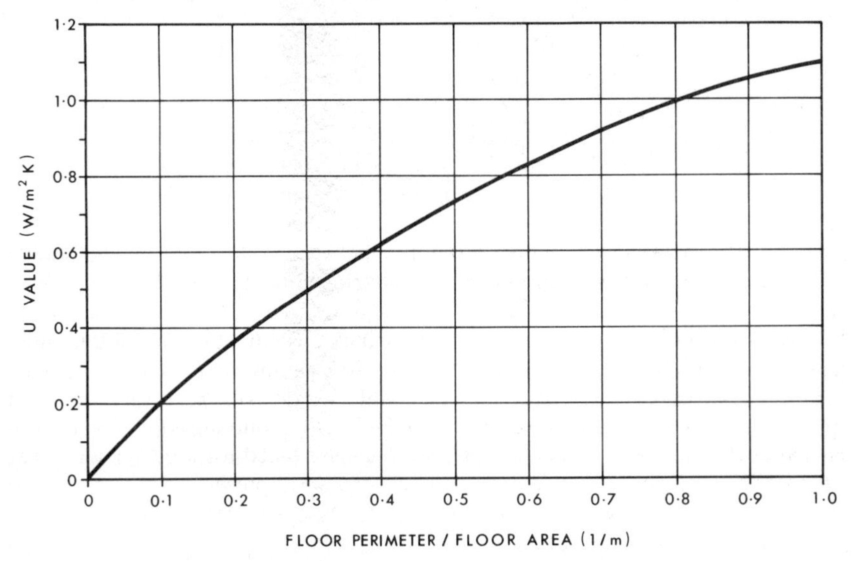

Figure 2.5 U values for floors

and dimensions, it has been shown by Anderson* that a more flexible alternative using the perimeter/area ratio of the floor has much to commend it (see Figure 2.5).

The boundary between the space to be heated and any unheated areas such as garages is included in the perimeter measurement but the unheated area itself is, of course, excluded. Where the perimeter/area ratio is less than about 0.3 m^{-1} then insulation may not be strictly necessary: however, the provision of a horizontal strip about 1 m wide around the exposed perimeter is good practice (Figure 2.8). The data plotted in Figure 2.5 have been adjusted to relate to the air temperature difference ($t_{ai} - t_{ao}$) used for the remaining surfaces.

Glass, glazing and windows

It will be appreciated from Figure 2.1 that, since the thermal resistance of a sheet of ordinary glass is negligible, the U value for single glazing is calculated very simply from the two relevant surface resistances, i.e. $1/(R_{so} + R_{si})$. Similarly, in the case of double and treble glazing, the U value may be calculated from those same surface resistances plus, as appropriate, one or two air gap resistances. Thus, for a normal exposure; a 12 mm wide sealed gap between panes and high emissivity surfaces, the notional values for glazing are:

$$\begin{aligned} \text{Single} &= 1/(0.06 + 0.12) = 5.56\ \text{W/m}^2\,\text{K} \\ \text{Double} &= 1/(0.18 + 0.18) = 2.78\ \text{W/m}^2\,\text{K} \\ \text{Treble} &= 1/(0.36 + 0.18) = 1.85\ \text{W/m}^2\,\text{K} \end{aligned}$$

These values, it will be noted, are not very different from those in the first column of Table 2.6. Both of the remaining examples in this same column relate to double glazing where that surface of the inner pane which faces the gap has been provided with a transparent low emissivity coating ($\epsilon = 0.2$). In the first case, a normal air gap is provided but, in the second, the cavity between the panes has been filled with the inert gas *argon* instead of air. The low emissivity coating and the argon fill each improve the thermal performance of the glazing considerably. A further advance, not yet fully available, is the double glazed window with an integral heating capacity, the inner pane being treated with a metal oxide which can be arranged to act as a resistance element.

The final four columns of Table 2.6 show practical values of the transmittance coefficient for real windows and differentiate between different materials and methods of mounting. The variation from the notional values is due not only to the considerable proportion of the total area available taken up by the frame (often 20%), but also to bridging effects to the structure.

It seems appropriate here to draw a clear distinction between double glazing and double windows. The principal difference is that double glazing mounts two sheets of glass, spaced and hermetically sealed in the works of the supplier in a single composite frame, whereas double windows normally consist of a secondary system of glazing, fitted quite independently, on the room side of an existing outer facade window. In consequence, since ready access to the space between the panes of double windows is necessary for good housekeeping, it follows that the air space between them will not be sealed, the resistance R_a will be reduced by about a quarter and the U values will be increased by about 0.5 $W/m^2\,K$.

* Anderson, B. R., *The U value of ground floors: application to building regulations.* BRE Information Paper IP90, April 1990.

Table 2.6 Thermal transmittance (*U* value, W/m² K) for glazing and windows with frames

	Glazing		*Windows (glazing in frames)*			
			Wood	*Metal*		*uPVC*
Type	*Vertical*	*Horizontal*		*Bare*	*Thermal Barrier*	
Single	5.6	7.1	4.5	5.7	5.4	4.7
Double	2.9	3.6	2.7	3.4	3.1	3.0
Triple	1.9	–	2.1	2.7	2.3	2.4
Double (low emissivity coating)	2.0	2.6	2.0	2.5	2.1	2.3
Double (low emissivity coating and argon fill)	1.7	2.2	1.9	2.4	2.0	2.2

Note: Values quoted are for double glazing sealed at works; frame area 30% for wood and uPVC, 10% for metal; low emissivity coating $\epsilon = 0.2$; normal exposure.

Windows which are double glazed, if of good quality, may be expected to retain their thermal characteristics whereas double windows, opened regularly for cleaning, are subject to misuse and thus a falling performance. As a contra argument, in the refurbishment of existing buildings, the capital cost of replacing windows *and* frames might be such that adding an additional pane is the only practical alternative.

Further, in terms of noise transfer from outside, double windows have an acoustic advantage arising from the wider air gap, provided that the reveals between the panes are treated suitably. A perfectionist might choose to make use of an expensive compromise and fit double glazing at the outer facade, to ensure the optimum thermal advantage, backing this up with acoustically treated reveals and a moveable inner pane!

Wall cavity fill

The thermal resistance of an unventilated air gap, as stated earlier and demonstrated in Table 2.1, is 0.18 m² K/W for building surfaces having a high emissivity, i.e. the majority. However, if that cavity were to be filled with an insulating material having a resistance of, say, $(1/0.035) \times (50/1000) = 1.43$ m² K/W, the *U* value of the structure would be greatly improved. Taking a conventional cavity construction having an outer skin of 105 mm brick and an inner skin of 100 mm lightweight concrete block, finished with 13 mm lightweight plaster, the thermal transmittance with an empty cavity would be 0.98 W/m² K, whereas with insulation therein it would be 0.44 W/m² K, i.e. less than half the former figure.

Features to be noted in respect of filling wall cavities, in this and other ways, Figures 2.6 and 2.7, are:

- Cavities in the walls of new buildings may, of course, be filled in a variety of ways during the construction period using materials such as expanded polystyrene sheet, glass or mineral fibre slabs and the like.
- Cavities in the walls of existing buildings, where these are sound and not overly exposed to damp, may be filled by the injection of materials such as either urea formaldehyde foam or beads of mineral wool or polystyrene, treated with a water

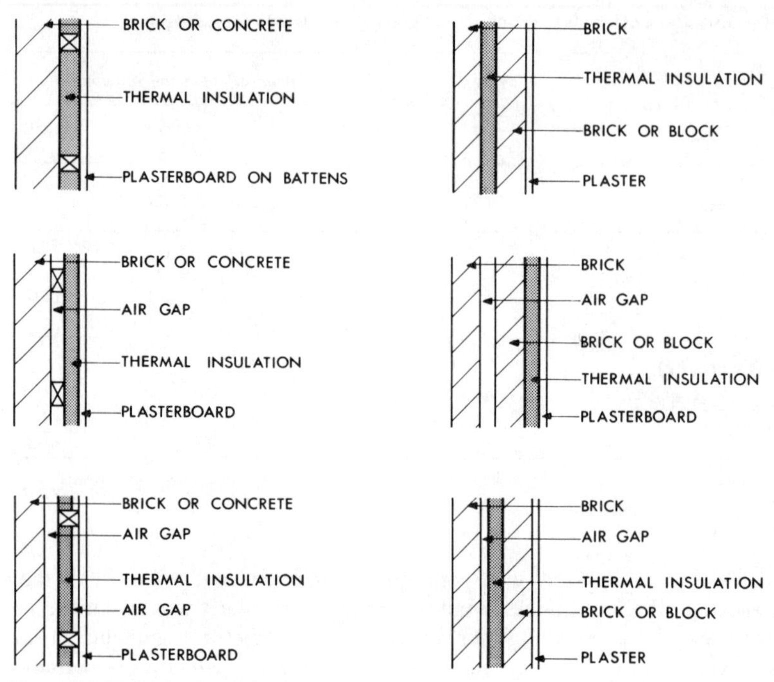

Figure 2.6 Methods of applying insulation to walls

repellant, blown into the cavity by air jet in a manner not dissimilar to that used with foam.

- Insulation will reduce the initial cost of any heating system.
- Cavity fill will reduce noise transmission from dwelling to dwelling as, for instance, at the point of intersection with a party wall.
- When a steady state condition is reached, the inner wall surface temperature will be at a higher temperature than would be the case with an unfilled cavity, thus reducing the risk of condensation.
- If heating is intermittent, the savings theoretically possible may not be achieved since all the heat which has been absorbed in the inner skin during the 'on' period may have been dissipated during the 'off' period.

Safeguards are necessary in adopting this method of treatment, owing to the fact that rain penetration through the outer skin will seek out any discontinuities in the injected or inbuilt material and will thus allow moisture penetration to the inner skin and plaster. It has always been a cardinal rule that mortar dropping onto wall ties and the like must be avoided in the building of cavity walls. It is, in consequence, recommended that application of this form of insulation be entrusted only to firms approved by the British Board of Agrément.

It is, furthermore, essential that the material used be water resistant and resistant to rotting, mould growth and attack by vermin. The principal danger is that of penetration, by driving rain, of the outer skin of the wall construction. Once a cavity has been filled, there is apparently no known method for clearing it completely!

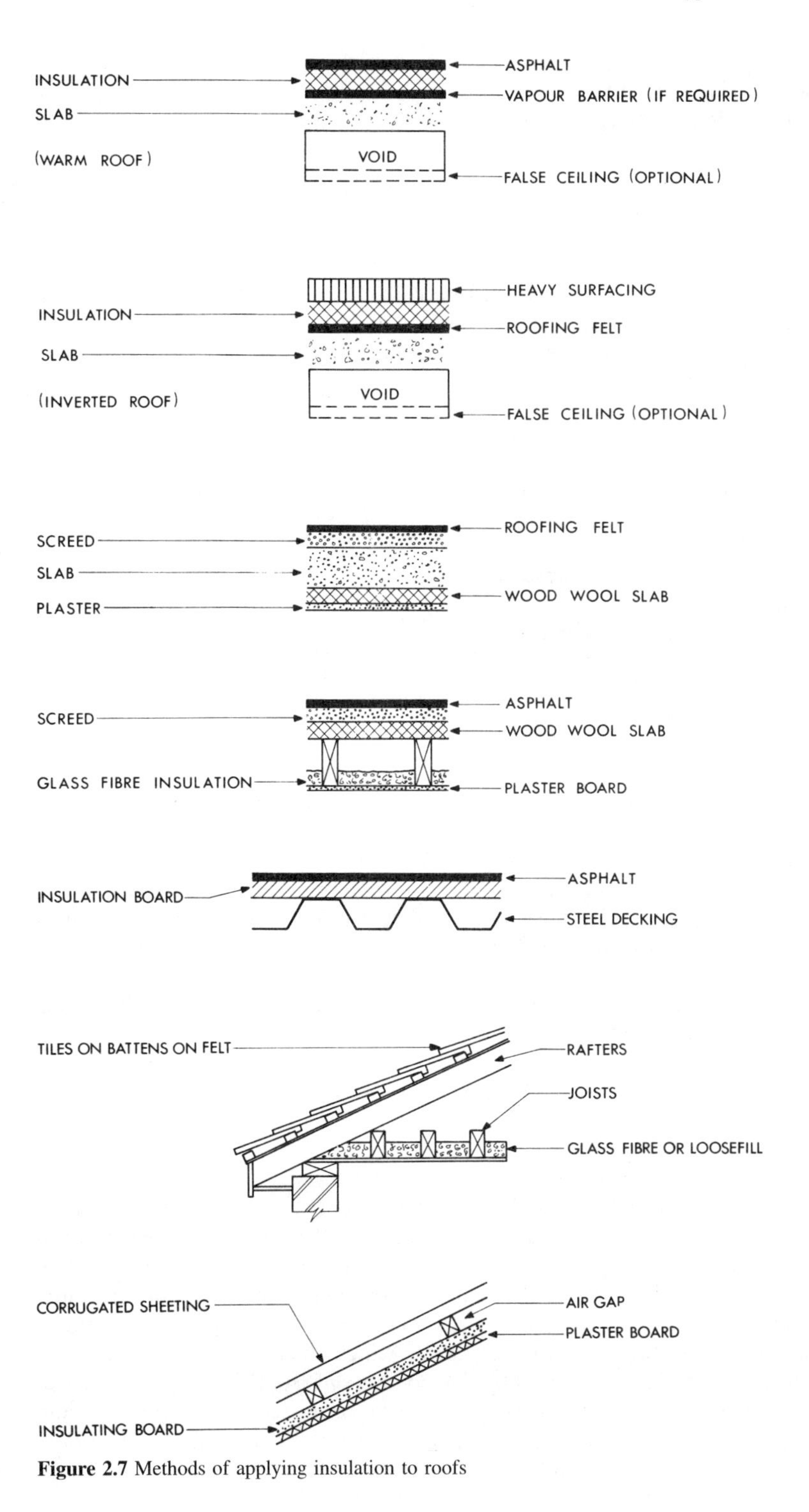

Figure 2.7 Methods of applying insulation to roofs

Flat and sloping roofs

As has been noted previously, attack upon the energy loss through factory roofs was the target of early legislation. Methods of insulation more appropriate to the present day are shown in Figure 2.7.

Floors

It has been noted earlier that the heat loss from a solid floor on earth will occur largely around the perimeter. It follows, therefore, that provision of insulation overall may not be necessary and that a horizontal strip about 1 m wide laid at the exposed edge of the floor, as Figure 2.8(a), will be adequate.

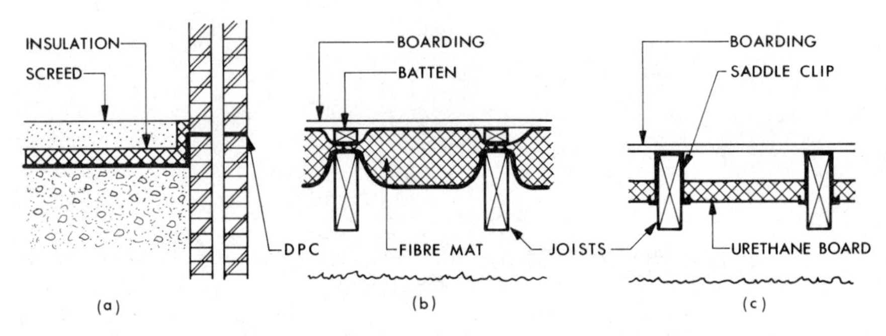

Figure 2.8 Methods of applying insulation to floors

For a timber floor suspended above an enclosed but ventilated air space, insulation is best applied not too far below the actual flooring. One method is to lay an insulating mat or quilt, over the joists and under battens, with a supporting membrane of plastic netting as Figure 2.8(b). Another uses rigid urethane boards, push-fitted between the joists and secured by means of saddle clips as Figure 2.8(c). Some means of access to any piping or wiring below the insulation must not be forgotten.

Application

It is now appropriate to reconsider the earlier example of conduction heat loss in the light of possible improvement to the thermal characteristics of the structural elements. For instance, the floor might be covered internally by laminated panels of 25 mm EPS and 9 mm ply and the roof might be covered externally with 65 mm high density mineral fibre finished with roofing felt and asphalt. The walls might be backed up with a 100 mm medium weight concrete block, thus forming a cavity which could contain a 50 mm glass fibre batt: the internal finish could be plasterboard on battens. As to the windows, those existing might be replaced with double glazing in a metal frame having a thermal barrier. Thus, assuming the same temperature difference as before (20 K), the calculation would be revised as follows:

Surface						*Area (A)* (m^2)			*U* (W/m^2 K)		*AU* (W/K)
Floor	=	14.5	×	8.5	=	123					
		6.0	×	8.5	=	51	174	×	0.40	=	69.6
Roof	=	6.0	×	8.5	=		51	×	0.44	=	22.4
Windows	=	12.0	×	2.0	=		24	×	3.1	=	74.4
Walls	=	48.5	×	4.0	=	194					
		less windows			=	24	170	×	0.43	=	73.1
							419				239.5

Thus, the conduction loss, air to air, is 240 × 20 = 4.8 kW and $\Sigma(AU)/\Sigma A$ = 240/419 = 0.57 W/m^2 K.

Selection of the means to be used in upgrading an existing structure or to produce an acceptable standard in a proposed building may follow the use of calculation routines which have been described earlier (p. 25). It may however be found convenient to make use of Tables 2.7 and 2.8 (which appear at the end of this chapter (p. 63)) when selecting the type and thickness of insulation to be used. The first of these lists U values and the corresponding resistances ($1/U$) for a limited selection of typical *but wholly uninsulated* constructions. It would, of course, be out of the question to provide a fully comprehensive version of such a table in view of the wide and ever-changing availability of composite constructions.

The complementary table (Table 2.8) provides values of resistances for a variety of insulation materials of standard thickness, as manufactured. The method of use of the two tables is best illustrated by a simple example:

> The walls of a new building are to be constructed as Item 7 in Table 2.7. The resistance is 0.549 m^2 K/W but Building Regulations (Table 2.5) require a U value of 0.45 W/m^2 K or a resistance ($1/U$) of 2.222 m^2 K/W.
>
> Resistance deficit = (2.222 – 0.549) = 1.673
> Deduct for an air gap (Table 2.1) = 0.18
> 1.493 m^2 K/W
>
> From Table 2.8, select insulation as, say:
> 40 mm of extruded polystyrene = 1.6 m^2 K/W
> or 35 mm of polyurethane (board) = 1.67 m^2 K/W
> or 30 mm of phenolic foam (board) = 1.67 m^2 K/W

Condensation

If the temperature at the internal surface of any element of the building structure falls below the dewpoint temperature of the air within the space, condensation of the water vapour in the air will take place on that surface. The problem is likely to occur as a result of dense occupancy, of the use of flueless heaters as shown in Table 4.2, and of domestic moisture-producing activities such as cooking, bathing, and clothes or dish washing. In commercial or industrial premises, further hazards exist as a result of steam or vapour producing activities and of the need for a humid atmosphere to suit certain processes.

In the case of windows, the problem may be ameliorated by either the introduction of double or triple glazing or by provision of a warm air convective current to 'sweep' the glass area. Where the building element is solid, however, such as a wall, a ceiling or a floor, the surface may be absorbent to a greater or lesser degree and thus condensation will be less visible although it will still exist. A concrete floor, suspended above a space open to the outside air, for example an office floor above an open car park, may be subject to condensation on the floor surface, even when insulation of apparently adequate quality and thickness has been incorporated in the structure.

Similarly, in multi-storey blocks of flats, conduction from the edge of an exposed balcony has been known to cause problems where the balcony and the slab forming the ceiling of the flat below have been constructed as a single unit. This type of structural bridging, epitomised by concrete or metal mullions and ribs formed across what should have been a weather barrier, without allowance for discontinuity in the thermal sense, has led to many cases of condensation on surfaces within the building.

The present drive for energy conservation and the consequent introduction of higher standards of structural insulation have led to problems arising from condensation actually within the structure. Perimeters used for the more traditional forms of building construction were homogeneous and to a large degree impermeable and had a wide margin of safety inherent in their character. More modern building structures with the required better level of insulation are, in effect, laminates of diverse internal and external finishes with layers of insulation etc. sandwiched between. Inevitably, some of the outer layers outside the insulation remain colder and the risk of intermediate or *interstitial* condensation then arises. A brief treatment of this aspect of the subject follows.

Interstitial condensation

The majority of the materials used in building construction and many insulating materials will allow the movement of water vapour through them by diffusion. If a higher vapour pressure exists on one side of the material than on the other, then movement of moisture will take place, subject only to the *vapour resistance* offered. Table 2.9 provides values of vapour resistivity for a limited range of building and insulating materials.

These values may be thought of as properties parallel to the values of thermal resistivity ($1/k$) listed with them for reference and, like them, to be multiplied by the material

Table 2.9 Vapour resistivities of some common materials

Material	*Density* (kg/m^3)	*Thermal resistivity* (m K/W)	*Vapour resistivity* [N s/(kg m × 10^9)]
Common brick	1700	1.19	35
Dense concrete	2100	0.71	200
Lightweight concrete	600	4.55[a]	45
Dense plaster	1300	2.00	50
Glass fibre slab	25	0.16	10
EPS slab	25	0.16	100

[a]Derived from 0.19 (Table 1.1) × 15% for sand cement joints = 0.22 W/m K and thus 1/0.22 = 4.55 m K/W.

Table 2.10 Vapour resistance of some common films

Material	*Thickness* (mm)	*Vapour resistance* (N s/kg × 10^9)
Polythene	0.05	125
Gloss paint	–	8
Varnish	0.05	5
Aluminium foil	–	>4000

thickness to provide individual resistances. Table 2.10 provides approximate values for the resistance of films. The individual vapour resistances may be added to provide a total for a building structure and, although the total should include for surface and air gap resistances, these are so relatively small that they may be ignored.

As to the rate of vapour transfer, by mass, this may be computed for either an element or a whole structure from:

$$m = \Delta p_v / G$$

where

m = rate of vapour transfer per unit area (kg/m^2 s)
Δp_v = vapour pressure difference across material or structure (Pa)
G = vapour resistance of material or structure (N s/kg)

Application

The application of these details in evaluation of a problem is best illustrated by an example and the curtain wall structure used earlier in Figure 2.2 is repeated here for convenience as Figure 2.9. All the temperature data given there are retained and expanded only to include values of percentage saturation, 100% externally and 58% within the room which were not relevant to the thermal calculation. For the necessary listings of vapour pressures,

Table 2.11 Selected vapour pressures

Temp. (°C)	*Saturated vapour pressure* (Pa)	*Temp.* (°C)	*Saturated vapour pressure* (Pa)	*% sat*	*Vapour pressure* (Pa) *19°C*	*Vapour pressure* (Pa) *20°C*
–4	437	8	1072	72	1591	1694
–3	476	9	1147	70	1547	1647
–2	517	10	1227	68	1504	1601
–1	562	11	1312	66	1469	1555
0	616	12	1402	64	1417	1508
1	657	13	1497	62	1373	1462
2	706	14	1598	60	1329	1415
3	758	15	1704	58	1286	1369
4	813	16	1817	56	1242	1322
5	872	17	1936	54	1198	1276
6	935	18	2063	52	1154	1229
7	1001	19	2196	50	1110	1182

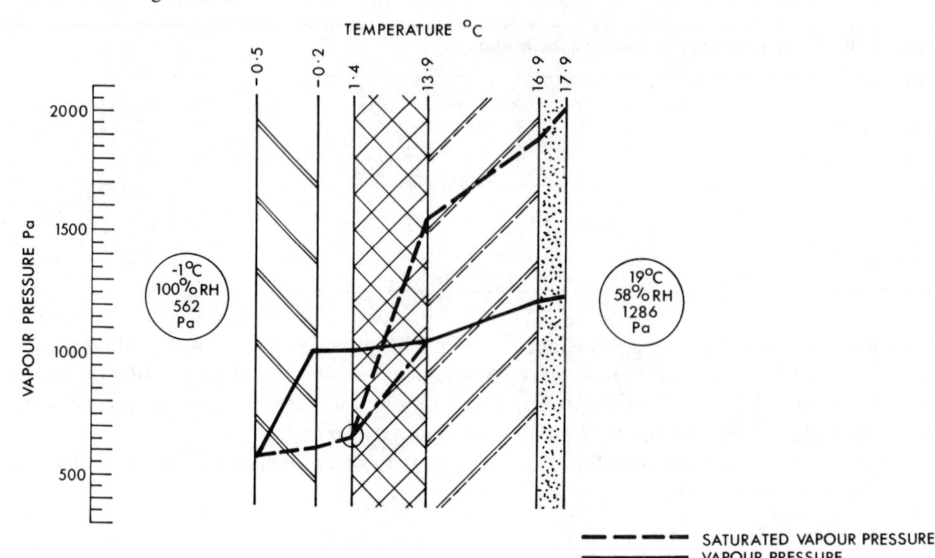

Figure 2.9 Condensation. Vapour pressure gradients

reference would be made in practice to the psychrometric tables in the *Guide Section C1* or some other source of tabulated data. To overcome this difficulty, Table 2.11 gives selected values and interpolation between these, where necessary, will provide figures of sufficient accuracy.

In exactly the same manner as that used to calculate the individual surface temperatures using thermal resistances, (p. 25), the vapour pressure gradient through the structure may be found using the vapour resistances from Table 2.9. The overall resistance will be made up thus:

Inside air to surface resistance		–
13 mm plaster	0.013 × 50 =	0.65
75 mm lightweight block	0.075 × 45 =	3.38
50 mm glass fibre slab	0.050 × 10 =	0.50
25 mm air gap		–
40 mm concrete panel	0.040 × 200 =	8.00
Outside air to surface resistance		–
		12.53 N s/kg × 10^9

The total change in vapour pressure across the structure will be, from Table 2.11, (1286 – 562) = 724 Pa and this must be allocated *pro rata* to the resistances as, for example, to the plaster layer 724 (0.65/12.53) = 38 Pa. This completed, the vapour pressure level at each surface or interface will be available:

Room side surface of plaster	1286 Pa
Interface, plaster to concrete block	1248 Pa
Interface, concrete block to glass fibre	1053 Pa
Interface, glass fibre to air gap	1024 Pa
Interface, air gap to concrete panel	1024 Pa
Outside surface of concrete panel	562 Pa

These values may now be plotted, as shown by the heavy full line in Figure 2.9, to illustrate the vapour pressure gradient through the structure.

To represent the dewpoints at the various surfaces and interfaces, saturation vapour pressure values may be read from Table 2.11 for the various temperatures noted at the top of Figure 2.9 and a gradient of saturation vapour pressure may then be plotted, as shown by a heavy broken line. As will be noted, the two lines cross over one another at vapour pressure values of 1040 and 580 Pa, suggesting that condensation will occur at these points. Since, of course, it is not possible for the vapour pressure level in the structure to exceed that of the saturation vapour pressure, the former will adjust to the gradient shown by the chain dotted line and indicate that excess moisture will condense out at the interface indicated.

The rate of vapour mass transfer from inside the space to the surface at which condensation occurs may be calculated from the sum of the vapour resistances along that path using the expression quoted previously, thus:

$$\begin{aligned} m &= (1286 - 1024)/(0.65 + 3.38 + 0.5) \times 10^9 \\ &= 57.84 \times 10^{-9}\,\text{kg/m}^2\,\text{s} \end{aligned}$$

and, similarly, from the condensation surface to outside:

$$\begin{aligned} m &= (1024 - 562)/(8.00 \times 10^9) \\ &= 57.75 \times 10^{-9}\,\text{kg/m}^2\,\text{s} \end{aligned}$$

The moisture deposited is the difference between these two figures, i.e.

$$m = (57.84 - 57.75) \times 10^{-9} = 0.09 \times 10^{-9}\,\text{kg/m}^2\,\text{s}$$

or, over an assumed period of 60 days as suggested by the Building Research Establishment:

$$\begin{aligned} m &= 60 \times 24 \times 3600\,(0.09 \times 10^{-9}) \\ &= 0.00047\,\text{kg/m}^2 \text{ of wall surface} \end{aligned}$$

As will be appreciated, the external conditions assumed for this example were severe but served to illustrate one method used to trace and evaluate interstitial condensation. For use at the design stage, inside and outside conditions of 15°C with 65° saturation and 5°C with 95° saturation are often recommended and these are rather less searching. For further guidance, the reader is referred to two excellent publications by the Building Research Establishment.*

Air infiltration

The subject matter of the last few paragraphs has related to conduction heat loss through the building fabric, a matter which is capable of examination on a rational basis. As was explained much earlier in this chapter, however, the matter of air infiltration from outside the building must be considered also. Leakage through windows and doors, an upward draught through an unsealed flue, and leakage through the structure itself, particularly in a factory-type sheeted building, will each have an influence.

The importance of air infiltration is that it may well account for as much as half or more of the total heat loss and yet it remains the least amenable to logical and systematic prediction. With improvement to the thermal properties of the building structure through

* Digest 110, *Condensation*, October 1969 (1972 Edition). Report BR 143, *Thermal Insulation: Avoiding Risks*, 1989.

added insulation, air infiltration has increasingly become the dominant component in heat loss. Consequently, temperature guarantees become more difficult to sustain or challenge since any performance test is as much related to the potential for faults in the building as to those in the heating system.

The heat needed to warm infiltration air is calculated using the specific heat capacity of air (at constant pressure) and the specific mass, both at 20°C. Thus, from Table 1.1, the quantity required to raise unit volume through one kelvin is $1.012 \times 1.205 = 1.219$ kJ/m^3 K.

There are two different methods of making an assessment of air infiltration. One is empirical and is based upon the number of times the air volume within a space will be changed in one hour,* this being referred to as the *air change rate*. The second and more specific approach is confined mainly to heavily glazed commercial buildings and, as will be explained later, relates areas of openings such as assumed lengths of cracks around windows and doors, etc., to rates of air flow.

Air change and ventilation allowance

For application to the air change concept and in order to work in units consistent with those used for conduction heat loss through the building fabric, the term *ventilation allowance* is now used, this being related to the air change rate (N):

$$N(1.219 \times 1000)/3600 = (0.339N)\ \text{J/s m}^3\,\text{K} \simeq (N/3)\ \text{W/m}^3\,\text{K}$$

Table 2.12 Natural air infiltration for heat losses (air change rate and ventilation allowance)

	Without weather stripping		*With weather stripping*	
Room or building	*Air change per hour* N	*Ventilation allowance* (W/m^3 K)	*Air change per hour* N	*Ventilation allowance* (W/m^3 K)
Large factory spaces				
Heavy construction				
300 to 3000 m^3	¾	0.25	–	–
3000 to 10 000 m^3	½	0.17	–	–
over 10 000 m^3	¼	0.08	–	–
Unlined sheet construction				
300 to 3000 m^3	1½	0.50	–	–
3000 to 10 000 m^3	1	0.33	–	–
over 10 000 m^3	¾	0.25	–	–
Living spaces and offices				
windows exposed on one side	1	0.33	¾	0.25
windows exposed on two sides	1½	0.50	1	0.33
windows exposed on more sides	2	0.67	1½	0.50
Miscellaneous				
assembly and lecture halls	½	0.17	–	–
circulating spaces	1½ to 2	0.50 to 0.67	–	–
laboratories	1 to 2	0.33 to 0.67	–	–
lavatories	2	0.67	–	–

Note: see DHSS and DES publications giving standards for Hospitals, Schools, etc.

* Although the hour is an unacceptable time interval for calculations made in strict SI units, the use of rates (which are at best no more than informed guesses) in multiples of 0.0003 air changes per second would not endear the concept to any practitioner!

Table 2.12 lists the commonly accepted rates of air change for various building types, together with the associated ventilation allowances. It should be noted that the values listed in this table are confined to the task of assessing natural air infiltration for heat loss calculations. They do not necessarily represent desirable ventilation rates for the comfort of occupants.

Infiltration through window cracks

Turning now to the more specific method of calculation referred to previously, it should be understood that this is not an alternative to the air-change method, which can be applied generally to buildings other than those for which it has been developed. The two influences considered in this method are that due to wind pressure and that due to the chimney or stack effect in a tall building: each of these effects will be dealt with in rather more detail in Chapter 13. In the present context, research* has concluded that wind pressure is the dominant factor, that due to stack effect being small in comparison except in very tall buildings and other unusual circumstances.

Air flow through cracks may be evaluated using the following general expression:

$$Q = C(\Delta P)^n$$

where

Q = air volume flow rate per metre run of window-opening joint (litre/s)
C = window air flow coefficient (litre/m s)
n = flow exponent, representing type of opening.
ΔP = pressure difference across the window (Pa)

The air flow coefficient depends upon the character of the window, being 0.1 where weather stripping has been applied and 0.2 where it has not. The exponent has been evaluated as 0.5 for large openings and 0.66 for cracks around windows and doors.

In strict terms, solutions from this equation are, however, applicable only to a building with an open plan form, air entering on one side having free access to a similar escape route on the other. In instances where the building has many internal partitions which will impede the cross-flow, then the *building infiltration rate* overall may be only 40% of the calculated value. A typical figure for the generality of buildings might be 70% of that produced by the equation.

The pressure difference (ΔP) is a function of the prevailing wind speed which will vary according to the terrain surrounding the building and the height of the windows above ground. Wind speed data published by the Meteorological Office relate to a height of 10 m above ground in open country but may be corrected for other situations by use of the expression:

$$V = V_{\mathrm{m}} k_{\mathrm{s}} z^{\mathrm{a}}$$

where

V = mean wind speed at height z (m/s)
V_{m} = mean wind speed at 10 m in open country (m/s)
z = height above ground (m)
k_{s} = a coefficient representing the terrain (Table 2.13)
a = an exponent representing height (Table 2.13)

* Air Infiltration and Ventilation Centre, *An Application Guide – Air Infiltration Calculation Techniques*, June 1986.

Table 2.13 Values for coefficient k_s and exponent a

Terrain	k_s	a
Open flat country	0.68	0.17
Country with scattered windbreaks	0.52	0.20
Urban	0.35	0.25
City	0.21	0.33

It is sensible, in establishing infiltration values, to adopt a datum wind speed at the higher end of the scale and, for the greater part of the British Isles, an hourly mean speed of 8 m/s is exceeded for only 10% of the time.

The pattern of wind flow over an exposed building takes a form such as that shown in Figure 2.10, but this, of course, is a generalisation since the effect of surrounding buildings and obstructions may well disrupt the pattern in an unpredictable manner, as may the aerodynamics of the building shape. However, it would appear that the sum of the positive pressure on the windward side and the negative pressure on the leeward side approximates to unity and that the pressure difference (ΔP) is thus numerically equal to the velocity pressure of the mean wind speed calculated as above, i.e. $p_v = 0.6\,V^2$ Pa.

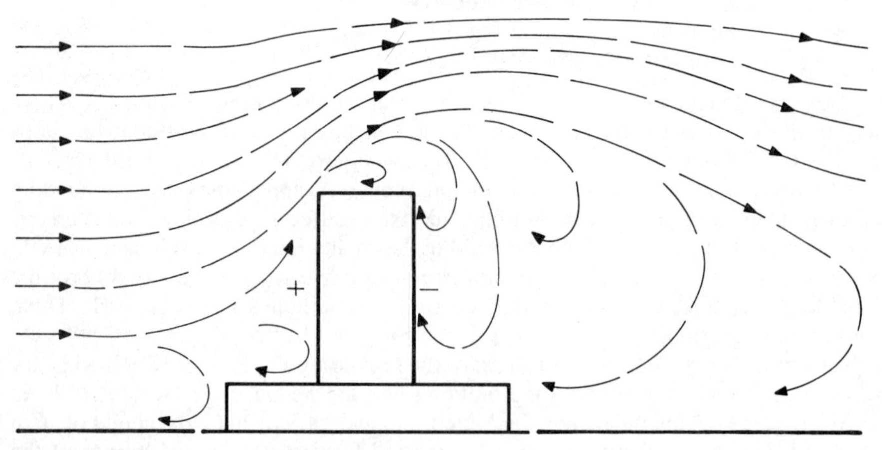

Figure 2.10 Wind currents about a tall building

The method may be summarised as shown in Table 2.14 where the individual values, for a limited range of building heights, are represented as the heat requirements per metre run of window opening joint, for an air temperature difference inside to outside of one kelvin. The table lists basic figures which may require adjustment to take account of:

- The reductions which may be appropriate in a building which is liberally compartmented by partitions, as mentioned above. It must be emphasized that these reductions apply to a whole building and *not* to individual rooms.
- The need in a corner room to make an addition of 50% to the tabulated figure to take account of cross-flow.

Table 2.14 Heat requirements for air infiltration through cracks around windows

Height of building (m)	*Heat requirement per unit length of opening window joint* (W/m K)							
	Without weather stripping				*With weather stripping*			
	Country		*Urban*	*City*	*Country*		*Urban*	*City*
	Open	*With wind breaks*			*Open*	*With wind breaks*		
6	2.43	1.84	1.22	0.75	1.22	0.92	0.61	0.38
8	2.60	1.98	1.35	0.81	1.30	0.99	0.67	0.40
10	2.71	2.10	1.49	0.94	1.36	1.05	0.72	0.47
15	3.00	2.33	1.66	1.12	1.49	1.17	0.83	0.56
20	3.19	2.52	1.82	1.27	1.59	1.26	0.91	0.64
25	3.35	2.67	1.96	1.40	1.68	1.34	0.98	0.70
30	3.49	2.80	2.08	1.52	1.75	1.40	1.04	0.76
40	3.73	3.03	2.29	1.72	1.86	1.51	1.14	0.86
50	3.92	3.21	2.46	1.90	1.96	1.61	1.23	0.95

Note: wind speed 8 m/s, at notional height of 10 m in open country, is exceeded for 10% of time only. Other wind speeds derived from this datum.

A further adjustment, which should properly be applied also to results derived from air-change calculations, makes a notional allowance for stack effect in tall buildings. In winter, the lower floors will have above average infiltration and although the upper floors will have less than the average, the *Guide Section A4* proposes that no allowance should be made, in the latter case, by deduction. The essence of the method is:

- At ground floor level, add the equivalent of 1% for each storey in the building; reduce this addition by 2% for each higher floor, i.e. for a 12-storey building add 12% at ground floor level, followed by 10% at first floor, decreasing to 2% at fifth floor with no addition thereafter.

This last adjustment results from research work* published in 1961 which concluded specifically that:

- Swinging door entrances – for a particular set of conditions – infiltrate about 25 m^3 through a single opening, per person entering or leaving a building, and 15 m^3 per person for a vestibule-type entrance.
- Revolving doors under similar conditions infiltrate about 2 m^3 per person, and motor driven doors about half that amount.
- Adequate heating in winter is required in entrances and vestibules at ground floor level by forced warm air heaters or air curtains, augmented by floor panel heating.

Reference must also be made here to significant problems which have been encountered with buildings clad in curtain walling. As may be seen from Figure 2.11, if particular care

* Min, T. C., 'Engineering concept and design of controlling ventilation and traffic through entrances in tall commercial buildings', *JIHVE*, 1961.

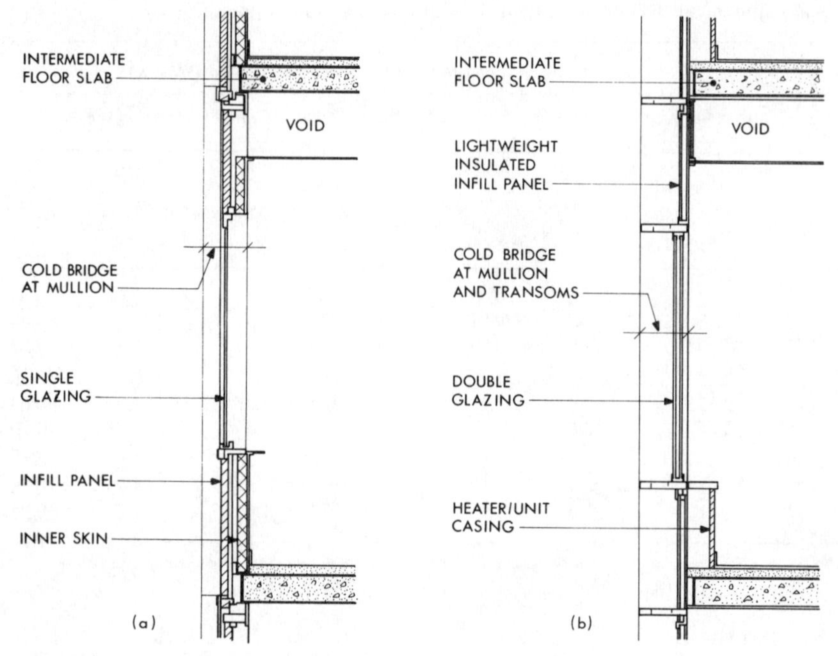

Figure 2.11 Examples of curtain walling

has not been taken to seal the structural joint where the cladding passes the edge beam of the floor slab, leakage can occur and stack effect will cause air to move upwards, floor by floor in cascade. A further example of air flow from an unexpected source can arise where wind pressure is applied to small openings in curtain wall cladding, such as drillings in members provided to drain away condensation, with uncontrolled infiltration to voids above suspended ceilings and consequent excess heat loss. The criteria for tightness of the building envelopes have to some extent been established and the conclusion reached that site workmanship is the major influence.

Application

Taking, once again, the simple space shown in Figure 2.3 and an air-to-air temperature difference of 20 K as before, the infiltration loss on an air-change basis would be:

Room volume

$$14.5 \times 8.5 \times 4 = 493$$
$$6.0 \times 8.5 \times 4 = \underline{204} \quad 697\,\text{m}^3$$

from Table 2.12, allowing for windows on two sides without weather stripping, take 1½ air changes or a ventilation allowance of 0.5 W/m^3 K. Thus,

$$697 \times 0.5 \times 20 = 6970\,\text{W}$$

and, in terms of floor area for unit temperature difference,

$$(697 \times 0.5)/419 = 0.83\,\text{W/m}^2\,\text{K}$$

If the windows were weather stripped, the air-change rate might be reduced to 1 with a ventilation allowance of 0.33 W/m^3 K. Thus,

$$697 \times 0.33 \times 20 = 4600\,\text{W}$$

and, as before,

$$(697 \times 0.33)/419 = 0.55\,\text{W/m}^2\,\text{K}$$

It is of interest to compare the magnitude of these two figures with the total conduction losses for the uninsulated and insulated versions of this building, as calculated previously, i.e. 14 840 and 4800 W respectively. If it be assumed that the windows without weather stripping are associated with the uninsulated building and *vice versa*, an overall saving of 57% is revealed. Furthermore, the infiltration losses are shown to be 32 and 49% of the two respective totals, not insignificant proportions!

As an example of application of the use of Table 2.14, take a typical private office, 5 m × 4 m × 2.5 m high on the third floor of a ten-storey city building (say 25 m high). The single outer wall has glazing on a 1 m module over the full width with a height of 1.5 m. The crack length of 14 m relates to two window modules, not weather stripped. The air-to-air temperature difference is 20 K.

From the table:

Basic heat requirement = 1.4 W/m K
Allowance for stack effect = add 4%
Thus, $14 \times 1.4 \times 1.04 \times 20 = 408\,\text{W}$

This requirement may be compared with that which would have resulted from use of an air-change basis:

Room volume of $5 \times 4 \times 2.5 = 50\,\text{m}^3$

from Table 2.12, take 1 air change or a ventilation allowance of 0.33 W/m K and add 4% for stack effect. Thus,

$$50 \times 0.33 \times 1.04 \times 20 = 343\,\text{W}$$

It will be appreciated that a comparison such as this has no particular significance since addition or reduction in the number of openable windows would not alter the room volume, nor would an increase or decrease in room depth affect the window crack length. In instances where some doubt may exist as to which method of assessment is the more valid, both should be evaluated and the higher result used.

Temperature difference

As will have been noted in the various examples so far provided, the total heat required to maintain a space at the chosen condition is calculated by multiplication of the conduction and air infiltration losses, both in W/K, by a 'temperature' difference between inside and outside. In each of these introductory examples, it was emphasised that air temperature had been used as a simplification.

Inside temperature

In recent years, following the work by Loudon at the Building Research Establishment to which previous reference has been made, the *Guide Section A5* adopted the concept of *environmental temperature* to represent the heat exchange between the surfaces surrounding a space and the space itself. Evaluation of this criterion is dependent upon the configuration of the surfaces and upon the convective and radiant heat transfer coefficients. For the conditions prevailing in the British Isles, environmental temperature may be taken as:

$$t_{ei} = 0.67\,t_{ri} + 0.33\,t_{ai}$$

where

t_{ei} = inside environmental temperature (°C)
t_{ri} = mean inside radiant temperature (°C)*
t_{ai} = inside air temperature (°C)

It is now a well-established convention that inside environmental temperature is used for the calculation of conduction loss and that inside air temperature is used for the calculation of infiltration loss. However, as was noted in Chapter 1, *dry resultant* temperature is that which best represents human comfort and a means to facilitate evaluation of the relationship between the various criteria is thus required. This matter will be discussed later in the present chapter.

Outside temperature

To represent conditions external to a building, outside environmental temperature is the appropriate standard. This is more usually known as the *sol-air* temperature which is a notional scale derived from the combined effect of air temperature and solar radiation, to produce the same rate of heat flow as that which would arise if these influences were considered separately. For winter conditions where overcast skies may be assumed to prevail, the outside environmental temperature will be equal to the outside air temperature.

With regard to the outside temperature adopted in the British Isles as a design datum, this will vary depending upon the precise location, the thermal mass of the building and the overload capacity of the heating plant. It may be of interest to follow the development of selection for suitable outside air temperatures for design purposes.

In 1950, a committee was set up by various institutions having an interest in this subject and the report *Basic Design Temperatures for Space Heating*† examined the frequency of cold spells and the days per annum when a specified *inside* temperature might not be met by designs based upon an external temperature of –1°C. One argument brought out in the report was that a heavily constructed building has sufficient thermal inertia to tide over a period of a few days of exceptionally cold weather, whereas a lightly constructed building has no such inherent ability. Conclusions drawn from the report were refined by Jamieson‡ and further emphasis given to the importance of thermal time-lag related to the overload capacity of the heating system.

* In this context, the various room surface areas times their surface temperatures and the product divided by the sum of the areas.
† Post-War Building Study No. 33. *Choice of Basic Design Temperatures.*
‡ Jamieson, H. C., 'Meteorological data and design temperatures', *JIHVE*, 1954, **22**, 465.

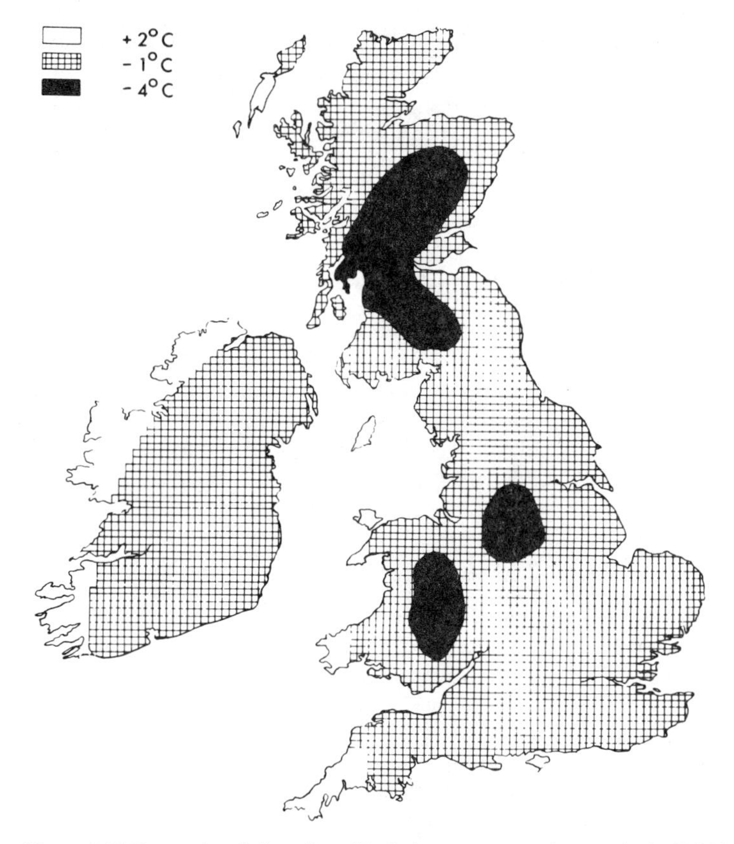

Figure 2.12 Proposed variation of outside design temperature by area in the British Isles

Arising from an analysis of background data to international norms, Billington* proposed that the British Isles be divided into three zones for the purpose of determining standards for structural insulation and for selection of external design temperatures. In place of the earlier single datum of −1°C, the alternatives shown in Figure 2.12 were proposed. A later study† analysed data for eight selected locations in order to establish the number of occasions per annum when the mean temperature falls below given levels. The results of this study are used in the *Guide Section A2* as a basis for recommended winter design conditions, as included here in Table 2.15. In order to take account of the '*heat island*' conditions in streets and around buildings in towns, adoption of a design basis one or two degrees higher than the levels listed in that table would seem to be reasonable for systems serving buildings in city centres.

* Billington, N. S., 'Thermal insulation of buildings', *JIHVE*, 1974, **42**, 63.
† Petherbridge, P. and Oughton, D. R., 'Weather and solar data', *BSER&T*, 1983, **4**, 4.

Table 2.15 Winter external design temperature (CIBSE recommendations)

Location	*External design temperature, t_{ai} (°C)*			
	Low thermal inertia		*High thermal inertia*	
	With overload capacity	*Without overload capacity*	*With overload capacity*	*Without overload capacity*
Belfast	–2.5	–5.0	–1.5	–4.0
Birmingham	–5.0	–7.5	–3.0	–5.5
Cardiff	–3.0	–5.5	–1.5	–4.0
Edinburgh	–4.5	–7.0	–3.5	–6.0
Glasgow	–4.0	–6.5	–2.0	–4.5
London	–3.0	–5.5	–2.0	–4.5
Manchester	–3.5	–6.0	–2.5	–5.0
Plymouth	–1.0	–3.5	0.0	–2.5

Temperature ratios

The relationship between the various temperatures noted under the two immediately preceding headings is a function of the thermal characteristics of the method of heating adopted. The inside air and mean radiant temperatures depend, for a given environmental temperature:

- For *convective heating,* upon the conduction loss alone.
- For *radiant heating,* upon the infiltration loss alone.

As may be appreciated, there are few methods and even fewer types of heat emitting equipment which fall wholly in one or other of these categories. It is therefore necessary that a means be provided which will facilitate the performance of routine calculations. The approach introduced in the *Guide Section A5* offers representative temperature ratios, F_1 and F_2 to group the variables, as follows:

$$F_1 = (t_{ei} - t_{ao}) / (t_{res} - t_{ao})$$
$$F_2 = (t_{ai} - t_{ao}) / (t_{res} - t_{ao})$$

Values of these ratios, as listed in Table 2.16 which is a summary of data included in the *Guide Section A9*, are entered into the composite equation:

$$Q_t = [(F_1 Q_u) + (F_2 Q_v)] (t_{res} - t_{ao})$$

where

Q_t = the total heat requirement, being the sum of conduction and infiltration losses (W)
Q_u = the conduction loss, per kelvin (W/K)
Q_v = the infiltration loss, per kelvin (W/K)

Table 2.16 Values for temperature ratios F_1 and F_2

System type and ventilation index		*Loaded average U value $\Sigma AU/\Sigma A$*							
Type of heat emitter (per cent convective)	*Ventilation index $0.33\ NV/\Sigma A$*	*0.6*		*1.0*		*1.5*		*2.0*	
		F_1	F_2	F_1	F_2	F_1	F_2	F_1	F_2
Forced warm air (100)	All	0.97	1.10	0.95	1.16	0.92	1.23	0.90	1.30
Natural convectors	0.1	0.97	1.08	0.96	1.13	0.93	1.20	0.91	1.26
Convector radiators	0.2	0.97	1.08	0.96	1.13	0.94	1.19	0.92	1.25
(90)	0.3	0.97	1.07	0.96	1.12	0.94	1.19	0.92	1.25
	0.6	0.98	1.07	0.96	1.12	0.94	1.18	0.92	1.24
	1.0	0.98	1.06	0.96	1.11	0.94	1.17	0.92	1.23
Multi-column radiators	0.1	0.98	1.06	0.96	1.11	0.95	1.16	0.93	1.21
Block storage heaters	0.2	0.98	1.06	0.97	1.10	0.95	1.15	0.93	1.21
(80)	0.3	0.98	1.05	0.97	1.09	0.95	1.14	0.94	1.20
	0.6	0.99	1.04	0.97	1.08	0.96	1.13	0.94	1.18
	1.0	0.99	1.02	0.98	1.06	0.96	1.11	0.95	1.16
Two-column radiators	0.1	0.98	1.05	0.97	1.08	0.96	1.12	0.95	1.16
Multi-panel radiators	0.2	0.99	1.04	0.98	1.07	0.96	1.12	0.95	1.16
(70)	0.3	0.99	1.03	0.98	1.06	0.97	1.11	0.95	1.15
	0.6	1.00	1.01	0.99	1.04	0.97	1.08	0.96	1.13
	1.0	1.01	0.98	0.99	1.02	0.98	1.06	0.97	1.10
Single panel radiators	0.1	1.00	1.01	0.99	1.03	0.98	1.05	0.98	1.07
Embedded floor panels	0.2	1.00	1.00	0.99	1.02	0.99	1.04	0.98	1.06
(50)	0.3	1.01	0.99	1.00	1.00	1.00	1.02	0.99	1.04
	0.6	1.02	0.95	1.01	0.97	1.00	0.99	1.00	1.01
	1.0	1.03	0.91	1.02	0.93	1.02	0.95	1.01	0.96
Ceiling panels	0.1	1.01	0.98	1.01	0.98	1.01	0.98	1.01	0.98
Wall panels	0.2	1.01	0.97	1.01	0.97	1.01	0.97	1.01	0.97
(33)	0.3	1.02	0.94	1.02	0.94	1.02	0.94	1.02	0.94
	0.6	1.03	0.91	1.03	0.91	1.03	0.91	1.03	0.91
	1.0	1.05	0.86	1.05	0.86	1.05	0.86	1.05	0.86
High temperature	0.1	1.02	0.94	1.03	0.92	1.04	0.89	1.05	0.86
radiant heaters	0.2	1.03	0.92	1.03	0.90	1.04	0.87	1.05	0.84
(10)	0.3	1.04	0.90	1.04	0.88	1.05	0.85	1.06	0.82
	0.6	1.05	0.85	1.06	0.83	1.07	0.80	1.08	0.77
	1.0	1.07	0.79	1.08	0.76	1.09	0.74	1.10	0.71

Application

Take the same building which has been used in preceding examples, as insulated and with an infiltration rate of 1½ air changes per hour. The internal dry resultant temperature is to be 19°C with an external air temperature of −1°C. Heating is to be by single-panel radiators.

From earlier calculations (p. 37)

$$\Sigma(AU)/\Sigma A = 240/419 = 0.57\ \mathrm{W/m^2\,K}$$

and (p. 46)

$$(0.33\,NV)/\Sigma A = 348/419 = 0.83\ \mathrm{W/m^2\,K}$$

By reference to Table 2.16, and interpolating,

$$F_1 = 1.025 \text{ and } F_2 = 0.93$$

thus

$$Q_u = 1.025 \times 240 \times 20 = 4920\,\text{W}$$
$$Q_v = 0.93 \times 348 \times 20 = 6480\,\text{W}$$

and hence

$$Q_t = 4920 + 6480 = 11.4\,\text{kW}$$

The various internal temperatures may now be established:

$$t_{ei} = (1.025 \times 20) - 1 = 19.5°\text{C}$$
$$t_{ai} = (0.93 \times 20) - 1 = 17.6°\text{C}$$

The last value above has particular significance since, for the building used as an example when heated by panel radiators, an inside air temperature of only 17.6°C is necessary to maintain a dry resultant temperature of 19°C. Had a warm air system been planned, reference to Table 2.16 shows that F_2 would be 1.1, with the result that an inside air temperature of (1.1 × 20) – 1 = 21°C would be required to maintain the same dry resultant temperature of 19°C.

Miscellaneous allowances

The methods outlined for the calculation of heat requirements, as set out in the routines described earlier in this chapter, do not take account of certain elusive factors which may affect the steady state load for a single space or for a whole building. These, in most part, cannot be evaluated other than empirically.

Allowance for height

It appears reasonable to make allowance for the height of a heated space, bearing in mind that warm air rises towards the ceiling, creating a temperature gradient actually within the space. Thus, when heat is provided to maintain a chosen temperature in the lower 1½–2 m of height, it follows that a higher temperature must exist near the ceiling or roof. In consequence, the conduction loss there will be greater, inevitably, through the surfaces of ceiling, roof, upper parts of walls and windows, etc.

This effect will be greatest with a convection type system, i.e. one which relies upon the movement of warmed air to heat the space, as is the case with forced and natural convectors and most forms of conventional radiators. Where the radiant component of system output is greater, as is the case with metal radiant panels, heated ceilings and floors, etc., a much more uniform temperature exists over the height of the space and, in the case of floor heating, there is virtually no temperature gradient whatsoever.

This situation is illustrated in Figures 2.13 and 2.14, the latter showing how a warm air system having a discharge at low level produces a temperature gradient which is much less

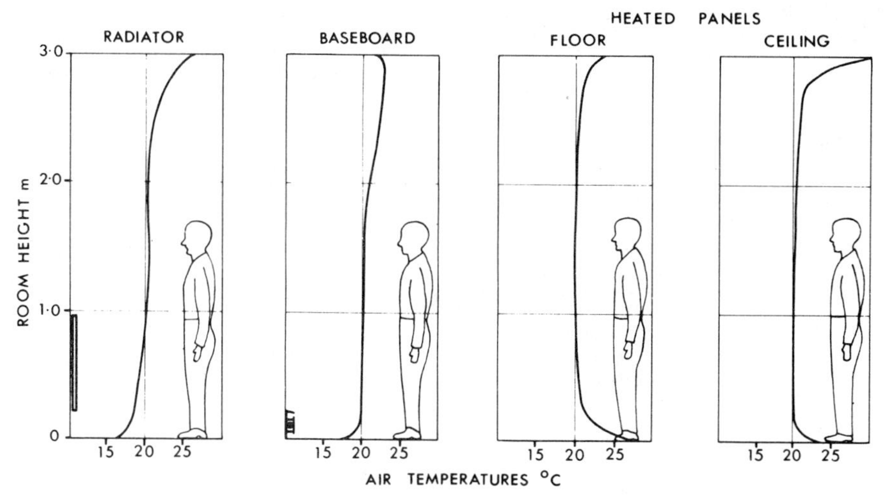

Figure 2.13 Vertical temperature gradients assuming heated rooms above and below

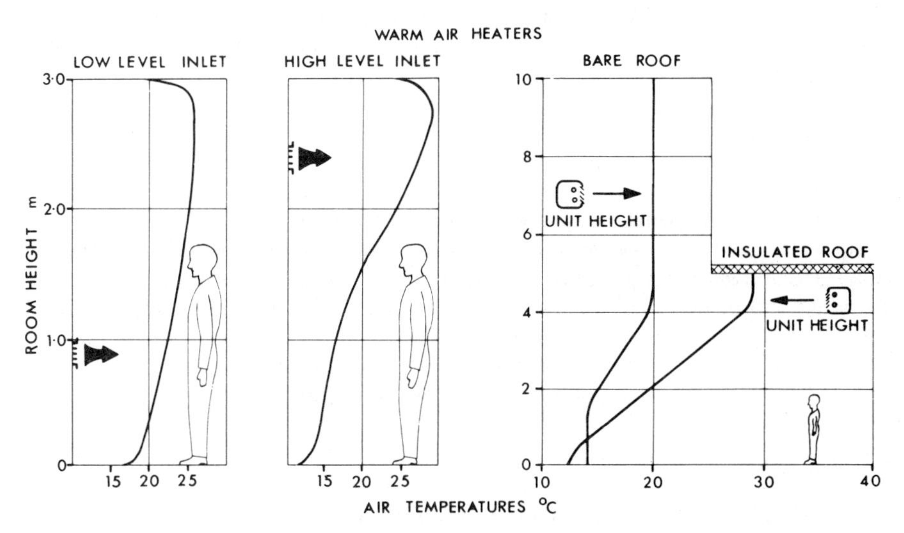

Figure 2.14 Vertical temperature gradients with warm air heating

than one discharging at high level due to better mixing with room air. Where radiators and natural convectors are to be used, much can be achieved by correct disposition of these items below large cooling surfaces such as windows. It might, in fact, be said that a system of this type which produces too large a temperature gradient has been badly designed. To make allowance for the effect of room height, the *Guide Section A9* proposes the use of multiplying *height factors*, as listed in Table 2.17.

Table 2.17 Height factors

Type of system	*Type and distribution of heaters*	*Percentage addition for height of heated space*		
		5 m	*5–10 m*	*over 10 m*
Mainly radiant	Warm floor	nil	nil	nil
	Warm ceiling	nil	0–5	–[a]
	Medium and high temperature cross radiation from intermediate level	nil	0–5	5–10
	Ditto, downward radiation from high level	nil	nil	0–5
Mainly convective	Natural warm air convection	nil	0–5	–[a]
	Forced warm air cross-flow from low level	0–5	5–15	15–30
	Ditto, downward from high level	0–5	5–10	10–20

[a] Not applicable to this application.

Whole building air infiltration

The air-change and infiltration data provided in Tables 2.12 and 2.14 relate to allowances appropriate to individual spaces. Where such spaces exist in a building more than one room 'thick', it follows that outside air entering rooms on the windward side is warmed there before passing through rooms on the leeward side and leaving the building. It may be argued, therefore, that although the listed figures should be applied to all individual rooms, the wind direction being unknown, a deduction could be made, in determining the *total* infiltration for the whole building, for this use of 'second-hand' air changes. The *Guide Section A4* includes a proposal which permits the deduction to be quantified but the validity of the assumptions made must be questioned and it is not included here.

In a different category, however, is the parallel proposal that, when a building is unoccupied, the allowance for air infiltration might be reduced. In circumstances where an office building will be occupied for working hours only during a five day week, plus an allowance for cleaning, etc., and furthermore where security is such that it may be assumed that both outer and inner doors will be closed out of those hours, then a reduction of air infiltration rates, by some proportion such as half, seems reasonable.

Internal heat gains

Passing reference has already been made to fortuitous heat gains and their potential to make a contribution towards keeping a building warm in winter and thus to reduce the capacity of any heating system which may be provided. Reports* by the Building Research Establishment provide a variety of interesting facts regarding the probable annual total of such input, as summarised in Table 2.18. The total of these items approximates to something over a third of the theoretical annual heat requirement of a dwelling so occupied.

In summer, of course, such gains are a penalty to be countered rather than a contribution to be welcomed and they will consequently be dealt with in more detail in the following

* *BRE Domestic Energy Model, Background, Philosophy and Description.* BRE Digest 94: 1985.

Table 2.18 Approximate heat input per annum to a dwelling from incidental sources

Source	*Heat input (GJ/annum)*
Two adults, one child (body heat)	5.9
Radiation from sun (15 m^2 glazing)	13.0
Cooking	
gas	4.3
electric	3.4
Electrical appliances (lighting; TV; washing machine, etc.)	3.4–8.2
Losses from water-heating pipes and appliances, etc.	10–30

chapter. For application to a commercial building, the level of reliance which may be placed upon availability may be used to separate winter heat gains into categories:

Reliable. Industrial or office machines operating permanently; electric lighting running permanently; occupied for 24 hours at a constant level.

Relatively reliable. Industrial and office machines that will operate continuously during working hours; electric lighting related to occupancy; permanent staff working to an established attendance routine.

Unreliable. Solar radiation and heat gains from industrial or office machines operated intermittently; electric task lighting; random occupation.

In the case of electric off-peak storage systems using room heaters, it has been suggested that full account should be taken of the 24 hour mean contribution from most of these sources on a 'design day', as related to the thermal capacity of the structure, and this proposition will be discussed further in Chapter 5. However, for conventional heating systems, total dependence upon other than the items listed in the first of these categories would be unwise. As to the second category, where a contribution may be absent over cold weekends and winter holidays, no allowance can be made if comfort temperature in the space is to be recovered in a reasonable time period prior to occupation. This is not to deny that use may be made of all such heat sources, but rather to suggest that a well-designed system will have sufficient capacity to cater for the full design load while retaining the ability to respond to any fortuitous contribution when this may be available.

Temperature control

Although this subject is dealt with in some detail in Chapter 22, it is apposite here to emphasise that the ideal system is one where separate automatic control of temperature is available in each room or space. With such a system, heat emitting elements having the capacity to bring the room up to temperature may be provided and then, as heat from occupants, lighting, machines and the sun begins to have effect, the output of the system is reduced to prevent overheating and save unnecessary fuel consumption. The availability today of inexpensive thermostatic radiator control valves has made room-by-room response an economic possibility.

Continuous versus intermittent operation

A building occupied for 24 hours each day, such as a hospital, a police station or a three-shift factory, will require that a continuous supply of heat is available. Most buildings, however, are occupied for a limited number of hours each day and, moreover, not for all days of the week: offices, schools and churches are examples of very different patterns of use. It is, in consequence, a matter of considerable importance as far as energy consumption is concerned to establish whether a pattern of intermittent operation will provide satisfactory comfort conditions for the whole of the occupied period. The manner in which proportions of total energy consumption vary with different operational régimes is shown in Table 2.19.

The *thermal response* or *time lag* of a building *vis à vis* a heat supply is perhaps best visualised initially by considering two extreme cases, the first being a lightly constructed building, insulated but overglazed, which will have minimal thermal capacity and hence a relatively short time lag. At the other extreme, a medieval or similar building with thick masonry walls, negligible window areas and massive construction throughout will respond very slowly to a change in external conditions, have a considerable thermal capacity and thus have an extended time lag taking as much as a week or more of heat input to recover from breakdown.

The first building would cool quickly overnight and, without heating, the internal temperature next morning would approximate to that outside. Furthermore, during a clear cloudless night, there might be a risk of condensation as the temperature of parts of the structure fell below the air temperature. Continuous heating would obviously be uneconomic in this case but, equally, a fully intermittent régime would only be adequate in the depth of winter if a lengthy pre-occupancy boost period were provided. The optimum solution might be a compromise with reduced temperature night-time heating, and a minimal pre-occupancy boost.

Conversely, the air temperature within the heavy building would fall only marginally during a winter night without a heat supply but there would be a considerable time delay

Table 2.19 Per cent annual fuel consumption for various operational regimes

	Internal temperatures maintained					
	20°C for 24 h		*20°C for 10 h 15°C for 14 h*		*20°C for 10 h 10°C for 14 h*	
Period	*Day*	*Night*	*Day*	*Night*	*Day*	*Night*
January	51	49	60	40	72	28
February	50	50	60	40	73	27
March	49	51	60	40	76	24
April	48	52	61	39	86	14
May	44	56	65	35	99	1
September	43	57	78	22	100	0
October	47	53	66	34	100	0
November	50	50	62	38	80	20
December	51	49	60	40	73	27
Average over season	48	52	64	36	84	16

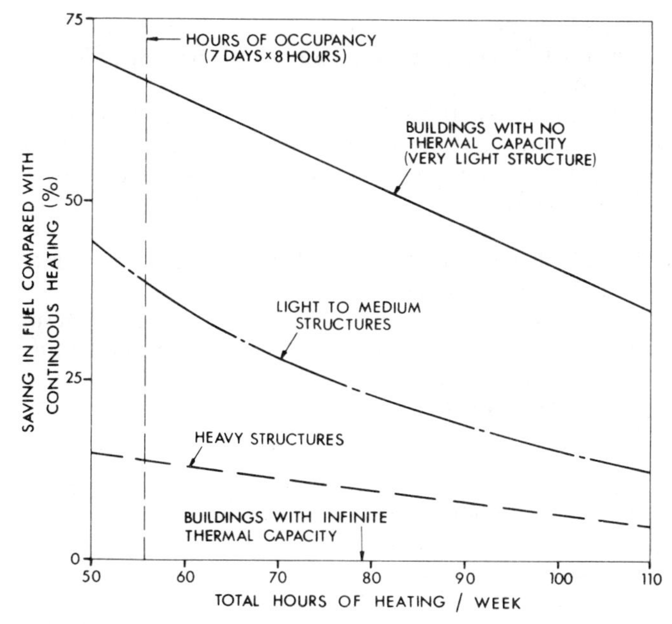

Figure 2.15 Comparative fuel savings for different weekly heating periods, indicating the effect of thermal capacity

the following morning before the daytime temperature was fully recovered. Intermittent heating should, however, provide a reasonably satisfactory performance, particularly if there were an insulating lining added to the walls. Continuous heating, at a marginally lower but constant level over 24 hours, would also meet requirements. Economically, there might be little to choose between the alternative methods in this case.

Between these two extremes lie all the buildings likely to be met in practice, be they of light, medium or heavy construction. Unfortunately, the choice between continuous or intermittent operation does not depend wholly upon the thermal response rate of the building and other factors must be taken into account. These include the proportion of the 24 hours (and perhaps the week also) during which the building is occupied, which may vary as a result of shift-work or flexible working hours in offices: further, the type and density of occupancy and, of course, the characteristics of the heating system will have an influence. Some of these effects have been the subject of a report by BSRIA* following extensive examination of a number of buildings of various types. Figure 2.15 provides a summary which illustrates the order of saving in fuel consumption which may be achieved as a result of intermittent operation.

Preheat capacity

When a heating system is operated intermittently, the temperature within the building will fall during the period of system shut-down and, in order to restore this to the proper level by the time the occupants arrive, a preheating period is necessary. Since the outside

* Billington, N. S., Colthorpe, K. J. and Shorter, D. N., *Intermittent Heating* HVRA (now BSRIA) Report 26, 1964.

temperature varies during the winter nights, so will the time required for preheat increase or decrease and control equipment which provides means for dealing with this variation is described in Chapter 22 (p. 620).

Some excess capacity in the heating system is desirable in order to reduce the length of the preheating period and, with the improved standards of insulation and lower rates of air infiltration common in modern buildings, larger plant margins are required than was the case hitherto. During the preheating period:

- The conduction loss is reduced by the use of curtains or shutters and closed windows and doors minimise the loss due to air infiltration.
- The effective output from the system is increased above the design capacity as a result of a higher output from heat emitting equipment due to the lower temperature of the inside air.
- The excess plant capacity allows operation at an elevated temperature.

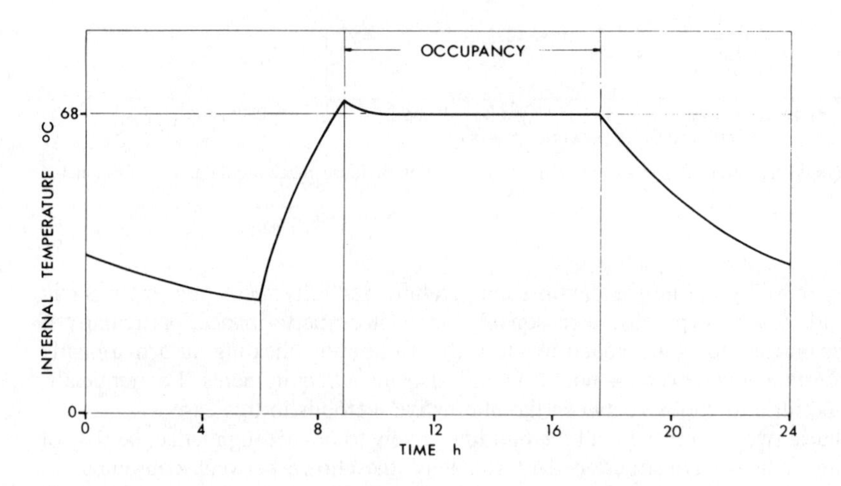

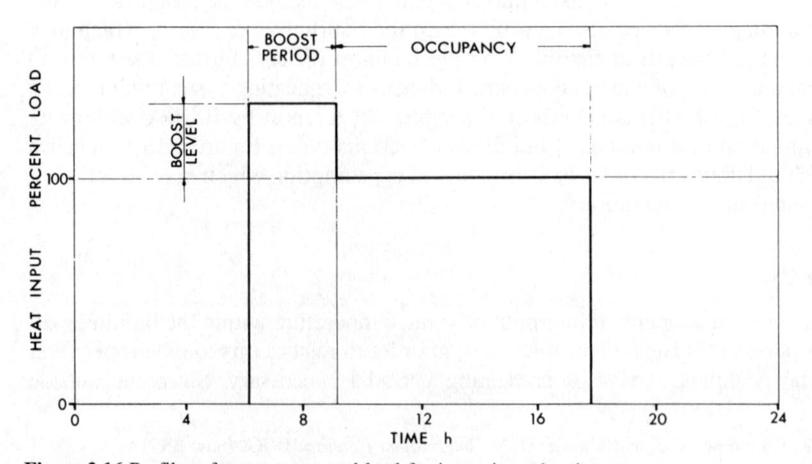

Figure 2.16 Profiles of temperature and load for intermittent heating

The combined effect of these items is illustrated in Figure 2.16, which represents a normal heating and cooling curve for a building heated intermittently. The effect of the changes noted in the first two groups can be quite significant, as illustrated by an example related, once again, to the simple space shown in Figure 2.3.

If shutters were used overnight at the windows and the air change reduced by half, then the heat loss might be reduced from 700 W/K to about 450 W/K. Taking an overnight inside temperature of, say, 10°C at the start of preheating, an average during that period might be 14°C; further, if the mean water temperature in the radiators were raised by as little as 5°C, then the dual effect would be to increase their output by about 20%. This is, of course, a considerable simplification of a complex problem but it serves to show that the overall result might be that a preheat capacity of more than twice normal system output could be available.

Preheat and plant size

It will be obvious from what has been said that the thermal characteristics of both building and heating system, the time lag of the former and the response rate of the latter, have a profound effect upon the preheat time required to restore conditions of comfort in a space after a period of shut-down. The *Guide Section A9* presents data which allow preheat time to be established relative to values of the other variables: a rounded up summary is presented in Table 2.20.

Table 2.20 Variations in preheat times

	Plant size ratio[a]			
	Heavyweight structure		*Lightweight structure*	
Preheat (h)	*Quick response system*	*Slow response system*	*Quick response system*	*Slow response system*
1	3.0	–	2.1	–
2	2.5	–	1.6	2.3
3	2.1	2.6	1.5	1.7
4	1.9	2.3	1.4	1.5
5	1.7	1.9	1.2	1.4
6	1.5	1.6	1.2	1.2

[a] Ratio of maximum plant output to design load at 20°C inside/outside. (Maximum plant output includes inherent capacity of system.)

It has been shown that there is little advantage to be gained by providing an oversized conventional system having a capacity much greater than that necessary to meet the calculated heat loss, with intent to reduce the preheat period. This is due to the fact that a system, made up of boiler plant, pipework and heat emitters, also has a time lag which increases with size. Margins on conventional heat emitters should be confined therefore to a nominal 10% at most in order to make allowance for the seemingly inevitable poor standard of building maintenance.

As to the influence of system type upon energy consumption with intermittent operation, it has been shown* that systems having a small thermal capacity and a rapid response rate, such as those using warm air, are able to achieve greater economies than those subject to time lag of any significance. This is an obvious conclusion and points to the generalisation that with buildings of light construction, greatest economy in energy consumption is achieved with intermittent operation of such systems, but only when the control arrangements provided are sufficiently responsive and are thus able to prevent over-run. Buildings of more solid construction are able to achieve reasonable economies with a steady supply of heat and a less sophisticated level of control. It follows, therefore, that while warm air, convector and radiator systems are well suited to intermittent operation, embedded panels in ceilings and floors are not so suitable.

Preheat for direct and storage heating

In the case of direct heating appliances having a fixed rate of input, some additional capacity is needed to provide for a rapid preheat and this may, to some extent, be available inherently since mass-produced equipment of this type is manufactured in multiples of 2 or 3 kW only, and selection on a generous basis should thus suffice. Heat output from all storage-type systems using room heaters is continuous, even though heat input is intermittent, and thus facilities for controlled preheat in the present sense need not be considered.

Steady state and dynamic response

This chapter would not be complete without a note to introduce the alternative and more thorough methods of establishing not only the room-by-room heat requirements of a building but also preheat times and other operational routines best suited to energy conservation.

Steady state calculations for winter heating, while adequate to produce results satisfactory for all but the most unorthodox forms of building construction, fail to provide data enabling full advantage to be taken of all the sophisticated control systems now available. The essence of that failure lies with the fundamental assumption that heat exchange is a function of values of temperatures inside and outside a building taken coincidentally in time. Except in the case of token weather barriers such as glass or thin sheeting, structural elements have both mass and thermal resistance which together introduce delay – the time lag which has been mentioned previously.

Computer based thermal modelling techniques enable transient heat flow to be analysed and the dynamic response of the building structure to be assessed, on an hour-by-hour basis, a task which would be infinitely tedious, if possible at all, using manual methods. For such non-steady state calculations, knowledge of other properties of the elements making up that structure are required, in addition to the U value. The more important of these are:

Admittance (Y value). This is a measure of the ability of a surface to smooth out temperature variations in a space and represents the rate of energy entry *into* a structure rather than that of passage *through* it. The units are $W/m^2 K$.

* Dick, J. B., 'Experimental and field studies in school heating'. *JIHVE*, 1955, **23**, 88.

Decrement factor (f). This represents the ability of a structural element to moderate the magnitude of a temperature change or swing at one face before this penetrates to the other. It is largely a function of the thickness of the element and is dimensionless.

Surface factor (F). This relates the admission and absorbtion of energy to the thermal capacity of a structural element. It is dimensionless.

For values of these factors and further details of these more probing calculations, reference should be made to the *Guide Section A3*.

Table 2.2 Typical transmittance coefficients (*U* values) for 'normal exposure' (including allowance for moisture content, as appropriate)

Construction (dimensions in mm*)*	*U value* (W/m^2 K)
External walls	
Solid walls, no insulation	
600 stone, bare	1.72
600 stone, 12 plasterboard on battens	1.35
105 brick, bare	3.27
105 brick, 13 dense plaster	3.0
220 brick, bare	2.26
220 brick, 13 dense plaster	2.14
335 brick, bare	1.73
335 brick, 12 plasterboard on battens	1.53
Dense concrete walls	
19 render, 200 concrete, 13 dense plaster	2.73
19 render, 200 concrete, 25 polyurethane, 12 plasterboard on battens	0.71
Pre-cast dense concrete walls	
80 concrete panel, 25 EPS, 100 concrete, 13 dense plaster	0.95
80 concrete panel, 50 EPS, 100 concrete, 12 plasterboard	0.53
Brick/brick cavity walls	
105 brick, 50 air space, 105 brick, 13 dense plaster	1.47
105 brick, 50 blown fibre, 105 brick, 13 dense plaster	0.57
Brick/dense concrete cavity walls	
105 brick, 50 air space, 100 concrete block, 13 dense plaster	1.75
105 brick, 50 EPS, 100 concrete block, 13 dense plaster	0.55
Brick/lightweight aggregate concrete block	
105 brick, 50 air space, 100 block, 13 dense plaster	0.96
105 brick, 50 UF foam, 100 block, 13 dense plaster	0.47
Brick/autoclaved aerated concrete block	
105 brick, 50 air space, 100 block, 13 lightweight plaster	1.07
105 brick, 25 air space, 25 EPS, 150 block, 13 lightweight plaster	0.47
Timber frame walls	
105 brick, 50 air space, 19 plywood sheathing, 95 studding, 12 plasterboard	1.13
105 brick, 50 air space, 19 plywood sheathing, 95 studding with mineral fibre between studs, 12 plasterboard	0.29

Table 2.2 *Continued*

Construction (dimensions in mm*)*	*U value* (W/m^2 K)
Party walls and partitions (internal)	
Brick	
13 dense plaster, 220 brick, 13 dense plaster	1.57
13 lightweight plaster, 105 brick, 13 lightweight plaster	1.89
Dense concrete block	
13 dense plaster, 215 block, 13 dense plaster	2.36
Lightweight concrete block	
12 plasterboard on battens, 100 block, 75 air space, 100 block, 12 plasterboard on battens	0.62
Flat roofs	
Cast concrete	
Waterproof covering, 75 screed, 150 concrete, 13 dense plaster	2.05
Waterproof covering, 100 polyurethane, vapour barrier, 75 screed, 150 cast concrete, 13 dense plaster	0.22
Timber	
Waterproof covering, 19 timber decking, ventilated air space, vapour barrier, 12 plasterboard	1.87
Waterproof covering, 35 polyurethane, vapour barrier, 19 timber decking, unventilated air space, 12 plasterboard	0.52
Pitched roofs	
Domestic and commercial	
Tiles on battens, roofing felt and rafters, bare plasterboard ceiling below joists	2.6
Tiles on battens, roofing felt and rafters, 50 glass fibre mat between joists, plasterboard ceiling below	0.6
Tiles on battens, roofing felt and rafters, 100 glass fibre mat between joists, plaster ceiling below	0.35
Industrial	
Corrugated double-skin decking with 25 glass fibre mat	1.1

Table 2.7 Transmission coefficients (U values) and resistances for a range of typical *uninsulated* constructions for 'normal' exposure (for use in conjunction with Table 2.8)

Construction (*dimensions in* mm)	U (W/m^2 K)	$1/U$ (m^2 K/W)
1. 19 render, 220 no fines concrete block : ■ : 12 plasterboard on battens	2.05	0.488
2. 19 render, 220 dense concrete block : ■ : 12 plasterboard on battens	2.40	0.416
3. 80 dense concrete panel : ■ : 100 dense concrete block, 13 dense plaster	2.68	0.373
4. 80 dense concrete panel : ■ : 100 dense concrete block, 13 dense plaster	3.09	0.324
5. 220 solid brick : ■ : 13 dense plaster	2.14	0.468
6. 220 solid brick : ■ : plasterboard on battens	1.93	0.517
7. 105 brick : ■ : 105 brick, 13 lightweight plaster	1.82	0.549
8. 105 brick : ■ : 105 brick, 13 dense plaster	2.0	0.499
9. 105 brick : ■ : 100 dense concrete block, 13 dense plaster	2.55	0.392
10. 105 brick : ■ : 100 lightweight concrete, 13 dense plaster	1.17	0.857
11. 105 brick : ■ : 100 autoclaved aerated concrete, 13 lightweight plaster	1.07	0.936
12. 105 brick, air gap, plywood membrane, 95 stud : ■ : 12 plasterboard on stud	1.44	0.696
13. 105 brick, air gap, plywood membrane, 140 stud : ■ : 12 plasterboard on stud	1.44	0.696
14. Waterproof covering, 75 screed : ■ : 150 concrete roof, 13 dense plaster	2.05	0.488
15. Tiles on battens, roofing felt and rafters : ■ : bare plaster ceiling below joists	2.6	0.385

Notes: if a better performance be required, subject to investigation of the risk of condensation, insulation may be placed as shown by the symbol ■.

Table 2.8 Resistances of insulation materials of standard thickness (m^2 K/W)

Material	Thickness (mm)											
	20	*25*	*30*	*35*	*40*	*45*	*50*	*55*	*60*	*65*	*70*	*75*
A.1 Expanded polystyrene (board)	0.54	0.68	0.81	0.95	1.08	1.22	1.35	1.49	1.62	1.76	1.89	2.02
A.2 Extruded polystyrene (board)	0.8	1.0	1.2	1.4	1.6	–	2.0	–	2.4	–	–	–
A.3 Glass fibre (rigid slab or batt)	–	0.81	0.97	–	1.29	–	1.61	–	1.94	–	–	2.42
A.4 Glass fibre (flexible and mat)	–	0.63	–	–	1.0	–	1.25	–	1.5	1.63	–	1.88
A.5 Mineral fibre (rigid slab or batt)	–	0.76	0.91	–	1.21	–	1.52	–	1.82	–	–	2.27
A.6 Mineral fibre (flexible and mat)	–	–	–	–	–	–	1.35	–	1.62	–	–	2.02
A.7 Phenolic foam (board)	1.11	1.39	1.67	1.94	2.22	2.5	2.78	3.06	3.33	3.61	3.88	4.17
A.8 Polyurethane foam, CFC blown (board)	1.05	1.32	1.58	1.84	2.11	–	2.63	–	3.16	–	–	–
A.9 Polyurethane foam, PHA blown (board)	0.95	1.19	1.43	1.67	1.9	–	2.38	–	2.86	–	–	–

Material	Thickness (mm)									
	50	*55*	*60*	*65*	*70*	*75*	*80*	*100*	*150*	*200*
B.1 Blown mineral fibre (for existing cavity)	1.25	1.38	1.5	1.63	1.75	1.88	–	–	–	–
B.2 Urea formaldehyde foam (for existing cavity)	1.61	1.77	1.94	2.1	2.25	2.42	–	–	–	–
B.3 Glass fibre mat (in roll)	–	–	1.5	–	–	–	2.0	2.5	3.75	5.0
B.4 Mineral fibre mat (in roll)	–	–	1.62	–	–	–	2.16	2.7	4.05	5.41
B.5 Vermiculite granules (loose fill)	–	–	–	–	–	–	1.23	1.54	2.31	3.08

Chapter 3

The building in summer

The traditional building techniques developed in the British Isles are those which have been described as 'carefully related to the climatic impacts'. The same authority* goes on to suggest that as a result of a poor understanding of solar effects, the environment within the generality of modern buildings has deteriorated. Identification of the climate referred to is less easy, as may be seen from Figure 3.1, which demonstrates that temperatures in the 'mid-seasons' of spring and autumn overlap with those of reputed summer and winter. A temperate maritime climate indeed!

It is impossible to separate those characteristics, desirable in a building envelope during winter, from those which are beneficial in high summer. Low *U* values, small window

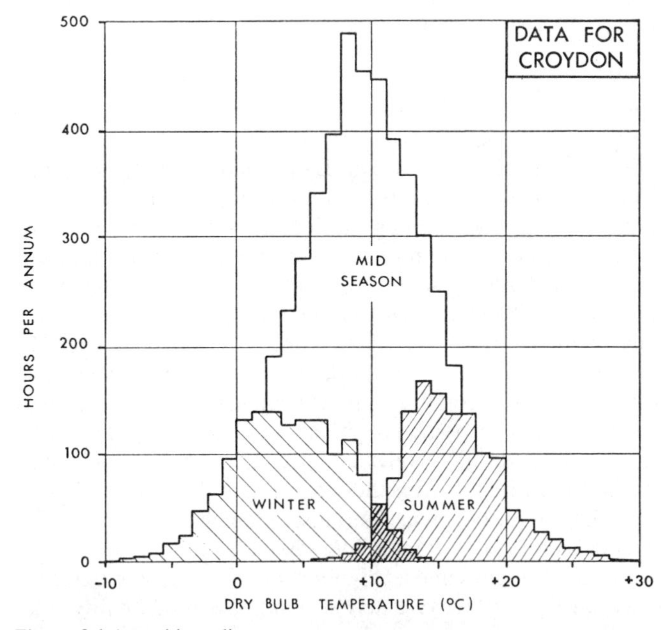

Figure 3.1 A maritime climate

* Page, J. K., *Sun in the Service of Mankind.* Session Report, UNESCO Conference, Paris, 1973.

areas, thermal insulation to a good standard and measures which ensure a reasonable rate of air infiltration are clearly beneficial in either season. It would be inappropriate in consequence to repeat here the comments made in the previous chapter. Nevertheless, the absolute significance of thermal transmittance for opaque building materials is, in the British Isles, less in summer than in winter simply because the temperature difference, inside to outside, is smaller in that season. Typical values might be 20 K (–1 to + 19°C) in winter and 7 K (28–21°C) in summer. Similarly, the effect of infiltration by outside air is less in summer, not only as a result of the reduced temperature difference, outside to inside, but also because wind speeds are usually less then than in winter.

Extraneous influences

The salient difference between the two seasons is of course that heat gains from people, artificial lighting and equipment or machinery combine with heat from the sun to create a penalty during the summer months, as far as human comfort is concerned. These heat gains may lead to high temperatures within a building and, in situations where the occupants are exposed to direct sunshine, there is the added effect of an unpleasantly high radiant intensity from surrounding glazed areas. The control of solar effects is critical to limitation of heat gains in summer and, in consequence, much of this chapter is devoted to that subject.

Not all of the gains noted above related wholly to *sensible heat*, i.e. that which leads to an increase in air temperature. Some part of the output may be in release of *latent heat* which leads to an increase in the amount of moisture in the air and, if the temperature remains constant, to an increase in the relative humidity. Sources of latent heat gains are building occupants, moist air infiltrated from outside and vapour from a variety of activities such as cooking and bathing, to name the simplest.

Conservation of energy

Note has previously been made of the requirements of Part L of Schedule 1 to the Building Regulations (1991), which came into operation in 1992, but these are directed principally towards limiting winter heat loss through the building fabric. Nevertheless, by imposing limits upon the maximum areas of windows and rooflights and prescribing maximum *U* values for the opaque elements of the building structure, some control of heat gains has been imposed. The shortcoming in the Regulations is that they do not include any really significant restraint upon solar gain through glazed areas since, in allowing twice the window area when double instead of single glazing is used, licence is given for solar gain to be increased by two-thirds. The reason for this apparent anomaly may be seen by reference to the respective exclusion characteristics of the alternatives (Table 3.3).

The Regulations could have been made more relevant to conditions in summer had some simple quantitative standard been imposed by the requirement that a solar control glass or, preferably, external shading of glazed areas from insolation be incorporated. For instance, if 35% of single glazing with internal light coloured venetian blinds were to be accepted as a datum, then maintenance of the same level of solar exclusion, with twice the area of double glazing, would require an effective external blind or a good-quality solar control glass such as heat reflecting gold. The energy savings produced by such means can be and are achieved and, failing sensible legislation, it is fortunate that the higher capital and recurrent costs inherent in failure to reduce solar gain encourage a degree of self-regulation by an informed design team.

Primary influences

The level and pattern of energy use in a building, particularly in summer, has its beginning when an architect accepts a brief from a client, on behalf of a full design team, and the initial composite sketch plans are developed. They are affected throughout the later detailed design and construction process (and until the day of demolition) by a host of influences including:

- *Building exposure*: orientation; shape; modules; mass; thermal insulation; glazing; solar shading; plant room siting; space for service distribution.
- *Plant and system design* to match the characteristics of the building and to meet the needs (known and unknown) of the ultimate occupants.
- *Commissioning and testing* of the completed plant and the adjustment to ensure that it operates as designed in all respects.
- *Operational routine* as adapted to match the building use in occupational pattern, working hours and the like.
- *Level of maintenance* provided to both building and plant; energy audits; preservation of records and updating.

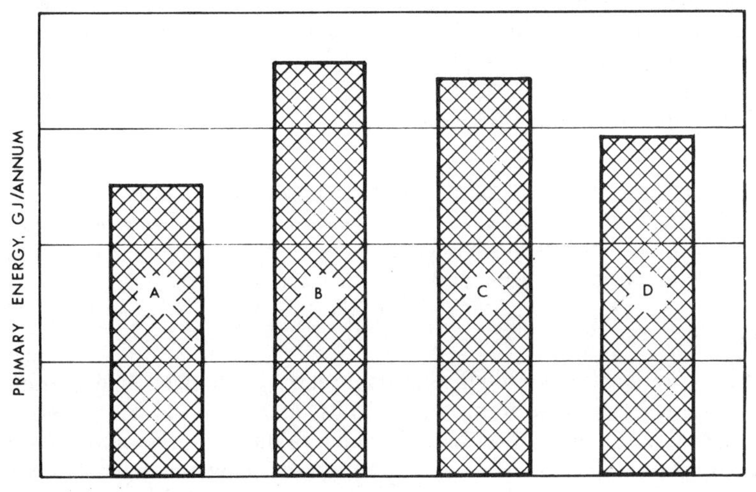

Figure 3.2 Primary energy consumption

Some of the characteristics which lead towards conservation of energy have been dealt with previously in Chapter 2 and will be explored further in this present chapter. In terms of primary energy, Figure 3.2 illustrates how building form can influence consumption, the three blocks in the histogram, identified as A, B and C, being taken from published work* and related to consumption per m^2 of floor area. If, however, consumption is related to *occupancy* and the advantage of open planning is taken into account, block D must be substituted for block C and it will be noted that the apparent penalty *re* block A (but for a greatly improved environment) is then halved. This example has been introduced here in order to illustrate that energy use per unit area may not necessarily be the appropriate criterion in all cases.

* Dick, J., *Ambient Energy in the Context of Buildings.* CIBS/RICS/DOE Conference, 1977.

Similarly, the *thermal balance point* of a building was a term which acquired a specific meaning in the very recent past. It was then defined as that outside dry bulb temperature at which, if all heat generated within the building from lighting, occupancy and machines, etc., were to be distributed usefully, energy from a subsidiary source would be unnecessary. A low balance point, then proposed as a standard of excellence, was later demonstrated to be quite the reverse since it implied a requirement for mechanical cooling at any higher outside temperature. The term has since fallen into disrepute.

The last three of the influences listed above are not altogether in the hands of the designers of the building. Commissioning and testing of plant is a matter of increasing importance, as components become more sophisticated, more 'packaged' and thus less susceptible to any level of repair. All too often, however, the contract period planned for these activities is curtailed due to over-run by the more barbaric building disciplines. It seems that little recognition is given to the fact that, although systems in any one building may be made up of standard production line components, the combination is a prototype in every case.

As to operational routines and maintenance, these are often given a low priority by building owners but this culpable neglect may result from the absence of a simple well-illustrated 'driver's handbook' aimed at Joe who used to stoke the old boiler. While the appointment of a specialist maintenance contractor may be recommended in instances where the building owner has no 'in-house' technicians in this field, that does not excuse the absence of adequate manuals.

Solar heat gains

It has already been stated that the most significant heat gain for the majority of modern buildings, although not always the greatest in terms of magnitude, is that from the sun. To understand the manner in which the sun has an influence upon a building, it is first necessary to have some appreciation of the principles of solar geometry.

Sun path diagrams

Tables providing values for the *altitude* (height as an angle above the horizon) and *azimuth* (compass bearing measured clockwise from the north) of the sun appear in the *Guide Section A2* for latitudes 0–55° and one date in each month of the year. This information is more easily understood by inspection of a diagram such as that shown in Figure 3.3, for a latitude of 50° north. In this example, the position of the sun has been computer plotted from sunrise to sunset on three discrete dates, against a background of altitude and azimuth. Similar profiles for other latitudes may be constructed and standard versions are available, if required, in the form of overlays.*

When a miniature plan of a building, correctly orientated, is placed at the centre of a diagram such as this, it is then possible to visualize the likely peak incidence of solar radiation on the various faces throughout the daylight hours of a chosen month. It will be clear that easterly faces will be subjected to solar influence first, followed by the south east and south, by which time the east face will be going into shadow again, and so on.

To make a preliminary assessment of the maximum solar heat gain for the building as a whole, a schedule of the various rooms and spaces may be constructed and, using values of intensity as described later for various sun altitudes, notes taken of the coincident load

* Petherbridge, P., *Sunpath Diagrams and Overlays for Solar Heat Calculations.* Research Paper No. 39, BRE, 1969.

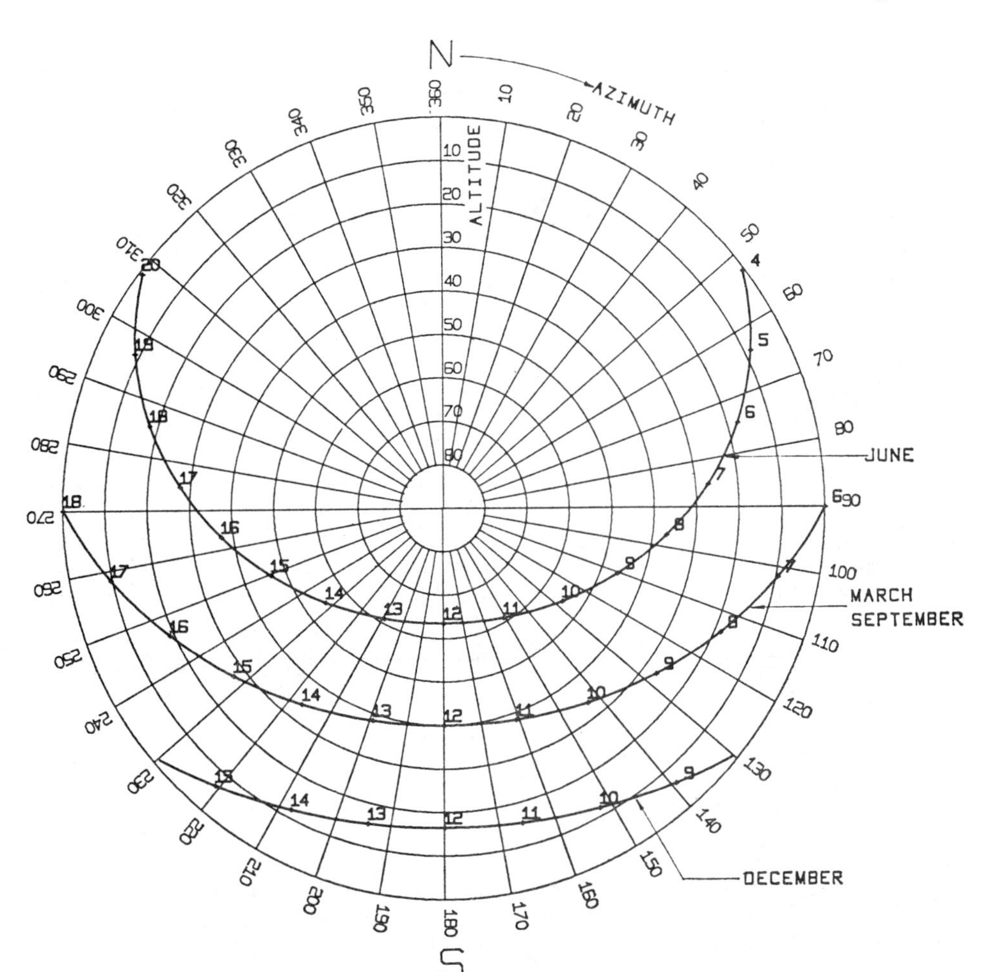

Figure 3.3 Sun-path diagram for latitude 50° north

each hour on each face. In addition, the peak load for each face or any other combination may be established. While this exercise is fairly straightforward in the case of a square or rectangular building it becomes more complex in instances where the plan has re-entrant angles or internal courts where other parts of the building considered cast moving shadows on adjacent faces. While a model can be helpful, extensive analyses of this nature are tedious if attempted manually and use of a suitable computer program is the obvious approach. Figure 3.4 shows a plot of the heat gain and cooling load for a whole building having a load of about 750 kW, drawn as program output by such means.

A second use for a sun-path diagram is in the design of shading devices taking the form of external sun-breaks. Their shape and angle of incidence may be set up on the plan and separate diagrams made for the various aspects; both vertical and horizontal devices may be so examined. This subject will be discussed in more detail later.

As may have been appreciated, sun-path diagrams provide a valuable aid in assisting visualisation of sun position *vis à vis* a building but, in terms of calculation, have been

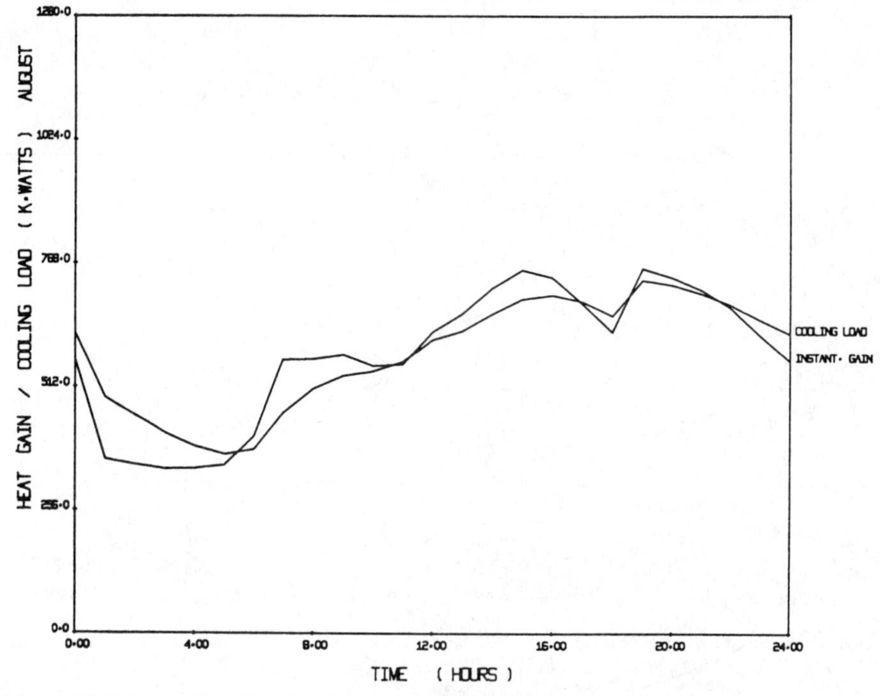

Figure 3.4 Computer plot of building heat gain and cooling load

superseded by the development and presentation of data which allow heat gain to be read directly for orientation and time of day without the intervention of a search for incident angles and basic intensities. In turn, computer use, which has opened up facilities for exploring coincident loads and more rigorous examination of time lag, etc., is well on the way to superseding the refined tabular data. In consequence, the reference here has been abridged and is included to indicate the process rather than the detail.

Incidence of solar radiation

The total of solar radiation to reach the surface of the earth has two components: *direct* and *diffuse*. Of the former, some 1% is ultraviolet, 40% is visible light and the remainder is infra-red. Diffuse, sometimes known as *sky* or *scattered* radiation, results from absorption by, and reflectance from, vapours and dusts, etc., in the atmosphere. It is at a maximum with cloud cover and a minimum with a clear sky.

For a latitude of 50° in the British Isles, the accepted value for the intensity of direct solar radiation on a horizontal surface at sea level (or up to about 300 m above) will be about 800 W/m^2 at noon in June under a clear sky. Diffuse radiation received on the same horizontal surface, again at noon in June, will be about 100 W/m^2 and 300 W/m^2 for clear and cloudy skies respectively. Intensities on surfaces other than the horizontal will be proportional to the *angle of incidence*, which is the angle between the direction of the sun's rays and the perpendicular to the surface.

As far as vertical surfaces (walls and windows) are concerned, some part of the diffuse radiation is received from the sky and the remainder by reflection from adjacent ground surfaces. The intensity of radiation received on the ground will be that for any horizontal

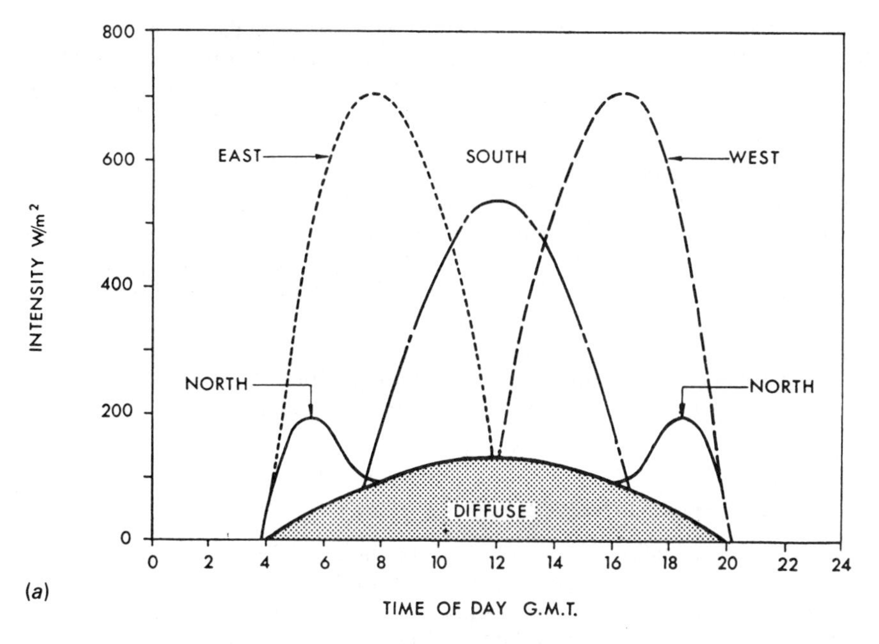

S. and N. faces 24 hour mean = 245 W/m².
E. and W. faces 24 hour mean = 390 W/m².

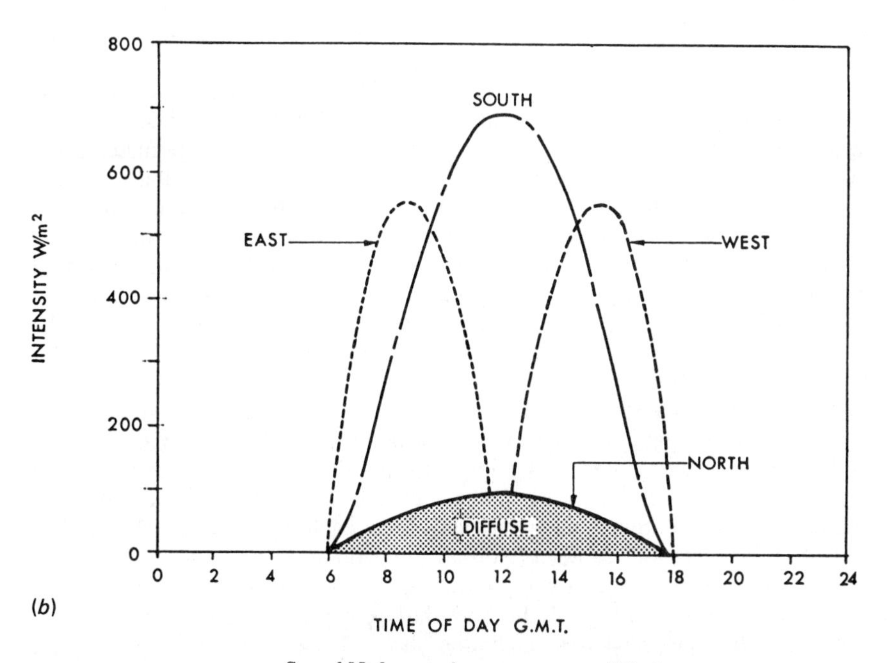

S. and N. faces 24 hour mean = 241 W/m².
E. and W. faces 24 hour mean = 236 W/m².

Figure 3.5 Solar heat on a vertical surface: (a) 21 June, (b) 22 March and 22 September

surface, as already noted, but the amount reflected will depend upon the nature and colour of the surface. In the past, correction factors related to grass, tarmac, water and snow were used but since these were too precise, the choice has been reduced in number to represent light and dark surfaces only. For the case of a city pavement, a convenient (albeit arbitrary) way to take account of *ground reflection* is to add a margin of 10% to those solar heat gains attributed to glazed areas.

The manner in which solar radiation varies throughout the day, for east, south and west faces at the summer solstice, is illustrated in Figure 3.5(a). It will be noted that since the sun altitude is high at midday (64°), the peak solar intensity on the south face is lower than that on the east and west faces which occur at times when the altitude is less. Figure 3.5(b) is a similar plot but shows the situation prevailing at either the spring or the autumn equinox: the intensity on the south face at a noon altitude of 40° now exceeds the peaks on the east and west faces. A measure of the relative intensities of radiation is given by the 24 hour mean values as stated below the two parts of the diagram.

Solar gain through opaque surfaces

When direct and diffuse radiation fall upon the opaque surfaces of a building, a proportion is reflected back into space but the greater part will warm the surface of the material. Of this latter component, some is lost by re-radiation and some by convection to the surrounding air but the remainder is absorbed into the material to a degree which depends upon the nature and colour of the surface. For example, a black non-metallic surface may retain as much as 90% of that remainder, whereas a highly polished metal surface will take in as little as 10%. The majority of building materials absorb some proportion within the range 50–80% and, of this, the greater part will be transmitted through the material by conduction at a rate which will depend upon the U value ($W/m^2 K$).

In instances where the building element has a negligible thermal capacity, as would be the case with thin unlined cement sheeting, then heat will be transferred at once but where there is mass of any real significance, then a time lag will occur. When the material has both sufficient thickness and substantial mass, this time delay will be prolonged to the extent that solar radiation which falls upon it during the early part of the day may not penetrate to the inside face until the outer surface is in shadow. The curious situation may then occur where the heat stored in the structure will be flowing in both directions, to outside and into the building. In some cases, even on east to south facades which receive solar exposure earlier in the day, the heat absorbed at the time of peak gain may not be released from the inner surfaces until after rooms there have ceased to be occupied.

As will be appreciated therefore, the temperature difference between outside air (t_{ao}) and inside air (t_{ai}), does not represent the situation properly since the outside surface temperature of the material will have been raised by the heat absorbed, as noted above. A convenient but approximate method of dealing with this situation is to make use of the concept of *sol–air temperature* (t_{so}), a hypothetical scale which takes into account not only the outside air temperature but also increments to it which represent the increase due to the effects of solar radiation. Tables of sol–air temperature, hour by hour during the critical months of insolation, are given in the *Guide Section A8* for a number of orientations and with respect to both 'light' and 'dark' coloured building surfaces.

The subject of dynamic response, including the influence of time lag (ϕ) and decrement factors (f), was discussed briefly at the end of the previous chapter. In summer, when heat gain is a penalty rather than a bonus, some attempt must be made in calculations performed manually to take these factors into consideration. The application of sol–air temperature tables to the problem takes account of the average of such temperatures over 24 hours as well as that relevant to the duration of the time lag, making use of the following equation to determine heat gain at the time taken for peak loading, θ hours:

$$Q/AU = (t_{som} - t_{ai}) + f(t_{so} - t_{som})$$

which may be transposed to:

$$Q/AU = (t_s - t_{ai})$$

where

Q/AU = heat flow into the space at the time, θ, which has been taken for peak loading (W/m^2).
t_{som} = mean sol–air temperature over 24 hours (°C).
t_{so} = sol–air temperature at time, θ – ϕ, i.e. prior to the appropriate time lag, ϕ hours (°C).
t_s = $t_{som}(1 - f) + f(t_{so})$ (°C).

Table 3.1 Selected notional temperatures (t_s°C) for a typical hot July day in south-east England, taking account of solar radiation and time-lag effects (British Summer Time)

Item		*BST taken for loads*	*Orientation*								
ϕ	*f*	θ	*Horizontal*	*N*	*NE*	*E*	*SE*	*S*	*SW*	*W*	*NW*
1	1.0	1100	38.0	26.5	33.0	43.0	43.5	35.0	27.5	27.5	27.0
		1300	47.0	31.5	32.0	38.5	45.5	45.0	36.5	32.5	32.0
		1500	48.5	33.0	33.5	34.0	38.5	46.5	47.0	40.0	39.0
		1700	44.0	32.5	33.0	33.5	33.5	40.5	49.0	48.5	39.0
4–6	0.47–0.53	1100	23.0	23.5	26.5	27.0	25.0	23.5	24.0	23.5	22.5
		1300	28.0	23.5	31.0	34.0	31.5	25.0	25.5	25.5	24.5
		1500	33.5	26.0	30.0	36.0	36.5	31.5	28.5	28.0	27.0
		1700	38.0	28.5	29.5	33.5	37.5	36.5	33.0	30.5	29.5
5–7	0.42–0.48	1100	22.5	21.5	22.5	23.5	23.5	23.0	23.5	23.5	22.5
		1300	22.5	24.5	29.5	31.0	28.5	24.5	25.0	25.0	24.0
		1500	30.5	24.5	30.5	34.5	33.5	28.0	26.5	27.0	26.0
		1700	35.5	27.0	29.5	34.0	36.0	33.5	29.5	29.0	28.0
6–8	0.37–0.43	1100	23.5	22.0	23.0	24.0	24.0	23.5	24.0	24.0	23.0
		1300	24.0	23.5	26.5	27.5	26.0	24.0	25.0	24.5	23.5
		1500	28.0	24.0	30.0	33.0	31.0	25.5	26.0	26.0	25.0
		1700	32.5	25.5	29.5	34.5	35.0	31.0	28.0	28.0	27.0
7–9	0.32–0.38	1100	24.5	22.5	24.0	25.0	25.0	24.5	25.0	25.0	24.0
		1300	24.0	22.0	23.5	24.5	25.0	24.0	25.0	24.5	23.5
		1500	26.5	24.5	29.0	31.0	28.5	25.5	26.0	25.5	24.5
		1700	30.0	24.5	29.5	33.0	32.5	28.0	27.5	27.5	26.5
8–11	0.22–0.33	1100	26.0	23.5	25.0	26.0	26.5	26.0	26.5	26.0	25.0
		1300	25.5	23.0	25.0	26.0	26.5	25.5	26.5	26.0	25.0
		1500	25.5	24.0	26.5	27.5	26.5	25.5	26.0	25.5	24.5
		1700	28.0	24.5	29.0	32.0	30.5	26.0	27.0	26.5	25.5

Note: based upon t_{ao} = 28°C at 16.00 hours with a 10 K diurnal range.

Selected values of this notional temperature (t_s) for a medium coloured building surface (i.e. one between 'light' and 'dark' and which absorbs 70% of total radiation) and for a number of combinations of time lag and decrement factors are listed in Table 3.1. These values apply strictly to south-east England in the month of July, but may be used for such latitudes over the months of May to September without too much loss of accuracy. Further, the temperatures listed may be applied, within the relevant margin of error, in conjunction with most published figures for time lag and decrement factor, including those given in Table 3.2 which represent these characteristics for a range of typical construction elements. Routine calculations are simple, as illustrated by the following example:

A wall facing south-east is constructed of 220 mm solid brick, with lightweight plaster internally. The internal air temperature is to be maintained at 21°C, the peak cooling load for the space concerned being taken at 16.00 hours BST in July.

Table 3.2 Representative design factors for solar gain through opaque materials (dimensions in mm)

Construction (dimensions in mm)	*U value* (W/m^2 K)	*Time lag,* ϕ (h)	*Decrement,* f
Walls			
105 brick, unplastered	3.27	3.0	0.88
220 brick, 13 lightweight plaster	1.9	7.0	0.46
220 brick, 50 air space, plasterboard on battens	1.43	7.1	0.39
19 render, 200 dense concrete, 25 polyurethane, 12 plasterboard	0.71	8.0	0.22
80 pre-cast concrete panel, 50 EPS, 100 dense concrete, 12 plasterboard on battens	0.55	8.2	0.29
105 brick, 50 air space, 105 brick, 13 dense plaster	1.36	7.9	0.4
105 brick, 50 EPS, 100 concrete block, 13 dense plaster	0.56	9.1	0.25
105 brick, 50 UF foam, 100 lightweight block, 13 dense plaster	0.47	8.4	0.44
105 brick, 50 air space, 19 plywood, 95 studding with mineral fibre between studs, 12 plasterboard	0.29	6.1	0.58
Roofs			
Light colour paving slab, 50 extruded polystyrene, waterproof membrane, 75 screed, 150 cast concrete, 13 plaster	0.51	10.0	0.17
Waterproof covering, 35 polyurethane, vapour barrier, 19 timber decking, unventilated air space, 12 plasterboard	0.52	1.9	0.93
Tiles on battens, roofing felt and rafters, 50 glass fibre mat between joists, plasterboard ceiling below	0.61	1.0	1.0
Tiles in battens, roofing felt and rafters, 100 glass fibre mat between joists, plaster ceiling below	0.35	1.0	1.0

From Table 2.2 and interpolating from Table 3.2:

U value = 1.9 W/m^2 K
Time lag = 7 hours
Decrement = 0.46

Thence, interpolating from Table 3.1:

$$t_s = (33.5 + 36.0)/2 = 34.8°\text{C}$$

and

$$\text{Heat gain at 16.00 hours} = 1.9\,(34.8 - 21) = 26\,\text{W/m}^2$$

It is appropriate to note that, when considered in relation to the customary internal design temperature range 21–23°C for summer, the data listed in Table 3.1 illustrate the point made earlier and show that there are numerous occasions when the heat gain from opaque surfaces may be negative. It is common practice to ignore any such small 'credits' in calculations for plant capacity, in the same way that small adventitious heat gains are ignored in winter calculations.

In the special case of lightweight curtain walling, as shown in Figure 2.11, the effect of solar radiation is to raise the outer surface temperature very rapidly when the finish is dark. It can be demonstrated that in the British Isles this temperature may be well over 50°C in the early afternoon on a south-west facade and, since the time lag of such a form of construction is an hour or less, with negligible decrement, it is inevitable that complaints of discomfort due to radiation will arise.

Flat roofs are subjected to solar radiation during the whole of the daylight hours and a relatively massive construction with a light coloured finish is advantageous. Roof insulation, as may be seen from Table 3.2, does little to increase the time lag but does of course reduce the U value. In the case of pitched roofs, the heat gain will depend upon the angle of the slope and may be more severe on one face than another, depending upon orientation. It is worth while making a check on the orientation of so-called 'north-light' roofs since it is not unknown for them to belie their name!

Solar gain through glazing

Windows are by far the most significant route by which solar heat enters a building, not least because that entry is without time lag. This is not to say, however, that the effect is instantaneous since this may well depend upon the nature and mass of the internal structure, furniture, carpets and other contents. The subject has been explored exhaustively by Loudon* at the Building Research Establishment in evolving a method whereby the excessive heat experienced in buildings without air-conditioning could be predicted.

As a result of this work and later refinements, a number of computer programs have been produced which, as exemplified by Figure 3.6, enable the relationship between inside and outside temperatures to be seen. The more sophisticated of such programs enable many parallel effects to be examined in detail and permit evaluation of structural forms and the dynamic response to these by engineering systems. The method involves the concept of environmental temperature, which has been referred to previously, thus taking account of both the mean radiant temperature and the air temperature within a space. Advantage may be taken of the same basic approach, including use of the admittance value of the various components of the building structure, to evaluate the extent to which

* Loudon, A. G., 'Summertime temperatures in buildings without air-conditioning', *JIHVE,* 1970, **37,** 280.

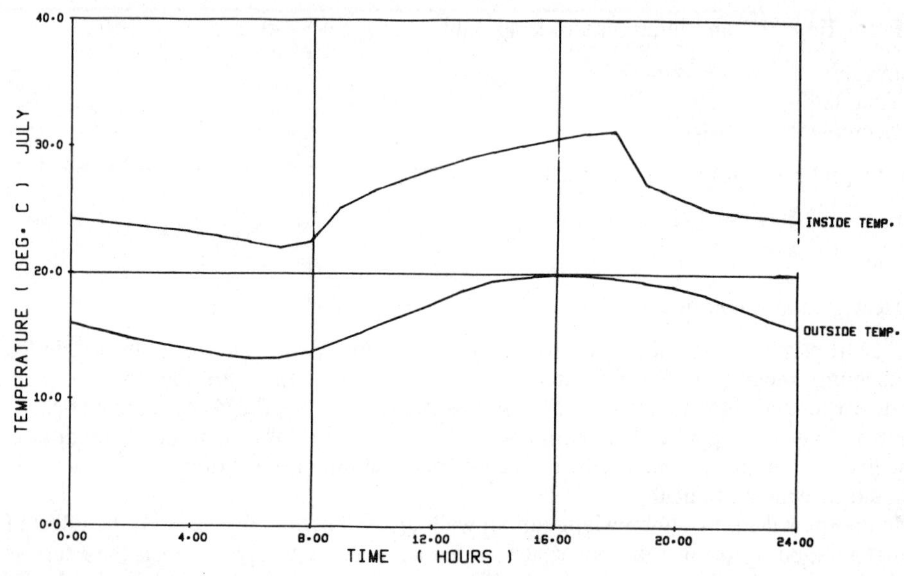

Figure 3.6 Computer plot of temperature without air conditioning

those components are able to smooth out diurnal temperature swings resultant upon intermittent solar heat gains.

Glass has certain unique characteristics as far as heat transfer is concerned in that it has different transmissivities at different wavelengths: it is, moreover, virtually opaque to radiation from any source having a surface temperature less than about 250°C. When a glazed surface is insolated, some part of the total incident radiation will be reflected away, some will be transmitted through the material, and some will be absorbed by it. This simple pattern of heat transfer is confused by re-radiation outwards and by other emissions in both directions, with the result that the final picture is quite complex.

It is of course much more effective to prevent solar radiation reaching glazed areas, in whole or in part, by making use of blinds, etc., fitted *externally*, than to permit exposure, allow penetration into the building, and then to search for means to curb the nuisance. Table 3.3 lists a number of combinations of glazing and shading devices and ranks these in order of efficacy. It will be noted that substitution of double for single glazing offers a much less dramatic improvement in this context than it does in reduction of *U* value. The computer based plots of Figure 3.7 provide parallel information and illustrate the effect upon peak air temperature within a south-west facing room which results from variation in types of protection and rates of air change.

Special solar control glasses reduce the total rate of heat transfer by increasing either the absorbed or the reflected component: a variety of different types is available:

Absorbing glasses. There are two main types here: body-tinted and metallic ion coated. The former type (BTG) is coloured throughout its thickness and, in consequence, the colour density and solar control properties increase with glass thickness: colourings available are commonly grey, bronze and green. The alternative variety has a layer of ions, which absorb incident radiation, injected into the surface of clear plate glass. Since the layer is below the surface and therefore protected, the glass may be used as a single pane.

Table 3.3 Relative efficiency of shading devices in reducing solar heat gain (plain single glazing = 0)

Type of solar protection	*Relative exclusion efficiency (%)*	
	Single glazing	*Double glazing*
External dark green miniature louvred blind	83	87
External canvas roller blind	82	85
External white louvred sunbreaker, 45° blades	82	85
External dark green open weave plastic blind	71	78
Heat reflecting glass, gold	66	67
Clear glass with solar control film, gold	66	67
Densely heat absorbing glass	–	67
Mid-pane white venetian blind	–	63
Internal cream holland blind	57	61
Lightly heat absorbing glass	–	50
Densely heat absorbing glass	49	–
Internal white cotton curtain	46	47
Internal white venetian blind	39	39
Lightly heat absorbing glass	33	–
Internal dark green open weave plastic blind	18	26
Clear glass in double window	–	16

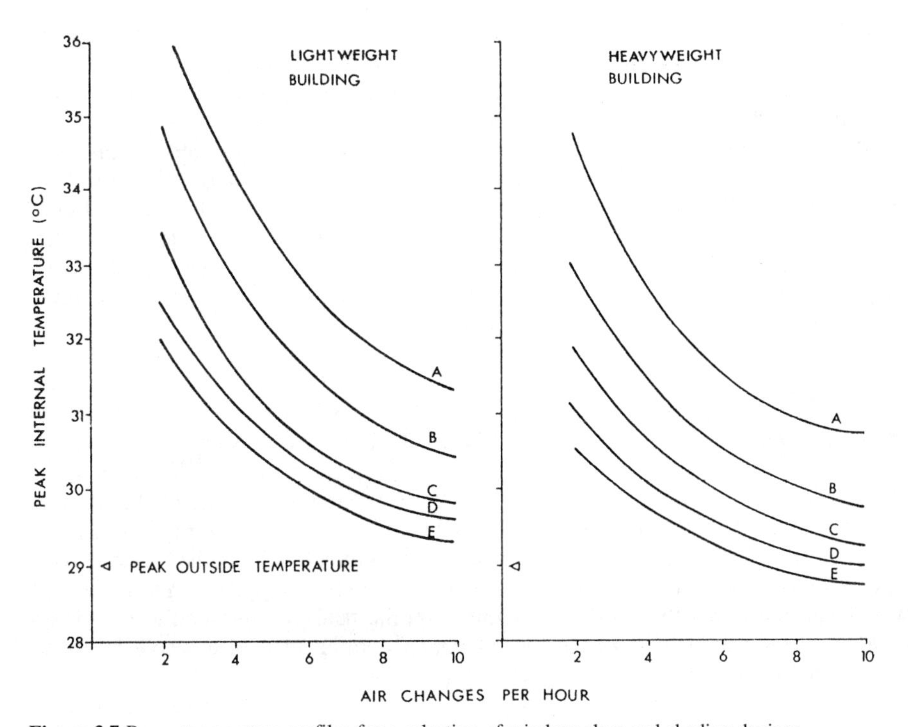

Figure 3.7 Room temperature profiles for a selection of window glass and shading devices

Reflective glasses. These have a metallic oxide or similar coating on one surface and are thus normally used in sealed double glazing units. They are manufactured in a standard colour range of silver, grey, bronze or blue but other colours may be obtained.

Solar control films. Solar control films are available in a wide range of colours, and are applied to the internal side of the glass. They produce an effect similar to that of reflective glass and can provide an economic solution as a retrofit measure to existing single glazed windows. A recent addition to the materials available is a film which acts like a photochromic glass, darkening as the incident sunlight increases its intensity.

'Smart' windows. These so-called materials are in the course of development for commercial use. Typically, the glass has four coatings, a transparent conductor, an electrochromic layer, an ion storage layer and a second transparent conductor. By the application of a voltage to the glass, the transmissivity property of the coating is changed and may be adjusted to control the amount of solar heat entering the space.

Low emissivity coatings. These are, strictly speaking, more appropriate to retention of solar heat within a space than to the control of its entry.

In order to take account of the surfaces and other features of a space within a building structure which absorb, to a greater or lesser degree, the solar heat admitted through glazing, the *Guide Section A9* includes a group of tables which list *cooling loads*, as distinct from *heat gains*. These apply to *lightweight* buildings, defined as having demountable partitions, suspended ceilings, and floors which are suspended or, if solid, have either a carpet or a wood block finish: a *heavyweight* building, conversely, has an overall solid construction with no soft finishes. Each such table applies to a single latitude and to a plant operational period of ten hours: correction factors are provided to cover some other circumstances. These data are represented here by Table 3.4,* which relates to unprotected clear glass, and by the parallel Table 3.6 which relates to a combination of glass with an internal white venetian blind. Correction factors for other glass and building combinations are provided by Tables 3.5 and 3.7 respectively.

In the case of the principal Tables 3.4 and 3.6, all the values listed assume that calculations for internal design conditions have been based upon dry resultant temperature. If air temperature were to be used, the correction factors at the ends of Tables 3.5 and 3.7 must be applied. Diffuse radiation from earth and sky is included in the data provided but, where the ground floor of a building abutts to a wide city pavement subject to solar radiation, an addition of 10% to the listed figure, as mentioned previously, may be justifiable.

Structural shading

In a deep plan building, it is possible that windows are provided more to encourage occupants to retain visual contact with outside rather than in competition with some subsidiary form of illumination. In consequence, the use of internal or external blinds may not be acceptable, despite the penalty paid in heat gain. It is therefore necessary to consider the extent to which vertical fins or horizontal projections may be of value. It is self-evident that, in the latitude of the British Isles, vertical fins are useful only on the east and west facades but that their influence there upon the peak cooling load to individual spaces may not be large. In terms of the *maximum* cooling load in a whole building, however, such fins may contribute to a reduction.

* Tables 3.4, 3.5, 3.6 and 3.7 appear at the end of this chapter (pp. 92–3).

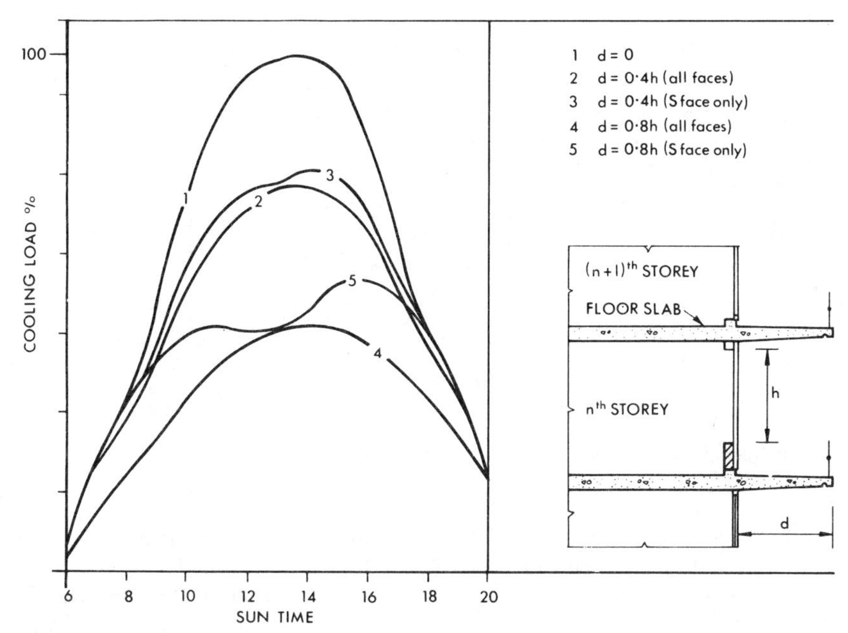

Figure 3.8 Effect of horizontal projections upon cooling load

Horizontal projections formed by outward extensions to floor slabs above windows may make a contribution if formed on the southerly facades, south east through to south west, as shown in Figure 3.8. The effect of such shading, for an air-conditioned commercial building, may be expressed also in terms of annual energy expended in cooling, as predicted for a south facing room using a variety of shading methods and dimensions, and listed in Table 3.8.

The diurnal cycle of the sun's motion subjects the various faces of a building to a pattern of shading which changes continuously and, as has been noted previously, if a

Table 3.8 Effect of structural shading upon annual energy consumption for cooling

Method of shading		
Type	*Depth* (mm)	*Annual index*
No shading	–	1.0
Window recess	150	0.94
	500	0.82
	1000	0.68
Vertical fins	150	0.97
	500	0.94
	1000	0.88
Horizontal overhang	500	0.87
	1000	0.76
	2000	0.57

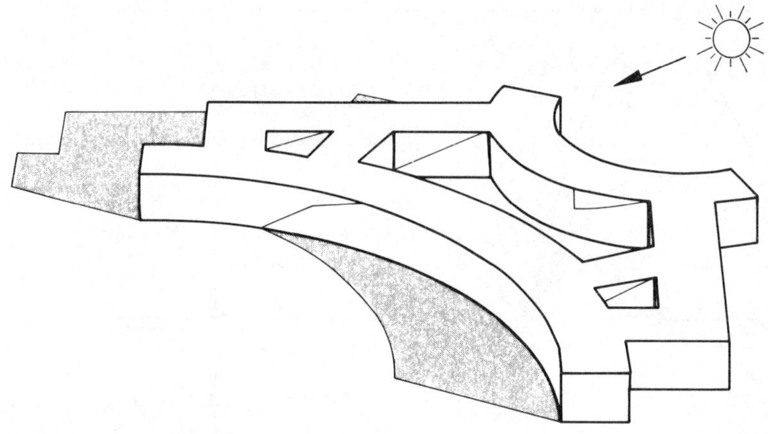

Figure 3.9 Computer plot of sun shading

building has wings forming an L or a T or some other more confused shape, the shading pattern may become complicated. In consequence of demands for detailed data, computer modelling programs have been produced and the form of output from one such is illustrated in Figure 3.9.

Building shape

Deep plan constructions, generally single-storey, have been a commonplace approach to industrial building for many years. In most cases room heights are generous, daylight is available from roof glazing and occupancy levels per unit of floor level are low. Simple heating and ventilating systems will thus, in the majority of cases, provide adequate service. As to solar gain through roof glazing in medium to heavy industrial buildings, experimental work at the (then) Building Research Station* demonstrated that application of two coats of whitewash, with a suitable binder such as tallow, to the outer surface would reduce solar overheating by about 70%.

Exceptions to this general statement, now tending to represent the rule, arise of course where an industrial process such as the assembly of microcomponents, etc., requires close control of temperature, humidity and airborne contaminants. Since requirements of this nature are usually accompanied by demands for a high and constant level of illumination, the logical outcome is the provision of an enclosed environment with full air conditioning.

In contrast to a deep plan arrangement designed for office use, as shown to the left of Figure 3.10, a not atypical high-rise building might be planned as on the right of the same diagram. For equal floor area overall and assuming reasonable proportions, length to width, the deep plan would have four storeys and the tower block twenty. In terms of exposure, the tall building would present 50% more surface to wind, rain, frost and sun at the perimeter but the need for artificial lighting in the core of the deep plan building would most probably lead to a requirement there for cooling throughout the year.

Once the size and character of an office building are such that use of natural ventilation, a simple heating system and natural lighting during daylight hours are inappropriate, the ground rules have changed and there is no absolute ideal for building shape. This is not

* Beckett, H. E., 'The exclusion of solar heat', *JIHVE*, 1935, **3**, 79.

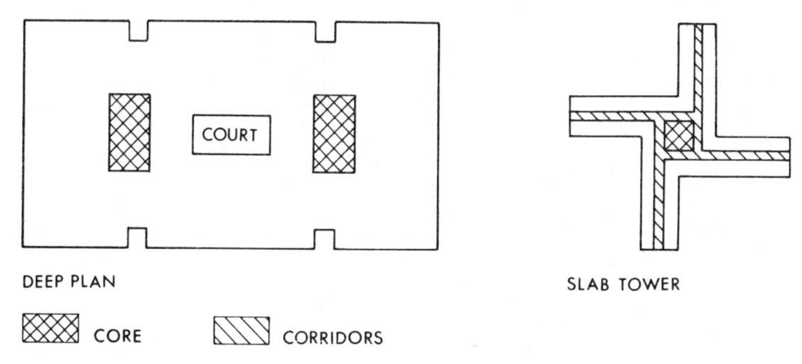

Figure 3.10 Arrangement of deep plan and tall building having the same area

to say that, for a given site and a given accommodation brief, a detailed analysis cannot produce positive advice as to the building form which will allow optimum system design and minimum recurrent cost to be examined. But rather that no more than 'brochure level' guidance, as exemplified by that given under the next heading, can be offered in the abstract.

Building orientation

For the particular case of the tall rectangular slab block, modelled to match a packet of breakfast cereal, and having a width/depth ratio of 3:1 as Figure 3.11, a simple calculation may be made to compare heat gains at two seasons for two different orientations. Using the 24 hour mean values for solar radiation from Figure 3.5:

Case 1, in June

S and N faces 245 × 3	=	735
E and W faces 390 × 1	=	390
		1125

Case 2, in June

S and N faces 245 × 1	=	245
E and W faces 390 × 3	=	1170
		1415

Case 1, in September

S and N faces 241 × 3	=	723
E and W faces 236 × 1	=	236
		959

Case 2, in September

S and N faces 241 × 1	=	241
E and W faces 236 × 3	=	708
		949

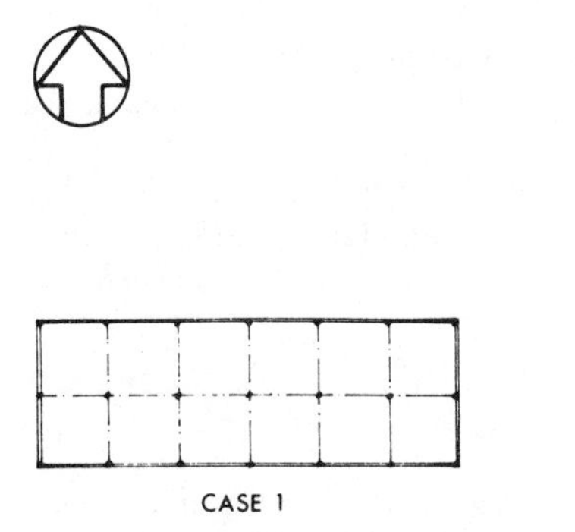

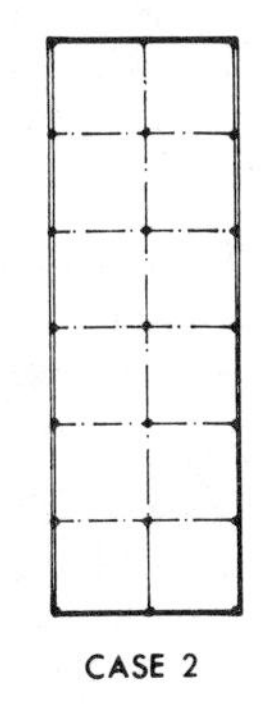

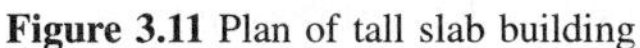

Figure 3.11 Plan of tall slab building

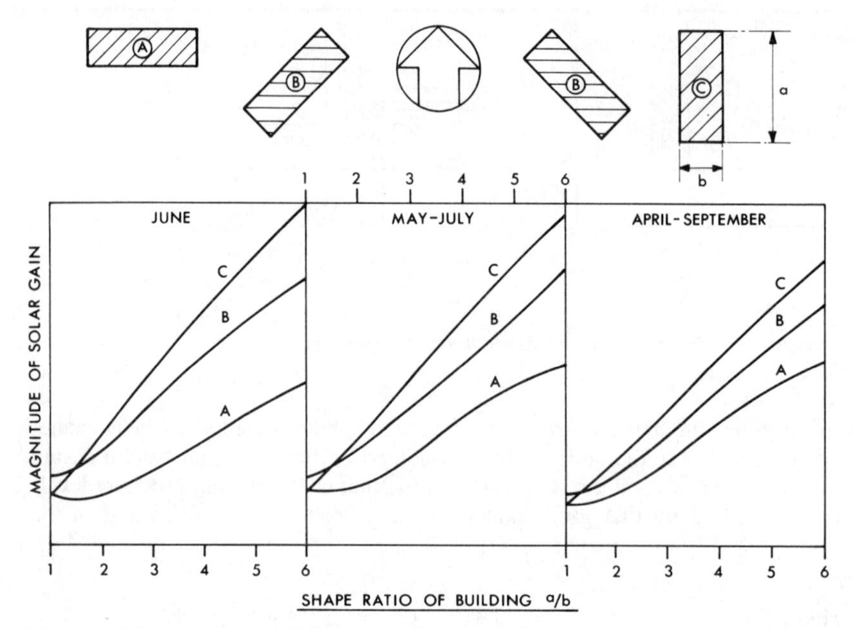

Figure 3.12 Solar incidence on buildings of different shapes and varying orientations

It is evident that, in mid-summer, the orientation of case 1 produces a lower heat gain than that of case 2 but that in spring and autumn the difference will be much less. In reality, with a slab block of this plan, the narrow ends would very probably be solid and windowless: in that situation, when due allowance is made for the substitution, case 1 shows up to even greater advantage. This rudimentary example serves to confirm the rule of preference for an east–west axis, now well established, and Figure 3.12 shows solutions for other orientations and for the three periods which represent the six warmer months.

Conduction

This subject has been explored in detail in Chapter 2 and the mechanism of heat transfer here is similar, being a function of the thermal transmittance coefficient (U) and the temperature difference, inside to outside. A factor may be applied in this instance to allow for the effect of the radiant component of heat transfer which arises from the adoption of dry resultant temperature as the criterion of comfort. However, for most practical applications, the value of the correction factor approaches unity and may be so taken. For those particular instances where the facade of a building has an extremely high proportion of glazing, correction is necessary and reference to the *Guide Section A9* is advised. The values for the coefficient U may be taken from Table 2.2, although a minor reduction could be made to allow for a higher surface resistance in summer when wind speeds are generally less.

Application

The manner in which these components of total heat gain relate to a building is best illustrated by application to the simple example shown in Figure 3.13: this represents an

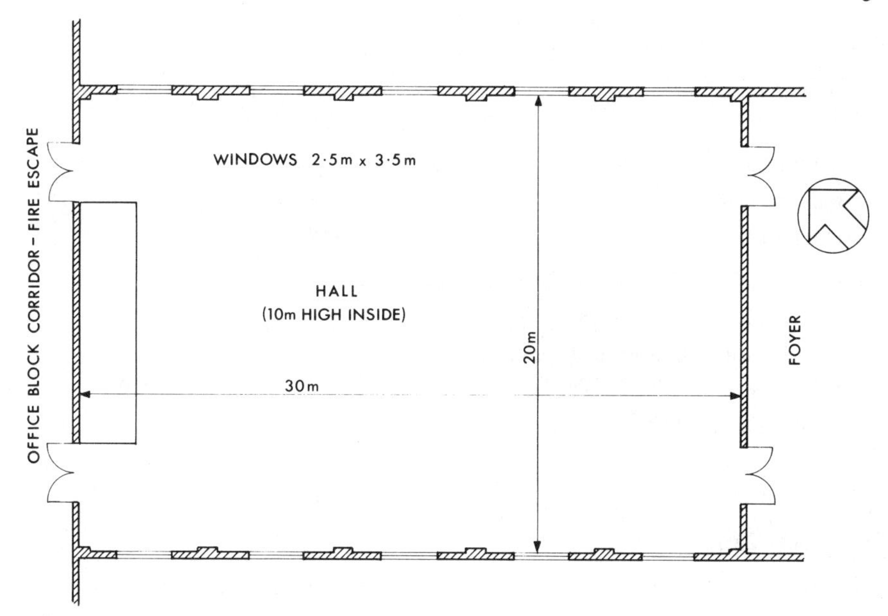

Figure 3.13 Heat gain example

all-purpose hall, in use for concerts and other activities, having the following constructional details etc.:

- The windows are single glazed in wooden frames and have light linen roller blinds mounted internally.
- The walls are 105 mm brick, 50 mm UF foam, 100 mm lightweight concrete block finished with 13 mm dense plaster.
- The roof is flat consisting of light coloured paving slabs, waterproof membrane, 50 mm extruded polystyrene, 75 mm screed, 150 mm cast concrete and 13 mm lightweight plaster.
- From the orientation shown, the building presents two faces to solar gain, north east and south west.
- For the purposes of the example an air temperature difference of 7 K (28°C outside and 21°C inside) may be assumed.

From examination of Table 3.6, it is clear that the south-west glass will be subject to the greatest heat gain at 15.00 hours BST in September, in the middle, perhaps, of an afternoon function. This time and date will therefore be the basis for the calculation which follows.

Glazing

Solar gain

Area:

SW exposure = 5(2.5 × 3.5) = 44 m^2

NE exposure = 5(2.5 × 3.5) = 44 m^2

From Table 3.6, the unit cooling loads are:

SW glazing	= 328 W/m^2
NE glazing	= 104 W/m^2

From Table 3.7, taking the building mass to be 'heavy':

Blind shade factor	= 0.63
Air temperature control factor	= 0.83

Thus, the cooling load arising from solar radiation:

SW glazing	= (44 × 328 × 0.63 × 0.83)	= 7546 W	
Add 10% for radiation from pavement		= 755 W	
NE glazing	= (44 × 104 × 0.63 × 0.83)	= 2393 W	= 10.69 kW

Conduction gain due to temperature difference:

Area:

Total	= 88 m^2

From Table 2.6:

U value	= 4.5 W/m^2 K

Temperatures:

Outside	= 28°C
Inside	= 21°C

Thus, gain = (88 × 4.5 × 7) = 2772 W = 2.77 kW

Walls

Combined solar and conduction gain

Area, wall less glazing:

SW exposure (30 × 10) – 44	= 256 m^2
NE exposure (30 × 10) – 44	= 256 m^2

From Table 3.2:

U value	= 0.47 W/m^2 K
Time lag	= 8.4 hours
Decrement	= 0.44

From Table 3.1, interpolating for t_s at 15.00 hours:

SW exposure	= 26.0°C
NE exposure	= 29.0°C

Thus, combined gain:

SW exposure	= (256 × 0.47) (26 – 21)	= 602 W	
NE exposure	= (256 × 0.47) (29 – 21)	= 963 W	= 1.57 kW

Roof

Combined solar and conduction gain:

Area:

(30 × 20) = 600 m^2

From Table 3.2:

U value	=	0.51 W/m^2 K
Time lag	=	10 hours
Decrement	=	0.17

From Table 3.1, extrapolating for t_s at 15.00 hours:

Horizontal = 26°C

Thus, combined gain:

Horizontal = (600 × 0.51) (26 – 21) = 1530 W = 1.53 kW

The total heat gain, carried forward to p. 507 = 16.56 kW

Air infiltration and ventilation

In addition to the heat which enters the building by direct transmission of solar radiation and by conduction, allowance must be made for that which enters with outside air by infiltration. This is not dissimilar to the interchange which takes place in winter except that heat is entering rather than leaving a building.

Air infiltration

This subject has been dealt with in some detail in the preceding chapter but, consequent upon the probability of lower wind speeds occurring during the summer months, a deduction, if thought worthwhile, may be made from the values appropriate to the winter: practice is to allow about half an air change per hour for the common run of buildings. When the designer can be sure that standards of construction for a particular project will be unusually high, with weather stripped or sealed windows and proper treatment of junctions between wall linings and false floors or ceilings, this allowance might be halved. When circumstances are appropriate, as was noted previously, the crack method of calculation for air infiltration may be used. In industrial buildings, where some part of the total volume only is air-conditioned, and goods must be transported away from or to a treated area, the use of air locks between the two areas is essential.

Ventilation

This matter has been dealt with in both of the preceding chapters and it is necessary only to emphasise the difference between *ventilation air* and, as far as system design may be concerned, what is known as *conditioned air.* The former is that quantity of outside air, normally quoted in terms of volume, necessary to dilute contamination from all sources to an acceptable level. The conditioned air quantity, however, normally quoted in terms of mass, is related to its capacity under certain limiting conditions to absorb whatever heat or moisture may be surplus within the space or, conversely, to supply heat or moisture when there is a deficit within the space. The conditioned air will, of necessity, be the means of introducing ventilation but since the whole quantity has been treated at the plant, unlike air which has entered by infiltration, no loss or gain will be involved *within* the space.

Miscellaneous heat gains

Activities which take place within a space produce heat and this, a possible although unreliable bonus in winter, is a certain penalty during summer.

Heat from occupants

The heat emitted by the human body depends upon the temperature, humidity and level of air movement in the occupied space and upon the level of activity: in an atmosphere of 20°C and 60% RH, it varies as shown in Table 3.9. For a man seated at rest, Figure 3.14 illustrates how the various outputs vary with the dry bulb temperature for an average level of humidity in still air. The sensible heat increases by about 20% with air movements up to 0.5 m/s but latent heat is affected only slightly by air movement for temperatures less than about 18°C, above which temperature the point at which the curve begins to rise is displaced, i.e. at 0.5 m/s, the rise begins at about 26°C.

It is the sensible heat alone which affects temperature within a space, that provided by one person seated at rest being sufficient to heat about 4 litre/s of air through 18 K.

Table 3.9 Heat emission by occupants

	Heat emission per occupant (W)		
Activity	*Sensible*	*Latent*	*Total*
Seated at rest	90	25	115
Sedentary worker	100	40	140
Walking slowly	110	50	160
Light bench work	130	105	235
Average manual work	140	125	265
Heavy work	190	250	440

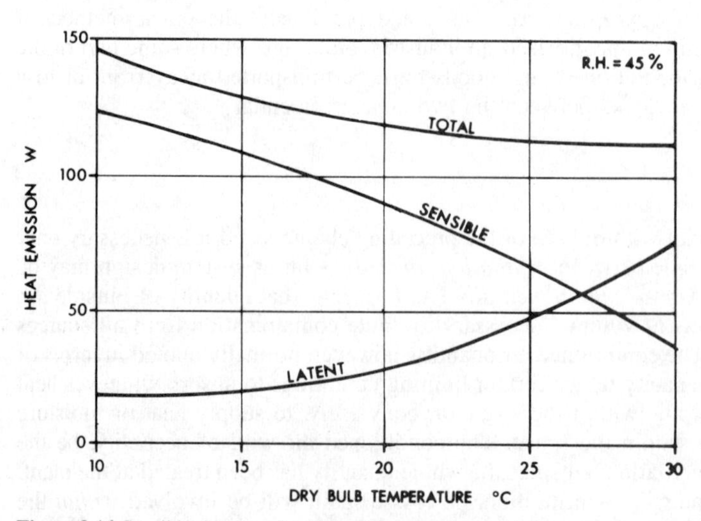

Figure 3.14 Sensible, latent and total heat for a person seated at rest

Heat from lighting

All the energy consumed in providing electricity for lighting is dissipated as heat, the rate depending upon the type of lamp and the design of the luminaire. For a given level of illuminance, tungsten lamps will emit four or five times as much heat as fluorescent tubes and the unit emission from the latter varies quite considerably with length and colour. Table 3.10 provides outline data for the electrical loading related to floor area but this type of information should be used for preliminary design only, pending details of the final lighting layout.

Table 3.10 Approximate heat gain from lighting (including control gear, if any)

	Approximate electrical loading (W/m²)						
Level of illumination (lux)	*Tungsten filament*		*White fluorescent*			*HP discharge*	
	Open	*With diffuser*	*Open*	*With diffuser*	*Recessed*	*Mercury*	*Sodium*
250	45	55	10	15	20	–	–
500	80	–	15	25	30	20	10
750	–	–	25	35	40	30	15

It has become common for facilities to be provided which permit extract ventilation from and around recessed luminaires. This has the effect of increasing luminous flux and improving efficacy (lumens/W). In terms of heat gain, however, as a generalization in the absence of specific details from a manufacturer, it may be assumed that the output *to the space* of 50% for a conventional luminaire is reduced to about 20% for a type having air-handling facilities. The balance of the output in the two cases, 50 and 80%, respectively, passes to the ceiling void and, although no longer a heat gain *to the space*, remains to be dealt with by the central cooling plant in circumstances where some or all of the air extracted is recirculated.

Heat from office machinery and computers

The rate of heat dissipated from these sources may be high, typically 22–30 W/m² and up to 40 W/m² in a commercial office accommodating a large number of items of electronic equipment. In areas devoted wholly to data processing, a peak of 70 W/m² may occur. Output is mainly convective and reference should be made to manufacturers' data to establish accurate information.

The level of heat gain from office equipment is likely to change in the foreseeable future under two opposing influences.* Over the next 5–10 years it is predicted that there will be a further increase in the use and that the machines will be more powerful. It is likely, however, that there will be significant improvements in the efficiency of machines thereby reducing energy consumption. The short-term effect is that power requirements may increase from the current typical maximum of about 280 to around 350 W per person,

* BSRIA Technical Note, *Small power loads*, TN 8/92.
Energy Efficiency Office, *Energy Efficiency in Offices – Small Power Loads*. Best Practice Guide No. 35.

but with a suggested fall in the longer term to 150 W per person, perhaps in the early years of the next century.

Heat gains in excess of 600 W/m^2 may be expected in computer rooms and those areas housing support equipment. Such spaces will almost certainly need to be air-conditioned in order to maintain the electronic equipment within the limits of temperature and humidity set down by the makers and to provide comfort conditions for operators. It is interesting to note signs of return to a practice adopted for early computers whereby certain processing units incorporated a cooling coil and a closed circuit air circulation within the casing. Manufacturers' literature in such cases requires only that a supply of chilled water, specified as to quantity (possibly in US gallons) and temperature, be provided.

Heat from machines and process equipment

Electrical energy provided to motor drives is for the most part converted wholly into heat at the time it is consumed, the exceptions being where work is done to produce potential energy, as in the case of water pumped to an elevated tank or goods raised in a hoist. In consequence, the energy input to machines constitutes a heat gain and may be evaluated by reference to the power absorbed and, where appropriate, any diversity of use. When both the motor and the driven machine are in the space, then the total power input must be considered, but if either are mounted elsewhere, then reference to Figure 3.15 will provide guidance as to the proportion of the total which must be allowed for.

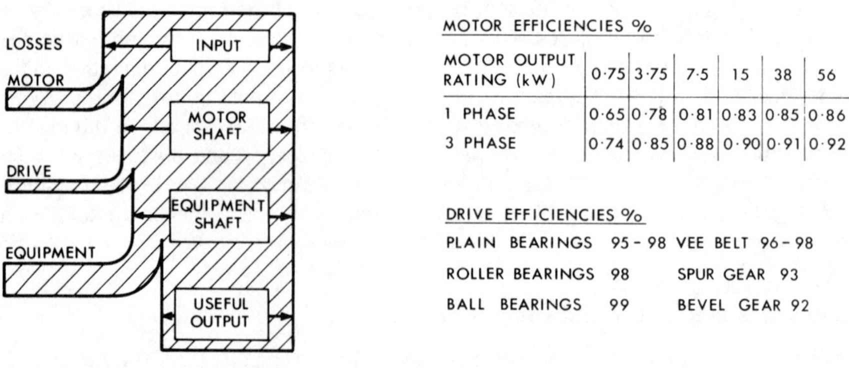

MOTOR EFFICIENCIES %

MOTOR OUTPUT RATING (kW)	0·75	3·75	7·5	15	38	56
1 PHASE	0·65	0·78	0·81	0·83	0·85	0·86
3 PHASE	0·74	0·85	0·88	0·90	0·91	0·92

DRIVE EFFICIENCIES %

PLAIN BEARINGS	95 - 98	VEE BELT	96 - 98
ROLLER BEARINGS	98	SPUR GEAR	93
BALL BEARINGS	99	BEVEL GEAR	92

Figure 3.15 Typical energy balance for an electric motor

No useful guidance can be given as to the amount of heat which may be liberated within a space by equipment such as steam presses, hot plates, drying ovens and gas or electric furnaces and reference must be made to manufacturers' literature. Similarly, for animal rooms, etc., standard authorities should be consulted, a wide ranging summary being included in the *Guide Section A7.*

Diversity

As has been mentioned in the two previous paragraphs, it may well be important that the matter of diversity in use be considered. In a workroom, for instance, there may be a variety of machines having 60 or so motors which together have a brochure rating of

300 kW but, at the same time, the total instantaneous loading may be only two-thirds of that total. While it would clearly be wrong to ignore this seeming diversity in use, the detail will often require careful analysis as to both timing and the physical positioning of the machines. At the risk of stating the obvious, a diversity of 66% does not mean that each motor is running at 66% full load (although this may be the reasoning of an ill-trained computer program) but more likely that of each 60 motors fitted, a random maximum of 40 are in use simultaneously. Those at one end of the space might, perhaps, all be idle at one time or, conversely, but one of each pair of adjacent machines might be running. Each case needs to be considered in detail.

Temperature control

The case made in Chapter 2 for temperature control room-by-room in winter, applies equally to buildings in summer. In the case of air-conditioning systems, the control philosophy requires very careful consideration since the plant configuration to deal with heating, cooling, humidification and dehumidification involves much more complex equipment than does a straightforward heating system. Furthermore, it is probable that some spaces within an air-conditioned building may require heating, coincidentally with a demand from others for cooling, as a result of different orientation, time of day and the level of miscellaneous internal gains. This situation will prevail in particular during the mid-seasons between summer and winter and it points to a very positive conclusion that system controls must be considered in detail right from the earliest stages of design.

Temperature difference

The comments made in Chapter 2, with regard to the use of dry resultant temperatures, are equally valid for summer conditions but, for building envelopes which meet with the current regulations as to their thermal properties (see Table 2.5), values from this scale may be taken as numerically equal to air temperature without any need to confuse the calculation process by application of correction factors.

Inside temperature

The only comment necessary under this heading is that it is usual for summer design purposes to make a distinction between continuous and transient occupancy, 21°C dry bulb and 23°C dry bulb being the commonly accepted levels, respectively, with 50% saturation in each case.

Outside temperature

As was the case for winter conditions, the design temperature datum adopted to represent outside conditions in the British Isles will vary according to the precise location. Table 3.11 lists the levels recommended in the *Guide Section A2* for the same eight locations given in Table 2.15. It will be noted that, in this instance, both dry and wet bulb temperatures are given, as will be required for the design of air conditioning systems.

Table 3.11 Summer external design temperature (CIBSE recommendations)

Location	Month	Temperature		
		Dry bulb[a] (°C)	Wet bulb (°C)	Diurnal range (K)
Belfast	July	23	17	8
Birmingham	July	26	19	9
Cardiff	August	25	18	7
Edinburgh	July	23	17	9
Glasgow	July	24	18	9
London	July	29	20	9
Manchester	July	26	18	8
Plymouth	July	25	19	6

[a] The peak dry bulb temperature will occur normally at 15.00 hours GMT.

When considering applications for buildings in town centres, it is good practice to use dry bulb temperatures one or two degrees higher than those listed in order to make allowance for the *heat island* effect created by traffic etc.

Isotherms (contours of equal temperature) showing those dry bulb and wet bulb temperatures which are exceeded for only 1% of the summer months in the British Isles are shown in Figure 3.16. These isotherms relate to sea level and the temperatures shown should be reduced by 0.6°C for each 100 m of height above that base. Other similar figures have been produced for temperatures which are exceeded for greater proportions of those months and, for instance, if a 2½% level were to be chosen, then temperatures would be about 1 K lower in each case.

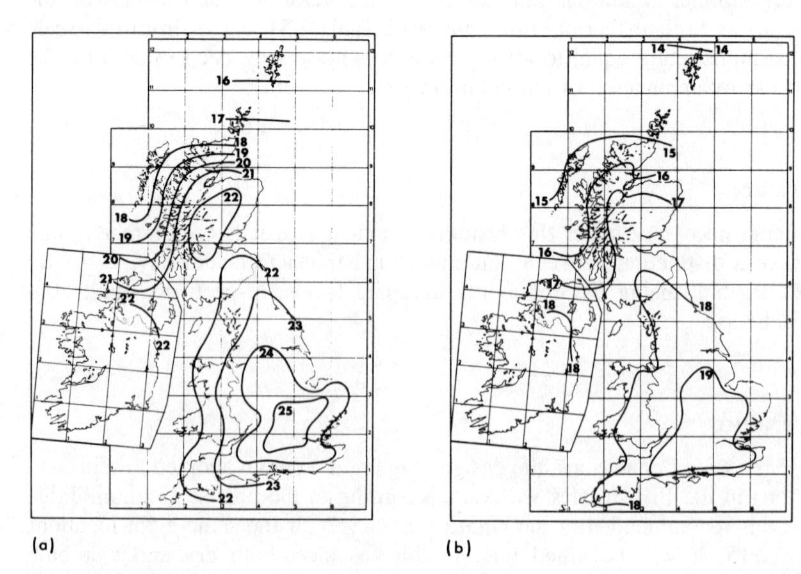

Figure 3.16 Dry bulb (a) and wet bulb (b) isotherms

Intermittent operation

In Chapter 2, when considering intermittent heating, the matter of excess plant and system capacity over and above that calculated for steady state heat losses in order to provide for morning preheating, was discussed in some detail. In the case of plant provided for cooling a building, the situation is not the same since the peak load (for the whole building) will normally occur somewhat later in the day, part way through the occupancy period after a steady rise in heat gain and requirements for cooling. In consequence, it is unlikely in most instances that any useful purpose would be served were the capacity of the plant to be increased above that required to meet the selected design condition.

It is nevertheless possible to go some way towards overcoming a deficit in plant size and to reduce running cost during warm daytime periods, by setting out deliberately to cool a building overnight when unoccupied. This may be achieved to some extent by circulating the relatively cold night air through the occupied areas with intent to cool the internal structure and thus create a reservoir against the demands imposed during the following day. This is not a routine to be recommended without a close study of potential problems such as the incidence of condensation, etc.

As to the calculation of summer loads generally, it must be borne in mind that the time at which the peak cooling load occurs in an individual space will vary with orientation. In consequence, the maximum simultaneous demand imposed upon a central plant is not normally the sum of the peak loads occurring in the individual spaces or zones of a building but almost certainly some lower figure. Calculation of the individual and simultaneous demands is further confused by the fact that the maximum cooling demand for a whole building is often displaced in time from what is the apparent peak of solar gain. This situation comes about as a result of an intricate pattern of interplay between fading instantaneous loads on one face and growing loads of similar type on another, all against a background of time delayed loads through opaque surfaces, to say nothing of lighting, occupancy and intermittent loads from machines. It is virtually impossible to make the necessary calculations using manual methods but computer programs exist which take the mixture in their stride.

Table 3.4 Cooling load (W/m²) due to solar gain through vertical glazing. (Latitude 51.7°N; lightweight building with 6 mm clear glass)

		British Summer Time (sun time + 1)										
Date	*Orientation*	*09.00*	*10.00*	*11.00*	*Noon*	*13.00*	*14.00*	*15.00*	*16.00*	*17.00*	*18.00*	*19.00*
June 21	N	89	104	122	137	147	150	146	135	119	103	92
	NE	358	307	220	159	165	169	164	153	138	120	102
	E	465	477	438	355	241	179	175	164	149	131	112
	SE	322	392	425	419	372	286	188	149	134	116	98
	S	89	127	208	289	342	359	337	280	196	116	83
	SW	102	121	138	158	202	300	381	423	424	386	313
	W	116	135	152	167	177	189	257	366	444	477	460
	NW	104	123	140	155	165	169	166	166	232	313	357
July 23	N	85	107	128	145	157	161	156	143	125	104	86
and	NE	314	276	205	162	173	177	172	159	141	120	98
May 22	E	419	440	412	343	244	188	183	170	152	131	109
	SE	309	381	418	419	380	304	209	162	144	123	101
	S	97	146	231	309	361	377	356	301	219	134	89
	SW	106	128	149	171	224	317	388	421	416	375	300
	W	114	136	156	174	186	197	258	353	417	439	413
	NW	102	124	144	162	173	177	173	168	214	281	312
August 24	N	59	82	102	119	131	134	129	117	100	79	58
and	NE	243	214	150	130	142	145	140	128	111	90	67
April 20	E	364	403	385	317	216	158	153	141	123	102	80
	SE	296	383	430	435	399	323	219	144	126	105	82
	S	90	167	264	346	398	415	394	338	253	154	80
	SW	88	110	131	156	235	335	406	437	427	376	285
	W	86	108	129	146	157	168	233	329	390	402	355
	NW	72	95	115	132	143	147	142	135	160	220	240
September 22	N	32	55	76	93	104	107	102	90	73	51	29
and	NE	143	142	91	96	107	110	106	94	76	54	32
March 22	E	263	371	375	306	191	122	118	106	88	67	44
	SE	245	396	474	494	459	376	254	136	101	79	56
	S	97	208	329	428	490	510	485	418	315	192	84
	SW	62	85	109	153	272	388	466	495	469	383	225
	W	50	73	94	111	122	134	209	318	377	364	243
	NW	36	59	80	97	108	111	107	98	97	146	133

Note: plant operation 10 hours. Glass is unprotected. Use correction factors from Table 3.5 only.

Table 3.5 Correction factors for shading (applying to Table 3.4 only)

		Glazing type		
Type of glass	*Building mass*	*Single*	*Double with 6 mm clear inside*	*Single with external light slatted blind used intermittently*
Clear 6 mm	Light	1.0	0.85	0.19
	Heavy	0.85	0.70	0.19
BTG 6 mm	Light	0.70	0.53	0.16
	Heavy	0.60	0.45	0.16
BTG 10 mm	Light	0.58	0.40	0.14
	Heavy	0.49	0.34	0.14
Reflecting	Light	0.49	0.37	0.13
	Heavy	0.42	0.31	0.13
Strongly reflecting	Light	0.25	0.16	0.09
	Heavy	0.21	0.14	0.10
Additional factor for air temperature control	Light	0.76	0.76	0.76
	Heavy	0.73	0.73	0.73

Table 3.6 Cooling load (W/m^2) due to solar gain through vertical glazing. (Latitude 51.7°N; lightweight building with intermittent blind use)

		British Summer Time (sun time + 1)										
Date	*Orientation*	*09.00*	*10.00*	*11.00*	*Noon*	*13.00*	*14.00*	*15.00*	*16.00*	*17.00*	*18.00*	*19.00*
June 21	N	81	96	114	128	138	142	137	126	111	95	122
	NE	223	169	81	144	151	154	150	139	124	106	87
	E	328	306	254	184	88	156	151	140	125	107	89
	SE	252	280	282	257	207	143	69	125	110	93	74
	S	102	124	179	220	238	230	197	145	89	39	61
	SW	69	87	105	184	183	239	273	281	261	217	152
	W	78	97	114	129	139	230	227	284	314	313	274
	NW	79	98	115	130	140	143	141	217	204	240	243
July 23 and May 22	N	85	107	128	145	157	161	156	143	125	104	86
	NE	205	161	80	155	167	170	165	152	135	114	92
	E	306	292	250	189	90	172	167	154	136	115	93
	SE	247	277	283	264	219	159	73	140	122	101	79
	S	75	209	194	234	252	244	212	161	59	112	68
	SW	74	96	117	211	196	247	275	278	256	210	144
	W	80	102	123	140	152	248	222	270	293	285	239
	NW	81	103	123	141	152	156	152	219	186	213	207
August 24 and April 20	N	59	82	102	119	131	134	129	117	100	79	58
	NE	159	118	57	126	137	141	136	124	106	85	63
	E	277	270	229	166	71	143	138	126	108	87	65
	SE	246	283	292	274	228	163	64	121	103	82	59
	S	118	153	212	254	272	264	231	177	111	36	52
	SW	56	78	99	205	204	257	285	286	259	203	120
	W	59	81	102	119	130	222	207	254	273	253	177
	NW	59	81	102	119	130	134	129	177	148	169	141
September 22 and March 21	N	32	55	76	93	104	107	102	90	73	51	29
	NE	205	39	89	94	105	109	104	92	74	53	30
	E	427	260	220	147	52	109	104	92	75	53	31
	SE	402	310	330	313	262	184	59	110	74	53	30
	S	137	189	262	312	333	324	285	219	137	36	48
	SW	31	53	78	224	237	298	328	322	274	175	37
	W	31	54	75	92	103	192	201	252	258	192	43
	NW	32	54	75	92	103	107	102	93	156	106	28

Note: plant operation 10 hours. Use correction factors from Table 3.7 only.

Table 3.7 Correction factors for light coloured internal blinds (applying to Table 3.6 only)

		Single glazing			*Double glazing*		
		Slatted blind			*Slatted blind*		
Type of glass	*Building mass*	*Open*	*Closed*	*Linen roller*	*Open*	*Closed*	*Linen roller*
Clear 6 mm	Light	1.00	0.77	0.66	0.95	0.74	0.65
	Heavy	0.97	0.77	0.63	0.94	0.76	0.64
BTG 6 mm	Light	0.86	0.77	0.72	0.66	0.55	0.51
	Heavy	0.85	0.77	0.71	0.66	0.57	0.51
BTG 10 mm	Light	0.78	0.73	0.70	0.54	0.47	0.45
	Heavy	0.77	0.73	0.70	0.53	0.48	0.45
Reflecting	Light	0.64	0.57	0.54	0.48	0.41	0.38
	Heavy	0.62	0.57	0.53	0.47	0.41	0.38
Strongly reflecting	Light	0.36	0.34	0.32	0.23	0.21	0.21
	Heavy	0.35	0.34	0.32	0.23	0.21	0.21
Additional factor for air temperature control	Light	0.91	0.91	0.91	0.91	0.91	0.91
	Heavy	0.83	0.83	0.83	0.90	0.90	0.90

Chapter 4

Survey of heating methods

Having arrived at the quantity of heat energy required for each space and thus for the building as a whole, as described in Chapter 2, it is now necessary to consider how this heat is to be supplied. It is thus appropriate to take a quick look at the whole range of available methods prior to examining these in more detail.

The number of heating methods and systems is almost unlimited if every combination of energy supply, method of conversion of that energy into heat, transmission medium and type of emitting element were to be considered. It is therefore quite useless to attempt to describe them at all clearly or systematically unless they are classified under the headings of the two main groups into which they all fall. These are:

- *Direct systems*, in which the energy purchased is consumed as required within the space to be heated.
- *Indirect systems*, in which the energy purchased is consumed at some more or less central point outside the space to be heated and then transferred to equipment in that space for liberation.

The *direct* systems may be divided into four primary categories, according to the type of fuel or energy source: solid; liquid; gaseous or electrical, as illustrated in Figure 4.1. Secondary sub-divisions relate to the characteristics of the terminal heat emitting elements i.e. whether these are *primarily radiant* or *primarily convective*.

In the case of *indirect systems,* the nature of the medium selected to distribute the thermal energy, from the point of generation or storage to the space to be heated, is of the greatest significance. This medium may be either a *liquid* (water), or a *vapour* (steam) or a *gas* (air) and each may be utilised over a wide range of temperatures and pressures. In most instances, the choice of fuel or energy source is quite independent of the distribution medium although there are rare circumstances where certain combinations would be inappropriate.

Consequent upon the scope thus presented for design development, the variety of system arrangements and terminal heat emitting equipment associated with each is considerable. The principal criteria by which indirect systems may be categorised are thus not easy to define since there are so many secondary, but not unimportant, sub-divisions within each that a diagram on the lines of Figure 4.1 would be confusing rather than helpful. It is more useful, therefore, to catalogue some of the options as shown in Table 4.1.

The text which follows concentrates upon particulars of the various types of equipment used with direct systems followed by no more than very brief notes covering the leading characteristics of indirect systems. The latter are dealt with in greater detail under specific headings in later chapters.

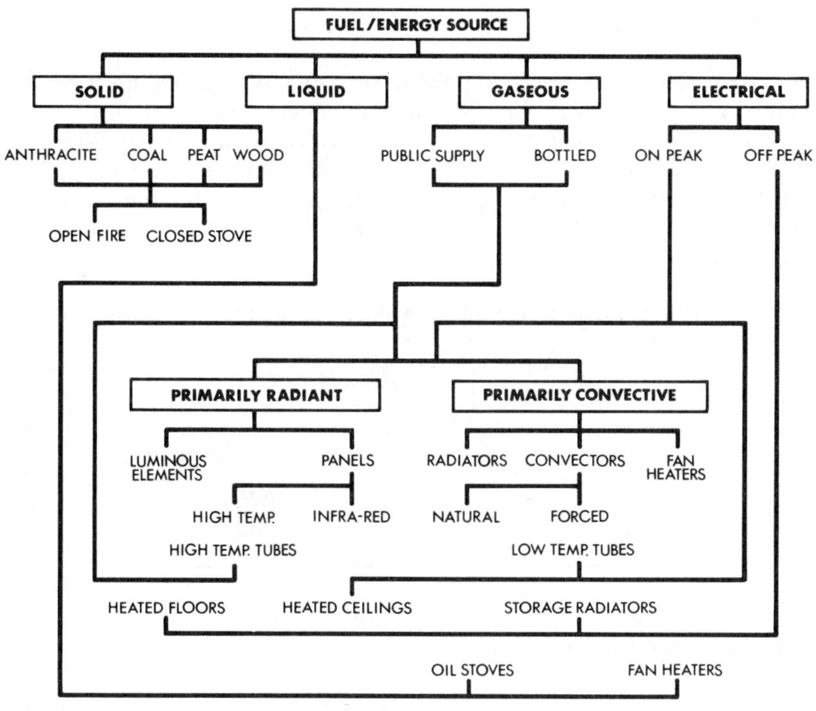

Figure 4.1 Categories of direct heating systems

Table 4.1 Options for indirect heating systems

	Distribution medium		
Item	*Water*	*Steam*	*Air*
Temperature or Pressure or Velocity	Warm or low temp. Medium temp. High temp.	Low pressure Medium pressure High pressure	Low velocity Medium velocity High velocity
Usage	Direct or at reduced level		
Arrangement	Visible in space or concealed		
Reticulation	Radial or concentric		
Pipework	Single or two-pipe Reversed return	Single run or ring main	
Motive power	Thermosyphon 'Accelerated' Pumped	Gravity return Pumped return Combined	Natural Mechanical
Terminals	See text for separation (primarily radiant or convective)		Grilles Diffusers Louvres
Plant	Dispersed, semi-dispersed or centralised		
Fuel or energy source	Solid, liquid, gaseous or electrical		

Condensation – a warning

Where a liquid or a gaseous fuel is used to provide an energy supply to a direct system of any description, it is important to appreciate that all equipment does not necessarily provide for the discharge of the products of combustion to outside the building. Where no flue is incorporated for this purpose, there will be a significant emission of water vapour to the heated space, dependent upon the type of fuel and this will in all probability lead to condensation on windows and cold walls. The data listed in Table 4.2 show the magnitude of this problem.

Table 4.2 Moisture released to the enclosure by combustion of fuels in direct heating appliances

	Rate of fuel consumption or energy use which will release 1 litre of water per hour			
Fuel	litre/h	kg/h	Therm/h	kW
Kerosene	1.0	–	–	10.2
Natural gas	–	–	0.23	6.6
Butane	–	0.65	–	8.8
Propane	–	0.62	–	8.5

Electrical off-peak storage systems

It seems open to question whether systems of this type fall within either of the definitions given previously and thus whether they may be described properly as either *direct* or *indirect*.

Since discussion of the techniques used in designs for heating a space using energy output from a thermal store, charged some hours previously, deserves a chapter to itself, it is not proposed to include any further mention here.

Direct systems – solid fuel

Open fires

These, man's primitive source of heat, were most economic in mediaeval times when the hearth was open to the heated space and gases escaped through a hole in the roof after cooling. With this arrangement little of the energy content of the fuel was lost but there were compensating disadvantages. Later, flues were invented whereby economy was sacrificed in the interests of cleanliness and easier respiration.

The modern version of the open fire is usually arranged for continuous burning, with means for controlling the air inlet and with a restricted throat to the flue, sometimes adjustable. It is designed to burn smokeless fuel and the efficiency may be of the order of 30–40% as compared with 15% or less for the pre-1939 pattern of grate. Some forms of modern appliance provide airways for a convection output thus increasing efficiency slightly.

Closed stoves

The large coal stoves so common in Europe and those in this country designed to burn anthracite are more efficient, labour saving and draught-reducing than the open fire and may be adjusted to keep alight overnight. Efficiencies of as high as 50% are claimed. Hopper-fed 'closable' anthracite stoves exist in many forms, free-standing and brick-set.

Examples of the wide variety of wood-burners which have been imported from Scandinavia, however pleasing aesthetically, are rarely either efficient or convenient in view of the need for frequent replenishment of the fuel charge. Logs, even when notionally dry, have a heat content only half that of other more compact solid fuels. Particular care is needed furthermore, but infrequently taken, to ensure that the condensation and tar products deposited in the chimney do not lead to inconvenient soot falls and fire hazards respectively.

Direct systems – liquid fuel (primarily convective)

Closed stoves

Commonly used in domestic premises for background heating in severe weather, such appliances have the merit of being portable and relatively cheap to run. Since the products of combustion are discharged into the space served, efficiency approaches 100% for ratings of about 2–5 kW. The twin hazards of fire and asphyxiation have, together with the problem of condensation of water vapour on windows and cold walls, led to a decrease in popularity of these units.

Industrial warm air systems

The type of unit illustrated in Figure 4.2 is used for industrial space heating and has the advantage of being simple in both construction and operation. It takes the form of heat transfer passages through which gases from an oil-fired combustion chamber are directed,

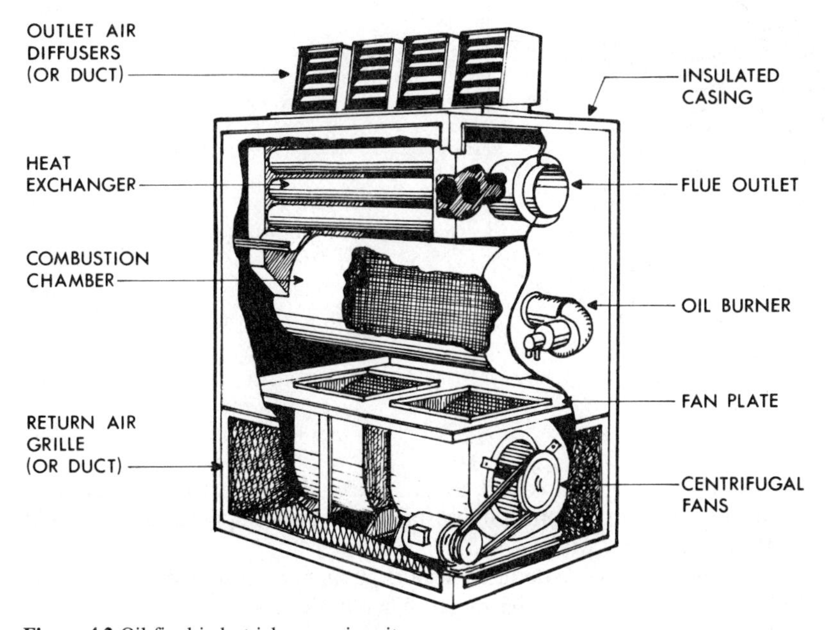

Figure 4.2 Oil-fired industrial warm air unit

and over which air is circulated by a fan or fans. Efficiencies of over 90% are claimed for a range of outputs from 30 to about 400 kW. The fuel supply to a number of such units may be piped from a central store.

The diagram shows a unit with a recirculated-air inlet and cabinet mounted outlet diffusers but an outside air supply and discharge ductwork can be fitted. It could be argued, however, that the addition of ductwork is not compatible with one particular advantage of such heaters, which is the relative ease with which they may be moved to suit changes in shop-floor layout, etc. Outlet flues are required in all cases.

Direct systems – gaseous fuels (primarily radiant)

Luminous fires

This type of equipment has been improved greatly over the last 30 years and, for connection to the public supply, it is now unusual to find a design which does not provide for direct connection to a flue. In addition to the radiant elements, additional heating surface is usually incorporated to provide further output by convection, as shown in

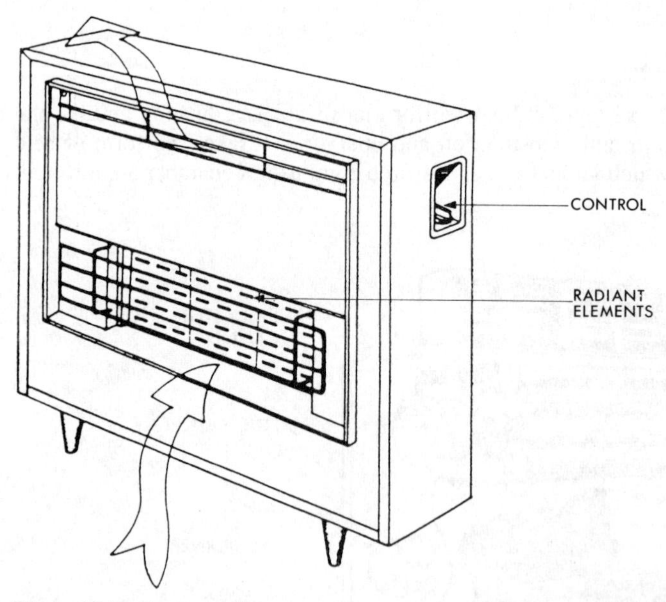

Figure 4.3 Luminous gas fire

Figure 4.3. Response to control is rapid but the intense radiant effect of the luminous elements may lead to discomfort in some circumstances. A total output of up to 3 kW is common but some patterns have a rating of 5 kW.

Also within this category fall those portable heaters having cabinets designed to house the butane cylinder which provides the fuel source. Such heaters, again generally rated to produce about 3 kW, cannot, by definition, be flued.

Infra-red heaters

Designed primarily for industrial applications in higher buildings (but often misapplied to others) heaters of this type are pipe-connected to either the public supply or to either butane or propane cylinders. For permanent installations when wall mounted or suspended from a roof, they may be rated at up to 30 kW but a range between 3 and 15 kW is more common.

Construction of one type, Figure 4.4(a), takes the form of a heavy duty rectangular reflector within which refractory elements and a burner array are fitted behind a safety guard. Another type, Figure 4.4(b), is cylindrical in plan with a shallow conical reflector over. Portable versions, as in Figure 4.4(c), are mounted on a telescopic stand which provides means to secure a propane cylinder as a fuel source.

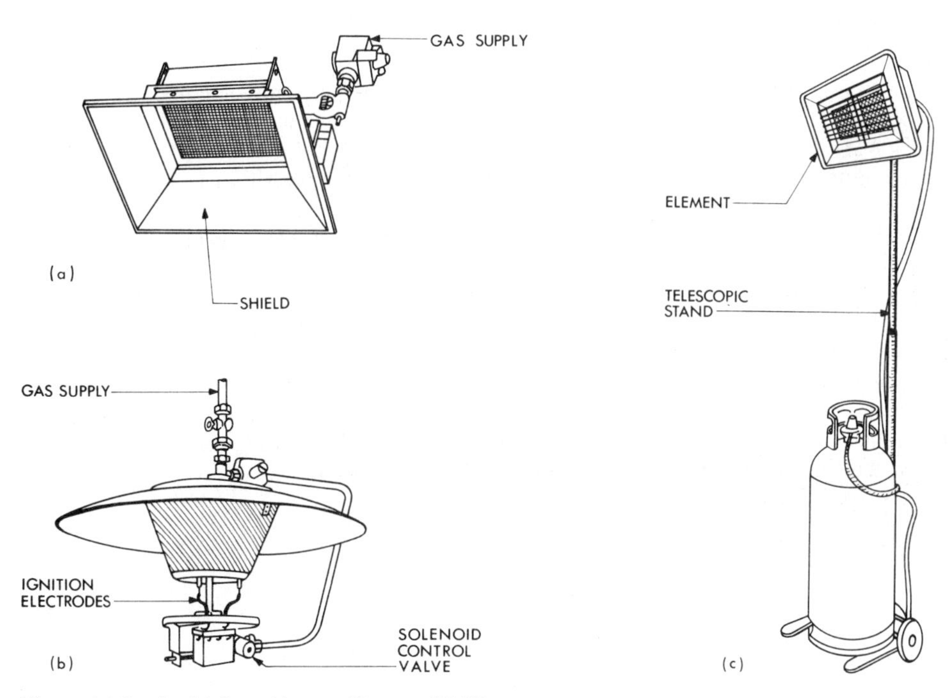

Figure 4.4 Gas-fired infra-red heaters (Spaceray/EMC)

Flue connections cannot usually be provided for heaters of this pattern and responsible manufacturers quote minimum rates of outside air ventilation necessary either per unit or per kW rating. In the case of portable units, these are usually sited in temporary positions within open buildings, stockyards and construction sites, etc., and the absence of a flue is not important.

Radiant tubes

For such systems, the principal elements are a burner; a steel tube of about 60–70 mm diameter; a reflector plate and a vacuum pump or extractor fan discharging to outside the building. The products of combustion are drawn through the tube which reaches a

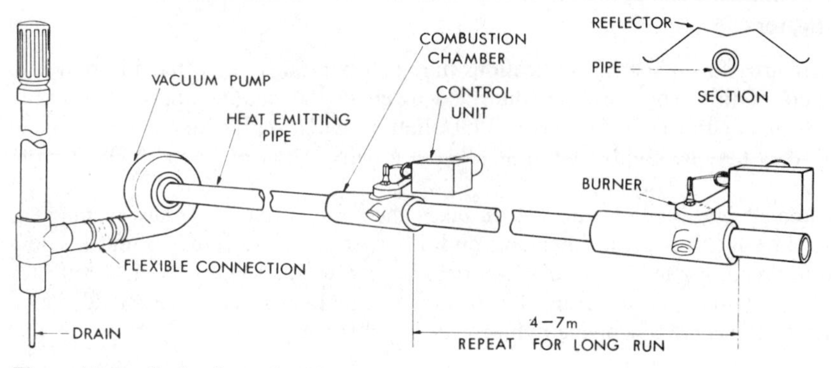

Figure 4.5 Gas-fired radiant tube (Nor-Ray-Vac)

temperature, below luminosity, of about 540°C. Air for combustion is drawn from the heated space, except in circumstances where danger exists as a result of an industrial process; in such applications a ducted air supply is provided. Efficiencies of between 75 and 90% are claimed by the manufacturers.

In one make, the units are in lengths of between 4 and 7 m with outputs of 12–14 kW and, where circumstances require, lengths are fitted in cascade with outlet gases from the first joining the second to be exhausted finally by a single vacuum pump: this arrangement is shown in Figure 4.5. Another type has a slightly different configuration in that each element, rated at about 15 kW for a 5 m length, is in the form of a 'U' tube with the suction fan adjacent to the burner box, as shown in Figure 4.6. The cascade arrangement is thus not practicable, the units being individual with a flue required for each.

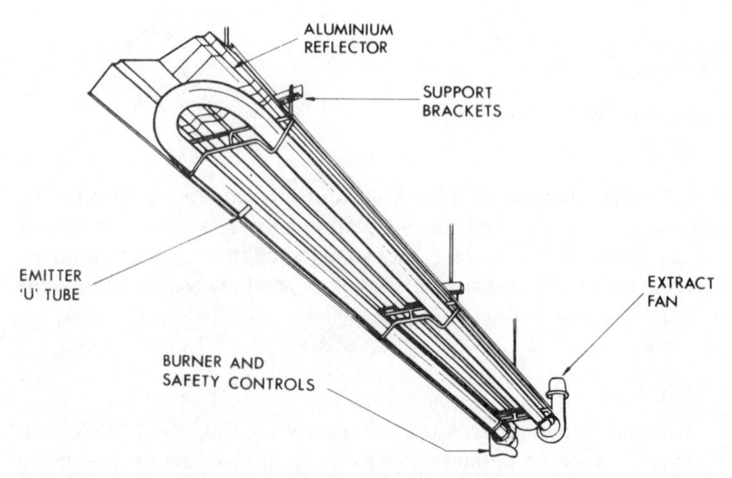

Figure 4.6 Gas-fired radiant tube (Gas-Rad)

Direct systems – gaseous fuels (primarily convective)

Natural convectors

As in the case of luminous fires, the better types of convector are designed to provide means for dispersal of the products of combustion via a conventional outlet or a balanced flue arrangement, as in Figure 4.7. In terms of construction, convectors consist of a heat exchanger so mounted within a casing as to provide a passage for the movement of convected air from a low level inlet to a top or top-front outlet. In most cases there will be some heat transfer to the front of the casing and a small radiant output may result. Designs aimed at the institutional building have ratings of up to 10 kW but 3 kW is a more representative average.

Convectors without flue connections are available but should not be used in domestic premises.

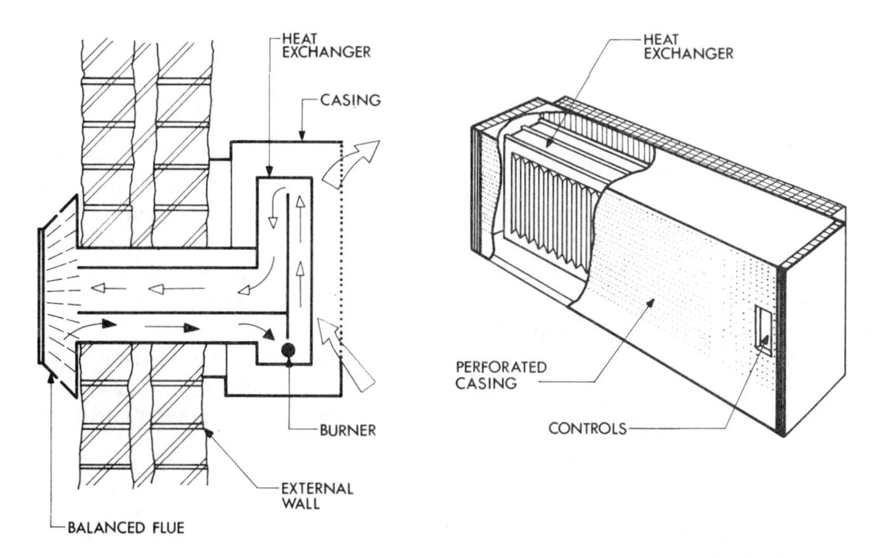

Figure 4.7 Gas-fired natural convector (Drugasar)

Forced convectors

A wide variety of units may be grouped under this heading, ranging from the robust portable blast heaters with axial flow fans used in warehouses and on construction sites, to permanently fixed unit heaters for suspension in industrial buildings. The former are often connected to an equally portable fuel supply in propane cylinders and are, of course, flueless.

The more permanent unit heaters should be connected to a properly arranged piped fuel system which may originate from either the public supply or an LPG tank source. Most manufacturers make provision for units of this type to have flued outlets to outside the building. Ratings vary with type (Figure 4.8) but most cover the range 30–120 kW.

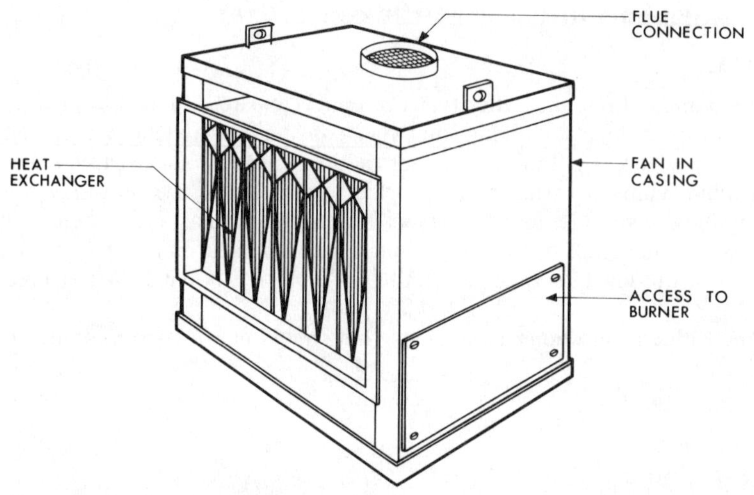

Figure 4.8 Gas-fired forced convector (Reznor)

Domestic warm air systems

A pattern of air heating unit developed for residential use is shown in Figure 4.9. It consists of a combustion chamber formed in some non-corrodible material, a fan for air delivery over that chamber and a casing into which air is drawn from the spaces served and to which it is returned after heating. Such units are made in a size range suited to small flats up to four-bedroom houses, i.e. about 8–24 kW.

For application to a flat, what are called 'stub-duct' connections are provided such that warmed air is delivered to the various rooms served, at low level, with return air at high level from a central hallway or from one of the rooms. When applied to a two-storey

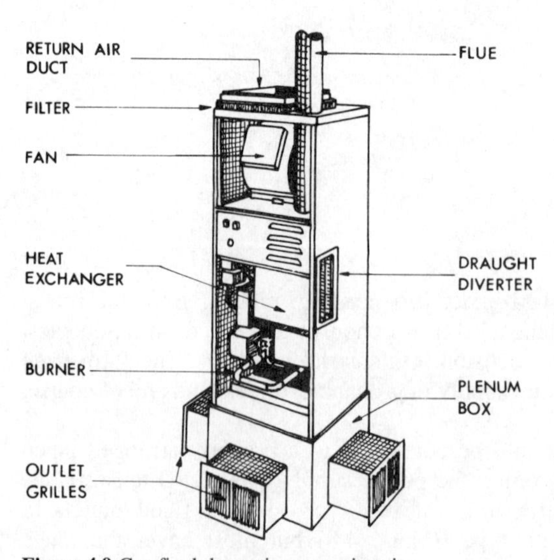

Figure 4.9 Gas-fired domestic warm air unit

dwelling, a ductwork arrangement carries a supply of warm air to the various rooms to be heated, return being to the unit which is usually sited in a hallway. To allow for this circulation, each doorway would have a gap below it or, preferably, be fitted with a return air grille.

Although popular with developers in the 1960s and 1970s, systems of this type have, in most cases, failed to give satisfactory service as a result of the wholly convective output provided. These systems are, however, economical and are suitable for total hand-over to occupiers with no subsequent landlord involvement.

Industrial warm air systems

For industrial application, the type of warm air unit illustrated in Figure 4.3, but with a burner suitable for gas rather than oil, is applicable. The absence of any necessity for energy storage by way of fuel is an advantage and, as in the case of the oil-fired version, some degree of mobility is retained.

While it is generally to be preferred that such equipment should be provided with a connection such that flue gases are discharged to the outside air, the advent of natural gas and the increasing use of bottled gas have led to the development of 'direct fired' air heaters, as shown in Figure 4.10.

Equipment of this type passes the air for circulation through or directly over an open combustion chamber with no separation or flue, the products of combustion passing out with the air into the heated space. The practicability of this arrangement rests upon two features:

- The fuels in question, with the consequent lack of any serious concentration of pollutants in the products of combustion (with the exception of water vapour).

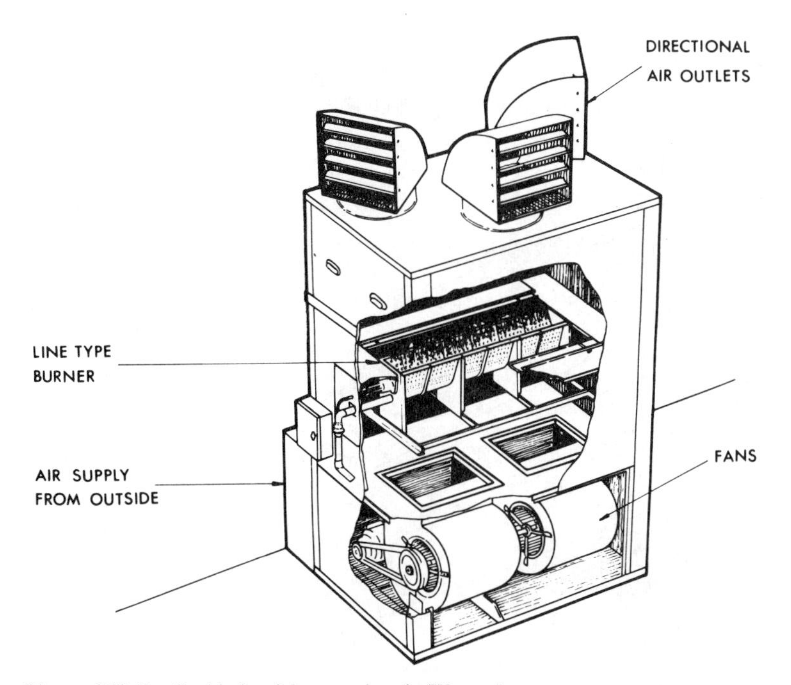

Figure 4.10 Gas-fired industrial warm air unit (Wanson)

- The method of application of the heater whereby the air drawn in for circulation is always *from outside the building* and not recirculated from the heated space.

In these circumstances, the purity of the discharge will be within the accepted threshold values laid down by the Health and Safety Executive. Unit ratings range up to about 650 kW and, since all the energy in the fuel passes to the air circulated, efficiency may be well over 90%.

Direct systems – electrical (primarily radiant)

Luminous fires

The earlier type of luminous fire, which consisted of an exposed coiled-wire element, mounted in some manner to a refractory block, is seldom seen today. Elements are now usually silica sheathed and are commonly mounted in front of a polished aluminium reflector having parabolic form. A great variety of types and designs is available with capacities ranging from 500 W to 3 kW. The effect of such heaters is localised and their use is generally confined to domestic premises, hotel lounges and the like.

Infra-red heaters

Wall or ceiling models of these are suitable for kitchens and bathrooms in a domestic context and more robust patterns may be used in commercial or industrial premises.

The elements used are similar to those fitted to luminous fires but, for a given rating, are commonly longer, as Figure 4.11, and arranged to operate at about 900°C. They are sometimes misapplied to churches where, at that temperature, the usual mounting position in the eaves is too high to provide effective radiant cover. Ratings are up to 3 kW per unit.

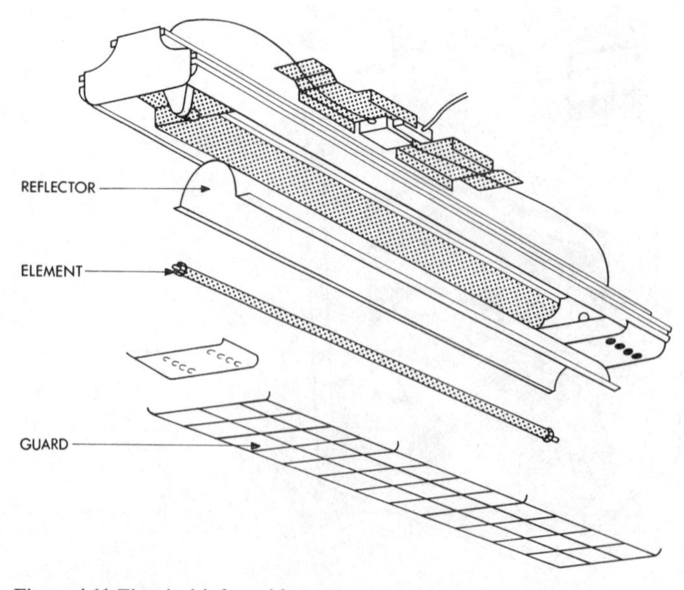

Figure 4.11 Electrical infra-red heater

Quartz lamp heaters

For application to large spaces either where the requirement is intermittent or where only localised areas require spot heating, quartz lamp heaters operate at a temperature of over 2000°C. The elements, each of which is rated at about 1.5 kW, consist of a tungsten wire coil sealed within a quartz tube containing gas and a suitable halide. As illustrated in Figure 4.12, a rather unlovely casing contains a number of elements (normally a maximum of six), each of which is mounted in front of a polished parabolic reflector.

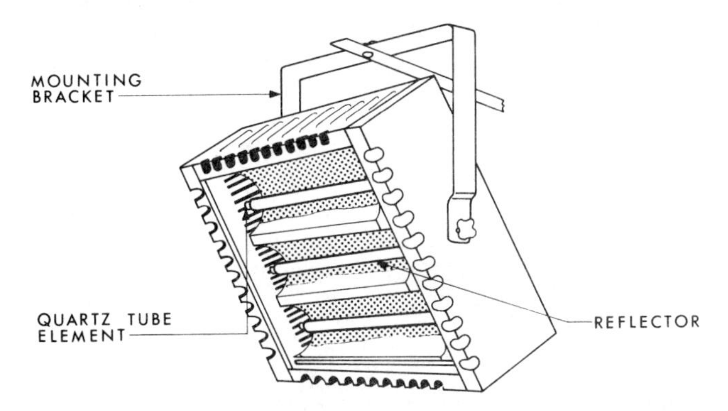

Figure 4.12 Electrical quartz lamp heater (Quartzray)

High temperature panels

Consisting of either a vitreous enamelled metal plate or a ceramic tile behind which a resistance element is mounted within a casing, panels of this type operate at a temperature of about 250°C and have ratings in the range 750 W to 2 kW. Although rather more sightly and easier to clean than infra-red or quartz lamp units, see Figure 4.13, application of these panels is usually confined to washrooms and the like in industrial premises.

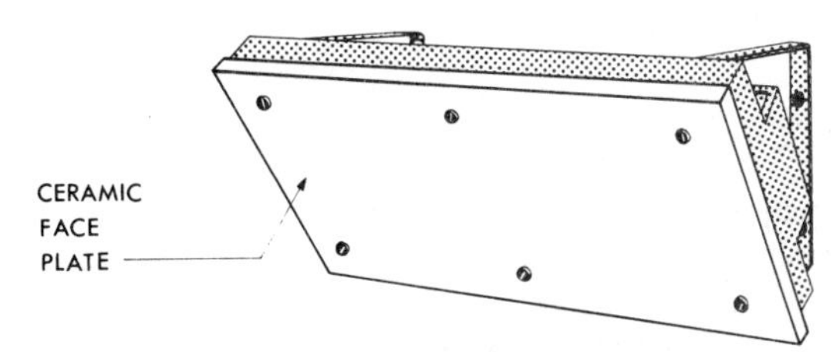

Figure 4.13 Electrical high temperature panel (Heatovent)

Low temperature panels

While equipment of this pattern should, strictly speaking, be listed alongside oil filled radiators and other convective heaters, it is convenient to deal with them here.

Panels of this type are in some cases faced with a metal plate having an enamel finish, other makes use resin impregnated hardboard or a plastic laminate. The heating element may be a conventional form of conductor within an insulating envelope or a polymer coated mesh of synthetic fibre sealed between sheets of polyester film. Many such panels are purpose made for special applications, one being church heating where a panel is fixed to the rear of each pew seat to provide localised warmth to members of the congregation without heating the church fully. Operating temperatures of 70–90°C are usual with ratings of 100–500 W.

Ceiling heating

It is necessary to make a clear distinction here between a purpose designed ceiling heating system and the fortuitous heat input from the underside of an intermediate floor slab which supports a floor heating system (possibly an off-peak storage type) serving the rooms above. It is the former alone which falls wholly within the definition of a *direct system*.

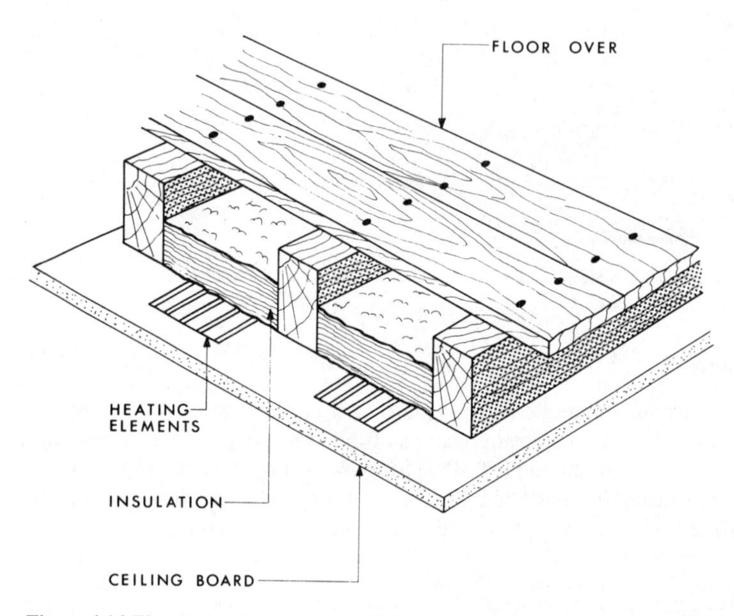

Figure 4.14 Electrical ceiling heating (ESWA)

The heating elements used for ceiling heating, which consist of waterproofed conductor strips protected by insulating plastic membranes, are installed immediately above and in contact with the ceiling finish, as shown in Figure 4.14. A layer of insulating material, as thick as possible (preferably 200 mm) is laid over the membrane. The heat emitting surface is, of course, the ceiling itself and some paints and other finishes are unsuitable. Electrical loadings up to a maximum of about 200 W/m^2 of the treated area may be made available. Ceiling heating takes up no floor or wall space and the attributes of an evenly distributed heat output may well be appropriate as a 'top-up' service, complementary to various forms of storage system.

Floor heating

Although the more common form of electrical floor heating is that which was designed to use energy provided during off-peak hours, an alternative direct method may be used, superimposed upon a solid ground floor. For this application, a variant of the type of element applied to the manufacture of low temperature panels is used, having a rating of up to about 150 W/m^2, laid close to the finished floor surface. Typically, the structural floor is covered with about 50 mm of insulating material and the heating elements follow prior to a final layer of chipboard with carpet tiles or some similar finish. Such an arrangement is particularly suitable for buildings which are used only intermittently and for relatively short periods such as churches, etc.

Direct systems – electrical (primarily convective)

Natural convectors

A considerable variety of patterns and sizes of such heaters is available, most now comprising a metal sheathed element or elements fitted within a rectangular casing. The enclosure is arranged to promote air flow over the elements via an opening at the base and a louvred outlet at the top. Convectors may, alternatively, be floor standing and portable or wall mounted: in the latter case, mounting brackets may provide means for easy access to the rear face for cleaning. Ratings range from 500 W to 3 kW, the larger sizes having multiple elements for control purposes. The temperature of the casing remains relatively low in most cases and output by radiation is minimal in consequence. Convectors of this and other types are particularly responsive to thermostatic control.

Skirting heaters

These are a variant of the convector type heaters noted above having a height of about 150 mm with a slotted casing, as in Figure 4.15, to promote air flow over the element. Most are arranged for either floor or wall mounting and have ratings of between 550 and 750 W per metre length. They may be fitted in small rooms or mounted below tall windows.

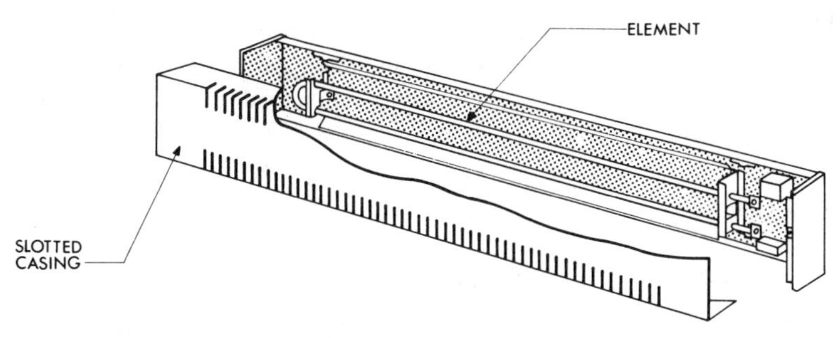

Figure 4.15 Electrical skirting heater (Dimplex)

Oil filled radiators

Similar in appearance to the pattern of steel panel radiators used in hot water heating systems, these have a light oil filling within which is an immersion heater element. The oil begins to circulate when heated and thus acts to transfer heat to the outer surfaces. With a maximum surface temperature of about 90°C, output ratings vary with size up to a normal maximum of 1.5 kW. Wall mounted or portable models on feet are available: the former are useful in circulation spaces.

Tubular heaters

As the name implies, these are steel or aluminium tubes commonly round or oval in cross-section, as shown in Figure 4.16, with no heat transfer filling other than air. The heating element extends from end to end to provide an even surface temperature of about 80°C. A single tube at 50 mm diameter has an output of about 180 W per metre length and tubes may be mounted in banks, one above the other, for a greater output per unit length. The diagram includes a convective type having a substantial casing with an output of about 400 W per metre run. This is not strictly a true tubular heater although it has a similar range of applications. It is smaller in cross-section than the skirting heaters of Figure 4.15 and rather more substantial.

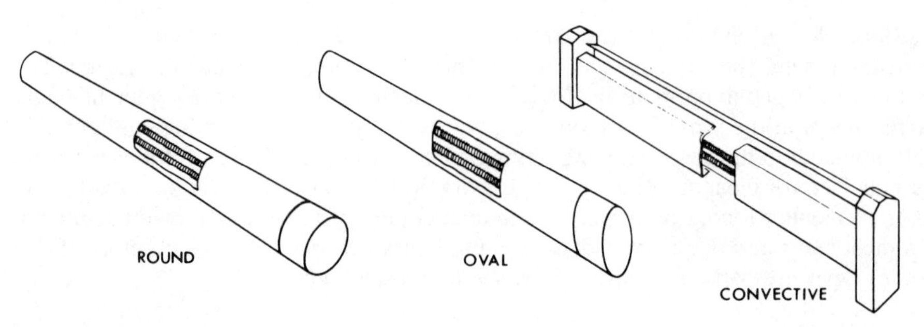

Figure 4.16 Electrical tubular heaters

Tubulars have been used in churches, placed under the pews, in which position they keep the lower air warm with no attempt to heat the whole enclosure. Fitted below clerestory windows in any tall building, they prevent down-draughts.

Forced convectors

This general heading covers a variety of items from the compact portable domestic fan heater rated at up to 3 kW at one end of the scale to large industrial units rated at 30 kW

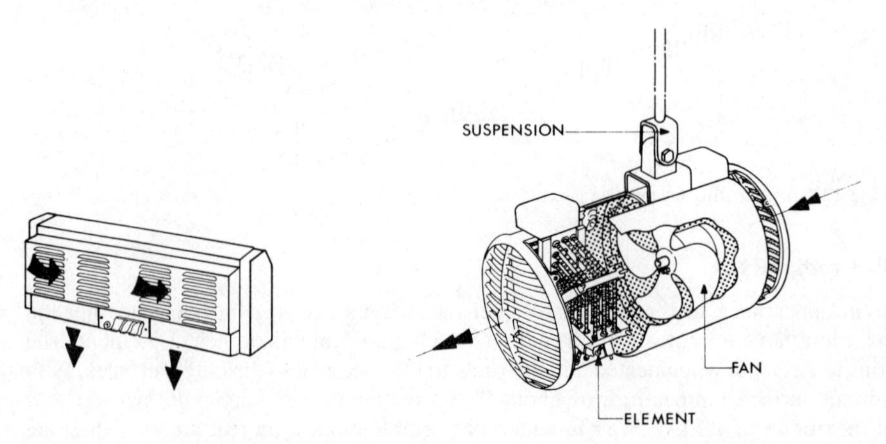

Figure 4.17 Electrical forced convectors (GEC-Xpelair)

or more at the other. Between these extremes are commercial type heaters, fitted with axial-flow fans, having ratings of about 3–6 kW and a range of cased tangential-fan units rated at up to 18 kW to meet the particular requirement of a source for warm air curtains at building exits and entrances. Figure 4.17 illustrates two such items.

Control of direct systems

The methods used to control direct systems and individual appliances are legion and range from simple on/off switches through to quite sophisticated systems for large installations. The following is a brief summary of methods available.

Individual domestic size units

- Simple switches providing 'on/off' action.
- Dial switches providing staged control, i.e. 1, 2 and 3 kW in the case of a gas or electric natural convector.
- Simple multi-switch control of any appliance which is fitted with a fan, i.e. a forced convector, including speed control if available.
- Automation of any of the foregoing via a thermostat which may be either integral to the unit or wall mounted as a remote room sensor.

Groups of units in larger systems

- Any of the above, applied to whole or part of the group, i.e. by zoning to suit aspect or sub-division.
- Further automation by sensing outside weather and use of time clocks to suit circumstances.

The above methods may be applied to liquid, gaseous and electrical equipment using self-acting controllers in the less complex applications and electric or electronic devices otherwise. Most equipment will be fitted with override or other safety controls as may be appropriate to the particular fuel or energy source.

Indirect systems – liquid media

Distribution of heat

Although non-toxic thermal fluids such as particular types of petroleum oil and a variety of synthetic chemical mixtures are available, the use of these is confined to high temperature process systems. Hot water, therefore, is the only liquid medium which needs to be considered under this heading.

Water has a high specific heat capacity per unit volume; it is non-toxic unless adulterated; it offers great flexibility in that the temperature may be adjusted easily to match the changes in demand imposed upon a heating system as a result of variations in the weather and it is extremely cheap. It may have the disadvantages of being corrosive to metals, of carrying mineral salts which produce scale deposits and of freezing at a relatively high temperature but each of these may be overcome without very much difficulty given an intelligent maintenance effort.

For the common run of building size in the domestic and commercial sectors, systems are usually designed to operate at atmospheric pressure. That is to say that they are commonly connected to a cistern, mounted not far above the highest point, the water content of which is open to atmosphere. In some circumstances a diaphragm expansion cylinder may be used instead of the open cistern but not, in this application, with intent to enable pressure to be raised much above atmospheric level. The operating temperature at the source of heat is thus normally controlled at a level below atmospheric boiling point and the system is described as being at *low temperature*.

In the case of very large individual buildings and of sites with several scattered buildings where considerable distances have to be covered by the pipework distribution arrangements, systems are unlikely to be open to atmosphere. By arranging that they are sealed, the pressure and thence the operating temperature of the water may be raised above atmospheric boiling point, so increasing the quantity of heat which is carried by a given size of pipe and reducing the physical dimensions of heat emitting equipment. Depending upon the level to which these operating criteria are raised, the description *medium temperature* or *high temperature* is applied to such a system.

Terminal equipment

All hot water heating systems, whatever the method or temperature of distribution, require some form of terminal heat emitting equipment for installation in the spaces to be served. A later chapter will consider this subject in more detail but, for the purpose of comparison in the present context, the following are the more common usages:

Low temperature systems only

- Exposed piping.
- Radiators.
- Pipe coils embedded in the structure.
- Metal panels in suspended ceilings.

Low, medium and higher temperature systems

- Metal radiant panels and strips.
- Natural convectors.
- Forced (fan) convectors.
- Skirting heaters.
- Unit heaters.
- Air/water heat exchangers for ventilation systems.

Indirect systems – vapour media

Distribution of heat

In earlier editions of this book, the use of steam as a distribution medium was considered in some detail. During the intervening period, in particular the last two decades, it has largely been superseded for that purpose by either medium or high temperature hot water. More recent practice, therefore, is to use steam only in circumstances where it is either available already on the site to be served or must be produced by a projected boiler plant to satisfy some process need. The upper limit of pressure at which steam is available will, in consequence of this practice, be determined by other than requirements for space heating. A supply at a reduced pressure may be provided but this is unlikely to be at the atmospheric or even sub-atmospheric levels which were popular in the past. This subject is considered further in a later chapter.

Terminal equipment

When steam is used this may be either as the primary medium in a calorifier (or heat exchanger) to produce hot water, or in heat emitting apparatus of one or more of the following types:

- Exposed pipes.
- Metal radiant panels or strips.
- Natural or forced (fan) convectors.
- Unit heaters.
- Steam/air heat exchangers for ventilation systems.

Indirect systems – gaseous media

Distribution of heat

Historically, the culture of the Romans developed a type of heating system which was primarily radiant, an art which was lost for some 1600 years. This system was centred about a furnace room below ground from which hot flue gases were conveyed, via ducts under the floor and flues in the walls, to emerge at various points about the building. Sometimes proper ducts were formed but in others the whole space under the floor (the *hypocaust*) was used. The wall flues took the form of hollow tiles, very similar to their modern counterparts. A similar but rather more sophisticated system was applied to the twentieth-century Anglican Cathedral in Liverpool using heated air, fan circulated through a pattern of structural ducts immediately below the floor finish.

Air, the only gas now in use as a medium for distribution, has a very low specific heat capacity per unit volume with the result that the distance over which heat can be transported is limited since duct sizes inevitably are large and the heat loss from them is thus disproportionately large also. This difficulty may be overcome in part by increasing the velocity within the duct but, by doing so, the power required to drive the fan may increase considerably.

The warm air equipment used in *direct* systems, whether oil, gas or electrically fired, could be considered as a source for an *indirect* air distribution system if any outlet ducting fitted to it were arranged to serve more than one room. In addition, for industrial or commercial applications, if the air circulated were to be drawn, in some part at least, from outside the building then the basic definition of what is called a plenum system would be met.

It is perhaps worth emphasis that there is a fundamental difference between a ventilation system and a plenum system. Although the components of each are similar, the air supply provided by the former to the rooms served will be at near-to-room temperature only, the building being heated by other means, whereas that provided by the latter will be at a much higher temperature in order to heat as well as to ventilate.

Factors affecting choice

The choice of system to be adopted can only be related to the type of building, since what may be suitable for a process plant would be out of the question for say a block of flats.

Home heating

Choice here may be influenced by personal preference, by the routine of daily occupation, by sales pressure from various fuel interests, by close regard for economy or, where a public authority is concerned, by what is permitted by various local or national regulations.

In an existing building, the choice of system may be limited by the facilities available, i.e. absence of a suitable flue, difficulties in arranging sensible and visually acceptable pipe routes, etc., and the availability, in rural areas, of a public supply of gas or electricity.

Direct heating systems using solid fuel fires and stoves involve labour, dust and dirt and, with modern habits of families being out all day, are often inconvenient. Gas and electric fires are then preferred, but tend to be expensive in running cost. Warm air systems, using off-peak supply where electricity is chosen, are thus often favoured since they may be controlled to give a set temperature at times when the house is occupied, with some minimal background warmth during the rest of the day to counteract, among other things, condensation. The absence of some form of radiant heat is a disadvantage during leisure hours.

Indirect systems for home heating can give whole-house comfort as well as hot water probably more consistently than any direct system. The capital cost may be higher but the running cost less, subject to the vagaries of the various fuel tariffs. The choice of fuel for a domestic indirect system, however, may not always be a matter of cost even supposing that relative prices were to remain stable. Space for a coal store or an oil tank may not be freely available and in rural areas having no access to a public gas supply, bottled gas or electricity may be the only two options available. To load and declinker a solid fuel system may be impossible if the tasks fall upon elderly people.

In a new house, some may prefer floor heating in order to avoid the loss of floor or wall space and where, in such circumstances, it has been possible to incorporate adequate insulation in the structure then off-peak electricity might be the chosen energy source. In an older house, skirting heating might be thought to be less obtrusive than radiators provided that furniture can be suitably disposed: radiators might be cheaper. Where a flat in an older property is leased then electrical storage radiators might be thought suitable by a tenant since this equipment could, if necessary, be removed to some other dwelling as furniture.

The pattern of home occupancy is an important factor as far as running costs are concerned but control systems suitable for domestic use have been developed to a high level of sophistication. Intending purchasers should not be misled by claims made by any one particular fuel or other interest since what can be done by one can probably be done equally well by all.

Thus, as may be seen, the choice for a home heating system is very much a matter of personal circumstances and personal taste. It is impossible to generalise, particularly as to comparative costs, in view of all the variables.

Flats – multi-storey

Systems of heating in common use and from which a choice would no doubt be made are:

- Electric storage radiators off-peak.
- Electric warm air.
- Gas warm air.
- Oil warm air from common storage.

- Central heating by hot water radiators.
- Hot water/warm air from a boiler plant per block.
- Group heating by hot water to a number of blocks.

The housing authority or estate developer in making a choice will be influenced by the capital cost, fuel cost, maintenance costs and labour to run. There is also the question of how the heat is to be charged – whether by the public utility reading its own meters (the consumer paying direct), or whether the landlord will be responsible for reading meters and collecting the money, or, again, whether the cost of heat is included in the rent or in a service charge.

All these matters have to be considered as well as the type of tenant and what kind of expenditure can be afforded before a recommendation can be made.

Commercial

Offices being usually in blocks are most economically heated by an indirect central system in some form. Choice is then confined to the kind of emitting surface, radiators, convectors, ceiling heating, etc., and to the kind of fuel.

For reasons already explained, an office block with large expanses of glass and probably having construction which could be characterised as lightweight, will be subject to rapid swings of temperature; hence a system which is quickly responsive to change of output is needed. Thus, embedded-coil systems, or electric-floor warming, are not to be recommended although they may find a place in buildings of heavy construction.

For deep planned buildings, it must be appreciated that a simple heating system cannot provide an adequate service in any circumstances. In the case of a tall building, where reliance upon opening windows for ventilation is inappropriate due to the increased wind effect, a mechanical ventilation system must be provided to introduce a supply of outside air. This supply, warmed in winter to room temperature or just below and supplemented by some means of air extraction, may suffice for that period of the year when outside temperatures are low enough to lead to a parallel heat loss, say 7°C and below. At any higher external temperature, heat *gains* from solar radiation plus those arising from occupancy and lighting, etc., will cause discomfort even when the heating system is shut off.

Very small blocks, where a central plant may be unsuitable for a variety of reasons, will probably best be served by one of the direct systems such as a large domestic type warm air unit or some type of electrical off-peak storage unit such as domestic warm air or individual radiators.

A study of probable temperature rise in modern office blocks reveals a universal tendency to overheating in summer, due to solar heat gains in addition to heat from office machinery and personnel. Thus it may be necessary to consider air-conditioning instead of heating alone, which matter is developed in a later chapter.

Public buildings and schools, universities, halls of residence, swimming pools, hospitals, hotels, etc.

This general class of substantial buildings will in most cases be in the hands of a consulting engineer, or public authority engineer, who can be expected to advise which system and fuel should be used. Matters to be reported on should cover, among other things, energy conservation, life-cycle costing, spatial requirements for plant, maintenance demands, amenity control, acoustic treatment and avoidance of pollution.

Industrial

Here the choice may be influenced by particular economic considerations related to the process or manufacture concerned, which might be short or long term, and which system will produce adequate conditions with minimum upkeep, fuel consumption and labour.

If steam were to be required for process work, the choice may fall in the direction of using steam for heating also, directly via unit heaters or some form of radiant surface or, alternatively, indirectly using a calorifier to produce hot water. If no steam should be required, hot water at medium pressure or high pressure is to be preferred. For heat emission, unit heaters may again be selected on account of low cost, but they involve maintenance which does not apply with radiant panel or strip heating.

Direct oil or gas fired air heaters meet certain cases where space limitations or other circumstances preclude the provision of a boiler plant or where extensions cannot be dealt with from existing boilers. Furthermore, such units can be moved to suit any change in building plan or workshop layout. Radiant tube systems, heated either by combustion products from gas burners or by hot air, should also be considered having regard to the high efficiency of heating effect and operating economy inherent to these arrangements.

Choice of fuel

There has in the past been complete freedom of choice but it is not known how long this may continue: any kind of control would necessarily limit this freedom. The predominant factor in the past has been determined by cost, but the instability of prices over the last few decades tends to make such comparisons unrealistic, however generalised they may be.

Thus, it is proposed here to do no more than indicate how the cost of various fuels may be compared, given the unit selling price or the basic tariff. The unit of heat for calculation purposes may be taken at 100 MJ since this is of a convenient magnitude and is roughly equal to the familiar *therm*. Thus, taking no account of any losses during conversion for use, the heat content of each unit sold would be:

Fuel	*Unit sold*	*Heat per unit sold*	*To produce 100 MJ*
Natural gas	kWh	0.036	27.77 kWh
Bottled gas	litre	0.256	3.309 litre
Solid fuel	kg	0.279	3.584 kg
Liquid fuel	litre	0.356	2.808 litre
Electricity	kWh	0.036	27.77 kWh

If unit prices, roughly representative of a domestic situation at the time of going to press, are put against these figures, then the unit costs in each case per 100 MJ of heat input potential are:

Fuel	*Unit price*		*Units per 100 MJ*		*Cost per 100 MJ*
North Sea gas	1.6 p	×	27.77	=	44 p
Propane	17.0 p	×	3.906	=	66 p
Coal (singles)	22.0 p	×	3.584	=	79 p
Oil (class D)	14.0 p	×	2.808	=	39 p
Electricity					
(on peak)	7.95 p	×	27.77	=	221 p
(off peak)	2.90 p	×	27.77	=	81 p

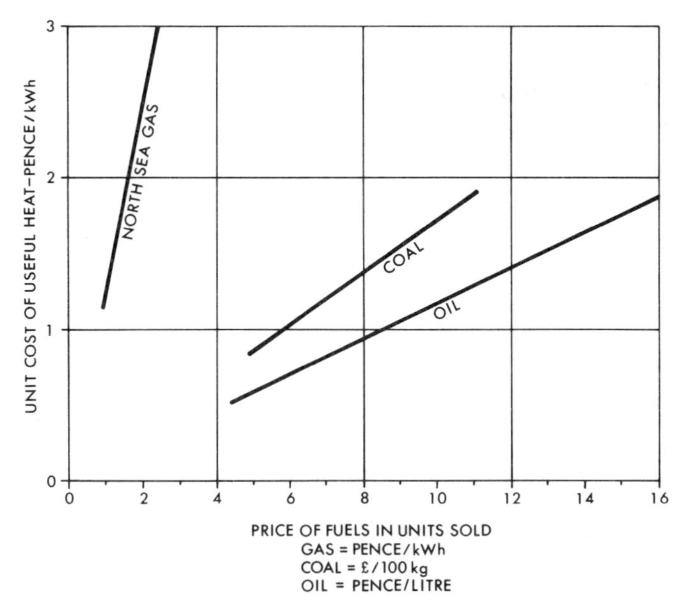

Figure 4.18 Fuel costs at normal utilisation efficiencies

Except in the case of direct electrical heating and, to a lesser extent direct gas heating, these figures do not represent the cost of the heat as delivered to the point of use. In an indirect system there will have been a loss in the conversion of raw energy during the combustion process, in the boiler and to the chimney, say 20% with solid fuel and 15% with oil or gas. In addition, some heat will have been lost from the pipework system en route to the space to be served – mains loss – which may amount to another 5% in certain circumstances.

Other factors affect fuel consumption such as controllability, no-load losses overnight and intermittent operation generally. For instance, an electrical storage system may have losses during the night as it charges but some part of these may be absorbed in the building structure; in contrast, boiler plant burning heavy fuel oil may carry an indirect 24 hour penalty as a result of electrical consumption for preheating or fuel pumping and, similarly, gas firing may carry the penalty of pressure boosting as well as that of a permanent pilot flame for ignition.

Each case merits separate evaluation. As a guide, a plot such as Figure 4.18 may be used to determine the cost of useful heat per 100 MJ arising from a range of unit selling prices for fuel, based upon the levels of efficiency previously quoted.

Chapter 5

Electrical storage heating

Direct in the sense that energy is consumed within the space to be heated but, conversely, indirect in the sense that consumption is remote (albeit only in time) from occupation of that space, any storage heating system must be considered as a hybrid.

The concept of storing heat energy may, of course, be applied in a variety of ways and to any type of fuel supply. For example, steam stored in an *accumulator* may be used in industrial plants to supply demands which are cyclic, or which fluctuate predictably, in order to provide a more constant load on boiler plant with consequent improved efficiency. Similarly, where a combination of circumstances precludes extension to a hot-water boiler plant, operation of that existing plant over the full 24 hours may allow an increased day-time demand to be met, in part directly and in part from water heated overnight and stored. In each of these instances, it should be noted, the unit cost of the fuel purchased would not vary.

Uniquely, and for the good reasons discussed in a later chapter, it is the policy of electrical supply companies in England and Wales to offer differential tariffs whereby current consumed during the seven hours after midnight is sold for rather less than half the price of that consumed during the remaining 17 hours. In Scotland the tariff is arranged slightly differently but a similar principle applies. The availability of such a reduction to consumers in general transforms the economics of electrical heating, provided that the various forms of storage equipment are applied to take maximum advantage of their distinctive characteristics.

Methods of storage

The methods commonly used for storage of heat taken during the hours of half-price supply, for use during the following day, may for convenience be grouped as being either room stores or central stores, thus:

Room stores

Storage radiators where heat is retained in a solid material contained within an insulated casing and emitted continuously with no more than marginal control of output.

Storage fan heaters where heat is retained similarly but where the addition of a fan, switched by a room thermostat, provides that a significant proportion of the heat from the store will be emitted only when required.

Warmed floors or *walls* where heat is retained in the building structure and emitted continuously with no control of output.

Central stores

Warm air units which are, in effect, large-scale storage fan heaters arranged as a central heat source for indirect warm air systems.

Dry-core boilers in which heat is retained as for storage fan heaters but arranged so that output is used to supply an air/water heat exchanger and, thence, an indirect system of piping, radiators, etc.

Wet-core boilers in which heat is retained in a hot water vessel, at atmospheric pressure, arranged so that the contents are used for circulation through an indirect system of piping, radiators, etc.

Thermal storage cylinders in which heat is retained in water, at elevated temperature and pressure, to serve indirect systems of all types in large commercial buildings. Electrode boilers, which are commonly used to charge such cylinders, are described in Chapter 10.

Capacity of the heat store

The amount of energy which must be stored in order to provide a satisfactory service will depend upon which particular system is used. In all instances, however, since heat will be discharged from the store over the whole 24 hour cycle, albeit perhaps at a low level during the period when output is not required, adequate capacity must be provided to include for the static heat output during the charging period. Some proportion of this output will, where the store is within the space to be heated, be absorbed in the surrounding building structure and thus not be wholly wasted.

Where heat output is uncontrolled, as in the case of all floor- or wall-warming systems and many *input-controlled* storage radiators, the night-time loss may be between 15 and 20% of the total input to storage. Where equipment such as some more modern storage radiators and most storage fan heaters is *output-controlled*, the situation is better in that night-time loss is a smaller proportion, about 10–12% of the total input. It must be added, nevertheless, that less than half of the energy which then remains in the store will be available for controlled output, the balance being a day-time static discharge. Those types of equipment which are designed to serve *indirect* systems are not subject to the same dimensional constraints as apply to room-sited units and thus can be not only better insulated but also better shaped as to their surface/volume ratio: static losses may thus be of the order of 5% or less of total input.

The foregoing comments are intended to emphasise that conventional methods of calculation for heat requirements must be modified to take account of the abnormal characteristics of storage systems. Research organisations in the electrical industry have produced a variety of application routines designed to provide the necessary adjustments: the notes which follow are a digest of these.

Criteria

In its simplest form, determination of the capacity of storage heaters would take account of the total design heat loss over 24 hours, calculated as described in Chapter 2 but making allowance for a reduced rate of air change during the night: Table 5.1 quotes recommendations in this latter respect for multi-storey office buildings. This 24 hour heat

Table 5.1 Recommended design values of rates for ventilation loss by natural infiltration in multi-storey office buildings

	Ventilation loss ($W/m^3 K$)			
	Height of building			
	15 m and under		*Over 15 m*	
Type of building	*Day*	*Night*	*Day*	*Night*
Buildings with little or no internal partitioning (i.e. 'open plan' buildings), or with partitions not of full height, or buildings with poorly-fitting windows and internal doors	0.5	0.3	0.7	0.3
Buildings with internal partitions of full height, without cross ventilation, and having self-closing doors to staircases, lift-lobbies, etc.	0.3	0.2	0.3	0.2

Note: The basic design temperature should take into account the fact that there is no inherent reserve and a basic temperature of −3.3°C for single-storey buildings, and −1.7°C to −1.1°C for multi-storey buildings is proposed.

requirement would then be divided by the number of hours (normally 7) during which energy input is available, the product being the appliance rating required. That is:

$$R_1 = (24\,Q/n)$$

where

R_1 = equipment rating (kW)
Q = design heat loss (kW)
n = available charging period (hours)

Such an approach does not, however, take account of a number of the other considerations which were discussed in Chapter 2, many of which are particularly relevant to a method of heating which is subjected to an intermittent energy supply associated with an output which, although at a level which varies, is continuous. The thermal response of

Table 5.2 Correction factors 'Z' for building type

Mean 24 hour heat gains (W/m^2)	*Correction factors 'Z' for following levels of design heat loss* (W/m^2)				
	< 30	*30–50*	*50–70*	*70–90*	*> 90*
4	0.55	0.70	0.75	0.80	0.90
8	0.45	0.60	0.70	0.75	0.80
12	[a]	0.50	0.60	0.70	0.70
16	[a]	[a]	0.45	0.55	0.60
Domestic	0.75	0.80	0.85	0.90	0.95

Notes:
1. Where windows are smaller than average, reduce by 0.05.
2. Where windows are larger than average, increase by 0.05.
[a] Premises not suitable for heating by storage methods.

Table 5.3 Correction factors 'Y' for night/day consumption

	Proportion of annual energy consumption used off-peak (%)										
	All	*99*	*98*	*97*	*96*	*95*	*94*	*93*	*92*	*91*	*90*
Correction factor 'Y' for design day rating	1.0	0.9	0.85	0.81	0.79	0.76	0.74	0.72	0.70	0.68	0.67

the building structure which is related to its mass; the level of insulation provided; the area of glazing; heat gains during the hours of occupancy from lighting and occupants are all relevant as is, where applicable, the degree of ouput control available. The complexities of these aspects are compounded in instances where output is primarily radiant and thus is interactive with other surfaces 'seen' by equipment or system. Calculations based upon consideration of all these variables would be out of the question for individual applications since, in any event, meticulous accuracy would be misplaced when applied to selection from a finite range of production equipment. Use is made therefore of correction factors, part theoretical and part empirical, which are applied to the simple equation noted above. These cover the following aspects:

- The order of design heat loss, the proportion of glazing, the thermal mass of the building structure and the level of heat gain from lighting and occupants. These are dealt with by what are called 'Z' factors, a digest of which is set out in Table 5.2.
- The proportion of design heat requirements to be met from storage in order to achieve a selected seasonal level of energy used at off-peak rates. This is dealt with by what are called 'Y' factors as listed in Table 5.3.

Capacity of room stores

Storage radiators

For these, the simplest of the various types of equipment available, the required rating (R_2) is determined generally from the basic calculation noted above by use of the correction factor 'Z' alone, such that:

$$R_2 = ZR_1$$

where

Z = factor from Table 5.2

To take account of those applications where it is accepted that some direct on-peak heaters will be fitted in parallel with the storage radiators, use is made also of the correction factor 'Y' such that the required capacity dealt with by storage (R_3) will be:

$$R_3 = YZR_1$$

where

Y = factor from Table 5.3

Storage fan heaters

Here, a new quantity has to be considered, the *active store* (R_4), which is a function of heater design but seems to be quantified in the publications of only a minority of manufacturers. This is the storage capacity remaining after the night-time static loss has been deducted from the total input, thus:

$$R_4 = XZR_1$$

where

X = an empirical factor (approximately 0.85)

Floor warming

In the heyday of electrical underfloor warming during the late 1950s and early 1960s, many authorities were offering a cheap off-peak supply not only for current used overnight but also, critically, for a 3 hour midday or afternoon boost. The success of this tariff was such that a temporary system peak arose during those daytime hours in 1969 and, as a result, it was not long before the facility for the low cost boost was withdrawn. This new situation had a very substantial effect upon the feasibility of providing adequate service to domestic premises from an electrical underfloor storage system. In fact it would not be an exaggeration to say that it was a death knell for this and similar applications.

It is common experience that the successful application of floor warming depends to a large extent upon the *swing* in space temperature during the discharge period or, perhaps more significantly, during the period when the heated space is occupied. This swing, which is a function of the thermal properties of the building and in particular those of the heated floor, should not exceed about 3.5 K. This means that if it is desired to maintain an average of 19°C, then the temperature will be up to 20.75°C in the morning and down to 17.25°C in the afternoon. It goes without saying that while the former might be just acceptable, it is unlikely that the latter would be tolerated. In an office building, heat gains from lighting, occupancy and other sources might be enough to 'top-up' the afternoon temperature to a barely satisfactory minimum but absolutely no margin would exist for any sub-normal condition. In domestic premises, some form of direct heating would certainly be required for evening use.

Similarly, if occupants of the heated space are not to experience discomfort as a result of hot and – consequently – tired feet, the maximum surface temperature of the floor must

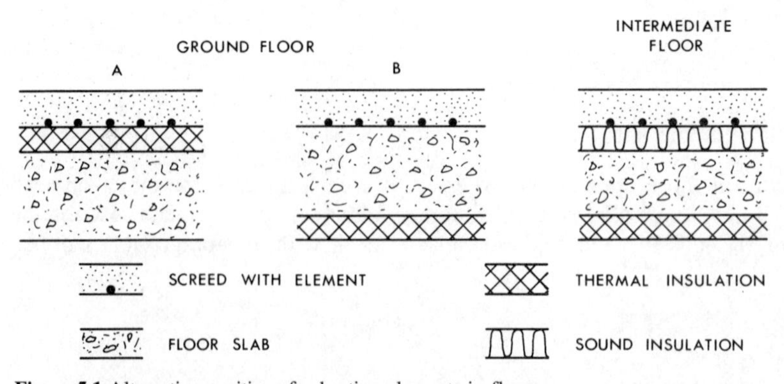

Figure 5.1 Alternative positions for heating elements in floors

Table 5.4 Maximum upward heat emission from floors charged over 7 hours

Position of floor (see Figure 5.1)	*Maximum upward emission* (W/m^2)			
	Bare floor		*With carpet*	
	50 mm screed	*100 mm screed*	*50 mm screed*	*100 mm screed*
Ground floor A	17	28	20	31
Ground floor B	31	55	42	68
Intermediate floor	23	39	30	48

be limited to 28°C. It is good practice to take a mean value of about 25°C for design purposes, recognising that while the surface temperature may reach the maximum tolerable level first thing in the morning at the end of the charging period, this is likely to be before occupancy begins.

Finally, it is necessary to consider the construction of the floor which is to be warmed, in whole or in part, and the floor finish which is to be applied. The thickness and position of any insulation are both critical to performance and affect directly the maximum upward emission which is tolerable. Table 5.4 lists these maximum values.

The unit emission from the heated floor surface is determined from the coefficient of upward heat transfer, a practical value for which is 10 W/m^2 K, and the difference between the mean surface temperature of the floor and the environmental temperature of the space to be heated:

$$E_1 = 1000(ZQ/A) \leqslant 10(t_s - t_{ei})$$

where

E = unit emission (W/m^2)
A = heated floor area (m^2)
t_s = temperature of floor surface (°C)
t_{ei} = environmental temperature of space (°C)

As will be appreciated, calculations based upon this relationship have to be solved by trial and error since neither the heated floor area nor the temperature of the floor surface are known initially other than by good practice in the latter case.

The environmental temperature must be calculated with some care, taking particular account of situations in multi-storey buildings where the space above may be provided with floor warming also. In such circumstances, the ceiling surface over the floor area being considered will be above room temperature to an extent determined by the finish applied to the floor of the space over. An approximation often adopted to cover this situation is to make a notional adjustment to the unit emission by use of a factor D, thus:

$$E_2 = 1000(DZQ/A)$$

where

D = 0.92 for a bare floor over
= 0.85 for carpet tiles over
= 0.78 for underlay and carpet over

Table 5.5 Correction factors '*F*' for floors having 50 mm insulation under

	Correction factors '*F*'		
Position of floor	*Bare floor*	*Carpet tiles*	*Carpet and underlay*
Ground floors, downward loss			
small area	1.19	1.25	1.30
large area	1.16	1.20	1.25
Intermediate floors, downward loss			
to heated room below	1.09	1.17	1.27
to unheated void below	1.25	1.35	1.40

Once the upward emission has been determined, the loading of the heating element may be established by adding an allowance for the edge loss around the perimeter and also for downward loss to earth in the case of a solid ground floor or downward emission to the space below in other circumstances. Each of these allowances may be calculated from first principles but, again, they are commonly dealt with by means of approximations; that for the edge loss being taken at 5 W/m run of perimeter and that for the downward loss or emission by use of a factor F selected from Table 5.5. Thus:

$$R_s = 24(FE_2A + 5p)/1000\,n$$

where

R_s = connected load (kW)
p = length of perimeter (m)

The design process is completed by calculations, first to ensure that the temperature at the plane of the heating element is not excessive bearing in mind the properties of the chosen element and second to determine the diurnal swing in room temperature in order to establish that this does not exceed 3.5 K, as discussed previously.

In conclusion, it is worth re-emphasis that since the operational characteristics of floor warming are relatively critical, it is necessary to pursue the full routine of design calculations although these are somewhat tedious, particularly for multi-storey buildings. The notes here, therefore, are no more than a summary of the necessary steps.

Capacity of central stores

The principal differences between the capacity required of a room store and that of a central store are that it may be necessary in the latter case to take account of distribution losses: this penalty may be off-set in some instances, however, by allowing for any diversity in the primary demand for output, room by room. A further penalty, of more significance in most cases, is that static losses by both night and day may be truly lost if the store is sited outside the spaces to be heated. It is of course not possible to generalise as to these negative or positive adjustments: they are functions of the application rather than the store.

Warm air units

The required capacity for such units is determined in the same way as that for storage fan heaters except in circumstances where a ventilation requirement exists and is met by adapting a unit to introduce an outside air supply. In those circumstances, the design heat requirement would be increased during the hours of occupation and the unit output rating would have to be adjusted to suit. Some units have elements rated in excess of the output required to charge the store: the additional capacity may then be used to provide preheat to the spaces served, still using the low-cost supply.

Dry core boilers

The particular characteristics of stores of this type, in conjunction with those of the connected indirect system, lead to a departure from the practice previously noted in that the capacity of the store is normally related to a charging period of about 5½ hours. This period is determined not by the time when a reduced-price supply is available, which is presumed to remain at 7 hours, but in consequence of the capability of the unit to produce a direct output to the connected system. This output is used during the last 1½ hours of the low-cost supply period to provide preheating to the space served without drawing upon the capacity of the store.

Since the 24 hour output of the connected indirect system is dissociated from the thermal cycle of the store, it may be assessed quite independently as a function of times of building use, etc. Thus, the required rating (R_s) is determined by:

$$R_s = (YHQ + L)/n$$

where

H = period of use or occupancy (hours)
Y = factor from Table 5.3 taken as 0.75 for this application.
L = 24 hour static heat loss from store when not useful emission (kWh)

Wet core boilers

As for dry core units, the characteristics of both store and system lead to special consideration of their application. The routine of calculation to determine the required rating is similar to that set out above except that the charging period is commonly reduced to 4½ hours in this case and rather more allowance is made for cold weather and evening requirements for heat to be dealt with by direct methods. Each of these approaches reduces the volume necessary for storage.

Thermal storage cylinders

In this case, where purpose designed cylinders are used to supply indirect systems in a large building, the method of calculation is a little different. A careful assessment of the true 24 hour load must be made, system by system and hour by hour. To the total thus established, it is necessary to add an allowance to take account of the 24 hour static heat losses from the vessels. The rating of the energy supply required is a simple application of the expression given above for dry core boilers, save that the factor 'Y' is not usually applied.

Determination of the net storage capacity required is based upon the assumption that *stratification* will exist within the vessels, temperature 'layers' remaining relatively undisturbed as discharge takes place. Hence, reference is made to the temperature at which the store is to be maintained and the volume-weighted average temperature of that returning thereto from the indirect system, thus:

$$S = 3600(HQ + L)/c\,(t_s - t_r)$$

where

S = water storage capacity (kg)
c = specific heat capacity (Table 5.13) (kJ/kg K)
t_s = temperature of stored water (°C)
t_r = temperature of return water (°C)

This capacity may then be converted to volume using the specific mass of water, again as read from Table 5.13. Additions are necessary to cater for the expansion of water when heated, as discussed later.

Equipment for room stores

It should be noted here that there have been very considerable advances over the last decade in the development of equipment for electrical thermal storage. Storage radiators ten years ago were clumsy and took up a disproportionate amount of room floor space. Figure 5.2 and Table 5.6, reproduced from the 1979 edition of this book, illustrate the equipment then available.

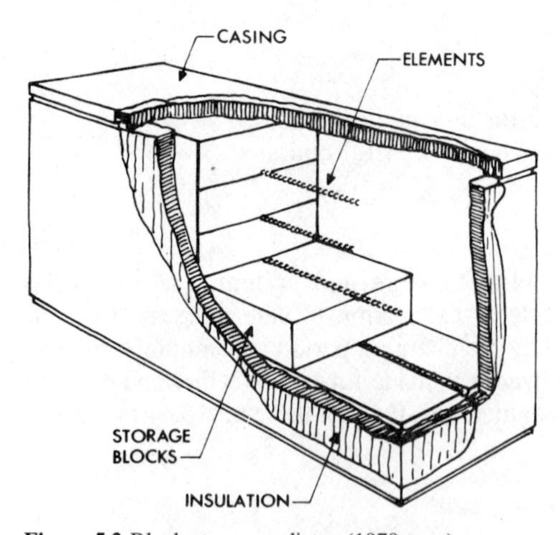

Figure 5.2 Block storage radiator (1979 type)

Two particular changes in practice have been adopted by most manufacturers of *dry core* storage equipment; the first relating to the material from which the pre-cast storage blocks are made, now either a high density refractory or an iron oxide (*Feolite*). Table 5.7 lists the properties of these and, for comparison, those of water and cast iron. The second is the use of opacified microporous silica panels (*Microtherm*) for insulation: this material has a thermal conductivity of only 0.030 W/m K at 800°C, and is thus about three times as effective as any material previously used.

Table 5.6 Block storage radiators. Typical ratings and dimensions in 1979

	Input rating (kW)		
Item	*2.0*	*2.6*	*3.3*
Energy (kWh)			
acceptance (8 hour)	16	20.8	26.4
active store	10.8	13.9	17.8
Output (W)			
maximum	1050	1345	1685
minimum	400	540	670
Temperature (K)			
front panel, above room	80	80	80
Dimensions (mm)			
height	614	699	699
length	735	735	960
depth (overall)	292	292	292
Weight (kg)			
total	123	150	199

Table 5.7 Typical heat capacities of storage materials

		Temperature during cycle (°C)		*Storage capacity*	
Storage material	*Specific mass* (kg/m^3)	Max.	Min.	kJ/kg	MJ/m^3
Water (at 100 kPa)	958	95	40	230	220
Water (at 450 kPa)	918	150	50	430	400
Cast iron	7900	750	450	140	1070
Refractory	2800	900	500	240	680
Feolite	3890	750	450	280	1100

Storage radiators

Units of this type enable energy to be stored in the space to be heated and in that respect could be said to be 100% efficient. Lack of control of heat output in most patterns, however, erodes this advantage. Such a radiator, as shown in Figure 5.3, comprises a number of sheathed elements enclosed within blocks of either refractory or Feolite, to form the heated core. This core is surrounded by insulation material which may be fitted in contact with the exterior casing or, in some more advanced patterns, held away from it to provide an airway.

The surface temperature of the casing reaches a maximum of about 80°C at the end of the charging period and this declines to about 40°C during the following day. Output is both radiant and convective in almost equal proportions and Figure 5.4 illustrates the

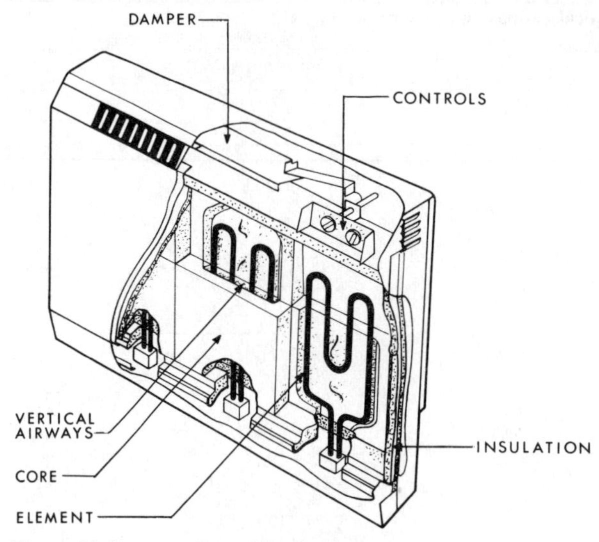

Figure 5.3 Storage radiator (Dimplex)

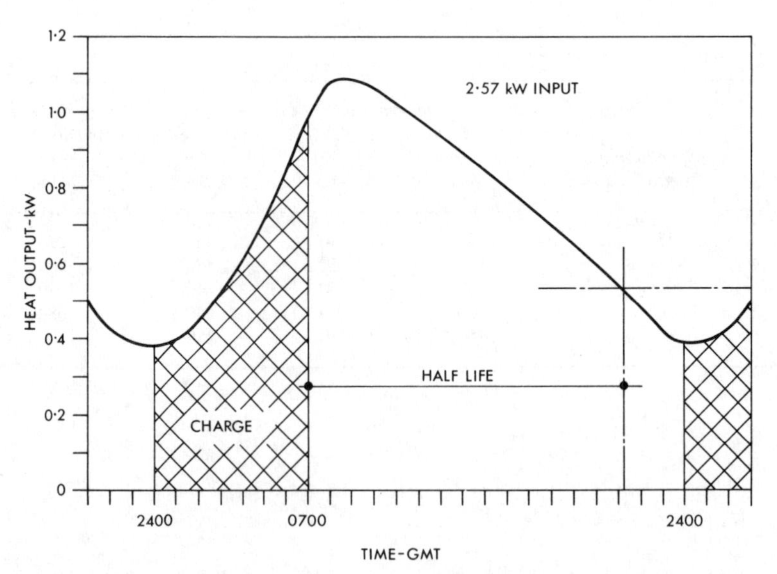

Figure 5.4 The 24 hour output pattern from a storage radiator (Unidare)

24 hour pattern of output, the *half-life* indicated being the point in time when output has fallen to 50% of the maximum.

Ratings vary with different makes but are usually 1.7, 2.55 and 3.4 kW, the seemingly odd figures being related to rounded 7 hour charge acceptances of 12, 18 and 24 kWh. Dimensions of typical radiators are listed in Table 5.8. Those types provided with airways and a damper may provide control of up to about 20% of the convective output.

Table 5.8 Storage radiators – typical particulars (Dimplex)

Item	*Input rating* (kW)			
	0.85	*1.7*	*2.55*	*3.4*
Performance				
charge acceptance (kWh)	5.95	11.9	17.85	23.8
half-life (h)	–[a]	15.4	15.0	15.2
Output				
maximum (W)	–[a]	775	1000	1350
minimum (W)	–[a]	250	320	470
Dimensions				
height (mm)	710	710	710	710
width (mm)	332	560	788	1016
depth (mm)	150	150	150	150
Weight				
in situ (kg)	41	77	110	145

[a] Not quoted by manufacturer.

Storage fan heaters

In most respects these are similar in construction to storage radiators, the differences relating to increased insulation, the design of air passages in the core and the addition of a small fan. In order to provide a uniform air-outlet temperature, irrespective of the store remaining, the casing may incorporate a chamber in which heated air from the core is mixed with room air via a thermostatically controlled damper. Output is about 75% convective.

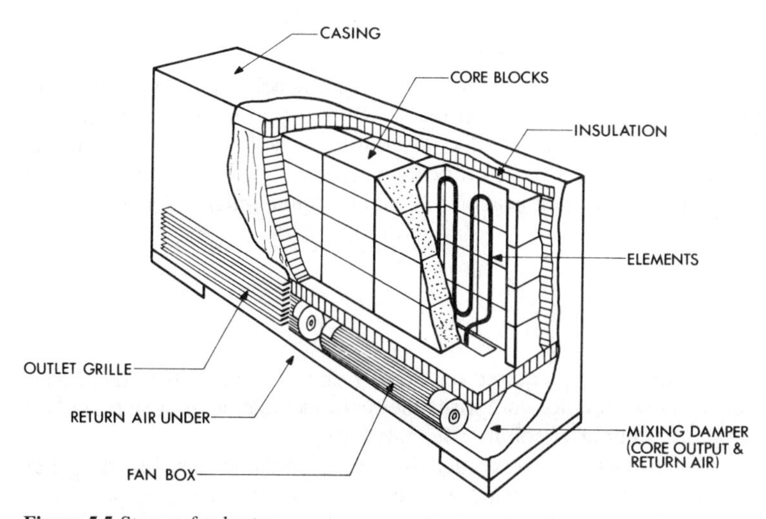

Figure 5.5 Storage fan heater

Table 5.9 Storage fan heaters – typical particulars (Stiebel Eltron)

	Input rating (kW)				
Item	*2*	*3*	*4*	*5*	*6*
Performance					
charge acceptance (kWh)	14	21	28	35	42
static loss during charge (kWh)	1.75	2.25	3.25	5.15	4.4
Output					
static loss by day (kWh)	6.25	9.85	13.35	14.9	18.6
available for control (kWh)	6.0	8.9	11.4	14.95	19.0
Dimensions					
height (mm)	640	640	640	640	640
width (mm)	584	775	965	1150	1340
depth (mm)	240	240	240	240	240
Weight					
in situ (kg)	90	128	165	200	239
Fans					
number	1	1	2	2	2
power, each (W)	20	20	15	15	15
Noise level					
rating (dB(A))	34	34	35	34	33

Ratings are in the range of 2–6 kW, from heaters of the type illustrated in Figure 5.5, and leading particulars of one make are listed in Table 5.9. Approximately 40% of total output may be controlled.

Booster heaters

It is of interest to note that, in recognition of the fact that the capacity of all types of room store is barely adequate for domestic use, most manufacturers now provide a facility for evening 'top-up'. This may take the form of either an additional element or of some means whereby the normal controls to one or more of the core elements may be overridden. The purpose in each case is to provide instant heat output which is quite independent of the off-peak charged supply.

Floor warming

An important part of the 'equipment' required for floor warming is the storage medium – the floor structure itself. The remainder comprises the resistance elements which are laid *in situ* within that structure. Output is about 50% convective.

The total length of cable elements must be such as to produce the required loading and one or more parallel circuits may be needed. Cables are laid in a grid-iron pattern at spacings which will vary with the loading and with the final floor finish: the centres may be between 100 and 300 mm and are sometimes varied across a floor, being closer at the perimeter than in the centre. Alternative methods of laying are used:

- *Rewirable* systems where the elements are laid within tubes which run across the floor between accessible trunking, using silicone rubber or glass fibre insulated cables.
- *Non-rewirable* systems where the elements are buried directly in the structure, using copper sheathed mineral insulated or cross-linked polythene insulated cables.

Cables are provided cut to length by the manufacturer to suit the required loading and are fitted with *cold tails* for connection to the supply.

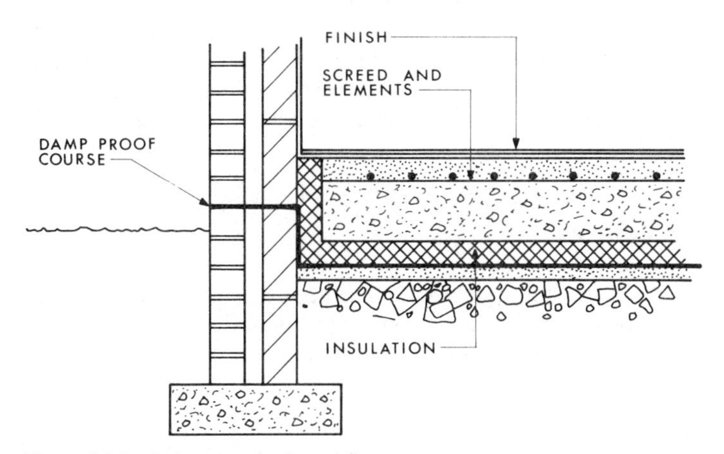

Figure 5.6 Insulation, etc., for heated floor

Of prime importance to the success of a floor warming system is the provision of a high standard of insulation below and around the floor structure, a 50 mm thickness of expanded polystyrene or other equivalent material is adequate and this should be placed as shown in Figure 5.6 wherever this is practical. It has been emphasised earlier that the positioning of the horizontal insulation layer has a significant effect upon the storage capacity of the floor.

Most types of floor finish have been used with warm floors but it is necessary to ensure that any adhesives involved when laying linoleum and some types of thermoplastic tile are suitable for the surface temperatures estimated.

Control of room stores

Room stores are normally provided with a *charge controller*, set during the previous evening to regulate the energy input overnight. Since, however, it is the weather of the morrow which needs to be sensed, the results produced are somewhat inexact. In its simplest form, as incorporated in most types of commercial storage radiator, such a controller is a manually set rotary switch having a scale representing the charging level. An intermediate setting delays the *start* of the charge period, thus varying its duration while avoiding post-charge losses. Automated versions of this arrangement, relative to night-time outside temperature, are available.

Heat output control of storage radiators is possible only when they are of a type which incorporates means to admit and regulate air flow over the core using, for example, a simple bi-metal operated damper reset by a cam from a rotary dial. The level of regulation available may be improved in the case of storage fan heaters, the power-assisted component of output then being controlled very easily via a room thermostat and the fan motor. A facility for multi-speed fan operation may exist in addition to provide 'low', 'normal' and 'boost' outputs.

Control of floor warming output, subsequent to the charging period, is not possible but, perhaps fortuitously, some inherent adjustment exists. For example, taking a mean floor surface temperature of 25°C, as previously noted, and an environmental temperature in the heated space of, say 18°C, the difference is then 7 K. If the room temperature were to rise by 2°C, then the notional temperature difference would be only 5 K and output from the floor would have fallen (theoretically) by approaching 30%. It is a difficult task nevertheless to explain this situation to a thin-soled secretary in an office which is over-heated by unseasonable solar radiation in January and yet retains a 'hot' floor!

Equipment for central stores

As was noted for room store units, the past decade has seen a number of significant advances in equipment development. Warm air as a heating medium in the domestic field no longer has the appeal it once enjoyed but some commercial interests have appreciated that it may have marginal advantages in application to spaces where occupation is either intermittent or transient.

In 1979, for domestic use, two quite different types of storage unit, aimed at provision of an indirect supply of hot water to piping and radiators were being developed. Both were noted as being of interest although, rather confusingly, they shared the single 'Centralec' name. One was a dry core unit based upon an adaptation of a large fan store and the other a prototype of the now well-established 'Economy 7 boiler'.

As to the first of these developments, it might seem odd at first sight that any application of the storage principle to a hot water heating system should be via three-stage heat exchange, first – energy to core, second – core to air; and third – air to water. The reasons are not hard to find. As to core capacity, Table 5.7 has shown that water at atmospheric pressure compares unfavourably as a storage medium, in terms of both mass and volume, with the other materials now used. As to heat transfer, it is possible to conceive an arrangement which would have water pipes embedded in a refractory core but, with temperatures there of 750°C, the prospect of any sort of control failure is uninviting quite apart from construction problems.

Warm air units

Extensively publicised under the 'Electricaire' label, such units are no more than large versions of storage fan heaters, arranged to serve as a heat source either to an extended space or to multiple rooms via an indirect system of air-duct distribution. Figure 5.7 illustrates one pattern where the heat store is mounted above the fan, mixing box and outlet plenum chamber. The reverse arrangement, with the heat store at the bottom, is preferred by some.

Typical equipment of this type will have the dimensions, etc., listed in Table 5.10 for a range of input ratings between 6 and 100 kW. The not inconsiderable floor loading

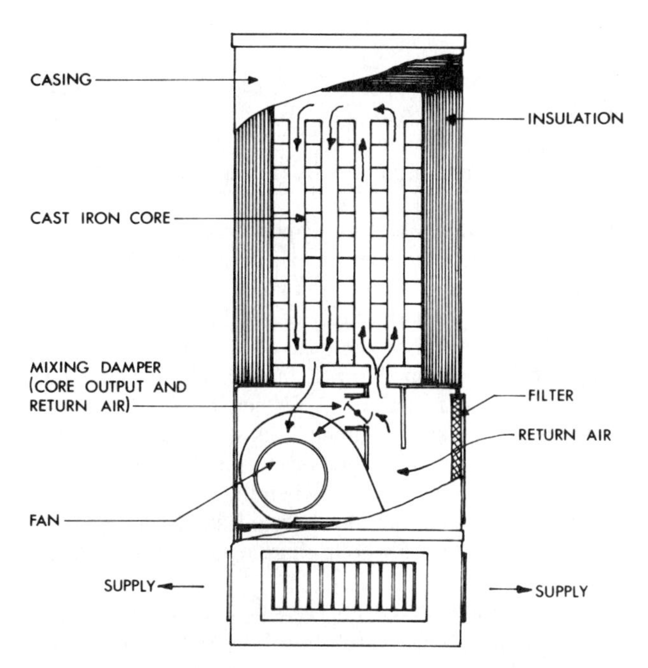

Figure 5.7 Warm air unit

Table 5.10 Warm air units – typical particulars

		Input rating (kW)					
Item		*6.79*	*9.86*	*14.79*	*30*	*72*	*100*
Performance							
charge acceptance	(kWh)	47	73	104	210	504	485
Dimensions							
height	(mm)	1300	1300	1750	2500	2178	2588
width	(mm)	630	630	630	860	2065	2228
depth	(mm)	610	610	610	775	1265	1156
Weight							
in situ	(kg)	370	484	754	1490	4400	4375
Fan							
duty	(litre/s)	116	175	244	–	1415	940
Manufacturer[a]		A	A	A	B	C	B

[a] A, Unidare Engineering; B, Chidlow and Co.; C, Powrmatic.

imposed by the larger sizes should be noted as should the fact that a three-phase power supply may sometimes be required. Output, usually arranged by admixture with recirculated air from the heated space in a similar manner to that described for the smaller equivalents, is normally at about 60°C and, as would be expected, is almost entirely convective.

Dry core boilers

Production-model equipment of this type ('Nightstor') is far removed in many details from the experimental units on test some years ago. The overall concept, however, remains the same in that the heat store is not dissimilar in principle from that provided in a warm air unit having a similar rating, although greatly refined in detail as to configuration.

The Feolite blocks which make up the storage core are disposed about a central *hot draught tube* and have vertical passages formed in them to accommodate the heating elements: they are contained within a casing which is lined with two layers of 'Microtherm' insulation. The core assembly stands upon a substantial layer of insulating blocks which isolate it thermally from the plenum chamber forming the base of the unit.

The plenum chamber is a box construction containing a fan, an air to water heat exchanger, a pump and the associated piping connections. A further insulated enclosure surrounds the complete unit and mounts the control equipment, as shown in Figure 5.8. The elements are arranged so that separate switching of one or more is possible in order to provide a facility for use as a daytime boost on occasions when the store is exhausted. Seven models are currently available of which the three listed in Table 5.11 are representative, the input ratings being related to a 5½ hour charging period. The larger units may require a three phase supply.

A dry core boiler introduced by another manufacturer provides usable stored energy of 52 kWh and operates on a different principle, with energy transfer from storage to the external space heating circuit via a steam/water heat exchanger. The steam is produced by a controlled 'boiling circuit' within the core, piped in stainless steel and topped up from a small de-ionised water tank. This alternative, rather less bulky than an air/water arrangement, allows the height of the boiler casing to be reduced to permit installation below a kitchen worktop.

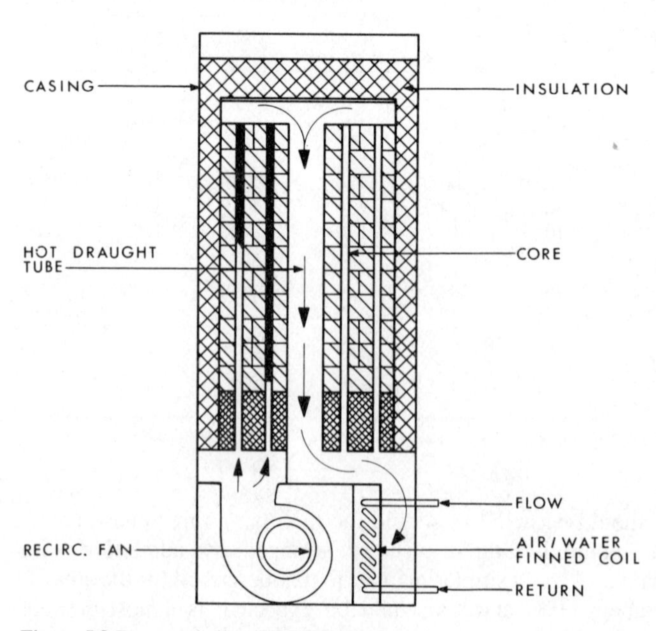

Figure 5.8 Dry core boiler (GEC Engineering)

Table 5.11 Dry core boilers – typical particulars (GEC Engineering)

Item		*Input rating* (kW)		
		12	*20*	*50*
Performance				
charge acceptance	(kWh)	60	100	250
recommended connected load	(kW)	6.7	11	35
Output				
nominal at full charge	(kW)	19	19	50
direct boost capacity	(kW)	6	8	17
case emission	(kW)	0.3/0.6	0.55/0.8	0.7/1.4
Dimensions				
height	(mm)	1317	1664	1664
width	(mm)	610	610	880
depth	(mm)	596	596	866
Weight				
in situ	(kg)	625	825	2000
Pump				
nominal duty	(litre/s)	0.23	0.23	0.64
minimum flow	(litre/s)	0.15	0.15	0.45

It should be noted that the characteristics of this equipment are such that water temperatures not very different from the traditional can be produced and that, in consequence, it may be used to replace a boiler connected to an existing hot water heating system. The proportions of radiant and convective output are, of course, determined by the type of indirect system connected.

Wet core boilers

It is difficult to imagine a simpler fundamental concept than that which is the basis for heat stores developed under the name 'Economy 7 boilers': some adjustment to the concept has, however, been necessary in order to produce a practical production unit.

Packaged equipment currently available consists of an insulated cylinder fitted with two banks of immersion heater elements, one near the bottom to produce the over-night charge and another near the top for use as a daytime boost on occasions when the store is exhausted. Water is stored at a temperature which is varied to suit the static head available from a roof mounted expansion cistern, 95°C being a not uncommon level.

The cylinder is piped to a simple three-way mixing valve set to produce a water outlet temperature of about 75°C. A pump is provided to circulate water round the heating system, the volume being restricted to ensure that water returns to the cylinder at not more than 40°C in order to take maximum advantage of the stored volume. Figure 5.9 illustrates the usual form of such equipment and Table 5.12 provides leading particulars of one make.

A unit of this type will have some advantages over the dry core equivalent, first since no heat exchangers are needed, the same water being used for both storage and distribution to the indirect system, and second because 24 hour standing losses are lower, the storage medium being held at a comparatively low temperature.

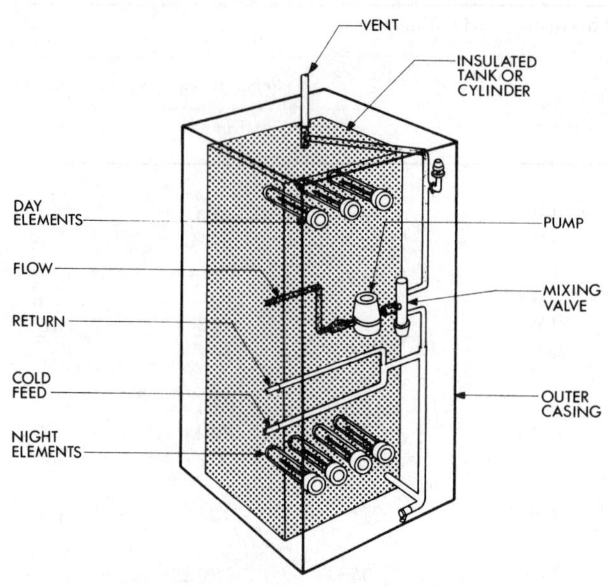

Figure 5.9 Wet core boiler

Table 5.12 Wet core boilers – typical particulars (Farnell Instruments)

Item		*Input rating* (kW)					
		6	*12*	*18*	*36*	*36*	*72*
Performance							
charge acceptance	(kWh)	23	60	90	135	180	303
recommended connected load	(kW)	2.5	6.67	10	15	20	33
Storage							
water at 98°C	(litre)	350	910	1360	2036	2720	4540
Output							
direct boost capacity	(kW)	3	9	12	18	24	36
Dimensions							
height	(mm)	2205[a]	2121	2121	2121	2121	2121
width	(mm)	750	1016	1219	1473	1372	1956
depth	(mm)	600	610	711	889	1219	1372
Weight							
in situ	(kg)	390	1174	1701	2642	3438	5766

[a] Includes height of integral expansion cistern.

Conversely, and probably much more importantly, it would seem that a wet core unit could not be used to replace a boiler without considerable alterations to the existing system connected to it. This situation arises since, with an outlet water temperature of say 75°C and the required return water temperature of 40°C, as mentioned above, the mean temperature at the existing radiators would be about 57.5°C. This is some 15–20 K lower than that at which they were, in all probability, designed to operate.

Thermal storage cylinders

These, of course, represent the traditional method of storing heat which has been made available at a cheaper overnight rate. Such vessels were used in systems designed and installed 60 or more years ago and their modern equivalents are very similar except in terms of quality. The principal differences between a store of this type and those which are outlined under the previous heading are first: the pressure and, thus, the temperature at which the hot water is stored; second the manner in which electrical energy is transferred to the water and third, a corollary, the overall size of the system. Figure 5.10 illustrates a cylinder arrangement suitable for a smaller installation, where banks of immersion heaters are used, and Figure 5.11 shows a complete installation incorporating cylinders in association with an *electrode boiler*.

Since, for a given heat output, the volume of storage required depends principally upon its temperature, it follows that the design criteria are interactive, i.e. the pressure which can be imposed upon the store and the space which can be made available for the vessels. Early systems made use of the height of the building, and the siting there of an expansion

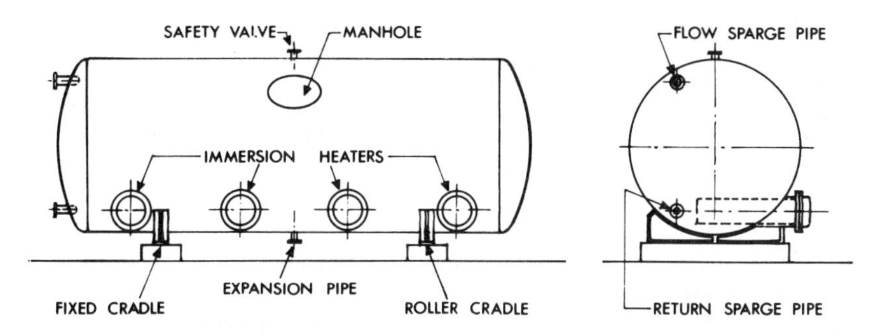

Figure 5.10 Immersion heater type storage vessels

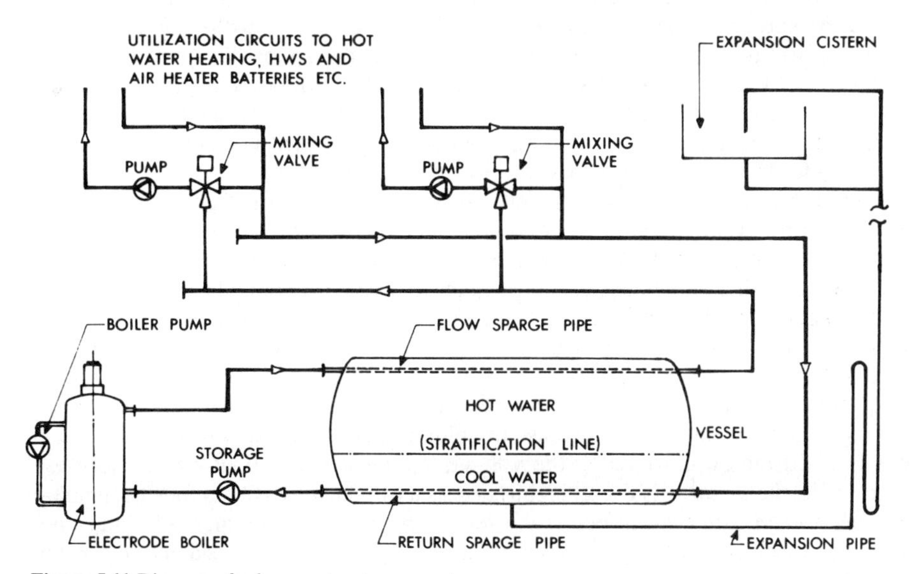

Figure 5.11 Diagram of a large water storage system

Table 5.13 Properties of water for thermal storage

Height of cistern	*Operating pressure (bar, gauge)*	*Properties of stored water (17 K anti-flash margin)*			
		Temperature	*Density*	*Specific heat capacity*	*Vapour pressure*
(m)		(°C)	(kg/m^3)	(kJ/kg K)	(kPa)
5	0.490	95	961.9	4.213	84.53
10	0.981	103	956.2	4.223	112.73
15	1.471	110	951.0	4.233	143.26
20	1.961	117	945.5	4.243	180.46
25	2.452	121	942.3	4.250	204.98
30	2.942	126	938.3	4.261	239.32
35	3.433	131	934.0	4.272	278.40
40	3.923	134	931.4	4.278	304.05
45	4.413	138	927.9	4.286	341.36
50	4.904	141	925.2	4.292	371.82
55	5.394	144	922.5	4.298	404.18
60	5.884	147	919.7	4.307	439.03
65	6.375	150	916.9	4.320	475.97
70	6.865	153	914.2	4.329	515.66
75	7.355	155	912.2	4.335	543.45
80	7.846	157	910.3	4.341	572.44

cistern to produce the necessary pressure, but more recent practice has been to use some external means (see Chapter 9) for this purpose. The temperature at which water is stored is normally about 10–17 K less than that at which steam would be produced. Table 5.13 provides data in this context.

Storage cylinders are almost always mounted horizontally, the size and shape being determined by site conditions. Diameters up to 4 m are common, with lengths of up to 10 m. Loss of heat is reduced as far as possible by insulation, up to 100 mm of glass fibre against the vessel being covered in some instances by 50 mm of dense material prior to a final casing with polished sheet metal. The cradles upon which vessels are mounted, fixed at one end and roller type at the other, stand upon a base which is insulated with hard material in order to minimise heat loss downwards by conduction.

Since effective use of storage depends upon minimising disturbance to potential stratification in the water stored, it is good practice to provide vessels with *sparge pipes* at both outlet and return connections. As shown in Figure 5.11, such pipes should extend over the whole length of the vessel, except for gaps to permit differential expansion. A rule of thumb suggests that the area of the holes in each sparge should equate to ten times the diameter thereof and be provided over the whole length in rows arranged to face the critical direction.

The quantity of water stored in cylinders of the size considered is such that, when heated to storage temperature, the increase in volume is considerable. For instance, when water is heated from 50°C to 120°C the volume is increased by about 4.5%, and when heated to 150°C the increase is 8%. It is thus necessary to make provision for this increase by allowing a *dead water space* below the return sparge pipe, equivalent to about one-eighth of the vessel diameter. It is to the bottom of this space that the feed and expansion pipe is connected.

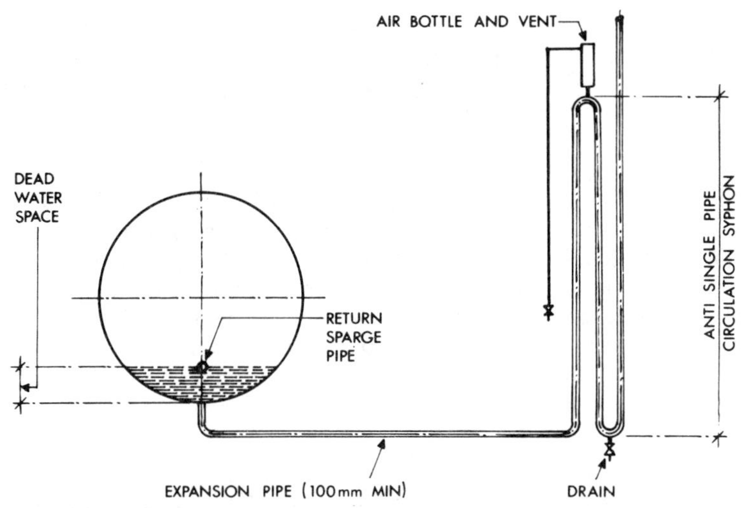

Figure 5.12 Expansion connection to storage vessel

Where pressure is applied by an elevated cistern, as in Figure 5.12, the syphon loop prevents circulation *within* the expansion pipe and thus, in theory, the dead water space stagnates and only cold water is allowed to pass to the high level cistern. In practice, some mixing will have taken place with the 'live' content of the vessel and there will also have been heat transfer by conduction. For installations where pressure is applied by external means, it is good practice to fit a simple heat exchanger in the feed and expansion pipe, rejecting to waste.

It is of interest to note that although most applications for thermal storage cylinders on any substantial scale relate to purpose-designed installation, one manufacturer produces a packaged version on the lines illustrated in Figure 5.13. This consists of a cylinder which is arranged to retain a space above the level of water fill in order to accommodate the volume of expansion and to provide a compressible pressure cushion. Unique, but inherent in the concept, is the introduction of a heat exchanger as separation between the water

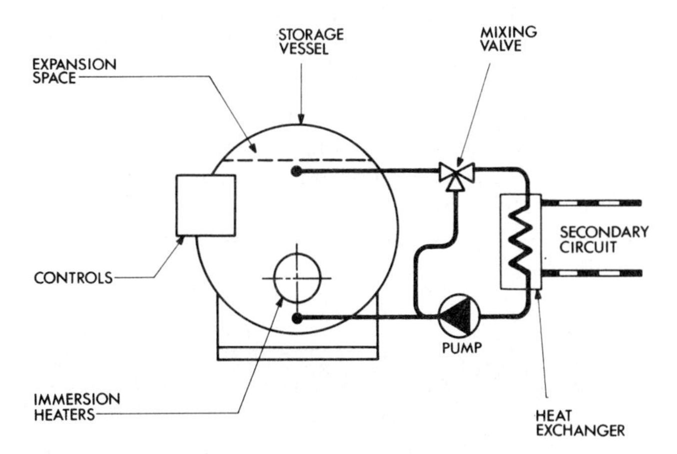

Figure 5.13 Packaged storage vessel and system (Heatrae Industrial)

store and the indirect system. The mixing valve shown, the variable speed circulating pump and all controls etc. are part of the package for a range of cylinder sizes from 2000 to 50 000 litre, with loadings of 36–720 kW. The standard operating criteria are 350 kPa and a charged temperature of 138°C.

Control of central stores

As with room stores, some form of *charge controller* for input is provided for central stores. In view of the increased size and capacity in this case, these are rather more complex and sophisticated and will in most instances incorporate some method for automatic sensing of outside temperature during the charging period, in anticipation that this will provide a closer approximation. It is not possible to generalise as to output control from central stores since this will almost always be related to the connected indirect systems.

Electrical supply to storage equipment and installations

This present volume does not pretend to deal with matters related to the design of electrical systems. The following notes are therefore included only as items for convenient reference.

Direct electric heating elements up to 3 kW (13 amp at 250 volt) are served by a two-wire circuit. Heavier duty elements are usually arranged in banks and the load divided over the phases of a three-phase supply if available.

For thermal storage with immersion heaters, the usual arrangement is balanced on a three-phase three- or four-wire supply. Thermal storage with electrode boilers is served invariably from a three-phase three-wire supply at medium or high voltage.

The current passing in the conductors may be calculated from the load and voltage. Taking as an example a heating load of 500 kW, the current passing through the circuit would be as follows:

for two-wire DC *or single-phase* AC at 240 volt, $500 \times 1000/240 = 2080$ ampere;
for three-wire AC *three-phase* at 415 volt between phases, load balanced,
$500 \times 1000/ (3 \times 415) = 700$ ampere per phase;
for four-wire AC *three-phase* at 415 volt between phases (240 volt phase to neutral), load balanced, the current would be the same as with the three-wire system, no current flowing in the neutral.

In the case of a number of small heaters, each will be served from a separate way on the distribution board. Having established the current flowing in the circuit supplying one heater, these are totalled up to give the current supplied by the main to the distribution board. From these currents the electrical distribution system is then designed. For a more complete study, reference should be made to one of the many textbooks on electrical engineering.

Chapter 6

Indirect heating systems

It has been demonstrated in an earlier chapter that, for consideration of indirect heating systems, the *media employed for distribution* provide the most convenient set of fundamental groupings. $500\times1000\ (\sqrt{3}\times415)=700$ amperes per phase. Further, it has been suggested that the abstract quantities of liquid, vapour and gas may, in practice, be represented by water, steam and air.

Water systems – characteristics

Water as a heating medium offers many advantages: among the various heat transfer liquids, it has the highest specific heat capacity and the highest thermal conductivity. Within the temperature range 20–200°C, although the specific heat capacity, kJ/kg, rises by about 8%, the specific mass, kg/m^3, falls by about 13% with the result that the *volumetric heat capacity* varies by only 5%. A digest of the properties of water over this temperature span is given in Table 6.1, which appears at the end of this chapter (p. 167).

Within this group, it is the temperature to which the water is raised in the boiler or other energy conversion plant which dictates the type of system which may be selected. Table 6.2 lists the four temperature ranges which are most commonly used and provides a name for each. It should be understood that such sub-divisions are quite arbitrary and have been evolved to suit either equipment and material usage or the physical properties of the water content within the range given.

Table 6.2 Design temperatures for water heating systems

	Design temperature	
System	*Flow at water heater* (°C)	*Differential at water heater* (K)
Low temperature warm water	40–70	5–10
Low temperature hot water	70–95	10–15
Medium temperature hot water	95–120	25–35
High temperature hot water	> 120	45–65

Low temperature warm water

Between the temperature limits quoted, warm water has been used in heating systems for three-quarters of a century or more. Until recently, the principal application has been confined to circulation through embedded panel heating coils where it is necessary that the flow temperature of the water supplied should not be more than about 53°C for ceilings and 43°C for floors. Since, as will be explained in Chapter 12, it is bad practice to operate normal boiler plant at such low temperatures, this warm water supply has been derived from a conventional hot water source, either by mixing boiler water output with that returning from the system or by temperature reduction via a heat exchanger.

The campaign for energy conservation which has developed during the last two decades has led to consideration being given to a variety of waste heat sources, many of which are either at warm water temperature when tapped or reach peak efficiency when producing output at such a level. Energy reclaimed from industrial processes exemplifies the former situation and condensing boilers or heat pumps fall within the latter. While each of these presents a practical and significant source, the problem of utilisation remains.

Obviously, not all new buildings are suited to the application of embedded panel heating and it is not a method which may be applied easily in refurbishment of the existing stock. While it is possible to design systems which are able to use warm water with conventional equipment and yet meet a mid-winter demand, these have yet to gain acceptance. There are two practical difficulties in this respect, the first being the significant increase in size of terminal heat emitting equipment, as Table 6.3. Of equal importance, however, is the reaction of occupants who cannot be convinced that a space is heated adequately by a radiator, however large, which is 'never more than just warm'!

Table 6.3 Variation in size of heat emitting equipment with water temperature

	Comparative length of heat emitting unit for following temperature differentials (K) *water to air at 20°C*				
Type of heat emitting unit	*55*	*45*	*35*	*25*	*15*
Primarily radiant	1	1.29	1.79	2.83	5.46
Primarily convective	1	1.35	1.95	3.31	7.14

As a result of these two similar but conflicting considerations, the concept of *bivalence* has emerged. This is a term adopted by the now defunct Electricity Council to represent the use, in parallel, of two energy sources to meet a common requirement. The principal application to date has been to demonstrate that a warm water source and a boiler plant may be arranged to work together throughout a heating season, the former carrying the load in mid-seasons and the latter taking over during periods of extreme weather. A heating analysis chart has been designed to provide a basis for representation of the proportions of the annual load which may be supported by either of the two energy sources and a simplified version is included here in Figure 6.1.

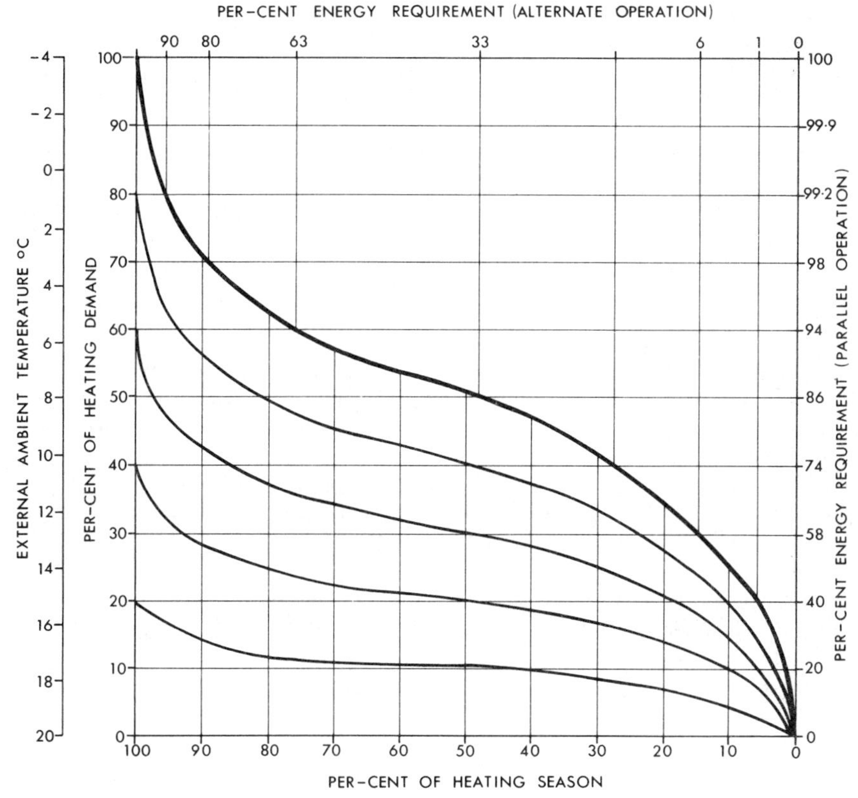

Figure 6.1 Analysis chart for application to bivalent energy systems

Low temperature hot water

Under this heading fall most of the systems fitted in domestic, commercial and institutional premises. The basis for the operating temperature is the traditional practice of providing that such systems should be connected above the highest point to a feed and expansion cistern open to atmosphere. Thus, a temperature of 82°C (180°F) was chosen for two reasons, first because it provided a reasonably wide margin with respect to boiling point at atmospheric pressure and second because it left some tolerance within that margin for elevation to perhaps 90°C under sub-normal weather conditions.

The water in a heating system expands when the temperature is raised and the purpose of the feed and expansion cistern is to receive the additional volume when the system is hot and return it when the system cools down. In following this cycle, the water content remains relatively unchanged and the scaling or corrosion which might arise from the admission and rejection of fresh water each time is largely avoided.

The volume of expansion over any temperature range may be found from Table 6.1 since the mass remains constant. Thus, between 10°C and 100°C for example, it is (999.7 −958.3)/958.3 = 4.32% or one twenty-third of the original volume. Manufacturers' literature must be consulted for the water content of equipment since this varies greatly with pattern: Table 6.4 provides that information for pipework.

Table 6.4 Water content of steel (BS 1387: medium) and copper (BS 2871: Table X) pipes

Nominal pipe size (mm)	*Water content* (litre/m)		*Nominal pipe size* (mm)	*Water content* (litre/m)	
	Steel	*Copper*		*Steel*	*Copper*
15	0.205	0.145	40	1.376	1.234
20	0.367	0.321	50	2.205	2.095
25	0.586	0.540	65	3.700	3.245
32	1.016	0.837	80	5.115	4.210

Of recent years, a small proportion of systems in this category has been provided with means, by way of a pressure cylinder as noted later in Chapter 8, to dispense with the open cistern. Nevertheless, although an output temperature of about 80°C may have originated in an era when hand firing with coke without any form of automatic control was the norm, no good reason to abandon it has yet been advanced. Furthermore, for domestic and most commercial premises, where operating skills and system maintenance are usually minimal, a safety margin of the order available is still to be preferred.

The difference of temperature, boiler outlet to boiler inlet, or flow to return in more common parlance, is a matter for the designer to determine. The differentials recommended are included in Table 6.2 although these are no longer listed in any section of the *Guide*. There is some evidence that wider differentials are used in continental practice but these relate to circumstances which have little relevance to good practice in the British Isles.

The types of heat emitting equipment (see Chapter 7) which are most suited to low temperature hot water are floor-panel coils embedded within sheaths (*Panelite*); heated acoustic ceilings; metal radiant panels; radiators of all types and natural or fan convectors.

Medium temperature hot water

It could be argued that systems of this type originated from those which enjoyed a vogue in about 1920, when fitted with a device known as a *heat generator*. This was an item of static equipment, containing a mercury column and reservoir, which was interposed between the feed and expansion cistern and the boiler. The effect was to multiply the pressure available by density difference and thus permit the working temperature to be elevated accordingly.

As now understood, however, systems of this type were so identified in the early 1950s and, as such, were advocated as being a compromise between low and high temperature practice. As will be noted later, high temperature systems at that time used steam boiler techniques to apply pressure, *split-casing* circulating pumps with water cooled glands and either high quality welding or flanged joints for pipework. These, needless to say, are correspondingly costly.

It was suggested that medium temperature systems could be constructed using what were, effectively, the best of low temperature equipment and methods. The use of screwed joints was proposed and adopted for valves and other fittings but, for most pipework joints, welding of an adequate commercial standard was becoming the more economic approach. A further concurrent development was the introduction of the factory made *gas pressurisation sets* which will be referred to later.

Taking mid-point temperatures from the ranges in Table 6.2, the differential, water to air, in a medium temperature system would be 105 – 20 = 85 K whereas that in a high temperature system would be 132 – 20 = 112 K. To achieve the same output, therefore, the medium temperature system would need to have over 40% more radiant heating surface. Nevertheless, there are many compensating advantages in terms of capital and running costs for smaller installations and, for these reasons, it has come to be accepted that medium temperature systems are best applied where the overall load is not greater than about 2.5–3.0 MW.

Most designers prefer to equip medium temperature systems with one or other of the packaged pressurising sets which are described in more detail later. However, it is possible in some instances to provide adequate pressure from a cistern elevated above the highest point of the system. A central boiler plant providing primary water to heat exchangers at basement or ground level in each of a group of buildings, might be so arranged. As in the case of pressure so applied to thermal storage vessels, Figure 12.9, it is necessary to take steps to ensure that heated water from the system does not circulate *within* the expansion pipe and rise to the cistern.

As an example of pressure so applied, consider a cistern mounted 17 m above any pipework carrying water at system flow temperature. A gauge pressure would be produced at that point of 17 × 9.81 = 167 kPa corresponding to 167 + 101 = 268 kPa absolute. From Table 6.1 it will be seen that an absolute vapour pressure of 270 kPa relates to a temperature of 130°C and thus, allowing for a 10 K safety margin, the system could be designed to operate with a flow temperature of 120°C.

In view of the elevated surface temperature, the types of heat emitting equipment which may be fitted are restricted to those which may be touched with safety by occupants. Thus, natural and fan convectors are used at working levels but unit heaters of various patterns, radiant strip heaters and radiant panels may be fitted at higher levels where they are out of reach.

High temperature hot water

The father of all high temperature systems was Perkins who filed a patent in 1831. In this system, the wrought iron piping was about 22 mm bore (32 mm outside diameter, jointed with right- and left-hand threads) and each circuit formed one continuous coil, part of it inside the boiler with the remainder forming the heating surface in the building, Figure 6.2: one such coil was measured recently as being over 180 m in length! Partial allowance for expansion of the water content was provided by means of closed vessels. Operating temperatures of 180°C were common and, firing being by hand with no real control, frequently reached 280°C (7 MPa!) near the boiler. Examples of the system remain in use, some converted to oil firing, in churches and chapels built towards the end of the last century and contractors still exist, in deepest Wales, who offer installation expertise.

The basic concept of water circulation at elevated temperatures and at pressures above atmospheric boiling point was revived during the years between the wars and applied, in particular, to large industrial premises. The essential differences between this application and the Perkins system were that circulation was by pump, through conventional pipe circuits arranged in parallel, and that the working pressure and temperature were controlled at the boiler plant to more specific levels.

Prior to the reintroduction of this system, the type of building to which it was particularly suited had been supplied by steam plant and arguments among advocates of the alternative methods, as to the superiority of each, continued for 20 years. The flexibility of the water system, both in the potential available to vary the output

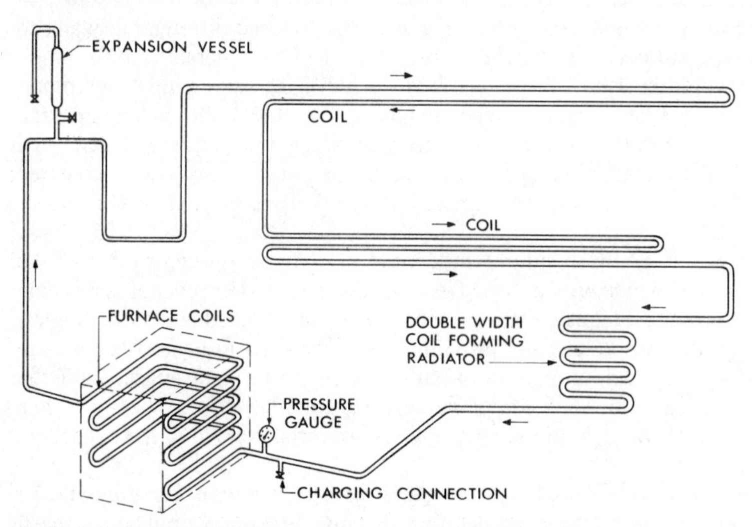

Figure 6.2 Diagram of Perkins' high temperature hot water system

temperature to suit weather conditions and in the physical freedom to route pipework largely irrespective of site levels, were among the considerations which led to its increasing popularity. In terms of operation and maintenance, virtually all equipment needing attention is, for the water system, concentrated in plant rooms. It has now come to be accepted that, for space heating, steam is not the preferred medium except in the particular circumstances which will be noted later.

In order to achieve the elevated temperatures associated with these systems, the practice adopted initially was as shown in Figure 6.3, and this arrangement still has its advocates. Steam is generated in a more-or-less conventional steam boiler and the water to be circulated is drawn from below the steam/water separation line, to be returned there also

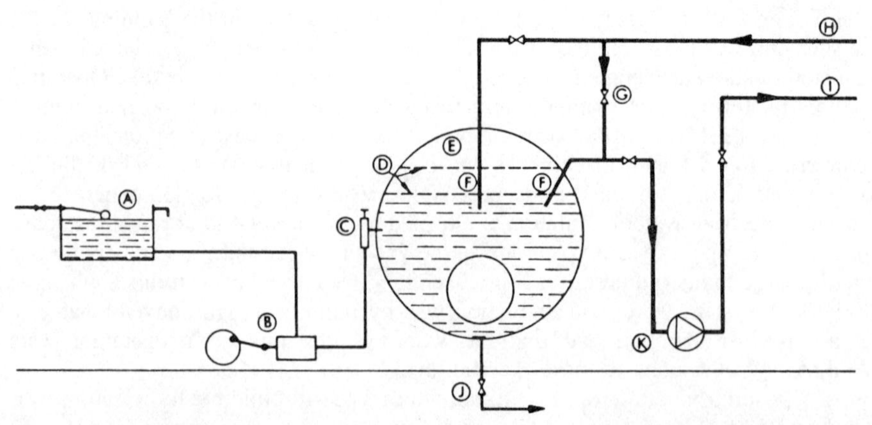

Figure 6.3 Steam pressurisation (shell boiler): A, feed cistern; B, feed pump; C, feed check valve; D, water line (upper and lower); E, steam space; F, dip pipes; G, cooling water by-pass; H, return; I, flow; J, blow-down; K, system circulating pump

after passing through the system. The connections are made in the form of *dip-pipes* from the top of the boiler in order to avoid draining the boiler, with consequent danger, should a serious leak occur in the system.

The temperature of both the steam and water are the same, at saturation level, and were pressure to fall in the water pipework without any parallel loss of temperature, the water would flash into steam and create an unstable condition. In order to prevent this happening, a small proportion of the water returning from the system is injected into the flow outlet as it leaves the boiler, so reducing the temperature there to below the point of ebullition. Although Figure 6.3 is no more than a diagram, it does illustrate a desirable arrangement whereby flow pipework is routed to below the steam/water interface in order to increase static pressure before connection to the circulating pump. Where the size of the installation requires that more than one boiler be installed, their respective water levels are kept uniform by use of balance pipes, one such joining the water spaces and another the steam spaces.

For much larger installations, much use has been made of high velocity water tube boilers with connections to a *steam drum*: where a number of boilers of this type are required, a common drum or drums are used, as Figure 6.4. The water of expansion which has to be dealt with as a result of diurnal temperature fluctuations may well be contained by variations in level within the drum but when the contents of a large system are heated up from cold, discharge either to a cistern or to drain is necessary.

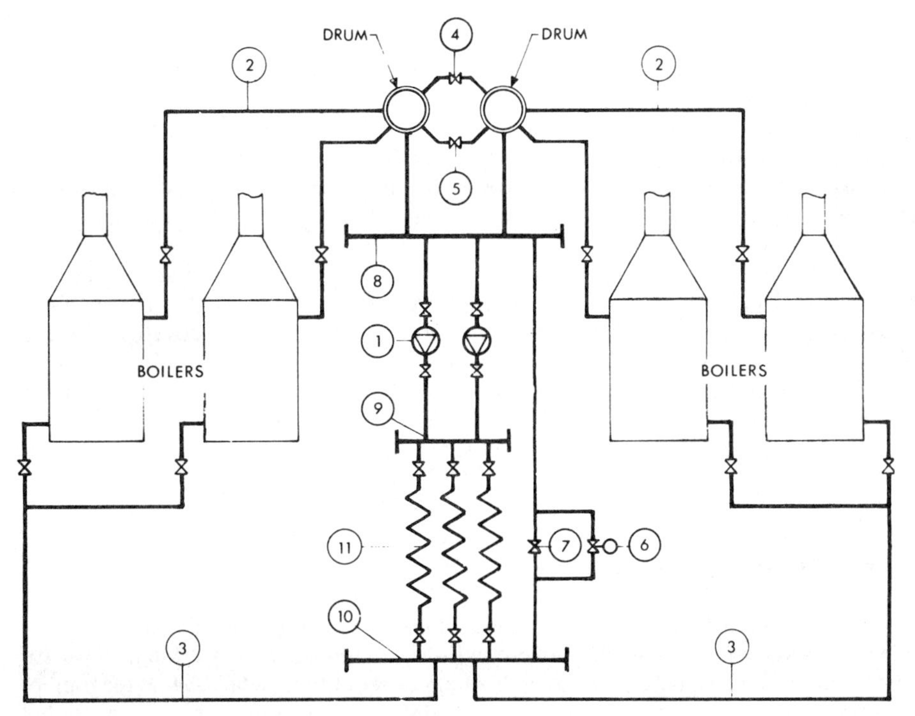

Figure 6.4 Steam pressurisation (water tube boilers): 1, system circulating pumps; 2, flows, boilers to drums; 3, returns, boilers to drums; 4, steam balance pipe; 5,water balance pipe; 6, mixing valve; 7, manual by-pass; 8, pump suction header; 9, flow header; 10, return header; 11, external circuits

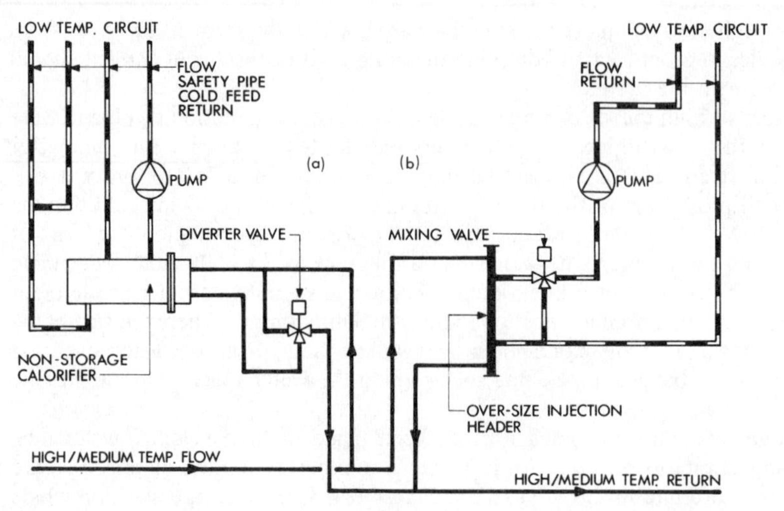

Figure 6.5 Low temperature circuits served from a high temperature system (isolating and regulating valves are omitted for clarity)

Make-up water for installations of this type, to replace the inevitable losses at valve and pump glands, is provided via a conventional boiler feed pump which draws its supply from a cistern containing treated water. The boiler water level is monitored and the make-up pump is usually controlled by hand.

The principal disadvantage of steam pressurisation is that the system is inherently unstable since the steam pressure applied is so closely associated with the temperature of the water circulated. Any minor variation in the applied load will be relatively critical to performance: experienced operatives and skilled maintenance personnel are necessary if satisfactory service is to be assured. As a result, most current high temperature installations use other methods to apply pressure, the equipment necessary being described in Chapter 8.

As in the case of medium temperature systems, it is the surface temperature of the heat emitting equipment which restricts selection of type. However, the higher temperatures available permit the use of most forms of radiant surface and these will show to advantage where they can be fitted at levels beyond reach. For a building or group of buildings where heating from an available high temperature source is unsuitable, one or other of the arrangements shown in Figure 6.5 may be used. The second of these, in (b), it should be noted, operates at the same *pressure* as the high temperature system.

Water systems – piping arrangements

Although the materials used and the techniques of construction span a wide range from the smallest domestic premises to the largest industrial building, piping arrangements for systems using water as a heat distributing medium differ only in detail. A line diagram of any system will reveal that the various circuits and sub-circuits form a network of parallel paths for water flow. If the arrangement of these is kept simple and they are arranged in a logical manner, a design has a sound basis from which it may be developed further.

Moreover, from such a basis, a system is likely to be produced which can be set to work, commissioned *and maintained* without undue difficulty.

The diagrams included under the subsequent headings do not show circulating pumps, feed and expansion pipes, valves, or connections to pressurising equipment, etc. These important aspects will be dealt with separately.

The single-pipe circuit

A piping arrangement of this type, as shown in Figure 6.6, is perhaps the simplest circuit possible, the various items of heat emitting equipment being connected to the same single pipe as shown. Each item of equipment is therefore in shunt with the short section of piping between the two junctions and, in consequence, the equipment must be of a type which does not offer a significant resistance to water flow. This means, in terms of current availability, that a simple radiator having liberally sized waterways is to be preferred.

The first radiator (A) will receive water at boiler output temperature and will reject it, somewhat cooler, after having extracted the appropriate heat quantity, back to the single pipe. Here the water rejected will mix with that which has bypassed the radiator, through the single pipe, and a mixture at less than boiler output temperature will reach radiator (B).

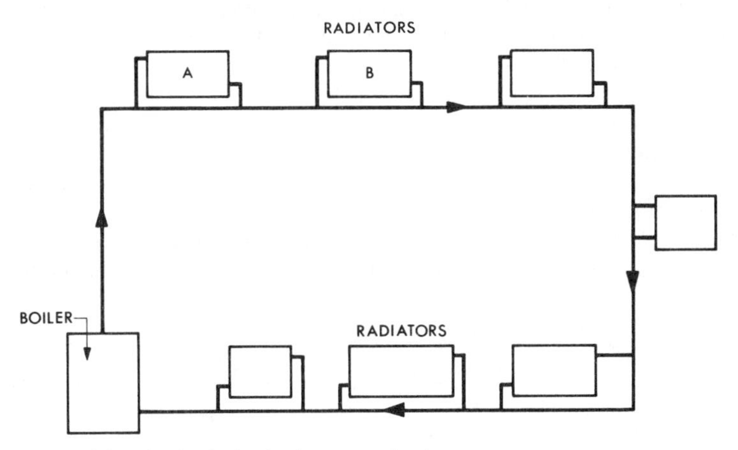

Figure 6.6 A simple single-pipe hot water circuit

Thus, the water temperature is progressively cooler as it reaches each following radiator and the last in the circuit is supplied at little more than boiler return temperature. In order to compensate for this decay, the size of successive radiators is increased and care is taken to select a suitable temperature drop across both them and the piping circuit. For example, consider a system having ten radiators, each of which is required to provide the same output. The temperature drop across the circuit might be chosen to be 10 K and that across each individual radiator to be 15 K. The water temperature in the single pipe would fall by an average of 1 K after each radiator and thus the mean temperature of the first and the last would be 72.5°C and 63.5°C, respectively.

The number of radiators which may be served by a single pipe is necessarily limited and thus a number of parallel circuits must be provided for in a system of any significant size.

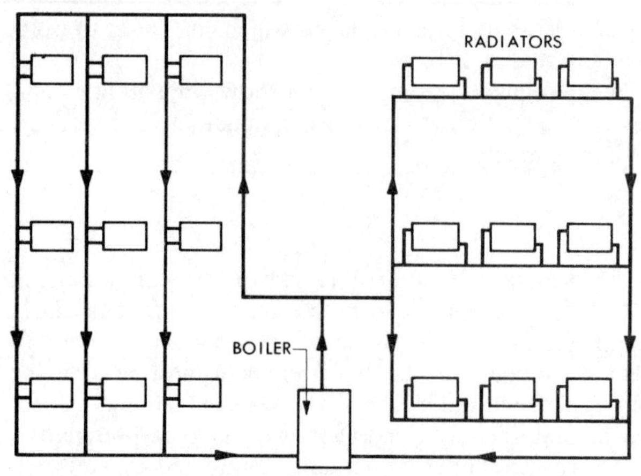

Figure 6.7 Multiple single-pipe hot water circuits in parallel

These may be arranged, as shown in Figure 6.7, either as horizontal pipes in a 'ladder' formation or as a series of vertical drops. There is of course an infinite number of ways in which parallel circuits of this type may be arranged but it is important to realise that the direction of flow through the radiators must correspond with that through the single pipe, i.e. it would generally be unworkable to serve radiators from a rising pipe.

The example used earlier was abridged by having radiators all of the same size but the calculations for any such circuits are not so much complex as tedious. They must be made, however, if a successful result is to be achieved. It seems probable that single-pipe circuits are currently out of favour simply because of failure in the past, by plumbers, to use simple arithmetic for domestic heating designs.

While the principles applied follow those noted above, *small-bore systems*, as tentatively introduced for commercial premises some 40 years ago, were later adapted to

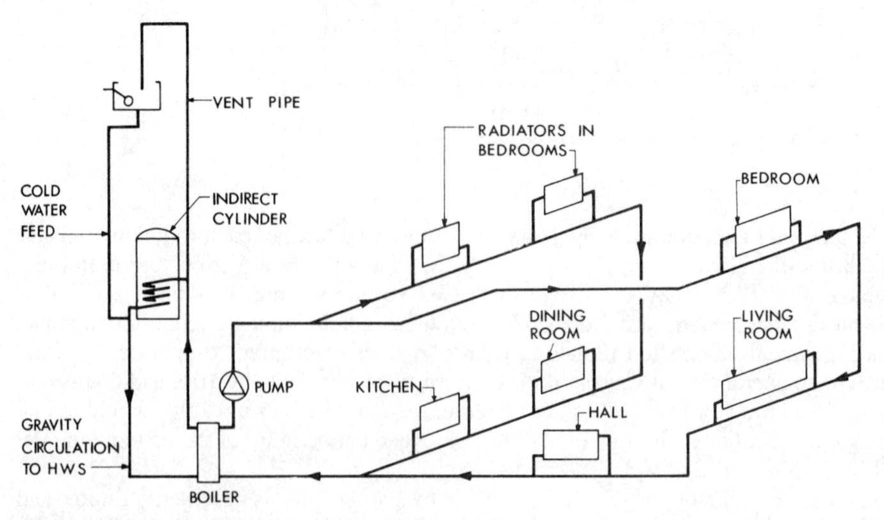

Figure 6.8 Small-bore hot water circuits (generally obsolete)

domestic use. The number of radiators which may be carried by a single circuit is limited by pump capacity and by the size of pipe used, commonly 15 mm copper. Where the system is to supply a large number of radiators, further ladder loops are added in parallel with the first, as shown in Figure 6.8.

Single-pipe circuits are generally confined to conventional low temperature hot water systems since the temperature decay would not be acceptable if the water were only warm at the point of origin. They have been applied to medium temperature systems using an unorthodox arrangement as referred to later under the *hybrid system* heading.

The two-pipe circuit

This arrangement has, in most respects, come to supersede the single-pipe configuration and has an inherent logic as far as parallel circuits are concerned, as shown in Figure 6.9. Flow and return mains originate from the boiler plant and each main or sub-circuit consists of branches from them. Each branch conveys an appropriate quantity of water to and from whatever heat emitting terminals are connected to it. In an ideal world, all circulating pipework would be insulated perfectly, and the water inlet temperature at each terminal would be exactly the same as that leaving the boiler. Similarly, the temperature of water returned to the boiler would be exactly the same as that leaving each terminal. In practice, heat output from the pipework – heat loss in this context – will reduce the temperature at the various inlets to a level which will vary, roughly, according to how distant each is from the boiler. Likewise, the water returned to the boiler will be at a lower temperature than that at which it leaves the outlets of the various terminals.

In terms of hydraulic balance, systems arranged as shown suffer from a number of difficulties. It will be obvious from the diagram that the most distant heat emitter is disadvantaged, by comparison with that nearest to the boiler, in this respect. The problem will be reduced if either the heat emitters, or the final sub-circuit pipework to them, offer a high resistance to water flow relative to that of the remainder of the system.

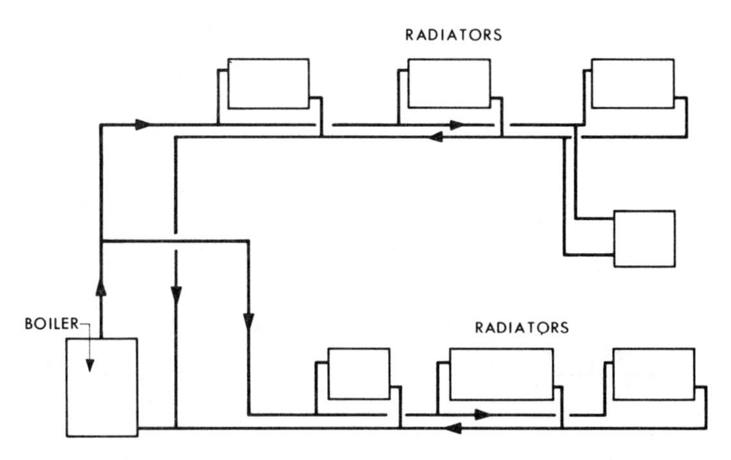

Figure 6.9 A simple two-pipe hot water circuit

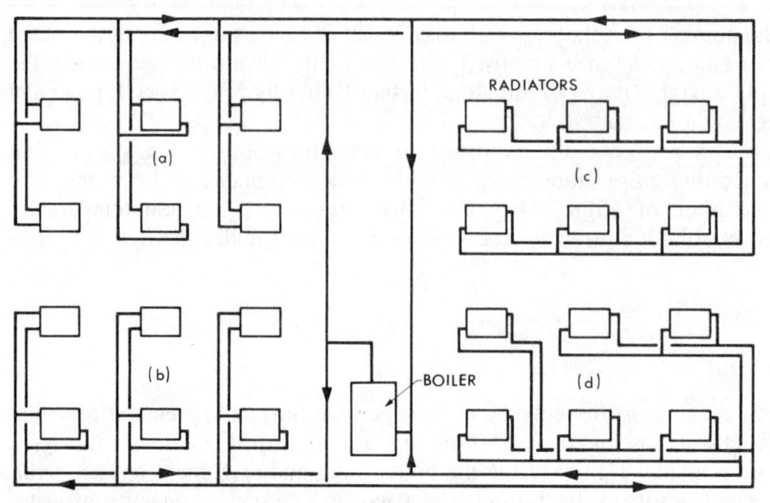

Figure 6.10 Multiple two-pipe hot water circuits: (a) drop feed; (b) riser feed; (c) drop feed to horizontals; (d) riser feed to horizontals

In terms of physical arrangements, two-pipe circuits are extremely flexible and may be set out to suit the building facilities available, as shown in Figure 6.10. They may be applied to all types of hot water systems, whatever the water temperature, and to supply all types of heat emitting equipment.

In the same way that the small-bore system was evolved to extend the scope of single-pipe circuits, the *micro-bore system* has been produced to further exploit the use of even smaller copper pipes, 6, 8, 10 and 12 mm in this instance. The two-pipe design concept, as Figure 6.11(a), differs somewhat from those described earlier under this

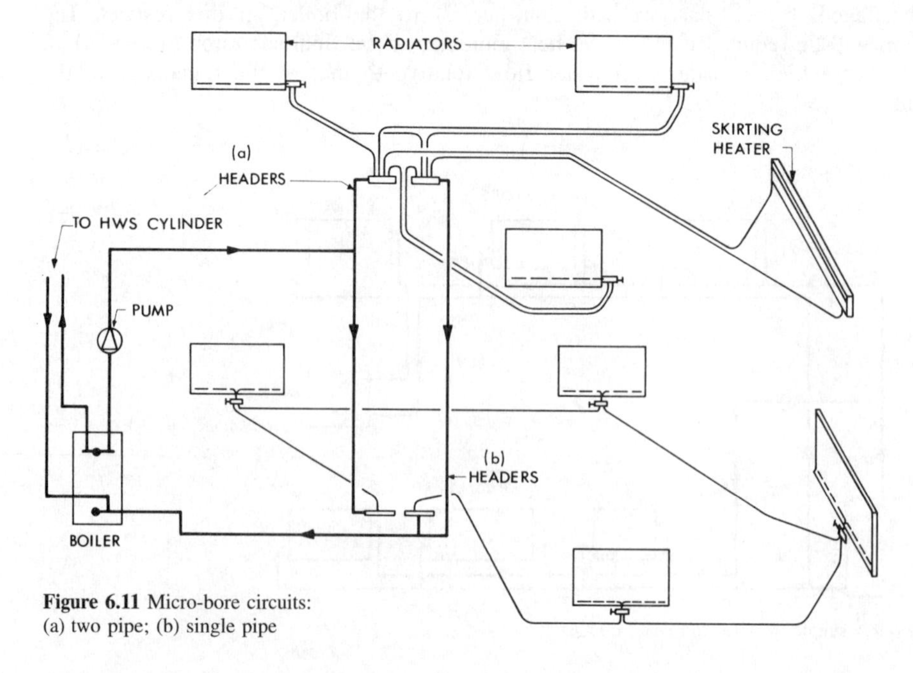

Figure 6.11 Micro-bore circuits: (a) two pipe; (b) single pipe

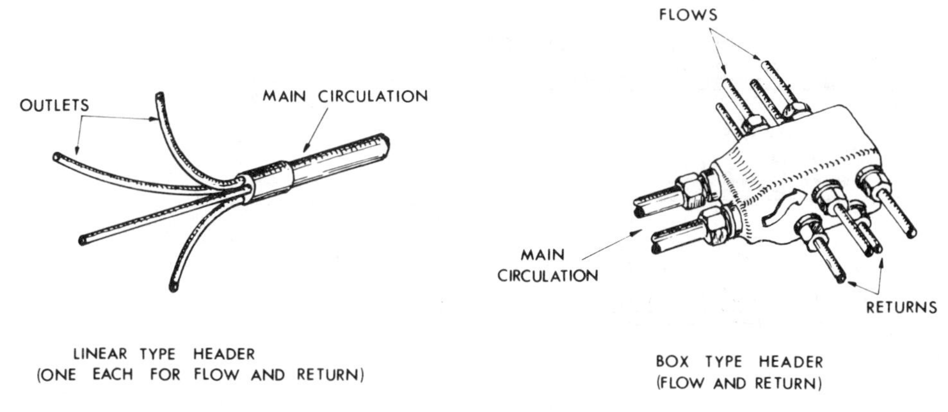

Figure 6.12 Alternative types of header for micro-bore systems

heading in that it is strictly radial, i.e. the flow and return connections to each heat emitter originate quite separately, each from a central manifold, two patterns of which are shown in Figure 6.12. Water authorities who represent the progressive minority have recognised that such connections are 'necessary for the efficient operation' of the system under *Model Water Byelaws* and have permitted them to be buried in the floor screed.

Use may be made of miniature valves, two per heat emitter in the normal manner, or of a special single valve, provided with flow and return connections, which carries an extension pipe to run through the waterway to the remote end. A derivative of this arrangement using proprietary radiators is a buried single-pipe circuit as Figure 6.11(b), the valve fitted at the bottom centre acting to divert water either through or around the radiator. Such systems have little to recommend them for other than the smallest domestic applications.

The reversed return circuit

It is almost always possible to arrange two-pipe circuits in a manner which will ameliorate the hydraulic problems mentioned above. This is achieved, in principle, by laying out the system so that the distance water has to travel, from the boiler to each individual terminal item and then back again to the boiler, is the same, as illustrated in Figure 6.13(a). Whether this is a practical proposition in terms of capital cost depends upon the building configuration and the ingenuity of the designer. The alternative layouts in Figure 6.13(b) and (c) show that it may in some cases be possible to produce a *reversed return* at minimal additional cost, whereas in others there are complications.

For a given heat output, the water quantity in circulation around a system is a function of the temperature difference, flow to return. Table 6.2 demonstrates that a low temperature hot water system is likely to have over four times more water flowing in each circuit and sub-circuit than would be the case for a high temperature hot water system having the same output. This comparison serves to show that the use of reversed return arrangements, always desirable, is much easier to justify on technical grounds for medium or high temperature systems.

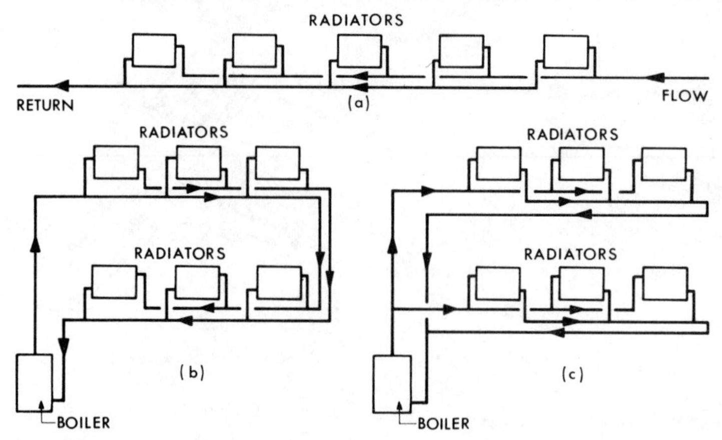

Figure 6.13 Reversed return circuits: (a) basic principle; (b) logical layout; (c) contrived layout

Hybrid circuits

The descriptions given of the three principal circuit configurations should not be read as suggesting that they are mutually exclusive. Provided that a clear design logic is followed, no harm is done by mixtures if they go to make up a total arrangement which suits the use and style of a building. Examples, not necessarily recommended for use, follow:

Ladder circuits. For high-rise buildings, a reversed return two-pipe system as Figure 6.14(a), serves single-pipe circuits at each floor. In contrast, a simple two-pipe system serves reversed return circuits at each floor, as in Figure 6.14(b).

Series circuits. Again for high-rise housing developments, a system much used in Europe has been applied to some buildings in the British Isles. It consists, as shown in Figure 6.15, of a medium temperature hot water circuit which, on a single-pipe basis, serves cased convectors (having 'hairpin' elements) in series: the return water therefrom passes to a two-pipe system serving radiators. By this means, the potential of the medium temperature water is exploited to the full.

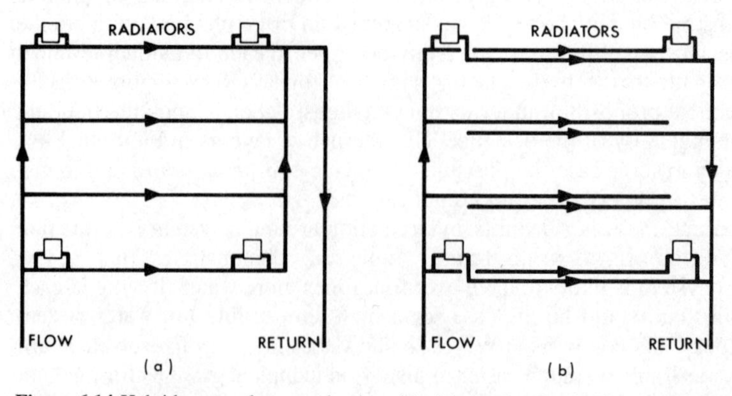

Figure 6.14 Hybrid reversed return circuits with: (a) simple sub-circuits; (b) simple supply mains

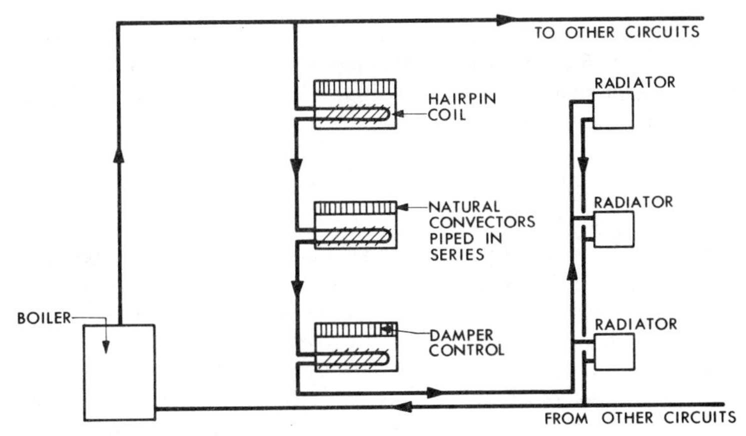

Figure 6.15 A hybrid series circuit

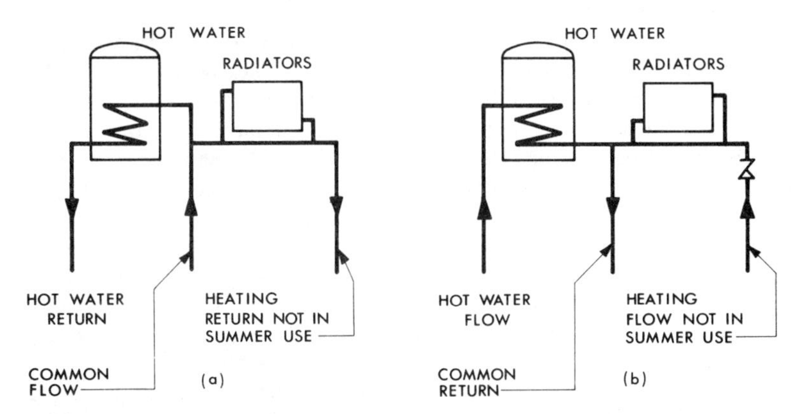

Figure 6.16 Three-pipe circuits (not recommended)

From time to time the concept of three-pipe arrangements emerges from hiding. It is then proposed that two circuits should be served, either by separate flows and a common return or by a common flow and separate returns. This arrangement is often considered, as an economy measure, particularly for group and small district heating schemes, as shown in Figure 6.16. On paper, such a piping system will allow each of the two circuits to be controlled separately: one, perhaps, being live in summer only and the other, or both, in winter. There is no published record that any such arrangement has been successful but ample evidence of failure: three-pipe circuits must be ranked with three-card tricks.

Steam systems – characteristics

Steam as a medium for heating radiators and the like is a thing of the past. The flexibility of hot water systems, with their ability to adjust temperature of output to suit weather conditions, coupled with an overall simplicity which has already been emphasised, has supplanted steam for most developments in residential and commercial buildings. Where steam is required for process in industrial premises or in hospitals for use in kitchens, laundries and sterilisers, etc., it may serve also as a primary distribution medium to heat exchangers. These are then used in the production of either low temperature hot water for heating systems or for hot water supply in a manner similar to that illustrated, for higher temperature water, as shown in Figure 6.5(a) (p. 146).

Generation of steam

When heat is applied to water in a closed vessel such as a boiler which is only partially filled, the temperature will rise to boiling point (100°C at atmospheric pressure): application of further heat will cause a change of state and water will become steam. The quantity of heat involved in this latter process – *the latent heat of evaporation* – is considerable, at 2257 kJ/kg, when this is compared with the 420 kJ/kg required to raise water from 0°C to 100°C. Since the vessel is closed, the steam has no means of escape and the addition of yet further heat will cause the pressure to rise and with it the temperature of the steam and of any water remaining.

Steam in contact with water is described as being *saturated*: if it carries some water droplets in suspension, then it is referred to as being *wet*. Saturated steam removed from the vessel in which it has been generated and subjected to further heat is said to be *superheated* but in this condition it is of little use for space heating purposes.

Utilisation involves the process of condensation, in which the latent heat is removed from the steam, by the use of heat exchangers of one sort or another, with the result that the steam reverts to water at the same pressure and temperature. This hot water or *condensate* must be removed from the heat exchangers as soon as it is formed in order to prevent them from becoming water-logged and thus useless. On release to atmospheric pressure, the condensate will fall in temperature since that part of the heat within it which is above atmospheric boiling point will act to re-evaporate a proportion of the remaining liquid. The vapour so produced is known as *flash steam* and this may be used in a variety of ways to economise in the use of what is known, conversely, as *live steam*.

Condensate is, for practical purposes, distilled water and will attract uncondensable and unattractive gases such as chlorine, carbon dioxide and oxygen. The resultant admixture will most probably be acidic and thus liable to cause corrosion in pipelines and elsewhere in the system.

Properties of steam

Comprehensive tables listing the properties of steam are published by HMSO for the National Engineering Laboratory and relevant extracts are provided in the *Guide Section C2*. Tables 6.5 and 6.6 which appear at the end of this chapter (pp. 167–8), are abridged extracts from the same source. Figure 6.17 provides a graphical representation of the principal characteristics from which the following may be noted:

Pressure. This is stated in absolute terms. Gauge pressure, as used for practical purposes, is the quoted value less the standard atmospheric pressure of 101.325 kPa.

Temperature. This is stated in relation to the rounded values for pressure. A water temperature of 100°C occurs at the standard atmospheric pressure of 101.325 kPa.

Enthalpy. These three columns of the tables give values for heat content or specific enthalpy, in terms of kJ/kg, for the water, for the latent heat and for the sum of the two (total heat of saturated vapour). The decline in latent heat content with pressure increase should be noted.

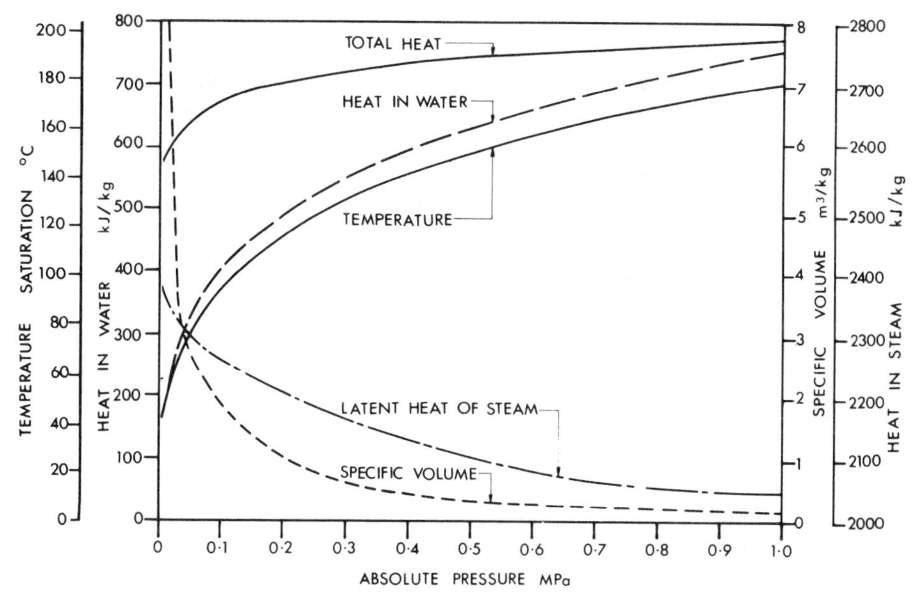

Figure 6.17 Properties of steam. Graphical representation

Volume. The last column in each table gives values for the specific volume of the steam. It is of interest to compare this, for any chosen pressure, with the equivalent for water at the same pressure, as read from Table 6.1.

System pressure

It will be clear from examination of the tables that high pressures are not necessarily desirable. For industrial purposes, quite aside from any question of supply to engines or turbines, there may be a need for high temperatures and, in some cases, superheat. While it is unsafe to generalise, it seems probable that an absolute pressure of about 1.2 MPa (12 bar) will be the highest level likely to be encountered in service to a hospital or a similar institutional building. However, it is most probable that steam generated at that pressure will be used only indirectly for heating service: this might be downstream of a pressure-reducing valve; as flash steam following a process use; or as hot water produced through a heat exchanger.

Removal of air

The presence of air and other incondensable gases is a problem in any steam system whatever the pressure. They exist as a result of aeration of condensate, decomposition of calcium bicarbonate to form calcium carbonate and CO_2 and the inevitable inward leakage through valve glands, etc., which follows any seasonal plant shut-down. Within any heat exchanger, a stagnant air film at the steam-side surface will act as an insulant and prevent condensation and the desired surrender of latent heat. At low pressures this effect is particularly persistent.

Flash steam recovery

Use may be made of Table 6.6 to determine the availability of flash steam, as best illustrated by the examples which follow. If steam at an absolute pressure of 400 kPa were to be used in some item of process equipment, the condensate temperature would be 144°C and the heat content would be 605 kJ/kg. If the discharge were to a vessel held at atmospheric pressure, the temperature could not then be more than 100°C, and thus heat must be abstracted from the condensate until the content reaches the level appropriate to the reduced pressure, i.e. 417 kJ/kg.

Table 6.7 Availability of flash steam

Initial absolute pressure (kPa)	*Per cent condensate available as flash steam at following reduced absolute pressures* (kPa)				
	100	*150*	*200*	*250*	*300*
1200	16.9	14.8	13.3	12.1	11.0
1100	16.1	14.1	12.5	11.3	10.2
1000	15.3	13.3	11.7	10.4	9.3
900	14.4	12.4	10.8	9.5	8.4
800	13.5	11.4	9.8	8.5	7.4
700	12.5	10.3	8.8	7.5	6.3
600	11.2	9.1	7.5	6.2	5.0
500	9.9	7.7	6.1	4.8	3.7
400	8.3	6.2	4.5	3.2	2.0
300	6.4	4.2	2.5	1.2	–

The difference of 605 – 417 = 188 kJ/kg would thus be available to evaporate a part of the remaining condensate and, since latent heat at atmospheric pressure is 2257 kJ/kg, the quantity of flash steam so produced would be 188/2257 = 0.083 kg/kg (or 8.3%). It is of course not necessary that the condensate be discharged to atmospheric pressure: flash steam could be produced at, say, 170 kPa in which case the availability from 400 kPa would be reduced to 0.054 kJ/kg. Flash steam, if not recovered, represents a loss in efficiency and the magnitude of this may be seen from Table 6.7.

Pressure reduction

Since a steam plant may well supply a number of different needs, the use of pressure reducing valves as described in Chapter 8 will often be a desirable feature. The process of pressure reduction involves what is known as *wiredrawing,* which again is best illustrated by an example. Taking the same absolute pressure as used in the preceding example, 400 kPa, the *total* heat content for saturated steam, as read from Table 6.6, is 2739 kJ/kg. If this steam were to be expanded through a reducing valve to provide an output at 150 kPa, the total heat would remain constant, since no work had been done. The temperature, in consequence, would be approximately 133°C and not the anticipated 111°C, as read from Table 6.6, which represents 22 K of superheat. Approximate values for superheat arising from wiredrawing are listed in Table 6.8; reference to the *Guide*

Table 6.8 Approximate superheat due to wiredrawing

Initial absolute pressure (kPa)	*Approximate superheat* (K) *at following reduced absolute pressures* (kPa)				
	100	*150*	*200*	*250*	*300*
1200	55	45	37	32	27
1100	53	43	36	30	26
1000	51	41	34	29	24
900	49	39	32	27	22
800	46	37	29	24	20
700	44	34	27	22	17
600	30	31	24	19	14
500	36	26	19	14	10
400	31	22	15	10	6
300	24	15	8	3	–

Section C for tables giving properties of superheated steam will enable similar values to be derived for other pressures.

Steam systems – piping arrangements

A much quoted sentence from the well-loved first edition of *The Efficient Use of Steam* postulates that 'A steam pipe should carry steam by the shortest route in the smallest pipe with the least heat loss and the smallest pressure drop that circumstances will allow.' As succinct guidance in relation to the design of distribution pipework, these few words gather together all the salient criteria.

Traditional arrangements

Disposal of condensate and elimination of air were the twin problems which dominated the design of steam pipework in early years, where system pressures adopted were in the range of 120 to 170 kPa absolute. Single-pipe arrangements were not uncommon, Figure 6.18(a), and these were featured in North American textbooks until 40 years ago. Horizontal pipework, where this occurred, was laid to a generous 'pitch' to encourage condensate return and vertical risers were oversized to allow steam and condensate to flow in opposite directions in the same pipe. Air vents were fitted at the end of each horizontal pipe run and at mid-height on each radiator. Water hammer was endemic.

Two-pipe gravity systems were classified as having 'wet' or 'dry' returns, dependent upon whether the condensate pipework was above or below the water level in the boiler, as in Figure 6.18(b). As in the case of the single-pipe system, radiators were either on, at temperatures between 105°C and 115°C, or off and cold; there was no intermediate level. Water make-up to both single and two-pipe systems was made from a cistern, fitted at a height sufficient to overcome the system pressure, to a 'boiler feeder' which consisted of a ball valve within a closed casing connected top and bottom to the steam and water contents of the boiler.

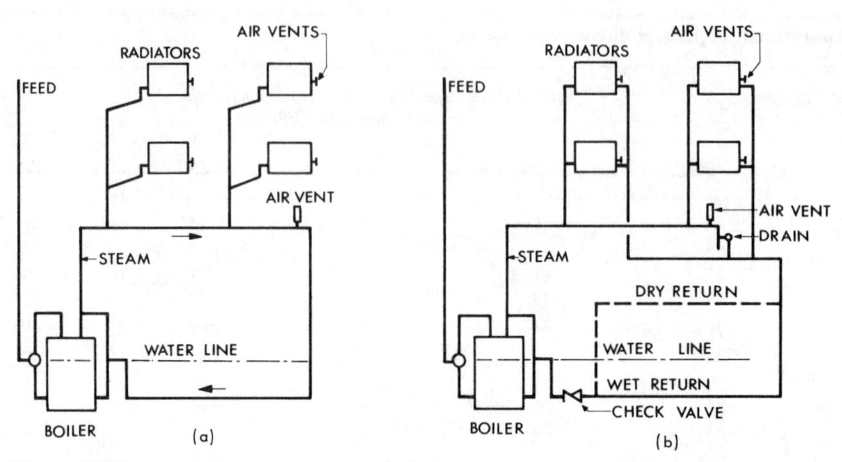

Figure 6.18 Low pressure steam systems: (a) single-pipe gravity return; (b) two-pipe gravity return

The subsequent *vacuum* systems where, as the name suggests, use was made of a vacuum pump to assist in removal of air and to return condensate to the boiler, were a substantial advance and permitted system size to be increased considerably. The later introduction of *sub-atmospheric* systems was perhaps the ultimate development since these permitted steam pressures to be both lowered and controlled such that temperatures could be varied between about 55°C and 105°C at source, according to weather conditions. Such precision was not unaccompanied by disadvantages, one of the more significant being the necessity to use specialised components such as glandless valves in an attempt to produce a piping system which remained absolutely 'airtight' both in service and after shut-down.

Current practice

Facilities for the removal of condensate so that the steam supply remains dry are among the more important features and supply pipework must be provided with *relay points*, as in Figure 6.19, to assist in this respect (see Figure 8.19 also, p. 217). Otherwise, pipework follows a two-pipe pattern very much as might be anticipated, incorporating specialised equipment as described in Chapter 8.

Condensate will, in most normal circumstances, be collected and returned to a tank or *hot well* near to the boiler. Where distance or site levels make direct return impossible, intermediate collection and pumping units may be required. At the hot well, to make up for any losses which may have occurred, a treated water supply will be provided and this may alternatively be directly 'on line' or via an intermediate store. The optimum size of a hot well will depend upon a number of factors such as the order of fluctuation in the steam demand, the corresponding – but not necessarily equivalent – rate of condensate

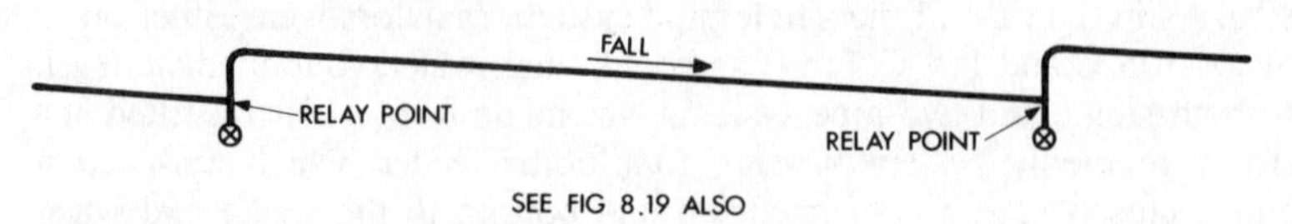

Figure 6.19 Steam main relay joints

return and the operational routine of the water treatment plant. It is good practice, nevertheless, to provide a minimum size related to the evaporation capacity of the boiler plant over 2 hours.

The temperature of the returned condensate is likely to be not much below 100°C but the hot well should nevertheless be provided with a steam coil to maintain that level. To avoid cavitation at the suction connections of the pumping equipment provided to return condensate to the boilers, the hot well should be elevated where this is practicable, to about 5 m above that level. Where packaged boilers are used, each is likely to be provided, at works, with a small cylindrical 'condense receiver' in association with the prewired feed pump and associated controls. The size of such receivers is usually quite inadequate to provide the necessary storage capacity but their volume may be deducted from the capacity of the hot well as proposed above.

Air systems – characteristics

Passing reference was made in Chapter 4 to the use of hot air as a heat transfer medium within floor ducts but such an application is a rarity and no purpose would be achieved by pursuing it further here. Table 6.9 lists the principal characteristics of dry air in context with its use for space heating and it is of interest to compare these with the equivalent values for hot water as listed in Table 6.1. The specific heat capacity of water, in terms of mass, is four times that of air at the same temperature and in terms of volume the difference is immensely greater.

Ventilation systems are dealt with later (Chapter 13) and it is necessary to draw a quite clear distinction between the function which they fulfil and that of the so-called *plenum* system, using air as a heating medium, which falls under this present heading. The essence

Table 6.9 Approximate properties of dry air (at atmospheric pressure)

	Re mass		*Re volume*	
Temperature (°C)	*Specific mass* (kg/m^3)	*Specific heat capacity* (kJ/kg K)	*Specific volume* (m^3/kg)	*Specific heat capacity* (kJ/m^3 K)
10	1.247	1.011	0.802	1.257
20	1.205	1.012	0.830	1.219
30	1.165	1.013	0.858	1.180
40	1.128	1.013	0.887	1.143
50	1.093	1.014	0.915	1.108
60	1.060	1.015	0.943	1.076
70	1.029	1.015	0.972	1.045
80	1.0	1.016	1.0	1.016
90	0.973	1.016	1.028	0.988
100	0.946	1.017	1.057	0.963
110	0.922	1.018	1.085	0.938
120	0.898	1.018	1.113	0.915
130	0.876	1.019	1.142	0.893
140	0.855	1.020	1.170	0.838
150	0.835	1.020	1.198	0.851

of the difference lies in the fact that an inlet ventilation system, properly designed, will do no more than introduce a supply of air into the space served at a temperature which is either at or just below that required in the space: heating is provided by some quite separate system.

The plenum system, conversely, introduces a supply of air at a temperature well in excess of that required in the space: ventilation, in effect, is a by product. In cooling to the space temperature, the air provided surrenders the heat it carried to windows, walls, floor and roof, etc., in just the same way that heated air convected from a hot water 'radiator' cools in circulation within a room. A further difference is that in the case of the plenum system, either all or a considerable proportion of the air provided is taken from outside with intent to produce a positive pressure differential, inside to outside, and thus reduce inward leakage through cracks around windows, doors etc.

The traditional plenum system

This is best discussed using, as a simple illustration, the building shown in Figure 2.3 (p. 26). As may be seen from Table 2.13, the component of the calculated heat loss associated with the building fabric is all that is required, in the case of a fully convective system, to determine the value of the two necessary temperature ratios. Thus:

From earlier calculations (p. 37):

$$\Sigma(AU/\Sigma A) = 240/419 = 0.57\,\mathrm{W/m^2\,K}$$

and by reference to Table 2.13:

$$F_1 = 0.97 \text{ and } F_2 = 1.10$$

thus

$$Q_u = 0.97 \times 240 \times 20 = 4\,656\,\mathrm{W}$$
$$Q_v = 1.10 \times 2091 \times 0.33 \times 20 = 15\,181\,\mathrm{W}$$

and hence

$$Q_{u+v} = 19\,837\,\mathrm{W}\ (\simeq 19.9\,\mathrm{kW})$$

The various internal temperatures may now be established:

$$t_{ei} = (0.97 \times 20) - 1 = 19.4^\circ\mathrm{C}$$
$$t_{ai} = (1.10 \times 20) - 1 = 21.0^\circ\mathrm{C}$$

The significance of this last value, for the building used as an example when heated by a plenum system, is that it shows that an inside air temperature of 21°C is necessary to maintain a dry resultant temperature of 19°C. This figure of 21°C would, taking a volumetric specific heat of air from Table 6.9, now be used to calculate the required air temperature reaching the room, thus:

$$t = (21 + 1) + [(4656)/(580 \times 1.18)] = 22 + 6.8 = 28.8^\circ\mathrm{C}$$

The heat load noted above is some 8.5 kW more than the total calculated on p. 52 for a panel radiator system, a substantial part of the difference coming about as a result of the necessarily increased rate of air change. The supply air will, of course, cool to only about 22°C in the space served and then be exhausted to outside at that temperature. Thus, about a third of the heat input is wasted unless either some form of heat reclaim equipment (pp. 490–500) is provided or a part of the supply air quantity is returned from the space via an extract system, for recirculation. The greater the proportion recirculated, however, the less the differential pressure and the greater the probability that strong winds and thermal forces will upset the theoretical pressure balance.

For commercial and institutional buildings having a multitude of compartments, the plenum system was misapplied since it could not be expected to provide satisfactory service in a situation where the requirements varied constantly from room to room. In terms of comfort, moreover, the elevated temperature of the air supply, often as high as 50°C, led to complaints that the atmosphere was oppressive. For low-cost installations in factories and buildings where the volume per occupant was relatively large and where floor space was valuable, application was more successful but was largely superseded when unit heaters became popular.

The high velocity system

The majority of the disadvantages associated with the traditional plenum system have been overcome by relatively recent developments in direct firing of central plant coupled with a high velocity air distribution system to specialist *terminal diffusers*. This system is directed principally at application to large industrial spaces, aircraft hangars, workshops, warehouses, etc.

The central plant consists of a forced convection air heater within which the products of combustion mix with the heated air which is provided for circulation to the space served. By this means, an outlet temperature of about 140°C may be achieved with a combustion efficiency which approaches 100%. This concept, which applies only when fuels containing minimum potential for pollution are used, is an extension of the principle illustrated in Figure 4.10. The basis of design used for the system here considered restricts the concentration of carbon dioxide, by volume, to 2800 parts per million in the air circulated, a quantity not much more than half the maximum permitted by the Health and Safety Executive.

The heated air, which represents only about half an air change within the space served, is distributed from the central plant through insulated sheet-metal ducts at a velocity of about 35 m/s, to an array of terminal diffusers. Each diffuser is equipped with one or more nozzles which *induce* a supply of air from the space into circulation via a venturi arrangement. The quantity of air induced, which may be drawn from an area within the building where temperature gradient has created a potential wastage, is about four times that of the supply from the central plant and the total is thus enough to allow control over the final distribution. The output discharge temperature is of the order of 50°C at a velocity of about 5 m/s. Figure 6.20 provides a diagram of the system arrangement.

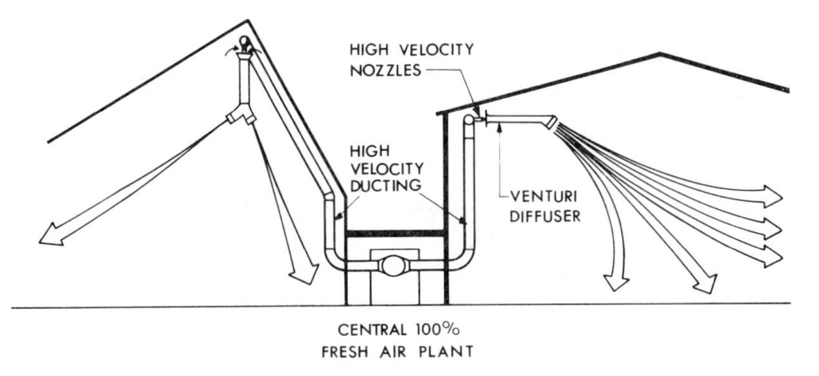

Figure 6.20 High velocity plenum system (Casaire)

Pipework heat emission

Whatever the heating medium may be; water, steam or air, the piping distribution arrangements have certain features in common. They will expand when heated, as described later in Chapter 8, and they will give out heat throughout their length whether this be wanted or not. The magnitude and effect of this second matter must be evaluated and dealt with by application of thermal insulation when necessary.

Pipework heat emission

Exposed pipework, as a form of heating surface, is rarely used in current practice but in any system of distribution, main and branch piping running through the spaces to be heated may offer a contribution. The theoretical emission from bare horizontal pipework may be calculated from expressions given in the *Guide Section C3* and extracts appropriate to a variety of water temperatures and steam pressures are listed in Tables 6.10, 6.12 and 6.14, which appear at the end of this chapter (pp. 169–70). Emission from pipes fixed vertically varies from the listed values and is about 25% less for small and 5% less for large pipes.

Pipework thermal insulation

In order to reduce the unwanted heat output from distribution mains running in trenches or ducts and through basements or any other spaces not requiring heat, thermal insulation is applied. The effect of such action depends upon the conductivity of the insulation material, Figure 6.21, its thickness and the surface finish. A bright metal cladding to insulation will reduce heat loss from that resulting from a dull painted finish by up to 10%.

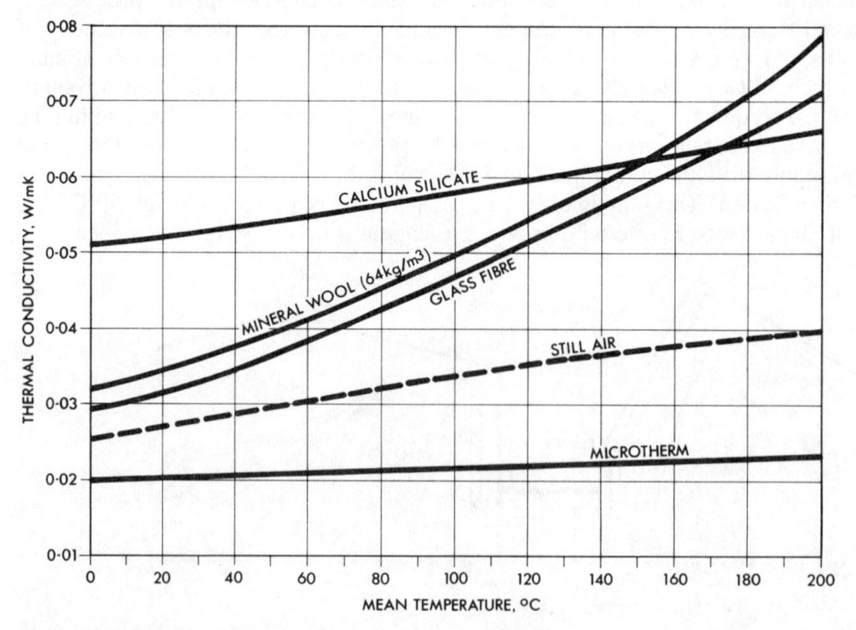

Figure 6.21 Thermal conductivity of pipe insulating materials

The thickness of the insulation which is economic for any given circumstance takes account of the capital cost of the material over its anticipated life equated to the saving in energy cost which will result.

As will be appreciated, the continuing trend towards increased fuel prices has led to a re-examination of many 'traditional' specifications for insulation thickness. In Tables 6.11, 6.13 and 6.15, companions to those referred to previously, comparable information as to emission from insulated pipework is provided. It will be noted that the larger the pipe, the greater the proportional saving for a given thickness of insulation, due to the increase in the outer radiating surface being relatively greater for small pipes than large. As to the materials available, wet applications are a thing of the past and the remaining options relate largely to the quality of the finish.

External distribution systems

It is not unusual that, in the design of a heating system for a large site, the matter of external pipe distribution arrangements has to be considered. There are three possible solutions to this problem:

- Arrange that piping be run overground, with particular care as to topography and any building features which may assist.
- Arrange that piping be routed through basements or, if none exist, through purpose built walk-in subways.
- Provide for excavation and either formation of underground ducts or direct underground burial of piping enclosures.

The last of these approaches will most probably be thought of as being the only practical solution in terms of first cost. The margin of advantage, however, will be less than first imagined if the whole of the site preparation and subsequent building construction methods are of the necessary quality. If they are not of that quality, excavation for and replacement of the pipework is not only inevitable in the long term but likely to be required within five years of installation.

Overground arrangements

On industrial and semi-industrial sites, such as dispersed hospitals, etc., where appearance is not a first-order priority, pipework may be run externally and be supported either at waist level on road verges or from gantries where vehicular traffic must have passage. The thermal insulation must be provided with a finish which is weatherproof and reasonably resistant to vandalism: supports, etc., must be arranged so that the finish is not punctured. Use of the roofing felt and wire netting combination so often seen is inadequate since it is not fully weatherproof with the result that the insulation material absorbs water and corrosion follows. Sheet metal cladding or a proprietary finish are to be preferred.

In some recent city centre and housing developments, there has been a tendency to separate pedestrian and vehicular traffic by use of bridges and overhead walkways. These offer the prospect of other routes for external piping, as shown in Figure 6.22, where protection to insulation by way of cladding may not be required.

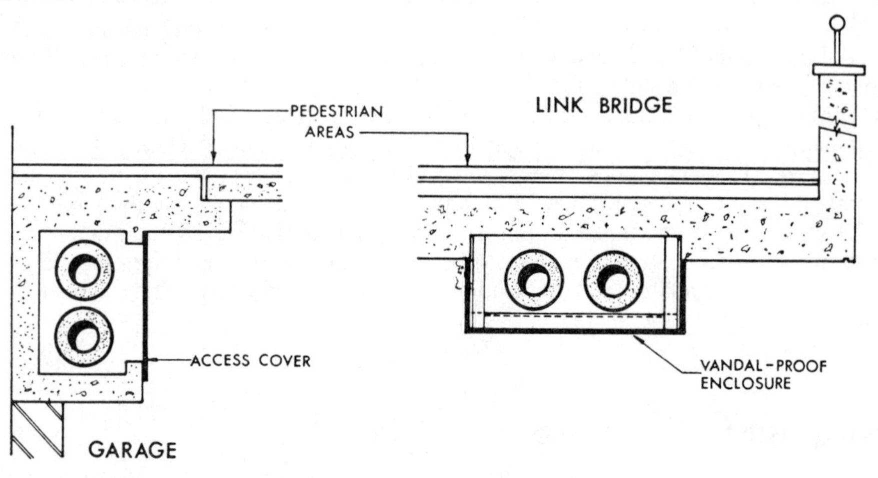

Figure 6.22 External pipework protected by overhead walkways

Basements and subways

In an era when the availability of skilled plant operators is limited, it is more than ever desirable that full consideration should be given to facilities for the day-to-day routine of preventative maintenance. The existence of a basement corridor through which pipework may be routed is a bonus in this respect, in particular where no suspended ceiling is fitted below the service route. Between buildings, a tunnel or subway of sensible size is needed: by the time that an excavation has been made and a waterproof construction formed, it is barely relevant in terms of real cost whether the cross-section is 1.5 m by 1.5 m or a metre bigger either way.

In terms of pipework installation, what little saving might have been made in constructing the small tunnel, rather than the larger subway, will have been swallowed up as a result of the sheer difficulty of working on hands and knees. As to subsequent maintenance, leaking valve glands, dripping joints, distorted flanges, corroded supports and damaged insulation would all go unnoticed were they to occur in the small tunnel.

Underground ducts, etc.

Traditional methods of pipe enclosure underground have of recent years come under scrutiny, particularly since practice in the USA and in Europe was known to be at variance with that in the British Isles. Following an expected pattern, developments in North America have been towards the production of relatively complex factory built units, with intent to reduce site works to a minimum. In contrast, European tendencies have been either towards simplification in choice of materials or to adapt and refine transatlantic usage.

The principal hazard to which underground enclosures are exposed is ground water since, when an excavation is made preparatory to the construction work proper, the natural water table is disturbed and any seepage tends to follow the boundary between the undisturbed ground and the later back-fill. Most of the pipeline enclosures used in the past have been site formed, as in Figure 6.23(a)–(e), and are thus vulnerable as far as site settlement is concerned. Ventilation ducts within buildings, constructed in a manner not

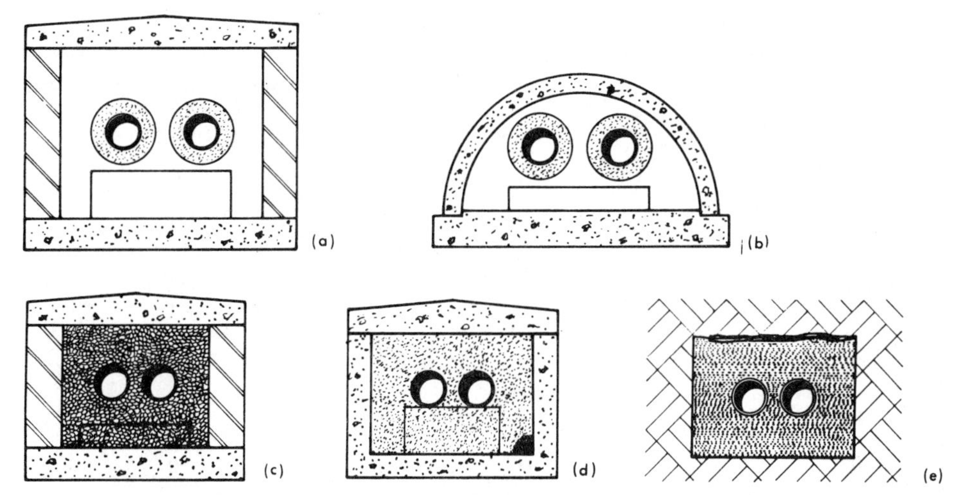

Figure 6.23 Underground pipe ducts. Past practice

unlike that of examples shown in the diagram, are rarely more than 50% airtight when new: it seems unlikely that a better result can be obtained under the more adverse conditions prevailing externally. Finally, many such ducts have been constructed within the constraints of an inadequate cost allocation, which has led to less than perfect detailing and workmanship. A brief description of the enclosures shown may, however, be useful, although none can be recommended:

(a) Insulated pipes laid within a duct having brick or concrete walls built as a preformed or *in-situ* channel and supporting a preformed concrete cover.
(b) Insulated pipes laid within a duct having a preformed half-cylindrical cover set on an *in-situ* concrete base.
(c) Construction as (a), but with a loose granular insulating fill poured round the bare pipes.
(d) Construction as (a), but with a fill of aerated concrete poured round the bare pipes.
(e) Raw excavation, sometimes lined with tar paper, where bare pipes are laid for insulation by a hydrophobic powder which cures when it is heated to form a protective skin on the pipe.

Instances of failure of each of the constructions shown have convinced many engineers that use of one of the factory formed *pipe-in-pipe* systems (Figure 6.24) is the way

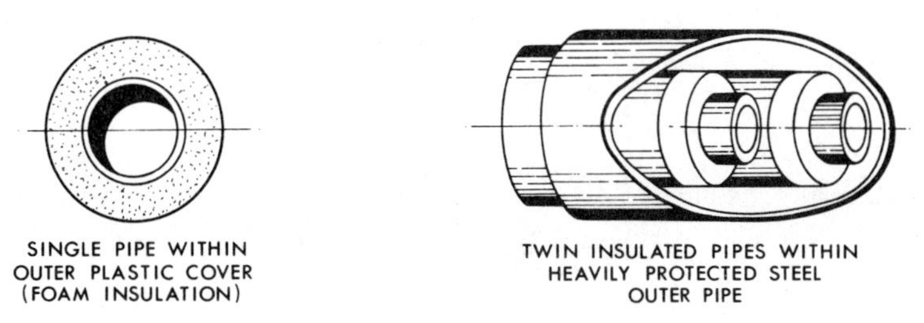

Figure 6.24 Pipe-in-pipe systems for underground heat distribution

Table 6.16 Heat loss from pipe-in-pipe system buried 600 mm below ground level

Size of service pipes (mm)	*Insulation thickness on pipes* (mm)	*Diameter of enclosing conduit* (mm)	*Heat loss (W/m run of enclosing conduit for following mean water temperatures)*		
			75°C	*100°C*	*125°C*
Two × 20	25	250	39	53	63
Two × 25	25	250	42	58	73
Two × 32	25	300	50	65	83
Two × 40	25	300	54	71	89
Two × 50	25	300	58	79	99
Two × 65	25	350	69	92	117
Two × 80	37	350	62	84	108
Two × 100	37	400	73	98	123
Two × 125	37	450	83	110	140
Two × 150	37	600	96	130	165

forward. Table 6.16 lists heat loss data for one such system. The Achilles heel of site application of these is the point at which they enter a valve pit or other site-formed structure, where the watertight continuity of the outer pipe is broken. A sceptic would emphasise that the better patterns of pipe-in-pipe are provided with devices of one sort or another, arranged to provide warning of the presence of vagrant moisture!

Table 6.1 Selected properties of water

Temperature (°C)	Absolute vapour pressure (kPa)	*Re mass* Specific mass (kg/m^3)	*Re mass* Heat capacity (kJ/kg K)	*Re volume* Specific volume (litre/kg)	*Re volume* Heat capacity (kJ/litre K)
10	1.23	999.7	4.193	1.000	4.191
20	2.34	998.2	4.183	1.002	4.175
25	3.17	997.0	4.180	1.003	4.168
30	4.24	995.6	4.179	1.004	4.160
35	5.62	994.0	4.178	1.006	4.153
40	7.38	992.2	4.179	1.008	4.146
45	9.58	990.2	4.180	1.010	4.139
50	12.33	988.0	4.181	1.012	4.131
55	15.74	985.7	4.183	1.015	4.123
60	19.92	983.2	4.185	1.017	4.115
65	25.01	980.5	4.188	1.020	4.106
70	31.16	977.7	4.191	1.023	4.098
75	38.55	974.9	4.194	1.026	4.088
80	47.36	971.8	4.198	1.029	4.080
85	57.80	968.6	4.203	1.032	4.071
90	70.11	965.3	4.208	1.036	4.062
95	84.53	961.9	4.213	1.040	4.052
100	101.33	958.3	4.219	1.044	4.043
110	143.26	951.0	4.233	1.052	4.026
120	198.53	943.1	4.248	1.060	4.006
130	270.12	934.8	4.271	1.070	3.992
140	361.36	926.1	4.291	1.080	3.973
150	475.97	916.9	4.322	1.091	3.961
160	618.05	907.4	4.349	1.102	3.947
170	792.03	897.3	4.387	1.111	3.930
180	1002.68	886.9	4.422	1.128	3.920

Table 6.5 Properties of saturated steam (below 1 bar)

Absolute pressure (kPa)	Temperature (°C)	*Specific enthalpy* (kJ/kg) *In water*	*Specific enthalpy* (kJ/kg) *Latent*	*Specific enthalpy* (kJ/kg) *Total*	Specific volume (m^3/kg)	Specific heat capacity (kJ/kg K)
10	45.83	191.8	2392.2	2584.0	14.673	1.89
20	60.09	251.5	2357.7	2609.2	7.648	1.91
40	75.89	317.7	2318.6	2636.3	3.992	1.94
60	85.95	359.9	2293.2	2653.1	2.731	1.96
80	93.51	391.7	2273.7	2665.4	2.087	1.98
100	99.63	417.5	2257.7	2675.2	1.694	2.01

Table 6.6 Properties of saturated steam (above 1 bar)

Absolute pressure (kPa)	*Temperature* (°C)	*Specific enthalpy* (kJ/kg)			*Specific volume* (m^3/kg)	*Specific heat capacity* (kJ/kg K)
		In water	*Latent*	*Total*		
100	99.6	417.5	2257.7	2675.2	1.694	2.01
120	104.8	439.3	2244.0	2683.3	1.428	2.03
140	109.3	458.4	2231.9	2690.3	1.237	2.05
160	113.3	475.4	2221.0	2696.4	1.091	2.06
180	116.9	490.7	2211.1	2701.8	0.977	2.07
200	120.2	504.7	2201.9	2706.6	0.886	2.09
220	123.3	517.6	2193.4	2711.0	0.810	2.11
240	126.1	529.6	2185.4	2715.0	0.747	2.13
260	128.7	540.9	2177.8	2718.7	0.693	2.14
280	131.2	551.5	2170.7	2722.2	0.646	2.16
300	133.5	561.4	2163.9	2725.3	0.606	2.17
320	135.8	570.9	2157.4	2728.3	0.570	2.19
340	137.9	579.9	2151.2	2731.1	0.539	2.20
360	139.9	588.5	2145.2	2733.7	0.510	2.21
380	141.8	596.8	2139.5	2736.3	0.485	2.22
400	143.6	604.7	2133.9	2738.6	0.462	2.24
420	145.4	612.3	2128.6	2740.9	0.442	2.25
440	147.1	619.6	2123.4	2743.0	0.423	2.27
460	148.7	626.7	2118.2	2744.9	0.405	2.28
480	150.3	633.5	2113.4	2746.9	0.389	2.29
500	151.9	640.1	2108.6	2748.7	0.375	2.31
520	153.3	646.5	2104.0	2750.5	0.361	2.32
540	154.8	652.8	2099.4	2752.2	0.349	2.33
560	156.2	658.8	2095.0	2753.8	0.337	2.34
580	157.5	664.7	2090.7	2755.4	0.326	2.35
600	158.8	670.4	2086.4	2756.8	0.316	2.37
620	160.1	676.0	2082.3	2758.3	0.306	2.38
640	161.4	681.5	2078.2	2759.7	0.297	2.39
660	162.6	686.8	2074.2	2761.0	0.288	2.40
680	163.8	692.0	2070.3	2762.3	0.280	2.41
700	164.9	697.1	2066.4	2763.5	0.273	2.43
720	166.1	702.0	2062.7	2764.7	0.266	2.44
740	167.2	706.9	2058.9	2765.8	0.259	2.45
760	168.3	711.7	2055.3	2767.0	0.252	2.47
780	169.4	716.4	2051.7	2768.1	0.246	2.48
800	170.4	720.9	2048.2	2769.1	0.240	2.49
820	171.4	725.4	2044.7	2770.1	0.235	2.51
840	172.5	729.8	2041.2	2771.0	0.229	2.52
860	173.4	734.2	2037.8	2772.0	0.224	2.53
880	174.4	738.4	2034.5	2772.9	0.220	2.55
900	175.4	742.6	2031.2	2773.8	0.215	2.56
920	176.3	746.8	2028.0	2774.8	0.210	2.57
940	177.2	750.8	2024.7	2775.5	0.206	2.58
960	178.1	754.8	2021.6	2776.4	0.202	2.60
980	179.0	758.7	2018.4	2777.2	0.198	2.61

Data for low temperature hot water pipework

Table 6.10 Theoretical heat emission from horizontal pipes (bare steel to BS 1387)[a]

Nominal bore (mm)	*Heat emission* (W/m run) *to ambient air at 20°C for following water temperatures* (°C)									Nominal bore (mm)
	40	*45*	*50*	*55*	*60*	*65*	*70*	*75*	*80*	
10	9	12	15	19	22	25	29	32	36	10
15	18	24	29	35	42	48	55	62	69	15
20	22	29	36	43	51	59	67	75	84	20
25	27	35	44	53	62	71	81	92	102	25
32	33	43	53	64	75	87	99	112	125	32
40	37	48	60	72	84	98	111	125	140	40
50	45	58	72	87	102	118	135	152	169	50
65	55	71	88	106	125	145	165	186	207	65
80	63	81	101	122	143	166	189	213	238	80
100	78	102	126	152	179	207	236	266	297	100
125	94	122	151	182	214	247	281	317	354	125
150	109	141	175	211	248	287	327	368	411	150

[a] May be used for tarnished, copper pipes to BS 2871 without significant error.

Table 6.11 Theoretical heat emission from horizontal pipes (with insulation of stated thickness)[a]

Nominal bore (mm)	*Heat emission* (W/m run) *to ambient air at 20°C for following water temperatures* (°C)									Insulation thickness (mm)
	40	*45*	*50*	*55*	*60*	*65*	*70*	*75*	*80*	
10	4	5	6	7	8	9	10	10	11	25
15	4	5	6	7	8	9	10	10	11	25
20	4	6	7	8	9	10	11	12	13	25
25	4	6	7	7	8	9	10	11	12	38
32	5	6	7	8	9	10	12	13	14	38
40	5	6	8	9	10	11	13	14	15	38
50	6	7	9	10	12	13	15	16	17	38
65	6	7	9	10	11	13	14	15	17	50
80	6	8	10	11	13	14	16	18	19	50
100	8	10	11	13	15	17	19	21	23	50
125	9	11	13	15	18	20	22	24	26	50
150	10	13	15	18	20	23	25	28	30	50

[a] Insulation conductivity = 0.04 W/mK.

Data for medium and high temperature hot water pipework

Table 6.12 Theoretical heat emission from horizontal pipes (bare steel to BS 1387)

Nominal bore (mm)	*Heat emission* (W/m run) *to ambient air at 20°C for following water temperatures* (°C)									Nominal bore (mm)
	100	*110*	*120*	*130*	*140*	*150*	*160*	*170*	*180*	
10	61	71	82	94	105	118	131	145	158	10
15	100	118	135	154	173	194	215	238	261	15
20	122	143	164	188	211	237	262	290	318	20
25	149	175	200	227	257	289	320	355	389	25
32	181	213	244	279	314	353	391	434	476	32
40	203	238	273	313	352	395	438	496	534	40
50	246	289	331	379	427	480	532	590	648	50
65	301	353	406	465	523	588	653	725	796	65
80	345	406	466	533	600	675	750	833	915	80
100	431	507	582	666	750	844	937	1039	1140	100
125	515	605	695	796	897	1009	1120	1245	1370	125
150	589	699	808	924	1040	1175	1310	1460	1600	150

Table 6.13 Theoretical heat emission from horizontal pipes (with insulation of stated thickness)[a]

Nominal bore (mm)	*Heat emission* (W/m run) *to ambient air at 20°C for following water temperatures* (°C)									
	Medium temperature hot water					*High temperature hot water*				
	90	*110*	*130*	*150*	*Insulation thickness* (mm)	*120*	*140*	*160*	*180*	*Insulation thickness* (mm)
10	15	18	21	25	38	18	21	22	24	50
15	15	19	23	27	38	19	23	27	30	50
20	17	22	26	31	38	21	25	29	34	50
25	19	24	30	35	38	24	29	34	38	50
32	19	24	30	35	50	27	32	38	43	50
40	20	26	32	38	50	29	35	41	46	50
50	23	30	36	43	50	33	40	46	53	50
65	27	34	42	49	50	35	42	49	56	65
80	30	39	47	56	50	39	47	55	62	65
100	32	41	50	59	65	45	54	63	72	65
125	36	46	56	66	65	51	61	71	82	65
150	40	51	63	74	65	57	68	80	91	65

[a] Insulation conductivity = 0.055 W/m K.

Data for steam pipework

Table 6.14 Theoretical heat emission from horizontal pipes (bare steel to BS 1387)

Nominal bore (mm)	*Heat emission* (W/m run) *to ambient air at 20°C for following absolute steam pressures* (kPa)									*Nominal bore* (mm)
	100	*200*	*300*	*400*	*500*	*600*	*700*	*800*	*900*	
10	61	82	94	105	118	131	138	145	151	10
15	100	135	154	173	194	215	227	238	250	15
20	122	164	188	211	237	262	276	290	304	20
25	149	200	227	257	289	320	338	355	372	25
32	181	244	279	314	353	391	413	434	455	32
40	203	273	313	352	395	438	462	486	510	40
50	246	331	379	427	480	532	561	590	619	50
65	301	406	465	523	588	653	689	725	761	65
80	345	466	533	600	675	750	792	833	874	80
100	431	582	666	750	844	937	988	1039	1090	100
125	515	695	796	897	1009	1120	1183	1245	1308	125
150	589	808	924	1040	1175	1310	1385	1460	1515	150

Table 6.15 Theoretical heat emission from horizontal pipes (with insulation of stated thickness)[a]

Nominal bore (mm)	*Heat emission* (W/m run) *to ambient air at 20°C for following absolute steam pressures* (kPa)									*Insulation thickness* (mm)
	100	*200*	*300*	*400*	*500*	*600*	*700*	*800*	*900*	
10	15	18	20	21	22	22	23	23	24	50
15	15	19	21	23	25	27	28	29	30	50
20	17	21	23	25	27	29	30	32	33	50
25	19	24	26	29	31	34	35	36	37	50
32	22	27	30	32	35	38	39	41	42	50
40	23	29	32	35	38	41	42	44	45	50
50	26	33	36	40	43	46	48	50	51	50
65	28	35	39	42	46	49	51	53	54	65
80	31	39	43	47	51	55	57	59	60	65
100	36	45	50	54	59	63	65	68	70	65
125	41	51	56	61	66	71	74	77	79	65
150	46	57	63	68	74	80	83	86	88	65

[a] Insulation conductivity = 0.055 W/m K.

Chapter 7

Heat emitting equipment

With the exception of certain rather specialist applications involving pipework embedded in the building structure, there are very few items of heat emitting equipment which could not be used in conjunction with the whole range of water and steam distribution media. That is not to say however that such use would be equally effective in all cases nor that it would always be acceptable from the point of view of avoidance of burns. Exposed heating surfaces must not be accesssible to touch if the temperature exceeds 80°C and in many circumstances a lower level of about 70°C is to be preferred.*

Principal criteria

Normal design temperatures

For water systems, recommended operating temperatures for an assumed external temperature of –1°C are given in Table 7.1, but this does not preclude the use of higher flow temperatures when external conditions are more severe than this datum level.

Fundamentals of heat transfer

The various types of heat emitting equipment which are the subject of this chapter are those fitted within or immediately adjacent to the space served. Output will be both by natural convection and by radiation, the proportion of each being determined by the form which the equipment takes. There are two empirical relationships applying to emission from plane and cylindrical surfaces which are used to put values to such output, these being:

Convection

$$h_c = C\,(T_s - T_a)^n$$

* DHSS Engineering Data, Ref. DN4 dated 1978 recommends that the average surface temperature in areas used by geriatric, mentally handicapped and paediatric patients should not exceed 42–43°C and proposes that a system flow temperature of 50°C should be adopted for these areas.

Radiation

$$h_r = 5.67\,\varepsilon\,(T_1 - T_2)$$

where

h_c = heat output by convection (W/m^2)
h_r = heat output by radiation (W/m^2)
C = a coefficient (Table 7.2)
ε = emissivity of the heated surface
n = an exponent (Table 7.2)
T_s = absolute temperature of the heated surface (K)
T_a = absolute temperature of room air (K)
T_1 = $(T_s/100)^4$
T_2 = $(T_a/100)^4$

Table 7.1 Design temperatures for hot water systems with various emitters

	Temperatures (°C) *at water heater*	
System	*Flow*	*Return*
Low temperature warm water		
Embedded pipe coils		
ceilings and walls	53	43
floors	43	35
Radiators (where acceptable)	60	55
Low temperature hot water		
Metal radiant panels and strips	83	73
Radiant ceilings		
metal	70	60
fibrous plaster	80	70
Sleeved pipe coils (Panelite)	70	60
Radiators and pipe coils		
gravity circulation[a]	80[b]	60
pumped single-pipe system[a]	75[b]	65
pumped two-pipe system	80[b]	70
Convectors		
natural, including skirtings	80	70
continuous	75	70
forced and unit heaters	83	73
Medium temperature hot water		
Metal radiant panels and strips	120	85
Convectors		
natural	110	85
forced and unit heaters	120	85
High temperature hot water		
Metal radiant panels and strips	150	95
Convectors		
forced and unit heaters	150	95

[a] Except where stated otherwise, all water circulation systems are pumped two-pipe arrangements.
[b] For domestic premises, best practice reduces these flow temperatures by 10 K.

Table 7.2 Values of coefficients and exponents for natural convection from plane surfaces

	Theoretical		Simplified	
Surface and aspect	*C*	*n*	*C*	*n*
Horizontal, facing up	1.7	1.33	2.5	1.25
Vertical	1.4	1.33	1.9	1.25
Horizontal, facing down	0.64	1.25	1.3	1.25

The coefficient and the exponent in the convection equation are varied, as shown in Table 7.2, in order to reflect the direction of heat emission, upwards or downwards, and the attitude of the surface, horizontal or vertical. These circumstances affect the nature of the air current pattern over the surface, which arises with buoyancy change, to the extent that movement may be either smooth or turbulent. In most practical cases of natural convection, where no fan or other external force is involved, flow will be in a transition stage, no longer smooth but not yet fully turbulent.

Use of the convection equation may be simplified in consequence, without significant error for this transition stage, by using a single value of 1.25 for the exponent and the alternative values, as listed also in Table 7.2, for the coefficient. As to the radiation equation, the value for emissivity may be taken as being between 0.8 and 0.95 for metal which is either tarnished or painted and also for most building surfaces.

Under the headings which follow, the various types of heating equipment are dealt with in terms of the predominant component of the output, radiant or convective. The suitability of each type, for use with the various distribution media, is discussed.

Radiant heating

As a result of the convection air movements noted earlier, no heated surface is able to provide a wholly radiant output, the nearest approach to that ideal being a heated ceiling in which instance the convective component is only about 30% of the total. It seems appropriate, however, to deal first with equipment manufactured away from the building site.

Metal radiant panels

Fabricated steel panels, of the type shown in Figure 7.1, are particularly suited to large factory spaces when supplied with high or medium temperature hot water or even with steam. With systems able to provide surface temperatures in the range 100–150°C, flat metal plates such as these produce a powerful radiation output which may be felt at some distance.

As may be seen from the diagram, the panels consist of a 15 mm sinuous pipe coil welded to a heavy gauge mild steel front plate. The rear of the panel may be treated in a number of ways as shown in section, i.e. (a) exposed coil, (b) double sided, (c) insulated

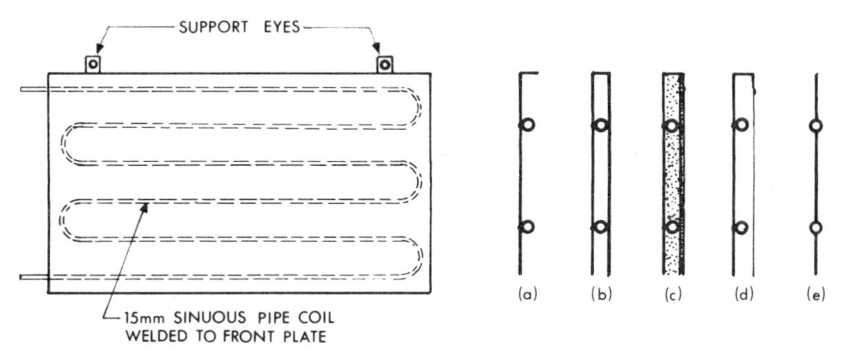

Figure 7.1 Metal radiant panels (medium and high temperature)

back, and (d) shielded back. In (e), an alternative form of exposed coil is shown where the plate is welded as a diaphragm on the centre line of the pipe loops. Panel sizes are commonly 2500 mm × 1200 mm or 1800 mm × 900 mm and approximate outputs for one size are given in Table 7.3. It will be noted that, although the rear is uninsulated in this case, the radiant output is 55–60% of the total quoted output.

Table 7.3 Emission from a radiant panel in free air: proportions radiant and convective

Hot water mean temperature (°C)	*Radiation*		*Convection*	
	kW	%	kW	%
100	2.87	0.55	2.32	0.45
150	5.78	0.57	4.30	0.43
200	9.90	0.61	6.53	0.39

Notes:
1. Values quoted are derived from Imperial data.
2. Panel was 2500 mm × 1200 mm without rear insulation.
3. Water temperature drop = 30 K. Air at 15°C.

It is usual to suspend such panels vertically along walls, or between intermediate columns in a large workshop, at a height of three or four metres above floor level. They may, as an alternative, be fixed horizontally or be inclined at an angle. In designing a system using such panels, it should be borne in mind that the convective component of output serves to counteract the cooling effect of the roof or roof glass but that an unnecessarily large allowance here will lead to an excessive temperature gradient. Panels fixed to external walls and those suspended horizontally are usually provided with insulated or shielded backs.

Another form of radiant panel, used with water at a lower temperature, may be fitted as a heated skirting or dado. Construction is from an all-welded waterway grid formed from 'D' section tubes at 150 mm centres with square section headers, the tubes being held against a substantial front panel by means of a profiled plate spot welded to it, as shown in Figure 7.2. This product, which is not mass produced but bespoke-made for particular applications, may also have expanded metal as a substitute for the front panel and be used

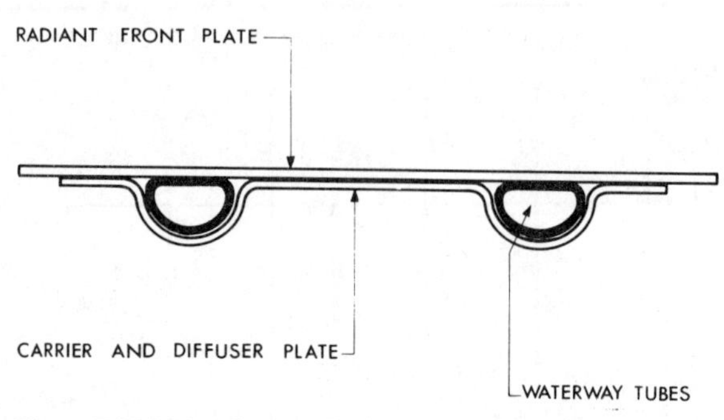

Figure 7.2 Metal radiant panels (low and medium temperature)

as the base for a floor finish at an exposed room perimeter. In the last application, only about 50% of output will be by radiation.

Past practice used another form of lower temperature panels, consisting of cast iron waterways to which a sheet steel front plate was fixed, for space heating in a variety of open areas such as department stores, lecture theatres and the like. Some of these applications were successful as a result of rule-of-thumb appreciation of the balance between radiant and convective effects.

Radiant strips

Again, for workshops, strip heaters of the type illustrated in Figure 7.3 may be applied with success when supplied with high or medium temperature hot water or steam. One type consists of a 32 mm pipe, to which a profiled aluminium plate is clipped, for running

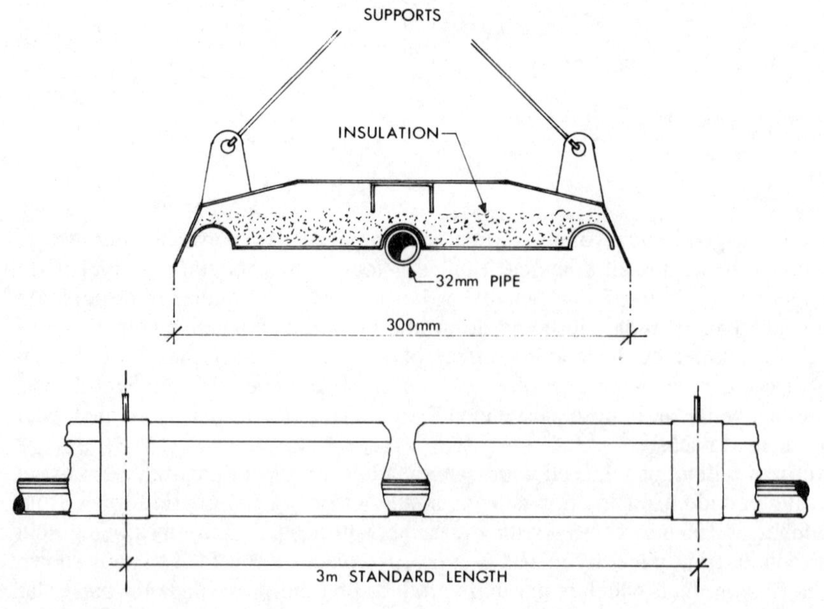

Figure 7.3 Radiant strip heating

in a continuous line overhead. The rear upward face may be insulated as shown or, unusually, left bare: provision may be made for lighting tubes to be contained below the aluminium plate.

Outputs to air at 15°C range from 0.5 kW/m run with a single pipe at 100°C to 3 kW/m run with four pipes at 150°C. The percentage radiant component of the total output is about the same, or slightly more than that of a panel operating under similar conditions. The manufacturers recommend that the water velocity should be maintained at a reasonably high level and this requirement may conflict in some cases with provision for thermal expansion. Figure 7.4 shows a layout in a long workshop which was conceived to overcome this problem.

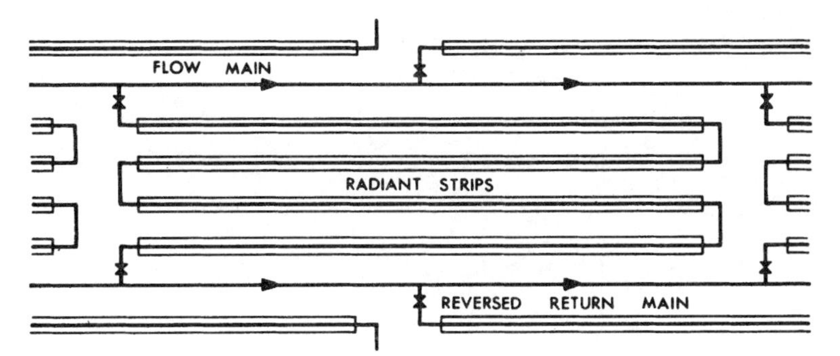

Figure 7.4 Application of strip heating to long workshop

Although not so described at the time, radiant strip heating was commonly provided to north-light factories some 50 or more years ago in the form of suspended rows of 100 mm pipe. A not unusual arrangement was to provide one such row below the south-facing roof and two rows below the glazing. Using low temperature hot water in each case, the total emission from a 100 mm pipe so suspended is not very different from that of a single-pipe radiant strip but the structural load imposed on the building structure is 13 times greater!

Metal radiant ceilings

The equipment to which reference has been made in previous paragraphs provides either a *linear* or a *point source* of radiation. Output from a metal-plate suspended ceiling, in contrast, originates from a relatively wide area and may thus operate at rather lower face temperatures. Such a ceiling will probably combine acoustical treatment with a radiant heat output and, in one proprietary make, consists of thin aluminium pans about 600 mm square having regular perforations. The pans are clipped to a piping grid at their junctions, and support a blanket layer of insulation material spread above them, as in Figure 7.5.

It is usual to arrange that the whole surface of the ceiling is so treated; those parts of the coil which are necessary for heating being supplied with a hot water circulation and the remainder used only as a means of suspension. This aspect is illustrated in Figure 7.6 which is a typical application layout serving a number of rooms. The total heat emission from a ceiling of this type, for a mean water temperature of 70°C and a room temperature of 20°C, would be about 160 W/m^2 downwards and 15 W/m^2 upwards to a heated room over. Two-thirds of the downward emission would be radiant.

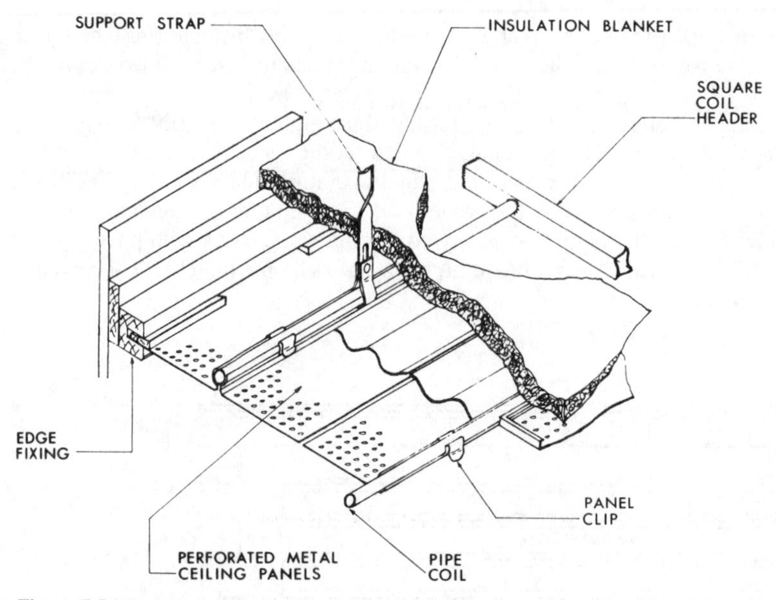

Figure 7.5 Metal plate radiant ceiling

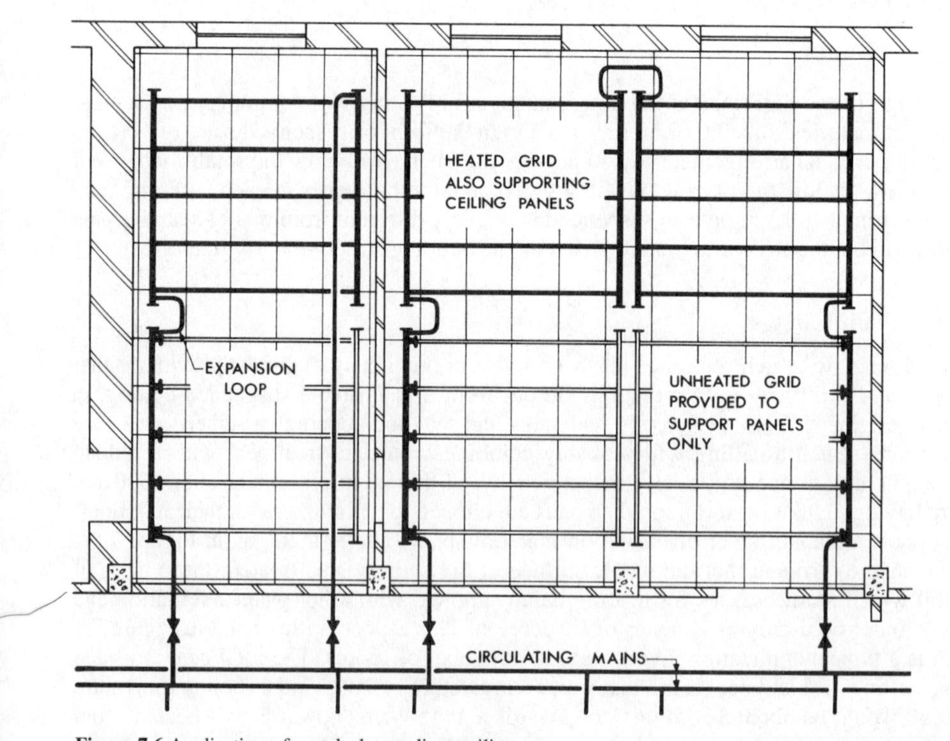

Figure 7.6 Application of metal plate radiant ceiling

Fibrous plaster radiant ceilings

In this case, an alternative form of suspended ceiling to that described above, a continuous low or medium temperature hot water pipe coil is supported above and independently of the fibrous plaster panels which provide the acoustical treatment and radiant surface, as in Figure 7.7. Above the pipe coil an insulating blanket is supported on wire mesh to contain the heat output. The extent and temperature of the pipe coil determine the level of heat output which, in round terms, is similar to that of the metal type described previously.

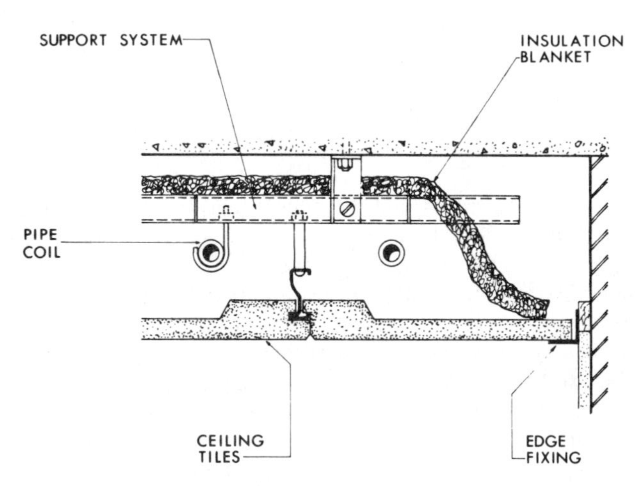

Figure 7.7 Fibrous plaster radiant ceiling

Embedded ceiling panels

The practice of heating rooms by means of embedded ceiling panels was common for the older type of building, pre about 1950, where a long time lag was inherent in the heavy construction and thus matched that of such a system. The forms of building framework which have become popular over the last 50 years, with extensive glazing and little thermal capacity, require systems able to provide a more rapid response to variations in outside temperature. The increasing use of suspended ceiling tiles, brought about to some

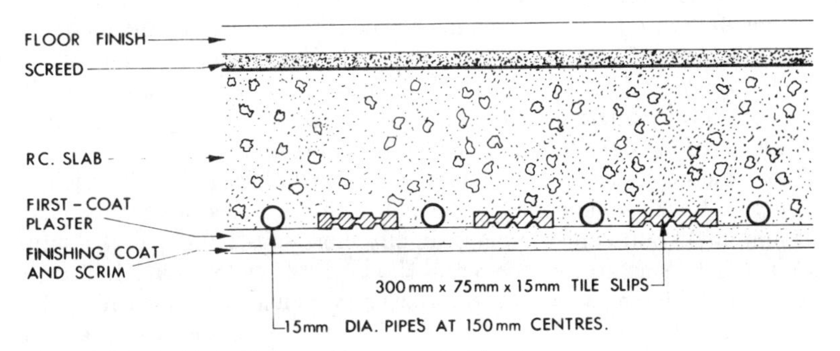

Figure 7.8 Embedded ceiling panel in structural slab

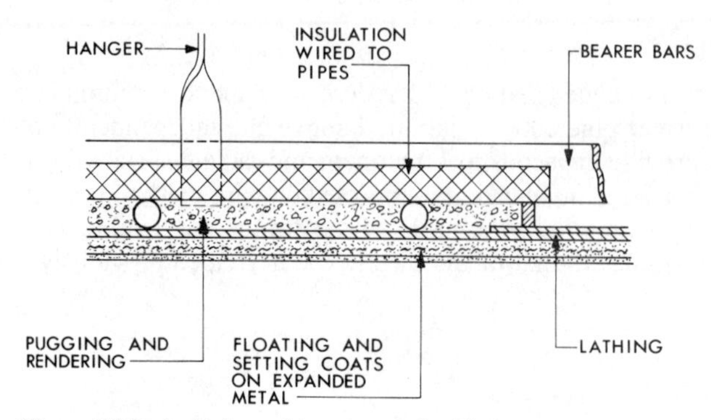

Figure 7.9 Embedded panel in suspended ceiling

extent by a desire to avoid 'wet' finishing trades and the consequent loss of plastering skills, has led to the obsolescence of the system. It is thus proposed to make no more than a brief reference to it here.

For supply with low temperature warm water, sinuous coils usually of 15 mm bore steel or copper pipe were laid at 150 mm centres on the shuttering for a concrete slab prior to the forming and laying of bottom reinforcement. To ensure that later *in situ* plastering would have a good key available, *slip tiles* were placed between the coils, as shown in Figure 7.8. Plastering was carried out to a special and quite rigid specification, currently administered by BSRIA. A technique for incorporating panel coils within suspended ceilings formed and plastered *in situ*, was similarly available and was used where required, as in Figure 7.9. It will be appreciated that access to the void was very limited and that subsequent maintenance or an addition to service facilities was difficult.

With a mean water temperature of 48°C and a room temperature of 18°C, a downward heat emission of 150–165 W/m^2 could be expected, with an upward emission to a heated room over of 50–75 W/m^2. The range of values in each case arises from the variation in resistance to heat flow upwards produced by alternative floor finishes. It has been noted previously that output from a heated ceiling is largely radiant.

Embedded floor panels

It will have been noted from the values listed in Table 7.1, that it is recommended practice for a low temperature warm water system serving embedded floor panels to be operated at a mean temperature some 10 K lower than that for equivalent coils in a ceiling. As a result, since the temperature difference between the heated floor surface and the room air is small (4–7 K), an inherent self-regulating potential exists. Thus, as the air temperature rises in a floor-heated room, so the output from the floor will fall away quite quickly. It is largely as a result of this characteristic that floor heating has retained popularity, *when properly designed and used for applications to which it is best suited.*

Traditionally, floor panels consisted of 15–20 mm bore copper pipes, laid without joints at 150–450 mm centres, above a solid concrete slab and within a graded floor screed not less than about 75 mm thick. Soft copper piping, slightly more substantial than that now available, was delivered to site in coils between 100 and 120 m long and laid to a predetermined pattern. After laying and prior to cover being provided by the screed, the piping was held in position by small daubs of mortar. There is no reason why this well understood technique should not continue, using the slightly inferior soft copper pipe to

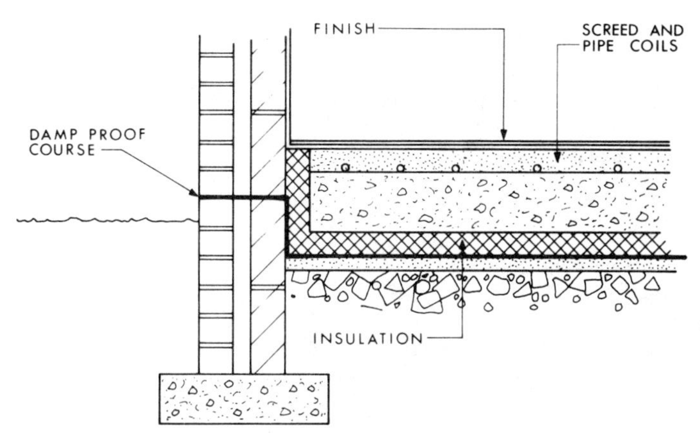

Figure 7.10 Embedded floor panel (note extent of insulation)

BS 2871: Table Y, also available in coils. Figure 7.10 shows the general arrangement, a prototype perhaps for the methods later adapted to electrical floor heating, Figure 5.6.

An alternative to soft copper piping is popular in Europe in the form of a *cross-linked polyethylene tube* of various proprietary types suitable for the temperature range involved. This is available for tubes having outside diameters of 17 and 20 mm, with 2 mm thick walls, in coil lengths of 120 m. The methods used in laying the coils are much the same as those previously noted except that metal strips holding plastic clips are laid on the base slab, at right-angles to the coil line, to form a locating grid. An emulsifying agent is added to the screed mix to improve contact with the coils. It appears that problems have arisen from *gaseous diffusion* inwards through the walls of the tube originally used, even under water pressure, leading to corrosion of metal parts elsewhere in the system. Substitute materials are now being provided, incorporating barrier films of either other plastics or metal foil.

A more familiar and well-established alternative to fully buried copper pipe coils is offered by 'Panelite', a proprietary arrangement now widely available, which consists of steel pipe coils encased in split asbestos cement covers or sheaths, as shown in Figure 7.11. This arrangement allows freedom for expansion and contraction and the air gap between the pipe and the covers acts as an insulant. As a result, higher water temperatures may be used, Table 7.1, than with the wholly embedded pipes: emission per unit area of floor surface is for practical purposes the same as that for the alternatives.

The two conflicting factors often encountered in designing floor heating coils are the necessity to avoid too high a temperature at the floor surface and the consequent marginal difficulty in obtaining a sufficient heat output. It is generally accepted that a floor surface temperature of 24°C should not be exceeded where occupants are static, 27°C where they are able to move about and about 30°C in corridors and halls, etc. It is usually necessary, in consequence, to provide cover to the whole floor of the area to be heated with a differential in the spacing between the coils. This may be 100–150 mm adjacent to the outer walls widening to perhaps 400–450 mm in the centre of the room with further variations in instances where the use of various parts of the area is well defined. The configuration of the pipe coils may be arranged such that the water temperature of adjacent pipes represents the mean of flow and return.

A variety of finishes may be used over heated floors and almost all types of hard material, marble, slate, stone, terrazzo and brick are suitable provided that provision is

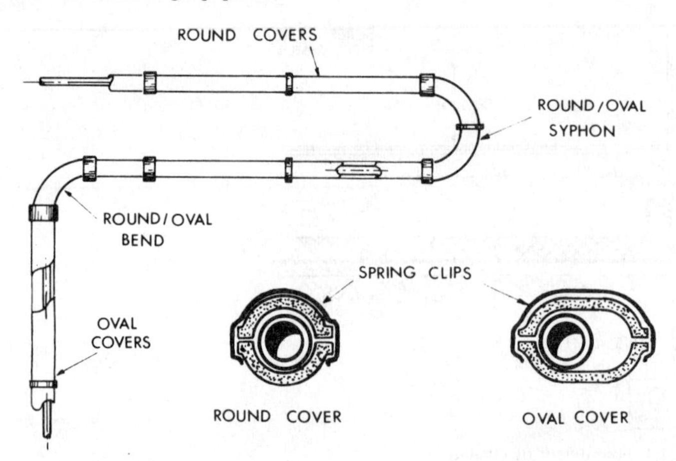

Figure 7.11 Details of sheaths for 'Panelite'

made for expansion and that laying is carried out in a manner which will not leave cavities within the finish to impede heat transmission. In the case of softer materials, wood blocks may be used if they are both properly seasoned and kilned and are laid in a mastic which will not soften or swell at the temperatures likely to be encountered. Cork tiles are a satisfactory finish but thermoplastic tiles and rubber sheet materials require individual investigation. Carpets may be used provided that they are not foam backed. Emission from heated floors, of which not much more than half is by radiation, depends greatly upon the finish and representative values are given in Table 7.4.

Table 7.4 Upward emission from embedded floor panel heating (floor slab on earth)

		Upward emission (W/m² *floor surface*) *for stated temperature differences, water to room air*					
	Centres of 15 mm bore pipes	*Embedded pipes*			*Panelite*		
Construction	(mm)	*23*	*28*	*33*	*40*	*45*	*50*
75 mm screed with hard 25 mm stone or equal finish	150	100	125	150	93	105	118
	225	88	110	132	81	92	103
	300	77	96	116	74	84	94
	375	67	84	101	67	76	85
	450	58	73	87	59	55	75
75 mm screed with 25 mm wood block	150	67	84	100	74	84	94
	225	60	75	90	67	76	85
	300	53	66	80	59	67	75
	375	47	59	71	54	61	68
	450	42	53	63	48	55	61
75 mm screed with felt underlay and used Wilton carpet	150	40	50	60	48	55	61
	225	36	45	54	44	50	56
	300	32	40	48	41	46	52
	375	29	36	44	37	42	47
	450	26	33	39	35	40	45

Note: for a floor having 50 mm insulation at the perimeter and below the heated area, the downward and edge loss will be approximately 20% of the upward emission as above.

The growing practice in commercial premises of providing a sub-floor void to accommodate communication cabling, etc., between various items of office equipment and, in some cases, a cooled air supply, rules out any question of embedded floor panel heating. But this restriction only serves to re-emphasise the comment made previously to the effect that this technique should only be adopted for those applications to which it is best suited, circulation spaces in commercial buildings, shopping malls, a variety of domestic areas and certain specialised industrial enclosures.

Combined radiant and convective heating

A classic paper by Peach* might well have been headed 'What's in a name' but in many respects the title chosen highlights the dichotomy which is enshrined in the language, too well established to be denied. The equipment which falls under this heading, although diverse in size shape and pattern, therefore consists very largely of what the public identifies as 'radiators'.

Pipework

The exception to that bulk inclusion is the earliest form of heat emitting equipment. Boiler manufacturers' catalogues, as recently as 1947, rated equipment in terms of unit emission from '4 inch pipe'! As a form of heating surface, exposed piping is rarely used in current practice but, in any system of distribution, main and branch piping running through the space to be heated will make a contribution. It is but one of the skills of the designer that he is able to make use of an output which might otherwise be wasted. Two examples will suffice to illustrate this point: in a factory mains pipework supplying either radiant panels or unit heaters will, if run below north light glazing, serve a more useful purpose than if routed elsewhere; sub-circuit pipe runs to radiators in, say, a school will contribute more usefully if routed on the surface than if buried in a water-logged trench outside the building.

Tables 6.12–6.17 provide extensive data regarding heat emission from both bare and insulated pipes. It remains only to note that where pipes are exposed in coil formation, vertically one above another, the reduction in emission will be:

2 pipes	5% reduction.
4 pipes	15% reduction.
6 pipes	25% reduction.

Cast iron radiators

The earliest point source heater which could, however loosely, be called a radiator was probably the double close-wound coil of the Perkins system: Billington† notes however that the first cast iron sectional radiator was produced in the USA in 1877 and that James Keith took out the first English patent in 1882. It is a matter for regret that, 100 years later, cast iron radiators are no longer manufactured in the British Isles and particularly so since, in European practice, there is a discernible trend, perhaps resultant from an increasing level of failure in other materials, to the use of cast iron. In contrast, a recent *Product*

* Peach, J., 'Radiators and other convectors', *JIHVE,* 1972, **39,** 239.
† Billington, N. S., 'A historical review of the art of heating and ventilating', *JIHVE,* 1955, **23,** 259.

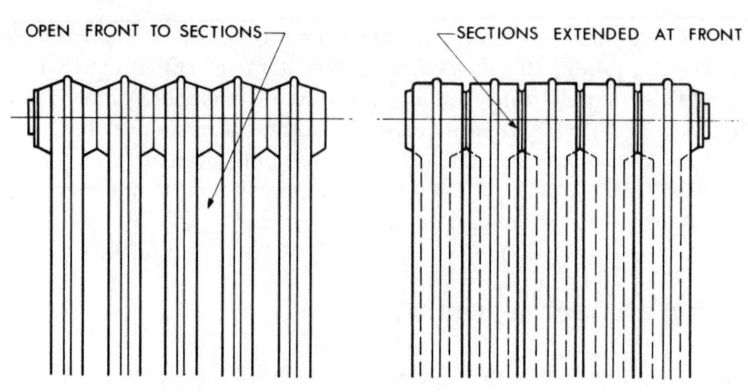

Figure 7.12 Elevations on two types of cast iron radiator

Profile produced by BSRIA states that, in 1987, aluminium and cast iron radiators represented less than 4% of the total sold in the home market.

There seems to be no modern substitute for the 75 mm deep cast iron 'Hospital'-type radiator which, in terms of neatness and an ability to withstand maltreatment by either physical impact or by untreated water, represented the ultimate of its type. Imported cast iron radiators seem generally to be of the sectional column type, not unlike the patterns made familiar in the British Isles during the 1930s. Erection of the sections is by means of left- and right-hand threaded nipples at the top and bottom ports. In some cases the profile of the sections is now arranged to present a front elevation which resembles a panel, thus increasing the output per unit area by about 10%. Figure 7.12 illustrates the difference between the two types, and a summary of typical ratings is included in Table 7.5* for a temperature difference, water to air, of 60 K which is the accepted standard for tests. Table 7.6 provides correction factors which enable these ratings to be adjusted for other conditions. The radiant component of output for column radiators varies between 30 and 17% as will be noted later.

Pressed steel radiators

The place of cast iron seems now to be taken by radiators fabricated from light gauge mild steel sheet pressings, welded together, a technique which originated in Scandinavia. Steel is more susceptible to corrosion than cast iron and thus some form of water treatment for the content of the heating system is prudent. The particular merits of steel radiators result from their small mass and their comparatively narrow waterways: they are light to handle on a building site and respond quickly to temperature control.

In consequence of the versatility of the manufacturing process, there is an almost infinite variety of shapes and patterns of steel radiator available. These fall, however, into a number of categories which divide into two clearly recognisable groups, the first of which is illustrated in Figure 7.13:

(a) The simple convoluted panel type which may be single, double or more. Each panel is normally fabricated from two identical sheets, seam welded at top, bottom and ends but some of the better patterns are made from a single sheet folded over to provide a rounded top.

* Tables 7.5, 7.6, 7.8 and 7.9 appear at the end of this chapter (pp. 197–8).

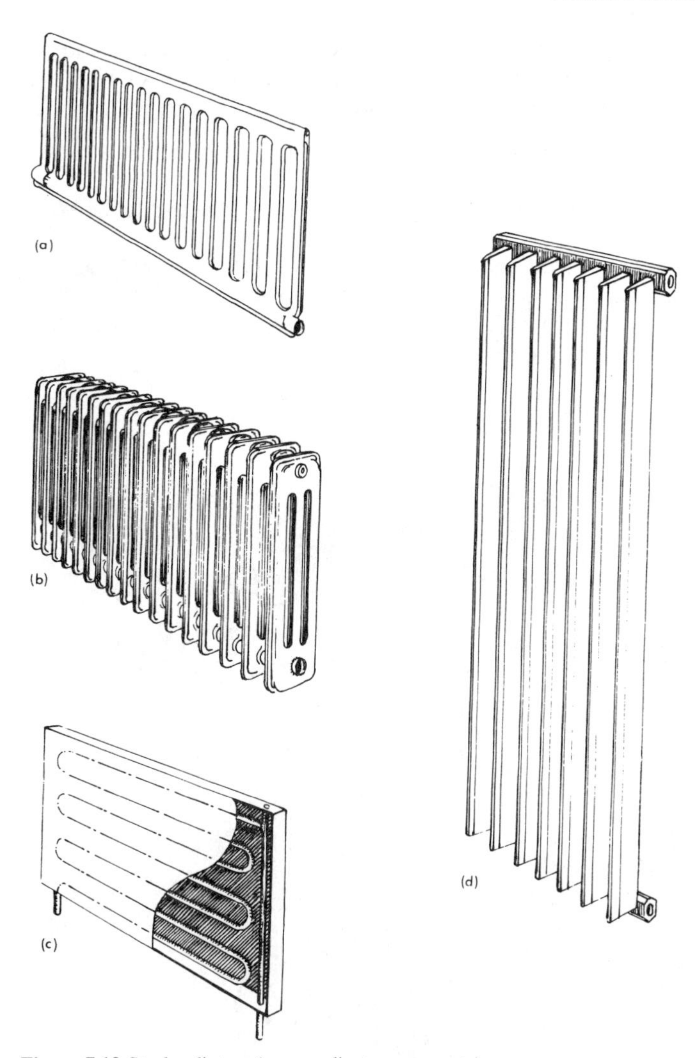

Figure 7.13 Steel radiators (some radiant components)

(b) The column type which, in effect, reproduces the appearance of a similar cast iron radiator.
(c) A sinuous coil panel type which is very similar to the radiant panels discussed earlier.
(d) A tubular type, using top and bottom headers with elegant pipe arrays between them: may be used as a room divider or barrier rail.

Radiators in the second group are provided with more secondary surface and thus will have a greater convective component in the total output. In at least two instances, the extent of the added surface and the minimal radiant output is such that the examples shown should properly be called convectors. It must be emphasised that although the extended surface provides additional output for a radiator having a given frontal area, it does so at the expense of cleanliness. The extended surface is usually hidden away at the rear and provides a series of small vertical air-ways which are both out of sight and notoriously difficult to clean. Figure 7.14 illustrates four of the more common types:

(a) Single convoluted panels with additional corrugated surface added at the rear, as a supplement.
(b) A flat plate panel front with additional waterway surface added at the rear to the extent that it predominates.
(c) A multiple panel with so much additional corrugated surface added that the panels are no more than incidental waterways, a convector in all but name.
(d) A tubular element attached to a front plate which acts as an enclosure and is provided with top and bottom louvres, a convector in all but name.

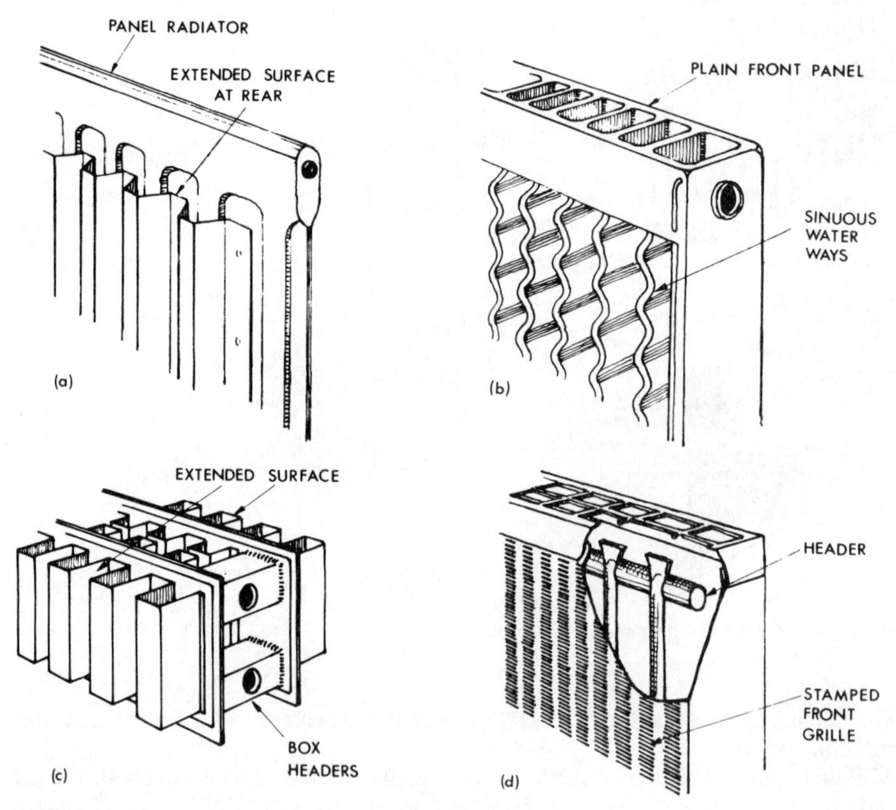

Figure 7.14 Steel radiators (principally convective)

A summary of typical ratings is included in Table 7.5, again for a temperature difference, water to air, of 60 K which may be corrected as before, using Table 7.6, for other conditions. Based on European research in an *isothermal enclosure* (a room having cooled external surfaces), Table 7.7 lists the relative proportions of radiant and convective output for a selection of radiator types.

Table 7.7 Emission from radiators in an isothermal enclosure: proportions radiant and convective

	Proportion (%)	
Radiator type	*Radiation*	*Convection*
Single panel	50	50
Double panel	30	70
2 column	30	70
4 column	19	81
6 column	17	83
Convector/radiator	10–15	90–85

Aluminium radiators

There are two quite clearly discernible patterns of aluminium radiator, the die-cast sectional type manufactured in Europe and the extruded high output type made in the British Isles. The former are very similar in most respects to the cast iron column radiators noted previously but the latter are provided with extensive finned surface to the extent that they should be categorised as convectors. Figure 7.15 illustrates the two types, (a) and (b) relating to the sectional arrangements and (c) to the convector. A digest of ratings is given in Table 7.5, for correction where necessary as before.

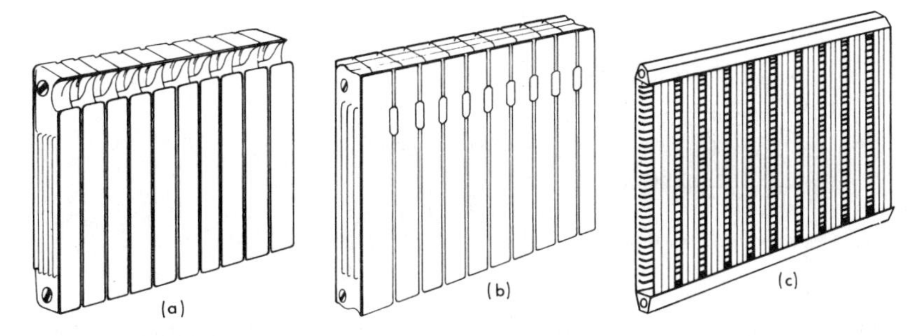

Figure 7.15 Aluminium radiators

The material and production techniques are such in each case that a clean smooth finish is provided but the problem of metal corrosion remains. One manufacturer provides an inhibitor which is inserted in capsule form during radiator construction, for release when installation is complete and hot water activates the curing process. Other manufacturers recommend the use of special additives to a system which has any mixture of metals.

Radiators generally

It is generally best to site radiators under windows in order to meet architectural, social and technical demands. Architectural, to avoid the presence of foreign objects on an otherwise unblemished wall, social to give freedom for furniture arrangement and to avoid the need for frequent redecoration to hide dirty marks from convection currents. In technical terms, siting under a window is preferred because:

- Heat is provided where most needed, at the point of maximum heat loss.
- Cold 'negative' radiation is countered at source and direction by hot 'positive' radiation.
- The temperature gradient in a room with radiators fitted below windows is less than if they were sited elsewhere.
- Marking of a wall surface by dust carried in rising convection currents is avoided.

In circumstances where it is not possible to site a radiator below a window, it is an advantage to fit a shelf above the heated surface. Provided that the jointing to the wall is sound and that adequate end shields are fitted, the shelf will reduce the wall stains caused by the dust entrained in rising convection currents. Such a shelf, if fitted not less than 80 mm above the top of the radiator, will reduce output by only about 4%.

It is important to remember that the length of a radiator should be matched to that of the window beneath which it is fitted. A narrow high output radiator beneath a wide window will produce an equally narrow 'fountain' of warm air moving upwards, with a cascade of cooler air falling at either end of the window (Figure 7.16).

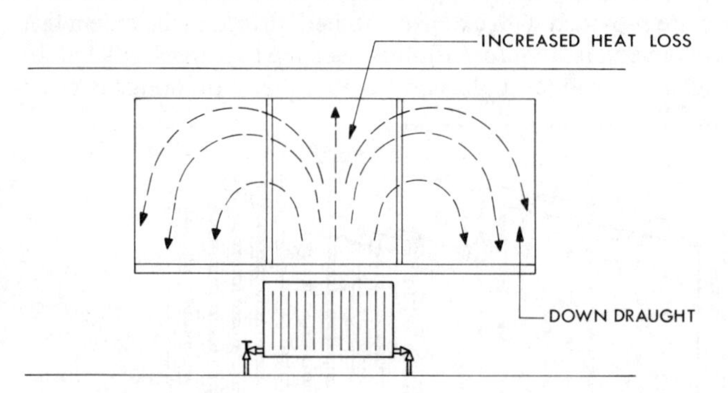

Figure 7.16 Mismatch of radiator to window

The construction of most types of radiator is such as to offer the facility for three different patterns of pipe connection. Two of these are shown in Figure 6.7, that at the left-hand side with flow and return top and bottom at the same end (TBSE); and that at the right-hand side with flow and return top and bottom at opposite ends (TBOE). The third alternative, probably now the most familiar, has flow and return both at the bottom at opposite ends (BBOE). Catalogue ratings are normally based upon the test routines required by BS 3528: 1986 which relate to the TBOE configuration: European laboratory investigations quoted by Peach suggest that variation in heat emission with the three connection patterns is:

TBSE = 1.00
TBOE = 0.97 to 0.95
BBOE = 0.78 to 0.84

Painting of radiators was a sore subject at one time as a result of the industrial and commercial practice of using metallic paints, aluminium or bronze. These reduced the emissivity of the surfaces, and in consequence the total output, by a significant extent. It was later found that a coat of clear varnish over the metallic paint resolved the problem. Attempts are made from time to time to market radiators which have been stove enamelled at works and are delivered to the site in a protective package. It is, however, a more common practice to provide no more than a primer coat in preparation for a site finish.

The concept of energy saving by provision of a reflective surface behind a radiator has been widely publicised and extravagent claims of economy made. Two methods have been advocated, first by the provision of a metallic-foil covering to the hidden wall surface and, second by attachment of polished metal strips to the rear of the radiator itself. The efficacy of either approach depends to a large extent upon how well the wall construction behind the radiator has been insulated (i.e. the U value). In the case of a solid 220 mm brick wall, the energy saving for a typical domestic living room might be of the order of 3% but for a 260 mm cavity wall with an effective insulant between the leaves, it would be less than 1%. There would of course be no saving at all if the radiator were fitted on a partition wall dividing two heated rooms.

Convective heating

While radiators and other similar surfaces are now used only with warm and low temperature hot water systems, earlier practice with steam and higher temperature water was to use such equipment but in a protective enclosure of some type. Billington notes the use of box and syphon coils and of finned pipes in the middle of the nineteenth century, enclosed in decorated openwork cases called *calorifères*. These may be thought of as ancestors of modern natural convectors ('indirect radiators' was the disparaging description used until relatively recently by other manufacturers!).

Natural convectors

Although a wide variety of equipment falls within this general category, it is used here to identify the cabinet type only, other patterns being dealt with separately. The principal components are a finned tubular element mounted near the bottom of a sheet metal casing such that a *chimney effect* is created and a rising column of warm air flows from the top, inducing an inlet of room air at or near the base. Figure 7.17 shows the general arrangement.

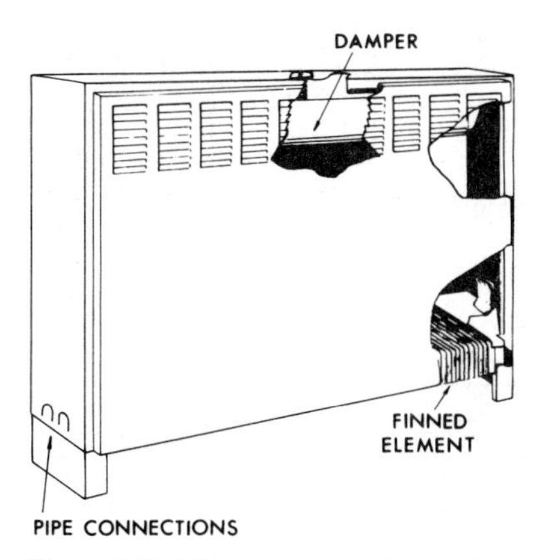

Figure 7.17 Cabinet type natural convector

The tubular element will normally span between two headers which accept the external pipe connections and, depending upon the required duty and the particular manufacturer, take a variety of forms. The relationships are complex since the tubes can vary as to number, size, and shape (round or oval) and the fins as to area, spacing, thickness, material and method of bonding to the tubes (solder or mechanical expansion). The casing height has an important effect upon output as may be seen from Table 7.8 which provides a digest of ratings for a temperature difference, water to air, as stated. Table 7.9 provides correction factors which enable these ratings to be adjusted for a narrow range of other conditions, the effect of water velocity in the tubes being the factor limiting further extrapolation.

Control of output may be either by adjustment of the temperature of the heating medium or, locally, by use of a damper fitted within the casing. Ideally, the damper should be mounted just above the element but it is more usually fitted at a convenient hand level as shown in Figure 7.17. Emission with the damper closed is by radiation from the casing, which then has an increased surface temperature, and amounts to about 20% of the normal output.

Continuous natural convectors

This application of the convector is particularly well suited to buildings designed, within severe cost constraints, on a modular basis with continuous glazing above sill level. The finned heating element is continuous from end to end of an elevation, subject to certain provisions for expansion, and the metal casing is continuous also, as in Figure 7.18. The louvred outlets at the top are in sections, arranged to suit the modules, and each is provided with a damper, no other form of control being suitable. The air inlet may be at an open base, as shown, or through a low level grille. In circumstances where the sill height permits, the base of the unit may be raised and the casing integrated with electrical distribution trunking. Where partitions occur, a soundproof barrier is provided within the casing.

There is, or should be, a limit to the length of any single run since output will decay

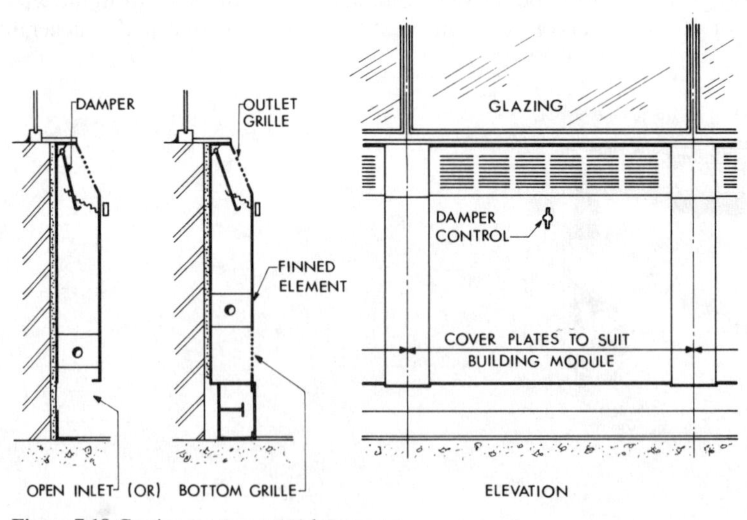

Figure 7.18 Continuous type natural convector

from module to module as the water temperature falls. A much better arrangement is to use the fundamental concept but adapt it such that only a limited number of modules (say two structural bays) is served in series from pipe mains running within the casing: the appearance remains the same but there is less variation in water temperature from the first to the last module. Table 7.8 provides some indication of the ratings available, using low temperature hot water and a 60 K differential to air.

Skirting (baseboard) convectors

This approach provides an unobtrusive method of heating for some parts of domestic premises and for halls and corridors etc. in commercial or institutional buildings. The best known pattern is as shown in Figure 7.19, the output being about 450 W per m run for a temperature difference, water to air, of 60 K, from a casing only 210 mm high and 65 mm deep. Accommodation is provided above the finned tube for a second pipe and, in consequence, full room-width runs may be arranged without other than the casing being visible. It is wise to consider the inhibition placed on furniture arrangements before arranging that the casing covers too many walls of a room.

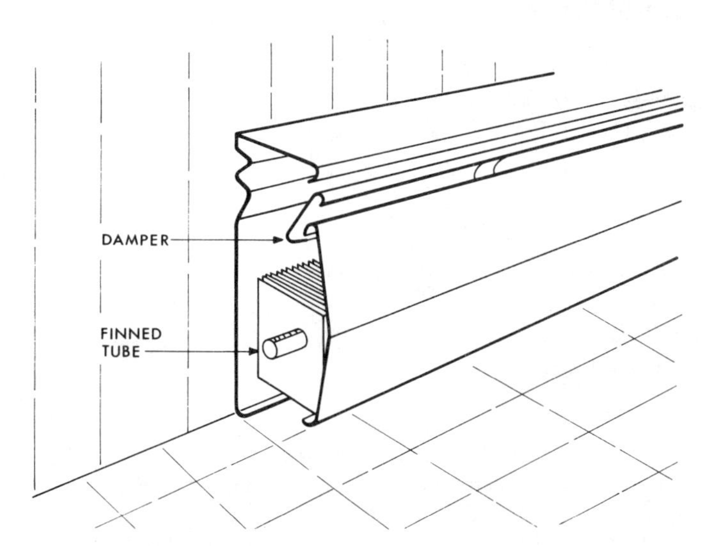

Figure 7.19 Skirting heater

Trench convectors

A modern version of an old church heating system, where cast iron pipes were run in shallow structural trenches half full of dust and covered by decorative cast iron grilles, is now available. Their use is primarily to counteract down draughts when recessed in the floor immediately adjacent to room height windows. The components usually include a metal channel section designed to accommodate a finned heating element, which may be fitted to a single pipe or a hairpin loop, and a transverse bar type grille. The grille is substantial and lies flush with the floor finish and may be either in sections or a full length ‘roll-up’ type. For a channel width and depth of about 150 mm × 75 mm, an output of 200 W/m run would be expected.

Forced (fan) convectors

Units of this type are best described by reference to an illustration such as Figure 7.20 which, for a commercial size of convector, shows the various component parts in separation. The finned tube element will probably differ from the pattern used in a natural convector in that the headers, to which pipe connections are made, are likely to be one above the other at the same end of the casing, the tubes being in hairpin form to make up a 'two deep' arrangement. In the earliest models made, propellor type fans were used but these have now, almost universally, been superseded by small quiet running centrifugal units having simple long-life motors mounted within the suction eye.

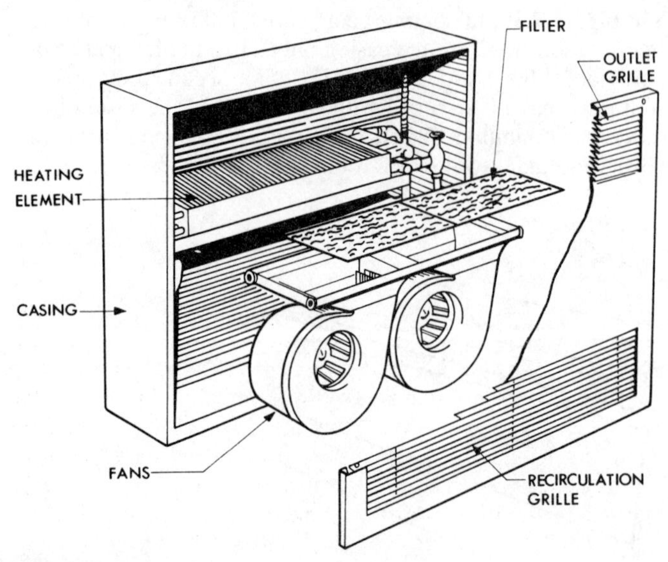

Figure 7.20 Forced (fan) convector, commercial type

The heating element and fans are contained within a casing which, where free-standing, is similar to that provided for a natural convector. For the type of unit illustrated, provision is made at the fan discharge for an air filter. This is often a washable type, consisting of no more than a thin mat of synthetic fibre stretched on a wire grid, and able to prevent carpet fluff and other airborne dust from fouling the heat exchanger. Such an arrangement does not keep the fan clean and a better alternative is to provide a simple modular throw-away type of filter immediately behind the recirculation grille.

Fan motors are usually provided with facilities for two or three running speeds and the lower or middle speed rating is that used for selection, operation then being acceptably quiet. The higher or boost speeds are available to provide an overheat facility after periods of shut-down when an increase in noise level may be tolerated for a short time. In some makes, the casing is provided with an acoustical lining in an endeavour to reduce the noise level. Use is often made of fan speed control by thermostat, timer or other device.

The size range available in the commercial pattern of unit is considerable and application is correspondingly wide, using hot water at low, medium and high temperature, or steam. In a block of flats, each separate dwelling may be provided with an *apartment heater* fed from a central boiler plant; in a school, one heater may be

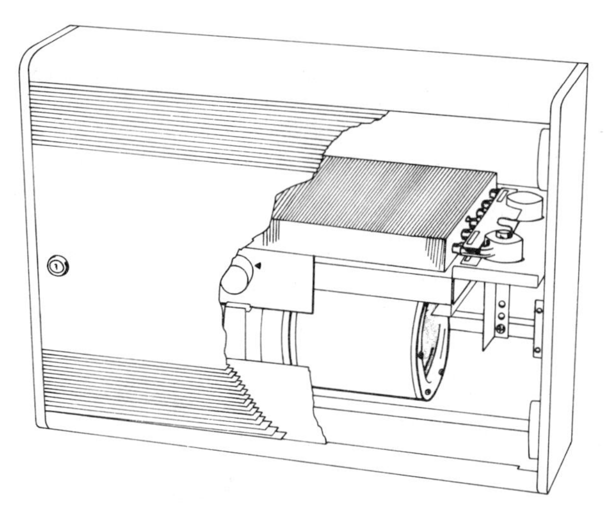

Figure 7.21 Forced (fan) convector, domestic type

provided for each classroom and several for service to a gymnasium or assembly hall. In lecture theatres and large spaces, which require mechanical ventilation when occupied but are not always in use, a fan convector may be used as a form of background heating. From time to time, circumstances arise where the provision of an outside air supply to a fan convector must be considered in an attempt to provide an elementary form of either plenum heating or ventilation. On a small scale, such as service to a single room, a satisfactory result may be obtained provided that the effect of wind pressure can be avoided; that protection against freezing of the heating element is incorporated; and that facilities for filter changing are assured.

For domestic applications, miniature units are available, as in Figure 7.21, with ratings of the order of 2–3 kW at normal speed using low temperature hot water. These are useful in circulation areas and in rooms where significant changes in incidental heat input occur and a quick response to control is required. They have been applied also, with some success, to study bedrooms in university halls of residence, the fan then being controlled by a time delay switch which isolates supply after, say, a 2-hour running period unless reset.

Unit heaters

Except for individual and special applications, the use of unit heaters for industrial premises seems to have given way to that of either radiant panels or an induction type high temperature plenum arrangement. This change has, no doubt, resulted from a realisation that the alternative methods are more energy economic. The most familiar type of unit heater is as illustrated in Figure 7.22, where the heating medium shown is steam but might equally well be low, medium or high temperature hot water. In addition, downward and horizontal discharge units have been used, the latter being floor mounted as shown in Figure 7.23. It is important, when applying the latter, to ensure that the discharge air pattern is unimpeded by partitions or other large objects.

Since the capacity range of unit heaters of the various types is from 10 to 300 kW, it

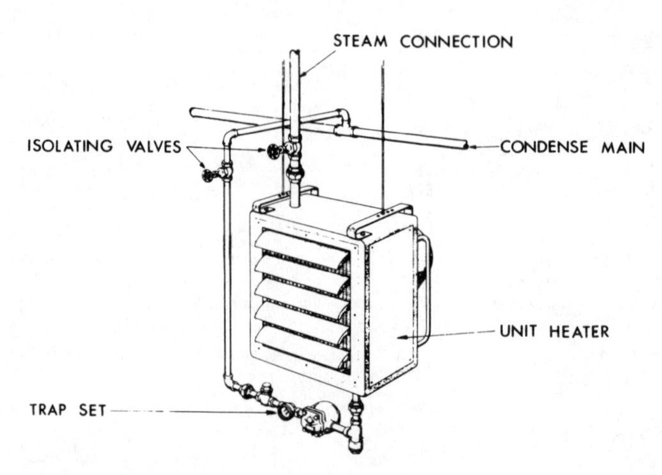

Figure 7.22 Suspended type unit heater, recirculating

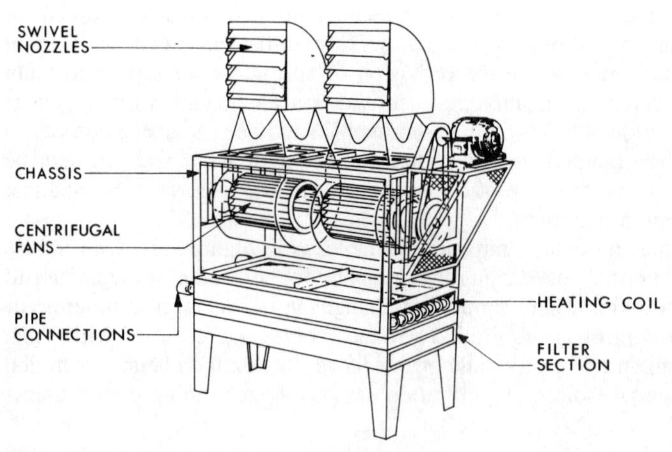

Figure 7.23 Projector type unit heater, recirculating

would serve no purpose to include a list here, particularly since dimensions are small and most are suspended rather than floor mounted. Selection must be from makers' lists, taking account of air volume, discharge temperature, mounting height and the *throw* which is produced. The principal criteria determining choice are that a relatively lower air discharge temperature (40–50°C) and a larger air volume are to be preferred over the converse since the buoyancy of air at higher temperatures will affect the throw available from a given mounting height as well as encouraging an adverse temperature gradient.

Unit heaters may be used to provide an outside air supply to industrial premises using

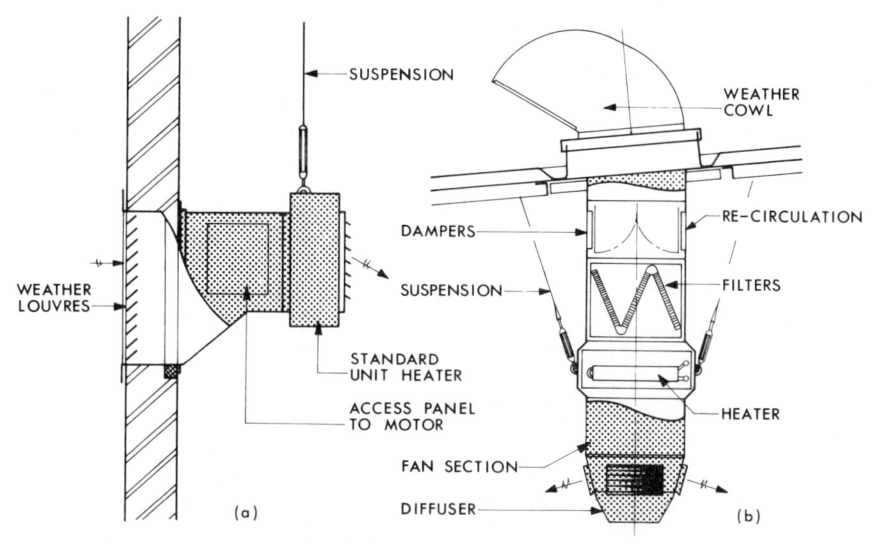

Figure 7.24 Unit heaters with outside air inlets

a ducting arrangement such as those shown in Figure 7.24, item (a) being a purpose made unit. In instances where the design incorporates some units having an outside air supply and others which recirculate, it is desirable that output ratings for the two different duties are so selected that all will provide the same discharge temperature. The recirculation opening and damper on the duct in example (b) are provided to allow for a boost discharge after a period of shut-down and also to provide a measure of frost protection. It should be noted that if the fan motor to such a unit were to fail, or be shut off by thermostatic or timer control, a reversed air flow will be induced by the heating element and, unless the damper is in the recirculating position, will then be discharged to outside. For this and other reasons, it is as well to provide for motorised operation of the damper, interlinked with the supply to the unit motor.

Convectors generally

In retrospect, and in the context of supply of medium and high temperature hot water and steam, it has not perhaps been emphasised adequately that it is only the odd isolated natural convector which would, in normal current practice, be connected to such systems. Offices and other subsidiary areas related to an industrial building so serviced would most probably be heated by water at a lower temperature, produced as was shown in Figure 6.5 (p. 146).

When control of natural convectors and similar equipment is to be by adjustment of the temperature of the heating medium, it should be noted that their output falls away more quickly than that from radiant equipment. This situation is reflected in the data which are included in Tables 7.6 and 7.9. It has been found in practice that the fundamental equations quoted at the beginning of this chapter must be adapted slightly to reflect the performance of actual equipment, output being proportional to the temperature difference, heating medium to air, as follows:

Radiant equipment
Proportional to $(t_m - t_a)^{1.3}$

Natural convectors, etc.
Proportional to $(t_m - t_a)^{1.5}$

It is, for this reason, inadvisable to supply equipment of the two categories from the same piping circuit but to provide for them separately using different rates of temperature adjustment. As to forced convectors and unit heaters, response to change in temperature difference is, in practical terms, directly proportional thereto. It would, however, be normal current practice to supply equipment of this type from a constant temperature circuit, control then being by either 'on/off' switching of the fan motor or speed change as mentioned previously.

In earlier paragraphs covering convective heating, no mention has been made of builders work enclosures for natural or forced convectors, skirting heaters, etc. These, of course, may be used in any situation where appearance is of importance and maltreated sheet metal is not acceptable. It is nevertheless important that the form and the dimensions of the enclosure, and the dimensions of inlet and outlet grilles be discussed with the heater manufacturer. Enclosures having grilles set into a flat top perform badly when covered with a variety of books and papers: a top angled at 45° is an encouraging first step in the movement of letters towards files (and files towards cabinets).

There are many further aspects in application of convective heating equipment which are unique to a given manufacturer and reference must in those circumstances be made to the technical data published. One single point, however, which seems to receive inadequate attention is that the heating elements of most natural and forced convectors are particularly susceptible to reduced output as a result of accumulation of air. As will be appreciated, there is little free space above the tubes in the element headers to accept air, in contrast with that available in a radiator. With those waters which tend to encourage the initial formation of gas and air mixtures, some engineers fit air bottles above the element header as a palliative.

Any form of convector, natural or forced, requires some level of maintenance to keep the finned heating element free from dust, etc., even if this is no more than good housekeeping: it is for this reason that particular attention to convenience of access is not only desirable but absolutely essential.

Table 7.5 Emissions from radiators for a temperature difference air to mean water of 60 K

Figure No.	Radiator: Type	Radiator: Pattern	Dimensions (mm): Depth	Dimensions (mm): Range of heights	Range of emissions (kW/m² of elevation)
7.12 (a)	Cast iron sectional (open)	2-column	70	430–980	1.87–2.10
		4-column	160	430–980	3.41–3.58
		6-column	250	280	5.43
7.12 (b)	Cast iron sectional (flat front)	2-column	71	430–980	2.29–2.32
		4-column	161	430–980	4.01–4.07
		6-column	251	280	6.01
7.13 (a)	Plain steel panel	Single	51	300–750	1.31–1.46
		Double	82	300–750	2.13–2.42
7.13 (b)	Steel sectional	2-column	110	450–1000	2.81
		3-column	160	450–1000	3.60–3.75
		5-column	250	300	5.94
7.13 (c)	Steel panel with coil	Figure 7.1 also	35	500–900	1.15–1.21
7.13 (d)	Steel tubular with headers	40 mm crs for	98	400–1000	2.93–3.12
		elements	166	400–1000	5.16–5.29
		60 mm crs for	98	400–1000	1.95–2.09
		elements	166	400–1000	3.43–3.53
7.14 (a)	Steel panel with surface added at rear	Single	30	320–720	1.62–1.79
		Double	64	320–720	3.01–3.46
7.14 (b)	Steel panel with added rear waterways	Single	40	300–1000	1.92–2.29
		Double	80	300–1000	2.95–4.07
7.14 (c)	Steel multi-panel with extensive added surface	Double	150	100–200	6.68–7.82
		Treble	225	100–200	9.88–11.08
7.14 (d)	Grid waterway in perforated case	Slim	70	370–1020	2.48–2.53
		Deep	172	370–1020	3.27–3.62
7.15 (a, b)	Aluminium sectional with flat panel front	Open top	95	430–690	4.52–4.65
		Closed top	160	285–435	6.68–6.73
7.15 (c)	Aluminium finned unit in casing	With damper at base	30	300–750	2.92–2.97

Table 7.6 Temperature corrections $(\Delta t/60)^{1.3}$

K	*0*	*1*	*2*	*3*	*4*	*5*	*6*	*7*	*8*	*9*
30	0.41	0.43	0.44	0.46	0.48	0.49	0.51	0.53	0.55	0.57
40	0.59	0.61	0.63	0.65	0.67	0.69	0.71	0.73	0.75	0.77
50	0.79	0.81	0.83	0.85	0.87	0.90	0.92	0.94	0.96	0.98
60	1.00	1.02	1.04	1.07	1.09	1.11	1.13	1.16	1.18	1.20
70	1.23	1.25	1.27	1.30	1.32	1.34	1.37	1.39	1.41	1.43
80	1.45	1.48	1.50	1.52	1.55	1.57	1.59	1.62	1.65	1.67
90	1.69	1.72	1.74	1.77	1.80	1.82	1.84	1.87	1.89	1.92
100	1.94	1.96	1.99	2.02	2.05	2.07	2.10	2.12	2.15	2.18
110	2.20	2.22	2.25	2.28	2.29	2.33	2.36	2.38	2.41	2.43

Table 7.8 Emissions from natural convectors for a temperature difference air to mean water of 60 K

Figure No.	*Convector*		*Dimensions* (mm)		*Range of emissions* (kW/m^2 *of elevation*)
	Type	*Pattern*	*Depth*	*Range of heights*	
7.16	Cabinet type with front outlet grille at top and bottom open for recirculation	2-tube element	110	400–700	1.46–1.89
		3-tube element	160	400–700	2.16–2.78
		4-tube element	210	400–700	2.73–3.50
7.17	Continuous type with outlet grille in sloping top front and bottom open for recirculation	Single tube	73	406–711	1.03–1.53
		Single tube	133	406–711	1.73–2.49
		Double tube	73	406–711	2.01–1.86
		Double tube	133	406–711	2.42–3.01

Table 7.9 Temperature corrections $(\Delta t/60)^{1.5}$

K	*0*	*1*	*2*	*3*	*4*	*5*	*6*	*7*	*8*	*9*
30	0.32	0.37	0.39	0.41	0.43	0.44	0.46	0.49	0.50	0.52
40	0.54	0.56	0.59	0.61	0.62	0.65	0.68	0.69	0.71	0.74
50	0.76	0.77	0.81	0.83	0.85	0.88	0.90	0.93	0.96	0.97
60	1.00	1.03	1.05	1.08	1.11	1.12	1.15	1.19	1.20	1.24
70	1.27	1.28	1.32	1.35	1.37	1.40	1.43	1.45	1.48	1.52
80	1.54	1.57	1.61	1.62	1.66	1.69	1.71	1.74	1.78	1.80
90	1.84	1.88	1.90	1.93	1.97	1.99	2.02	2.24	2.08	2.12
100	2.16	2.18	2.21	2.25	2.28	2.32	2.36	2.37	2.42	2.46
110	2.48	2.52	2.56	2.58	2.62	2.66	2.69	2.72	2.77	2.79

Chapter 8

Pumps and other auxiliary equipment

The principal component parts which must be brought together to make up an indirect heating system are the energy source, the distribution pipework and the heat emitting equipment. There are, however, a number of important auxiliary items which each make a contribution to the functioning of the whole. The subject matter of the present chapter is devoted to consideration of these enabling items, some understanding of which is fundamental to appreciation of system operation.

Pumps

There are two basic categories of pump used in connection with indirect heating systems, *positive displacement* and *centrifugal.* Of the former, the rotary gear type is used exclusively (in the present context) for liquid fuel handling, as described in Chapter 11. Direct acting positive displacement pumps were applied to early heating systems, but they are now rarely used, even for those boiler feed duties for which they were once popular as a result of their robust reliability. The wide use of the centrifugal type for other applications is such that no further introduction is required.

Reciprocating pumps

As the name implies, pumps of this type produce a discharge as a result of the axial reciprocating movement of a plunger within a cylinder, displacing fluid from suction to delivery. Pumps may be arranged vertically or horizontally and be either single acting with drive from a piston rod and crank or double acting with a direct drive through a shaft which is common to the plunger and to a steam piston and cylinder. A twin cylinder balanced action is usual where a pump is single acting. Pumps of this type operate at low speeds and are particularly suited to providing output against high pressures in circumstances where an inherent minor intermittency in delivery is not of consequence.

Rotary gear pumps

These take the form of two interlinked and contra-rotating gears, set with close clearances within a single casing. On the suction side, as the gears disengage, fluid fills the spaces between the teeth and is conveyed round the periphery of the casing. It is then discharged as the gears re-engage, to provide a practically constant level of delivery against any chosen pressure.

Centrifugal pumps (general)

Some form of centrifugal pump, be it single-stage or multi-stage, is now used to meet most duties including that of boiler feed against pressure. In principle, such pumps consist of an impeller, having backward curved blades, which rotates within a scroll casing. This casing, more properly called a *volute*, has a profile which is so formed that the flow, discharged at high velocity by the impeller to the circumference, passes smoothly through an increasing area. As a result, when the point of discharge is reached, most of the velocity energy has been converted into pressure. Multi-stage units consist of a grouping of such volutes, arranged in series such that the discharge from the first passes to the inlet of the second and so on.

A number of simple relationships shows how, for a given diameter of impeller, the performance of a centrifugal pump will change as the speed is varied, as follows:

- Volume varies directly with speed.
- Pressure varies with the square of the speed.
- Power absorbed varies with the cube of the speed.

To these may be added a further series of similar relationships which show that, for a given speed, the performance of a pump will alter as the diameter of the impeller is changed. To expand further upon these and other similar matters is outside the scope of this book.

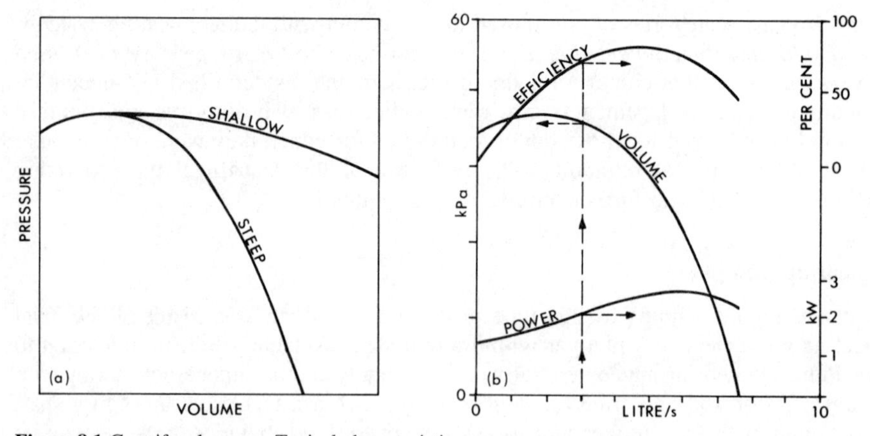

Figure 8.1 Centrifugal pumps. Typical characteristic curves

The performance of a typical centrifugal pump may be represented by *characteristic curves* as illustrated in Figure 8.1. Part (a) of this diagram shows alternative relationships between the volume flow and the pressure developed. It will be seen that, dependent upon the pump design, the shape of the curve may be either steep or shallow to suit the particular application. Part (b) of the diagram repeats the steep curve and adds further characteristics which illustrate how *efficiency* varies over the operating range and, consequently, the *power absorbed*. It will be noted that, in this instance, near-to-peak efficiency is retained over a relatively wide range of volumes.

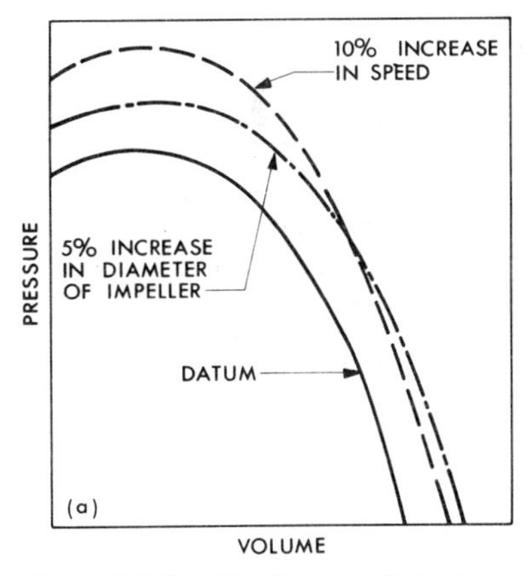

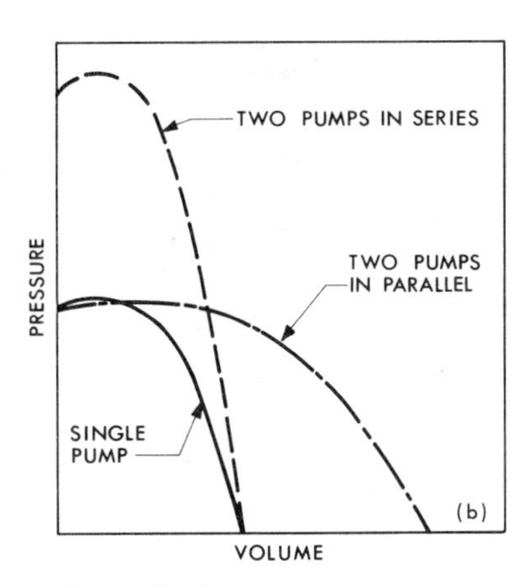

Figure 8.2 Centrifugal pumps. Duty change and alternative applications

The manner in which the volume/pressure characteristic may be adjusted either by varying the speed or fitting an impeller of a different diameter is shown in Figure 8.2(a). Centrifugal pumps may be arranged to operate either in series or in parallel and the outputs which result from such arrangements are shown in Figure 8.2(b), taking the same volume/pressure characteristic as was used in part (b) of the preceding diagram, but drawn to a smaller scale. It will be seen that two *identical* pumps which are operating in parallel will deliver twice the original volume for any given pressure and, similarly, that two *identical* pumps which are operating in series will produce twice the original pressure for any given volume. In instances where two *dissimilar* pumps are arranged to operate in either parallel or series, the result may be unsatisfactory and a detailed examination of the characteristic performance of each must be undertaken.

Centrifugal pumps for water systems

Passing from the general to the specific, there are a number of types or styles of centrifugal pump available for application to water heating systems, and these fall into groups which may be listed as follows:

- Form of volute – split casing or end suction.
- Drive arrangement – close-coupled, direct or belt.
- Shaft axis – horizontal or vertical.

Dealing with these in turn, the split casing type, as in Figure 8.3(a), is used only for the largest installations where access to the impeller or casing is necessary for cleaning purposes. End suction is the most common arrangement as shown in Figure 8.3(b).

A close-coupled drive, where the pump is attached to a flange on the motor casing and the impeller is mounted on an extension to the motor shaft, is a compact arrangement well suited to industrial use or pressure development but is often too noisy for use as a circulating pump. Direct drive, where motor and pump are mounted on a common baseplate, with drives coaxial one to the other but joined only by a flexible coupling, is

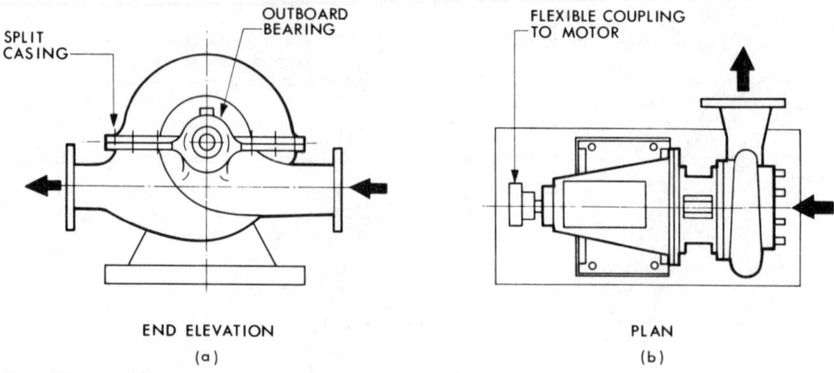

Figure 8.3 Centrifugal pumps. Forms of volute

a convenient arrangement and will provide long service without problems provided that the pump and motor shafts are aligned at manufacturers' works. Overall, however, drives which employ generously proportioned vee belts and pulleys, as in Figure 8.4, are those preferred by many designers. Here, the pump shaft runs at a speed which is quite independent of that of the motor and water-borne noise is reduced by the separation of the two components. Adjustments to the duty of the pump may be achieved by a simple change of pulleys and belts.

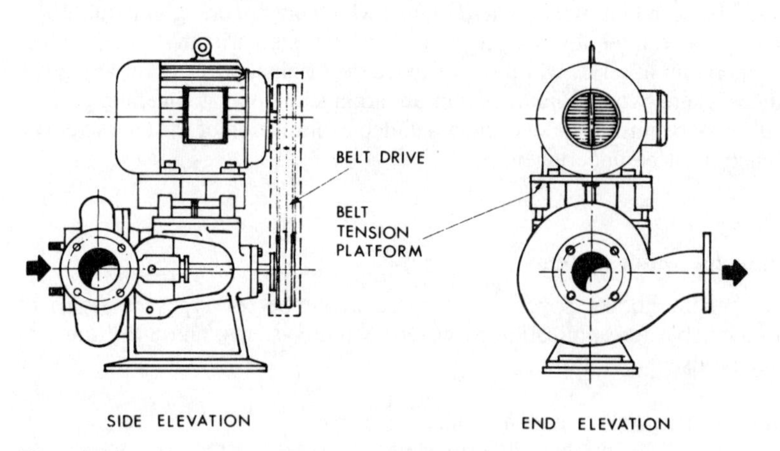

Figure 8.4 Centrifugal pumps. Belt driven (Pullen)

The axis of the driving shaft between motor and pump is most commonly arranged to be horizontal and supported by intermediate bearings: there is in consequence much to commend this arrangement. As an alternative, which may be more economic of space than the horizontal direct drive arrangement, pumps having vertical shafts (sometimes known in this context as *spindles*) are available, as in Figure 8.5. Some, as part (a) of the diagram, are provided with stools or feet for floor mounting but smaller sizes, as part (b), are sold as being suitable for pipeline installation. It is wise to establish that

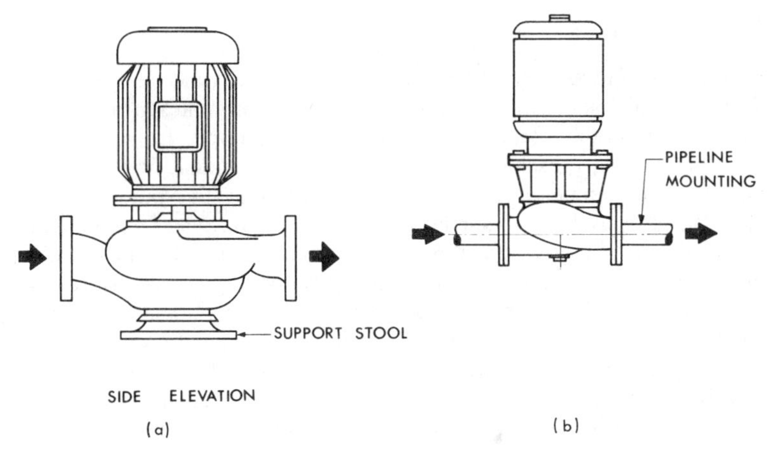

Figure 8.5 Centrifugal pumps. Vertical spindle types

the latter have thrust bearings and shaft seals which have been designed to suit a vertical configuration.

The types and styles noted above do not include that which is probably best known: the glandless submerged (canned) rotor pattern as fitted in most domestic systems. This is effectively an end suction type, and has the advantage of being without a gland or seal since the driving shaft does not pass through the casing. Figure 8.6 provides a cross-section through two pumps of this pattern. The penalty paid for this considerable advantage is that the clearances between the stator and the rotor can are necessarily small and, in consequence, any foreign matter in the water circulated may lead to seizure. Most such pumps incorporate devices whereby output may be varied: by an electrical adjustment to motor speed; by use of a hydraulic *spoiler by-pass* or by some combination of these two methods. Most manufacturers now advise that pumps of this type should in no circumstances be mounted with the shaft in a vertical plane.

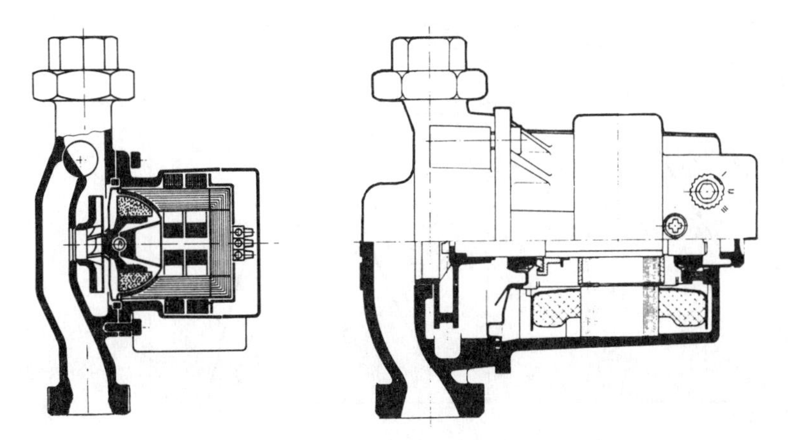

Figure 8.6 Centrifugal pumps. Submerged (canned) rotor type (Brefco)

Centrifugal pumps for condensate

A separation of condensate pumps from those handling water circulation may seem artificial but there are several important differences in the present context. First is the matter of pressure differential in that condensate is most usually pumped from suction at a near-to-atmospheric condition to discharge at a much higher potential. Second is the difficulty of quantifying the temperature/pressure relationship at the pump suction connection where condensate may be less 'stable' than is the case for water circulated in a closed system subjected overall to pressure above ebullition point. Third is the question of performance, since practice is to select a condensate pump having a capacity several times the calculated mean demand, in the realisation that performance must meet peaks well in excess of that level. Lastly, since condensate pumps are in most cases installed adjacent to plant items producing a much higher noise level, this will affect choice.

In consequence of these circumstances, the centrifugal pump most commonly used to handle condensate is, in comparison with those used for circulation purposes, designed to produce an output against a significant pressure differential, with good access for cleaning and without much regard for noise production: close coupled drive is the type most usually adopted and the volute is often split at right-angles to the shaft as an aid to cleaning. Reference will be made later in this chapter to factory made condensate receiver sets incorporating such pumps.

System characteristics

The operation of a centrifugal pump cannot be considered in isolation from the distribution system with which it is associated, and which system is made up of a pipework reticulation and various single items such as boilers and terminal equipment. As water flows through such a system, surface friction and other parallel effects will cause a loss of pressure which, for practical purposes, may be considered as varying in proportion to the square of the water velocity and hence of the quantity flowing. Taking any given system therefore, once the pressure loss has been determined for one flow quantity (see Chapter 9), other values may be produced as the basis for plotting a curve which represents the pressure reaction of that system to varying rates of water throughput.

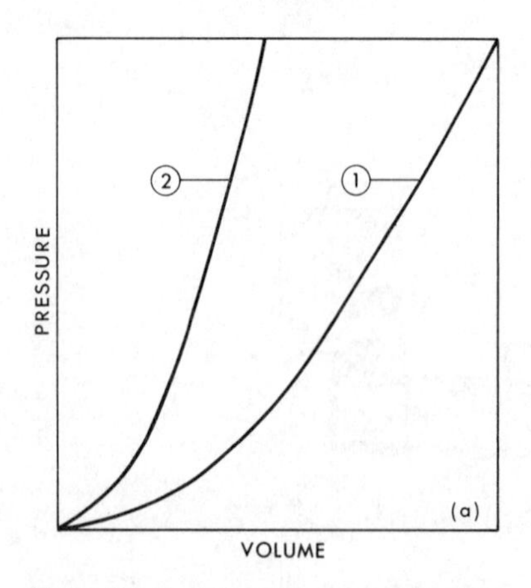

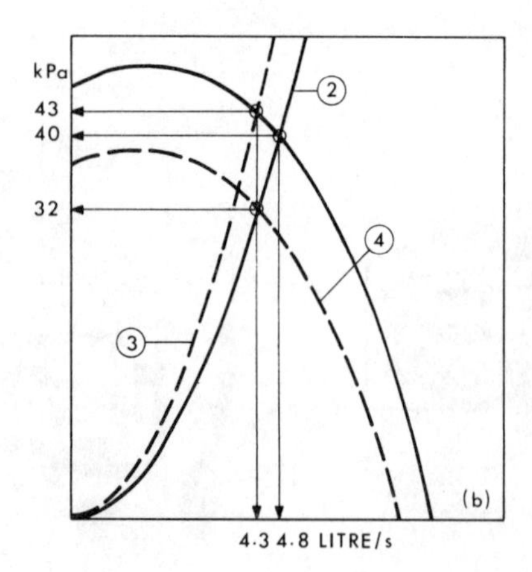

Figure 8.7 Centrifugal pumps. Application to systems

Two characteristic curves produced in this manner are plotted in Figure 8.7(a), that marked '1' being for a *system* having a relatively low resistance in comparison with that marked '2'. The use of such curves is illustrated in Figure 8.7(b) where that having the higher resistance is superimposed upon the pump volume/pressure characteristic reproduced from Figure 8.1(b). The point at which the system curve intersects with the pump curve represents the duty at which that particular combination would operate, i.e. 4.8 litre/s against a pressure of 40 kPa. As an example of the type of adjustment to such an intersection which may be required in practice, let us assume that a flow rate of 4.3 litre/s is critical to the operation of the system. There are two obvious ways in which this may be produced, as shown in the diagram:

- By adding resistance to the system (part closing a valve or some other similar action) such that a new *system* characteristic is developed, as '3', where 4.3 litre/s will be delivered against a pressure of 43 kPa.

or

- By reducing the speed of the pump by 10% (changing a pulley or an impeller) such that a new *pump* characteristic is developed, as '4', where 4.3 litre/s will be delivered against a pressure of 32 kPa.

The second alternative would be preferred in this case since less energy will be expended.

Pump application

The duties required of centrifugal pumps may be related to the heating capacity of a system, these then being translated to litre/s. Mass flow of water is thus, in this respect, the energy output in kW divided by the specific heat capacity of water and by the temperature differential across the system, flow to return. For all practical purposes, therefore, over the temperature range 10–180°C which is encountered in practice, the flow rate in litre/s = 0.238 × kW/K.

To limit a circulating pump to this exact duty would mean that the water flow through the many parallel circuits and, in effect, each terminal fitting, must be precisely the calculated quantity. This is obviously impracticable bearing in mind that the pipework system will be built up on site from commercially available materials under less than ideal working conditions and may, in any event, deviate slightly from the design. In consequence, it is usual to make the best calculation possible at the design stage and, this having been done, to add a margin to the calculated pump volume requirement. The size of the margin, which will vary between 10 and 20%, depends upon the complexity of the system arrangement: for either a simple single-pipe layout or for a reversed return system, 10% would be adequate. There is a difference of opinion among designers as to the validity of adding some similar margin to the calculated pressure loss: experience suggests that such an addition should be made only after careful study of the pump characteristics.

For larger systems it is usual to provide a duplicate pump for each circuit, in order to provide some insurance against failure. The two pumps are piped to the system in parallel and each is fitted with isolating valves on the suction and delivery connections. Where pumps are provided in duplicate for a high temperature system, it is good practice to arrange for a small circulation to be maintained through the stand-by pump so that it may be brought into service at the system working temperature, thus avoiding thermal shock.

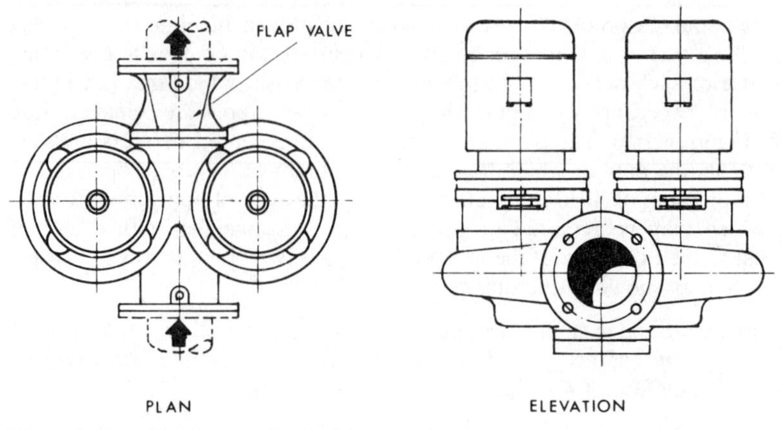

Figure 8.8 Centrifugal pumps. Dual vertical spindle type

It used to be thought good practice to provide a non-return valve on the delivery connection from each pump so that change-over could be a simple matter of electrical switching, without making use of the isolating valves. Such arrangements resulted, however, in the pumps and the valves being totally neglected, there being no need for the plant operator to visit them. In consequence, non-return valves are now rarely provided except in the case of packaged twin pump sets, as Figure 8.8, where a single flap is integral to the construction. The particular merit of providing duplicate pumps in this way is that the combination is very compact: some such dual sets are made with pumps of different sizes in order to provide for day/night or winter/summer duties.

Pump construction

Casings for centrifugal pumps are commonly manufactured in close-grained cast iron although a copper alloy may be used in particular circumstances. Impellers are of cast gunmetal, machined and balanced to close tolerances, and are mounted on stainless steel shafts. A problem which is common to almost all such pumps, no matter how arranged,

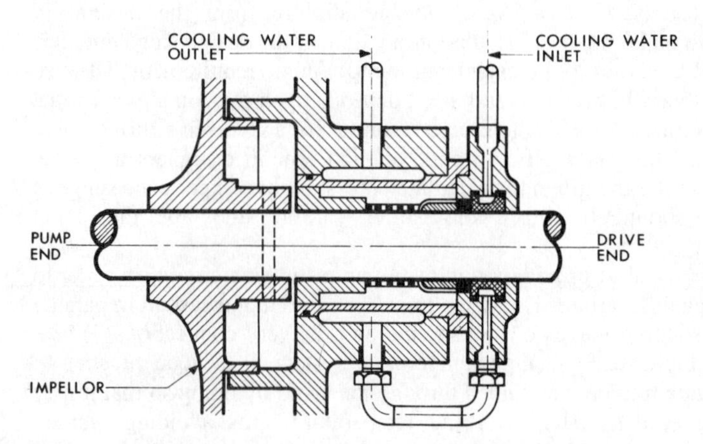

Figure 8.9 Centrifugal pumps. Water cooled bearing for high temperatures

is that the rotating shaft must pass through some form of gland or seal in the casing to whatever form of drive is to be used. This gland must, of course, allow for free rotation of the shaft with the minimum practicable leakage of the fluid pumped. Mechanical seals are generally preferred to packed glands for most applications but special designs are necessary for higher pressures and for temperatures above atmospheric boiling point. In the latter instance, water cooling arrangements as in Figure 8.9 are required.

Pump mountings

In addition to the isolating valves mentioned previously, each pump, or pair of pumps where in duplicate, should be provided with facilities for establishing the suction and delivery pressures. These may take the form of a pair of pressure gauges, a single differential gauge which is connected to both suction and delivery or, at the very least, a pair of pressure tapping points. Pressure readings will allow the performance of the pump to be checked against the characteristic curve when the system is commissioned and will provide a means for monitoring any fall off in performance during the life of the plant.

Some designers provide a small bore (say 25 mm) cross-connection between the suction and delivery pipework at the pump and fit this with isolating valves and a strainer. The pump capacity is increased by a small percentage and the arrangement ensures that a proportion of the system water content is being filtered continuously.

System pressurisation

Low temperature hot water heating systems, where the operating temperature is held below atmospheric boiling point, derive an adequate working pressure as a result of connection to an open feed and expansion cistern fitted above the highest point. Nevertheless, a cistern so sited suffers, particularly in domestic situations, from being out of sight and out of mind with the result that odd noises may be the first sign that it is empty, following seizure of the ball valve. Furthermore, the very requirement that the cistern be fitted high may well lead to it being positioned where frost is a potential hazard in exceptionally severe weather.

It has been explained previously (Chapter 6, p. 144), that the early high temperature hot water systems were pressurised by means of a steam space which was either, in the case of shell boilers, within the shell or, in the case of water tube boilers, within a separate steam drum. The disadvantage of this method, as explained, was the inherent instability resulting from the close association between the pressure applied and the temperature of the water circulated, with the result that skilled operators were required.

At either end of the temperature spectrum, therefore, the adoption of some alternative method of pressurisation needed consideration.*

Pressurisation by expansion

This, at the simplest domestic level, involves little more than the addition of an unvented expansion vessel to a heating system which is then charged with water and sealed. The function of the vessel is to take up the increased volume of the water content of the system as it is heated and, by so doing, apply additional pressure. In practice, proprietary type vessels are used which incorporate a flexible rubber diaphragm separating the water

* Kell, J. R., 'A survey of methods of pressurization', *JIHVE*, 1958, **261**, 1.

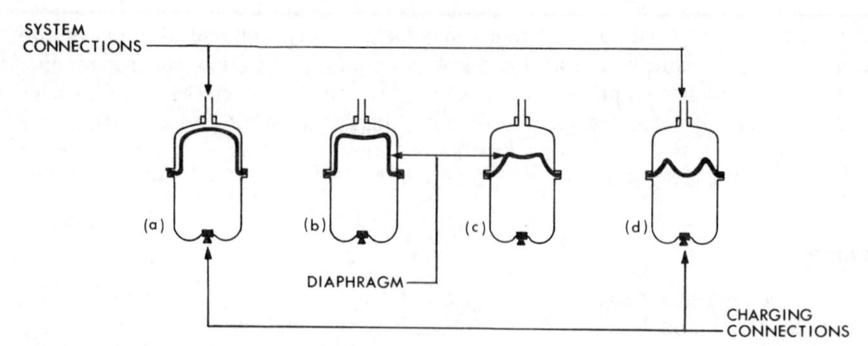

Figure 8.10 Application of a simple expansion vessel

content of the system on one side, from a factory applied charge of nitrogen on the other. The size of vessel required is a function of the initial and final pressures and the water capacity of the system: suppliers rate a standard range accordingly.

The way in which such a vessel performs is shown in Figure 8.10, the sequence being (a) before connection, with the diaphragm held to the vessel wall by the nitrogen charge; (b) connected to the system which has been filled; (c) during heating, as the water expands; (d) at system working temperature, the water content now fully expanded. In addition to the vessel, other fittings necessary are a safety valve fitted to the boiler (of a rather better quality than is normally provided for domestic systems), a fill/non-return valve which will accept a temporary hose connection from a water supply and an automatic air release valve, possibly associated with a centrifugal air separator.

For larger systems operating at low temperature, the principles of operation remain the same. The expansion vessel will increase in size and may even be duplicated. A small filling unit is usually provided for 'topping up' purposes, consisting of a cistern with a ball valve for water supply and an electrically driven pump all as illustrated in Figure 8.11. A low pressure switch fitted to or near the boiler will control the operation of the pump to ensure that a minimum water pressure exists and a parallel high pressure switch may be incorporated to stop firing of the boiler in the event of over pressure, at a level below safety valve operation. The various components of a system of this type may be built up into a complete packaged unit.

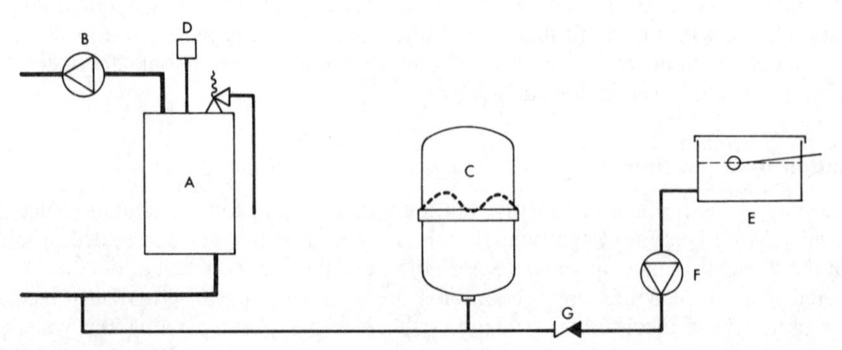

Figure 8.11 Components of a simple packaged expansion unit. A, boiler; B, system circulating pump; C, expansion vessel; D, topping-up control; E, feed cistern; F, pressurising pump; G, non-return valve

Pressurisation by pump

For medium temperature hot water systems, operating at temperatures up to about 110°C and at pressures of about 400 kPa, an alternative approach to pressurisation has led to the introduction of units of the type shown in Figure 8.12. These rely principally upon pressure-pump operation in conjunction with a *spill valve* and a cistern fitted at low level to take up the water of expansion. Some units of this type are factory assembled and are provided with duplicate pumps and controls.

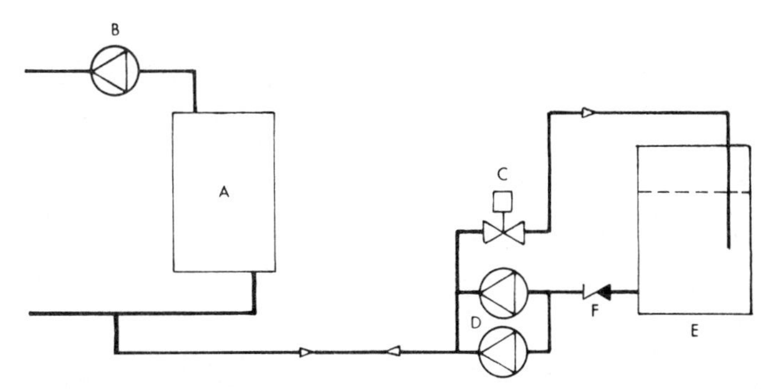

Figure 8.12 Components of pump-pressurising packaged unit. A, boiler; B, system circulating pump; C, water pressure spill valve; D, pressure pumps; E, expansion water spill cistern; F, non-return valve

Starting a system so pressurised from cold, the spill valve allows the water of expansion to escape into the cistern once a preset pressure is reached. While the system remains at the design working temperature, and thus at constant pressure, the spill valve remains closed and the pump is idle. When the system temperature and pressure fall as a result of decreasing demand or boiler cycling, the pump will start and the preset pressure will be restored. Since the *rate* of fall in pressure may vary considerably, the pump does not run continuously on demand but is arranged to cycle 'on/off' by means of timed interruption of the electrical supply. This arrangement provides, with admirable simplicity, the equivalent of a variable pump output. Arrangements such as this are sometimes criticised as being too reliant upon the availability of an electrical supply to maintain the necessary working pressure. Since automatic boiler firing depends also upon an electrical power source, avoidance of an unbalanced condition may be overcome by prudent design and the provision of self-actuating override controls.

Pressurisation by gas

This was, in chronological sequence, probably the first independent method to be introduced. It was conceived as a result of the difficulties experienced in application of steam as a pressurising agent for high temperature hot water systems, as noted previously. Early versions were bespoke to suit the characteristics of particular installations and were built up on site from standard components to the requirements of the system designer. A range of packaged units, rather more sophisticated as to detail, soon followed but these retained the original principles of operation.

The logic of the design sequence was that a pressure cylinder would be connected to the pipework arrangements in a chosen position, generally in the main return near the boiler where the water is coolest. This cylinder would be filled in part by water and in part by air or an inert gas, the initial supply of which derived from a small air compressor or from a gas bottle. By these means, an initial pressure could be applied at a level suited to the limits imposed by the system construction but in any event well above the boiling point of the water content. The water of expansion would be discharged from the system through a spill valve to a cistern open to atmosphere and, as the system cooled and contracted in turn, a pressure pump would draw water from the spill cistern and return it to the system.

The basic elements of the arrangement are shown in Figure 8.13, one of the more important being the pressure controller which regulates the admission of water from the pump or its expulsion through the spill valve. The use of an inert gas such as nitrogen as the cushion, in place of air, offers the advantage that it is less soluble in water: at a later stage in the development of packaged units, the spill cistern also was provided with a cover and was pressurised very lightly with nitrogen for the same reason. In most units, two pumps are provided, to run in parallel when required to meet an unusual demand, and the water of expansion passes through some form of heat exchanger in order to lower its temperature and, if possible, prevent flash.

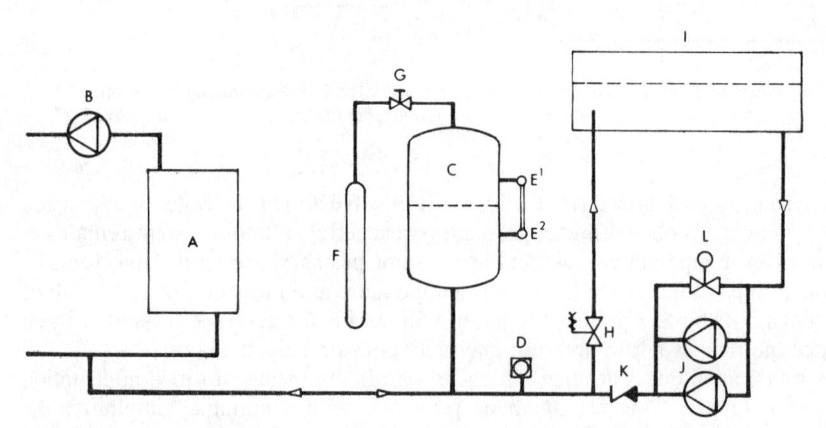

Figure 8.13 Components of gas-pressurising packaged unit. A, boiler; B, system circulating pump; C, pressure cylinder; D, pressure control; E_1 and E_2, high and low level cut-outs; F, gas cylinder (or compressed air supply); G, manual top-up valve; H, water pressure spill valve; I, expansion water spill cistern; J, pressure pumps; K, non-return valve; L, pump pressure relief valve

The calculation of the working pressure, to which components of the system and of the pressurising set are exposed, starts from a decision as to the flow temperature required at the highest point in the system, such as a unit heater or a high level main pipe. This, when established, is used with values interpolated from Table 6.1 as follows:

Assume that the boiler flow temperature is 160°C and that, at a high point 10 m above the pressure cylinder water level, the temperature in the pipework is 5 K less, i.e. 155°C.

Adding 15 K as an anti-flash margin to the water temperature at the high point, the vapour pressure equivalent to 170°C = 792 kPa

Allowing for:

a 10 m rise to the high point, add (10 × 9.81)	=	98 kPa
the differential on the pressure controller	=	50 kPa
Normal system operating pressure	=	940 kPa

And, hence

Pressure pump
starts at 890 kPa
stops at 940 kPa

Pump relief bypass
opens at 990 kPa

Spill valve
starts to open at 955 kPa
and fully open at 965 kPa

Safety valve settings
on cylinder 1015 kPa
on boiler 1040 kPa

It will be noted that an anti-flash margin of 15 K has been added for the purpose of this example: this would be varied to suit the particular circumstances in each case. The pressure exerted by the circulating pump will, in normal working conditions, augment this anti-flash margin but it is almost inevitable that at some time during the life of the system there will be a pump failure of some sort: it is usual therefore not to rely upon this added margin.

The required volume of the cylinder, if left uninsulated, is a function of the polytropic expression where PV^n is a constant. It has been found by experiment that $n = 1.26$ approximately and thus the movement of the water level may be calculated from:

$$a = h\,(1 - e^{0.794})$$

where

a = movement in the water level (m)
h = height of gas space prior to the start of compression (m)
e = P_1/P_2
P_1 = initial absolute pressure in gas space (kPa)
P_2 = final absolute pressure in gas space (kPa)

Control of the pump and spill valve is greatly simplified if the movement of the water line in the pressure cylinder is at a maximum: this movement is independent of the diameter of the cylinder but that dimension must be considered in conjunction with the duty of the pressure pump which should have a realistic running time of at least two minutes between start and stop when restoring the water level. The pump duty is a function of the rapid contraction which would take place if the boilers were to be shut down in emergency while the circulation continued.

Application of pressurisation

Although the equipment and packaged units which have been described under the preceding three headings were suggested as being appropriate to low, medium and high temperature heating systems in that sequence, this was merely to note that they were often so applied. There is of course no reason why packages for pumped or gas pressurisation should not be applied to any category of system provided that they have been designed to suit the pressure and volume conditions obtaining and that their use can be justified in economic terms.

Non-storage calorifiers

A calorifier is a heat exchanger, i.e. an item of equipment used to transmit heat from one fluid, at a higher temperature, to another at a lower temperature. The *non-storage* pattern, which is the subject of the following paragraphs, is so called to distinguish it from those related to hot-tap supplies which are large by comparison and provide a reservoir of hot water. While non-storage calorifiers are manufactured to suit either steam or water as the higher temperature source, the latter are now less commonly used for duties such as that which was shown in Figure 6.5(a), the function they fulfilled now being dealt with by injection circuits as Figure 6.5(b). They are, however, still required for particular applications, such as the heating of chlorinated swimming pool water, where separation between the higher temperature (primary) and lower temperature (secondary) contents is necessary. Subsequent references here are directed principally to steam-to-water types but apply with little revision to water-to-water equipment.

Simple non-storage calorifiers

A considerable number of patterns of non-storage units has been produced but all have had four principal components: an *outer shell*; a tubular *heater battery*; one or more *tube plates* and a *steam chest* for pipe connections. The primary heating medium is passed through the tubes with the secondary contents in the surrounding shell, this disposition reducing heat losses with the lower temperature fluid against the outside surface. The various patterns of calorifier have generally evolved from the manner in which the tubes were arranged, in hairpin or 'U' form connected to a single tube plate or as straight runs between tube plates at each end of the shell.

In most cases tubes have been plain but one manufacturer produced an indented type which offered a high rate of heat exchange per unit length and an ability to shed scale with movement. For most applications, however, the pattern shown in Figure 8.14 is suitable but an alternative vertical form, with the steam chest at the bottom supported clear of the floor, is sometimes used. This vertical form is economic in plan space but height is required to allow the shell to be lifted from the 'U' tube battery.

The outer shell is often constructed in cast iron but may be of welded mild steel. Tubes may be formed in steel or copper, the latter being the more usual material, and are individually expanded into the tube plate which is commonly of brass, the formation in

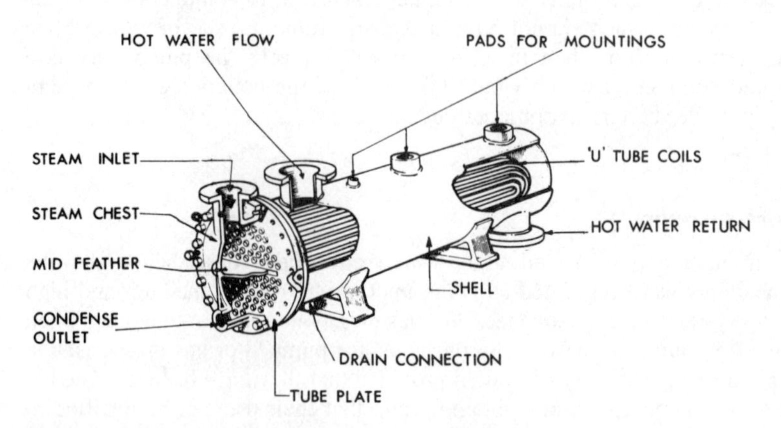

Figure 8.14 Horizontal 'U' tube steam-to-water calorifier

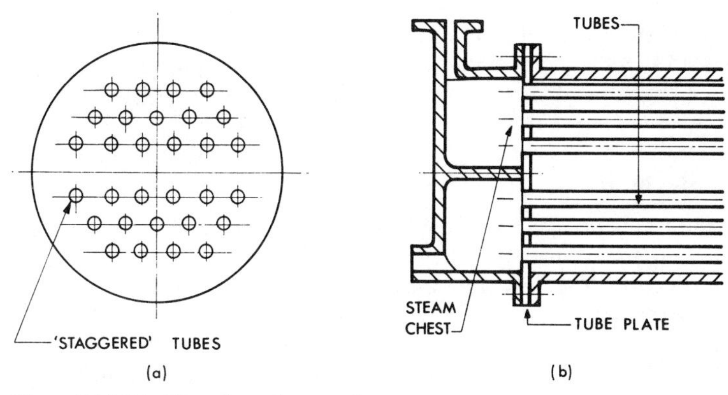

Figure 8.15 Calorifier tube and steam chest arrangements

end view being staggered as shown in Figure 8.15(a). The steam chest, usually of cast iron, mounts the tube plate to the shell and provides facilities for steam and condensate pipe connections as Figure 8.15(b). Since the tubes and tube plate are a heavy assembly, they are sometimes provided with a trolley type runway within the shell to facilitate removal for inspection: the pipe connections should be arranged to make this convenient.

Calorifier-condensate cooler units

Reference has been made previously (p. 156) to the recovery of flash steam. In circumstances where non-storage calorifiers are supplied with other than low pressure steam, discharge of the resulting condensate may give rise to problems and, in any event, a potential for energy saving will exist. Equipment suited to flash steam recovery is available in the form of two heat exchanger shells close-coupled, as illustrated in Figure 8.16. The upper unit is a normal steam-to-water calorifier equipped with a trapped outlet

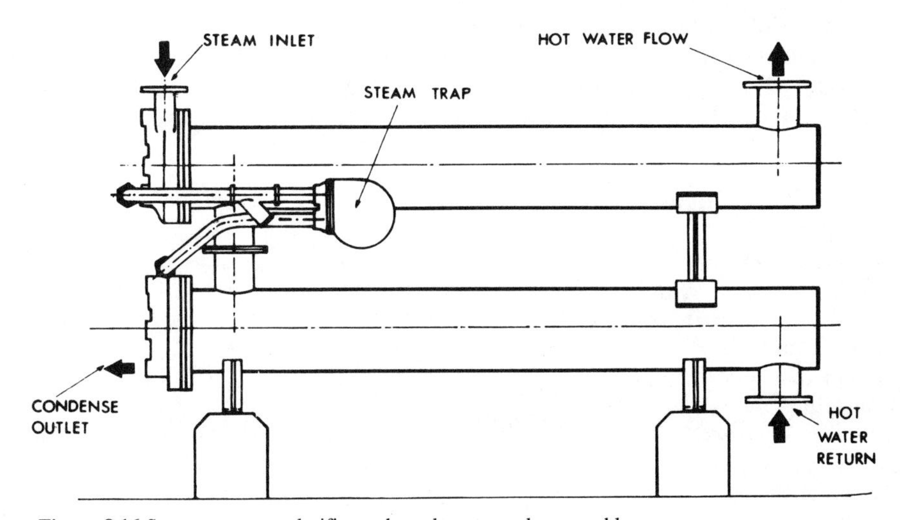

Figure 8.16 Steam-to-water calorifier and condensate cooler assembly

which discharges a mixture of flash steam and condensate to a tubular heater battery in the lower unit. The hot water circuit is directed through the two heat exchangers in series and in counterflow.

Calorifier ratings and mountings

Since, in most respects, a non-storage calorifier is a 'boiler substitute', calculations to determine the rating and number of units required should follow almost the same routine as that used for boiler selection. The only difference of any significance is that low load efficiency need be considered only as a function of the standing heat loss from a calorifier which is heated when output is not required. As in the case of a boiler, reference should be made to manufacturers' lists for actual outputs and sizes particularly since the use by some of controlled flow paths and high velocities within the shell produce an enhanced performance.

As to mountings, those required for a non-storage calorifier are precisely the same as the items which would be fitted to a boiler having an equivalent rating, as noted in Chapter 10, pp. 270–1.

Condensate handling equipment

When steam has given up its latent content in heating equipment of any type, the condensate remains at the temperature of the steam and it would be inexcusably wasteful to discharge it to drain since it might well represent 20% or so of the energy input to the boiler plant. The condensate must, however, be removed as soon as it is formed, in order to prevent water-logging, and returned to the boiler plant for re-evaporation.

Steam traps

At the exit point of the heating equipment it is necessary to introduce a device which will allow condensate to pass but not steam: this is a function fulfilled by the *steam trap*. There are various types of trap and these fall into three broad categories, identified by the means adopted to distinguish and separate condensate from live steam, as follows:

- *Mechanical, incorporating*
  Open top bucket
  Inverted bucket
  Ball float.
- *Thermostatic, with various elements*
  Balanced pressure
  Liquid expansion
  Bi-metallic.
- *Miscellaneous, including*
  Labyrinth
  Thermodynamic
  Impulse.

Some of these types are illustrated in Figure 8.17 but specialist texts* and manufacturers' literature should be consulted for specific details. It is usual to add auxiliary components to the trap proper in order to make up what is called a *trap set* and these will probably include some or all of: a strainer at entry; a check valve at exit; a sight glass and isolating valves when appropriate. The choice of a trap type suitable for any particular application depends upon a number of factors; load characteristics (constant or fluctuating); inlet and outlet pressures; associated thermostatic or other controls to the steam supply and the relative levels of trap and condensate piping, to name but a few. While it is not possible to generalise for all varieties of application, the trap patterns listed in Table 8.1 are generally suited to equipment within the scope of this book.

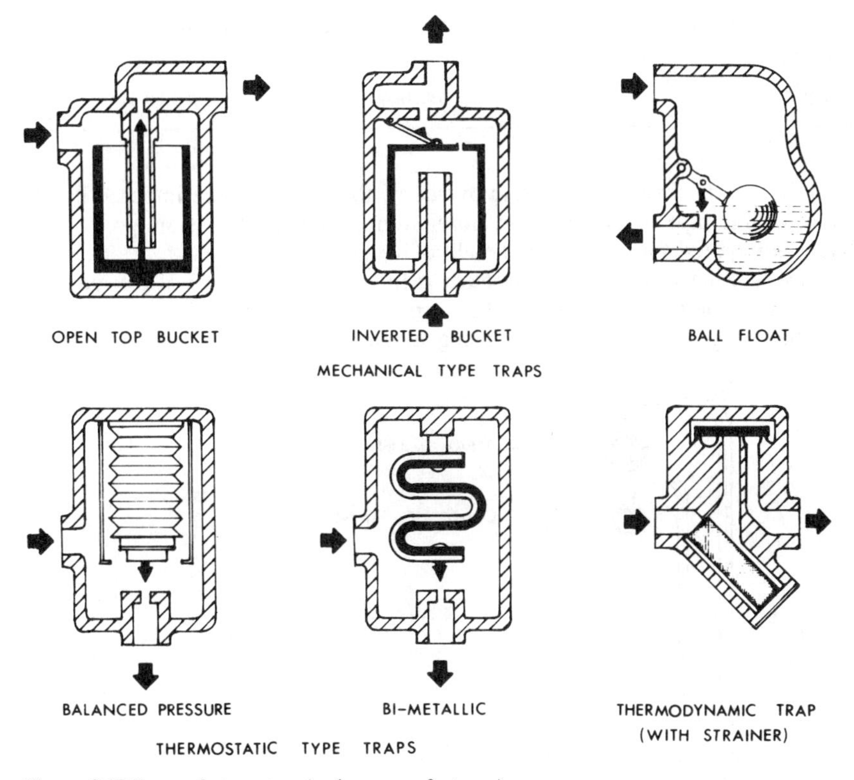

Figure 8.17 Types of steam trap (various manufacturers)

In an ideal situation, a trap would discharge into a condensate main run below it and the main would then fall in level back to the boiler plant. Such an arrangement may sometimes be possible in an industrial building but it is very often the case that a trap must discharge into a main above it; this is quite practicable *provided that* the steam pressure

* Northcroft, L. G., *Steam Trapping and Air Venting*. Hutchinson, 1945.
Spirax–Sarco. Various excellent instructional courses and texts.

Table 8.1 Steam traps: general application

Application	*Type of steam trap*
Drain points on steam mains	Mechanical open top, inverted bucket or thermodynamic
Fan convectors (large)	Mechanical ball float or inverted bucket
Natural convectors and radiators	Thermostatic balanced pressure
Oil storage tank coils and outflow heaters	Mechanical open top or inverted bucket
Oil tracing lines	Thermostatic bi-metallic or thermodynamic
Plenum heating or air-conditioning heater batteries	Mechanical ball float or inverted bucket: may be multiple for large units
Storage and non-storage calorifiers	Mechanical ball float or open bucket
Unit heaters (small)	Thermostatic balanced pressure
Unit heaters (large)	Mechanical ball float or inverted bucket

at the inlet to the trap is always adequate to overcome the back pressure imposed by the water in the discharge pipe, Figure 8.18. In simple terms, an available steam pressure of at least 10 kPa is required for each metre height of vertical 'lift'.

In instances where the steam using equipment is fitted with some type of thermostatic or other automatic control which takes the form of a throttling valve, it must be remembered that such a valve acts by causing a *reduction in steam pressure* to the

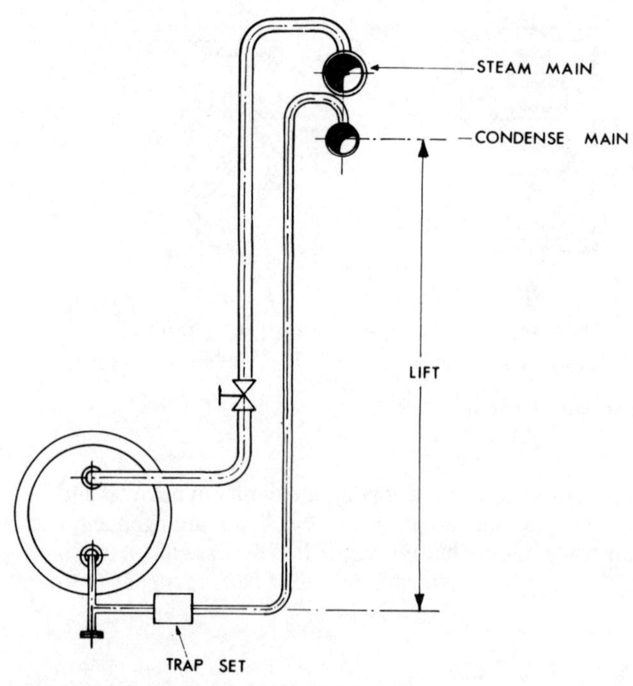

Figure 8.18 Lifting condensate to a higher level

equipment (and thus to the steam trap). In consequence, although the initial steam supply pressure to the control valve may be adequate to provide the required lift, this will not be the case at low loads. It is *always* good practice, therefore, to avoid a situation where condensate has to be lifted from a trap serving equipment fitted with automatic control of steam input.

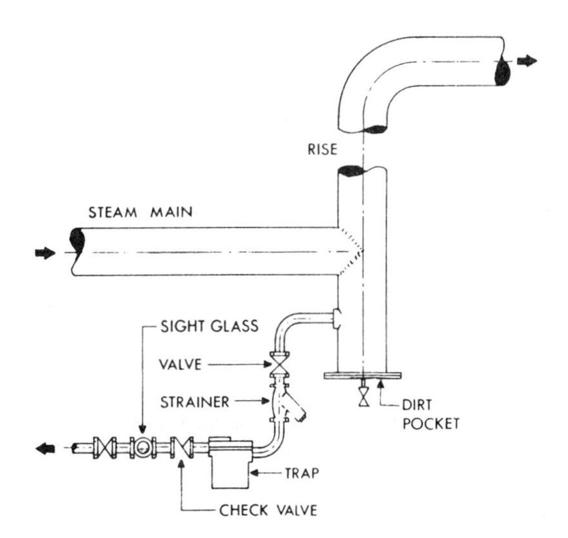

Figure 8.19 Steam main relay (to a higher level) and drain (drip) points

Condensate which is formed at individual items of equipment is easy to identify as to source and, hence, not difficult to deal with. It must not be forgotten, however, that steam mains, however well insulated, emit heat and that condensate will thus be formed along their run. Mains should always be arranged to fall in the direction of steam flow and arrangements made to provide drain (or 'drip') points at intervals along the runs, as in Figure 8.19.

Condensate return pumping units

In instances where either distance or site levels make it impossible for condensate to be returned to the boiler plant by gravity flow, it may be delivered either piecemeal or in quantity to some convenient central point where a built-up pumping unit of the type shown in Figure 8.20 is sited. Such *condensate receiver sets* are made to a range of manufacturers' ratings and consist of a cylinder fitted with a float operated switch which is mounted above a pump or pumps. The cylinder is sized to accommodate about one and a half to twice the maximum quantity of condensate returned per minute and each pump is rated to empty the cylinder in about two minutes. In order to avoid any possibility of pressure build up, the cylinder must be provided with an open vent to atmosphere, terminating in what is known as an *exhaust head* from which a drain pipe must be run to a safe position.

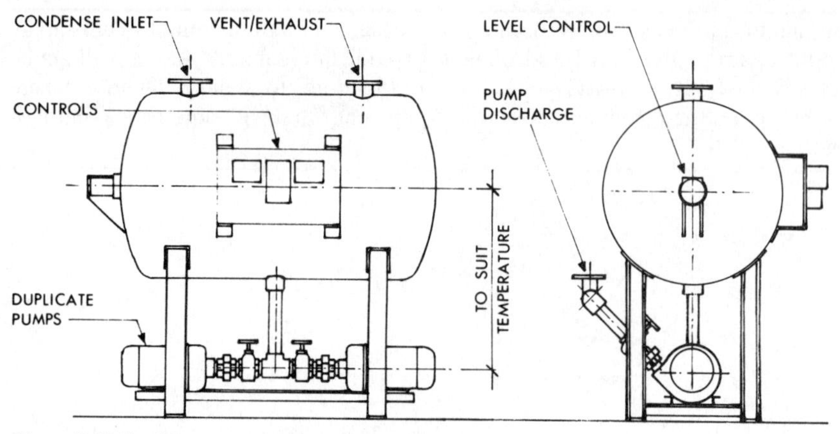

Figure 8.20 Condensate pump and receiver unit (Dunham)

Super-lifting steam traps

Where it is difficult to provide for a pipe run from isolated items of equipment to convey condensate back to a condensate receiver set, use may be made of a *super-lifting trap* which is, in effect, a steam-pressure-operated pump, Figure 8.21. A small vented receiver is required to collect incoming condensate which then feeds the trap, in preparation for intermittent discharge.

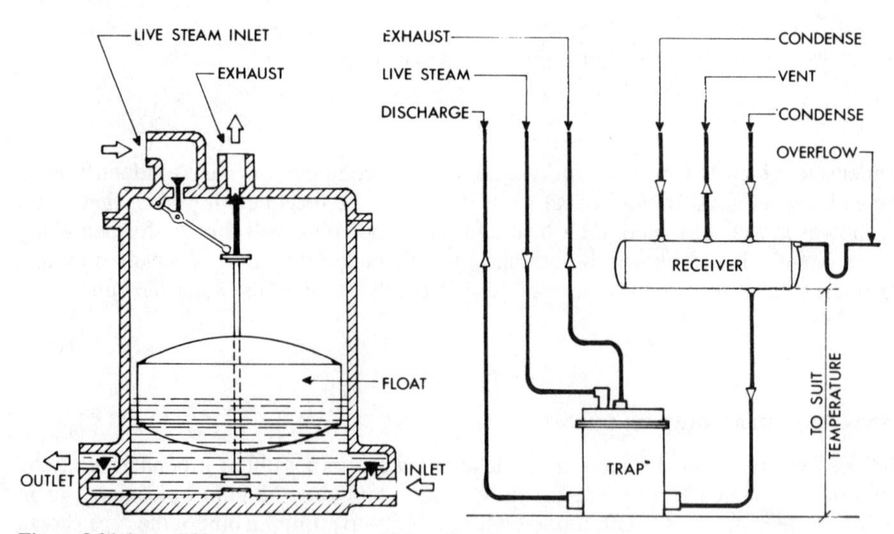

Figure 8.21 Super-lifting or pumping trap (Spirax-Sarco)

Steam pressure reduction

Where, for some industrial process, it may have been necessary to generate steam at a pressure quite unsuitable for heating purposes, some means of reduction will be necessary. For this purpose, pressure-reducing valves are available, as in Figure 8.22. It has been

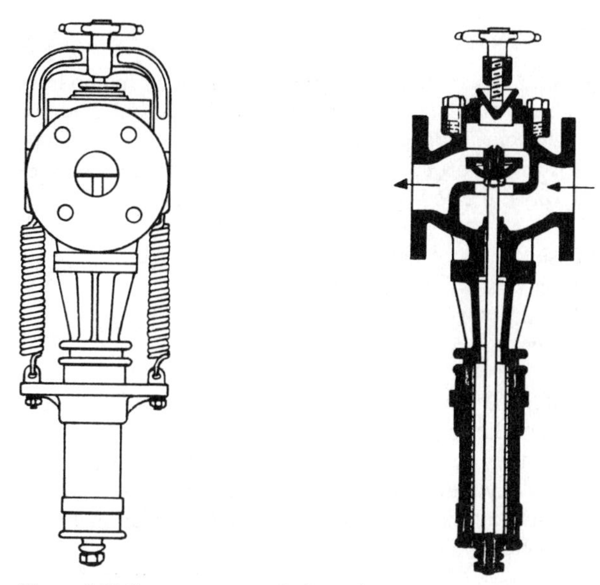

Figure 8.22 Steam pressure reducing valve

explained earlier, on p.156, that dry-saturated steam entering a valve of this type will produce a super-heated output. However, if there were some level of wetness in the entering stream, as is probable at the end of a long pipe run, and some losses following pressure reduction then the condition at the reduced pressure may approach that of saturation.

As in the case of steam traps, the provision of a pressure reducing valve involves installation of a number of auxiliary components as shown in Figure 8.23, the complete array being known as a *pressure reducing set*.

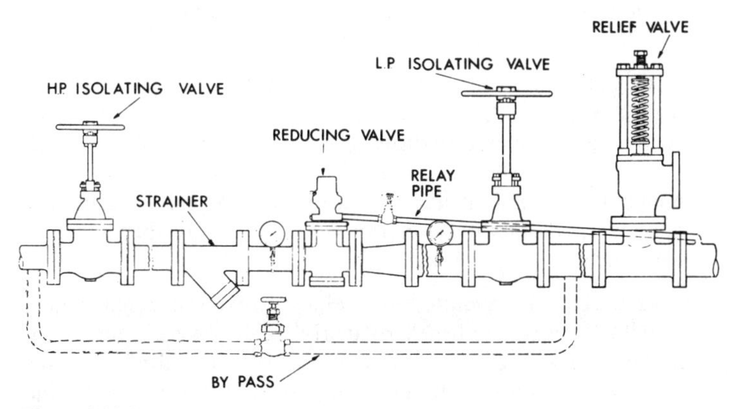

Figure 8.23 Steam pressure reducing 'set'

Heat meters

Circumstances arise from time to time where it is necessary to consider methods of making charges for heat supply. These may relate to internal audits within a single industrial organisation, to multiple tenancies on an industrial estate served from a single boiler plant or to households supplied from a group heating system. Such charges may be assessed and levied in a number of ways, the four principal methods being by:

- A flat rate proportional charge based on floor area.
- A flat rate charge based upon an agreed estimated usage.
- A fixed charge plus a unit rate based upon metering.
- A unit rate based wholly upon metering.

The first two of the alternatives offer the advantage that they may be assessed before the event and thus included in any periodic exercise of forward planning. In an era of rising fuel and labour costs, however, they must, in equity, be subjected to a periodic review and may be thought not to reflect any individual efforts made towards conservation of energy. In consequence, the situation overall must be considered in relation to the practicability of using some form of meter to determine heat consumption.

Metering of steam systems

In a context which is wholly industrial and where steam is the medium for heat distribution, the boiler plant will most probably be equipped with an accurate *inferential type* steam flow meter. Although it would be possible to use similar equipment at the remote points of use, the application of a mechanical meter of some sort to condensate flow would be a more economic solution. Rotary piston, rotary vane and rotary helix patterns are available and have an accuracy of about plus or minus 2%. The rotary vane type has generous clearances and is thus particularly suitable for use with condensate.

Heat meters for hot water systems

Complex hot water meters have been produced but few have performed up to expectations for any length of time. An acceptable level of accuracy in the integration of the two variables, flow quantity and temperature difference, has been the problem requiring an economic solution. Three quite different methods have been adopted by various manufacturers as follows:

Mechanical. Where the drive between components of a conventional water meter is adjusted mechanically by temperature sensitive elements.

Electrical. Where both flow rate and temperature difference are measured and integrated by electronics.

Inferential. Where a measured but very small quantity of return water is heated electrically back to the supply temperature, as sensed by thermostat. The energy then taken is measured by a simple kWh meter.

All such meters, as shown in Figure 8.24, are quite large, relatively expensive and subject to error if not maintained with care and returned regularly to the manufacturers for recalibration. Their application is, in consequence, best confined to central plant rooms where they may be used to provide records of system output overall. For use at remote points of heat consumption it is necessary to fall back on a range of simpler devices.

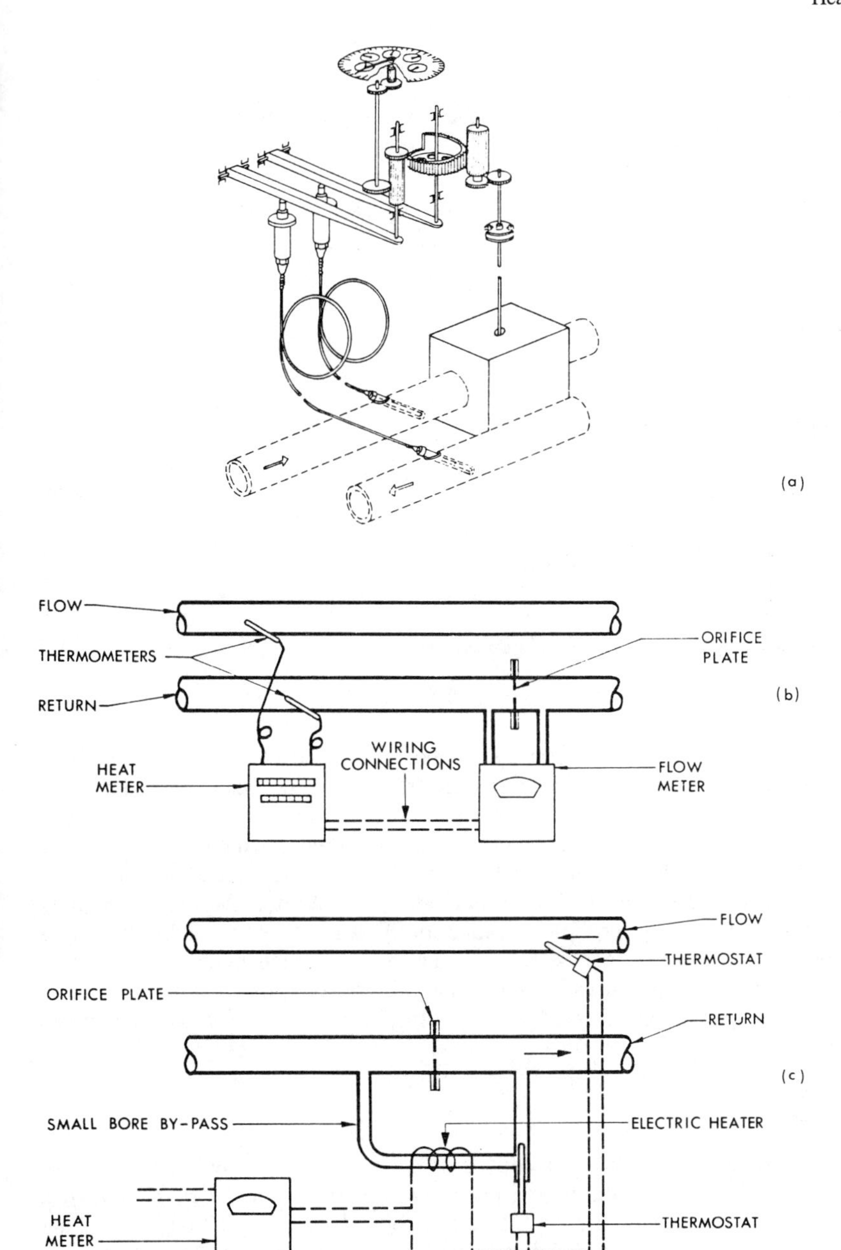

Figure 8.24 Mechanical, electrical and 'shunt' heat meters for hot water

Apportioning meters

For radiator systems, the type of device shown in Figure 8.25, one of which must be clamped to each radiator, was used in Europe for many years. It consisted of a housing bearing a scale, to the rear of which was a *replaceable* open phial containing a volatile

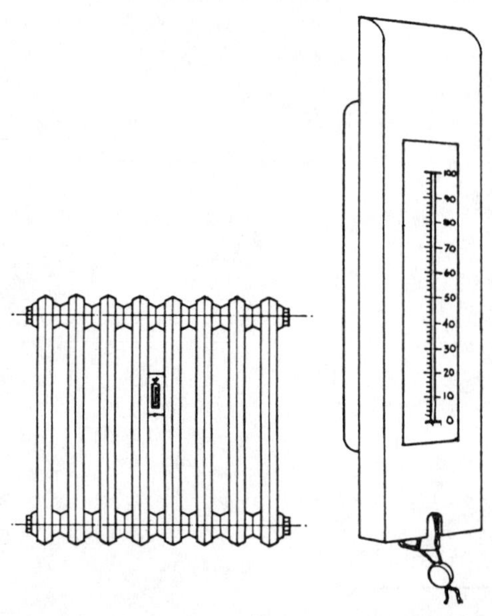

Figure 8.25 Evaporative proportioning meter for radiators

liquid. The amount of liquid evaporated by the heat from the radiator was a broadly approximate measure of usage and this was read from the scale each season or other period. This type of device was subject to vandalism and has now been been superseded by a relatively simple and secure electronic counter which integrates time and temperature and provides a numerical LED display. As before, readings are taken and totalled for the whole system and used as a basis to apportion the overall running cost to individual radiators or groups of radiators. The more obvious disadvantages include the necessity to provide access for a meter reader to each radiator, not always achieved easily in domestic premises. It is understood, however, that the manufacturers continue to offer an all-in meter reading and accounting service in given circumstances.

Water flow meters

An alternative method, for radiator systems, uses a simple rotary water meter, as for condensate measurement, associated with a *temperature limiting controller*. This device, when fitted to the return pipe of the circuit being metered, acts to maintain a constant outlet temperature and, by doing so, adjusts water flow in proportion to heat demand.

Hours-run meters

For any part of a system which requires a power supply to an individual fan or circulating pump, it is possible to arrange for a cyclometer type *hours-run meter* to be wired in parallel therewith and sited so that it may be read from outside the premises. The products of the reading taken, and factors representing the heat capacity of the individual units powered, are totalled as for other apportioning meters and used to determine the overall running cost. Allowance has to be made of course for any useful heat output from the metered components, by radiation or natural convection, which may take place when the fan or pump is not running.

Air venting, etc.

Mention has been made in Chapter 6 (p. 155) of the adverse effect of a stagnant air film upon steam-side heat transfer. As far as water systems are concerned, there will be an initial presence of both dissolved and free air and, since the solubility of air in water falls with increasing temperature, the free component will increase rather than decrease as the system is brought into service. The water content of an open system, or of one pressurised by equipment incorporating an open spill cistern, will be subjected to a process of continual re-aeration as a result of expansion and contraction. Furthermore, small leakages at pump and valve glands etc. and evaporation from almost any type of cistern will require that raw water be admitted regularly to make up the deficit.

If the presence of air is not to impede pipe circulation; reduce heat output from radiators, etc.; lead to cavitation difficulties in low pressure areas such as pump suction connections and, primarily in domestic systems, produce 'kettling' noises in boilers; then active measures must be taken towards its elimination. The matter of corrosion arising from air in the water content will be considered separately.

Air venting

The traditional approach to the disposal of air from low temperature hot water systems was to route a full-bore flow pipe *immediately* from the boiler plant to the highest level in the building and, at that point, to provide an open air vent to above the feed and expansion cistern. This approach, although a counsel of perfection in some respects, is rarely practicable in modern buildings and in any event would be a considerable impediment to design development.

For low temperature hot water systems, it is current practice to take advantage of any sensible opportunity to dispose of air without using components which require maintenance. A simple open vent taken straight from the boiler plant, as shown in Figure 10.13 (p. 271), although provided for a different purpose, is favourably placed to carry away air coming out of solution in the boiler. Elsewhere, if high points in the system cannot be provided with air bottles having small bore vents to outside, then robust and simple automatic air valves, as shown in Figure 8.26, must be fitted. Centrifugal and other types of air separators are sometimes added to domestic systems but there should be no need for such devices in a system which has been designed properly.

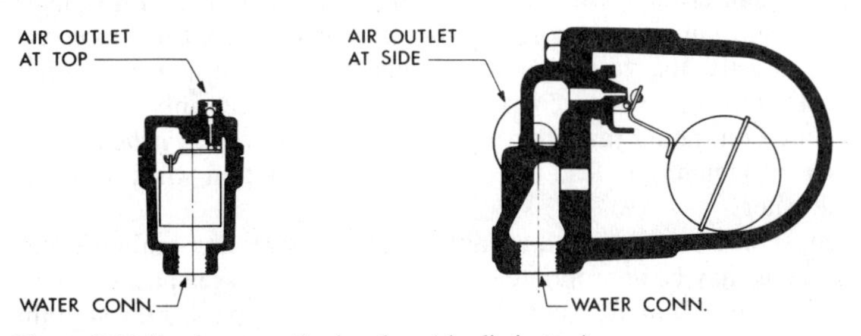

Figure 8.26 Simple automatic air valves (air eliminators)

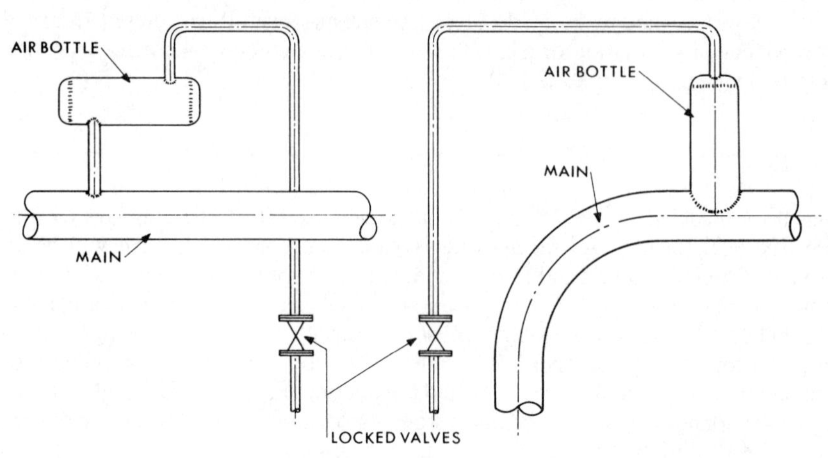

Figure 8.27 Air bottles for high temperature systems

For medium and high temperature water systems, where air release will – inevitably – be accompanied by some flash steam, full bore air bottles as shown in Figure 8.27 must be used with the discharge pipe led down to a position clear of any traffic. The needle valve fitted to the discharge pipe should be provided with a lock-shield type terminal or other means to avoid unauthorised use.

Corrosion

A full scale treatment of this subject would be quite beyond the scope of this book as would be any attempt at discussion in depth of water treatment and other ameliorative measures. It would, however, be wrong to ignore the subjects altogether in view of their importance: a reader in search of further information might refer either to the *Guide Section B7* or, better still, consult a professional specialist adviser. The necessity to treat feed water to steam systems has been well known and well explored for many years and it is not necessary to deal with that aspect of the matter here, other than by reference to Table 8.2 which summarises the additive processes commonly used.

As far as water systems are concerned, a radical change in the availability of equipment and consequently in design practice has taken place during the last 20 years. Boilers of solid construction with wide waterways, whether cast iron sectional or welded steel shell-type, have given way generally to higher-efficiency units of smaller size manufactured to designs meeting quite different criteria. Similarly, heat emitting equipment is no longer made in the British Isles in substantially proportioned cast iron but in light gauge pressed-steel sheet or aluminium. Lastly, the various new-type components are now very often piped together with small-bore light-gauge copper, rather than with sensibly thick mild steel pipes. The former are commonly made by a process which leaves a *carbonaceous* film on the internal surfaces which may lead to accelerated deterioration when these are exposed to certain hard waters.

Ideally, a water system should be filled with a stable, non-corrosive, non-scaling water, softened or de-mineralised as may be appropriate. No engineer who has seen a swimming pool filled by water straight from a mains supply, without first being passed through the filtration plant as is good practice, will remain under any illusion as to the condition of

Table 8.2 Water treatment additives for low pressure steam and hot water systems

Treatment	*Process/additive*	
Elimination		
Dissolved oxygen	Hydrazine	N_2H_4
	Sodium hydroxide	NaOH
	Sodim sulphate	Na_2SO_4
Inhibition		
Corrosion	Sodium compounds	
	benzoate	
	borate	$Na_2B_4O_7$
	nitrite	$NaNO_3$
	phosphate	Na_2HPO
	silicate	
	Tannins	
Scale	Lignins	
	Organic polymers	
	Phosphates	
	Phosphonates	
Neutralising CO_2	Filming amines	
pH control	Alkalis	
	Neutralising amines	
	Sodim hydroxide	NaOH

Note: this list is for information only: specialist advice should be sought for all water treatment problems.

water from such a source! During the construction of the heating system, mill scale and foundry sand will have accompanied material deliveries and welding beads, cement, plaster and other site debris will have been introduced: not all of this will have been removed by any flushing process.

The mixture of diverse materials noted above, when brought into contact with mains water, lays the foundation for ultimate disaster if no positive remedial action is taken. Dependent upon the composition of the particular water supply, corrosive or scale forming or both, some level of treatment will be necessary. The symptoms of a problem relate to evidence of:

- Corrosion at exposed parts.
- Deposition of salts at valve glands etc.
- Gassing at any tested air vents (hydrogen).
- Sludge formation in the system (magnetite).
- Bacteriological growth (anaerobic).

Corrosion, deposition and gassing, although apparently diverse, may well all result from the decomposition of the calcium bicarbonate ($Ca(HCO_3)_2$), common in all raw waters, to produce calcium carbonate ($CaCO_3$) and carbon dioxide (CO_2). As has been noted under the previous heading, the presence of air within a hot water system is inevitable at some time: the oxygen content of that air is a significant source of corrosion problems, in combination with the differing electrical potentials of the various metals involved. The formation of *magnetite sludge* (Fe_3O_4) results from oxygen combination with ferrous components. The deposition of a hard calcium carbonate scale on the internal surfaces of pipes may provide some measure of protection against corrosion but where such a coating breaks down due to thermal movement, attack may then be concentrated upon a small area of the pipe wall thus exposed. The presence of *anaerobic* bacteria within the quiet dark warmth of a pressurising vessel, where they ingest sulphates to produce corrosive sulphides, may generally be disturbed by raising the temperature of the water content to as far above atmospheric boiling temperature as the design of the system will permit.

As far as most conventional domestic and commercial systems are concerned, failing a detailed expert analysis of the situation by an independent professional adviser, the application of one of the commercially available corrosion inhibitors must be recommended. These take the form of chemical additives to the water content of the system at a calculated concentration. In the case of an existing system which may already have been so treated, it is imperative that details of any previous inhibitors be identified since the various specialist suppliers approach the subject in different ways and a disastrous colloidal interaction might result if two separate chemical compounds, however dilute, were to be mixed. Additives having reliable antecedents are available, based upon a selection from the processes listed in Table 8.2. or their equivalents.

Chapter 9

Piping design for indirect heating systems

Having selected the distribution medium, the general arrangement of the system and the type, position and size of the heat emitters, the next logical stage in the design process is to select, from the range of commercial pipes sizes available, those which will provide the most economical and effective pattern of connection. The principal factors which may influence this choice may be considered, simply as a matter of convenience, as falling under two headings:

- *External constraints*
  appearance;
  noise due to water turbulence or equipment;
  space occupied (i.e. pipe shaft size);
  force imposed on building components by mass;
  physical strength of pipe between supports.
- *Technical constraints*
  velocity *re* erosion, turbulence, air entrainment;
  dynamic pressure and pump availability;
  total pressure *re* cavitation;
  profile of hydraulic gradient for water flow.

Superimposed upon these aspects is the matter of achieving a long term balance between capital and recurrent costs in that a large pipe will be expensive to install initially and will emit heat (probably unwanted) in quantity. The energy required to induce fluid to flow through it, however, will be low. A small pipe will require a greater energy to induce flow but will emit less heat and be cheaper to install.

Water systems – principles

Circulation of water in a piping system may be maintained either by a thermo-syphon effect or by means of a pump. A thermo-syphon or *gravity* circulation is produced by a temperature variation, the water in the hot flow pipe being less dense than that in the cooler return pipe. The consequent difference in mass between a rising hot water column and a falling cold water column stimulates movement throughout the pipework system. As to circulating pumps, the centrifugal types which are most used were discussed in Chapter 8 (p. 200).

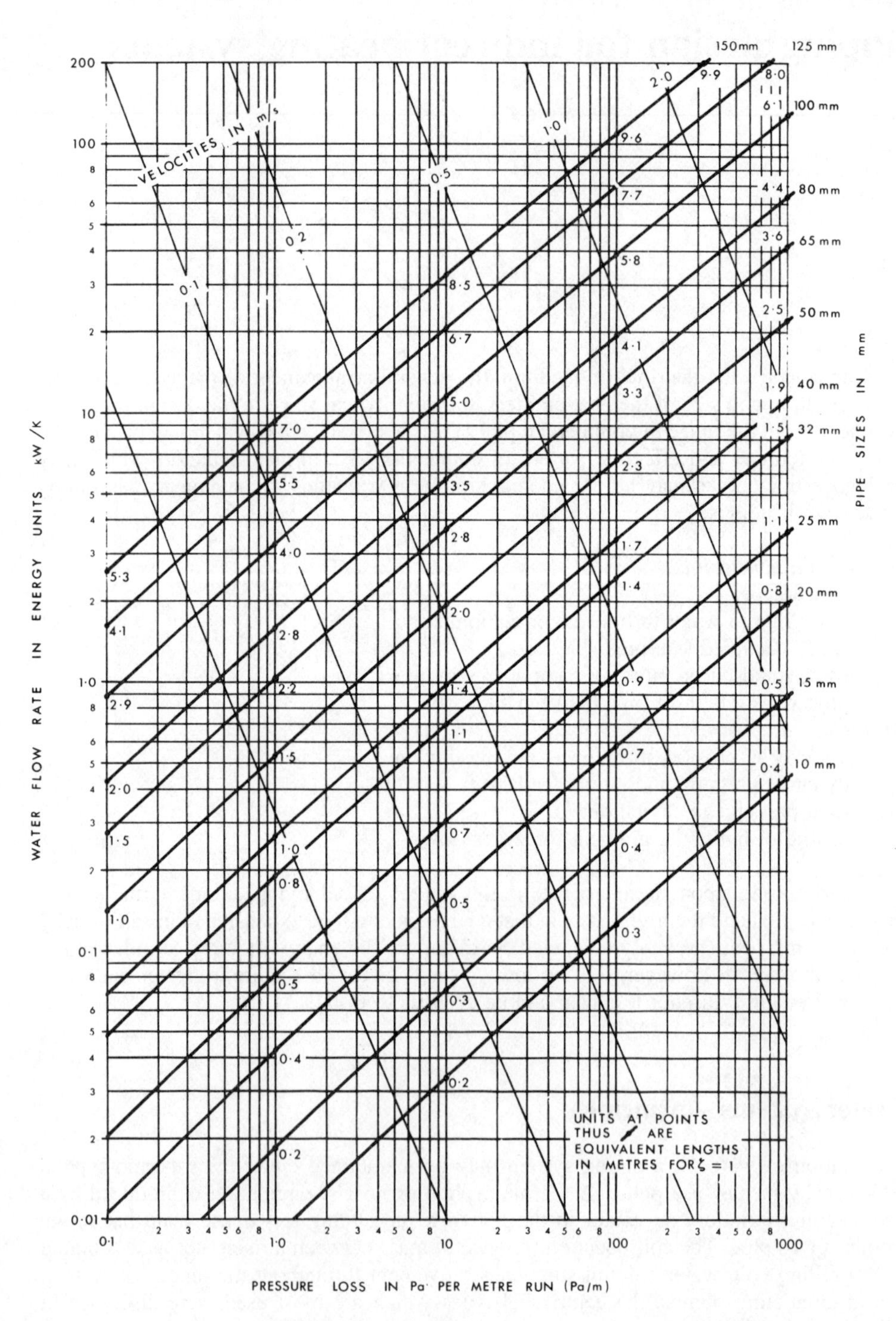

Figure 9.1 Sizing chart in energy units (kW/K) for water flow at 75°C in steel pipes (BS 1387: medium)

Flow of water in pipes

It would be inappropriate to discuss this wide subject in any depth here: a considerable literature in terms of text books and technical papers is available and an adequate bibliography is presented in the *Guide Section C4*. For present purposes it is enough to note that the relationship between mass flow rate and loss of pressure is complex and depends upon the velocity of flow; the diameter of the pipe; the roughness of the internal wetted surfaces and certain temperature-dependent characteristics of the fluid flowing such as specific mass and viscosity. The *Guide Section C4* provides a selection of tables showing pressure loss and mass flow of water at several temperatures and for a range of steel and copper pipes.

The mass flow of water required to convey a given quantity of heat from a central heat source to the various heat emitters, in unit time, is a function of the temperature difference

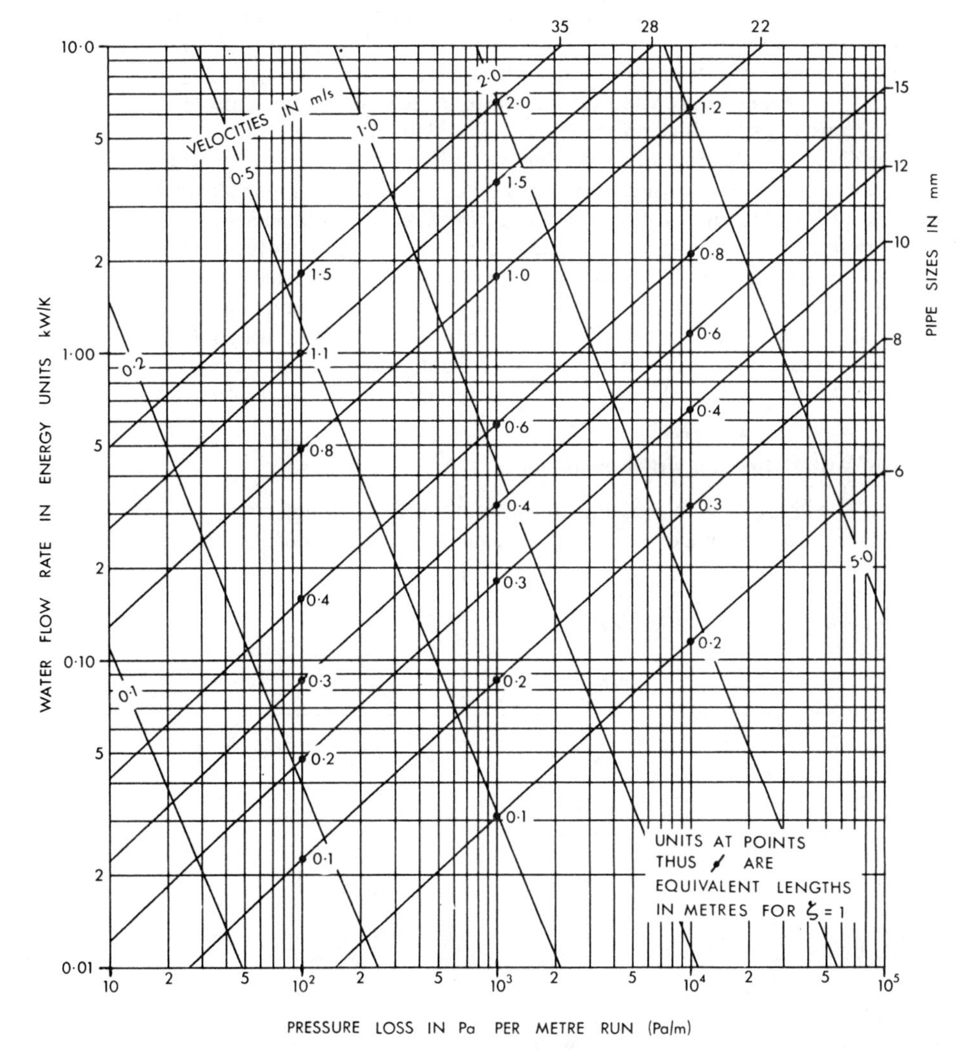

Figure 9.2 Sizing chart in energy units (kW/K) for water flow at 75°C in copper pipes (BS 2871: Table X)

(Δt) between the flow (t_f) and the return (t_r) and of the specific heat capacity of water. This latter may, for practical purposes over the range of temperatures considered here, be taken as constant as 4.2 –kJ/kg K. Mass flow may thus be calculated from:

$$\text{kg/s} = (\text{kJ/s})/(4.2\,\Delta t)$$

In a complex system where heat loads are known only in energy (kW) terms, this is a tedious calculation to work out a great many times for each and every pipework section, as required by the tables included in the *Guide Section C4*, and one which may lead to nostalgic recollections of the Imperial system where units of heat flow had a calorimetric base. However, since 1 kW = 1 kJ/s, it is easy to produce data in more direct terms (kW/K) by simple substitution in the expression set out above. By adoption of this approach, it is possible to retain the kilowatt as the base unit throughout all calculations: heat loss; selection of heat emitting equipment; piping design and, finally, the overall totals for boiler power. It must be remembered always that the kilowatt is no more than a measure of energy flow and use of it in this manner is an aid towards understanding the thread of continuity which persists from first to last.

Figure 9.1 has been prepared on this basis to give a vertical scale of energy flow in kW/K against a base scale of pressure loss in Pa/m run, for water flow at 75°C in steel pipes. Similar data could be provided for medium and high temperature hot water systems but the temperature ranges at which these operate are less well established: since they are not, use is made of correction factors and the *Guide Section C4* includes a table listing those appropriate to a water temperature of 150°C. Since the error involved in using Figure 9.1 without correction is only about 3%, it would be out of place to pursue this added complexity here. Where small bore copper tube is to be used for low temperature systems, however, it is necessary to use alternative data since both the roughness of the wetted surfaces and the internal pipe diameters vary from those of steel. Reference to Figure 9.2 provides the alternative information, again presented on the basis of energy flow.

Flow of water through single resistances

In parallel with the loss of pressure consequent on flow through straight pipes, additional losses arise at bends, branches, valves, etc., and at major components such as boilers, heat

Table 9.1 Pressure loss factors for single resistances

Item	ζ	*Item*	ζ
Elbows		Valves (open)	
90°	0.75	globe	10
45°	0.5	angle	5
welding	0.35	non-return	0.1
copper	0.75	gate	0.3
Bend	0.45	Column radiator	5
Return bend	0.8	Panel radiator	2.5
Reducer		Boiler	
area ratio 0.1	0.6	cast iron	2.5
area ratio 0.8	0.05	shell	1.5

Take through tees as 0.5 plus reduction or enlargement.
Take branch tees as 0.2 plus equivalent elbow or bend.

emitters and so on. These come about as a result of disruption to the velocity profile which is developed in flow along straight pipes, the turbulence created at the *single resistance* persisting for a considerable distance downstream. The pressure loss characteristics of most types of single resistance have been determined experimentally, as a function of flow velocity, and are represented by the symbol ζ.

A convenient way of making allowance for single resistances is to add to the measured length of straight pipe an *equivalent length* (*EL*) for each such item. Values for unit equivalent length are marked on the charts in Figures 9.1 and 9.2 and these, multiplied by the appropriate value of ζ, provide the actual equivalent length to be added. The *Guide Section C4* lists approaching a hundred values of ζ for various types of pipe fitting, etc., but Table 9.1 provides a short summary adequate to cover most normal situations.

Velocities

Experimental data regarding acceptable maximum and minimum water velocities in pipes are meagre but there seems to be general agreement that erosion is unlikely to become a problem at less than about 2.5–3 m/s: similarly, it has been suggested that velocities of less than 2–3 m/s in straight pipes are not likely to give rise to a noise level which is unacceptable. As to minimum velocities, these are of importance only as far as air entrainment is concerned and a value of 0.5 m/s has been proposed although this would seem to be rather on the high side for small pipes. The ranges of values listed in Table 9.2 represent normal practice for pumped circulations.

Table 9.2 Water velocities in pipework

Pipes nominal bore (mm)	*Application group*	*Velocities in normal use* (m/s)
10, 15, 20	Small domestic	0.25–0.75
25, 32, 40	Domestic	0.5 –1.0
50, 65, 80	Small commercial	0.75–1.75
100, 125, 150	Commercial	1.25–2.5

Water systems – applications

In applying theory to practice, it is as well to remember that a system made up from commercial pipework bears little resemblance to a laboratory test rig. Manufacturing tolerances for internal diameters vary with the material, being about ± 10% for steel, and other variables such as internal roughness with ageing, thickness of scale deposition, etc., add to the uncertainty. Furthermore, for one reason or another, site work may not proceed exactly to the design: additional bends may be introduced, tee branches may be arranged in a different way and welded joints in awkward positions may not be quite as free of protrusions into the bore as they should be. These comments are not intended to suggest that pipework sizes should be selected solely by a plumber's rheumy eye nor that data included in the *Guide Section C4* are too exact for practical use. They are offered rather to emphasise that calculations made with a slide rule are more pertinent in this context than results displayed on a pocket calculator tuned to provide eight or more decimal places.

Except for the smallest heating systems and for primary circuits to some indirect hot water supply systems (p. 579), water circulation by gravity has now been supplanted by use of centrifugal pumps. The origins of this change were the sponsorship of domestic small bore systems in the mid-1950s by a solid fuel research association and the coincident introduction of the small relatively inexpensive submerged rotor pumps noted in the preceding chapter. In these circumstances, it is logical to give precedence here to consideration of pumped systems and to refer to gravity circulation subsequently.

A single-pipe circuit

To consider the simplest of circuits first, Figure 9.3(a) shows a single-pipe system serving six radiators. The pipe must carry sufficient energy, via the water flowing in it, to cater for the emission from all the radiators plus that of the piping. This total will determine the net requirement imposed upon the boiler.

Assume that the radiators emit 6 × 4 kW	=	24 kW	
and that pipe emission (taken as 32 mm)	=	3 kW	
Total load	=	27 kW	
If the temperature drop, flow to return, Δt = 10 K, then energy flow = 27/10	=		2.7 kW/K
Assume that the piping measured length	=	30 m	
and that the equivalent length	=	5 m	
Total length	=	35 m	

Referring now to Figure 9.1, a choice is available:

40 mm pipe with a pressure loss of 65 Pa/m, thus 65 × 35	=	2.3 kPa
32 mm pipe with a pressure loss of 120 Pa/m, thus 120 × 35	=	4.2 kPa
25 mm pipe with a pressure loss of 550 Pa/m, thus 550 × 35	=	19.3 kPa

Both pressure loss and velocity (0.45 m/s) are low in the 40 mm pipe and pressure loss is a little on the high side in a 25 mm pipe: thus, it would seem that the original assumption of 32 mm was correct. The net water duty required would be (2.7/4.2) = 0.65 litre/s

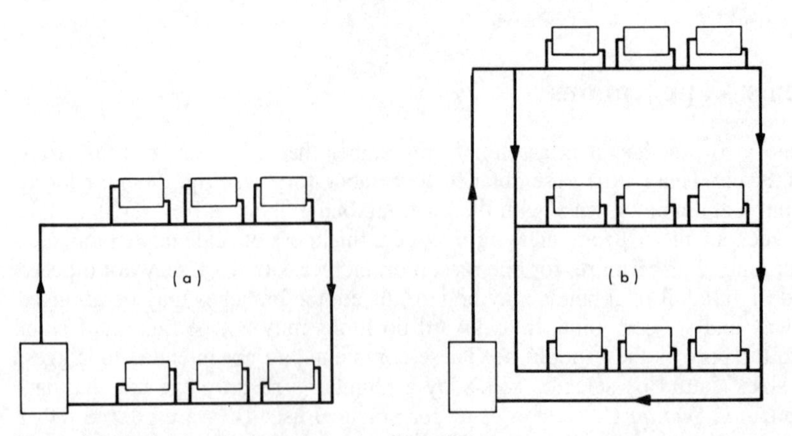

Figure 9.3 Single-pipe circuits: (a) simple, (b) multiple

against a pressure of 4.2 kPa and a pump rated at perhaps 0.7 litre/s against 5 kPa would be chosen. The piping connections to individual radiators would be sized separately as will be shown later.

For reasons which will become apparent, parallel single-pipe circuits are dealt with later.

A two-pipe circuit

A simple example of a two-pipe circuit is shown in Figure 9.4 and this will serve to illustrate how the pipe sizes might be selected in an approximate way, perhaps to produce a preliminary cost.

Assume that each of the heat emitters		
is a 20 kW fan convector, 5 × 20 kW	=	100 kW
and that pipe emission is say 10%	=	10 kW
Total load	=	110 kW
If the temperature drop, flow to return,		
Δt = 12 K, then energy flow = 110/12	=	9.1 kW/K
Measured length of piping, from Figure 9.4	=	200 m
assume that equivalent length	=	20 m
Total length	=	220 m

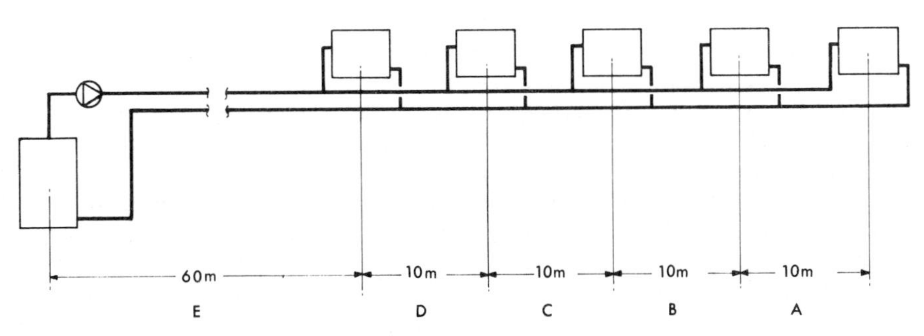

Figure 9.4 Example of a two-pipe circuit

If we assume that a pump pressure of say 20 kPa would be appropriate, then the unit pressure drop for the longest (*index*) pipe run, from the boiler to the most distant fan convector, would be 20/220 = 90 Pa/m. From Figure 9.1, using this value, pipe sizes for energy loads between 1 and 10 kW/K may be read off and the following schedule constructed:

Pipe size	kW/K	kW *for 12 K* Δt
25 mm	1.1	13.2
32 mm	2.2	26.4
40 mm	3.1	37.2
50 mm	6.3	75.6
65 mm	10.2	122.4

Assuming that the pipe emission is constant at 10% of the net energy load, a second schedule may be constructed:

Pipe section	*Total load*	*Pipe size*
A	22 kW	32 mm
B	44 kW	40 mm
C	66 kW	50 mm
D	88 kW	50 mm
E	110 kW	65 mm

It will be noted that the loads on pipe sections B and D, 44 and 88 kW, are slightly above those listed in the last column of the first schedule for 40 and 50 mm pipes respectively. By reference to Figure 9.1, however, it will be apparent that the unit pressure loss for these sections will be increased only slightly if these sizes are selected: the next pipe size up in each case would be much too large. The short connections to the individual fan convectors would, continuing with the same approach, be 32 mm in each case but, alternatively, might each be considered separately and used to balance the system a little better. For example, since the total length related to the first item served is not much more than half that of the index circuit, 25 mm pipe might be used for the connections.

Parallel single-pipe circuits

By inspection of Figure 9.3(b), it will be apparent that each single-pipe circuit here may be treated as if the points of connection to pipework common to all circuits were flow and return connections to a boiler. The size of the single pipe may then be selected on that basis, as discussed previously. Thereafter, each single-pipe circuit may be considered as if it were a compact heat emitting unit connected to a two-pipe circuit and sizes for the common pipework then selected accordingly.

The pressure loss which is used to select the pump will be that of the longest or *index* circuit and the volume handled will be the sum of the individual requirements. If a single value of unit pressure loss were to be used then those circuits which are shorter or carry less load than the index run will lose less pressure and it will be necessary to take account of this when sizing the common pipework and, in all probability, also necessary to provide regulating valves to overcome out-of-balance conditions.

As in the case of the example for a simple two-pipe arrangement, the process outlined above is approximate only but would serve also as a basis for making a preliminary cost estimate.

Calculations for final design

Moving now from approximations to the calculation routine required to complete a design, the system shown in Figure 9.5 will serve as a basis to illustrate the necessary steps which are as follows:

- The *iterative* process of allocating guessed preliminary pipe sizes to a system in order to establish the amount of heat emitted by each section (guessed or estimated as a percentage in the previous examples), is tedious and use may be made of Figure 9.6 as a short cut. From the curves included, a near approximation to heat emission from a pipe

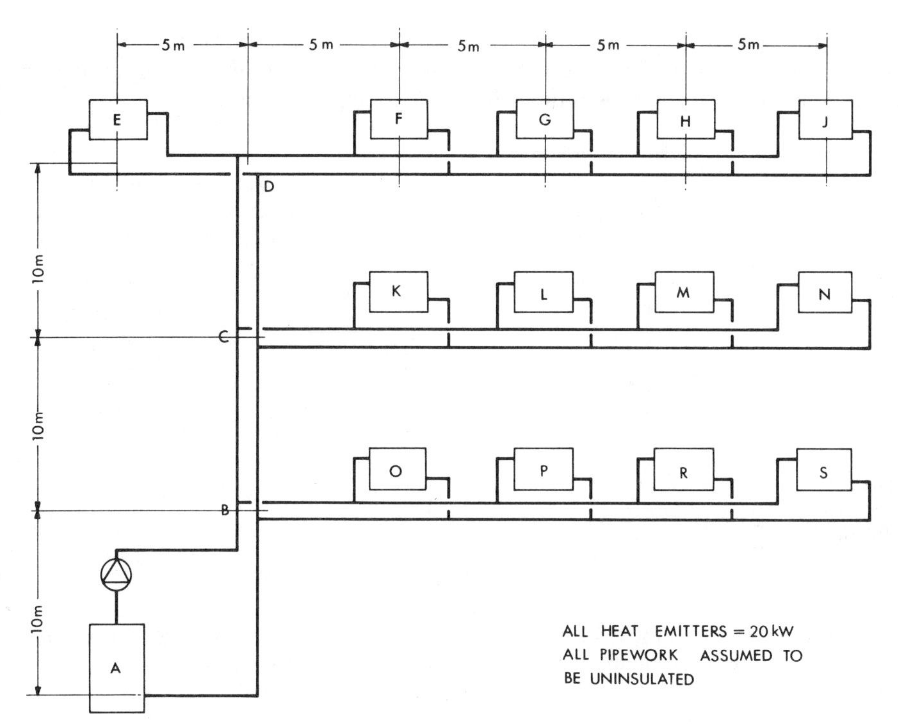

Figure 9.5 Example of two-pipe sizing exercise

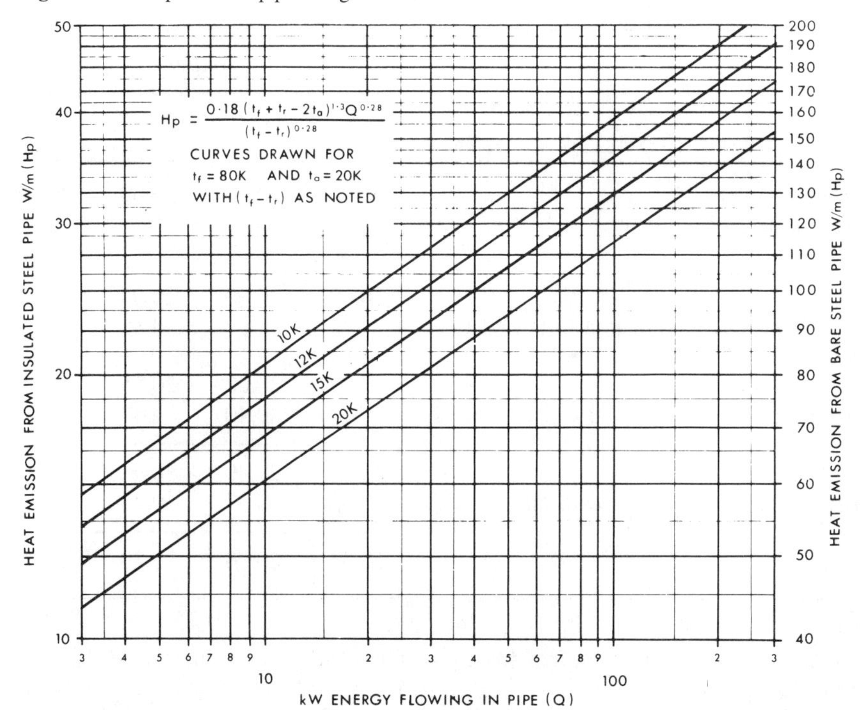

Figure 9.6 Piping heat emission related to energy flow

may be read from one or other of the vertical scales for a chosen temperature drop Δt, flow to return, against the energy flow rate shown on the bottom scale. Thus, for the system in Figure 9.5, a list may be made as follows for a 12 K temperature drop:

Pipe section key	*Net energy flow* (kW)	*Pipe heat emission* (W/m)	*Length* (m)	*Total* (kW)
AB	260	188	30	5.64
BC	180	178	20	3.56
CD	100	148	20	2.96
DE	20	92	10	0.92
DF, CK, BO	80	135	10	1.35
FG, KL, OP	60	125	10	1.25
GH, LM, PR	40	112	10	1.12
HJ, MN, RS	20	92	10	0.92

- The apportionment of these heat emissions, section by section, must now be carried out systematically on a 'compound interest' basis. A first step would be to consider the three identical main circuits on the right-hand side of the figure, taking the upper identifications as follows:

	Heater F	G	H	J
Rated emission (kW)	20.00	20.00	20.00	20.00
Heat from pipe HJ is allocated to J alone				0.92
	20.00	20.00	20.00	20.92
Heat from pipe GH is allocated to H and J			0.55	0.57
	20.00	20.00	20.55	21.49
Heat from pipe FG is allocated to G, H and J		0.40	0.41	0.44
	20.00	20.40	20.96	21.93
Heat from pipe DF is allocated to all	0.32	0.33	0.34	0.35
Final total	20.32	20.73	21.30	22.28

- Proceeding similarly, the heat loss from pipe CD is allocated to heaters E, F, G, H and J, that from pipe BC to heaters E, F, G, H, J, K, L, M and N, and lastly, that from pipe AB is allocated to all the heaters shown. This process produces the total loadings shown below and it will be seen that, although the *average* mains loss is just over 10%, that for heater J is 18.9% while that for heater O is only 3.5%.

Heater	*Gross load* (kW)	*Mains loss* (%)
O	20.71	3.5
P	21.15	5.7
R	21.74	8.7
S	22.73	13.6
K	21.09	5.4
L	21.53	7.1
M	22.12	10.6
N	23.14	15.7
E	22.35	11.7
F	21.69	8.4
G	22.14	10.7
H	22.76	13.8
J	23.79	18.9

The effect of the mains losses upon the heat emitters is that the full *system* temperature drop of 12 K will not be available at the individual flow and return connections to those heaters. Taking the two extreme examples:

*Heater O, temperature drops**

in flow main = $(12 \times 0.35)/20.71$ = 0.25 K

across heater = $(12 \times 20)/20.71$ = 11.5 K

in return main = $(12 \times 0.35)/20.71$ = 0.25 K

*Heater J, temperature drops**

in flow main = $(12 \times 1.89)/23.79$ = 0.95 K

across heater = $(12 \times 20)/23.79$ = 10.1 K

in return main = $(12 \times 1.89)/23.79$ = 0.95 K

The object of this calculation is to produce energy loadings (and hence water quantities) which will lead to the *mean* temperature at each heat emitter being the same throughout the system. Naturally, this small simple example does not show up the relative importance of mains loss as would be the case in an extensive system, but no doubt it will serve to show the principle.

It will be seen that, for a large installation involving a great many branches of different lengths and sub-circuits of varying size, the apportionment of mains losses, if pursued to an ultimate refinement, can be a very laborious process. Various methods have been devised to simplify this task, such as have appeared in previous editions of this book. By

* These are not strictly correct since the flow pipe is hotter than the return, and will thus have a slightly higher heat loss.

way of rough compromise, if the total mains loss of one circuit from the boiler be calculated and divided by the number of branches on that circuit, even if not of uniform load, some attempt at apportionment can generally be made by sight to achieve a percentage basis which is probably not far from reality.

However arrived at, the mains loss for each section is added to the emitter load of the branch, and these are added progressively back to the boiler, or to the headers if there are several main circuits.

- The loads which each section of main must carry are now available and the size can be judged from a starting basis of say 100 Pa/m unit pressure drop. The length of each section, flow plus return, plus single resistances can be set down and a table prepared, thus, for heater J in the example:

Section of pipe (1)	*Load carried* (kW/K) *(2)*	*Total length, (L + EL)* (m) *(3)*	*Pipe size* (mm) *(4)*	*Unit pressure loss* (Pa/m) *(5)*	*Section pressure loss,* (3) × (5) *(6)*
AB	23.85	41.1	80	140	5760
BC	16.70	23.6	80	70	1652
CD	9.37	21.3	65	58	1235
DF	7.52	11.1	50	120	1332
FG	5.72	11.2	50	70	784
GH	3.88	11.1	40	120	1332
HJ	1.98	14.0	32	60	840
Emitter	1.98	–	–	Catalogue	5000
				Total for circuit =	17 935 Pa
				Say, 18 kPa	

- Next come the other branches and sub-circuits. At each off-take from the index circuit there will be some surplus pressure available: this must be dissipated in the branch connections, otherwise short circuiting will occur.

Each branch taken in turn then becomes a fresh exercise to be retabulated as above, sizes being adjusted to absorb surplus pressure. There is of course a limit to the practicability of such adjustments since the range of commercial pipe sizes is not infinite and, because pressure loss varies with the square of the velocity, 'steps' between pipe sizes are quite large. For example, take an energy flow rate of 3 kW/K and note from Figure 9.1 that, whereas unit pressure loss in a 32 mm pipe is 140 Pa/m, it is *five times* greater in a 25 mm pipe.

- Having established the total pressure loss of the system based upon a number of initial assumptions, it is necessary to consider whether the result achieved has invalidated any of these to the extent where it might be necessary to repeat the calculation routine using corrected figures. Furthermore, the question arises as to whether, by selecting a different temperature drop or by varying the pump pressure, a more economical solution might have been produced.

Hydraulic gradients

Although, as noted above, it is rarely possible to select pipe sizes for intermediate circuits such that the pressure loss there is in exact balance with that of the index circuit, it is possible to reduce the disparity by careful manipulation of the *hydraulic gradient*. Figure 9.7 shows, at the top, a string of seven heat emitting elements served by a two-pipe arrangement. Below this are plots of three alternative hydraulic gradients against a base scale of pipe length and a vertical scale of pressure loss. The vertical lines at the bottom of the plot represent pressure loss through the actual heat emitters. It will be noted that the total pressure loss represented by each curve is the same.

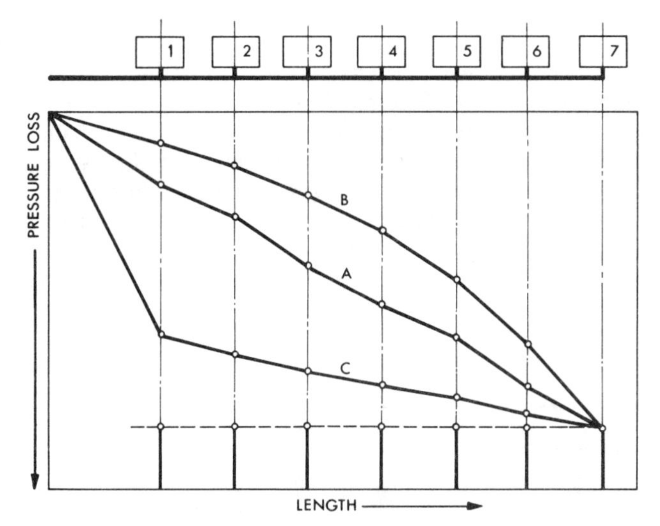

Figure 9.7 Hydraulic gradients

From the point of origin at top left, curve A is the profile of a pipe system chosen on the basis of constant pressure drop: it will be seen that a significant amount of valve regulation will be required to balance all the circuits. The concave-down curve B is a profile where unit pressure drop is low in the early sections of the system and becomes higher towards the index heat emitter: an even greater out of balance situation exists here. Curve C in contrast, which is concave-up, is the profile of a system which is all but self balancing.

The lesson to be learnt from this simple example is that, for a given *overall* pressure loss through a multi-section pipe run, neither constant unit pressure loss nor constant velocity are suitable design criteria: ideally, pipe sizes should be such that the velocity in the index circuit will *decrease* stage by stage. That said, many of the constraints which have been mentioned earlier in this chapter may militate against consistent adoption of this principle.

Reversed-return systems

It will be obvious that with this arrangement the loads to be carried by the piping are added progressively forwards for the return and backwards for the flow. Branches, however, are dealt with in the same way as for a two-pipe sytem. Although, in theory, there is no index circuit, the old bugbear of commercial pipe sizes confuses matters and it is sometimes found that an energy load in the *centre* of the string of circuits is disadvantaged. Moreover, a further word of caution is necessary since a condition can arise at an intermediate branch where a greater than average pressure loss has occurred in the flow sections and a less than average in the return sections. As a result, the pressure differential at the branch connections may even be negative! A careful check of pressures is required where loads or lengths of sections of pipe run vary greatly from their average.

Multiple zone systems

Previous reference has been made to those low temperature hot water systems which are fed from a high or medium temperature source (Figure 6.5, p. 146). The same principle may be applied in circumstances where it is necessary to provide individual temperature or other control to a number of circuits all fed from a single low temperature heat source as Figure 9.8. This application is mentioned here, in the context of pipe size selection, simply because the water quantity circulating in the ring main must be slightly more than the total of that handled by the zone-circuit pumps: a margin of the order of 8–10% has been found adequate.

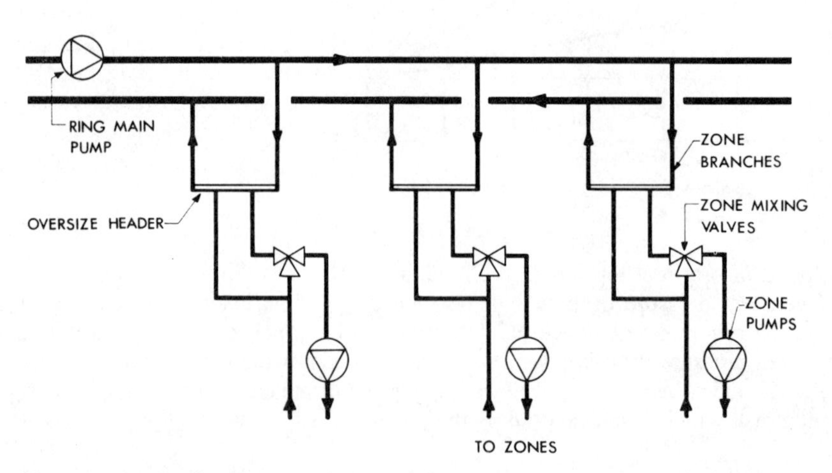

Figure 9.8 Header arrangement from ring main to multiple zones

The connections to the zone circuits are often taken from an oversize section of pipework fed from the ring main: some engineers will select a pipe one size greater than the zone connections and others prefer to use a standard size of, say, 100–150 mm in all cases. Neither practice is right (or wrong) since, were it possible to be absolutely sure that the water quantities circulating in both the zone circuits and the ring main loop were always in correct balance, there would be no need for any enlargement at all!

Embedded panel systems

The particular circumstances which are unique to this type of system simplify the routine of pipe size selection since:

- The design temperature drop is usually small with the result that the water quantity in circulation is correspondingly large.
- The pressure loss in the heat emitting elements (the pipe coils) is high relative to that in the main pipework.
- The pipe coils are, as a result of care in design, of much the same length and thus carry similar water quantities with similar pressure loss.

In consequence of these characteristics, such systems are very nearly self-balancing. In determining the volume which is to be handled by the circulating pump, note must be taken of both the upward and downward components of emission.

Water systems – footnotes

'Hydraulic' design

This ill-named concept, which may have originated from study of the self-balancing characteristics of the embedded panel system, is worth noting as an oddity if nothing else. The basis of the routine is that the system pipework of the index circuit alone is selected in the normal way and the pressure loss through it established for the required water quantity. Balance is achieved for other circuits not by adjustment of pipe size but by selection of a water quantity which will produce equal pressure loss. With this approach, the temperature drop across the various circuits will differ but the argument is that this, in terms of emission (i.e. the difference between mean water temperature and air temperature in the space to be heated) will be so small as to be irrelevant. Application of the concept means, of course, that the volume duty of the circulating pump will be greater than necessary, as will be the energy which it consumes. This approach is not recommended for normal usage.

Mechanised design

A totally refined solution to routine calculations directed at selection of pipe sizes would be a lengthy task using long-hand methods, having regard to the large number of interdependent variables. Traditional methods have followed principles and limitations which, in the main, have worked in practice. With the advent of the computer, new techniques have been developed to assimilate all the variables and to produce iterative solutions which would be too time consuming to deal with by any other means.

A number of computer programs are now available, some written by engineers, and experience suggests that a project of suitable size may be processed in one-quarter of the time and at one-third of the cost which would apply to an equivalent manual exercise. In this context, however, the limitations imposed by commercial pipe sizes and other imponderables apply equally to both computer and manual solutions. In the end, despite the means, circuits must be balanced one with the other by Fred's skill in adjustment of regulating valves!

Regulating valves

Reference has been made here, in several earlier paragraphs, to the virtual impossibility of achieving an exact balance between various branch circuits by selection of pipe sizes alone. In order to overcome this problem, branches are provided with regulating valves which may be adjusted to take up over-pressure. Traditionally, gate valves were used for this purpose but they, in common with most other commercially available angle radiator valves, were virtually useless for this application since the relation between control of

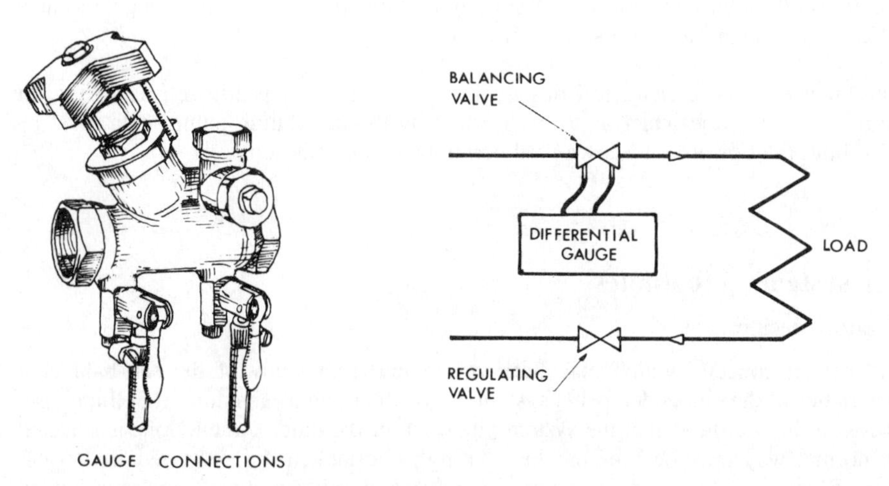

Figure 9.9 Special circuit balancing valve

water flow and spindle rotation was far too critical. Valves having a performance which is nearer to requirements are now available and, in addition, it is now becoming accepted practice to provide major branches with balancing valves of the type shown in Figure 9.9. These have connections to which a sensitive manometer or other test instrument may be fitted to provide data for comparison with curves or tables provided by the manufacturer to show pressure loss and water flow characteristics.

System pressure condition

The point at which the feed and expansion pipe (or the equivalent influence from a pressurisation unit) is connected to a piping system is of some significance as far as boiler and pump operation are concerned. This point of connection is the only position at which any external pressure is applied and is known as the *neutral point* of the system. With respect to the boiler or other energy source, this neutral point could be connected either to the flow or to the return but common sense dictates that it should be at the position where the water of expansion will be at the lower of the two temperatures, if this be practicable. As to position *vis à vis* the circulating pump, the alternatives are only that it should be on either the suction side or the delivery side.

Since, as will be obvious, the pressure at the neutral point can be no greater or less than that imposed by the external influence, it follows that connection to pump delivery will result in the whole of the piping system being under *less* than external pressure whereas

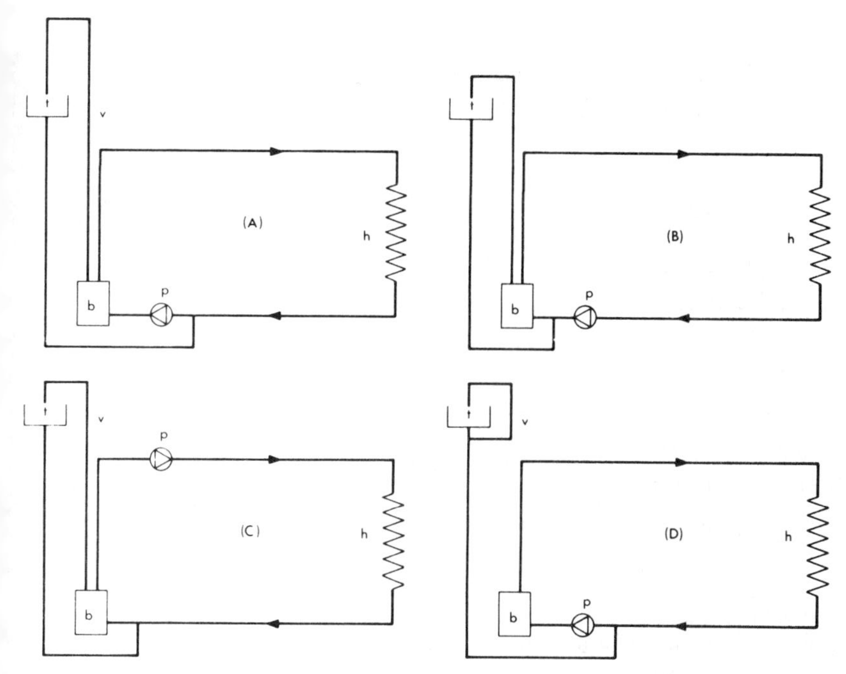

Figure 9.10 Alternative arrangements for cold water feed and expansion pipe

connection to pump suction will place the system under external pressure *plus* whatever residual pump pressure is available at any given point.

Four situations are shown in Figure 9.10:

- In case A, the vent pipe must be carried above the water level in the cistern, to a height exceeding the pressure exerted there by the pump, if water discharge is to be avoided in normal operation. With a high pump pressure this may not be possible.
- In case B, the vent pipe is not subject to pump pressure and no additional height is necessary but there is some tendency to draw air into the system at any air vent or other open high point.
- In case C, with the pump in the flow pipe, the vent and the feed and expansion pipe are in balance. Since virtually the whole system is under pump pressure, air release will present no problems.
- In case D, the feed and expansion pipe is combined with the vent. This is a solution shown to be very dangerous more than 50 years ago but still sometimes used.

Water systems – gravity circulation

A simple single-pipe system is shown in Figure 9.11 and, with a gravity circulation, the force creating water movement will be that due to the difference in mass between the positive column P_1 at temperature t_2 and the negative column N_1 at temperature t_1. The

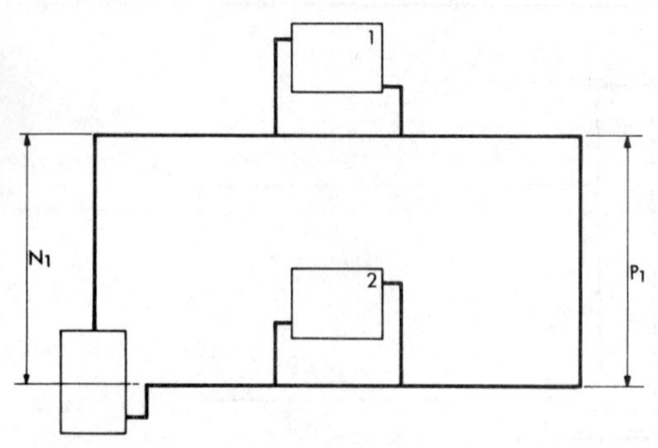

Figure 9.11 Simple example of single-pipe gravity circulation

force available will be that of unit mass, subject to acceleration due to gravity, which may be taken as $9.81\,m/s^2$, and thus:

$$CP = 9.81\,(p_2 - p_1)$$

where

CP = circulating pressure per m height (Pa/m)
p_2 = specific mass of water at temperature t_2 (kg/m^3)
p_1 = specific mass of water at temperature t_1 (kg/m^3)

As a matter of convenience, values of *circulating pressure* (*CP*) have been pre-calculated and are listed in Table 9.3 for a range of flow temperatures and temperature drops, flow to return.

Table 9.3 Circulating pressures for gravity hot water systems

Flow temperature (°C)	*Circulating pressure* (Pa) *per metre height, for the following temperature differences* (K) *flow to return*						
	10	*11*	*12*	*14*	*16*	*18*	*20*
60	47.4	56.1	64.6	72.8	80.7	88.4	95.7
65	50.3	59.7	68.8	77.6	86.2	94.6	103
70	53.1	63.1	72.8	82.3	91.5	101	109
75	55.9	66.4	76.7	86.8	96.6	106	116
80	58.5	69.6	80.5	91.1	102	112	122
85	61.1	72.7	84.1	95.3	106	117	128

It will be obvious that the water temperature at any point around the piping circuit may be determined knowing the ratio between the heat emitted between the boiler and the point in question and the heat emission overall. For example, if the output of radiator '1' and the upper horizontal pipework were 25% of the overall emission, and the temperatures at the boiler were 80°C flow and 60°C return ($\Delta t = 20$ K), then the temperature of the water at the top of the drop P would be $80 - (0.25 \times 20) = 75$°C. The circulating pressure available may be determined by use of this simple relationship.

Equilibrium conditions

For a given quantity of heat to be transmitted from the boiler to radiators '1' and '2', water must flow through the pipework, the mass depending upon the temperature drop. The higher the temperature drop, t_1 to t_2, the more heat will be carried by each kilogram of water in circulation. On the other hand, the greater the difference between t_1 and t_2, the greater the circulating pressure and, in consequence, the greater the mass flowing. In practice, with constant heat output from the boiler, these effects fall into equilibrium such that the temperatures t_1 and t_2 adjust themselves to produce precisely that circulating pressure necessary to maintain a mass of water in motion which is sufficient, in turn, to release the quantity of energy necessary to create the appropriate temperature difference between them.

Thermal centrelines

For the purpose of calculation, heights are measured from the *thermal centre line* of boilers and radiators, that is, the mid-point between the flow and return connections. For a system having more than one circuit, for example the two-pipe arrangement illustrated in Figure 9.12, it is necessary to calculate the circulating pressure for each individual circuit and to divide this by the appropriate total length (measured plus equivalent length) in order to obtain a unit of available pressure (*CP*/metre length) and thus determine the least favourably placed, or index circuit. This, when identified (for example that serving radiator '3' in the diagram), is used to select sizes for all common pipework, those for the remaining circuits then being chosen to absorb the residual of their individual values of *CP*/length.

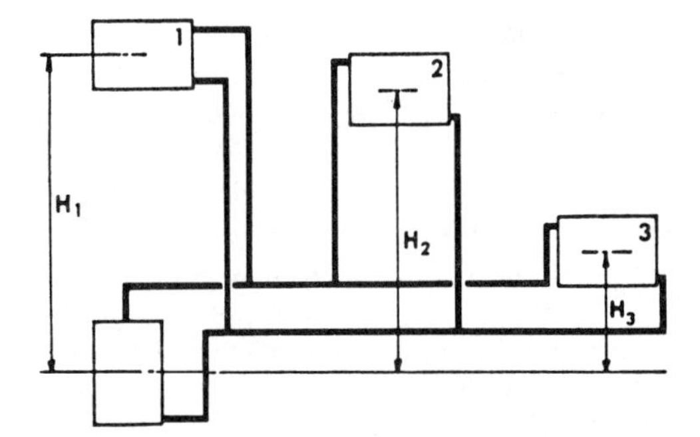

Figure 9.12 Simple example of two-pipe gravity circulation

Radiator connections

It is necessary, in conclusion, to mention a particular example of gravity circulation still in use to augment pump pressure. This occurs in the case of any radiator fed from a single-pipe system. The gravity circulation through the radiator is determined as though there were a boiler in the centre of the single pipe, Figure 9.13. The circulating pressure available may thus be calculated from the temperature difference across the radiator, which will be other than the system temperature drop, and this may then be added to the pressure differential between the two 'T' junctions resultant from any pumped circulation through the single pipe.

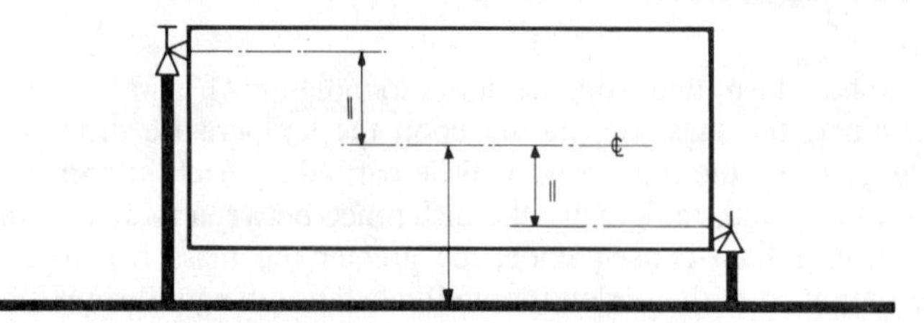

Figure 9.13 Connections to a radiator from a single-pipe circuit

These few notes are intended to do no more than provide a brief insight into the principles adopted to design gravity systems. Those readers who wish to have more knowledge of the genuine 'art' involved may consult earlier editions of this book or other works produced before 1950. The very fact that systems having connected loads of up to 3 MW were designed on this basis may come as a surprise when viewed in the context of pressure availabilities of not much more than 4 or 5 Pa/m in comparison with the 100–500 available today with pumped systems. It is no wonder that pipes of 200 mm diameter were required in boiler houses serving loads of this order and that the mass of calculations necessary was equally formidable!

Steam systems – principles

As will be obvious, no external motive force is required in any normal circumstances to produce steam flow since pressure is inherent in the state. An argument can be developed as to the practical unit to be used for steam mass, as distinct from that which best suits the purist approach. The most recent edition of the *Guide Section C4* has adopted kg/s in substitution for the far more useful unit of g/s listed in the 1970 edition. The latter equates to an energy supply unit of 2.25 kW and, moreover, permits working in whole numbers rather than with three or four decimal places.

While reasonable care must be taken in the calculations leading to selection of pipe sizes for steam service, it must be remembered that a number of imponderables exist in that heat losses from the pipework, whether insulated or not, will lead to formation of condensate. In consequence, the steam pipe will be carrying a mixture of vapour and liquid, the proportions of the two components varying from time to time dependent upon the external ambient air temperature and fluctuations in the rate of mass steam flow. The variables relating to flow of condensate are no less obscure since air and flash steam will accompany the boiling water in pipes which are often only half full and are under a changing mixture of pressure influences.

Calculation routines

For selection of steam pipe sizes, the following information must be available:

- The mass flow required, in g/s. For heating service this will equate to the energy requirement in watts divided by the latent heat at the pressure of utilisation. For kitchen or other equipment, the manufacturers' data must be consulted.
- Where the steam main pipework is comparatively short, heat loss from it may be ignored but, for extensive runs, use must be made of Tables 6.14 and 6.15 (p. 171) and the net requirement then adjusted to suit.

- The initial pressure available at the steam source.
- The minimum pressure required at the point of consumption.
- Lengths of the various pipeline sections, with details of all single resistances.

Selection by steam velocity

Main headers in boiler houses and other plant rooms, off-takes from them and actual connections to principal items of steam consuming plant are most conveniently sized on the basis of velocity. Volume flow may be calculated from the energy load using the data listed in Table 6.6 and, from that, velocity using the internal cross-sectional areas of pipes, in m^2, given in Table 9.4. Velocities which are generally recommended for use in various applications are quoted in Table 9.5.

Table 9.4 Cross-sectional areas of steel pipes (BS 1387: medium)

Pipe size (mm)	*Area* (m^2)	*Pipe size* (mm)	*Area* (m^2)	*Pipe size* (mm)	*Area* (m^2)
15	0.000 175	40	0.001 272	100	0.008 38
20	0.000 326	50	0.002 070	125	0.013 05
25	0.000 518	65	0.003 530	150	0.018 70
32	0.000 927	80	0.004 905		

Table 9.5 Conventional velocities for saturated steam

Position	*Velocity* (m/s)	*Position*	*Velocity* (m/s)
Boiler outlet connections (LP)	5–10	Trunk steam mains	30–50
Boiler outlet connections (HP)	15–20	Unit heater connections	25–30
Boiler headers (HP and LP)	20–30	Calorifier connections	25–30

Selection by pressure drop

Elsewhere in the piping system, selection of size is usually made on the basis of overall pressure drop. If no particular criteria were imposed and depending upon the extent of the distribution system, it is normal to allow for a pressure drop of betwen 5 and 10% of the initial pressure, subject to a limiting maximum velocity of 50 m/s. Since steam is compressible, both the specific mass and the viscosity change with pressure: as a result, presentation of data relating pressure drop to flow is more complex than is the case with water. Bearing in mind the many extraneous influences mentioned earlier, an approximate approach to the relationship is required and in the context of the subject matter of this book, one which will provide adequate accuracy for velocities between 5 and 50 m/s and pressures between about 100 and 1000 kPa.

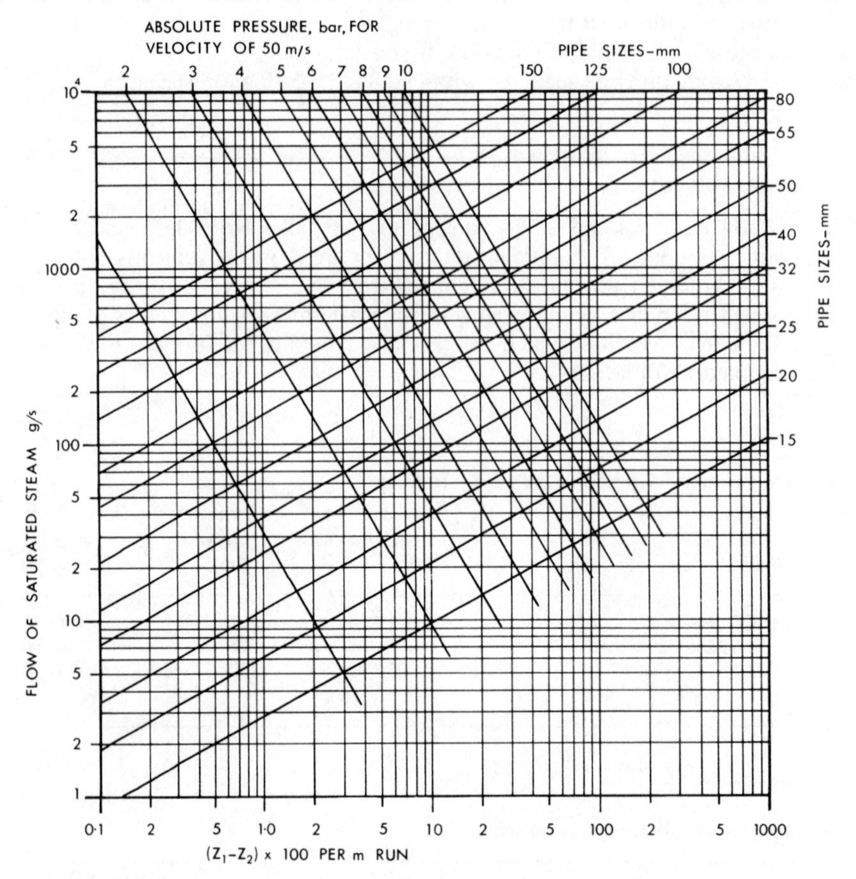

Figure 9.14 Sizing chart for flow of saturated steam in steel pipes (BS 1387: heavy)

A method which was developed originally by Rietschel* in 1922, and used in early editions of this book, fulfilled this specification but became slightly dated and, in any event referred to a different quality of pipe than that available in the British Isles. For the 1965 edition of the *Guide*, the basis for the expression was re-examined and a slightly modified version was used to produce the data included both there and in later editions including the current *Section C4.* This approach defines the relationship between steam flow and pressure loss as:

$$Z_1 - Z_2 = 3.648\ (M^{1.889} \times L)/(10^6 \times d^{5.027})$$

where

Z_1 = initial pressure factor = $P_1^{\ 1.929}$
Z_2 = final pressure factor = $P_2^{\ 1.929}$
P = steam pressure, absolute (kPa)

* Rietschel, H. and Brabbee, K., *Leitfaden der Heiz-und Lüftungstechnik,* Julius Springer, 1922, **2**, 28.

M = mass of steam flowing (kg/s)
L = length of pipe run (m)
d = diameter of pipe (mm)

Solutions may be presented in tabular or graphical form as in Figure 9.14 where the *difference* between the pressure factors Z_1 and Z_2 is plotted on the base scale against the mass flow of steam on the vertical scale. It should be noted that, to avoid decimal places and use whole numbers, mass flow in the figure is plotted in terms of g/s and the pressure factors listed in Table 9.6 are actually 100 Z. The values within the figure and table, nevertheless, have been adjusted to be compatible. An example best illustrates the method of use:

Initial steam pressure = 600 kPa
Pressure loss, say 10% = 60 kPa
Final steam pressure = 540 kPa

Energy to be transmitted = 1000 kW
Latent heat, final (Table 6.6) = 2099 kJ/kg
Steam flow thus is $10^6/2099$ = 476 g/s

Z for 600 kPa (Table 9.6) = 3170
Z for 540 kPa (Table 9.6) = 2587
Length including single resistances = 55 m

Thus

$$(Z_1 - Z_2)/L = (3170 - 2587)/55 = 10.6$$

Reading this last value on the base scale of Figure 9.14, against 476 g/s on the vertical scale, it will be seen that a 65 mm pipe will carry about 510 g/s. By reference to Tables 6.6 and 9.4, the velocity of flow will be $0.476 \times 0.349/0.003\,53 = 47$ m/s and if drainage of the main presents any problem, it would be better to select an 80 mm pipe since the 10% pressure drop assumed is near the top of the preferred range of velocities.

Table 9.6 Values of Z for use with Figure 9.14 (P = absolute pressure in kPa)

P	Z	P	Z	P	Z	P	Z
100	100	240	541	460	1899	740	4751
110	120	250	586	480	2061	760	5001
120	142	260	632	500	2230	780	5258
130	166	270	679	520	2405	800	5522
140	191	280	729	540	2587	820	5791
150	219	290	780	560	2775	840	6066
160	248	300	832	580	2969	860	6348
170	278	320	943	600	3170	880	6636
180	311	340	1060	620	3377	900	6930
190	345	360	1183	640	3590	920	7230
200	381	380	1313	660	3810	940	7536
210	418	400	1450	680	4036	960	7849
220	458	420	1593	700	4268	980	8167
230	499	440	1743	720	4506	1000	8492

Note: for convenience in using whole numbers, the values given are actually (100Z). Figure 9.14 has been adjusted similarly.

Condensate pipe sizes

Pipes carrying condensate fall into three classes as far as selection of size is concerned:

- Pump discharges following condensate collection in a receiver which is so vented to atmosphere that the majority of the air and flash steam has been removed. These may be considered as flowing full and may be sized using the water flow data of Figures 9.1 and 9.2, plus a 25% margin on the pressure drop read therefrom.
- Gravity flow at low pressure where extraneous means have provided that the majority of the air and flash steam has been discharged. A pressure drop three times that due to water flow alone may be anticipated.
- Gravity flow immediately following trap discharge where air and flash steam will be present in quantity. A drop in pressure ten times that due to water flow alone may be anticipated.

For these three very different situations, Table 9.7 provides values for condensate flow in g/s based upon a pressure drop of 40 Pa per m run of condensate main. It will be appreciated that these values are not the result of a deeply theoretical analysis: the same could be said, however, of the proven traditional approach which was to use condensate pipes one size smaller than the associated steam pipe! This somewhat intractable problem may be summarised by advocating that a well designed system of condensate collection will provide for flow by gravity, using consistent grading, to an adequate number of pump and receiver units. In no circumstances should a trap discharge be connected to the pump delivery pipe from such a unit.

Table 9.7 Approximate data for condensate flow in steel (BS 1387: heavy) and copper (BS 2871: Table X)

Piping nominal bore (mm)		*Flow of condensate* (g/s) *for listed conditions and pressure loss of 4 mm/m run* (40 Pa/m)					
		Liquid condensate without air or vapour		*Liquid condensate with some air and vapour*		*Mixture of liquid condensate with air and vapour*	
Steel	*Copper*	*Steel*	*Copper*	*Steel*	*Copper*	*Steel*	*Copper*
10	12	15	10	10	10	5	–
15	15	30	20	20	15	10	5
20	22	65	65	40	40	20	20
25	28	120	135	75	85	35	40
32	35	260	240	170	155	80	75
40	42	400	410	260	265	125	125
50	54	765	835	500	540	240	255
65	67	1560	1500	1030	975	500	460
80	76	2420	2130	1600	1380	780	655
100	108	4920	5580	3260	3640	1590	1740

Provision for thermal expansion

While the coefficients of linear expansion for steel and copper are low in comparison with those for other metals and only about a tenth of those for pipeline plastics, this does not mean that their effects may be ignored. Table 9.8 shows the extent of the growth in length which arises, relative to temperature increase from an assumed datum of 20°C. Provision to accommodate these changes must be made in the physical layout of systems and, as is obvious, the seriousness of the matter increases for those which operate at higher temperatures.

Table 9.8 Expansion of pipework for temperature difference (K)

Temperature difference (K)	*Expansion* (mm/m) *for the following materials*		*Temperature difference* (K)	*Expansion* (mm/m) *for the following materials*	
	Steel	*Copper*		*Steel*	*Copper*
50	0.567	0.846	110	1.247	1.861
60	0.680	1.105	120	1.361	2.030
70	0.794	1.184	130	1.474	2.200
80	0.907	1.354	140	1.588	2.369
90	1.021	1.523	150	1.701	2.538
100	1.134	1.692	160	1.814	2.707

In all instances, short rigid connections between the various system components, boilers, pumps, etc., should be avoided since situations may arise where a temperature differential exists between piping and component, positive or negative as the case may be. Dependent upon the pipework configuration, the shape change of such a connection may impose a torsional movement upon materials not best suited to the resulting stress. Given a fair length and plenty of bends, a pipework system can often accommodate itself to cope with some movement without recourse to specialist measures.

Expansion joints and loops

Pipework expansion may be provided for in a number of ways:

- By length and changes in direction as commented upon above. An example is provided in Figure 9.15(a) and it will be noted that certain points are designated as anchors or guides.
- By purpose made *horse-shoe* or *lyre* expansion loops set into the pipework in lieu of the fabricated rectangular shape shown in Figure 9.15(b).
- By *bellows* inserts which may be manufactured either in heavy pipe quality material or in a laminated stainless steel which is quite thin in section, Figure 9.15(c).
- By sliding *telescopic* type joints, Figure 9.15(d), which rely upon a packed gland for integrity and are restrained by bolts to prevent separation of the parts under pressure.

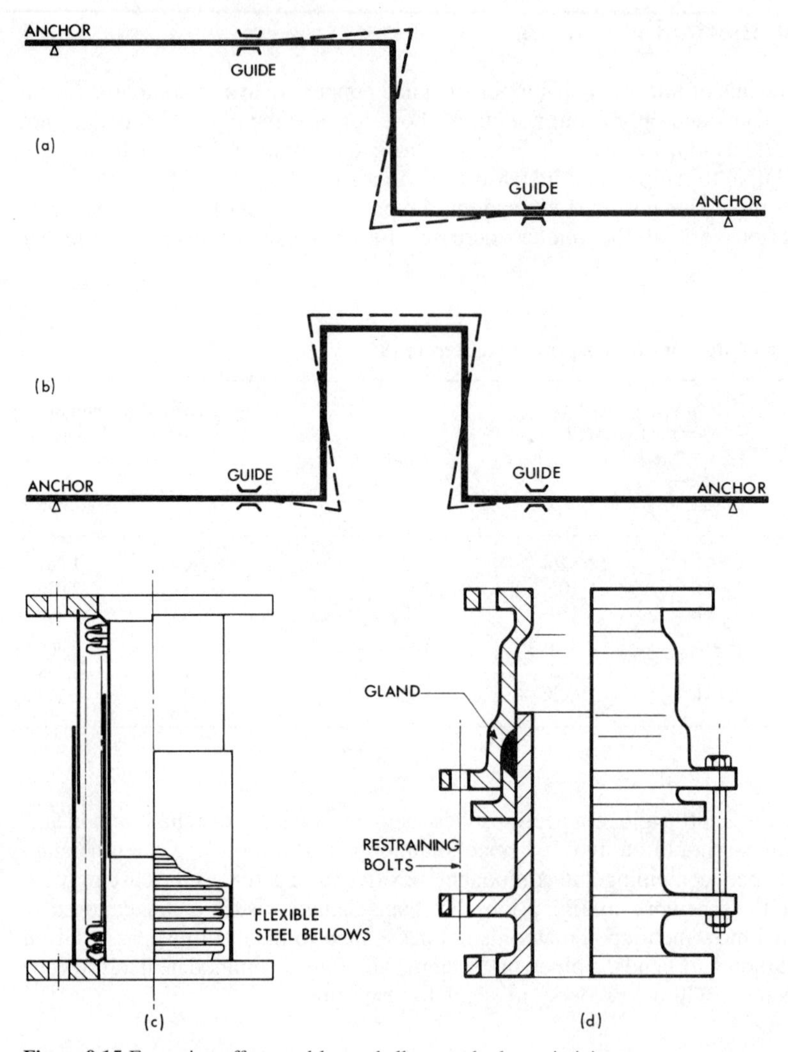

Figure 9.15 Expansion offsets and loops: bellows and telescopic joints

Of these four methods, the last is no longer in common use since it could accommodate only a completely axial movement and, in any case was prone to persistent leakage. Similarly, the simple bellows is now rarely used since, when internal pressure is applied, the convolutions attempt to open out and thus apply a considerable unnecessary force upon the anchors. Horse-shoe and lyre loops are much used in steam process and chemical engineering but, perhaps because they may be made up on a building site from available materials to suit particular dimensions, rectangular loops and offsets are the most commonly used arrangement within the context of this chapter. It must be added, however, that the flexibility of these arrangements is greatly assisted when bellows of the *articulated* type are introduced into the offset or loop, as Figure 9.16, these acting then in flexure rather than in compression.

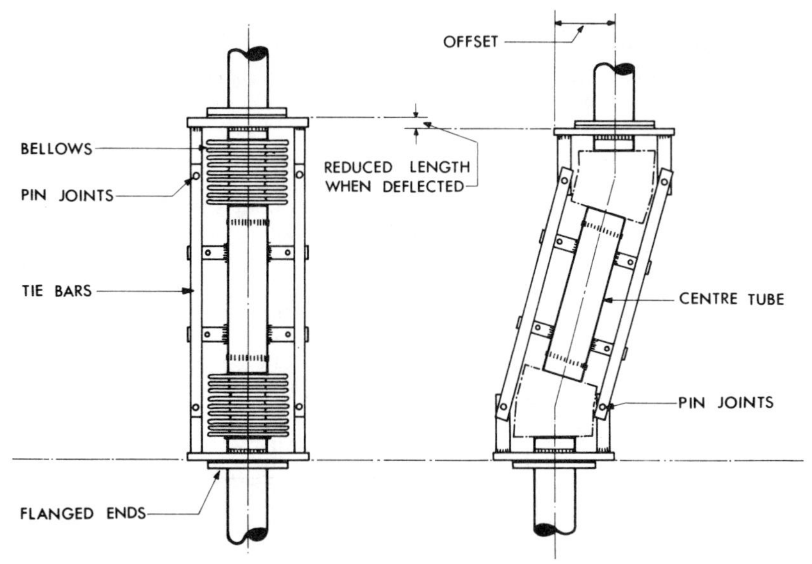

Figure 9.16 Articulated bellows and their application (Engineering Appliances)

Cold draw

It is usual, in making calculations for dealing with the results of thermal expansion, to take account of what is called *cold draw.* In effect, this represents action taken during erection whereby the loop or offset is stretched to take up a proportion (often half) of the anticipated compression in service. By this expedient, the force on the anchor points is reduced *pro rata.*

Offsets and loops

The *Guide Section B16* presents a collection of data which enable the dimensions of offsets and rectangular loops to be evaluated and the thrust upon anchors determined. The following five salient points, not in order of merit, require greater emphasis:

- Contrary to previous usage, it is the distance between any *guides*, as distinct from *anchors*, in the pipework which is the critical dimension.

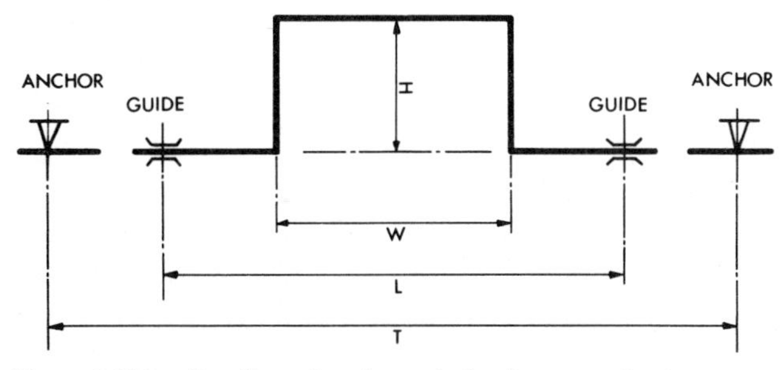

Figure 9.17 Leading dimensions for analysis of an expansion loop

Table 9.9 Fabricated expansion loops (see Figure 9.17 for symbols)

	Pipework dimensions (m)				*Total thrust on anchors* (kN)		
Nominal bore (K)	*Pipe run*		*Loop*		*System temperature* (°C)		
	T	*L*	*W*	*H*	*80*	*110*	*140*
25	21	4	1.00	0.66	0.34	0.49	0.63
32	24	5	1.25	0.71	0.50	0.71	0.91
40	27	6	1.50	0.75	0.72	1.01	1.29
50	30	8	1.75	0.88	1.02	1.41	1.80
65	34	11	2.00	1.10	1.54	2.10	2.65
80	38	14	2.25	1.16	2.16	2.90	3.65
100	43	19	2.50	1.58	3.15	4.16	5.16
125	48	25	2.75	1.79	4.54	5.87	7.20
150	52	30	3.00	2.14	6.03	7.69	9.36

- A *symmetrical* offset or loop, between guides, produces the least thrust at the anchors.
- Anchor points in long runs of pipework may be subjected to equal thrust from each direction and thus to a balanced force of compression.
- Least thrust is imposed upon anchor points when the height of a rectangular loop is both twice its width and equal to the distance between the associated guides.
- When loops are fabricated, short radius elbows should be used rather than wide bends. The latter offer stiffness, and consequent increased thrust on anchor points, in their tendency to assume an oval cross-section.

The leading dimensions necessary to initiate calculation of the forces and moments acting as a result of thermal expansion are identified in Figure 9.17 and, from these, results such as those listed in Table 9.9 may be produced. These latter are not, it must be emphasised, recommended values but no more than a collection made to illustrate the order of the numbers which may be expected. In practice, it is usual to use a computer to undertake what is otherwise a laborious and time consuming exercise.

Chapter 10

Boilers and firing equipment

In the two subsequent chapters, the properties of available fuels or energy sources and the chemistry of combustion will be examined. As an introduction to this subject, however, we now turn to the plant items employed in practice to burn those fuels and to transfer the energy so liberated to a heating medium. Features which are particular to boilers designed to serve domestic hot water systems are described elsewhere (Chapter 20, p. 555).

Basic considerations

Boiler power

The load imposed upon a boiler plant is derived from knowledge of not only the sum of the heat losses calculated for the individual spaces under basic design-temperature conditions, but also of any interaction or diversity between the components thereof. As has been mentioned earlier, the total of the heat losses may well exceed the actual peak demand due to the fact that infiltration air entering rooms on one side of the building may leave via rooms on the other side. As a result, an adjustment is necessary in order to avoid making allowance twice for the same air-change. In addition, where heating is continuous, it is known that further diversities exist and use of the following correction factors has been proposed:

Single space	1.0
Single building or zone, controlled centrally	0.9
Single building or zone, controlled per room	0.8
Group of buildings with similar pattern of use	0.8

The appropriate corrections having been made and the peak load thus estimated, an arbitrary allowance or a calculated estimate must be added to cover unwanted heat losses from distribution pipework. It is then necessary to consider by how much the boiler rating should exceed this corrected total in order to make sure that the design temperatures within the various heated spaces may be reached in a reasonable time. When heating is continuous, of course, the excess will be a minimum; the more intermittent the usage, the greater the excess capacity required.

There is a further case for some margin of capacity, where a boiler plant is thermostatically controlled, in order to provide what might be called *acceleration* – the ability of the plant to surmount the load under peak conditions while remaining still under control. In addition, it is wise to remember that boiler plants are generally not well maintained. The result is that, especially where solid fuel is used, output is likely to decline from the rated level due to irrational attention to combustion equipment and consequent fouling of heat transfer surface by soot or other deposits.

To some extent the various corrections cancel out and in the past these refinements in calculation were often ignored, a blanket margin of 25% being added to the total of heat losses to deal with such adjustments, to make allowance for distribution heat losses and to cover a multitude of other sins. More recent thinking has concluded that much smaller allowances, of the order of 10–15%, are adequate but there is no proper alternative to a painstaking analysis of the various aspects and the addition then of no more than a notional 5% or so to the calculated total. By such means, the designer will know just what allowance he has made for what eventuality.

By way of example, to illustrate the associated problems, if the total heat loss for a building, after adjustment for diversity, etc., were 490 kW and a calculated addition of 25 kW were made for heat losses from piping outside the occupied space, then a net boiler capacity of 515 kW would be needed. Adding an arbitrary margin of 10% to that figure would produce a gross requirement of about 570 kW. Boilers, as will be appreciated, are not purpose made to size and a side issue might be that for physical or other reasons one particular make was preferred. The catalogue range then examined would perhaps offer a choice between output ratings of 550 or 600 kW and since these would, respectively, represent margins of either 6.8 or 16.5%, the former would probably be chosen as being the more suitable.

In days when hand-stoking with coke was the norm, a generously oversized boiler with a large firebox would allow longer periods of running without attention on one charge of fuel. With automated firing however, and in particular with oil fuel or gas, it is a positive penalty on running cost to select a single boiler with too great a margin. The consequent extended periods of firing at low load will lead to frequent cycling of the firing mechanism and a reduced annual average combustion efficiency will result, to say nothing of smoke and other nuisances.

Boiler efficiency

Heat *input* to boilers of all types must not be confused with their rated *output* although it is rare for a manufacturer to quote other than an output rating. *Input* is measured in a rational manner best suited to the fuel: coal is weighed and the volume of liquid or gaseous fuel is metered. Detailed procedures for input tests are set out in various British Standards, the duration usually being a 6-hour test period with pre- and post-control periods. From such a test and taking account of an associated analysis of the fuel, a heat balance may be struck. The complete routine is extended but the final summary may be brief, for example:

		(%)
Heat content of fuel (net calorific value)		100.0
Loss of sensible heat in flue gases	11.4	
Loss due to unburnt carbon monoxide	1.9	
Combustible matter in ash or clinker	1.4	
Radiation losses (calculated)	4.2	18.9
Overall thermal efficiency		81.1

Heat *output* in the case of hot water boilers is troublesome and costly to measure on site since simultaneous integration of both water quantity and temperature difference, inlet to outlet, is needed over an extended period of steady state conditions. In contrast, for steam boilers, the water pumped in and then evaporated may be measured easily and conveniently over the test period using a simple turbine-type water meter. Thence, from the inlet water temperature and the outlet steam pressure, the total heat added per unit quantity of water may be calculated, which represents the output.

Quoted figures for works test efficiencies for most modern boilers range between 85 and 90% and in the particular case of condensing boilers, referred to later, may be somewhat higher. In daily use, however, efficiencies may be expected to average at least 5 and more probably 10% below those recorded under controlled conditions on a test bed.

Criteria for boiler selection

Hot water boilers are normally rated in kW and some manufacturers of steam boilers use this same basis in parallel with the more conventional mass approach of *kg/s from and at 100*°C. This alternative rating denotes the output which would be achieved if steam were being generated at atmospheric pressure from boiling water, i.e. it assumes that latent heat only is added. Such a rating is of little practical use since the feed water supplied is seldom at 100°C and the pressure is usually above atmospheric level. For any set of conditions other than those listed, the actual performance may be calculated from:

$$E_a = 2258\,(E_r)/(H_t - H_w)$$

where

E_a = actual evaporation (kg/s)
E_r = rated evaporation (kg/s)
H_t = total heat at required pressure (kJ/kg)
H_w = heat in water at feed temperature (kJ/kg)

Ratings fall broadly into the following three ranges:

Small 10–50 kW (mainly domestic)
Medium 50–500 kW
Large 500 kW and above

Boiler margins

In any sizeable installation of over say 300 kW, there is a case for providing more than one boiler. If there are two boilers then one will suffice in mild weather, working near to its full output which is advantageous in terms of efficiency and avoidance of corrosion. The second boiler is then brought into use during cold weather and can act as a standby for a large part of the year against breakdown or outage for maintenance. In the past it was routine to select each of two boilers to have two-thirds of the total required capacity but more modern practice, with which the authors do not wholly agree, suggests that each should meet no more than half of the total requirement. For larger installations still, three, four or more boilers may be used, giving greater flexibility still since several boilers which each have a small margin may then give almost a complete one-boiler standby.

Selection of the size of individual units for a multi-boiler plant, nonetheless, is a matter for compromise and cannot be determined wholly by a statistical approach. On the one hand, limitation of the number of spare parts to be stocked suggests the use of equally sized units, e.g. three boilers at 167 kW to meet a gross total of 500 kW. Conversely, to hold maximum combustion efficiency while meeting a load which varies with the external temperature, a case might be made for selecting units of unequal size, e.g. for the same gross total of 500 kW, one boiler at 100 and two at 200 kW which would provide five steps in output instead of three.

The first of these two loading arrangements would require no more than a simple sequence controller, manually reset at intervals to change the order of firing (1–2–3, 2–3–1, 3–1–2) in order to avoid undue use of any one boiler. The second example, however, would need a very much more complex sensing system with automatic though limited potential (and this beyond the ability of an operative to monitor) for change of sequence. The end result might be that the owning cost of the electronics would outweigh any saving resulting from better annual combustion efficiency!

Packaged modules

The modular approach to the provision of boiler power differs from the conventional in that it represents the provision of an array of small boiler-burner units, each of which has a rating in the range of 40–70 kW dependent upon the maker. The units are assembled with their water ways connected in parallel, as Figure 10.1, so that, for instance, an array of ten 40 kW units would substitute for a single large boiler rated at 400 kW. The principal gain arising from such an arrangement is in flexibility to meet variable loads at high efficiency. For the example quoted, if only one of the ten units were to be fired then the *turn-down ratio* would be 10:1 at the same efficiency as if all ten units were in use.

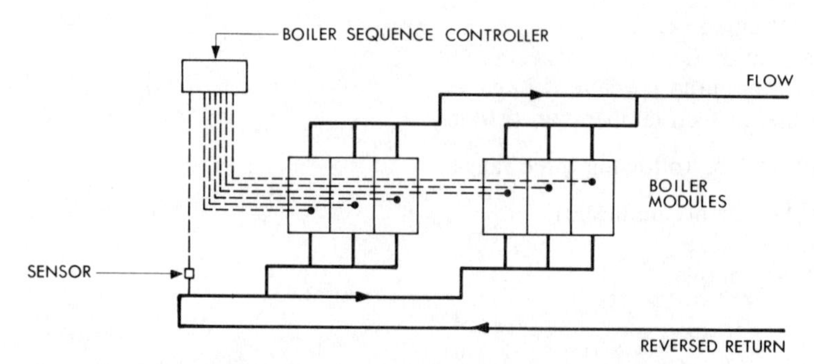

Figure 10.1 Typical pipework arrangements for modular boilers

Essential to the modular concept is the in-built system of control which, via a single monitor, will arrange for the optimum number of modules to be fired to meet the imposed thermal load. Facilities are provided in the control arrangements to allow changes in the order in which the modules are fired. Care must be taken in the design of the associated water circuits with particular reference to possible interaction between the boiler load controller and any diverting or mixing valves fitted there: some means to maintain constant water flow is required.

The advantages claimed for the concept, which is, as will be noted, no more than a works designed package echoing the *ad hoc* arrangement for load matching 'steps' described under the first heading here, include the low thermal capacity of the boiler modules and the consequent quick response to both firing and control.

Thermal storage

An alternative approach, which is in some respects diametrically opposite to that noted above, takes account of not only the cold weather peak load but also the extent by which this will be reduced on even the most severe winter day by lighting, solar radiation, occupancy and other heat sources. As may have been noted from Chapter 5, such aspects are taken into account as a matter of course when electrical off-peak systems are designed.

In essence, the heat storage method which is noted here consists of a package made up from a well insulated water vessel interposed between the boiler unit and the distribution system. The boiler charges the store and maintains the water temperature in it, quite irrespective of the concurrent output of the system. By this means, it is claimed that the frequency of the on/off cycling of the firing arrangements is greatly reduced, particularly under the less than full load conditions which persist for most of the year, thus increasing annual efficiency. It is further claimed that as a result of adoption of such an arrangement, the capacity of the boiler plant may be reduced. Figure 10.2 shows the arrangement of such a system.

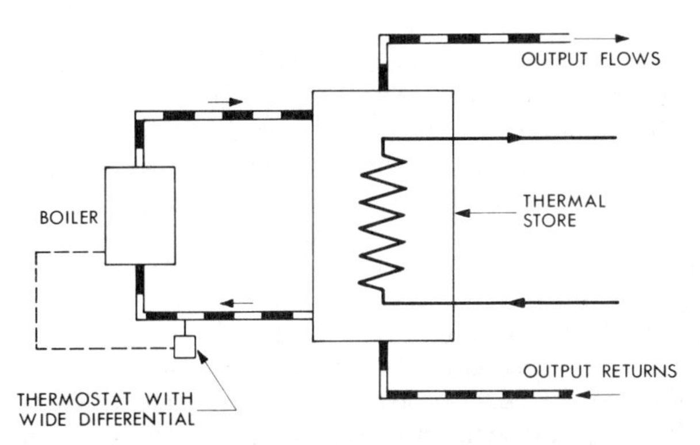

Figure 10.2 Pipework arrangements for heat storage

Certainly, on a domestic scale, this approach has proved effective and further reference will be found in Chapter 20 to the situation where a so-called *combination boiler* is used and both heating and domestic hot water supply systems are served from it. Field trials have been carried out with a larger installation but no definitive guidance as to plant or vessel sizing is yet available. Obviously, the routines developed to suit the off-peak storage systems described earlier would not be appropriate.

As a word of warning, however, it is worth remembering, when considering this or any other design hypothesis (however promising in theory) which relies for success upon a reduction in plant potential, that building occupiers are inclined to deplore ingenuity when full capacity is unavailable on a cold winter morning.

Condensing

An explanation will be given later, Chapter 12 (p. 313), why fears exist that a corrosion problem may arise at boiler heat exchange surfaces when flue gases containing water vapour are allowed to cool below about 250°C when in contact with them. It has been

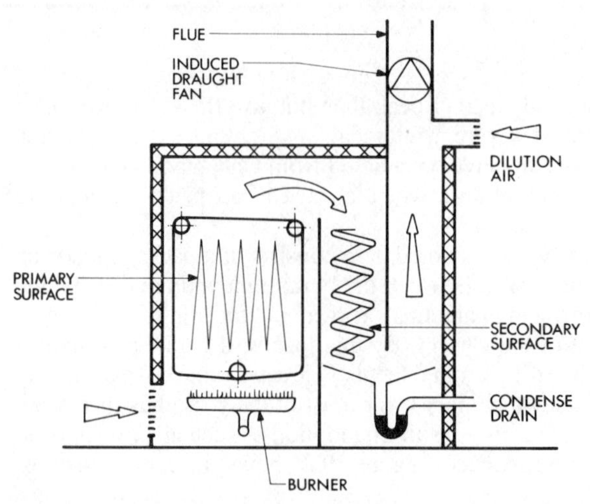

Figure 10.3 Principle of condensing boilers

appreciated, nonetheless, that a very considerable quantity of heat is lost as a result, in particular that within water vapour which has not been condensed (i.e. the difference between the gross and the net calorific values of the fuel, see Chapter 12). Over the years, various items of *add-on* equipment have reached the market in an attempt to reclaim this wastage but none has seemed to endure. The advent of fuels having negligible or no sulphur content has led to further research.

First developed in the domestic range of sizes but now increasingly available for commercial use with hot water systems, the principle underlying the design of this type of boiler is the introduction of a second heat exchanger, of stainless steel or a protected material, through which the flue gases pass after leaving the boiler proper. As a result of the additional resistance to gas flow so caused, it is usually necessary that forced or induced draught be provided. The outline arrangement of such a boiler, now made for loads between 12 and 800 kW, is illustrated in Figure 10.3.

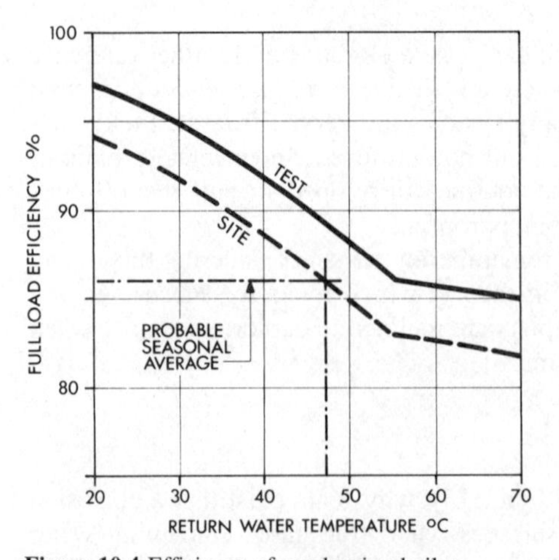

Figure 10.4 Efficiency of condensing boilers

The full potential of the *condensing boiler* is, of course, available only when the flue gases are reduced in temperature to that of the ingoing combustion air, ideally about 15°C. In practice, the factor which determines performance is the temperature at which water returns to the boiler from the associated system since the heat exchanger surfaces upon which condensation takes place are normally about 5 K above this. The practical order of efficiencies which may be obtained, with various return water temperatures, is shown in Figure 10.4.

Balanced flue

In the sense that a balanced flue relates to the provision of combustion air to a *boiler house*, this subject is dealt with in Chapter 12. However, the market now provides for individual boiler variants in the output range of 12–60 kW which incorporate integral

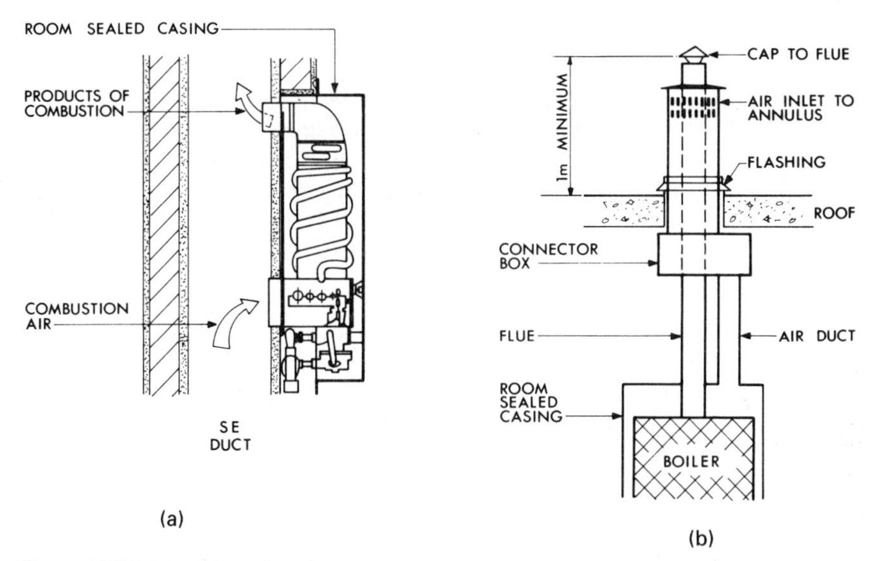

Figure 10.5 Balanced flues for boilers: (a) horizontal; (b) vertical

purpose-designed balanced flues, as shown in Figure 10.5. It must be noted that Building Regulations require that a relaxation be sought in instances where it is proposed to install a horizontal flue discharge.

Hot water supply

It sometimes happens that it is convenient to provide service to a hot water supply system, during the heating season, by use of a combined boiler plant serving a heat exchanger or calorifier. Since demand for hot water is in most circumstances spasmodic, it is usually possible to provide that service by 'robbing' the heating system for short periods. In other words, when selecting the boiler power for such a combined system, it may not be necessary to add the whole or even a part of the heat exchanger rating to that of the heating system. This approach is not valid, of course, where a very large hot water load occurs in a building having a small heating load, e.g., for a large kitchen or laundry, etc. (although the converse might possibly apply in these circumstances!)

In summer, when the heating plant is not in use, service to the heat exchanger might still be dealt with in the case of a single boiler if this were not disporportionately large. Similarly, where a multi-boiler plant is envisaged, it could be that the selection overall would offer the opportunity to include one boiler which was small enough to be kept in use during the summer.

Boiler houses – size and location

Various attempts have been made to provide guidance for provision, in a new building, of boiler and other plant rooms of adequate size and suitable location. In general terms, it is easier to suggest that the site should be at the thermal centre of the load than to define the size in either area or length × width.

Size

It is often the case that information in this respect is requested long before adequate details of the building structure are available with the result that the required boiler capacity cannot be determined to greater accuracy than plus or minus 25% or more. Furthermore, there are so many types of boiler and kinds of arrangement that advice at this stage of building development cannot be more than in very general terms. Some boilers are long and narrow; some are short and tall; some are sectional whereas others are factory packaged with the result that, in that instance, availability of access for delivery may be a critical factor.

As good a rule of thumb as any is that a boiler house should have an area of never less than 35 m^2 plus a further 35 m^2 per MW of boiler capacity. The aspect ratio should not exceed 2½ to 1 and additions must be made for fuel storage, etc.: height should never be less than 4 m. Any lesser dimensions may present an escape hazard and fall foul of a local authority for that reason alone.

Location

Past practice was to house the boiler plant in a basement, under the stairs in a position useless for any other purposes, a relic of the days when coke was delivered by horse and cart. Construction below ground is relatively expensive and accommodation through the building, to roof level, must be found for a flue. In many ways preferable from the point of view of operatives and such maintenance staff as may be available, a purpose-built detached plant house may be sited at ground level, particularly where several buildings are served from a central plant.

Alternatively, where planning permission and practicality permit, there is sometimes a case for siting a boiler house on the roof. With mechanical draught, the flue may be quite short and both ventilation and access are easy. Roof-top boiler houses are really only sensible when firing is by oil or gas, although prototype installations exist where small coal has been delivered to roof level plant by vehicles equipped with pneumatic blowers. It is, perhaps, cynical to suspect that this last application is no more than an attempt to 'keep up with the Joneses' and ignores problems of ash and clinker disposal, etc. Oil is pumped up to a daily service tank, as shown in Figure 11.6, but design considerations in all instances must include provision against any noise or vibration which might, via the structure, affect occupants.

Types of boiler

Cast iron

By far the most familiar type of heating boiler is the cast iron sectional pattern which, in one form or another, has been in use for more than a century. Originally designed to burn coke, by hand stoking on to a set of fixed firebars, it has been developed and refined for use with solid fuel, automatically fired, and for application of oil or gas firing at a much improved efficiency. Sizes range from the smallest up to 3 MW.

Generally, the boiler construction consists of individual sections connected together by use of *machined nipples*, usually three in number, all pulled together and held water-tight by steel tie-bars externally, Some of the larger sizes have half-sections, with multiple nipples but pulled together as before. Front and back sections differ from those in the centre as they make provision for firing and cleaning at the front and for a flue outlet at the back. Intermediate sections may be added for extensions or as replacements on failure.

Figure 10.6 shows a type of cast iron boiler which has been designed especially for oil or gas firing and in which the flue passages have been arranged for higher gas velocities and hence greater output for a given physical size. This type has a *water-way* bottom instead of being open and hence the structural base, in brick or concrete, on which it stands is subject only to relatively low temperatures, thus avoiding the complexities of floor insulation.

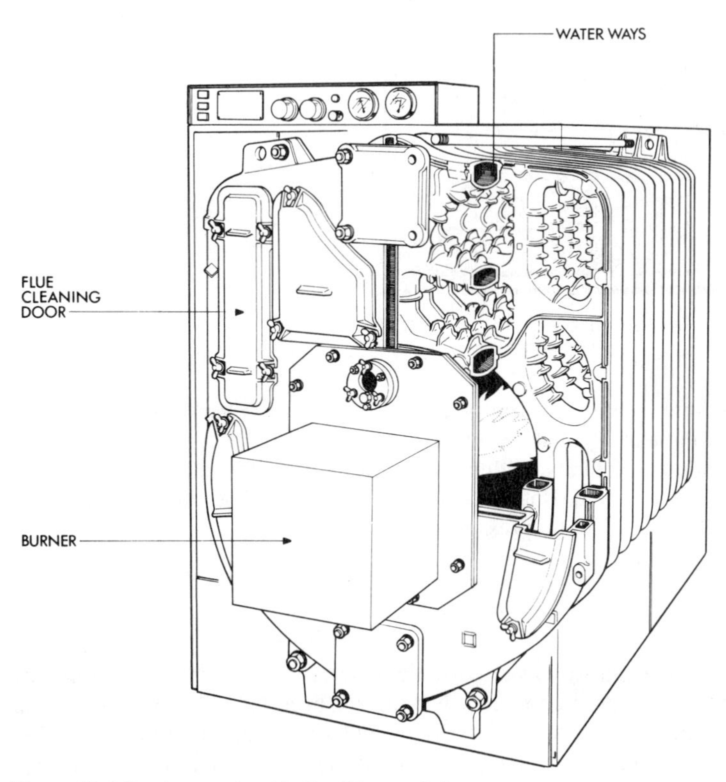

Figure 10.6 Cast iron sectional boiler (Hamworthy)

For use with atmospheric gas burners, as described later, a special type of cast iron boiler is produced, as illustrated in Figure 10.7 and, in this case, the sections are built up side by side. The outer surfaces of each casting, facing on to the internal flue-way, have integral nodules or studs cast on to them to provide additional surface for heat transfer. Sizes range from 20 kW to 1 MW per boiler unit and combustion efficiency may be about 80%.

The normal range of cast iron boilers is designed for operation under water pressures of up to about 400 kPa (40 m head of water) which covers most building sites except in high rise city centre developments. Certain manufacturers, however, do produce higher rated boilers using a spheroidal graphite grade of cast iron which will withstand water pressures of up to 1 MPa (100 m head of water).

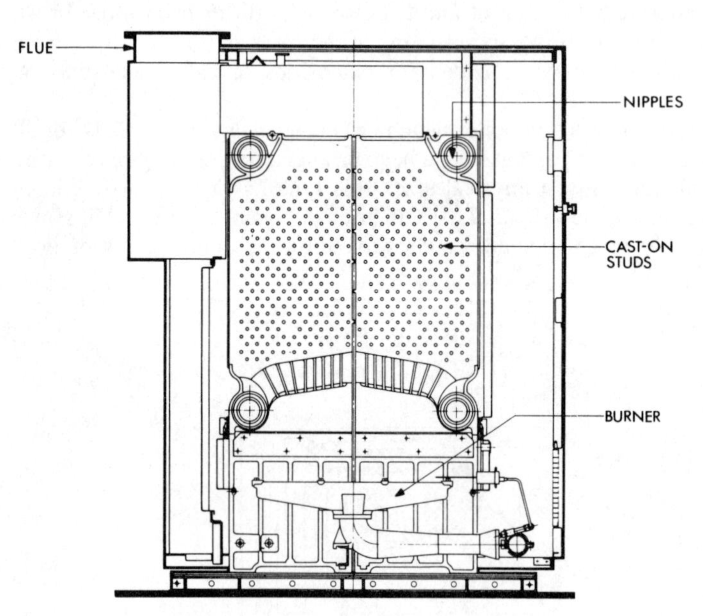

Figure 10.7 Specialist gas boiler (Broag)

To supply low pressure steam, cast iron sectional boilers were very commonly used during the era when this heating medium was popular. The earlier types had a limited steam space made available actually within the sections but later patterns mounted a horizontal steam drum above them. Ratings available were up to 750 kW for working at not much more than 200 kPa absolute pressure (2 bar).

Steel – sectional

The place once held by wrought iron as a favourite material for boiler construction, due to its ductility and ability to resist corrosion, has largely been taken by mild steel as a result of the disappearance of the manual skills in *puddling*. Steel is homogeneous in structure, in contrast to the more laminar nature of wrought iron, and is very liable to attack by sulphurous corrosion products.

For installations of modest size and too small to require the capacity of a simple shell boiler of *Cornish* or similar type, the majority of early heating boilers were constructed of wrought iron in semi-circular saddle section and set in brickwork. Subsequently, with the introduction and success of cast-iron sectional designs, patterns of fabricated steel boilers were produced which echoed the sectional form but offered advantages in their ability to withstand higher pressures.

Since the formation of convolution to provide additional heating surface is difficult in welded steel-plate sections, the physical size of a mild steel boiler of sectional type is generally greater than the equivalent rating in cast iron. For service to hot water systems, mild steel sectional boilers may be obtained in a variety of forms to cover a wide range of duties from about 30 kW to 1.5 MW.

Several patterns of steel sectional boiler were at one time made complete with a horizontal steam drum, as for the cast iron equivalents noted previously. These were often used to supply low pressure steam to kitchens and the like. Ratings up to about 1 MW were available for an absolute working pressure of about 300 kPa (3 bar). It would seem that such equipment is no longer made.

Steel – reverse flow

For long life, a mild steel water boiler requires that the system which it serves be designed to operate at a temperature outside the region where severe corrosive attack may be anticipated. Given these conditions, steel has advantages over cast iron in that it is more versatile: the more modern developments of the last 30 years demonstrate this versatility as shown in Figure 10.8.

In this type, a pressurised combustion chamber is provided in the form of a welded cylinder having a blind rear end. In consequence, the burner discharge is reversed to provide, *in counterflow*, a second pass of flue gases within the chamber. This reversal causes considerable turbulence in the flame zone before the gases enter the third and final pass to a rear smoke hood through secondary surface arranged as a circumferential ring of fire tubes around the combustion chamber. The front access door has a double skin and is water cooled. A range of ratings from 100 kW to 3.5 MW is available for operation at absolute pressures up to 1 MPa (10 bar).

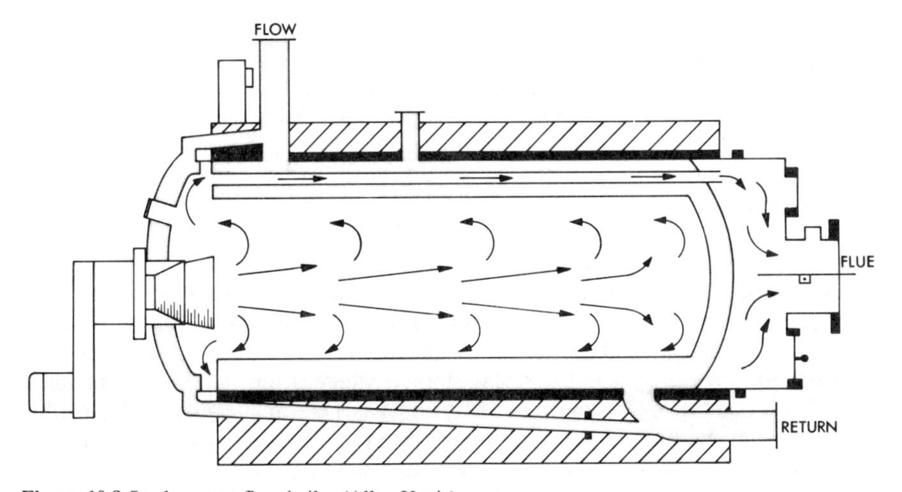

Figure 10.8 Steel reverse flow boiler (Allen Ygnis)

Shell type

It is convenient to separate boilers falling into this category from those others which are manufactured from the same material, namely mild steel, and to make brief mention of earlier patterns from which the present range has evolved. Single-flue *Cornish* and twin-flue *Lancashire* boilers, both types brick-set, are now rarely seen. Nevertheless, in their day, they had the merits of sturdiness, extreme simplicity and vast thermal storage capacity. All these were useful attributes, when hand fired using coal of indifferent quality to meet fluctuating loads, but none of them seem relevant today when viewed in conjunction with a combustion efficiency, on test, of only about 60%.

For both water and steam service, with ratings between 1 MW and 10 MW, shell boilers of both *economic* and *super-economic* type are available, one being as illustrated in Figure 10.9. Without any brick setting, either type may have one or two furnace tubes within the

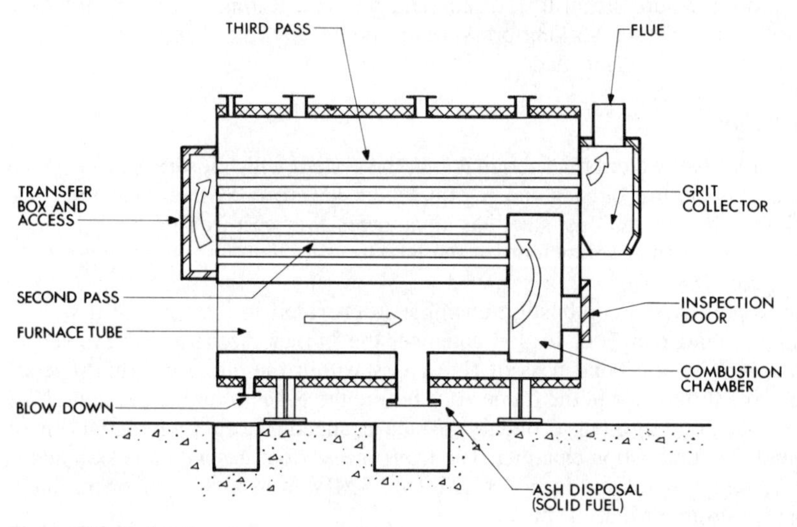

Figure 10.9 Shell boiler of super-economic type

pressure shell, from which the flue gases pass to a combustion chamber at the rear of the boiler. In the *two-pass* economic design, the gases return through a secondary array of fire tubes to a smoke hood at the boiler front from which they are discharged. In the *three-pass* super-economic design, a transfer box takes the place of the front smoke hood and a further array of fire tubes conveys the gases back again to the rear of the boiler for collection prior to dispersal.

The manner in which the combustion chamber is constructed, for either type, as a refractory lined external box at the back of the boiler or as a water-immersed pressure vessel within the main shell, types the design as being either *dry back* or *wet back*.

Apart from the different patterns of mountings fitted, as described later, the principal difference between the steam and hot water variants of this type is that the former provide a space within the pressure shell for steam storage whereas the latter are commonly *drowned*, i.e. are completely water filled. Mention has been made in Chapter 6 of a method adopted in some high temperature hot water systems whereby pressurisation by steam was employed: with this technique, the steam storage space was retained even though the boiler served a hot water system.

For applications where a substantial, large capacity boiler is required, this type will provide a combustion efficiency of the order of 80%. A further advantage is that the type is not only capable of burning either solid, liquid or gaseous fuel but also of site conversion once or more during its working life to a fuel other than that for which it was initially equipped.

Packaged

As boiler design has grown in sophistication and the associated firing equipment, controls, etc., have become increasingly complex, a market has grown for a range of factory-built boiler–burner units, incorporating all necessary working parts and pre-commissioned at manufacturers' works. The *packaged boiler* may be no more than an assembly of production line components but with the very important difference that all those components are, as it were, hand picked for compatibility, Figure 10.10.

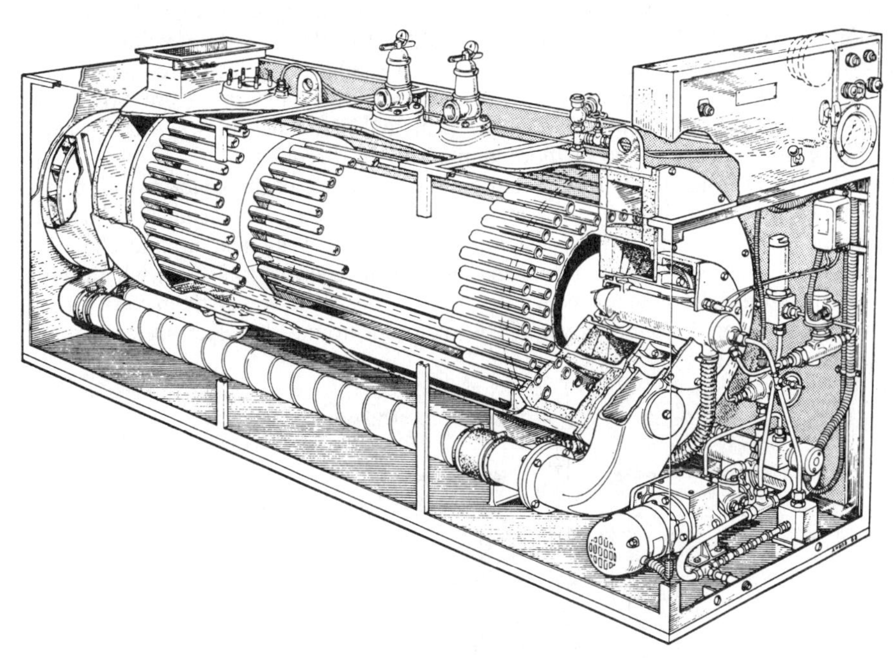

Figure 10.10 Packaged hot water boiler

The firing equipment will have been chosen to give the optimum flame shape for the combustion chamber; the forced or induced draught arrangements will be matched to provide the most suitable gas flow pattern and the control system will be fitted, wired and proven. Site tasks after delivery, and time for erection and commissioning should be reduced in consequence.

In some respects it is unfortunate that the introduction of fire-tube shell boilers in packaged form coincided with a trend towards smaller furnace volumes, reduced diameter fire tubes and a general move to squeeze the last gram of heat transfer from each kilogram of metal. The consequent need for greater care in water treatment and greater skill in maintenance has coincided with a period when both care and skill are in short supply. The package concept has acquired some quite undeserved disrepute in consequence.

Water tube

Boilers of this type, while available in sizes down to as little as 500 kW, are commonly used only for much larger duties of up to about 10 MW. The most typical application has been for service to high temperature hot water plants for very large industrial sites. Since fire-tube boilers are now available in larger sizes than was the earlier practice, for absolute pressures up to about 1.5 MPa (15 bar), a tendency now exists to use these for such systems.

Electrode

For use with off-peak current and the large-scale thermal storage cylinders described in Chapter 5, a heating source may be banks of immersion heaters but is more likely to be one or more electrode boilers. These may be connected to either a medium voltage supply (up to 650 V) or to high voltage (3.3, 6.6 or 11 kV), the latter being the more common for installations of any significant size. Such boilers, as illustrated in Figure 10.11, are available in ratings up to about 2.5 MW, the principal difference between the various designs being in the detail of the method adopted for load regulation.

Current is passed from electrode to electrode, the resistance of the water in which they are immersed acting as the heating element. The load is varied by increasing or decreasing the length of the path which the current has to take between and around the non-conducting *neutral shields* which shroud the electrodes. The material from which these shields are made and the actual mechanism used to raise and lower them varies from maker to maker. A brisk water movement is required around the electrodes and most

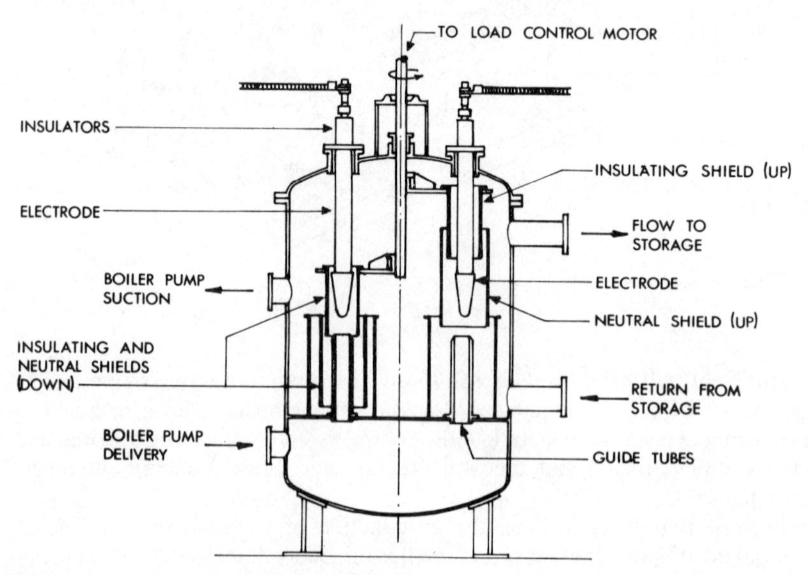

Figure 10.11 High voltage electrode water heater

manufacturers mount a centrifugal pump to the boiler shell for this purpose. This pump must not be confused with that which must be provided to circulate water, during the charging period, between the boiler and the storage cylinder or cylinders.

Where electrode boilers are used for steam generation, at ratings between 20 kW and 2.5 MW, it is most probable that they will operate 'on-peak' although some form of thermal storage may still be used to meet short fluctuations in demand. The arrangement of the electrodes in a steam boiler is usually relatively simple since the imposed load depends upon how much of the electrodes is immersed. Load regulation is thus achieved by adjustment of the water level within the shell. A typical form for electrodes is as three interleaved *scrolls*, as shown in Figure 10.12, one per phase within a single neutral shield.

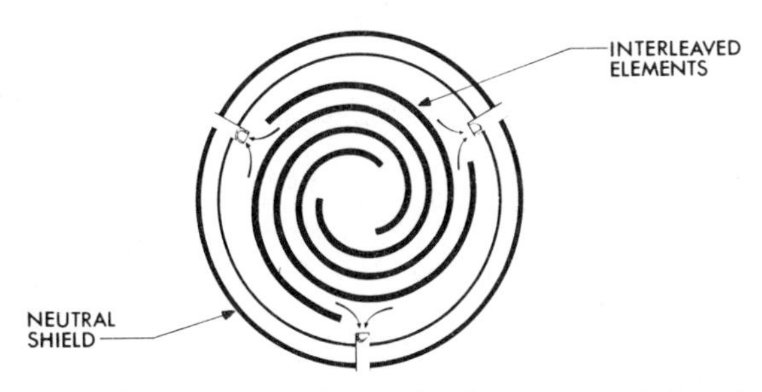

Figure 10.12 Interleaved elements for electrode steam boiler (plan view)

For both steam and water operation, the conductivity of the water may have to be adjusted from time to time by the addition of soda, or other salts if need be, in order to preserve the required electrical resistance.

Flow heaters

Packaged electrical units, for service with a variety of loads, are made under this description to distinguish them from those which serve a static store. Most are designed for use during the hours when 'off-peak' current is available at reduced cost, to act as a substitute for a daytime source of energy not then available, e.g., an industrial boiler plant which is shut down overnight. In contrast, some units are sold in the domestic sector as a substitute for a conventional boiler, to use daytime current at 'on-peak' prices. It is difficult to imagine what purpose this second type can serve which could not be provided more economically by individual room heaters.

Miscellaneous

Experiments were made in France in the 1960s with concrete as a material for sectional type boilers. A special quartz aggregate was used and each section was cast in halves to a form which included flueways, etc., for the passage of combustion gases. Within the mass of this concrete a sinuous coil of steel pipe was embedded with ends brought to the side of each half section at top and bottom, for connection to headers. Advantages, apart from the high resistance and low water content, were that no refractory was required internally and, as the fire-box surfaces were able to reach red heat, smuts and ash were consumed and flue cleaning thus reduced to virtually nothing. Such boilers, however, do not seem to have reached any commercial market.

Boiler fittings and mountings

The much-loved first edition of *The Efficient Use of Fuel* provided a definition which distinguished between the two classes of attachments to boilers. 'Generally speaking', the author wrote, 'the term *mounting* implies that the equipment is mounted on a pad or stool riveted on to the fabric of the boiler as distinct from a *fitting* which may or may not be attached to the boiler, but for which there is neither pad nor stool.' This distinction served for an era when boilers were purchased 'raw' and both mountings and fittings were selected and bought quite separately: it seems less appropriate to the complete packages now sold.

Hot water boilers

The principal fittings and mountings in this case must include:

- Relief valve.
- Altitude gauge.
- Thermometer.
- Drain cock.
- Feed and expansion pipe.
- Open vent pipe.
- Control devices.

The first four of these items require no comment other than emphasis that relief valves should be of a type approved by the building owner's insurance company and purpose set for the particular installation. Similarly, the altitude gauge and thermometer should be of a pattern having large legible figures and the former should be calibrated in kPa, m head of water or other accepted SI units. It is not possible to generalise as to provision for control devices since these may be either rudimentary or complex.

The feed and expansion pipe and the open vent pipe are, in many respects, a part of the connected system but both are very closely associated with the boiler plant. The former dictates the pressure at which the plant will operate and the latter is a safety feature paralleled only by the relief valve. Table 10.1 lists sizes for these pipes and for relief

Table 10.1 Feed and expansion pipes, open vents and relief valves

	Pipe size (mm)		*Relief valve, min. clear bore* (mm)	
Boiler rating (kW)	*Feed and expansion pipe*	*Vent pipe*	*Solid fuel firing*	*Oil or gas firing*
up to 74	20	25	20	25
75 to 224	25	32	20	25
225 to 349	32	40	25	32
350 to 399	32	40	32	40
400 to 449	40	50	32	40
450 to 499	40	50	40	50
500 to 749	40	50	50	65
750 to 900	50	65	65	80

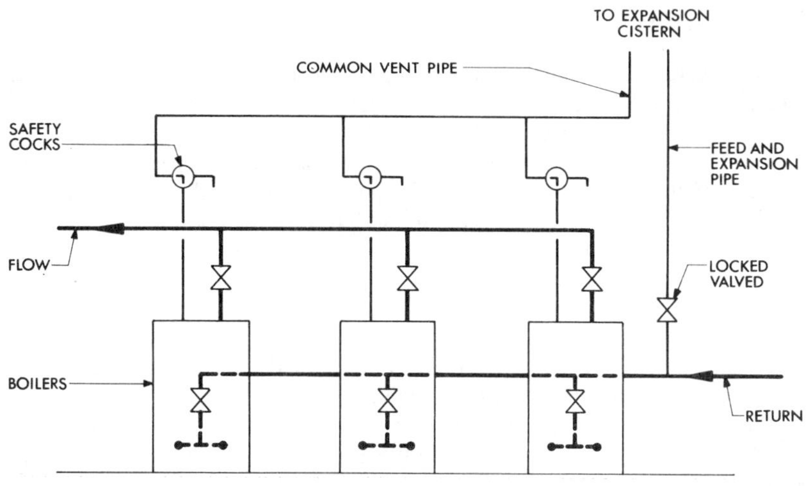

Figure 10.13 Open vent arrangement for multi-boiler plant

valves. It has already been noted in preceding chapters that the open vent pipe is *not* a means for dealing with the increased water volume of expansion following heating.

Where a plant has only a single boiler, an open vent pipe may be routed from it to a point above the expansion cistern. Where more than one boiler is provided, it is necessary to fit a special type of 'safety cock' (approved by insurance companies) so that individual boilers are, first, open to atmosphere when working and, second, may be isolated from the system for service. Figure 10.13 illustrates this arrangement.

Steam boilers

As far as fittings and mountings for steam boilers are concerned, stringent requirements for these are set out in various of the publications of the Health and Safety Executive. Certain of the mountings are required to be duplicated and the following is no more than a list of the principal items:

- Safety valves.
- Pressure gauge.
- Water level gauges.
- Steam stop valve.
- Auxiliary steam stop valve.
- Feed and check valve.
- Blow-down valve.
- Alarm devices (high and low water, etc.).
- Control devices.

Some packaged boilers are provided with an electrically driven boiler feed pump supplied from a frame-mounted condensate cylinder. This latter must not be confused with the large-scale hot well required in the case of most systems to collect the intermittent flow of returning condensate (see p. 158).

Boiler firing – solid fuel

At the present time, there is little interest in the use of solid fuel for firing either large or small boiler plants but, as has happened in the past, economic or other pressures may lead to a resurgence of activity at some time in the future. Hand firing, with its human failings and other vagaries, is unlikely to return but a variety of mechanised equipment, admittedly not yet wholly automatic, is widely available. Properly applied and maintained, such equipment is able to produce an adequately high combustion efficiency.

Magazine boilers

The boiler illustrated in Figure 10.14 is a domestic pattern unit designed to burn anthracite grains but equipment of very similar type is available for commercial applications up to ratings of about 600 kW. The principle common to all is top charging of the fuel into a hopper whence it descends by gravity to the burning zone. Ash and clinker are removed by hand from the base.

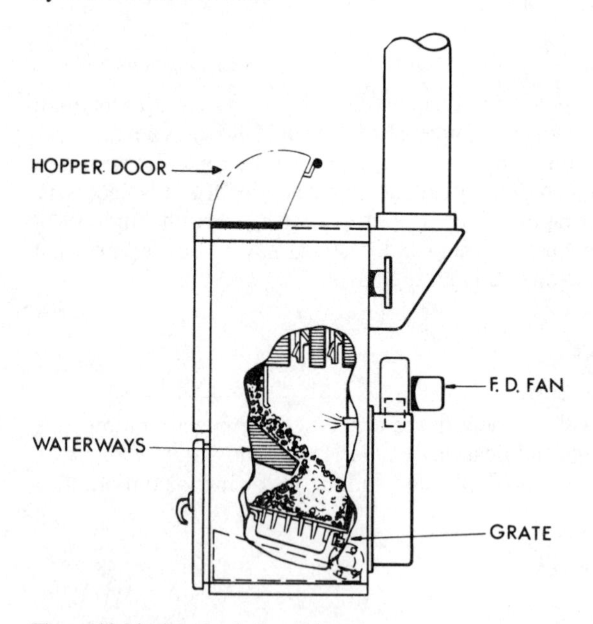

Figure 10.14 Magazine boiler (Trianco)

Automatic stokers

Of the overall choice available, there are several specific types of automatic stoker which are particularly appropriate for use with the size range of boiler which falls within the scope of this book. These include those which are most commonly used, namely the various patterns of underfeed stoker and also, as shown in Figure 10.15, sprinklers, coking types and moving beds.

All modern types of stoker may be arranged for some form of semi- or fully automatic control whether this be variable through speed control of the mechanism or merely ‘on/

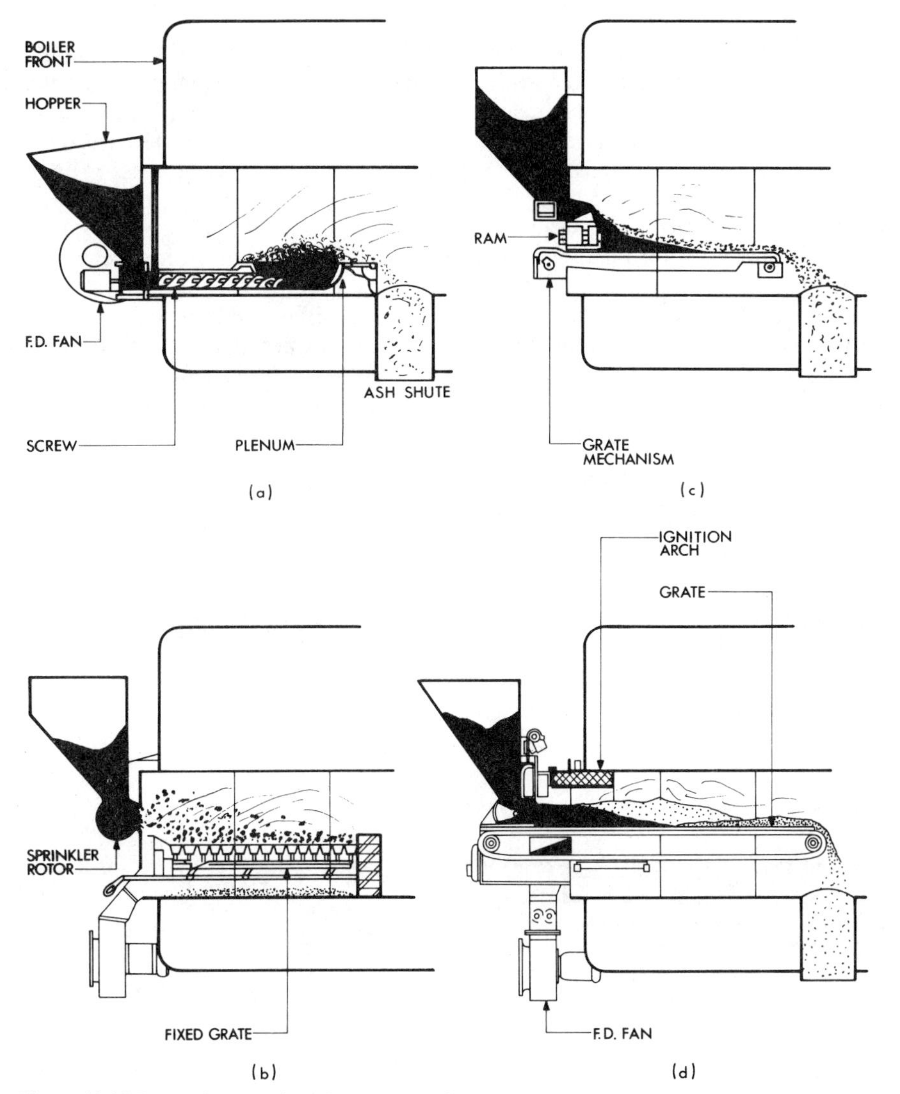

Figure 10.15 Types of automatic stoker: (a) underfeed; (b) sprinkler; (c) coking; (d) moving bed

off'. In the latter case a '*kindling*' control is a necessary feature, the stoker operating for a few minutes in each and every hour in order to keep the fire alive. A recent development, for use when starting the fire, is a form of electric ignition.

It must be emphasised that no single type of automatic stoker is suited to all types and sizes of coal. It is fundamental to selection of the appropriate type of machine that the characteristics of the fuel available are examined, in consultation with the local solid fuel agency. Table 11.2 lists published coal Rank numbers and other details.

The principal design features of the various generic types of stoker may be summarised as follows:

Underfeed (Figure 10.15(a))
This has always been the most popular type of stoker for application to boilers serving heating systems since it is simple in construction and application; tolerant of a low level of intelligence for maintenance and is relatively inexpensive. Requirements as to fuel are that coal be weakly caking, low in ash content and of even size at not much over 20 mm (coal Rank 600–900). More details of this type are set out in later paragraphs.

Sprinkler (Figure 10.15(b))
The original arrangement for this type was conceived early in the nineteenth century to imitate the spreading motion achieved by hand-firing using a shovel. Several basic patterns of equipment have since been developed. One, as illustrated, spreads fuel via a *rotary distributor* mounted at the mouth of an overhead hopper. A second type sprays pneumatically conveyed fuel from a pipe terminal fitted to the furnace front plate. Yet a third, which is integral to the boiler design, also sprays fuel on the fire-bed but in this case

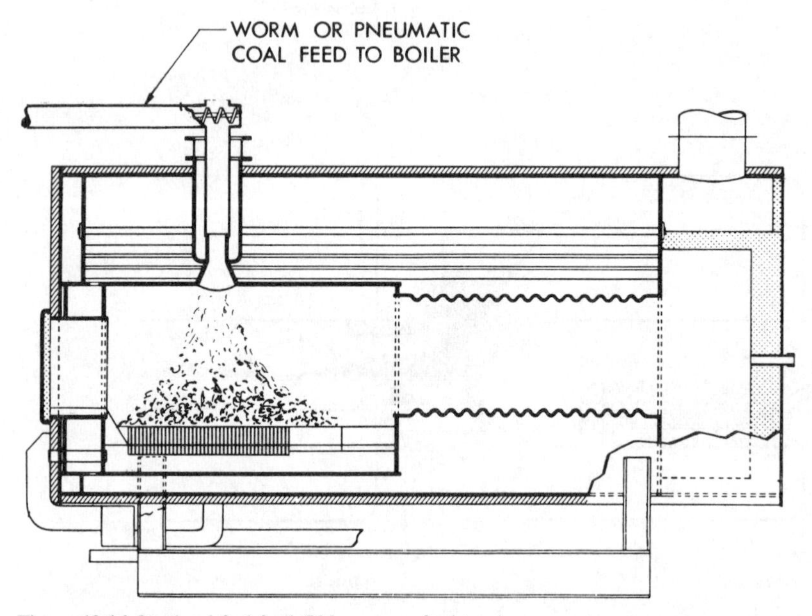

Figure 10.16 Overhead fuel feed (Vekos worm feed type)

from a vertical manifold penetrating the pressure shell as shown in Figure 10.16. Sprinklers will burn most types of small washed coal except anthracite: a low ash content is preferred (coal Rank 600–700).

Coking (Figure 10.15(c))
The operating principle in this equipment rests with the action of a reciprocating ram mechanism which pushes fuel forward from an overhead hopper at the boiler front and deposits it on a coking plate in the hottest area of the furnace. At that point, the volatiles are consumed and the residual coke is then passed towards the boiler rear by the action of a moving grate. Early designs used a deep fuel bed and were, in consequence, unsuited to any type of shell boiler: this problem was later overcome when *'low-ram'* patterns

appeared. Singles and doubles are the appropriate sizes of a coal having slight caking properties and an ash content of not more than about 8% (coal Rank 700–900).

Moving bed (Figure 10.15(d))
A chain grate stoker, as the name suggests, has a fire grate made up of a continuous chain of links or bars which run over driving sprockets at the boiler front. Fuel is fed to the grate by gravity from a hopper, the depth of the bed being controlled by a guillotine plate mounted above the furnace entry. The position of an '*ignition arch*' behind the guillotine is arranged to expose it to maximum radiation so that it may raise and maintain the temperature of incoming fuel. Combustion takes place progressively, downwards from the ignited upper layer of fuel towards that in contact with the grate. Wet smalls is the preferred fuel (coal Rank 700–900).

Hopper or bunker underfeed stokers

In the simplest form of such equipment, fuel is fed from a hopper down to a worm or screw at its base: this is rotated at slow speed through reduction gearing from a motor drive. Beyond the feed point the worm is enclosed within a tube and serves to convey the fuel into a fire-pot which is built into a refractory base inside the boiler, as shown in Figure 10.17. Safeguards are provided, in the form of *shear-pins* or a *slipping clutch*, to prevent damage to the worm if jammed by a foreign body or by oversize coal.

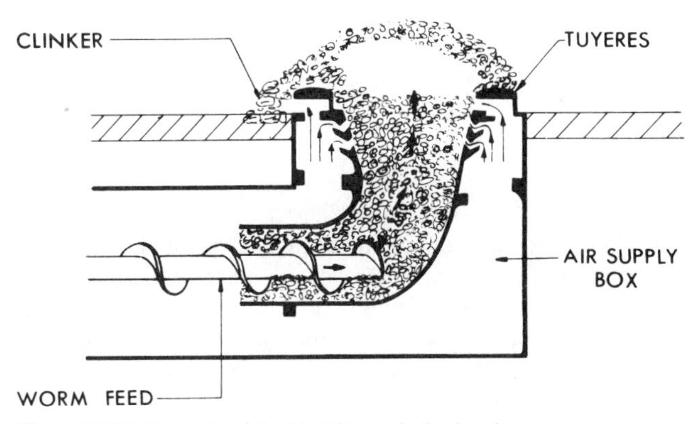

Figure 10.17 Fire-pot and fuel bed for underfeed stoker

A second tube delivers air to the fire-pot from a fan driven by the same motor that powers the worm. This air is discharged through a series of slots or openings, disposed around the fire-pot to form *tuyères* which will not be fouled by fuel or ash. Secondary air is supplied by the same fan and preheated in order to control smoke emission.

Thus, forced draught is provided and a very high combustion rate is possible, so high in fact that no grate is necessary. The fuel is burnt as it passes over the edge of the fire-pot and all the ash is reduced to clinker in the process. Any accumulation of clinker, as an impediment to combustion, is prevented as the fresh coal brought in by the worm pushes it to one side from which position it is removed at intervals.

An alternative method of feeding the fuel is direct from a fuel bunker. In this application, either the firing worm or a subsidiary extends into the main fuel store and

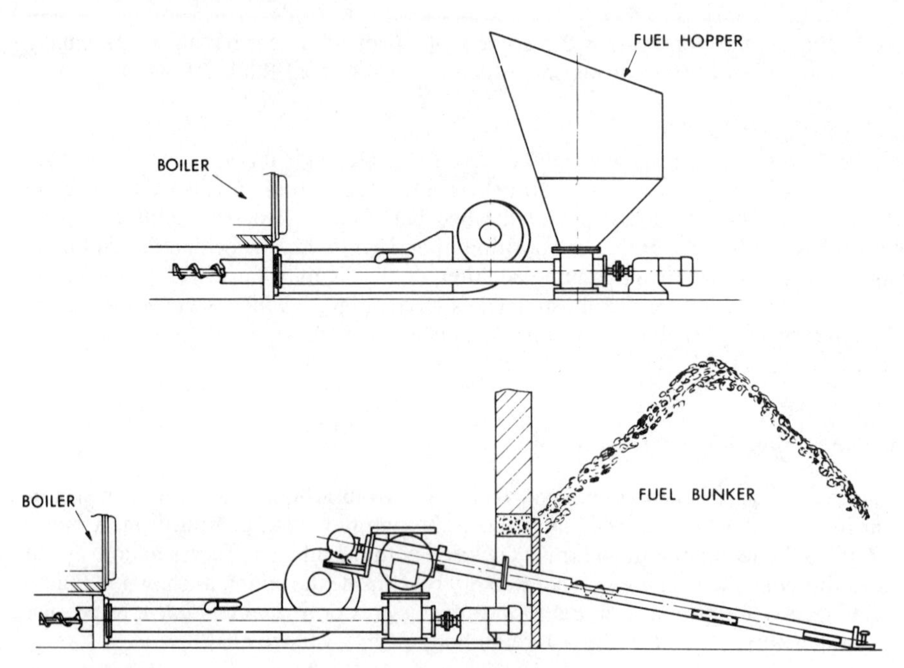

Figure 10.18 Hopper and bunker supply to underfeed stoker

the hopper is thus eliminated. Both hopper and bunker feed methods are illustrated in Figure 10.18.

Holden Heat-House

A range of coal fired boiler-burner units was introduced in the mid 1980s as illustrated in Figure 10.19. The boiler has a vertical multi-tubular heat exchanger which is fitted with a cleaning mechanism designed to operate automatically as required. Other unique features include an integral worm feed supplying fuel to a rotating grate and a second worm arrangement for extraction of ash direct to an external container. The firebed is provided with forced draught for primary and secondary air.

Present availability is for units rated at 29 and 45 kW but a further model rated at 90 kW was under development. The fuel used is described as 'pearls', 6 mm × 12 mm, delivered under a 'Coalflow' contract service which also undertakes annual maintenance.

Boiler firing – oil fuel

Achievement of clean and efficient burning with oil fuel rests almost entirely with the matter of atomisation, that is to say the intimate mixing on a molecular scale of the carbon in the fuel with the oxygen in the air supply. All manner of methods have been used over the years in attempts to produce an ideal solution but, for the size and type of boilers which are chosen to serve heating plants, it is necessary to consider only the following:

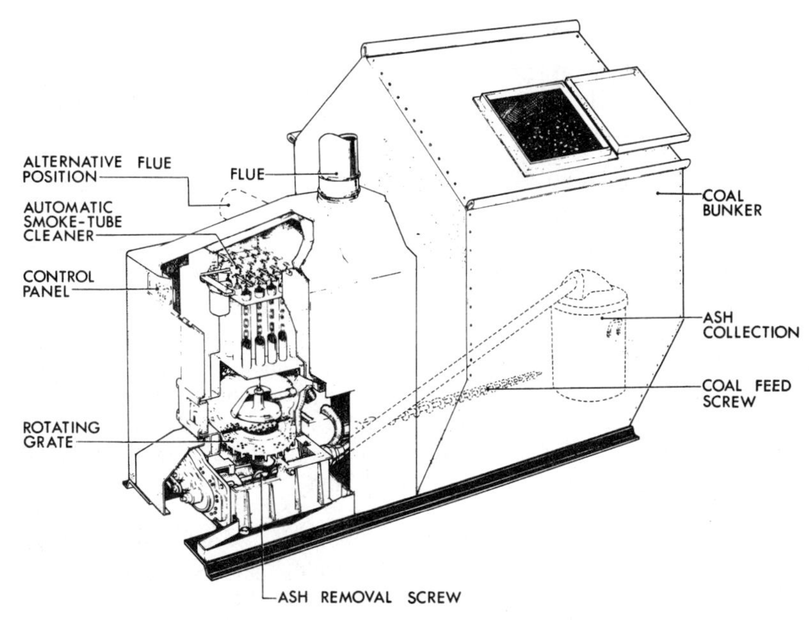

Figure 10.19 Self-cleaning underfeed stoker (Holden Heat)

Vaporisation

This may be compared with the principle used in a blow-lamp or a primus stove and is applied only to very light oils, such as kerosene. On start-up, the oil is preheated, often electrically, to form a vapour which is then ignited: subsequent vaporisation is produced by heat from the flame. Pot-type burners, now probably obsolete, which used this principle were notoriously unstable.

Pressure atomisation

In this arrangement, the oil supply is fed at a controlled rate to a nozzle the discharge from which meets a stream of air (or sometimes steam) at high, medium or low pressure. The primary air quantity so provided is less than that required for combustion and acts principally to atomise the fuel.

Mechanical atomisation

Oil is fed, at a controlled rate, typically to the inside surface of a conical cup which is rotated at a high speed. As the oil leaves the edge of the cup under centrifugal force, it is atomised by a primary air stream supplied concentrically around the cup and contra-rotating *vis à vis* the oil supply. More details of this type are given later.

Pressure jet

At a relatively high pressure, oil is supplied to a fine nozzle which is so designed as to apply a swirling motion to the spray of droplets discharged. An air supply, primary or total, is provided in a contra-rotating swirl. Additional information regarding this type is given later.

Emulsification

This process may be applied to the principle of either pressure atomisation or pressure jet, the difference being that the oil is pre-mixed with a controlled quantity of primary air before it is delivered to the nozzle in the form of an emulsion.

Combustion air

As noted above, the total air supply to an oil burner falls into two categories: the primary air which is intimately involved in the atomisation process and the secondary air which makes up the balance necessary to complete the combustion reactions, including any excess air which may be required.

The secondary air may be induced to flow through pre-set registers at the burner front, either by natural draught from the chimney or as a result of a mechanical *induced draught* fan at the boiler exit, in which latter circumstances the boiler combustion chamber is under suction. Alternatively, and as is now more usual practice for applications of substantial size, the secondary air is supplied by a *forced draught* fan into what then becomes a pressurised combustion chamber.

It will be seen that there is a variety of possible combinations of atomising methods with systems of primary and secondary air supply and hence many possible variations in burner design. When fully automatic operation is required, as in the present application, the choice is generally limited to either pressure jet or rotary cup equipment.

Gun-type pressure jet burners

By far the most adaptable type of burner, the gun-type unit, is available in sizes suitable for boiler outputs ranging from 10 kW to 2.5 MW. Burners of this pattern consist, in the simplest form, of a direct electrical drive to a centrifugal fan which produces the whole of the air required for combustion, including any necessary excess. A positive

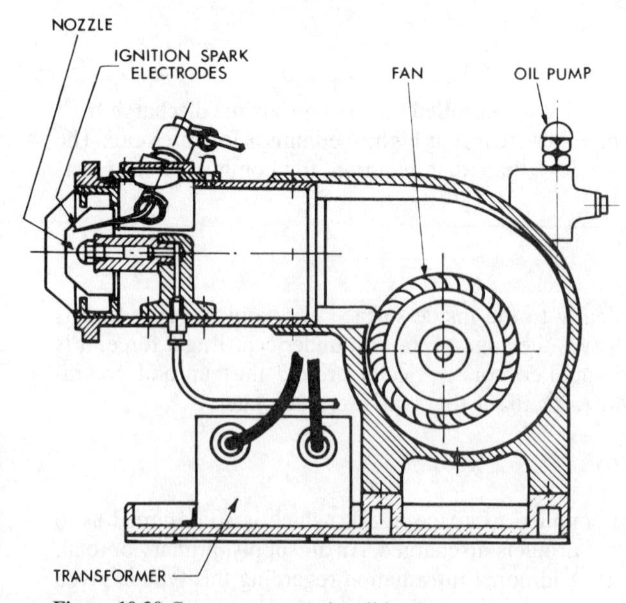

Figure 10.20 Gun-type pressure jet oil burner

displacement oil pump is coupled to the fan drive and supplies oil at an absolute pressure of up to about 1 MPa (10 bar) to a fine-calibrated jet. Air from the fan is delivered via swirling vanes disposed round the nozzle following the general pattern shown in Figure 10.20.

The air quantity supplied by the fan is set by slots or dampers at the inlet and the oil quantity delivered is adjusted by control of the pump output pressure and the size of the jet orifice. The flame shape produced by the burner may be varied to suit the geometry of the boiler combustion chamber by changing the angle of the swirling vanes and by selecting the angle of divergence of the jet orifice.

When the fuel used is light oil (class C2 or class D), there is no requirement for preheating. For heavier grades, an electrical heater is incorporated as part of the burner unit to raise the oil temperature to between 60 and 70°C. Ignition is by spark between two electrodes located near the nozzle tip which have a high tension feed from a transformer. These suffice for smaller size burners and lighter oils but, in other instances, a two-stage process for ignition is used via a gas supply, the electrodes igniting a gas pilot and the pilot igniting the oil.

Rotary cup burners

The principal components of this type of burner are illustrated in Figure 10.21, the secondary air *quarl* shown being at the point where the burner is mounted to the front of the boiler combustion chamber. Drive from the motor is not only to the primary air fan and the oil pump but also, via a hollow shaft, to the spinning cup atomiser. Oil is delivered

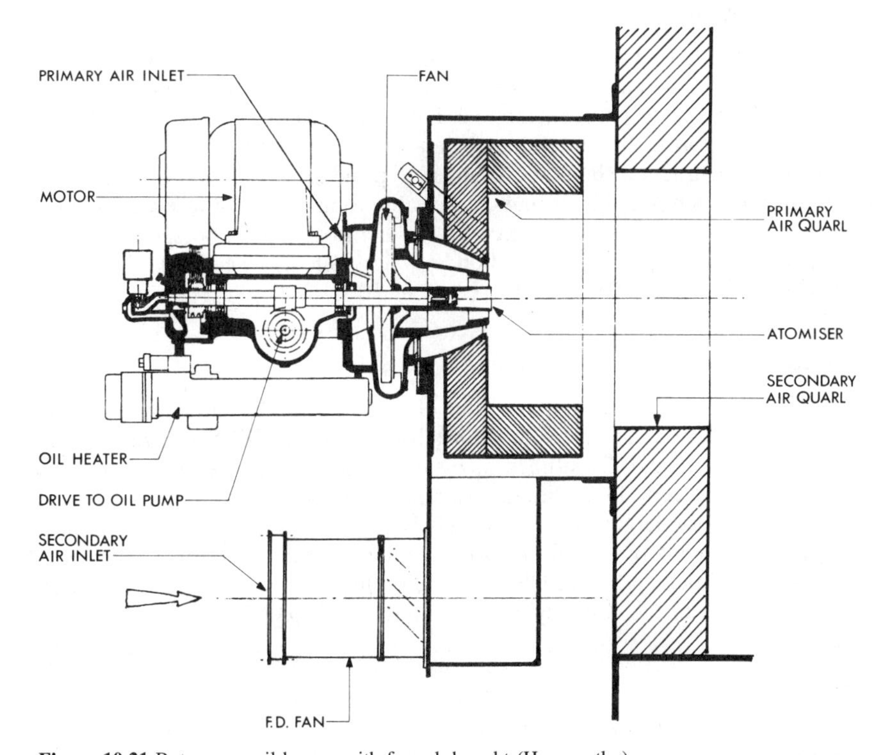

Figure 10.21 Rotary cup oil burner with forced draught (Hamworthy)

through this shaft, which rotates at about 100 rev/s, to the cup. The primary air supply from the fan passes through an annular chamber having internal swirling vanes to be delivered by nozzles circumferential to the cup.

The secondary air required for combustion is introduced at a lower pressure by a further forced draught fan, usually an axial flow type, into the burner box where it can be pre-heated to some extent by radiation from the quarls. Characterised dampers are provided for both primary and secondary air and are linked to a metering valve in the oil supply.

As noted above for larger gun-type burners, ignition here is two stage, using a gas pilot ignited by electrodes. Temperature control of the fuel supply, which will usually be a heavier grade of oil for the 150 kW to 5 MW range of this type of burner, is provided for at the boiler front but topping up of temperature is all that is necessary since fuel of these grades must be supplied from a heated oil ring main.

Furnace linings

The fire-brick construction of a combustion chamber subjected to oil firing, where this is provided, needs to be in a material which will withstand temperatures of between 1400 and 1600°C without fusing or premature disintegration. Linings must not be carried solid up to the metal of the boiler construction but must allow an air gap of 15 mm or so for brickwork expansion. Many boilers of the packaged type dispense with any brickwork apart from the quarl.

Under-hearth insulation

Where a boiler does not have a water-way below the combustion chamber, it is necessary to provide a layer of insulation above the floor in order to prevent damage to the structure and, in particular, to any damp-proof membrane. For boilers up to a size of, say, 600 kW, a 150 mm layer of *moler* insulating brick may be enough but for larger boilers some form of open honeycomb construction should be arranged to provide additional cooling through circulation of air. To encourage this, it may be possible to arrange that a forced draught fan draws its air supply via passages formed within the honeycomb, so serving a dual purpose by preheating this supply by conduction. It is desirable that a concrete slab below a boiler should not reach a temperature of over about 70°C.

Regulations, etc.

Since oil storage and oil firing both impose some elements of fire hazard, various bodies have drawn up regulations relating to these installations. A British Standard *Code of Practice* (BS 5410: Parts 1 and 2) outlines many of the hazards and proposes ways in which they may be reduced. In addition, insurance companies and the fire service departments of some of the larger local authorities have their own regulations and these should be ascertained when any new installation is under consideration. Local authorities have certain responsibilities under the Clean Air Act which may affect the grade of oil used and the dimensions of the chimney, as referred to in Chapter 12.

Burner controls

The few notes which follow are not intended to be a full and complete dissertation covering the whole subject of automatic control to oil burners, particularly since many of the primary devices may be applied with equal validity to the control of automatic firing equipment for other fuels. Primary functional control is certainly common to all, but unique requirements as to safety and function, particular to individual fuels, are subjects which cannot be dealt with satisfactorily here.

Fully automatic types of oil burner have been referred to and it will be obvious that these, by definition, incorporate means for self-igniting and self-extinguishing. Primary control of these functions is normally by either two separate devices, be they thermostats or pressure sensors, or one single dual-service device. One such item will be the basic controller and the other, normally a hand-reset pattern, will be a high-limit safety override.

Whether the firing rate is capacity controlled between the extremes of on and off depends upon the method of operation and, very largely, the size of the fired boiler: small units normally operate in the '*on/off*' mode whereas larger units incorporate some means of varying or modulating output. Apart from the nuisance which would arise resultant upon the cycling of a large burner to meet a small load, '*on/off*' operation in those circumstances would create thermal shock to the boiler.

Obviously, both fuel and combustion air must be controlled in parallel, which tends to involve complexity if adjustment is envisaged right across the spectrum of capacity. This difficulty is commonly overcome in one of two ways, the simpler being to arrange for two preset levels of operation, low and high, i.e. perhaps a sequence of 40 and 100% capacity. More complex, but still simpler than the ideal, is an '*on/off*' operation up to one third load and full modulation to match the load thereafter.

With a self-contained burner in which one motor drives both fan and fuel pump, there is little likelihood of fuel being delivered into the boiler without a parallel supply of air. In types where separate drives are necessary, suitable interlocks are available. Residual problems arise however when a flame is not established due to failure of ignition or, having been established, is later extinguished for some reason such as an intermittent stoppage in the fuel supply.

A variety of flame failure devices has been used in the past but it is now almost universal practice to incorporate a photo-electric cell either within the burner housing or at some other site where the flame may be viewed. Arrangements are incorporated in the control sequence such that this cell is out of circuit for a predetermined few seconds on start-up but that, thereafter, the burner will cut out if no flame is sensed. In some instances the control circuits provide a facility which will allow a second attempt at ignition to be made after a suitable time delay but will ensure that if that second attempt fails, the burner is deactivated until reset by hand.

With larger and more sophisticated burners, a commonplace addition to the control system is a purging sequence which arranges for the air supply fan to run for a period before fuel is made available to the burner head and also to run for a similar period after supply has been shut off. This sequence is, of course, interlocked with the flame failure circuit and will be initiated whether or not a flame is established.

Boiler firing – gaseous fuel

A gaseous fuel discharged at a low velocity from a simple nozzle will burn with a soft lazy flame at the interface between the gas envelope and the surrounding air. Such a flame is unstable with a low heat flux and if the velocity be increased with intent to produce a more useful combustion characteristic, then there will be a tendency for the flame to lift from the nozzle and extinguish. A variety of complex jets has been produced to substitute for the simple nozzle and other methods, principally involving some process of aeration, have been evolved to address the problem. Those most commonly used in boiler firing are:

Neat flame burner. As the name suggests, these are naturally aerated burners having shielded jets and their use is generally confined to conversion exercises for boilers rated at not much more than 50 kW. Such a burner is illustrated in Figure 10.22.

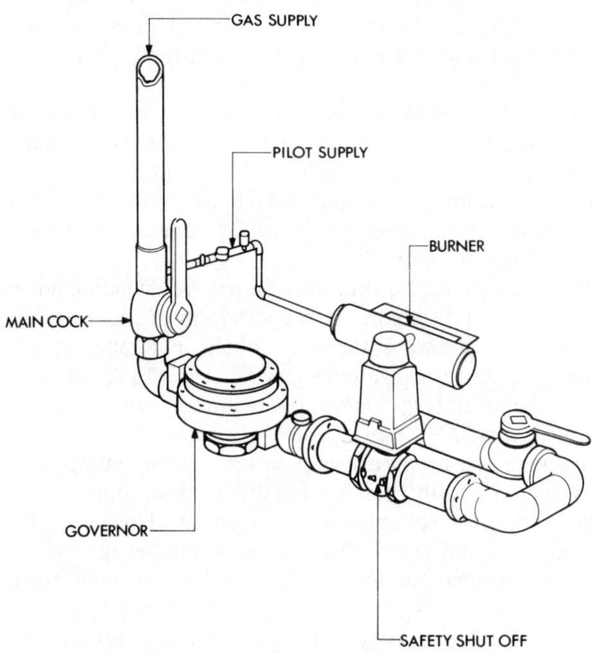

Figure 10.22 Neat flame gas burner

Atmospheric burners. Since a fuel gas reaches the point of use under pressure, it is logical to make use of this potential energy to induce a proportion of the air required for combustion into a mixing tube. This, the principle used in the Bunsen burner, is adapted also for the atmospheric burner.

Packaged burners. The introduction of a fan powered air supply is the principal distinguishing feature of this general type. Discharged at the burner face, the air supply mixes there with the fuel which is admitted via control ports. In many respects the concept here may be compared with that of gun-type oil burners.

It should be noted that whilst the general thrust of the comments made above and of those which follow is directed to the burning of North Sea gas derived from the public supply, this is not to suggest that the equipment described is exclusive to that fuel. On the contrary, much if not all of it may be used equally satisfactorily with LPG when appropriate adjustments have been made to jet sizes and to air supplies.

Atmospheric burners

These represent a type of equipment which follows the traditional pattern associated with the firing of cast-iron sectional boilers purpose designed for this fuel, as illustrated in

Figure 10.7. They are extremely quiet in operation and, as multiple units, may be obtained in ratings up to about 1 MW. A gas pressure of about 1.5–2 kPa is required at the burner, down stream from the governor.

Gas is supplied to a calibrated nozzle fitted to a manifold and flows through a venturi to induce a supply of primary air. The resultant mixture is delivered by the manifold to a series of outlet ports where the flames entrain secondary air from the surrounding space at the boiler base.

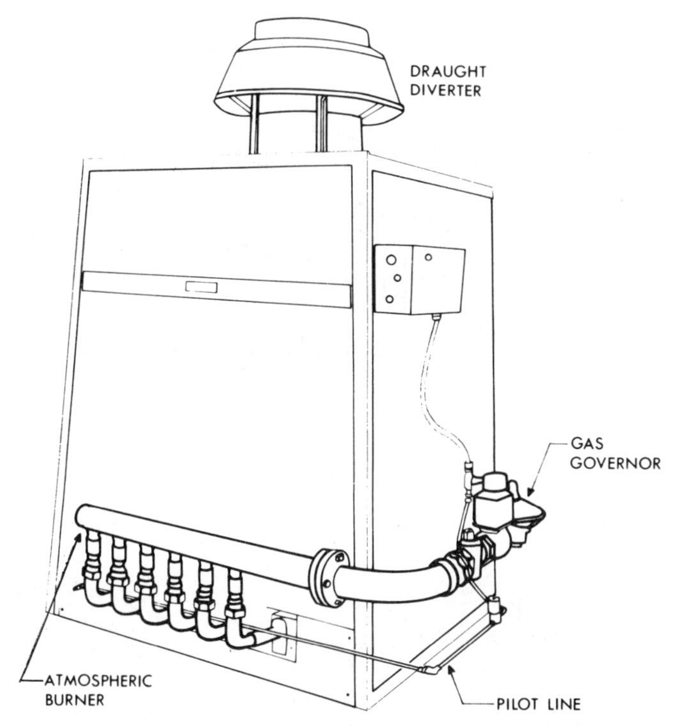

Figure 10.23 Atmospheric gas burner and boiler

Combustion gases rise by convection through channels between the heat exchange surfaces, boiler height providing an adequate motive force. As illustrated in Figure 10.23, it is necessary that the flue terminal of the boiler incorporates a *draught diverter* and gas dilution device in order to prevent any interaction between that small natural draught and the effect of a connected chimney. An atmospheric burner must not be used in any situation where draught is not so stabilised.

Packaged burners

Known also as forced draught burners, units of this type are built up around an integral fan which is arranged to provide, under pressure, all the air required for combustion. In most models, such fans operate with direct coupled drives at relatively high speed and are, as a result, objectionably noisy.

As may be seen from Figure 10.24 which illustrates one such burner, both gas and air under pressure are delivered to the combustion chamber via a complex arrangement of ports in a nozzle plate, having been pre-mixed in the fan discharge air box. The precise design arrangement adopted to ensure that gas and air are properly mixed varies between manufacturers, as does the geometry of the burner head.

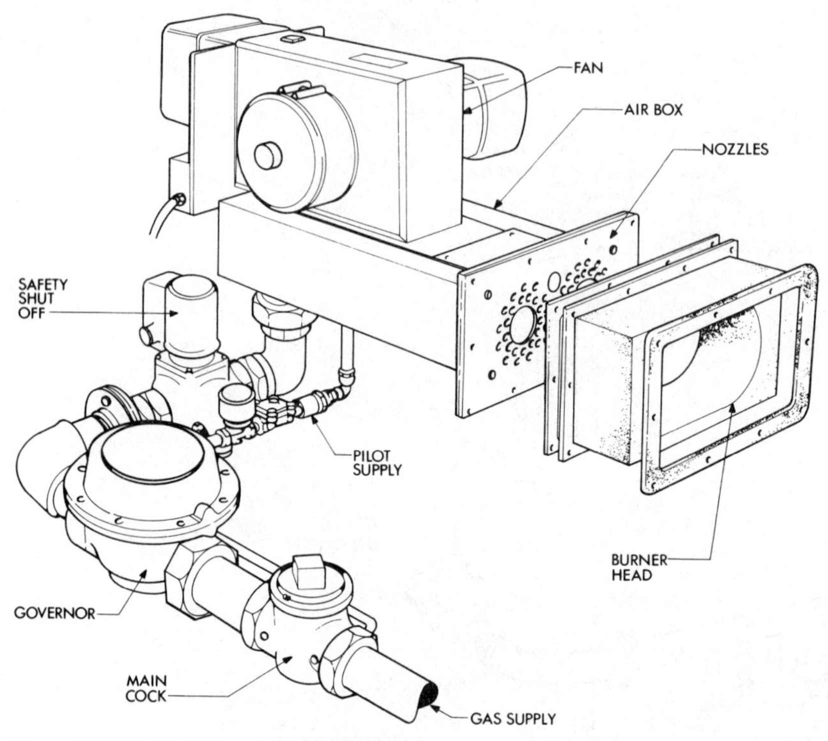

Figure 10.24 Packaged forced draught gas burner

Since the boiler combustion space is pressurised by the forced draught air supply, the flue connection may be made without regard to the limitations existing in the case of an atmospheric burner and no draught diverter is required. In order to overcome the noise nuisance mentioned previously, some manufacturers provide insulating shrouds tailored to allow for relatively easy removal and consequent access for burner maintenance.

Burner controls

In the case of atmospheric burners, the control systems are comparatively simple, usually consisting of an electrical ignition device which operates in conjunction with a gas ladder and a flame failure sensor connected to a safety shut-off valve, which will be activated also by sub-normal gas pressure. Basic and high-limit controllers are either separate fittings or a single dual-function type with the high limit arranged such that hand reset is necessary.

As in the case of oil burner controls, the operational sequence will vary with the size of the burner from simple '*on/off*' for the smaller range to more complex systems for larger sizes. Some makes of burner are now provided with a further safety thermostat which is fitted to sense flue down-draught.

For packaged burners, the same principles apply but the details are somewhat more complex to take account of fan controls and, where larger sizes are involved, the sequencing arrangements will probably provide for purging periods during both start-up and shut-down cycles. The method used for detection of flame failure is either by probe or by ultra-violet scanner, the luminosity of a gas flame having been found insufficient to operate the type of light sensitive cell used to view oil combustion.

Dual-fuel burners

Brief mention has been made already of equipment which offers the facility of use with more than one fuel. Combination dual-fuel burners, capable of firing oil or gas are as

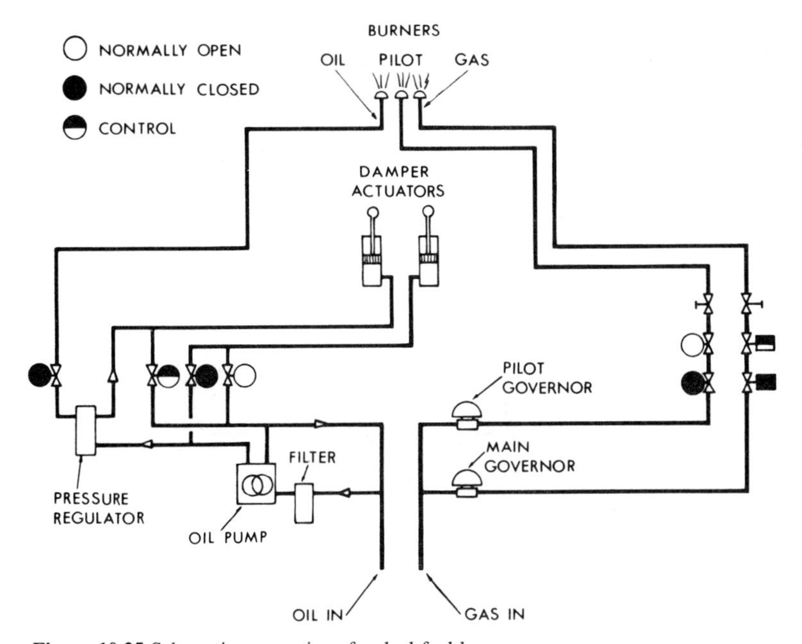

Figure 10.25 Schematic connections for dual-fuel burner

shown in Figure 10.25. It is usual for both supplies to be connected and change-over is thus a relatively simple matter of shut-down; isolation of one supply; activation of the alternative and start up.

Gas boosters

In certain circumstances it may be found necessary to provide pressure boosting equipment to serve burners which require a fuel supply at a pressure higher than that available at the meter. Such a situation may arise in particular where a dual-fuel burner is selected and the pressure at which the combustion air is provided exceeds that at which gas is available. Pressures higher than normal offer advantages in that piping, automatic control valves, etc., may be of smaller size but it would rarely be economic to introduce boosting equipment for such reasons alone.

Booster equipment may take the form of one centrifugal unit per burner or a common unit serving a range of burners. Since operation must be fully automatic and cross-linked with the burner control systems, physical booster/burner connections to provide standby support may be difficult to arrange and may dictate the overall arrangements for a given situation. The circuit arrangement illustrated in Figure 10.26 applies to a single booster serving two burners.

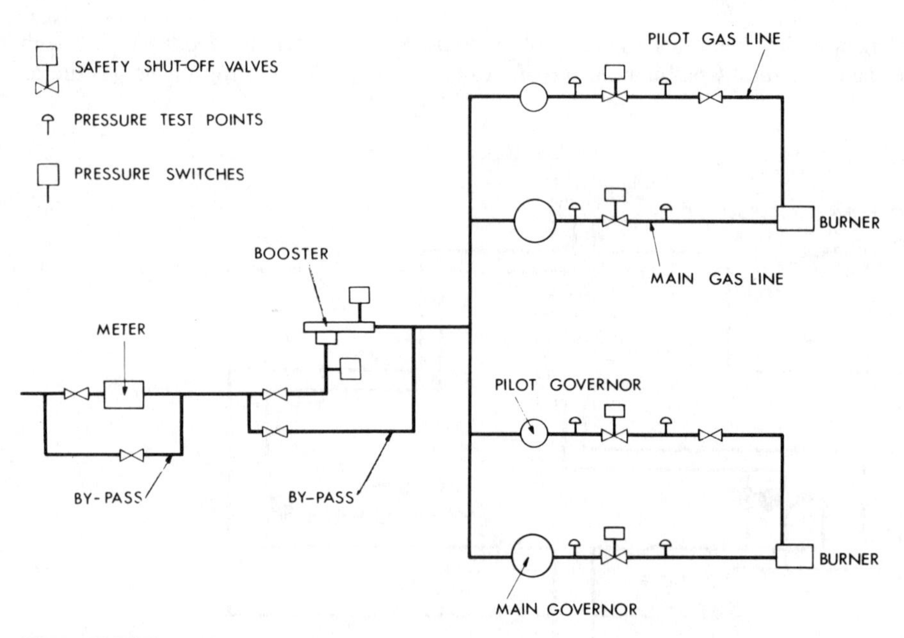

Figure 10.26 Pipework arrangements for gas booster

Safety precautions

In addition to the various devices incorporated in burner control systems, a number of further items must be considered. These apply equally to oil and gas fired plants, as follows:

- Plant operators should be certificated as competent to undertake the tasks expected of them. A *Tech. Eng.* qualification should be the minimum level acceptable.
- Boiler dampers should be either removed, locked open with an indicating plate showing the position of the vane, or provided with an interlock with the burner control.
- Both boiler house and fuel tank room should be provided either with piped foam connections for use by the fire brigade or with chemical fire extinguishers of appropriate type.

Miscellaneous burner equipment

Successive energy crises have led to interest in the consumption of a wide variety of waste products for heat production and to the development of specialist boilers and burners suitable for their disposal. This subject covers too wide a field to be dealt with here with all the attendant ramifications of incomplete incineration, furnace fouling by non-degradable material and atmospheric pollution by exhaust gases. Two items, from opposite ends of the technological spectrum, may be noted as being of interest.

Straw boilers

For disposal of a rural waste product, straw in bales, a number of purpose-made boilers have appeared on the market, together with a variety of ancillary equipment (Figure 10.27). Bales are produced in two sizes, the small rectangular size which measures 350 × 450 × 900 and weighs about 18 kg and the large cylindrical sizes some of which measure up to 2 m in diameter and weigh upwards of 250 kg. Moisture content is quoted at 17% and the net CV appears to be about 12 MJ/kg.

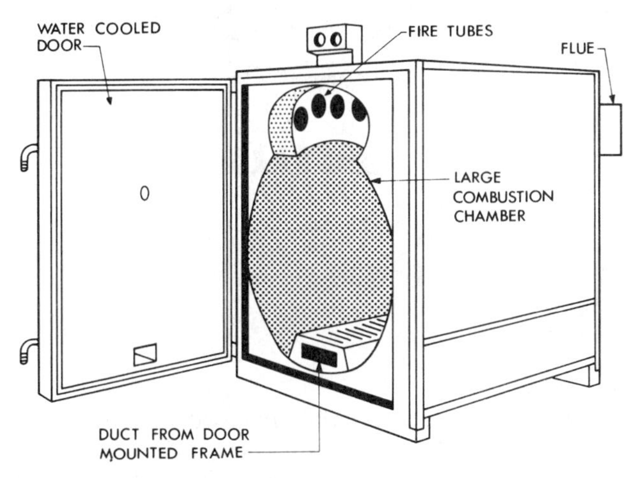

Figure 10.27 Straw burning boiler (Ranco)

The original ranges of boilers were designed to burn the small bales and had ratings of between 20 and 100 kW based upon a 4 hour loading cycle, the combustion chamber of a 100 kW unit, for instance, being large enough to accept seven small bales. Later models designed to burn the larger cylindrical bales are rated at between 100 and 250 kW based upon a 6 hour loading cycle. The loading doors for all sizes are, of course, extremely large and seem to be arranged in most cases for water cooling.

To overcome loading difficulties which, with the larger bales, must be considerable, automatic feed systems are available. These use hydraulic power to divide the bales into smaller compressed portions and to provide a piston feed to the boiler via a conveyor tube. This tube is fitted with a water drenching system and a spring-return fire guillotine.

Fluidised bed boilers

The dramatic effect upon the use of low grade fuel and upon boiler design, which it is expected may result from progress in the development in fluidised bed firing, has yet to be fully appreciated. Whereas, for instance, it is current practice to halve the rated output of a given boiler when it is converted to firing by solid fuel from either gas or oil, this derating would not be necessary if the solid fuel firing were via a fluidised bed.

The concept of combustion within a fluidised bed is so different from conventional practice that it deserves further explanation of the principles involved. If a bed of sand, or a similar inert material such as a crushed refractory, be mounted over a plenum box then, when a critical air velocity is reached through the bed, it will behave very much as if it were boiling, with the bed particles mixing rapidly throughout the depth. If such a bed be heated to a temperature of say 750°C and fuel be then added, combustion will be self-sustaining and the entire bed will become incandescent. The principles are illustrated in Figure 10.28.

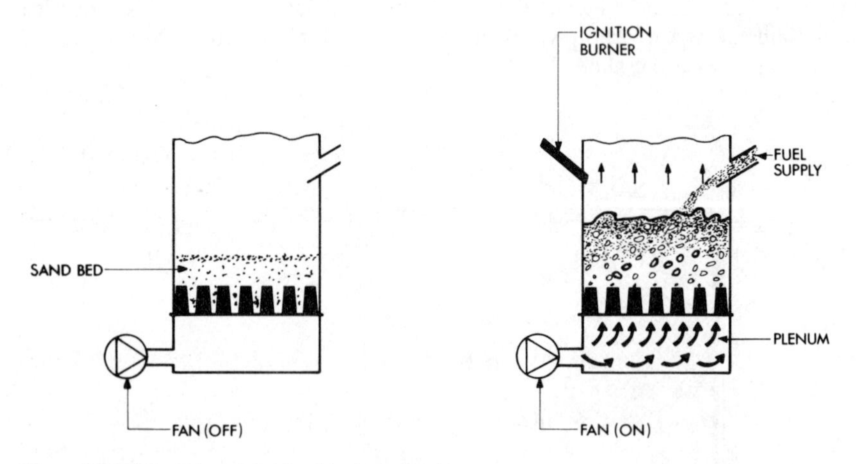

Figure 10.28 Principles of fluidised bed combustion

Heat exchange within a combustion chamber provided with a fluidised bed is extremely complex, including radiant and convective transfer to the surrounding surfaces and direct transfer to any surfaces immersed in the bed, the latter being particularly significant. What is important, however, is that combustion is highly efficient and almost complete with minimal ash residuals.

Developments up to the present time have been directed particularly to the large water-tube boiler field for two reasons. Firstly, if problems in this area can be overcome, then very low grade fuel and waste products of all types may become acceptable for large plants. Secondly, problems in shell boilers arising from lack of available bed depth and from the restricted combustion chamber volume have yet to be completely resolved. A semi-prototype application to a heating installation, rated at 1.7 MW, was installed as a retro-fit in 1984 but a detailed report covering difficulties encountered in operation has yet to be published. It would be a mistake to believe that commercial availability of fluidised bed boilers is imminent, particularly in sizes appropriate to run-of-the-mill plants. But this is not to say that long-term prospects may not be encouraging, once details are made available for analysis.

Instrumentation

The theoretical background to analysis of the products of combustion, as a measure of the 'efficiency' of the combustion process, will be dealt with in Chapter 12. There is now available a number of sophisticated instruments which enable the necessary measurements to be taken but, for site use, the more robust and simple devices of the past may continue to be of value.

Flue gas analysis

The standard apparatus is the *Orsat* with which a sample of the gas is first dried to remove water vapour and is then subjected to an adsorption process by three liquids in turn. The first is a solution of caustic soda which adsorbs CO_2 and the second an alkaline solution of pyrogallic acid plus caustic soda which adsorbs O_2. The third is a solution of cuprous chloride in hydrochloric acid which reacts with CO. Measuring burettes enable the proportion of gas adsorbed in each case to be assessed with comparative ease and the volumetric content of the whole established.

Flue gas temperature

Gases at boiler exit are measured as to temperature by means of either pyrometers or high temperature thermometers which latter may be either nitrogen filled, mercury-in-glass, mercury-in-steel or some form of thermocouple. A variety of devices has been developed to ensure that the gas flow sensed is representative.

Smoke indication

The *Ringlemann* charts are a series of grids of increasing darkness which provide a scale against which smoke emission from a boiler plant may be compared. The charts are enshrined in legislation in that the Clean Air Act prohibits emission of smoke darker than Scale No. 2 except in smokeless zones where no levels on the scale are acceptable. Another scale is the *Bacharach* which uses a filter paper through which a controlled volume of flue gas is passed. Discoloration is matched against standard shades to which are allocated numbers in a scale of 0–9.

Boiler plant is often fitted with smoke density indicating or, in some cases, recording equipment. These devices rely upon an arrangement whereby a light source on one side of the flue relates to a light sensitive cell on the other. In practice, deposition of soot on either the viewing or the transmitting ports may distort the readings.

Chapter 11

Fuels, storage and handling

Except where circumstances require that energy cannot be provided by other than either electricity or gas from a public supply, provision must be made on site to accommodate fuel storage. The correct position for the fuel store can be decided only after taking into account a number of determining factors including aesthetics, reasonable proximity to the boiler plant and convenience of access for deliveries. It is a mistake to agree upon a position, however well suited to the fuel chosen to fire the plant initially, which may then inhibit absolutely any change to an alternative fuel at some time in the future.

Storage capacity

It is usual to provide facilities to store fuel in a sufficient quantity to enable the boiler plant to run at full load for a period of three weeks. Except in special cases, such as a hospital unit, the full load figure would not apply to a 24 hour period but to say 12 hours per day. This rule is one of convenience and is related to an arbitrary estimate of duration for exceptionally severe weather, for an industrial dispute or some other cause for cessation of supply.

For a plant which is quite critical to an industrial process or to, say, a computer building and also in rural situations remote from trunk roads, it may be wise to provide storage for a longer period. In terms of economics, seasonal price differentials may make it attractive to stock-pile fuel in summer. In the special case of small installations in domestic premises, the fuel supplier may propose a storage capacity suited to his delivery routine and offer a reduced unit cost if this be provided.

Solid fuel

An explanation is provided in the following chapter of the various types of analysis to which solid fuel is subjected. A *proximate* analysis will provide information relevant to the practical aspects of combustion, the most important items being the proportion of *volatile* matter, the *caking qualities* and the properties, etc., of the *ash*. Volatiles, a grouping of tars and gases produced as the fuel decomposes in burning, affect not only the rate of combustion but also the length of flame produced. After the volatiles have been burnt off, the solid residue (less any ash) is known as the *fixed carbon*. The cohesive nature of the fixed carbon is an indication of the caking and swelling properties which are important since they affect the porosity of a fuel bed to air flow. The ash content has an indirect influence upon the caking properties and upon the formation of clinker.

Now that coke is, for practical purposes, no longer available for use as a boiler fuel and since, except as a matter of passing interest, we are not concerned here with the incineration of waste products, the principal solid fuel for consideration is coal. Characteristics of the various categories (or Ranks) which are available are summarised in Tables 11.1* and 11.2, the details given being from typical analyses. Size nomenclature is given in Table 11.2, the more common grades being Singles and Smalls. The former is the more consistent as to quality and size mixture and is thus the more conveniently handled by mechanical equipment.

Delivery

Coal delivery, except to very large industrial users able to take a 400 tonne train load, will normally be by road either by a 5–20 tonne tipper or by a 15–20 tonne special purpose vehicle. The latter will be equipped with either a belt conveyor or with facilities for pneumatic discharge. It should be the aim of delivery methods to avoid degradation and segregation of the fuel, i.e. not to break up the larger pieces and not to separate out the large from the small within the mix.

Delivery by tipper vehicle is more appropriate to larger plants, rated perhaps at above 8 MW although there can be no absolute rule as to minimum rating, and to those which are sited away from occupied buildings. Tipping might be through a road grid into an underground store or into a captive tippler hopper as described later. For a smaller plant or one sited near to areas sensitive to noise and dust, totally enclosed pneumatic handling is particularly suitable.

Storage

Solid fuel may, of course, be stored in the open without cover and for very large plants in industry this may be the method adopted. In other circumstances, the site and arrangements selected for storage may very well depend upon the economics of excavation for underground bunkers, the practicability of arranging overground silos, the necessity for mechanised fuel handling and the method of boiler firing to be adopted. An example of the last influence is the use of automatic underfeed stokers, particularly where these are to be of bunker-to-boiler type (see p. 275).

It is necessary in designing all types of silo or bunker to take into account the natural angle of repose which the fuel will adopt when heaped and the consequent necessity to slope the sides of the store to encourage movement, both as shown in Table 11.3. When silos or bunkers are constructed on a permanent basis in concrete, either underground or at surface level, it is wise to provide flexibility for a possible future change in use (to house oil tanks for example) by arranging for the structural walls to be vertical and forming the internal sloping faces in falsework.

An overground silo may be built in any of a number of materials and in some cases may be only a semi-permanent structure built up from pre-fabricated sections. Ferrous sheets of cast iron or mild steel, unless provided with very substantial protection, are not a suitable material for use in construction of storage for a wet and acidic material such as coal.

* Table 11.1 and all ensuing tables dealing with fuel properties are grouped together at the end of this chapter (pp. 307–11).

Handling

The range of equipment available for conveying coal between a central store and the point of combustion is wide. In the case of a large plant, handling may be dealt with in more than one stage, possibly making use of several different methods. The principal criteria to be met are cleanliness, quietness and amenability to automatic control, all achieved with an acceptable level of efficiency. Traditional methods, using open draglines, cables, belts, rollers, buckets and other medieval devices are now acceptable only on remote industrial sites.

A prime requirement for any mechanical handling system is that it should be fully enclosed, a prerequisite that is often incompatible with an adequate level of maintenance. Two methods alone seem to meet with this demand, screw conveyors and what are known as the *en masse* systems. The former operate using the principle of an Archimedean screw. The latter transport the fuel by means of a chain which moves slowly through an enclosed casing. Attached to the sides of the chain are 'flukes', several times wider than the driving links, which take advantage of cohesion within the fuel mass to induce surrounding material to flow as a column.

Alternative to mechanical means are pneumatic systems of conveying by either what are known as *lean phase* or *dense phase* methods. In the former case, the solids are conveyed at a relatively high velocity (20–30 m/s) in a consistent but dilute fluidised mixture with the transporting air volume exceeding that of the fuel moved by as much as 300:1. Using the dense phase approach, the solids are conveyed in compact 'slugs' at a velocity of only about 3 m/s with an air/fuel ratio of perhaps 25:1. The latter method is to be preferred since it is less demanding as to energy and causes less equipment damage by abrasion. An important component in the dense phase system is the shutter or valve which admits the fuel intermittently into the conveying pipeline.

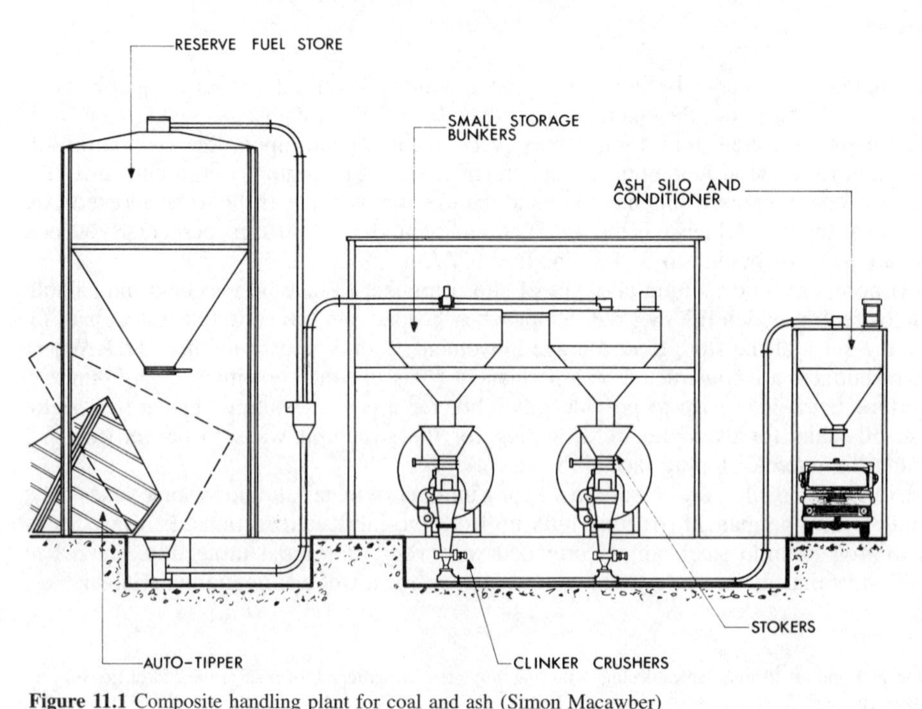

Figure 11.1 Composite handling plant for coal and ash (Simon Macawber)

A complete handling system, from fuel intake to stoker hoppers is illustrated in Figure 11.1. Delivery is to tippler hopper and thence either direct to service bunkers feeding the stoker hoppers or to an intermediate storage silo. Once fuel has been unloaded into the tippler hopper, it is distributed thereafter within a completely enclosed system under automatic control. The fuel levels in the various bunkers and hoppers are sensed by a variety of probes, either rotating paddle or self-cleaning capacitance type, and the appropriate section of the system is energised. The various components illustrated may be disposed to suit other configurations or methods of delivery, intermediate storage and boiler firing.

Ash and clinker handling

While, strictly speaking, the treatment and handling of ash and clinker falls outside the subject matter of this chapter, it is convenient to deal with it here. It will be noted that the overall arrangement illustrated in Figure 11.1 includes a pneumatic system of waste disposal, integrated with that for fuel supply. Use of the word 'waste' may perhaps be an error in this context since ash and clinker have a value and may be sold in some circumstances to offset the cost of collection.

Clinker, of course, must be crushed before any sensible action can be taken to deal with its disposal. In some applications, crushing equipment may be incorporated adjacent to each individual boiler but such an arrangement is not always possible. In many respects, a pneumatic lean phase arrangement has advantages as far as handling ash and clinker may be concerned but each boiler plant will have individual characteristics which will dictate the system to be adopted. In a fully automated system, final disposal of ash and clinker will be via a cylcone hopper to a suitably enclosed vehicle.

Liquid fuel

Of the two principal types of liquid fuel, petroleum oil and the coal tar series, it is the former only which need concern us here. Crude petroleum oil, a complex mixture of paraffins, aromatics and naphthalenes, is subjected at the refinery to heat treatment and other processes leading to progressive decomposition. The various 'families' of components are segregated and subsequently re-grouped into commercially usable products, bitumen being the ultimate residual. It will be appreciated, therefore, that the five grades of petroleum oil which are now available, as listed in Table 11.4, are not unique compounds but products created by selective blending. Rising or falling demand for a given product in the chain may upset the viability of the whole and lead to a quite significant reappraisal of unit costs overall.

At one extreme, kerosene (grade C2) is rarely fired to other than small domestic boilers rated at not more than about 50 kW. Of the remainder, the heavier the grade, the less the unit volume cost but the greater, and more costly, the facilities required for storage, handling and burning. A total economic judgement has to be undertaken, taking all these aspects into account before a selection can be made. For instance, light fuel oil (grade E), was for 20 years or more a much used technical/economic compromise, but can no longer be so classified and thus is now little used.

As far as the various characteristics of the oils listed in Table 11.4 are concerned, the following should be noted:

Flash point (closed). This is the minimum temperature at which, when the oil is heated, a flash may be obtained within the apparatus. Whilst the flash point is a limiting factor upon the amount of low boiling-point material which may be incorporated in the oil, it is not very significant from the point of view of combustion.

Viscosity. This quantity, which is comprehensible when given as a Redwood number (seconds taken for a given quantity to flow through an orifice of given size), is important in so far as it affects the ability of oil to flow in delivery and from storage. It is usually determined via a 'U'-tube viscometer. Values for kinematic viscosity are given in centistokes.

Pour point. This generally follows viscosity and is the temperature at which the oil ceases to run freely. For practical purposes, this must be below that of the normal temperature of a storage vessel.

Sulphur content. Very significant as far as the potential for corrosion is concerned and is also a determinant of statutory requirements for chimney height.

Ash. Although present in small quantities, this has little significance.

Delivery

Except in the case of very large installations, oil delivery is made by road tanker. Thus, site facilities must be provided for tanker parking and the position of the fill terminal be within an agreed distance from the park. The fill terminal should be a standard 50, 65 or 80 mm male gas thread, the size being determined by the length of the fill pipe and the grade of oil. The thread should be suitable for hose coupling and fitted with a non-ferrous cap as protection. As may be seen from Table 11.4, the heavier grades of oil are delivered already heated and for such usage the fill pipe should be insulated and provided with electrical trace heating.

Where more than one tank exists, filling should preferably be through separate pipes, one per tank, although a single fill pipe and an arrangement of three-way cocks or valves at the tanks may be used. Fill pipes should, where practicable, be laid to drain into the tank and be fitted with an isolating valve at the terminal end. Although the tanker driver should purge the fill pipe with air on completion of each delivery, the provision of a permanent drip-tray arrangement is a prudent step.

Storage

Since the storage of a quantity of oil in or near a building provides the elements of a fire hazard, various bodies have drawn up regulations covering the installation of storage tanks and other aspects of oil firing. A number of British Standards has been published and reference to the latest (currently BS 5410) is advisable, always bearing in mind that these publications provide no more than minimum specifications. Local authorities, fire authorities and insurance companies have their own requirements as far as oil storage is concerned and these too should be noted.

At domestic level, a tank capacity of not less than about 3500 litres makes sensible allowance for the minimum delivery of 2500 litres preferred by oil companies. If a facility exists whereby a metered 'milk round' service is offered, then a tank holding as little as 1300 litres may be installed to serve a house built in an area which is fully guaranteed to be frost, ice, snow and strike free.

In larger installations, it is frequently necessary to provide two or more storage tanks. This has the advantage that filling and usage may be rotated, thus allowing any sludge or water to settle out and maintenance to be effected. Table 11.5. gives the consumption of

Table 11.5 Approximate requirements for oil storage

Boiler rating (kW)	*Storage for 3 weeks* (litre)	*Boiler rating* (kW)	*Storage for 3 weeks* (litre)
20	700	200	6720
40	1390	300	9870
60	2080	400	12880
80	2760	500	15750
100	3430	750	22310
150	5090	1000	28000

Assumptions:
1. 21 days × 12 hours.
2. Boiler efficiency = 75%.
3. Oil CV = 41 MJ/kg.
4. Error at 1000 kW = 10% and pro rata below.

boiler plants of various sizes based upon a three week operation at full load, 12 hours per day.

Tanks used for oil storage are most often of welded steel construction and may be either rectangular or cylindrical. A generality might be that vessels mounted in the open air are more commonly cylindrical, horizontal or vertical, and that tanks housed within a building are more commonly rectangular. Construction may be prefabricated completely or carried out *in situ* from steel plate or formed sections. The latter method is obviously necessary for confined situations. Capacities and sizes, cylindrical and rectangular, are given in Table 11.6.

Cylindrical vessels for mounting in the open should, where horizontal, be supported on sleeper walls topped by steel cradles. Vertical cylinders should be constructed in a manner which keeps the bottom plate free from overall contact with the supporting base. Rectangular tanks sited in the open should have top plates hipped to reject rainwater and, wherever sited, be supported on sleeper walls topped by steel joists. Cradles and joists should both be bedded on bituminous felt or lead packing. All tanks must be arranged such that water, sludge and other accumulations may be drawn off at a convenient low point.

A domestic range of oil tanks manufactured in medium density polythene is now available. The material is rotationally moulded in one piece to provide a stress-free container of uniform wall thickness which is resistant to cracking. The rectangular tanks, with domed top, and the vertical cylindrical pattern must be mounted on a flat surface and not piers. The capacities and dimensions are listed in Table 11.6.

Tank rooms

Where within a building, a tank room to house oil storage must be separated from a boiler house by brick or concrete construction. The floor and lower walls must be treated with an oil-proof render to contain the whole capacity of storage as a precaution against leakage. An access opening, in consequence, must have the sill at an elevated level with ladders inside and outside and be provided with a fire resisting door all as illustrated in Figure 11.2. Arrangements for inlet and outlet ventilation must be made, separate from all other systems.

There are three methods used for external underground storage, firstly by literally burying a cylindrical vessel in a prepared pit. Substantial protection of the external steel surface is essential and the bottom of the excavation should be drained. In waterlogged ground, it is generally necessary to anchor the vessel to a heavy block of concrete to overcome buoyancy when the tank is less than full. This, a not very satisfactory expedient, is nowadays little used.

Table 11.6 Oil tank capacities

Diameter (m)	*Length* (m)	*Gross capacity* (litre)	*Net capacity*[a] (litre)
Cylindrical			
1	2	1 570	1 300
	2.5	1 960	1 600
1.5	2	3 530	3 200
	2.5	4 420	4 000
	3	5 300	4 800
2	3	9 430	8 800
	3.5	11 000	10 200
	4	12 570	11 800
2.5	3.5	17 180	16 200
	4	19 640	18 400
	4.5	22 090	20 800
3	4	28 280	26 900
	5	35 350	33 600
	6	42 420	40 200

Length (m)	*Width* (m)	*Depth* (m)	*Gross capacity* (litre)	*Net capacity*[a] (litre)
Rectangular				
1	1	1	1 000	700
1.5	1	1	1 500	1 100
2	1	1	2 000	1 500
3	1.5	1	4 500	3 300
3	1.5	1.5	6 750	5 600
3	2	1.5	9 000	7 500
4	2	2	16 000	14 000
4	3	2	24 000	21 000
4	4	2	32 000	28 000
6	4	2	48 000	42 000

Length (m)	*Width* (m)	*Depth* (m)	*Diameter* (m)	*Net capacity* (litre)
Plastic				
1.37	1.06	1.25	–	1 250
1.8	–	–	1.1	1 350
2.02	1.36	1.36	–	2 500
1.35	–	–	1.61	2 600

[a] Allowance 150 mm up to outlet and 100 mm ullage (at top), rounded to nearest 100 litres lower.

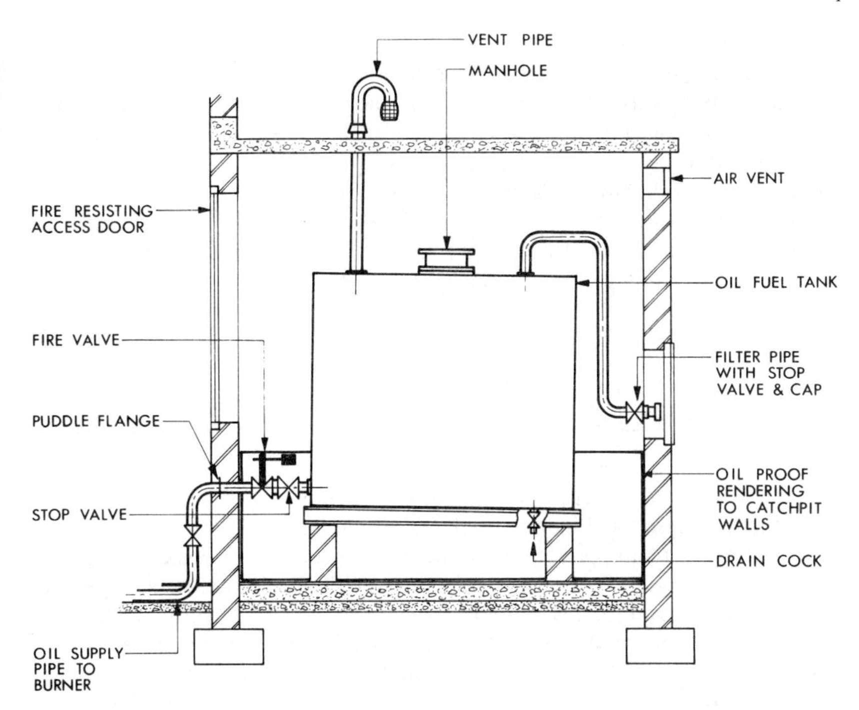

Figure 11.2 Oil tank and connections, above ground

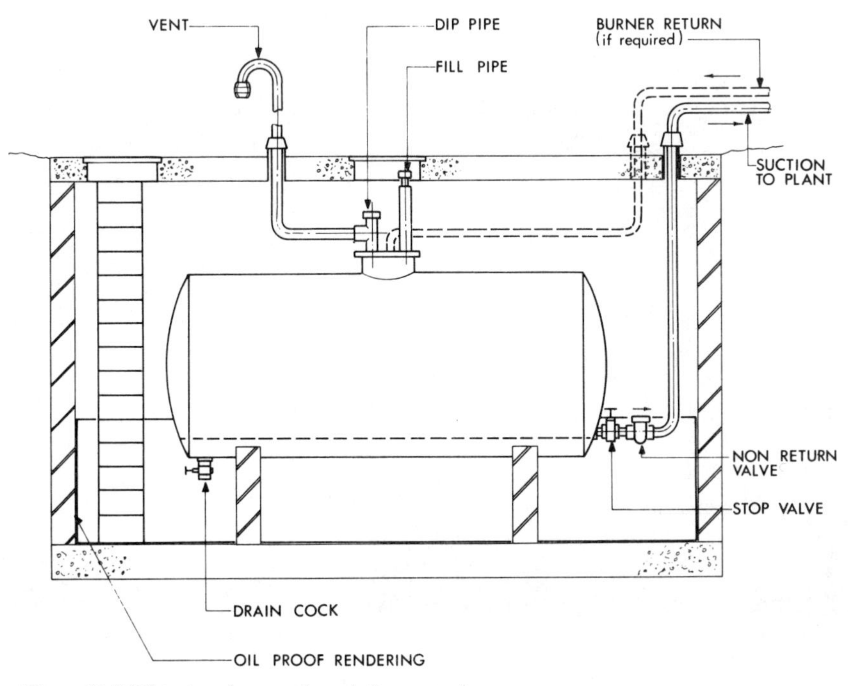

Figure 11.3 Oil tank and connections, below ground

A second method involves the provision of an underground chamber, as illustrated in Figure 11.3. Construction must be such that there cannot be either ground water leakage inwards or oil leakage outwards. Manholes for access must be substantial and convenient if maintenance is to be adequate. After allowing for a walking space around the tank and for shape factors, the ratio of excavation to oil storage volume will be less than 2:1.

The third method is based upon a radically different approach in that the tank shell is itself a concrete structure lined with special tiles. This concept, which is a proprietary design, has much to commend it, not least being the fact that the whole of the available cube internal to the underground structure is used for storage.

Tank fittings

Storage tanks must be provided with a variety of fittings, some as part of the tank and others related to filling. The following is a brief summary of the principal items:

Manhole. Except in the case of small domestic tanks, a manhole with an air-tight cover joint must be fitted. Internal and external ladders should be fitted.

Vent pipe. A vent from each tank, preferably of the same size as the fill pipe, is required. It should be carried up to some point where any fumes will not be troublesome, perhaps to the roof of an adjacent building. A bird guard should be fitted at the terminal. Oil may rise in the vent pipe when a tank is over-filled and thus impose an excessive pressure on the shell. An alarm device of some sort is desirable to give warning of this condition and an oil-seal trap or one of the proprietary unloading devices should be fitted.

Sludge valve. Water, being denser than oil, collects at the bottom of a tank together with sludge and other solid matter. This accumulation must be removed at intervals through

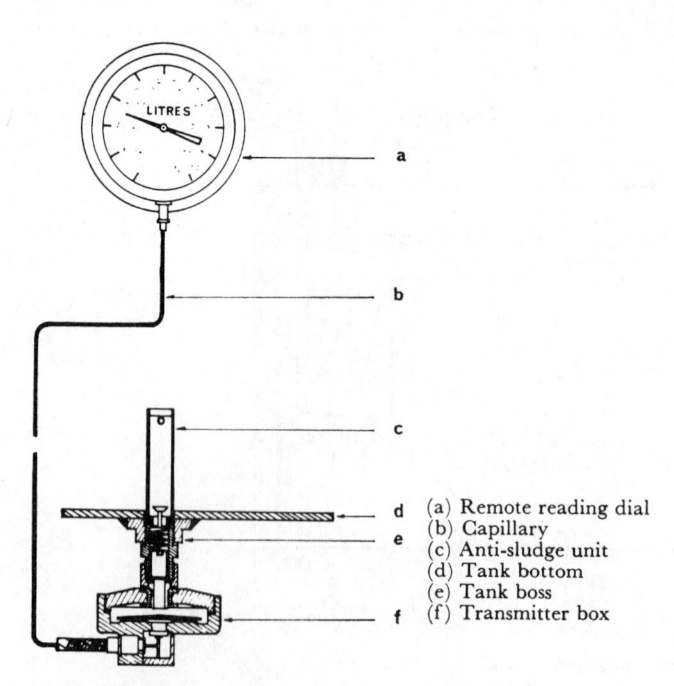

Figure 11.4 Oil tank contents gauge

a sludge valve which should be easily accessible. In the case of buried or underground tanks, this can be a messy business.

Level indicator. Gauge glass indicators, although positive, are not very durable; dip-sticks are inconvenient; direct mounted dial gauges are subject to damage and what are known as 'cat and mouse' float levels are thought by some to be fallible. There is a variety of hydrostatic remote reading gauges available, as Figure 11.4, and electrically or (better) pneumatically operated versions may be obtained.

Heaters. In the case of heavy oils, grades F and G, it is necessary to provide means of maintaining a suitable minimum temperature in storage, see Table 11.4, either by pipe coils or by electrical heaters. These are commonly of very simple construction and provided with a coarse control of temperature.

Outlet valve. For isolation, the outlet from each tank must be provided with a simple and reliable stop valve.

Handling

The treatment required to handle the flow of oil between storage and the point of utilisation, and the associated equipment, varies greatly with the grade to be used. Necessary in all cases is an oil filter, which needs no further comment, and a fire valve mounted preferably externally, but always as near as possible, to the oil pipe entry to the boiler house.

In its crudest form, a fire valve consists of a lever type isolator, heavily weighted, which is kept open by a taut wire stretched across the boiler house and has a fusible link of low melting-point alloy over the oil burner with a hand operated release/test point at the boiler house door. Such devices are almost useless since they rely for operation upon the flexibility of the wire which, inevitably, becomes corroded or deformed where it passes over pulleys and, in any event, is rarely if ever tested for operation. It is now usual to provide a much superior system where the weight-closed valve is kept open by a solenoid operated catch, current to the circuit being maintained through a heat sensitive device (thermal fuse) mounted above the burner. Alarm bell or other contacts are easily added to such a circuit which inherently 'fails safe'.

For heavy oils, grades F and G, it is necessary to provide heating and pumping equipment to take fuel from storage and deliver it to the burners. At the tank, an outflow heater may be provided, as in Figure 11.5, designed to preheat oil flow at the point of exit rather than in store. The principal item of equipment may be a prefabricated pumping and heating unit as shown in Figure 11.6 which incorporates not only the necessary positive

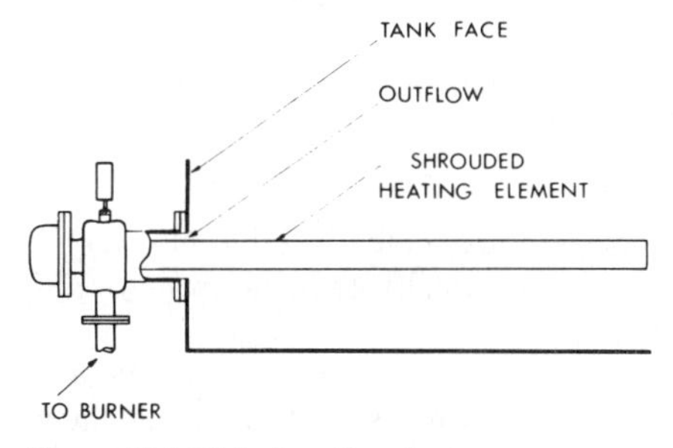

Figure 11.5 Oil tank outflow heater

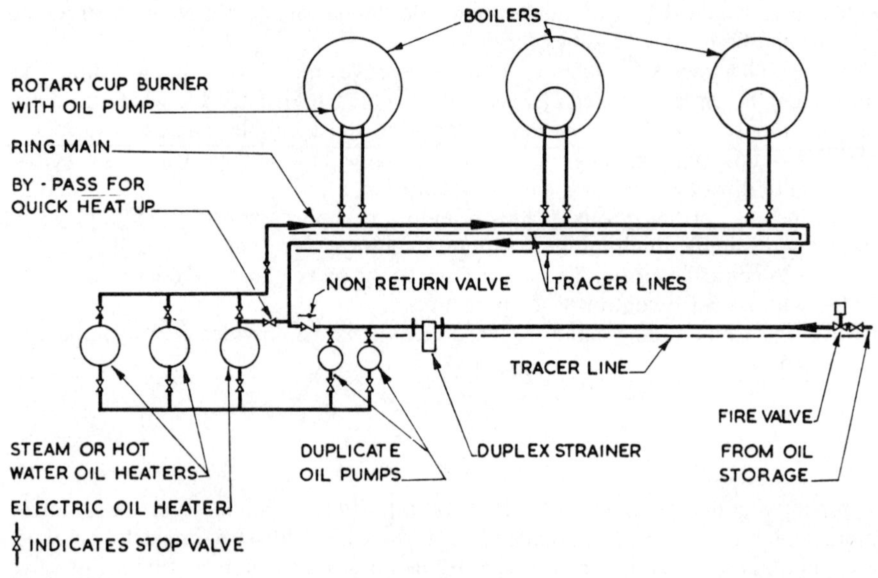

Figure 11.6 Pumping and heating set with oil ring main

displacement pumps but also hot water or steam heaters with electrical back-up for use when starting the plant from cold.

From the pumping and heating set, the hot oil circulation is extended right up to the burner in order to avoid any problems of smoke on start up but variations in the actual arrangement may be found to be necessary with burners which do not incorporate integral

Table 11.7 Flow of oil in steel pipes (BS 1387: medium)

	Value of factor θ for pipes having stated nominal bore (mm)								
Grade of oil	*10*	*15*	*20*	*25*	*32*	*40*	*50*	*65*	*80*
D	0.56	0.19	0.06	0.02	0.01	–	–	–	–
E	–	7.69	2.43	0.96	0.32	0.17	0.07	0.02	0.01
F	–	–	3.04	1.19	0.40	0.22	0.08	0.03	0.02
G	–	–	4.86	1.91	0.64	0.34	0.13	0.05	0.03

Notes:
Factors apply to oil at handling temperature, Table 11.4.
Single resistances:
For approximate purposes, equivalent lengths of bends, tees, etc., may be ignored.

pumping equipment. Pipework carrying oil at the temperatures noted in Table 11.4 should be insulated and be provided with *trace heating*, either electrically or by hot water/steam as may be available.

The viscosities and velocities encountered in oil distribution piping are normally such that the flow is streamline or laminar. The consequent loss in pressure may be calculated from a simplified approximate equation:

$$\Delta p = \Theta \cdot Q$$

where

Δp = pressure drop per m run (Pa/m)
Θ = a factor read from Table 11.7
Q = boiler power, at 75% efficiency (kW)

Subsidiary storage and handling

The subject of roof-top boiler houses has been referred to in the preceding chapter, particularly in the context of multi-storey buildings. In such cases, it is usual for the main oil store to be retained at ground level and for a supply to be pumped to a small *daily service tank* in or near the boiler house, as shown in Figure 11.7. The size of the service tank is dictated by regulations as are the methods of control for the transfer pumping equipment. Pumps are usually provided in duplicate and arranged to be started by hand but stopped by a level control fitted in the service tank.

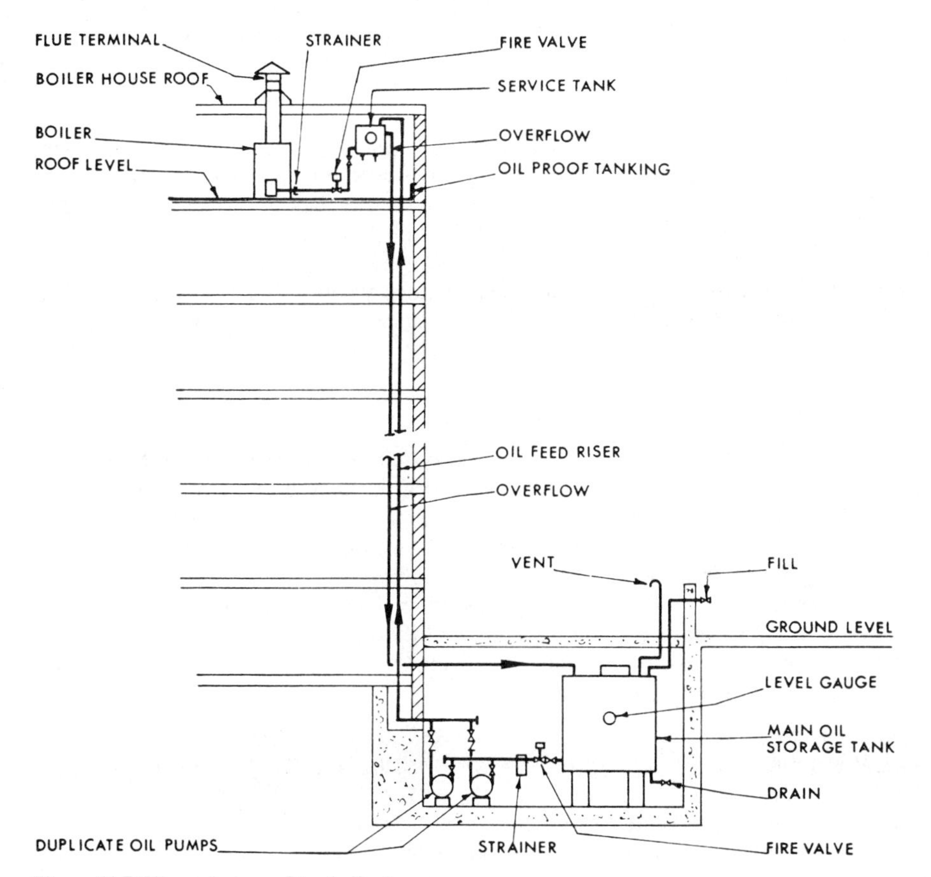

Figure 11.7 Oil supply to roof-top boiler house

North Sea natural gas

Current statistics suggest that natural gas from the North Sea is providing more than 98% of the national '*mains*' supply in the British Isles and, in consequence, manufactured gas need be considered only in the context of history and as a basis for comparison. For distribution from import terminals, a high pressure pipe network has been created to serve the various areas of the mainland. Some of the terminals are capable also of receiving deliveries of liquified natural gas supplied under long term contract from North Africa. To cater for the inevitable diurnal fluctuations in demand, the not inconsiderable volumetric capacity of the primary high pressure distribution pipework is supplemented at national level from storage underground and, for local low pressure storage, by the familiar above-ground gasometers.

In order to make provision for exceptional peak demands upon the system, special '*interruptible*' tariffs have been offered to large industrial users, the arrangement being that their supply may be cut, after due warning, for specified periods. In such instances an alternative fuel supply is needed in the form of individual bulk stores, on site, of either liquified petroleum gas (LPG) or fuel oil. The boiler plants concerned must, of course, be equipped to utilise both natural gas and the alternative fuel. In general terms, however, natural gas offers the advantage of requiring no storage facilities at the point of use.

As may be seen from Table 11.8, which lists the properties of both natural and manufactured gas, the former is a methane-rich mixture of a number of hydrocarbons.

Delivery

A statutory requirement exists whereby the natural gas suppliers are obliged to provide and maintain gas to premises sited within 23 m of a distribution main. In principle, the cost of making a connection to that main and of running the first 9 m of service pipe is met by the supplier and the remainder is a charge to the building owner. The service pipe will terminate in an isolating valve or cock and be followed by a meter which may be either a *positive displacement* type or, for large installations where space is at a premium, an *inferential* (shunt) unit. The latter are inherently suspect as to accuracy and should be checked at frequent intervals.

Pipework

From the meter outlet, the gas supply distribution system becomes the responsibility of the building owner. The *Guide Section C4* provides data in tabular form for flow of gas in pipes and Table 11.9 is a digest from this, for steel pipes. Retaining the kilowatt as a basis of calculation, a unit flow of natural gas (1 litre/s) is equivalent to an energy supply of 39 kW (strictly 38.6). Thus the input requirement, in kW, of any boiler or other appliance may be converted into gas flow by using this equivalence. Alternatively, makers' lists sometimes give input gas rates volumetrically and these may be used.

In terms of pressure loss, it is usual to limit this to between 75 and 125 Pa from the meter to the point of use. Thus, for example:

Input rating to boiler		= 750 kW
Natural gas rate	= 750/39	= 19.2 litre/s
Pressure available, say,		= 100 Pa
Pipe run from intake to boiler		= 30 m
Pressure available/m run	= 100/30	= 3.3 Pa
Thus, from Table 11.9, pipe size		= 65 mm

Table 11.9 Flow of natural gas in steel pipes (BS 1387: medium)

Pa per m run	*Litre per second in pipes having stated nominal bore* (mm)										
	15	*20*	*25*	*32*	*40*	*50*	*65*	*80*	*100*	*125*	*150*
0.5	0.07	0.24	0.47	1.02	1.6	3.0	6.1	9.5	20	35	57
1.0	0.14	0.37	0.71	1.54	2.4	4.5	9.1	14.1	29	52	84
1.5	0.21	0.48	0.90	1.96	3.0	5.7	11.5	17.9	37	65	106
2.0	0.25	0.56	1.08	2.31	3.5	6.7	13.6	21.0	43	76	124
2.5	0.29	0.64	1.23	2.64	4.0	7.6	15.4	23.9	49	86	140
3.5	0.34	0.78	1.50	3.20	4.8	9.2	18.6	28.9	59	104	169
5.0	0.43	0.96	1.83	3.92	5.9	11.2	22.7	35.1	72	127	205
7.0	0.51	1.17	2.23	4.75	7.2	13.6	27.4	42.3	86	152	246
10.0	0.64	1.44	2.72	5.81	8.8	16.6	33.4	51.5	105	185	298
15.0	0.81	1.82	3.43	7.30	11.0	20.7	41.8	64.3	130	230	370
20.0	0.95	2.14	4.04	8.57	12.9	24.3	48.9	75.3	152	269	432
25.0	1.09	2.43	4.58	9.71	14.7	27.5	55.2	84.9	172	303	487
	0.5	0.5	1	1	1.5	2	2.5	3.5	5	7	8

Note: single resistances.
For approximate purposes, equivalent lengths may be taken as length in metres as listed in the bottom line of the table for each bend, tee, etc.

The pressure available for distribution of gas within a building is bound up with governor settings, requirements for burner intake pressure and with demands for boosting equipment. Such matters need investigation in the light of manufacturers' application data in individual cases and with the gas suppliers.

Liquified petroleum gas

Only 78% of households in Great Britain (55% in Scotland and 64% in Wales) are listed as being consumers of gas from a public '*mains*' supply. Liquified petroleum gases (LPG) are convenient alternatives in circumstances where a supply of natural mains gas is not available. They may be used not only for the familiar portable heaters and other domestic appliances but for firing to static boilers, etc., also.

Whereas natural gas from the North Sea is a mixture of various hydrocarbon compounds, liquified petroleum gases are available as commercially pure butane and propane. Use of butane is normally confined to the smaller applications. Table 11.10 lists the properties of both gases and, for comparison, those of natural gas.

Delivery and storage

LPG is stored on site, as a liquid, in pressure vessels which are fitted by the gas suppliers and remain their property. They are refilled as may be necessary by delivery from a road tanker. It is a general recommendation that vessels for larger installations be fitted in duplicate so that inspection and any maintenance may be carried out without interrupting output. Dimensions and other leading particulars of storage vessels are given in Table 11.11.

Table 11.11 Storage of liquified petroleum gases (LPG)

	Cylinder		*Capacity*			*Size* (mm)		
Gas	*Colour*	*Pressure* (kPa)	*kg*	*Liquid*[a] (litre)	*Gas* (m^3)	*Height or length*	*Diameter*	*Normal offtake* (litre/s)
Butane	Blue	172	4.5	7.8	1.8	340	240	0.04
			7	12.2	2.8	495	256	0.06
			15	26.1	6.1	580	318	0.08
Propane	Red	690	13	25.4	7.0	580	318	0.16
			19	37.2	10.2	810	318	0.2
			47	92	25.2	1290	375	0.35
	White	690	615	1200	330	2000	1000	1.6
	or		1020	2000	550	3000	1000	2.0
	Green		1735	3400	930	3800	1200	3.7
	White	690	715	1400	380	Sphere	1500	1.8
			2040	4000	1100	3800	1200	4.4

[a] Notional capacity 87% fill at 15°C.
Note: larger bulk storage tanks of 24 m^3 are available but no standard sizes exist.

Regulations exist as to the siting of storage vessels and these are set out in a variety of publications, some by HMSO, some by British Standards and others (*the most useful*) by the distributors' Trade Association. A prime requirement is that vessels must not be installed within buildings and particularly in basements, LPG being heavier than air. According to the capacity of a store, the distance by which it is separated from sources of ignition (such as balanced flue boilers having low level terminals), site boundaries and buildings is laid down, and varies from 3 m for a small domestic size vessel to 15 m for the largest. Access by tanker must not be more than 30 m distant from the store.

Storage pressure for propane is 690 kPa and this is reduced to about 75 kPa at a first stage regulator which supplies the primary distribution main running up to the outside of a building. A second stage regulator fitted there will have an output at 3.7 kPa to suit the standard low pressure input rating for which most utilisation equipment is designed.

Pipework

From the second stage regulator, pipework for LPG should be sized using the same principles as have been set out previously for natural gas but using the data in Table 11.12. A pressure drop of about 250 Pa between the second stage regulator and the point of use is the accepted criterion. Once again, retaining the kilowatt as a basis for calculation, a unit flow of propane (1 litre/s) is equivalent to an energy supply of 95 kW.

Electricity

The latest statistics available show that in the British Isles, a large proportion of the output of the major electricity generating companies (almost 55%) was produced by conventional coal fired/steam turbine power stations. Oil and natural gas fired stations contributed about 14% and 3% respectively and, of the remainder, 18% was produced by nuclear

Table 11.12 Flow of LPG in steel and copper pipes

Pa per m *run*	*Litres per second in pipes having stated nominal size* (mm)								
	Medium steel (BS 1387)				*Copper* (BS 2871 (Table X))				
	8	*15*	*20*	*25*	*6*	*10*	*12*	*22*	*28*
4	0.030	0.23	0.46	0.93	0.006	0.044	0.075	0.39	0.78
5	0.034	0.26	0.53	1.11	0.007	0.049	0.083	0.42	0.89
7	0.039	0.31	0.63	1.32	0.008	0.059	0.103	0.54	1.06
10	0.047	0.38	0.75	1.61	0.010	0.072	0.122	0.66	1.31
15	0.058	0.46	0.92	2.00	0.012	0.091	0.156	0.83	1.64
20	0.068	0.53	1.08	2.33	0.015	0.107	0.183	0.99	1.94
25	0.076	0.60	1.22	2.61	0.017	0.121	0.208	1.11	2.19
30	0.083	0.66	1.33	2.89	0.019	0.135	0.230	1.25	2.47
40	0.097	0.77	1.56	3.33	0.022	0.158	0.272	1.47	2.86
50	0.108	0.86	1.74	3.75	0.025	0.178	0.306	1.67	3.31
70	0.119	0.95	1.93	4.17	0.027	0.201	0.342	1.86	3.64
80	0.139	1.11	2.25	4.86	0.032	0.236	0.403	2.17	4.31
	0.3	0.5	0.7	1.3	0.1	0.3	0.4	0.7	1.0

Notes: single resistances.
For approximate purposes, equivalent lengths may be taken as lengths in metres as listed in the bottom line of the table for each bend, tee, etc.

power; 7% by hydro-electric plant and the remainder by gas turbines and oil engines. Each of these main types has a different capital/running cost relationship and, in consequence, each is best suited to contribute to a particular pattern of loading. Nuclear stations, for instance, are more economical when meeting the base load and gas-turbine stations for dealing with a short term peak load. Furthermore, stations of the same type will have differing costs of production depending upon age, design, etc., but the National Grid allows the most efficient stations to be run at any given time to match the load at that time.

A typical load curve has a morning peak which usually extends from 8 a.m. onwards until about noon with a lesser peak in the late afternoon. Such changes in demand, insofar as they relate to motive power and lighting, etc., are inevitable but nevertheless create problems for the supply industry. While electricity, uniquely among energy supplies, cannot be stored as such it can be converted without loss into heat which may then be stored using both old and new techniques. Those most relevant to the present subject are described in an earlier chapter (p. 116).

Tariff structures

Because plant of overall higher efficiency operates at night, generation costs are lower then than during the day. If the costs of metering were of no consequence either to the supply industry or to the consumer, then a price structure which took account of both demand pattern and energy use might be the universal and preferable tariff practice.

For the larger consumer, a tariff including separate components for maximum demand and energy use is commonly used and a variety of such arrangements exists to suit different circumstances. In each case, the charge for energy used may be at a single rate or be biassed as to time of use, by day or by night. The family of tariffs used in the domestic sector take the form of a standing charge which reflects the cost of making the supply available, plus unit rates set at a level appropriate to recovery of both the demand- and energy-related costs. The latter may, again, be as a single rate for all units no matter when consumed; as a dual rate where units consumed for a specific purpose during the night are charged separately or, more usually, in simple day/night form where one rate applies during the day and a lower rate is charged for all use during the night. For domestic consumers, the off-peak rate is commonly 50% or less of that charged for daytime use.

Since the various supply companies throughout the country are independent organisations, the details of their tariffs are related to the particular supply and distribution circumstances arising within the geographical area which they serve. In general terms, however, roughly seven hours of off-peak supply are available each 24 hours, commonly from midnight (24.00 GMT) until 7 a.m. (07.00 GMT). During the summer months, when British Summer Time is in force, the availability of the off-peak supply remains related to GMT and is thus out of phase.

Space heating

When a daytime tariff applies, electricity is an expensive source of energy but, in spite of this, it is used extensively as a result of its almost universal availability and the flexibility it offers. It is, however, as a supply taken off-peak that electricity has rather more to offer the user. At any time this supply offers:

- Transmissibility to any point regardless of physical levels or similar limitations.
- Absence, local to the point of use, of dust from fuel and ash, fumes, etc.
- Reduction in on-site labour for plant operation.
- Availability at actual point-of-use.
- Relative ease of control for energy input.

It is often argued that to produce electricity from a raw fuel involves much wastage since more than three-quarters of the calorific value of that fuel, as supplied to the generating station, is wasted in cooling towers, rivers and canals or is lost in the distribution system. The contra-argument was that the grade of raw fuel used in generating stations is so low that it would be difficult if not impossible to burn it elsewhere and, whilst this statement is valid as advanced, the value of that low grade fuel as a chemical feed-stock cannot be ignored. The construction of a generating station designed to burn natural gas alone is attractive in the sense of capital expenditure. Properly equipped and sited near to a demand for heat supply (see Chapter 24), a case just might be made for such usage. Otherwise, the profligate consumption of such a refined source of energy (and one which may well be in short supply before the end of the century) is incomprehensible.

Miscellaneous fuels

In addition to the more usual fuels and sources of energy, others are in use for a variety of reasons. To give but two examples:

Timber, as logs or off-cuts

If burnt in a properly designed stove with an integral boiler, timber may in some circumstances provide primary or back-up facilities at domestic level for heating tap water or a few radiators. When sold as fuel, timber may be either waste off-cuts or logs purpose-cut from stick which is either diseased or otherwise unsuitable for use as a raw material in manufacture. It may originate from a wide range of growth, imported or home produced, and be *green, air dried, seasoned* or *oven dried.*

The principal characteristics of timber are set out in Table 11.13 together with, for comparison, those for a medium volatile coal. Moisture content, as would be expected, has a very significant effect upon the calorific value, of the order of plus or minus 30%, and any manufacturer's performance data are likely to be rather higher than results achieved in practice due to the firing problems which arise from the high volatile content.

Waste products

These, from a very wide variety of sources in manufacturing industries and from municipal collection, may be burnt at a relatively low efficiency in large scale specialist furnaces equipped with accessible heat exchangers. The characteristics of such waste material vary so widely that no sensible generalisations are of use.

The approximate data listed in Table 11.14 illustrate the changing nature of what might be supposed to be a representative mixture. In contrast to urban waste, however, baled straw where surplus to requirements as an animal fodder may be available in rural areas for use as a fuel.

Table 11.1 Properties of coal (proximate analysis)

	Volatiles			*Air dried, moisture*	*CV* (MJ/kg)	
Rank No.	*%*[a]	*Type*	*Caking properties*	(%)	*Gross*	*Net*
100	< 10	Very low	Non-caking	2	32.4	31.6
201	12	Low	Non-caking[b]	1	33.0	32.2
202	15	Low	Weakly[b]	1	33.2	32.3
204	19	Low	Strongly[b]	1	33.2	32.6
300	25	Medium	Weakly[c]	1	33.2	32.3
301	25	Medium	Strongly[c]	1	32.9	31.8
400	> 30	High	Very strongly[c]	2	32.3	31.2
500	> 30	High	Strongly[c]	2.5	31.5	30.3
600	> 30	High	Medium[c]	4	30.1	29.1
700	> 30	High	Weakly[c]	5	29.1	27.9
800	> 30	High	Very weakly[c]	8	27.9	26.6
900	> 30	High	Non-caking[c]	10	25.4	24.1

[a] Average for singles. [b] Sub- and semi-bituminous. [c] Bituminous.

Table 11.2 Properties of coal (ultimate analysis)

	Content (%)						
Rank No.	*Moisture*[a]	*Ash*	*Carbon*	*Hydrogen*	*Nitrogen*	*Oxygen*	*Sulphur*
100	3.5	5	84.5	3.0	1.1	1.8	1.1
201	3	5	83.7	3.7	1.3	2.2	1.1
202	3	5	83.4	3.8	1.3	2.4	1.1
204	3	5	83.1	4.1	1.3	2.4	1.1
300	3	5	82.1	4.4	1.4	2.8	1.3
301	3	5	81.0	4.5	1.4	3.8	1.3
400	4	5	78.5	4.7	1.7	4.2	1.9
500	5	5	76.3	4.8	1.6	5.4	1.9
600	6.5	5	73.9	4.7	1.5	6.5	1.9
700	9	5	71.1	4.5	1.5	7.1	1.8
800	11	5	67.8	4.4	1.4	8.5	1.9
900	16	5	63.0	3.9	1.3	9.0	1.8

[a] As fired.

Table 11.3 Properties of coal (physical data)

		Size range (mm)		Angle to horizontal (deg.)		
					Hopper sides	
Group name	*Bulk density*[a] (kg/m^3)	*Upper*	*Lower*	*In repose*	*Steel*	*Concrete*
Trebles		90–64	50–40			
Doubles	580–700	57–45	40–25	40	40	47
Singles		40–25	25–13			
Peas		20–13	13–6			
Grains	700–800	11–6	6–3	55	55	60
Smalls[b]		50–25	None			

[a] Dry coal loosely packed. [b]Smalls are not a graded group.

Table 11.4 Properties of fuel oils

Description	Grade of oil				
	C2 Kerosene	*D Gas oil*	*E Light*	*F Medium*	*G Heavy*
Specific mass (*re* water)	0.789	0.834	0.929	0.949	0.969
Flash point (closed) (°C)	38	55	66	66	66
Viscosity at 38°C					
Redwood No 1	28	34	250	1000	3500
kinematic (cSt)	1 to 2	4	62	247	864
Pour point (°C)	–	–18	–7	21	21
Calorific value (MJ/kg)					
gross	46.4	45.5	43.4	42.9	42.5
net	43.6	42.7	41.0	40.5	40.0
Impurities (% mass)					
sulphur	0.2	0.75	3.2	3.5	3.5
ash	–	0.01	0.05	0.12	0.2
Temperature (°C)					
storage (minimum)	Atmospheric		10	25	35
handling (minimum)	Atmospheric		10	30	45

Table 11.8 Properties of natural and manufactured gas

Item	*Natural gas*	*Manufactured gas*
Specific mass		
kg/m^3 at 15°C	0.78	0.6
re air = 1	0.6	0.48
Calorific value (MJ/m^3)		
gross	41.6	20.9
net	38.7	18.6
Wobbe No.	49.9	27.0
Burning velocity (m/s)	0.35	0.8
Ignition temperature (°C)	650	600
Operating pressure (kPa)	2	1
Toxicity	Nil	Toxic
Constituents (% volume)		
methane	92.6	33.5
ethane	3.6	–
propane	0.8	–
butane	0.2	–
pentane and above	0.1	–
hydrogen	–	47.9
nitrogen	2.6	–
CO_2	0.1	4.9
CO	–	13.7

Table 11.10 Properties of natural and liquified petroleum gas

		Commercial LPG	
Item	*Natural gas*	*Butane*	*Propane*
Specific mass			
as gas at 15°C (kg/m^3)	0.78	2.45	1.85
as gas at 15°C (*re* air = 1)	0.60	2.0	1.5
as liquid (kg/litre)	–	0.575	0.512
Volume of gas			
per volume of liquid (m^3/m^3)	–	233	274
per mass of liquid (m^3/kg)	–	0.41	0.54
Calorific value (gross)			
re mass (MJ/kg)	53.3	49.2	50
as gas (MJ/m^3)	41.6	121.5	95
as liquid (MJ/litre)	–	28.2	25.5
Ignition temperature	650	510	510
Operating pressure (kPa)	2	3.7	3.7
Toxicity	Nil	Nil	Nil

Table 11.13 Properties of typical timber

		Timber	
Description	*Medium volatile coal* (30 lb)	*Average hardwood*	*Average softwood*
Typical ultimate analysis (%)			
carbon	81.0	45	44
hydrogen	4.5	6	5
nitrogen	1.4	1	1
oxygen	3.8	38	37
sulphur	1.3	–	trace
ash	5.0	2	1
moisture[a]	3.0	8	12
Calorific value (MJ/kg)			
gross	32.9	19	17
net	31.8	17	16
Bulk density[b] (kg/m^3)			
green	–	600	450
air dried	750	300	250

[a] Moisture quoted for air dried timber.
[b] Based on half value of solid specific mass.

Table 11.14 Properties of municipal waste

Item	*% by mass*		
	Early 1960s	*Early 1970s*	*Early 1980s*
Constituents			
cinders and dust	42	12	3
paper, card, wood	28	37	42
textiles	1	2	2
vegetable matter[a]	2	5	7
unclassified combustible	1	5	4
incombustibles	12	17	19
moisture	14	22	23
Bulk density (kg/m^3)		120–300	
Gross calorific value (MJ/kg)		6–8	

[a] Vegetable and other decomposing matter contains approximately 75% water.

Chapter 12

Combustion and chimneys

As an extension to the understanding of fuel characteristics and of the methods used for storage, handling and burning, some knowledge of the chemistry of combustion is useful. In addition, this has a bearing upon the selection of and design for the chimneys necessary to disperse the products of combustion. The text which follows is no more than an introduction to the subject, with particular reference to those fuels which have been noted in the preceding chapter as being commonly used for plant of the scale necessary to service individual buildings or relatively compact groups of buildings.

Fuels encountered in practice consist of carbon, hydrogen and oxygen combined to form relatively complex mixtures or compounds; the hydrogen may be uncombined and will be found present in the free state in most gaseous fuels. In addition, there are generally small quantities of sulphur, nitrogen and – in the case of most solid and liquid fuels – ash. The moisture content of solid fuel takes two forms, the superficial or *free* moisture resulting from either pit-head water-screening or subsequent open-air storage, and the *inherent* or air-dried content.

Combustion processes

The combustion of fuel is an *oxidation* process which is accompanied by the liberation of heat energy. The reactions occur as the carbon, hydrogen and, when present, sulphur combine with the oxygen in the air supplied. These reactions can take place only at a relatively high temperature, known as the *ignition temperature*, which varies between 400°C and 700°C according to the fuel. If an adequate supply of air has been made available, then the carbon will burn completely to form *carbon dioxide* but, when combustion is incomplete due to shortage of air, *carbon monoxide* will be formed. The hydrogen will burn to form water vapour and any sulphur present will burn to produce *sulphur dioxide* which may perhaps later combine with more oxygen to form *sulphur trioxide*.

On average, by mass, air contains 23.21% oxygen and 75.81% nitrogen, plus traces of argon, helium, krypton and other inert gases: without sensible error, therefore, the proportions may be considered as 23.2% oxygen and 76.8% incombustible. By volume, the proportions are 20.9% oxygen and 79.1% incombustible. The various elements previously mentioned combine with oxygen in proportion to their respective molecular weights which are: oxygen (O_2 = 32), carbon (C = 12), hydrogen (H_2 = 2), nitrogen (N_2 = 28) and sulphur (S = 32). Table 12.1 lists the fundamental combustion equations.

Table 12.1 Combustion reactions (elements)

	kg/kg *of combustible*							
	Requirements		*Products of combustion*					*Heat liberated (MJ/kg combustible)*
Reaction	O_2	*Air*	CO_2	CO	H_2O	SO_2	N_2	
$C + O_2 = CO_2$	2.67		3.67	–	–	–	8.84	33.6
$12 + 32 = 44$		11.51						
$2C + O_2 = 2CO$	1.33		–	2.33	–	–	4.40	10.1
$24 + 32 = 56$		5.73						
$2CO + O_2 = 2CO_2$	0.57		1.57	–	–	–	1.89	23.5
$56 + 32 = 88$		2.46						
$2H + O_2 = 2H_2O$	8.0		–	–	9.0	–	26.48	142.7
$4 + 32 = 36$		34.48						
$S_2 + 2O_2 = 2SO_2$	1.0		–	–	–	2.0	3.31	9.2
$64 + 64 = 128$		4.31						

Fuel analysis

The *ultimate analysis* of a fuel gives the percentage by mass of the various elements or compounds contained in a sample. Results from this type of analysis are used in the majority of calculations made in combustion problems. The *proximate analysis* of a fuel (specifically a solid fuel) gives the percentage by mass of certain characteristic groupings of elements which, as a result, indicates the probable physical nature of the combustion performance.

Excess air

The requirements for oxygen, and thus of air, which are listed in Table 12.1 are those calculated from the equations to be precisely the quantities necessary for the various reactions to take place. They represent the ideal situation or what is known as the *stoichiometric* condition. In practice, complete combustion cannot be achieved unless more air is provided than that which is theoretically required.

This situation comes about for very practical reasons related to the parallel difficulties of ensuring that the combustible elements will be mixed intimately with the air and that the consequent reactions will be completed before the products of combustion are discharged. The additional or *excess* air should not, however, be more than is necessary to prevent the discharge of unburnt fuel since the presence of a surplus will result not only in unnecessary energy losses, as noted later, but also in corrosion hazards associated with the formation of sulphur trioxide.

Flue gas temperature

The temperature at which the products of combustion are discharged finally raises difficulties since, if this be allowed to fall to too low a level, condensation will take place. Any water vapour in the gases, resultant from the combustion of hydrogen, will be deposited at what is called the *water dew point,* approximately 60°C. In the case of sulphur bearing fuels, there is the *acid dew point* also, at which the water vapour combines with any sulphur trioxide present to form sulphuric acid. This occurs when temperatures fall much below about 130°C.

In order to reduce the prospects of corrosion resultant upon presence of sulphur trioxide, practice is to design for discharge temperatures of 250–270°C (some 240 K average above the surrounding atmosphere). Hence, the combustion products hold a considerable amount of sensible heat in addition to the latent heat within the uncondensed water vapour which has been referred to previously. This sensible heat may, in turn, be quantified as being proportional to the mass of the total products *including any excess air* carried over which serves to re-emphasise the importance of setting and maintaining an optimum quantity of excess.

Flue gas analysis

Carbon being one of the principal components of most practical fuels, knowledge of the proportion of carbon dioxide present in the products provides a useful indication of completeness of combustion. Dilution of these products with excess air leads to a parallel decrease in the proportion of carbon dioxide present and thus measurement of the CO_2 content provides a relatively simple criterion which indicates not only the degree to which the combustion reactions have been completed but also the proportion of excess air admitted.

Taking the simplest case of pure carbon as an example, it may be seen from Table 12.1 that, for the ideal state of complete combustion, the CO_2 content of the products by volume is:

$$\begin{aligned} CO_2 &= 100\ (3.67/44)/[(3.67/44) + (8.84/28)] \\ &= 8.3/(0.083 + 0.317) = \underline{21\%\ (\text{max})} \end{aligned}$$

If 50% excess air were admitted, the CO_2 content would then be:

$$\begin{aligned} CO_2 &= 8.3/[0.4 + 0.5\ (2.67/32 + 0.317)] \\ &= 8.3/(0.4 + 0.2) = \underline{13.8\%} \end{aligned}$$

In the case of plants larger than those within the scope of this present chapter, it is necessary to be aware not only of the carbon dioxide content of the products of combustion but also that of other components such as oxygen, carbon monoxide and the sulphur oxides, etc.

Calorific value

The calorific value of a fuel is the quantity of heat energy released as a result of the complete combustion of unit mass. Two values are normally quoted, the *gross* or higher and the *net* or lower. The *gross calorific value* includes the latent heat within any water vapour formed as a result of the combustion of hydrogen and the *net calorific value* excludes this constituent.

It is not normally practicable to recover this latent heat by condensing the water vapour and thus the net calorific value is the more useful figure. Since, again as may be seen from Table 12.1, the mass of water produced is 36/4 = 9 times the mass of hydrogen burnt, it follows that the latent heat within the uncondensed vapour may be calculated as:

$$\begin{aligned} \text{Latent heat } re\ 15°\text{C} &= 2450\ \text{kJ/kg} \\ \text{Thus } CV_g - CV_n &= H(9 \times 2450)/(100 \times 1000) \\ &= \underline{(0.221\text{H})\ \text{MJ/kg}} \end{aligned}$$

Since the heat liberated as a result of the oxidation of various elements is known, a calculation may be made using a suitable analysis to derive a theoretical calorific value. For solid fuels in particular, the result of such a calculation will lack accuracy as a result of the presence of a variety of extraneous elements or impurities. Practical evaluations are thus made using a *calorimeter.*

Chimney loss

The chimney loss, as the name suggests, is a summation of all those losses which have accumulated by the time that the products of combustion reach the chimney. They are directly proportional to the mass gas flow, to the temperature at which discharge from the boiler occurs and to the specific heat capacity of the gases. Further, the total includes also the latent and sensible heat content of water vapour arising from the combustion of hydrogen. The inter-relation of the various components of the total is complex and it is convenient to make use of diagrams as Figures 12.1–12.3 which are abridged versions of those included in the *Guide Section C5.*

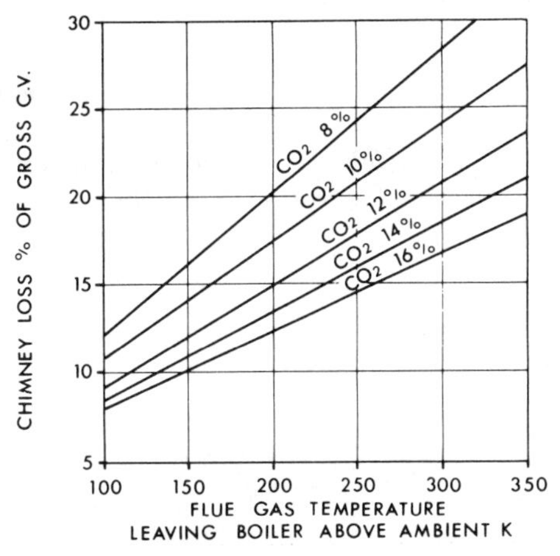

Figure 12.1 Chimney loss for coal firing

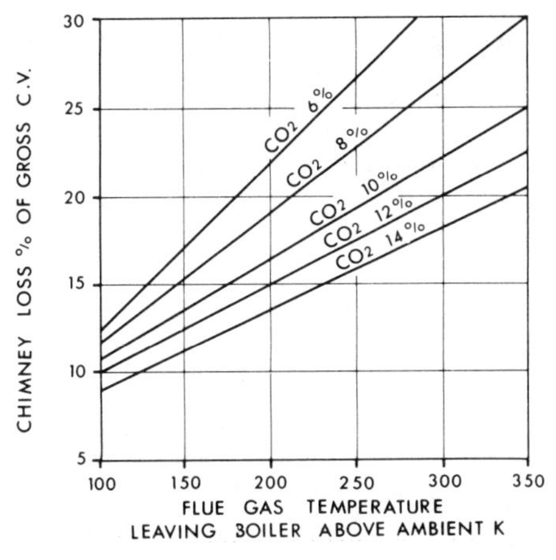

Figure 12.2 Chimney loss for oil firing

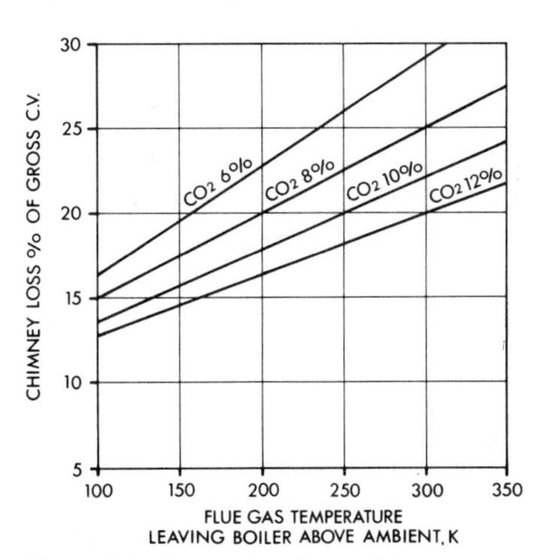

Figure 12.3 Chimney loss for gas firing

Sample calculations

Solid or liquid fuels

A single example common to both these fuels may be taken to introduce the use of Table 12.1 by assuming that a sample has, by mass, the following ultimate analysis:

carbon 80%, hydrogen 4%, oxygen 1.5%, sulphur 2%, ash and moisture 12.5%

The air required for stoichiometric combustion would be:

By mass

C to CO_2	$= 11.51 \times 0.80$	$= 9.21$
H, O } H_2O	$= 34.48 \times (0.04 - 0.015/8)$	$= 1.32$
S to SO_2	$= 4.31 \times 0.02$	$= 0.09$
∴ Air required, kg/kg		$= 10.62$

By volume (air at 15°C $= 0.816\,m^3/kg$)

∴ Air required, m^3/kg $= 8.66$

Note: the oxygen content of 1.5% combines with one-eighth of its weight of hydrogen. An external supply of air is required to burn the remaining hydrogen content.

The CO_2 present in the combustion products, with 75% excess air would be:

Combustion products, by mass:

CO_2	$= 3.67 \times 0.80$	=		2.94
H_2O	$= 9.0 \times 0.04$	=		0.36
SO_2	$= 2.0 \times 0.02$	=		0.04
N_2	$= 8.84 \times 0.80$	=	7.07	
	26.48×0.04	=	1.06	
	3.31×0.02	=	0.07	8.20

(check: [10.62(76.8/100)] = 8.16

N_2 in excess air 0.75×8.20 = 6.17
O_2 in excess air $6.17 \times 23.2/76.8$ = 1.86

The water vapour may be ignored since it will have been condensed out before the CO_2 is measured.

Hence, by volume:

CO_2	= 100(2.94/44) =	6.68	
SO_2	= 100(0.04/64) =	0.06	
N_2	= 100(14.35/28) =	51.25	
O_2	= 100(1.86/32) =	5.81	63.80

Thus,

$CO_2 = 6.80/63.80 = 10.5\%$.

If the flue gas temperature were to be 300°C with a CO_2 level of 10.5%, the chimney loss in the case of solid fuel firing, as read from Figure 12.1, would be 23.5%.

Table 12.2 Combustion reactions (gases)

Reaction	Requirements kg/kg O_2	Requirements kg/kg Air	Requirements m^3/m^3 O_2	Requirements m^3/m^3 Air	Heat liberated (MJ/m^3 combustible)
Methane					
$CH_4 + 2O_2 = CO_2 + 2H_2O$	4		2		40
16 + 64 = 44 + 36		17.24		9.57	
Ethane					
$2C_2H_6 + 7O_2 = 4CO_2 + 6H_2O$	3.73		3.5		69
60 + 224 = 176 + 108		16.08		16.75	
Propane					
$2C_3H_8 + 10O_2 = 6CO_2 + 8H_2O$	3.64		5		95
88 + 320 = 264 + 144		15.69		23.92	
Butane					
$2C_4H_{10} + 13O_2 = 8CO_2 + 10H_2O$	3.59		6.5		121
116 + 416 = 352 + 180		15.47		31.1	
Pentane and above					
C_5+	3.1		8.3		164
		13.36		39.7	

Gaseous fuels

In this case, although the principles remain exactly the same as before, it is convenient to approach any calculations by making use of the data for the various hydrocarbon compounds listed in Table 12.2. Hence, we consider a fuel having the following composition, by volume:

methane 90%, ethane 5%, propane 1%, nitrogen 3.5%, carbon dioxide 0.5%

The air required for stoichiometric combustion would be:

By volume

Methane	0.9×2	= 1.8
Ethane	0.05×3.5	= 0.175
Propane	0.01×5	= 0.05

∴ Oxygen required (m^3/m^3) = 2.025
Thus air required (m^3/m^3) = 2.025/(20.9/100)
= 9.7

The CO_2 present in the combustion products, with 25% excess air would be:

Combustion products, by volume

$0.9\,m^3$ of methane
combines with 0.9×2 = $1.8\,m^3$ of O_2
produces 1.8(79.1/20.9) = $6.8\,m^3$ of N_2
produces 0.9×1 = $0.9\,m^3$ of CO_2

$0.05\,m^3$ of ethane
combines with 0.05×3.5 = $0.175\,m^3$ of O_2
produces $0.175(79.1/20.9)$ = $0.66\,m^3$ of N_2
produces 0.05×2 = $0.1\,m^3$ of CO_2

$0.01\,m^3$ of propane
combines with 0.01×5 = $0.05\,m^3$ of O_2
produces $0.05(79.1/20.9)$ = $0.19\,m^3$ of N_2
produces 0.01×3 = $0.03\,m^3$ of CO_2

Thus, volume of CO_2 in products (m^3)
$= 0.005 + (0.9 + 0.1 + 0.03) = 1.035$

and volume of N_2 in products (m^3)
$= 0.035 + (6.8 + 0.66 + 0.19) = 7.685$

and excess air volume (m^3)
$= 0.25[7.685/(79.1/100)] = 2.429$

Thus,

$$CO_2 = 1.035(1.035 + 7.685 + 2.429) = 9.3\%$$

Chimneys

The purpose of a chimney is to provide a means whereby the products of combustion may be exhausted in such a manner as either to avoid pollution or, at worst, to provide means of dispersion such that pollution is diluted to an acceptable level. Obviously, the nature of the pollutants will vary with the nature of the fuel used. Particulate matter may be carried over in some cases, acid vapours in others and in some the discharge may be no more than a thermal carrier-plume.

Chimney design is a task which may involve a number of disciplines in the solution of problems related to the aesthetic, structural and functional aspects. Very many chimneys are concealed for much of their height within the massing of a tall building but this approach to concealment may lead to local difficulties as a result of lee-side downwash or vortex effects. It is not possible to discuss these matters here and, in any event, they are best resolved through model tests in a wind tunnel using techniques which are now well established.

There are certain principles applying to all chimneys, regardless of fuel type, which are now recognised as being fundamental to good design. These are:

- The velocity of the gases through the stack should be as high as possible commensurate with the draught available.
- The temperature of the gases should be maintained, throughout the stack, at near to the entry level by effective insulation.
- The use of draught stabilisers (as distinct from draught diverters) and leakages which admit cold air should be avoided.
- The inner surfaces of the flue construction should be smooth and changes in cross section, bends, etc., should be designed aerodynamically.
- The use of a single flue-way to serve more than one boiler should be avoided wherever possible. In the rare case where a common flue for a multi-boiler plant is genuinely unavoidable, means should be provided to isolate the outlets of idle boilers and to maintain a constant efflux velocity at the chimney terminal.

Chimneys for solid or liquid fuels

For the purpose of this present text the design process may be simplified by reducing the number of variables which relate to the particular cases of solid fuel and oil. Since the remainder, including the calorific value of the fuel envisaged for design purposes, will almost certainly vary during the life of the chimney, further precision will serve little purpose.

- *Products of combustion, quantity*
  Assume excess air provided to be 75%
  Assume flue gas temperature, on average, to be 200–300°C
  Assume boiler efficiency to be 75%

For these conditions, it may be shown that the products of combustion per MJ of boiler output are between 1.1 and 1.3 m^3/MJ. It is convenient to take a mean of 1.2 m^3/MJ. The temperatures assumed cover the range mentioned previously.

- *Time*

Bringing a time scale into consideration, the volume of the products as above will derive from combustion associated with 1 MJ/s, i.e. 1 MW.

- *Velocity of flue gases in the chimney*

It may be assumed that the range of velocities appropriate to natural and to mechanical draught are, respectively, 4–8 m/s and 10–12 m/s.

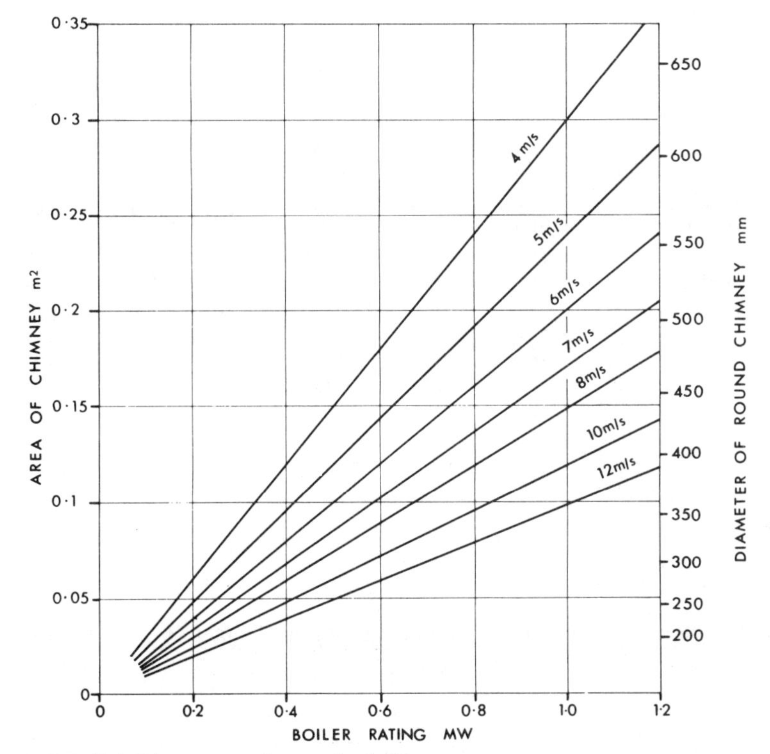

Figure 12.4 Chimney areas for stated velocities

- *Area*

Figure 12.4 gives areas and diameters for round chimneys direct, having selected a velocity. The *equivalent diameter* of a square or rectangular chimney, in mm, is 1000 × the square root of the area as read from the figure.

The cross-sectional area of the chimney may then be derived according to boiler duty, fuel and type of draught. Thus, if the boiler duty were 1 MW, oil-fired, with natural draught (5 m/s) and a chimney temperature average of 250°C,

$$1.2/5 = 0.24\,\text{m}^2 = 550\,\text{mm in diameter}$$

- *Efflux velocity*

In order that the plume of flue gas should rise clear of the chimney top and not flow down the outside (*down-wash*), the diameter at the top should be reduced so as to maintain as high a velocity as practicable. For small boilers with natural draught, a velocity of 6 m/s is advised. Larger boilers with mechanical draught should achieve 7.5–15 m/s.

The velocity pressure corresponding to these rates may be read from the bottom line of Table 12.3 as appropriate to a flue gas temperature of 250°C.

Table 12.3 Pressure loss per metre height of smooth chimney (multiply values by 4 for brick or rough cement rendering)

Effective chimney diameter (mm)	*Pressure loss,* (Pa per m height or run) *for gas flow at the following velocities* (m/s)								
	4	*5*	*6*	*7*	*8*	*9*	*10*	*11*	*12*
200	0.45	0.71	1.02	1.39	1.81	2.29	2.82	3.41	4.07
250	0.34	0.52	0.76	1.02	1.34	1.70	2.10	2.53	3.04
300	0.27	0.41	0.60	0.81	1.06	1.35	1.66	2.00	2.39
350	0.22	0.35	0.49	0.68	0.88	1.12	1.38	1.67	1.99
400	0.18	0.29	0.41	0.56	0.73	0.93	1.14	1.38	1.65
450	0.16	0.25	0.36	0.49	0.64	0.81	1.00	1.21	1.44
500	0.14	0.22	0.32	0.44	0.57	0.72	0.89	1.07	1.28
550	0.13	0.20	0.28	0.38	0.50	0.64	0.78	0.95	1.12
600	0.11	0.18	0.25	0.35	0.45	0.57	0.71	0.85	1.02
650	0.10	0.16	0.23	0.32	0.41	0.52	0.64	0.78	0.92
700	–	0.14	0.21	0.28	0.37	0.47	0.57	0.69	0.83
750	–	0.13	0.19	0.26	0.34	0.43	0.53	0.64	0.76
Velocity pressure	5.4	8.4	12.1	16.5	21.5	27.3	33.6	40.6	48.4

- *Draught required*

The following may be calculated from the data on duct sizing in Chapter 16, adjusted for temperature:

At boiler exit, consult makers' data but

Oil-fired boilers vary from	7 to 50 Pa
Solid fuel fired, if burning rate is 5 kg/m^2 grate area	70 Pa

Flue connections, boiler to chimney, depending on number of bends and other losses, average — 15 to 30 Pa
Efflux velocity pressure, e.g. for 6 m/s — 12.1 Pa

The new total for an oil-fired boiler may then be between 50 and 100 Pa and for a solid fuel boiler 100 to 150 Pa.

- *Draught produced by chimney*

The theoretical draught of a chimney at the two temperatures named varies with the external ambient temperature. Assuming that this is 20°C in summer and 0°C in winter, unit values are:

Winter

per metre height at 300°C	6.7 Pa
at 200°C	5.5 Pa

Summer

per metre height at 300°C	5.8 Pa
at 200°C	4.5 Pa

Figure 12.5 is drawn on this basis and thus a chimney of 30 m height will in summer produce, theoretically, a draught of 155 Pa with flue gases at 250°C.

- *Draught loss in chimney*

The pressure loss per metre height may be taken from Table 12.3. Values for other velocities may be interpolated. Figure 12.6 gives the pressure loss for chimneys of given heights, the loss per unit length having been determined from Table 12.3. The pressure loss arising from the flow of gases in rough brickwork or concrete chimneys will be as much as three or four times greater than that for relatively smooth sheet steel.

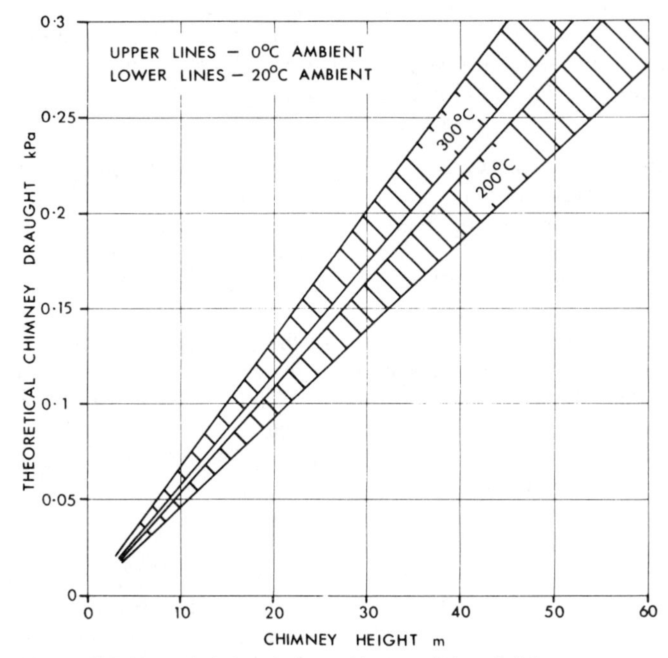

Figure 12.5 Theoretical draught for a chimney of given height

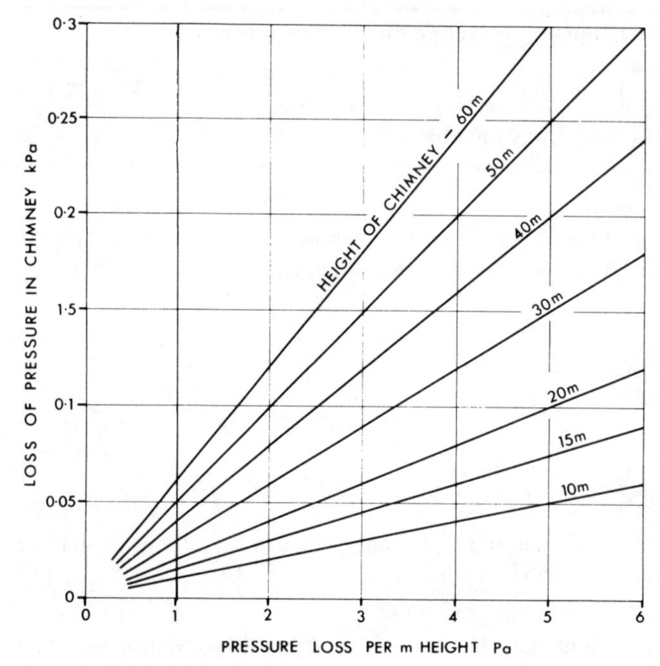

Figure 12.6 Draught loss for a chimney of given height (smooth surface)

- *Rectangular equivalents*

For square or rectangular chimneys, the effective areas are those of the circle or ellipse which may be inscribed within them. The *equivalent diameter* of such flues is therefore the square root of the square or rectangular area. If the flue must be rectangular, it is a general rule that a ratio of sides of 3:1 should not be exceeded.

In conclusion:

- Using Figure 12.4 select a velocity according to whether draught is natural or mechanical and find chimney area and diameter.
- From Table 12.3 determine pressure loss per metre of height for this diameter.
- Make an assumption as to chimney height and hence from Figure 12.6 note pressure loss in chimney. Add for loss through boiler and flue connection and for efflux velocity.
- Using Figure 12.5, the available draught may be found for the same assumed chimney height. If this were equal to or in excess of the sum of the losses, the assumption as to height may stand. If the available draught were insufficient, either the height must be increased or the velocity reduced, or both. If the calculated draught were much in excess of requirements, a smaller chimney or less height might be the solution, or if neither were possible, or desirable, a damper may be used.

Clean Air Act

The chimney height determined by following the routine just described is that which is required for combustion. It is however also necessary to consider the height in relation to the mandatory requirements of the Clean Air Act of 1968. For this purpose, reference should be made to the third (1981) edition of the document *Memorandum on Chimney Heights*.

Memorandum on Chimney Heights
The purpose of the *Memorandum* is to provide a basis for limiting the pollution due to sulphur dioxide (SO_2) near to the chimney by increasing the height as the rate of emission increases. In addition, the relationship between chimney height and building height is taken into account as well as the type of locality, here classified in précis as follows:

A An undeveloped area where background pollution is low with no other emission nearby.
B A partially developed area with low background pollution and no other emission nearby.
C A built-up residential area with only moderate background pollution and no other emission nearby.
D An urban area of mixed industrial and residential development, with considerable background pollution and with other emission nearby.
E A large city, or an urban area of dense residential and heavy industrial development with severe background pollution.

The concentration of SO_2 is determined from equations quoted in the *Memorandum* which require knowledge of the maximum rate of fuel consumption, the sulphur content of the fuel and the boiler efficiency. For calculation from first principles, the required information is given in Tables 11.2 and 11.4 but, as an approximation for the coals most in use and the heavier oils, emission of SO_2 in g/s may be approximated by multiplying the boiler rating in MW by either 1.7 for coal firing or 2.2 for oil firing.

If the calculated SO_2 content is less than 0.38 g/s then the chimney height need be only 3 m higher than that which has been calculated as necessary for combustion. For SO_2

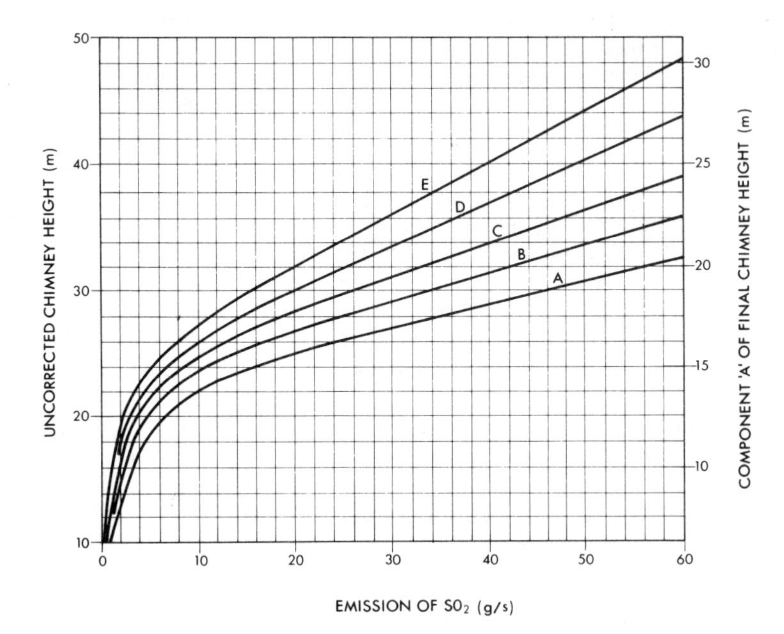

Figure 12.7 Emission of SO_2 and chimney height

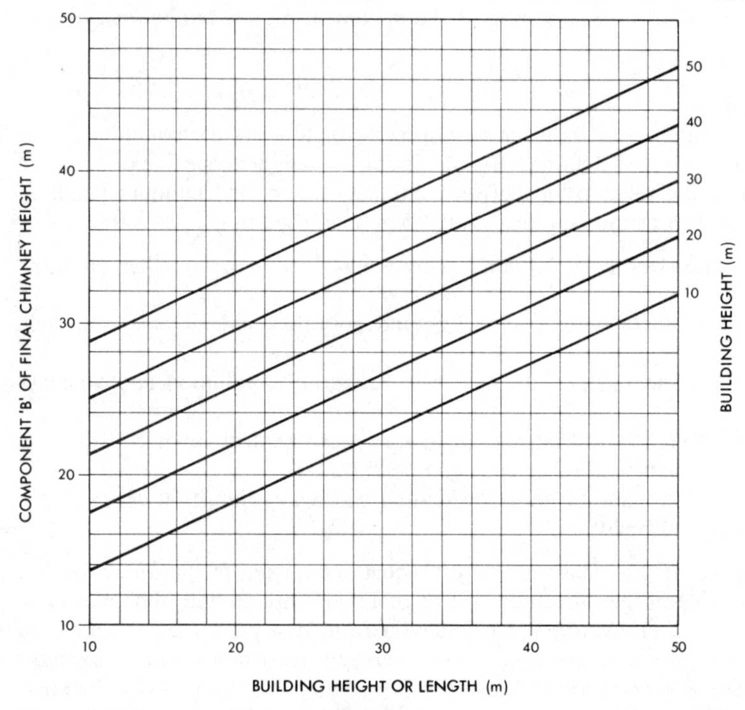

Figure 12.8 Component 'B' for chimney height

contents greater than this, reference must be made either to the *Memorandum*, which contains a set of ill-conceived nomograms, or to Figure 12.7 (plotted for the *Guide* by the present authors from the same basic data), from which an 'uncorrected' chimney height may be read, taking account of the location, from the left-hand scale. Where the fuel has a sulphur content of more than 2% as in the case of most oils, a primary correction must then be made by adding 10% to the height read from the diagram. If, after this increase, the dimension is more than 2½ times the height of the building or any other building in the immediate vicinity, then no further correction is necessary.

In other circumstances, it is necessary to take the dimensions of the building into account. A convenient way of dealing with this is to consider the final chimney height as having two separate components, the first of which (A) is read from the right-hand scale of Figure 12.7. The second (B) is read from Figure 12.8 by entering building height or length, whichever is the greater, to the base scale. The sum of the two components, the final chimney height, must be corrected as before by the addition of 10% if the sulphur content of the fuel is greater than 2%.

It should be noted that the use of the short cuts given here does not wholly invalidate the final result produced since the basic processes laid down in the *Memorandum* are, at best, only pseudo-scientific approximations. Nevertheless, the rule book should always be followed when making the necessary applications to a local authority.

Mechanical draught

Any solid fuel or oil-fired boiler may be fitted with an *induced-draught* fan for exhausting the products of combustion and discharging them up the chimney. Many boilers are obtainable with such a fan fitted as part of the standard unit. In others, the fan supplying combustion air has sufficient power to expel the products under pressure, i.e. using *forced draught*.

The advantages of mechanical draught are: first, that the boiler can be designed for higher velocities over the heating surface, so giving a higher rating for a given size; second, that the natural draught produced by stack height is no longer of importance, and the chimney may thus be short (subject to the Clean Air Act), or, if the boiler is on the roof, notional only.

Induced-draught fans usually take in some diluting air and makers' data should be consulted for the volume to be handled by the chimney, the area of which will then be determined by the velocity decided upon, such as 10–12 m/s. The resistance would be calculated as before. It is usual, where the fan is an integral part of the boiler package, for about 60 Pa surplus pressure to be made available for chimney loss and efflux velocity.

Chimneys for gaseous fuels

The requirements for chimneys for gas firing differ from those previously discussed in cases where the air admitted by the *draught diverter* acts as a dilutant to the combustion gases. The CO_2 content of the latter, at the boiler discharge, may be 9% with a gas temperature of 240°C, but the normal for the 'secondary' flue after the draft diverter is about 4% with a gas temperature of about 120°C. Curves for sizing and data as to resistance factors are given in the *Guide, Section B13* and a British Gas publication* provides comprehensive tables for a wide range of circumstances. In brief, the required chimney height is that which will provide for adequate dispersion of the products of combustion such that their concentration at ground level does not exceed a critical value.

The negligible sulphur content of natural gas means that the provisions of the *Memorandum on Chimney Heights* (p. 323) apply only to the very largest installations. In terms of height, the flue terminal need be only about 2 m above roof level of a building, or 4 m above ground level where free standing, for outputs of up to 5 MW. For such natural draught burners with gas dilution at a diverter, the area of the flue for each unit may be read from Figure 12.9 where the value of the factor F has been calculated from:

$$F = H_n/(S + 2 + b/2)$$

where

H_n = height of flue above boiler outlet (m)
S = two-thirds of any suction required at flue base (Pa)
b = number of bends in the route of the flue

For a forced draught burner, the design technique necessary does not vary greatly from that described earlier for solid fuel and oil equipment. The data provided by Figures 12.5 and 12.6 are equally valid. Some manufacturers list technical information regarding performance of burners and of draught requirements at the point of connection to the chimney: most do not. These data must be obtained before design of the chimney can be tackled with confidence.

* *Technical Notes on the Design of Flues for Larger Gas Boilers*, Dec., 1971

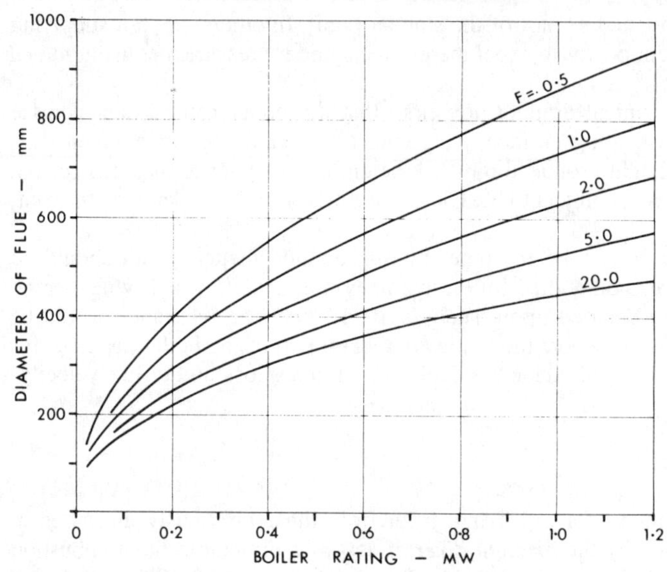

Figure 12.9 Chimney areas

Chimney construction

Domestic chimneys

Building Regulations, 1991, require that all domestic chimneys be lined with some impervious materials such as tile. For advice on details of construction, reference may be made to the Building Research Station's *Digest No. 60.*

Larger installations

The importance of keeping the products of combustion in the chimney warm has been emphasised earlier. With this in mind some form of insulation is a necessity and various forms of construction are illustrated in Figure 12.10. They are:

A An outer stack in brickwork or concrete enclosing an independent lining of firebrick or moler brick with an air gap between.
B An outer stack in brickwork enclosing a moler brick lining which may be bonded in.
C An outer stack in brickwork lined with fire clay tiles.
D An outer stack in reinforced concrete sections lined with insulating moler brick or concrete.
E A welded steel flue within aluminium cladding. The construction may be guyed or self supporting. The annular space between flue and cladding may be packed with insulation material.
F One of a family of prefabricated chimneys manufactured in lengths of about 1 m with socket-and-spigot or twist-lock joints. A wide variety of constructions is available, insulated and un-insulated.

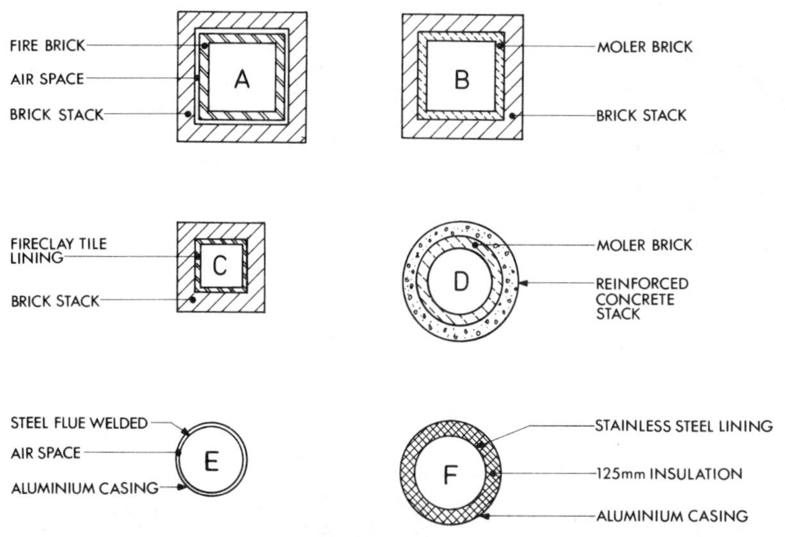

Figure 12.10 Methods for single-flue construction

The relative heat losses from these constructions may be calculated as for *U* values – the inner surface resistance being assumed as zero. Cost, permanence and appearance if free standing, will always be determining factors.

Multiple flues

It will be apparent that, where more than one boiler occurs and there is a common chimney, it would be impossible to maintain the design efflux velocity with anything less than the full number of boilers in use. Furthermore, where mechanical draught is used and the flue is under pressure, back draught might occur to any boilers which are not being fired. Hence, it is now advocated, as has been noted previously, that each boiler should have its own individual flue connection and chimney, and that they should not be combined into one large stack as was the practice in the past. The separate vertical flues may be grouped into one stack, as in Figure 12.11.

Smoke problems

The corrosive effect of sulphurous gases has been alluded to previously. If the products of combustion, on entering the chimney, are allowed to cool to the region of the acid dew point, a new hazard is set up with oil firing, namely acid smuts.

It appears that the minute unburnt particles of carbon always present in flue gases are liable to form nuclei on which condensation takes place, and they then agglomerate into visible black oily specks. In a chimney they are particularly liable to collect on any roughnesses, or at points of change of velocity. In so doing, on starting up, the sudden shock may cause them to be discharged from the top – often giving rise to complaints from surrounding property.

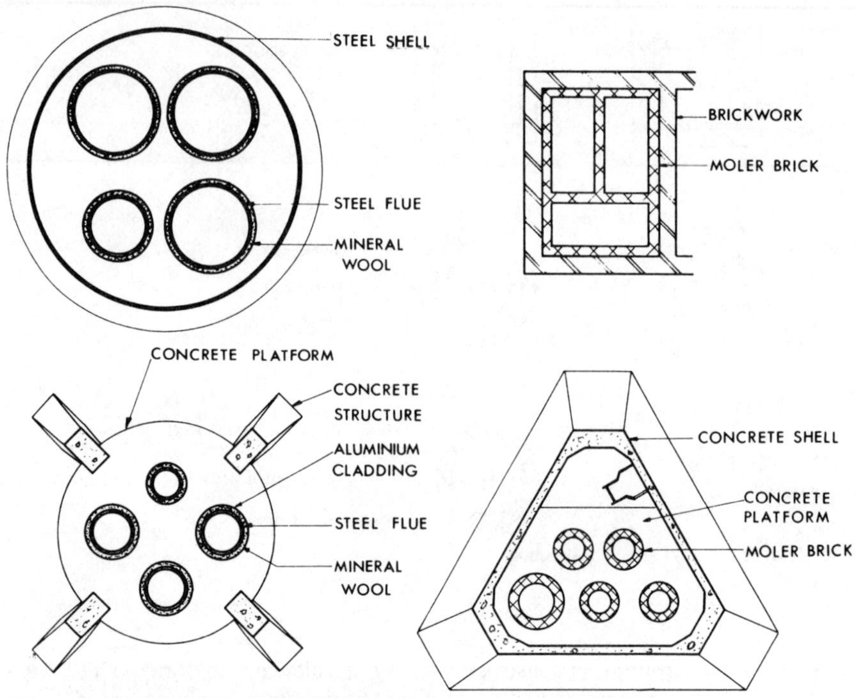

Figure 12.11 Methods for multiple-flue construction

Acid smut formation is less of a problem with a low sulphur oil such as Class D, even though draught stabilisers are often part of standard boilers on a domestic scale.

For the control of draught on plants burning the heavier oils, in lieu of a draught stabiliser admitting cold air, some form of damper control is necessary which may be automatic, of a type such as Figure 12.12.

Treatment of flue gases by means of a proprietary additive to the fuel or in the hot-gas outlet may be considered where smut nuisance is particularly troublesome.

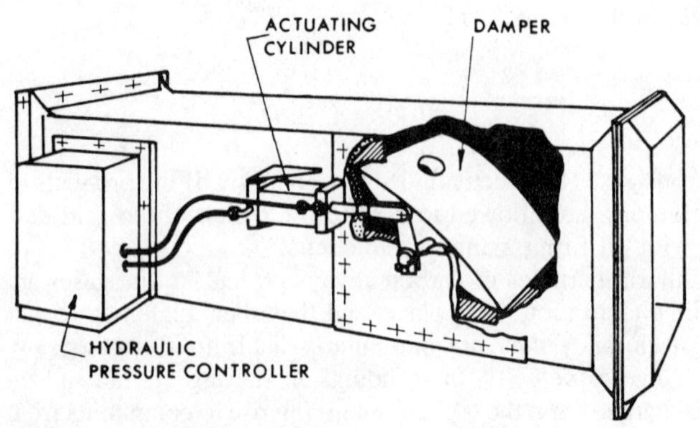

Figure 12.12 Damper for automatic draught control

Materials for chimneys and flues (gas fired boilers)

For each unit of heat released, combustion of natural gas will produce between one and a half and two and a half times as much water vapour as will combustion of oil or coal. Hence, condensation is likely to occur in gas flues and chimneys, particularly on start-up. They should be lined with an impervious and acid-resisting material such as glazed stoneware or asbestos cement. It is desirable to keep the chimney warm to assist draught, either by enclosing it in brickwork or concrete, or by insulation. Connecting flues between such boilers and vertical chimneys are usually of asbestos cement or double-wall stainless steel.

Special methods for flue gas disposal (Natural gas)

Apart from the conventional chimneys mentioned, a number of special methods for flue gas disposal have been developed, principally by the various research bodies within the gas industry. Some of these have been adapted for use with equipment burning the lightest grades of fuel oil but such applications apply only on a domestic scale.

Fan diluted draught

In this system, one form of which is shown in Figure 12.13, the combustion products, together with a quantity of diluting air, are exhausted to the atmosphere by a fan. By this means it is possible to dispense with a chimney entirely, for a gas fired boiler plant, by discharging through the wall of the boilerhouse. The diluting air quantity must be such as to bring the CO_2 content of the mixture down to 1%, which involves a fan to handle approximately 100 m^3 per m^3 of natural gas burnt. Included in the system is a fan failure device to shut off the burners in the event of draught failure. The duct sizing with this system should be on the basis of a gas velocity of 3–5 m/s.

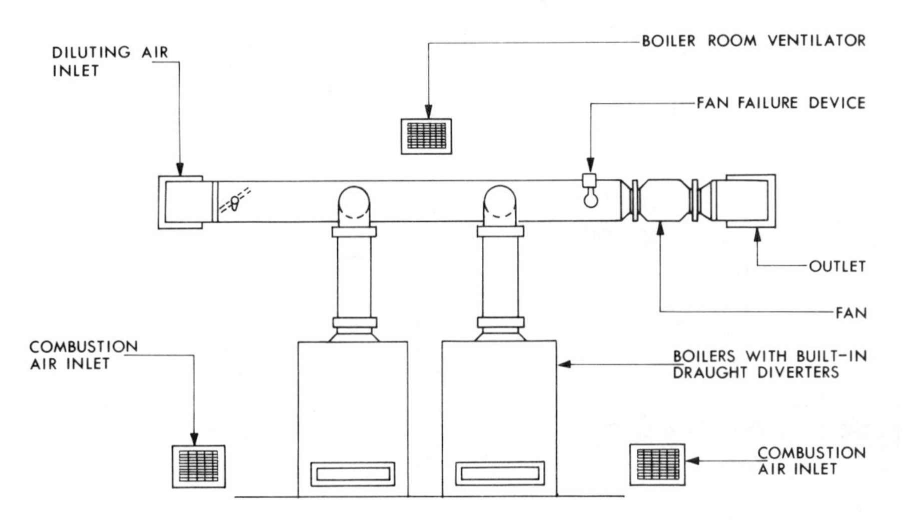

Figure 12.13 Fan diluted draught

Domestic multi-appliance flues

Another flue system is the SE, as shown in Figure 12.14(a). This is suitable for multi-storey buildings and, as will be seen, air enters at the base and the products of combustion from the various appliances discharge into the same duct. Thus the gas combustion space is virtually sealed off from the occupied space. An alternative, for use where a bottom inlet is not possible, is the 'U' duct shown in Figure 12.14(b). Fresh air taken in at the roof is conveyed down a shaft running parallel with the rising shaft, which acts as a shared flue as in the SE duct, the products of combustion being exhausted at the roof from a terminal

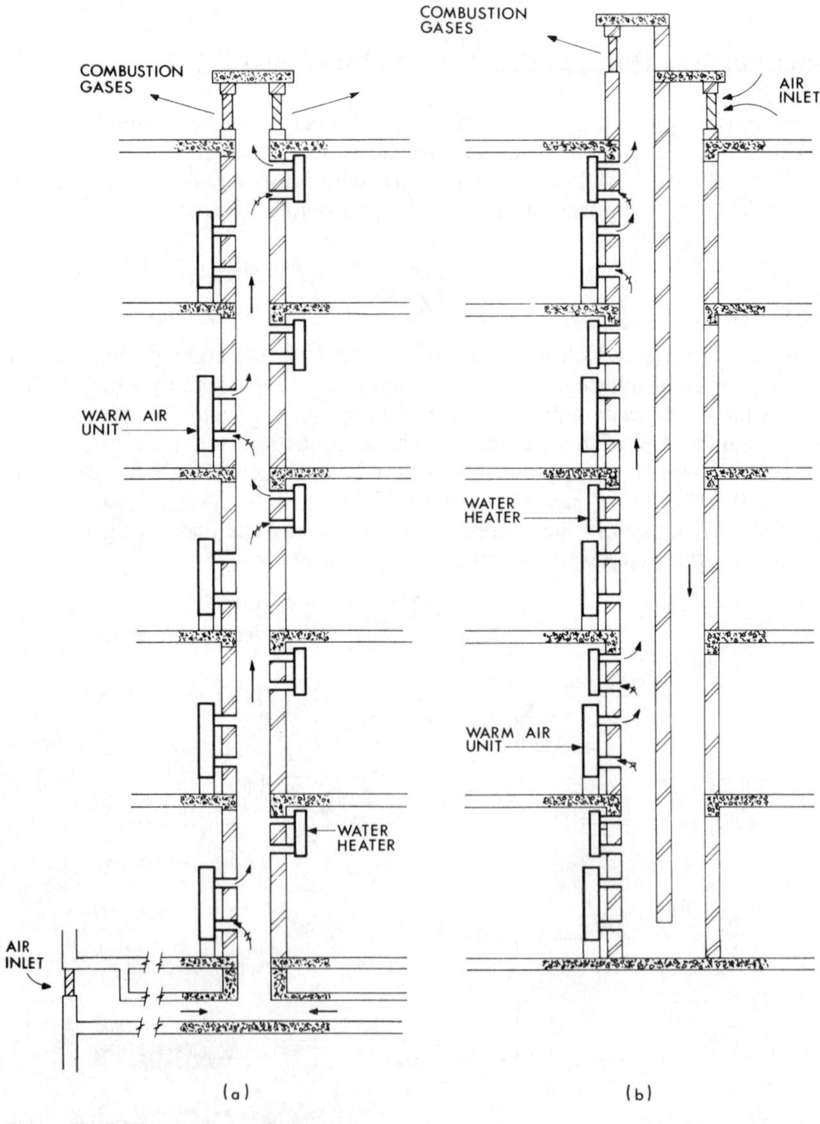

Figure 12.14 SE and 'U' duct flue systems

adjacent to the intake, but at a slightly higher level. All gas burning appliances take their combustion air from, and return the products of combustion to, the rising duct.

Balanced flues

In difficult situations where the flue terminal cannot be carried above roof level, but must terminate in some position such as in an internal light well, differences of pressure are found to cause back draughts. These may be overcome by the provision of a balanced flue, as in Figure 12.15, which comprises an inlet duct similar to the rising flue except that it is carried down to near floor level in the boiler chamber. The sizing of this duct must be generous, since it has to convey not only the air for combustion, but also that drawn in by the draught diverter where fitted; and the resistance should, in any event, be kept as low as possible.

Terminals

Care must be taken to see that flues discharge the products of combustion freely to the atmosphere. In buildings with return walls, areas, etc., peculiar atmospheric pressure

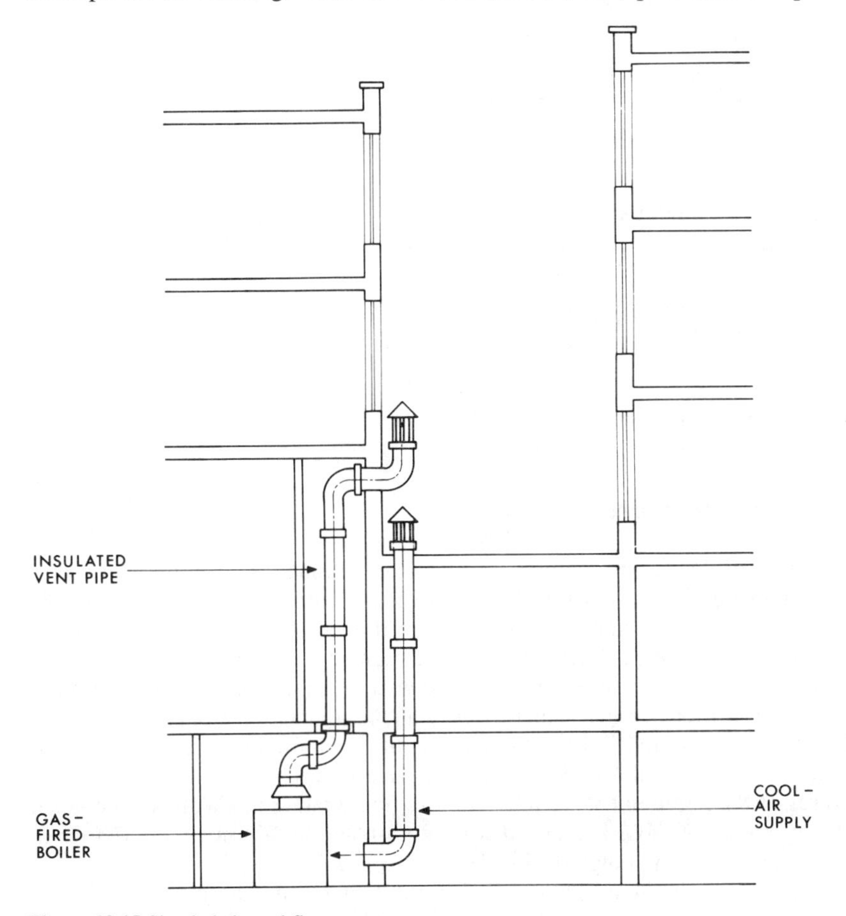

Figure 12.15 Simple balanced flue arrangement

conditions occur and it is advisable to extend the flue to a height of at least 500 mm above the eaves of the roof. The possibilities of downdraught are then much reduced.

Every flue should be fitted with a terminal of approved design to prevent birds nesting, etc. Bafflers, or more correctly 'draught diverters', should be fitted to the flues of all natural-draught gas boilers and heaters if these are not incorporated in the design of the heater.

Roof-top boiler house

The flue problem is avoided altogether in a boiler installation if the boilers are on the roof. Figure 12.16 shows such an arrangement with short outlets from the boilers carried through the roof of the penthouse to the atmosphere. Gas boilers on the roof are obviously simpler to deal with than oil – there is no delivery pumping and negligible fire risk. Thus, in many examples of multi-storey buildings, the use of gas in this location for boilers has much to commend it.

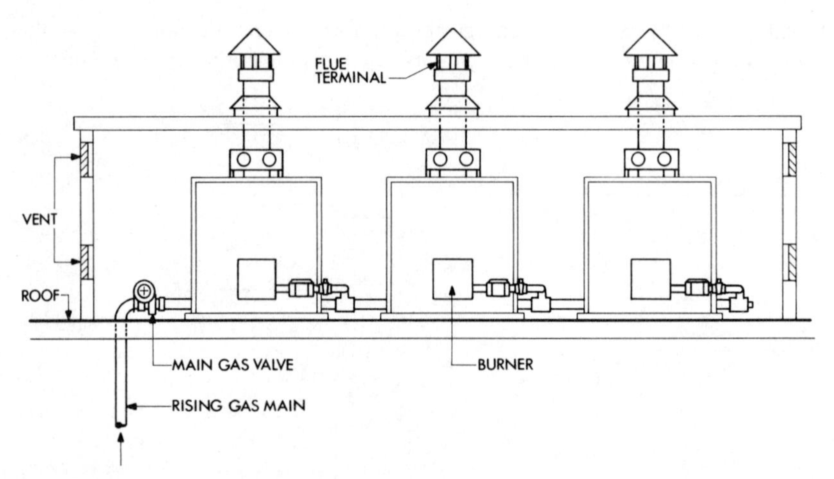

Figure 12.16 Roof-top boiler house

Air supply to boiler houses

The manner in which calculations are made in order to determine the quantity of air which is required theoretically for combustion has been set out earlier in this chapter. In the design of a boiler house, provision must be made for this air to enter, together with any excess which may be necessary. Where atmospheric burners are used with gaseous fuels, the associated boilers must, as a result, be equipped with draught diverters and yet further air will be necessary to replace that which is used as the dilutant (see p. 283).

Table 12.4 sets out, in round figures, the air quantities required for various fuels and situations. The last column of this table makes use of the respective calorific values of the fuels and expresses the requirements as a function of boiler rating in kW. It will be noted that, with the exception of those associated with atmospheric gas burners, the figures in that last column all fall in the range 0.44–0.56 litre/s per kW.

As far as is possible, air should be supplied to the boiler house by natural induction through door or wall louvres and other openings. A velocity of between 1 and 2 m/s is

Table 12.4 Air required for combustion of various fuels

Fuel and burner	Theoretical		Excess air (%)	Practical		
	m^3/kg	m^3/m^3		m^3/kg	m^3/m^3	litre/s per kW[a]
Bituminous coal	8	–	50	12	–	0.53
Fuel oil	12	–	30	16	–	0.51
Natural gas						
atmospheric burner (primary)	–	10	60	–	16	0.56
atmospheric burner (total)	–	10	180	–	28	0.98
packaged burner	–	10	30	–	13	0.46
Propane						
atmospheric burner (primary)	–	24	60	–	38	0.53
atmospheric burner (total)	–	24	180	–	67	0.94
packaged burner	–	24	30	–	31	0.44

[a] litre/s per kW boiler rating, at 75% efficiency.

commonly assumed in calculating the free area of opening needed. Taking the higher figure of 0.56 litre/s per kW (= 0.56 m^3/s per MW), and a face velocity of 1 m/s, we arrive at a free area for inlet louvres of, say, 0.6 m^2/MW of boiler rating.

In parallel with the need to supply combustion air, it is necessary to consider the extent to which the boiler house may require ventilation in order to prevent overheating. A separate lightweight building at ground level would very probably not require any such treatment but, at the other extreme, a basement enclosure might be intolerably hot during some seasons.

A calculation may be made, using data published in a *Practice Note* by *CIBSE*, to estimate the heat gain and thus establish the necessary ventilation rate. However, a time honoured rule of thumb suggests that if the area of the combustion air louvres be doubled, this will be adequate as an inlet area to serve the dual purpose. Thus, we arrive at a total of about 1.2 m^2/MW of boiler rating which is, by coincidence, the figure recommended in the *Guide Section B13*. This volume covers the case of the atmospheric burners since, of course, the additional dilutant air acts to ventilate before it is induced into the flue.

In addition to the inlet louvres, which should be sited at low level, it is necessary to provide for disposal of the surplus ventilating air. As this is warmer than the supply, it will collect near to the roof or ceiling and any louvres provided to outside should be sited as high as possible in the boiler house. The possibility of noise nuisance through any openings, inlet or outlet, to the boiler house must not be overlooked and special louvres having acoustic linings may be required.

Where the boiler plant is sited remote from the building perimeter, it may be necessary to provide mechanical systems for both inlet and extract ventilation. Practical difficulties are inherent in such systems since boiler loading and consequent combustion air requirements will vary during the year, leading to imbalance between inlet and extract unless some form of capacity control be provided. The associated fans must have their drives interlocked electrically with the fuel burner arrangements.

Chapter 13

Ventilation

The act or art of ventilating is, according to the *Oxford Dictionary*, to 'expose to fresh air' and to 'cause air to circulate freely in an enclosed space' (*L. ventilatio*). As now understood, the purpose may be either to supply the untainted air necessary for human existence or to provide for dispersal of smells, noxious gases and dangerous concentrations of fumes and smoke. It has become practice to avoid the use of the definition *fresh air*, with its connotation of purity, and to refer instead to *outside air* leaving open the possibility that pre-treatment may be necessary.

The need for a supply of outside air as a contribution to human comfort, the parallel requirement for the removal of contaminants and some of the limiting considerations have been touched upon in the first chapter of this book. The subject matter which follows is concerned primarily therefore with quantification of both need and requirement followed by consideration of some of the methods used to meet them.

Section L2 of the revision to the *Building Regulations* dated 26 January 1993, currently in circulation as a draft and discussed earlier in Chapter 2, refers to a draft report prepared by the Building Research Establishment*. This latter document emphasises the penalty, in terms of additional energy consumption, which arises from the use of mechanical ventilation or air-conditioning systems and proposes a requirement that such systems be installed only where it is reasonably necessary. The text emphasises that natural ventilation may, in certain circumstances, provide tolerable comfort conditions in the climate of the British Isles for office and other buildings provided that:

- There will be, in the vicinity of the site, no excessive atmospheric or noise pollution which will require the building to be sealed.
- The purpose of the building is such that there will be no activities or processes which require close control of the internal temperature or humidity.
- The design of the building is such that excessive solar gains will not occur.
- There will be no excessive heat gains arising from office machinery, etc.

It is further suggested that adequate ventilation may be provided through windows in either a single sided building having offices 10 m deep or a double sided building 22 m wide with offices either side of a 2 m corridor. The reader is advised to refer to this report when published (and to any subsequent revisions) to confirm the final recommendations.

* Grigg, P. F., *The Use of Air-conditioning and Mechanical Ventilation in Non-domestic Buildings*. Draft BRE Report, 26 January 1993.

Air supply for human emissions

The volume of air necessary to provide for human occupancy may be considered under the following principal headings:

- Provision of oxygen for respiration.
- Removal of products of exhalation.
- Removal of body odour.
- Removal of unwanted heat.
- Removal of unwanted moisture.
- Removal of contaminants.

Respiration and exhalation

At rest, the normal adult inhales between 0.10 and 0.12 litre/s of air and of this only about some 5% is absorbed as oxygen by the lungs. The exhaled breath contains between 3 and 4% of carbon dioxide (CO_2) which amounts to about 0.004 litre/s.* The accepted level for a maximum concentration of CO_2 within an occupied space is 5000 parts per million, or 0.5% by volume, for an exposure of 8 hours. The outside air requirement, at an equilibrium condition, to restrict the level of concentration to the maximum level permitted, may be found by using the contaminant expression given on page 339: this indicates a very low rate of 0.847 litre/s per person.

Table 13.1 lists ventilation rates for concentrations other than the maximum and for other levels of activity but there are parallel aspects of human occupation of a space which are more important criteria for the need for ventilation.

Table 13.1 Ventilation rates required to limit CO_2 concentrations

	Minimum ventilation required (litre/s per person)		
Activity	*0.1% CO_2*	*0.25% CO_2*	*0.5% CO_2*
At rest	5.7	1.8	0.85
Light work	8.6–18.5	2.7–5.9	1.3–2.8
Moderate work	–	5.9–9.1	2.8–4.2
Heavy work	–	9.1–11.8	4.2–5.5
Very heavy work	–	11.8–14.5	5.5–6.8

As a result of these parallel effects, a room may feel fresh and pleasant with a CO_2 content higher than that noted above and yet feel stuffy with a much lower content. In consequence, it has in the past been considered that it is an unsatisfactory datum for assessment of ventilation quality. However, the results of current research at BSRIA and elsewhere suggest that it might well be used as a means for ventilation control.

* *Note:* this is sometimes quoted as 0.004719 litre/s which is no more than an exact metric conversion of 0.01 ft^3/minute which, in turn, is an average between values of 0.513 and 0.684 ft^3/hour. Since these last two figures were derived from an estimate of the CO_2 content of the average exhalation from average human lungs, 3–4% of 17 ft^3/hour, the authors are inclined to believe that a somewhat rounder figure is to be preferred.

Body odour

One of the essentials of good ventilation is the removal of odours arising from human occupation which problem generally becomes serious only in crowded places. A supply of outside air at a rate of at least 5 litre/s per person has been found to be the minimum quantity necessary to be reasonably sure that no trouble will arise from this source but a rate of about 8 litre/s is to be preferred. This quantity should be adjusted upwards in instances where occupation is particularly dense and, for factory canteens and the like, a rate of 10–15 litre/s per person may be appropriate.

Unwanted heat

A sedentary worker, as may be noted from Table 13.2, will emit sensible heat at a rate of about 0.1 kW. If it be assumed that ventilation air is provided to the workplace at the rate of 16 litre/s per person, double the minimum quantity required to combat body odour, then this sensible heat will lead to the air temperature rising by $(0.1 \times 1000)/(16 \times 1.205 \times 1.012) = 5.1$ K, assuming that there is no heat loss from the room.

Table 13.2 Heat and moisture from occupants

	Heat emission per occupant (W)		
Activity[a]	*Sensible*	*Latent*	*Total*
Sedentary worker	100	40	140

[a] See Table 3.9 for a range of activities.

Unwanted moisture

Similarly, Table 13.2 shows that the same sedentary worker will produce latent heat at a rate of about 0.04 kW. This represents a moisture output of $0.04 \times 3600 \times 1000/2450 =$ 59 g of water vapour per hour or, with respect to the ventilation air quantity of 16 litre/s per person noted above, 1.02 g/m^3. Table 13.3, last column, shows the moisture produced by un-flued direct heating appliances.

Table 13.3 Contaminants released to an enclosure by the combustion of fuel in a direct heating appliance

	Rate of release of contaminants (g/h per kW)		
Fuel	CO_2	SO_2	*Water vapour*
Kerosene	244	0.155	151
Natural gas	187	–	98
Butane	220	–	113
Propane	216	–	117

Contaminants

Much recent research has been directed to assessment of the ventilation rate required to deal with pollution in offices and work rooms arising from tobacco smoke. The results are not easy to quantify in simple terms since the statistical relationship between density of occupation and number of smokers must be taken into account. In a large open office it might be assumed that only a quarter of the occupants will smoke whereas in a small private office the ratio of smokers to abstainers may be much higher. Indications are that, for a minimum outside air supply of 8 litre/s per person with no smoking, one and a half times that quantity is required where *some* smoking takes place and twice to four times that where the concentration is *heavy to very heavy.*

Air supply for other reasons

As in the case of emissions from occupants, those from other sources may be grouped as being unwanted heat, unwanted moisture and contaminants, as follows:

Unwanted heat

Incidental sources of heat have been referred to in Chapter 3, up to 40 W/m^2 being quite normal in a commercial office, and where these occur it may be possible to introduce a sufficient quantity of outside air to prevent the internal temperature from rising above some predetermined level. That quantity may be found from the expression:

$$Q = (1000\,\mathrm{H})/\rho c$$

where

Q = outside air quantity for 1 K rise (litre/s)
H = heat emission (kW)
ρ = specific mass of air (kg/m^3)
c = specific heat capacity of air (kJ/kgK)

Similarly heat gains from the sun, particularly during the summer months, may in some cases be removed by a simple ventilation system but in many instances the magnitude of such a gain is such that a very large air volume would be needed which may be undesirable from other points of view. Where the building structure is heavy, it may be possible to improve daytime conditions during the summer by use of a high ventilation rate during the night, when the outside air is cooler, and a lower rate during the day.

In general terms, however, it must be remembered always that the removal of heat by simple ventilation is achieved by raising the temperature of the air introduced. In any normal circumstance therefore, with reliance upon ventilation alone, the result may well be that the temperature within the space is higher than that prevailing outside. If such a situation were considered to be unacceptable then it would be necessary to consider the application of air-conditioning, as dealt with in later chapters.

Unwanted moisture

In certain special circumstances, the need to combat condensation may be the criterion for ventilation rate. One such instance is that of a swimming pool hall where, of the requirements to supply air to the occupants – removal of the familiar 'chemical' smell and action to combat condensation – the latter is by far the more important. It has been shown that, with double glazing, the outside air supply rate should not be less than 15 litre/s per m^2 of the water and wetted surround surface, i.e. about 18 litre/s per m^2 of water surface.*

In housing, it has been shown that a mean ventilation rate of one air change per hour (see later comment upon this quantity) is required to prevent condensation on single glazed windows in a dwelling which is heated uniformly.† In domestic kitchens, the ventilation requirement to avoid condensation is about 100 litre/s for electric cooking and half as much again for gas cooking. Packaged units of continental manufacture which provide for whole-dwelling supply and discharge ventilation, including energy recovery, appear to provide for an outside air volume of 70 litre/s or about 1.5 air changes per hour.

Contaminants

Under this heading, a whole range of industrial hazards arises in addition to the particular requirements for medical buildings, laboratories, animal rooms, horticulture, etc., which

Table 13.4 Limiting values of some common contaminants

	Exposure limit (ppm)			*Exposure limit* (ppm)	
Contaminant	*8 hour*	*10 min*	*Contaminant*	*8 hour*	*10 min*
Acetone	750	1500	Hydrogen peroxide	1	1.5
Ammonia	25	35	LPG	1000	1250
Bromine	0.1	0.3	Methanol	200	250
Butane	600	750	Nitrous oxide	100	–
Carbon dioxide	5000	15 000	Ozone	0.1	0.3
Carbon monoxide	50	300	Sulphur dioxide	2	5
Carbon tetrachloride	2	–	Trichloroethane	100	150
Chlorine	0.5	1	Turpentine	100	150
Formaldehyde	2	2	White spirit	100	125

Note: these data are presented to illustrate the wide range of limiting values. Reference in practice should be made in all cases to the current edition of the Health and Safety Executive publication *Occupational Exposure Limits, EH 40/–* which is published annually.

are too diverse and too specialised to be dealt with here. It is not possible to offer any generalised observations in this respect other than to refer to the equilibrium expression which follows here. This may be used when details of the process and the permissible‡ concentration of the contaminant are known in any set of consistent units: a brief list of common contaminants is given in Table 13.4.

* Doe, L. N., Gura, J. H., and Martin, P. L., 'Building services for swimming pools', *JIHVE,* 1967, **35**, 261.
† Loudon, A. G., and Hendry, I. W. L., *Ventilation and Condensation Control.* IHVE, RIBA. IOB. Conference 1972.
‡ Health and Safety Executive. *Occupational Exposure Limits.* Guidance Note EH 40/1993, HMSO.

Table 13.5 Ventilation based on air change rates

Type of enclosure	*Air changes per hour*	*Ventilation Allowance* (W/m^3 K)
Assembly halls	3–6	1.0–2.0
Bedrooms	1–2	0.33–0.66
Boiler rooms and engine rooms	10–15	3.33–5.0
Class rooms	3–4	1.0–1.33
Corridors	2–3	0.67–1.0
Entrance halls	3–4	1.0–1.33
Factories, large open type	1–4	0.33–1.33
Factories, densely occupied workrooms	6–8	2.0–2.67
Foundries, with exhaust plant	8–10	2.67–3.33
Foundries, without exhaust plant	10–20	3.33–6.67
Hospital operating rooms	20	6.67
Hospital treatment rooms	10	3.33
Kitchens above ground	20–30	6.67–10.0
Kitchens below ground	40–80	13.33–26.67
Laboratories	10–15	3.33–5.0
Laundries, dye houses and spinning mills	10–20	3.33–6.67
Libraries	3–4	1.0–1.33
Living rooms	1–2	0.33–0.67
Offices above ground	2–6	0.67–2.0
Offices below ground	10–20	3.33–6.67
Restaurants and canteens	10–15	3.33–5.0
Rolling mills	8–10	2.67–3.33
Stores and warehouses	1–2	0.33–0.67
Workshops with unhealthy fumes	20–30	6.67–10.0

$$Q = P[(1 - C_1)/(C_1 - C_2)]$$

where

Q = rate of supply of outside air
P = rate of contaminant release
C_1 = permissible concentration in room
C_2 = contaminant concentration in outside air

Air change rates

Where the occupancy is unknown or variable, an arbitrary basis for the ventilation rate must be taken. Table 13.5 may be used as a guide but the cubic content of the space and the length of time for occupation may have an effect upon the rate of air supply necessary. For example, the air volume of a very large hall might be considered as holding a 'store' of air which could be used to reduce the quantity provided over a short period of occupancy.

Legislation and other rules

No précis of current Acts of Parliament or the Regulations and Guidance Notes published (often separately for England and Wales; Scotland; and Northern Ireland) under those Acts can be either complete or up to date; nor can a digest of local authority bye-laws or regulations purport to be comprehensive. The notes which follow therefore are no more than an alphabetical listing to illustrate some of the less specialised regulations:

- Auditoria. The requirements of the now defunct GLC were that an outside air quantity of 7.8 litre/s per person be provided by a ventilation system. Half of that quantity was permitted if cooling were added.
- Car parks. *Building Regulations*, etc., require, for natural ventilation, that openings to outside should have an area of 5% of floor area. Between 6 and 10 air changes are required, dependent upon circumstances, if ventilation is to be mechanical.
- Dwellings. *Building Regulations* require that habitable rooms, naturally ventilated, must be provided with an opening to outside with an area of at least 5% of floor area. When ventilation is by mechanical means, one air change per hour must be provided to habitable rooms and three air changes per hour to bathrooms (or 15 litre/s) and one air change per hour to kitchens (or either 60 litre/s generally or 30 litre/s via a cooker hood).
- Factories. Since almost all work places fall within this one category, specific Orders under the Factories Act, the Offices, Shops and Railway Premises Act and the Health and Safety at Work Act will apply. An Order published by the Health and Safety Executive suggests that a minimum of 6 litre/s of outside air should be provided per person.
- Hospitals. Publications by the Department of Health and Social Security cover requirements for Buildings over a very wide field of activities.
- Kitchens. School kitchens are required, in a Building Note published by the Scottish Education Department, to have an outside air supply of not less than 17.5 litre/s per m^2 of floor area nor less than 20 air changes per hour.
- Schools. The Department of Education and Science requires that working areas should be provided with an outside air supply of 8.5 litre/s per person.
- Theatres. These, with other places of public entertainment such as dance halls, fall under the same category as auditoria.
- Toilets. Many local authorities (and 1993 draft *Building Regulations*) require three air changes per hour or 6 litre/s of outside air per WC pan. (For public use, it is good practice to provide double this air quantity.) Duplicate fans and motors are commonly required.

Readers who require detailed information upon a particular area of legislation would be well advised to consult BSRIA since this organisation publishes a wide range of bibliographies.

Criteria for air supply to occupied rooms

Good ventilation cannot be defined in simple terms which can be quantified as hard and fast rules. Reference must be made to conditions which have been found in practice to give reasonably satisfactory results, as outlined in the following few paragraphs. It may be of interest to compare these with the fundamentals of natural ventilation given by Walker* in 1850:

- Windows are to admit light and not air; ventilation should be catered for separately.
- Both inlets and outlets are necessary.

* Walker, W. *Useful Hints on Ventilation.* Parkes, Manchester.

- Incoming air should be warmed to avoid draughts.
- Inlets and outlets should be well distributed.
- Ventilating openings should be permanent, realising that once closed, they will remain closed.

Distribution and air movement

The admission of outside air into an occupied space is discussed at some length in Chapter 15, the cardinal principles being that:

- It is evenly diffused over the whole area served, particularly at breathing level.
- It should not strike directly upon the occupants.
- It should, nevertheless, provide a feeling of air movement and not allow any areas of stagnation.

The method of distribution adopted often determines the volume of air to be circulated. Though a small quantity of outside air may be all that is required for a sparse population in a large room, it may be impossible to diffuse this evenly over the whole area and to avoid pockets of stagnation. It will be necessary in consequence to increase the volume in circulation either by adding further outside air or by arranging to add a proportion of room air as recirculation.

Conversely, in a small crowded room, distribution problems may make it impossible to introduce the required quantity of outside air without causing draughts. In this case, the volume could be reduced which might lead to an unavoidable temperature rise which may not be acceptable: in this situation, the only solution would be to cool the air before it is admitted.

Given a room of reasonable proportions, the limits of effective distribution consistent with the maintenance of comfort appear to range between a minimum of about four and a maximum of up to 20 air changes per hour. Obviously, the higher rates cannot be achieved unless mechanical means are employed to provide the necessary propulsion.

The extraction of vitiated air from a space is not to be relied upon to create a directional effect. Movement upwards, downwards or sideways will, in a general sense, be towards the point of removal but with no positive momentum.

Temperature

If the temperature of the air admitted were too much below that of the room, it would fall to the floor without proper mixing and might cause cold draughts. On the other hand, the temperature must not be too much above that of the room since in that case the air would rise to the ceiling with the result that stagnation could occur in the breathing zone.

In the case of a straightforward ventilation system, which is under discussion here, it is assumed that the heating of the building is dealt with by some separate radiator or other similar plant. This will have been designed to balance the heat loss through the building fabric and to maintain an internal temperature of, say, 20°C during winter weather. In these circumstances, the ventilation air supply should be delivered to the room at about 17°C, or 3 K lower than the desired temperature, so that it has the potential to reach 20°C as it absorbs unwanted heat prior to leaving the room.

Under summer conditions, without provision for mechanical cooling, no control can be exercised over the temperature of the air delivered to the room other than by the small reduction made possible by a process of evaporation, referred to later.

Humidity

Ventilation air admitted from outside will have the same moisture content as that prevailing at the source. In cold weather, as was mentioned in an earlier chapter, this moisture content may be low and although the quantity will not alter in the *absolute* sense when the air is heated to say 17°C, the *relative* humidity will have fallen. This situation may be corrected, where mechanical inlet plant is used, by incorporating humidification equipment in the form of water sprays or some other device for producing water vapour. By such means, the amount of water added may be controlled to give a desirable relative humidity of 40–60%, at room temperature.

In the normal mid-season mild weather of spring and autumn in the British Isles, the humidity prevailing outside will generally be at a level which will permit some temperature adjustment without any necessity to take corrective action. The human body, as was explained earlier, is not too critical of humidity variations at the temperature levels then prevailing. There is, in addition, a reservoir effect in most buildings as a result of the hygroscopic retention of fabrics, timber, paper, etc., which serves to steady changes in humidity taken over a long period.

In summer, when the humidity outside is high, there can be no complete control of internal humidity without full air-conditioning which, by definition, includes means for dehumidification. Humidifiers, of whatever type, can only *add* moisture but, in doing so, some may cool the air by evaporation to a temperature approaching that of the wet bulb.

Thus, with outside air at, say, 24°C dry bulb and 18°C wet bulb, (relative humidity = 52%), a humidifier may be able to reduce the temperature to 19°C, but at the same time increasing the relative humidity to 90%. This second condition may well be more oppressive than the first and it is often found that humidifiers, where provided, are shut off in hot weather for that reason. Reference to Figure 1.2 (p. 10) will show that the first condition noted above is just inside the summer comfort zone, but that the second condition is well outside.

It is necessary to add a note of caution here since, as a result of the investigations into the incidence of *Legionnaires' disease* which are discussed more fully in both Chapter 14 and Chapter 20, the use of water spray types of humidifier in air streams is now looked upon with suspicion. For the use here mentioned, it is probable that the critical temperature band 25–45°C will be avoided but the margin is too narrow to be ignored.

Air purity

Most buildings requiring some form of ventilating system are in the centre of towns or cities where atmospheric pollution is at the highest level normally encountered. As a result of the action which followed the implementation of the 1968 Clean Air Act, the soot, tar, ash and sulphur dioxide which had previously been a major problem as a result of indiscriminate use of coal and heavy oil, are no longer the principal pollutant. Their place has largely been taken by the effluents produced by motor vehicles, in contrast to those complained of by Florence Nightingale in 1860, when she referred to

> Dirty air coming in from without, soiled by sewer emanations, the evaporation from dirty streets, bits of unburnt fuel, bits of straw and bits of horse dung.

When air is introduced into a building by ventilation, it brings with it a proportion of whatever pollutants exist outside. If inlet is to be by natural means, i.e. through open windows and the like, then there is very little to be done to prevent this incursion. Where the supply of air is to be by mechanical means then a whole variety of types of filters is available. Some of these are discussed later in Chapter 17.

The use of *activated carbon* filters will revive vitiated air by adsorbtion of many impurities such as body odours, sulphur dioxide, petrol fumes, and other noxious products of civilisation. Filters of this type are, however, high in first cost and have a relatively short service life before return to the manufacturers for re-activation becomes necessary. Nevertheless, use of this equipment is growing for certain specialised installations and, with the growth, costs may fall and usage spread to more mundane applications.

Methods of ventilation

Where air movement is induced either by wind or by the effect of temperature difference, ventilation is termed *natural*. On the other hand, where air movement results from power drive applied to a fan or fans, the arrangement is described as being *mechanical*. Since inlet and extract have to be considered separately, there are four possible combinations as covered under the sub-headings which follow.

Natural inlet and natural extract

Arrangements of this sort are used in buildings which seem to fall into three categories which, for want of better descriptions, might be called traditional, commercial and industrial.

Traditional

This category includes most school and university buildings, hospitals, shops, office buildings constructed between the wars, almost all domestic premises and other rooms and buildings with low levels of occupancy. Conditions internally depend upon clean outside air and upon other external features which permit windows to be opened.

Where an open fireplace exists, the flue provides a route for exhausting air from the room. Further, in instances where a fireplace is in use for burning solid fuel, the flue serves

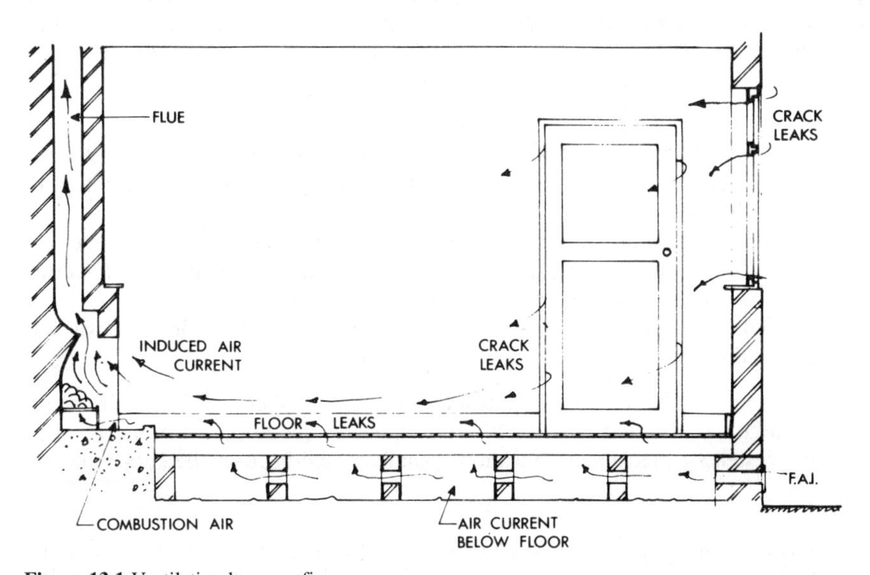

Figure 13.1 Ventilation by open fire

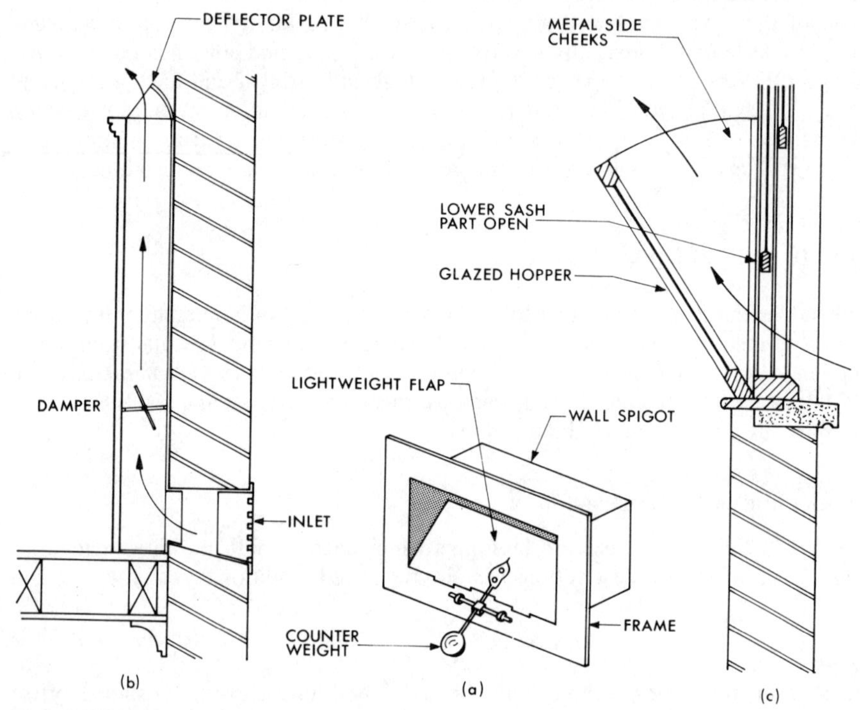

Figure 13.2 Natural ventilators

the dual role of not only carrying away products of combustion but also of inducing a flow of room air much greater in quantity than is normally required, as illustrated in Figure 13.1: an adjustable restriction at the throat of the flue is advised in order to ameliorate this problem. It must be added that a gas fire connected to a flue will have the same effect but at a much reduced level. Various methods of providing natural ventilation to rooms have been devised as shown in Figure 13.2:

(a) The Dr Arnott ventilator of the 1850s which, built into a chimney breast at high level in the room, acted to remove 'used' air as induced by the flow of products of combustion in the flue.
(b) The Tobin tube of the same era which sometimes had a water tray fitted at the low level inlet 'to cleanse the entering air from smuts or organic impurities' or a wetted fabric 'sock' suspended from the outlet as a terminal filter.
(c) The hopper window of Edwardian school rooms which, when the lower sash was opened, diverted entering air to a higher level in the room.

Many modern windows of better quality, whether metal or timber, provide some means for 'trickle' or 'night' ventilation. This feature is sometimes incorporated in the window proper or in the fastening mechanism but more usually at the head of the sub-frame.

Commercial

For larger rooms, assembly halls, workshops, etc., simple high level outlet ventilators, combined with low level inlets for outside air, provide a solution which is low in capital

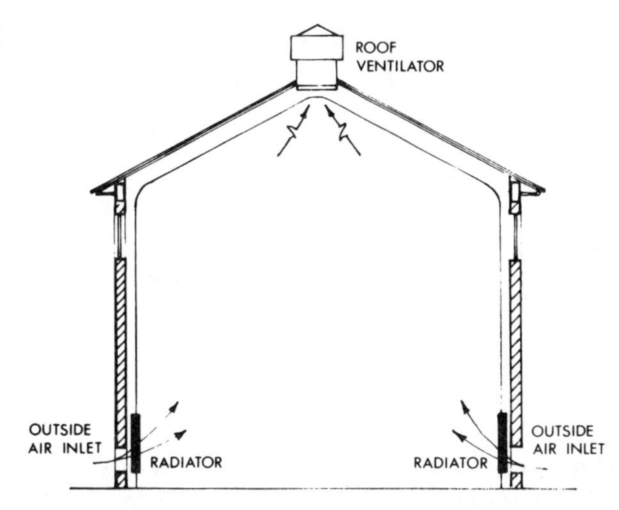

Figure 13.3 Natural inlet and natural extract

cost but unreliable in operation. A simple system is illustrated in Figure 13.3 where air inlet is through grilles fitted behind radiators, an arrangement much favoured in the past. The grille openings, inside and outside, soon become dirt and insect traps and are often sealed by occupants in order to avoid winter draughts. In hot weather, when ventilation may be most needed, little appreciable air movement will occur in consequence.

Industrial

This, in effect, is an extension to the previous category but with the clear distinction that the type of building is different as is the equipment used. There are a great many heavy industries where natural ventilation has been, and will be applied with success to single-storey shed type constructions. These industries are those in which there is a considerable

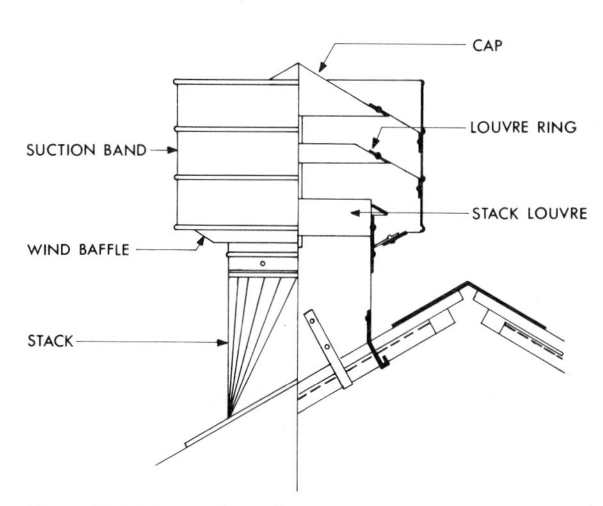

Figure 13.4 Robertson's ventilator

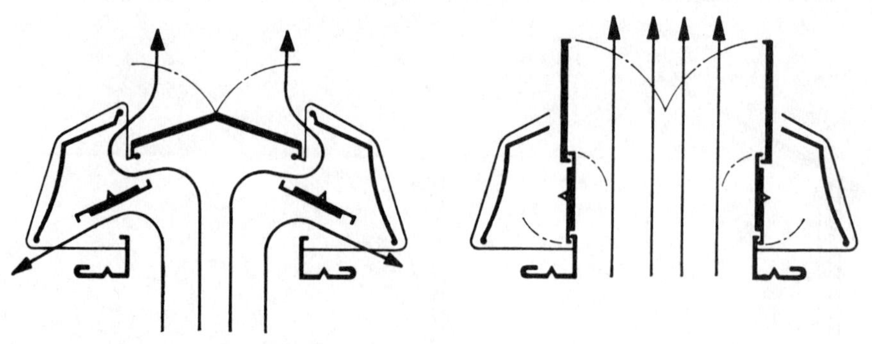

Figure 13.5 Colt type MF ventilator

release of heat or steam, such as foundries, rolling mills, welding shops and dye-houses, etc. The quantity of heat to be removed is significant and a temperature rise of 2 K per metre height may often be encountered. The equipment used in this application has been evolved over the years with a proven capability, an example being the well-known Robertson's roof extract unit shown in Figure 13.4.

An alternative form of roof unit is illustrated in Figure 13.5, this being a multi-function device, having side dampers which may be closed by remote operation at times when no ventilation is required, and a weather cap, shown in the right-hand diagram, which may be opened similarly when a completely unobstructed outlet is necessary. Such units are manufactured to provide outlet areas of between about 0.5 and 1.3 m^2. The development of versatile roof ventilators of this type offers considerable advantages from the point of view of smoke venting since they encourage rapid clearance and prevent spread.

On an industrial scale, inlets for natural ventilation may well present a problem because of their size and possible proximity to those operatives who may not be affected directly

Figure 13.6 Louvred wall panel for natural inlet

by excessive heat. Wall openings, provided with louvres as Figure 13.6, may be acceptable and be arranged for automatic opening in parallel with the associated roof outlets. Air admitted by such louvres is unlikely to be heated at the point of entry in winter, since the volume required may well be too large to pass over a conventional heater without mechanical assistance.

Temperature difference effects

In some respects, the motive force arising as a result of temperature difference is similar to that utilised in a gravity hot water heating system. This so-called *stack effect* relates a column of air within a building to one of similar cross section outside and considers the average specific mass of each, together with the acceleration effect of gravity ($9.81\ m/s^2$). The mass varies inversely with absolute temperature, approximately as $353/(t + 273)$, and thus the following expression may be derived:

$$\Delta P = (9.81 \times 353 \times H)[1/(t_1 + 273) - 1/(t_2 + 273)]$$

and thence, for the range of temperatures encountered in practice

$$Q = (268 \times A)[H(t_1 - t_2)]^{0.5}$$

where

ΔP = pressure difference (Pa)
Q = air volume (litre/s)
A = free area of inlets or outlets (m^2)
H = height, centre of inlets to centre of outlets (m)
t_1 = average indoor temperature over height, H (°C)
t_2 = average outdoor temperature over height, H (°C)

Table 13.6 lists values calculated from this expression, including a resistance factor as noted later, for a range of heights and temperature differences.

Table 13.6 Air volumes for natural ventilation due to stack effect

Effective height (m)	*Air volume* (litre/s per m^2) *free opening for temperature difference* (K)								
	2	*4*	*6*	*8*	*10*	*15*	*20*	*25*	*30*
2	240	340	420	480	540	660	760	850	930
4	340	480	590	680	760	930	1070	1200	1310
6	420	590	720	830	930	1140	1310	1470	1610
8	480	680	830	960	1070	1310	1520	1700	1860
10	540	760	930	1070	1200	1470	1700	1900	2080
15	660	930	1140	1310	1470	1800	2080	2320	2550
20	760	1070	1310	1520	1700	2080	2400	2680	2940
25	850	1200	1470	1700	1900	2320	2680	3000	3290
30	930	1310	1610	1860	2080	2550	2940	3290	3600

Notes:
1. Areas of openings are assumed to be equal, i.e. upper = lower.
2. Openings are assumed to be 45% effective.

Wind effects

Natural ventilation resulting from wind is unpredictable and any attempt to quantify the effect should be based upon data which give some indication of the frequency of occurrence. Statistics which are reproduced in the *Guide Section A2* relate to the *meteorological wind speed,* i.e. as measured in open country at a height of 10 m. For use in practice, factors are published which enable this speed to be adjusted to suit other heights and alternative topographies. In addition, *isopleths* (contours of hourly mean wind speeds) are available together with multipliers which allow frequency of occurrence to be estimated.

In the context of ventilation, the design criterion must be aimed at the low end of the scale and, for a speed which will be exceeded for 80% of the time, a figure of about 3 m/s would appear to be a reasonable choice. Table 13.7 has been prepared, from the simple relationship $Q = A \times V$, using this wind speed for the columns headed 'open country' and adjusted values for other locations. A resistance factor, as noted in the following paragraph, has been applied to produce the volumes listed.

Table 13.7 Air volumes for natural ventilation due to wind

Effective height above ground (m)	*Air volume* (litre/s per m²) *free opening with notional wind speed of* 3 m/s							
	Openings normal to wind				*Openings at angle to wind*			
	Country		*Urban*	*City*	*Country*		*Urban*	*City*
	Open	*With wind breaks*			*Open*	*With wind breaks*		
2	1260	990	690	440	690	540	380	240
4	1420	1130	820	550	720	620	450	300
6	1520	1230	900	630	830	670	490	340
8	1610	1300	970	690	880	710	530	370
10	1660	1360	1040	740	910	740	570	400
15	1780	1470	1140	850	970	800	620	460
20	1870	1560	1220	930	1020	850	670	510
25	1940	1630	1290	1000	1060	890	700	550
30	2000	1690	1350	1060	1090	920	740	580

Notes:
1. Areas of openings are assumed to be equal, i.e. inlet = outlet.
2. Wind speed is 3 m/s (exceeded for 80% of time).
3. Wind speed is at notional height of 10 m in open country: other speeds are derived from this datum.

Resistance factors

Theoretical air flow quantities due to temperature and wind effects are not achieved in practice due to flow resistances and the physical arrangement of inlet and outlet openings. For the flow due to temperature effects, a correction factor of 0.45 was used in preparing Table 13.6 and, for Table 13.7, correction factors of 0.55 and 0.3 were used for the first and second sets of columns, respectively. These factors are probably appropriate to buildings where simple louvred ventilators are used but should not be confused with the comprehensive performance data published by specialist manufacturers of ventilators for the industrial category.

Other considerations

The combined results of temperature difference and wind effects may be estimated by their addition, with respect to sign: that is to say, with the assurance that both are acting to produce air flow in the same direction (which is by no means always the case!). Similarly, the matter of a positive wind effect upon an exposed face of a building and a negative effect upon the lee side must be considered, particularly in circumstances where openings may not be of the same size. The entire subject cannot be dealt with in the abstract since the variables are complex and, to some extent, depend upon the shape of the building. The essential factors are given in the *Guide Section A4*.

Natural inlet and mechanical extract

A mechanical extract system will function irrespective of wind and temperature difference and will be positive in action. Since the air to be extracted from the space must be replaced and the means provided for admission of outside air will present some resistance to flow, leakage inward from surrounding spaces is more likely than leakage outward. In consequence, escape of steam, fumes and noxious vapours generated within the ventilated space is less likely than would be the case if reliance were placed upon natural extract alone.

A difficulty arises however in providing a satisfactory means of admitting the bulk of the air required to balance the extract volume, and of heating it in winter. Fresh air inlets behind ranges of gilled tube or some other form of natural convector, as in Figure 13.7, need regular cleaning and have only a limited application.

For summer use only in a small building, infiltration via the type of window ventilator ('trickle' or 'night') referred to on p. 344 may suffice. In some circumstances, replacement air may be drawn from another part of the building to serve a dual purpose; a good example being extract ventilation from a small kitchen which, by drawing air through serving hatches, serves to dilute the spread of cooking smells to the associated canteen.

For industrial applications, the natural roof ventilator shown earlier might be replaced by some form of fan-powered extract unit as Figure 13.8. Units of this type are made in a variety of patterns with mounting arrangements and weathering covers to suit most types of roof construction, pitched and flat. The air extracted by such units is discharged at a

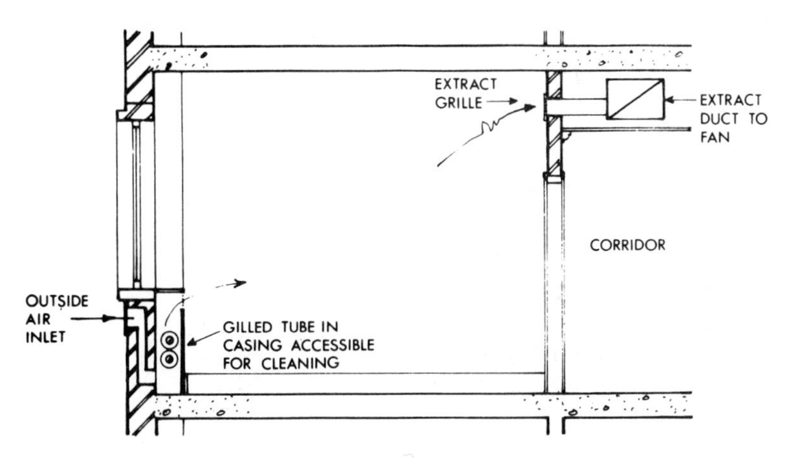

Figure 13.7 Natural inlet and mechanical extract

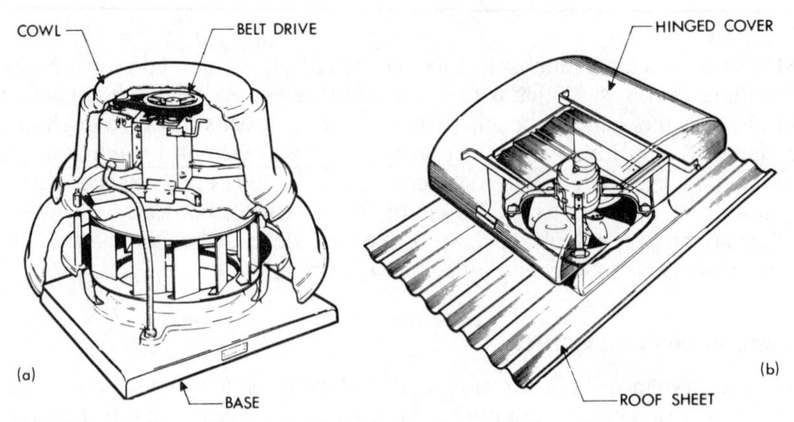

Figure 13.8 Fan powered extract units: (a) centrifugal; (b) propellor

relatively low velocity and has a tendency to hug the roof profile, thus, where fumes and other pollutants are carried in that air, it is better practice to use a *vertical jet-discharge* unit as Figure 13.9. Axial-flow fans, which are commonly more noisy, are often fitted to such units, together with hinged vanes or dampers, for rain protection when the fan is idle. These units may be connected to either horizontal or vertical ductwork.

With the evolution of well detailed weathering methods for roof extract units, the previous practice of mounting fans to factory gable walls has lost favour. It is necessary that fans so arranged be fitted with baffles or discharge bends in order to avoid the effect of wind blowing in opposition to the discharge. Self-closing dampers are required, in addition, to prevent unwanted inlet or discharge when the fan is not in operation. For domestic use and indeed for application to small rooms generally, the familiar window or wall mounted extract units manufactured in moulded plastic, Figure 13.10, may well

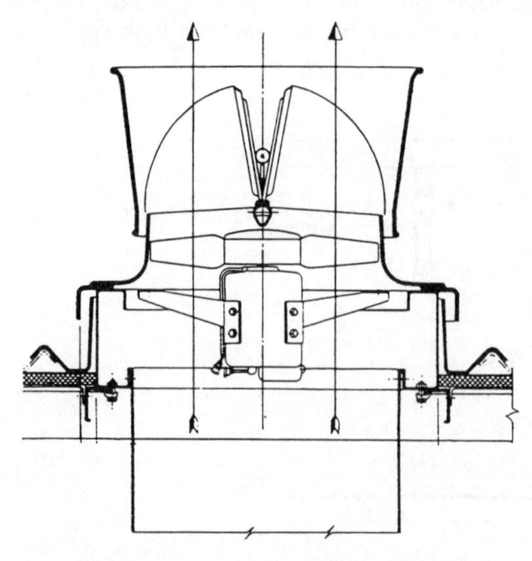

Figure 13.9 Vertical jet-discharge roof unit

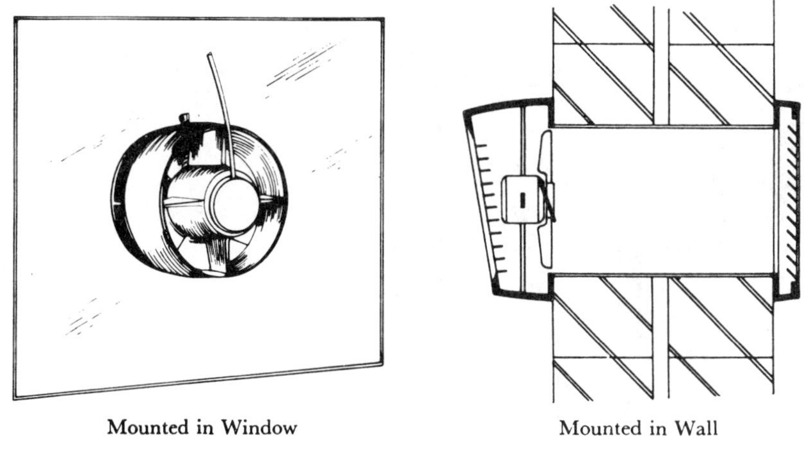

Figure 13.10 Domestic window or wall mounted fans

provide a solution to a local ventilation problem which is adequate for most of the time. Such units may be provided with speed control and, in some instances, are reversible so that they may, alternately, be used for either an inlet or an extract duty.

A specific area where local authority requirements bear upon the provision of mechanical extract with natural inlet relates to internal toilets and bathrooms, particularly in multi-storey housing. An investigation which was reported 30 years ago remains valid,* the principal conclusion being that a ventilation volume of 5.5 litre/s should be provided for each fixture, i.e. per WC pan or bath, and thus 11 litre/s for a bathroom with WC. In order to overcome stack effect and other chance pressure differences due to open doors, etc., a duct system design as Figure 13.11 was recommended; this has the added advantage

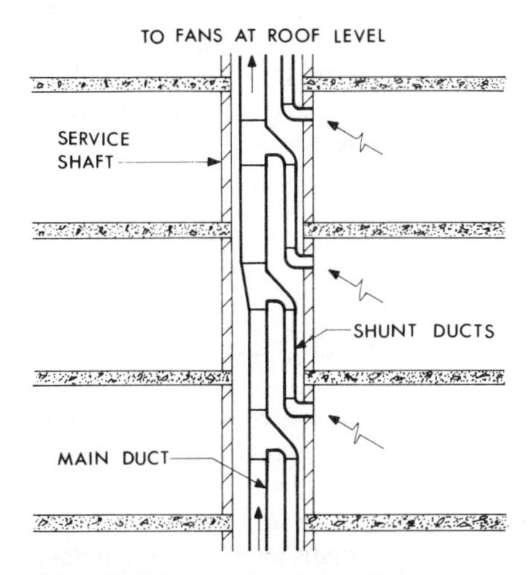

Figure 13.11 Extract duct from toilets in multi-storey flats (fire barriers not shown)

* Wise, A. F. E. and Curtis, M., 'Ventilation of internal bathrooms and water closets in multi-storey flats,' *JIHVE*, 1964, **32**, 180.

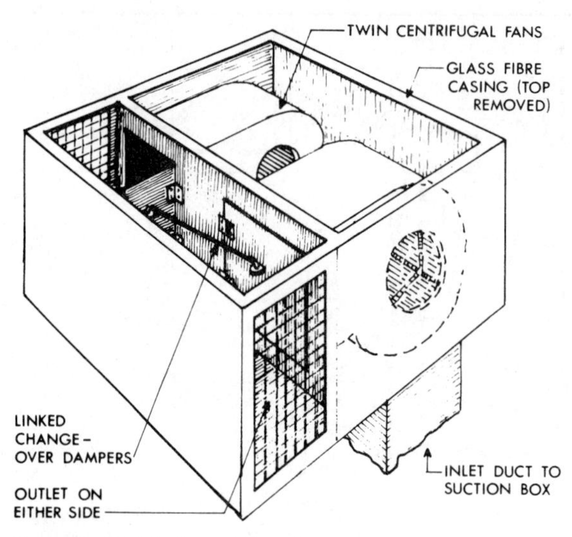

Figure 13.12 Twin fan unit for toilet extract

of providing an extended travel for voice transfer between dwellings. In the case of office blocks and other large buildings, an extract rate of 10–12 air changes per hour is desirable.

In this context, most authorities require that the exhaust fan must be duplicated, with arrangements for automatic change-over in the event of failure. Conventional fans may be fitted in parallel but it is more convenient to fit a purpose-made twin fan unit of the type shown in Figure 13.12.

As in other instances, provision must be made for a supply of inlet air from outside and, in a block of flats where toilets and bathrooms are likely to be arranged one above the other, inlets may connect to a rising shaft with, perhaps, some form of heater at the base. Subject to decision by the fire authority, it may be possible to use the shaft provided for rising water pipes for ductwork but not, of course, if space be allocated also to wastes from sanitary fittings. Some authorities require that a lobby be introduced at the entrance to toilets in public buildings and that this be provided with a supply of outside air by an inlet plant serving no other duty.

Mechanical inlet and natural extract

This proposition should not be confused with the plenum heating system described in Chapter 6. It is typified, for application to an office building, by the arrangement shown in Figure 13.13. As may be seen, an air supply from some central source is ducted through a corridor ceiling void to individual rooms; the vitiated air escapes therefrom, through a low level register, to the corridor proper and thence to outside. It is now unlikely that a fire authority would permit the construction of such a system since it has the potential to endanger the atmosphere of an escape route. A single room, perhaps an office, might have an outside air supply from some form of reversible fan unit as described earlier but this hardly falls within the context of the present heading.

For an industrial application, such a system might consist of a number of individual unit heaters, each with an outside air inlet as described in Chapter 8, and working in parallel

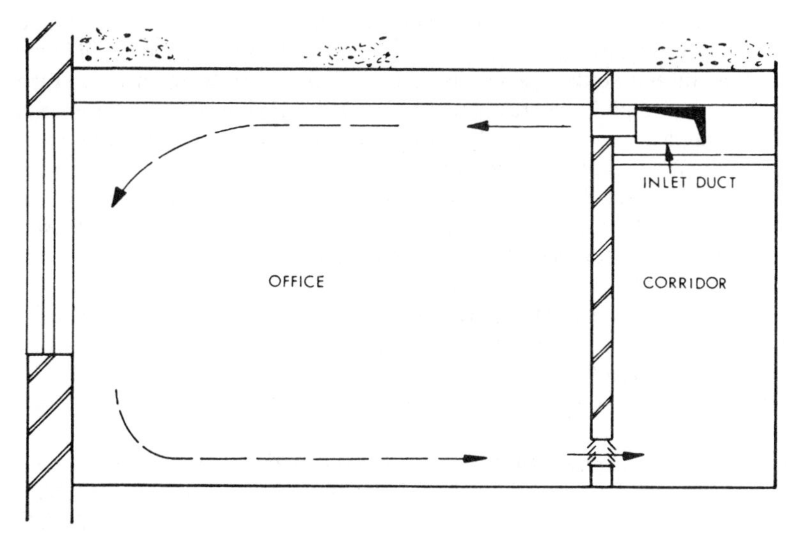

Figure 13.13 Mechanical inlet and natural extract

with natural roof ventilators. But this arrangement is barely a credible proposition for present-day requirements.

The ventilation of boiler houses is a special case as a result of the air consumed in combustion. The volume required may be estimated, as was described in Chapter 12 (p. 332), and, where the boiler house is above ground, this quantity may be adequate without any augmentation. Where a boiler house is sited in a basement, or otherwise remote from outside air, the amount of heat emitted by the plant may produce conditions which are unacceptable to maintenance staff and thus be the criterion for determination of the ventilation air quantity. As a notional figure representing emission by a heating boiler plant in winter, a supply of between five and six times the combustion air quantity is a suitable volume for preliminary design purposes.

It is most desirable that this outside air supply be provided by mechanical means in order to maintain a positive pressure to the boiler plant. The siting of the louvres which admit outside air and any parallel openings for exhaust to atmosphere should be chosen with care since they will act also as a route for the escape of noise. The provision of louvres which have been designed to absorb this noise is a prudent step towards avoidance of what may be a major nuisance and source of complaint.

Mechanical inlet and extract

This final combination is that which must be noted as not only having the widest application but also as providing the greatest challenge since the air distribution, in terms of quantity, pressure and temperature, is wholly in the hands of the designer. It may be applied to all manner of spaces and is greatly to be preferred to any of the compromise systems discussed earlier in this chapter. In application, the ratio between the air volume duties of the inlet and extract systems must be selected with care in order to suit the particular application.

For instance, in normal living and working spaces where no noxious fumes are generated, the extract volume should be arranged to be slightly less (by, say, 10–20%)

than that provided by the inlet system: any air movement will, in consequence, be outward rather than inward. Conversely, in cases where fumes of any sort might be generated in the working space and should not be distributed, the balance would best be reversed and the inlet volume arranged to be slightly less (by, say, 10–20%) than that handled by the extract system. These two examples serve only to show that each case must be considered on its merits and that there is no single perfect solution.

It will be obvious from what has been said earlier that there is a great number of combinations of inlet and extract arrangements, with and without ductwork, etc., to suit various purposes. It is now commonplace to provide balanced inlet and extract systems for office and other commercial or recreational accommodation but, possibly because of the large air volumes involved, a similar approach is not always applied to industrial buildings. Wherever any form of mechanical extract is used, it is a parallel necessity that replacement air be introduced in a suitable way and at an appropriate temperature. In the most general terms, a satisfactory solution can be produced only by the addition of some form of mechanical inlet.

In this context, the supply arrangement need not necessarily be a complex system and one solution is to provide unit heaters having an outside air inlet, suitably filtered, as Figure 7.24(b). These may serve also as a means of preheating the building for occupation, when fitted with a damper mechanism which allows recirculation. An alternative method of warming replacement air is the use of direct fired oil or gas heaters as illustrated in Figure 4.10.

Kitchens

The removal of cooking odours from a kitchen and prevention of spread to adjacent rooms is most desirable in dwellings but essential as far as hotels, restaurants and institutions are concerned. The ventilation rate must be high if the system is to be a success and air change rates per hour from a minimum of 30 to as many as 100 are not unusual. In order to prevent the spread of odours from the kitchen, the replacement air delivered by a supply

Table 13.8 Nominal exhaust rates for kitchen appliances

	Air extraction rates (litre/s)	
Equipment	*Unit*	*Per* m^2 *net area of appliances*
Roasting and grilling		
ranges, unit type, (approximately 1 m^2)	300	300
pastry ovens	300	300
fish fryers	450	600
grills	250–300	450
steak grills	450	900
speciality grills	450	900
Steaming and vapour producing		
boiling pans (140–180 litre)	300	600
steamers	300	600
sterilising sinks	250	600
bains-marie	200	300
tea sets	150–250	300

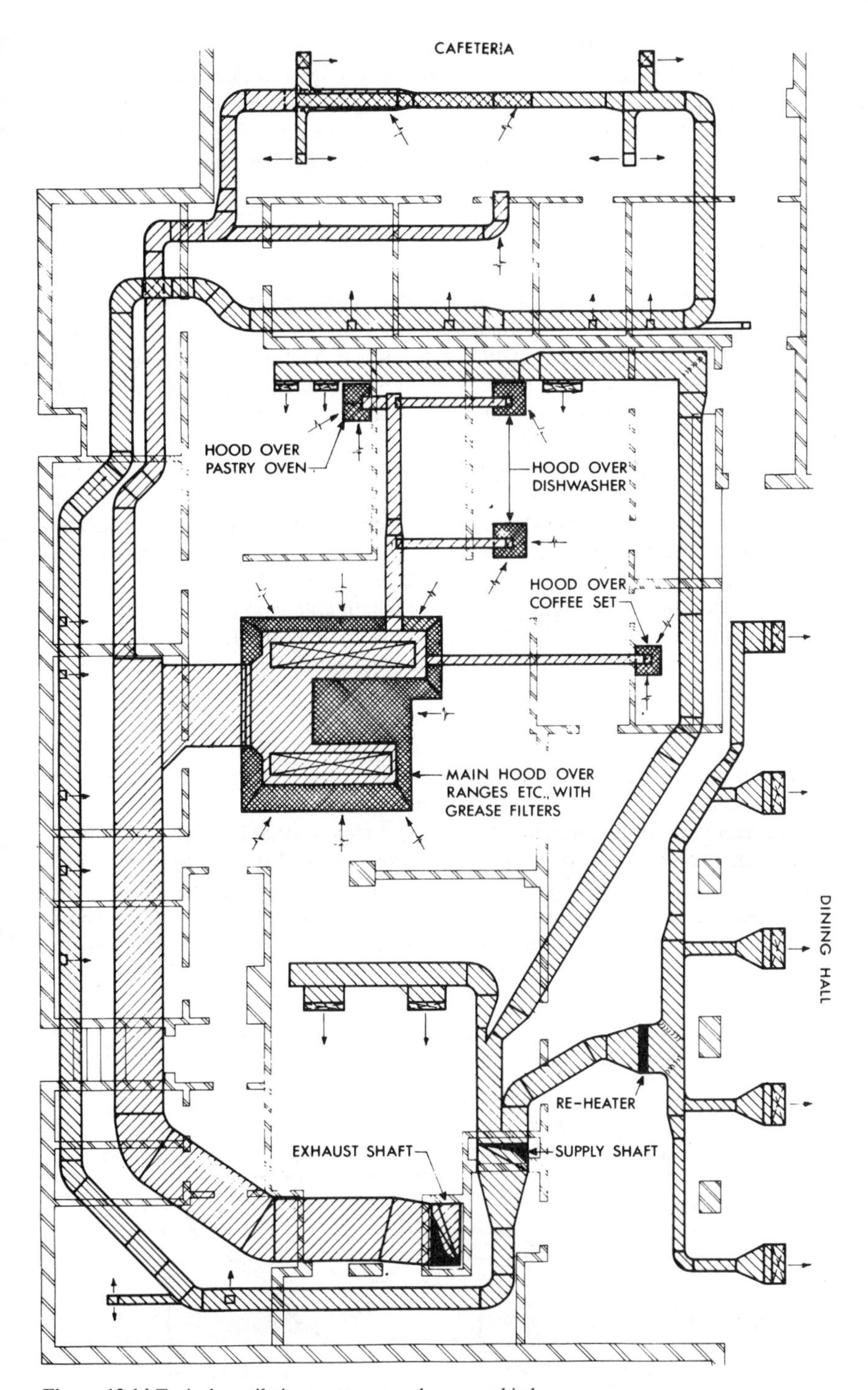

Figure 13.14 Typical ventilation systems to a basement kitchen

plant should generally be 15–20% *less* than the volume extracted, the balance being drawn either from the restaurant or some other intervening area through serving hatches and transfer grilles associated with them. Air velocities through hatches should not be allowed to exceed about 0.2 m/s if complaints are to be avoided. Table 13.8, taken from the *Guide Section B2*, lists the more common appliances and the notional exhaust air requirement per unit and per m^2 area of appliance.

Ventilation by canopy

An extract ventilation system, using collecting hoods, or more properly canopies, over all the principal items of equipment is the method most commonly adopted, as shown in Figure 13.14, which illustrates a large institutional kitchen. The design and construction of canopies, commonly finished in polished aluminium or stainless steel, is now usually dealt with by specialist manufacturers but was, in the recent past, in the hands of contractors fabricating and installing ductwork.

If canopies are to be effective, their size in plan should be such that they overlap the area of the block of appliances which they serve by about 300 mm to 400 mm on all open sides. The *capture velocity* of the air extracted over this plan area should be not much less than 0.4 m/s. Provision must be made for drainage of the condensate which will form on the inside surfaces, by means of a perimeter channel, and grease filters must be fitted, at the point of air exit into ductwork, from any canopy which collects fumes from a process generating oily vapours. Cleanliness in a kitchen is a prime requirement and all exposed surfaces, including the not inconsiderable internals and vertical outside enclosures of canopies, need regular cleaning *in situ* with detergents.

Energy recovery
Bearing in mind the large volume of air extracted from a kitchen through conventional canopies and the consequent need to supply replacement air which must not be much below kitchen temperature which, inevitably, is high, it follows that an appreciable waste of energy will occur. An energy recovery system of conventional plate type may be used and Figure 13.15 shows how a compact arrangement may be mounted to the roof of a small kitchen to operate in conjunction with a simple canopy arrangement. It is of course necessary in all cases to provide a high level of secondary filtration to the exhaust air.

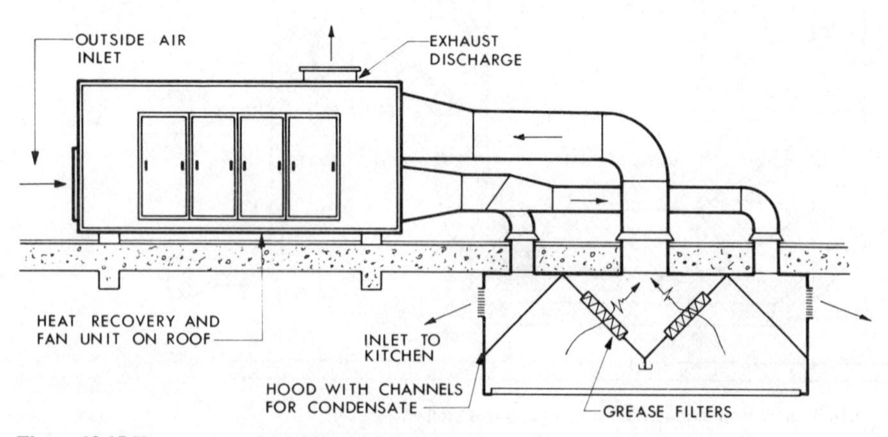

Figure 13.15 Heat recovery from kitchen extract system

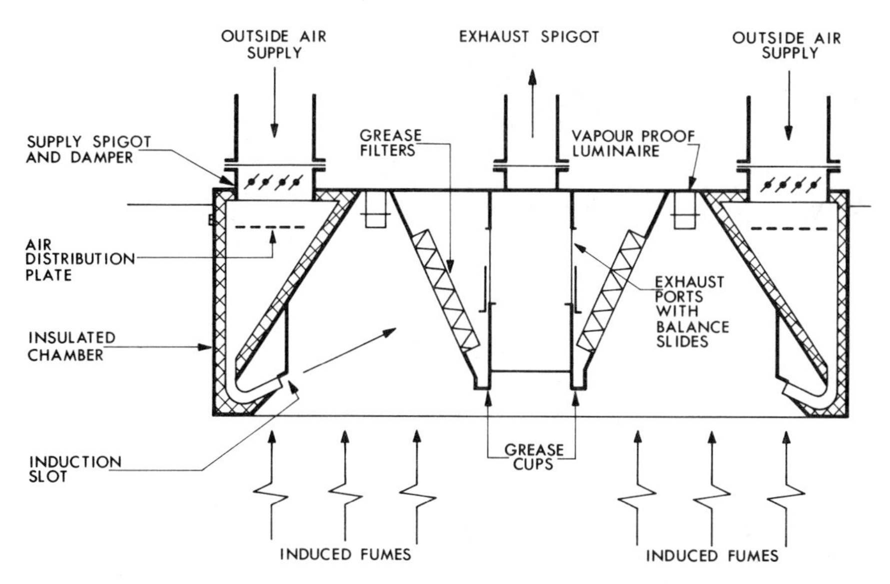

Figure 13.16 Energy saving canopy (Stott-Benham)

Carry-over of grease deposits which will occur if maintenance effort is neglected will soon lead to a commensurate reduction in the efficiency of recovery.

Energy saving canopies

A relatively recent development has gone a long way towards overcoming the problems of energy recovery with the introduction of canopies which use supply air at outside temperature ducted to a venturi slot or slots within the canopy to produce high velocity jets which induce kitchen air and fumes into the canopy prior to exhaust therefrom. With this arrangement, as shown in Figure 13.16, only about one-third of the total supply air quantity provided to the kitchen needs to be raised or cooled to near kitchen temperature. An added advantage is that the resultant air mixture passing through the grease filters is at a lower temperature than it would be in a conventional canopy and the efficiency of filtration is thus improved.

It would be wrong, nevertheless, to pretend that the use of canopies of any design in a kitchen is not without problems, particularly in spaces where the floor to ceiling height is, as so often seems to be the case, less than it should be. The effective form of a canopy catchment, a hollow pyramid or cone, is usually concealed within an enclosure having vertical sides, supposedly in the interests of cleanliness, but the visual effect of such large vertical plane surfaces suspended from the ceiling down almost to head height, is claustrophobic.

Ventilation by other means

To provide ventilation without the use of canopies, a variety of patented and purpose designed ceilings has been produced. One type uses 500 mm square modular ceiling shells, manufactured in stainless steel, to serve as terminals, some to exhaust contaminated air from appliances sited below them and others to introduce replacement supply air. As shown in Figure 13.17(a), the former have two shells, an inner and an outer, each with a

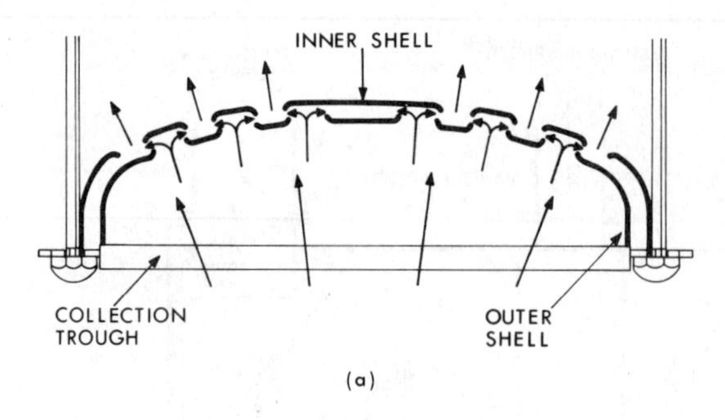

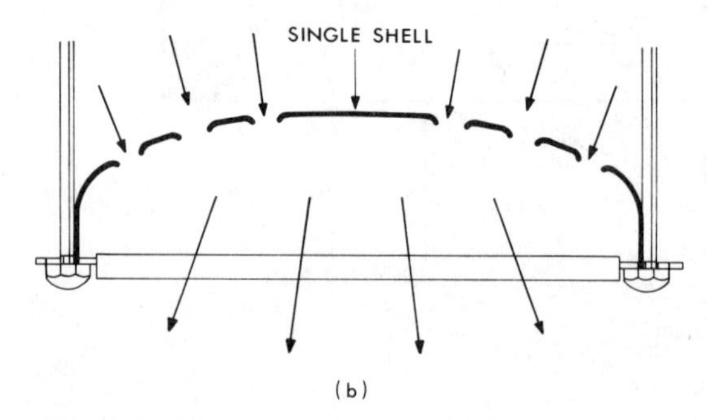

Figure 13.17 Modular ceiling shells (Heydal, OSC Process Engineering)

parabolic profile and holes, which are offset in one shell with respect to those in the other. The latter, as is shown in Figure 13.17(b), has a single shell only.

The twin shell arrangement of the exhaust module serves to separate out grease, moisture and suspended particles from the exhaust air, as a result of passage through the offset holes and the change of direction thus induced, and allows either deposition on the shell surface or flow to collection troughs mounted below each module: in either case, the profile of the shell is such that drips are avoided.

As might be expected, a number of relatively critical design parameters are inherent in such a system which include the following:

- The exhaust air volume must be within the limits of 55 litre/s minimum and 130 litre/s maximum per m^2.
- The number of supply air modules must be determined according to room height and the kitchen space/supply air temperature differential, typically within the range of 30 to 90 litre/s per module.
- A minimum clearance space of 350 mm above the modules and below the structure is required in order to provide for easy removal of the shells.

As in the case of ventilation by canopy, cleanliness is a prime requirement. The manufacturers state that the shells and collection troughs may be dismounted without tools and then cleaned as required in a catering type dishwasher. It would seem sensible for the cleaning staff to carry out preventative maintenance using a rotating stock of spare components, say, 50%.

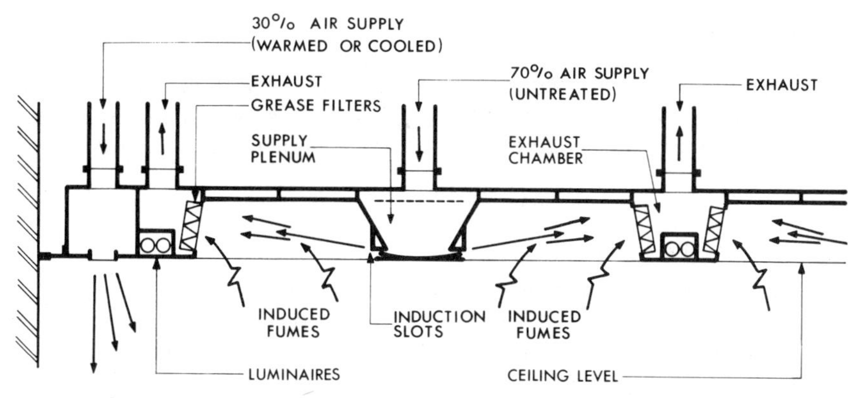

Figure 3.18 Energy-saving ceilings (Stott-Benham)

Energy saving ceilings

An alternative form of 'ventilating ceiling' is available which offers the advantage of energy economy in that it makes use of venturi slots within a series of shallow concavities in the ceiling profile to echo the performance of the previously described energy saving canopies. Figure 13.18 shows this arrangement. The whole of the ceiling is manufactured in stainless steel or anodised aluminium and is insulated to avoid condensation. It should be noted that, as a result of the direction of a large proportion of the air supply to the venturi slots within the ceiling troughs, the air change rate in the occupied area is reduced *pro rata*. This reduced level will lead to a reduction in air velocity through servery hatches and, in the majority of cases, provide a more comfortable working environment for the kitchen staff.

Ductwork connections

Ductwork generally is dealt with in Chapter 16 (p. 423) but that for kitchens, whether supply or exhaust, is a special case. In many instances, with a view to cleanliness, modern kitchens with canopies are provided with a false ceiling to conceal down-standing beams, ducting, pipework and electrical conduits, etc. It is often argued that the resultant void could, if adequate in depth, be used as a plenum for supply air, with ducting reduced to short lengths and bell mouths, or as a suction chamber at negative pressure for exhaust air. Poor site workmanship by building contractors and the natural permeability of most structures and materials should be taken into account and also the inevitability of a long-term cleaning problem.

As far as supply air is concerned, use of such a ceiling void would be practicable only if there were absolutely no possibility of fouling by a noxious leak from, say, a waste pipe or a gas service or, furthermore, that there would be no possibility that a requirement for access, perhaps for maintenance, would ever occur.

It goes without saying that canopies should always be connected directly by sheet metal ducting to the suction side of the exhaust fan. Similarly, in the case of exhaust from a kitchen through a ventilated ceiling, sheet metal suction chambers above the ceiling and sheet metal ducting to the fan suction therefrom is much to be preferred. In the interests of cleanliness and as a precaution against fire, facilities for cleaning, by way of access holes and covers, should be provided at regular intervals along all exhaust ductwork. As a counsel of perfection, similar access on supply ductwork is desirable.

Special applications

It is not possible within the scope of this present chapter to deal with the whole range of demands for ventilation which are imposed by specialist applications. The notes included in the following paragraphs are therefore no more than an introduction to the wide variety of those demands.

Smoke control in escape routes

With the advent of tall buildings and, in their wake, a greater necessity for mechanical air movement in the occupied areas, some fire authorities realised that the traditional approach to smoke clearance throughout escape routes required re-examination. As a result, the concept of air pressurisation to those routes has received much attention. In view of the importance of the subject and the absence of any significant engineering analyses of the solutions offered, it is necessary to refer the reader to BS 5588: Part 4: 1986, *Code of Practice for Smoke Control in Protected Escape Routes Using Pressurization*, which provides an empirical approach to the problem.

In essence, the Code designates lift and other lobbies, with associated corridors in certain circumstances, as being areas in which air pressure should be maintained at levels in excess of that which obtains in surrounding accommodation zones. It is proposed that this situation should be achieved by the admission of an outside air supply to the designated areas, via an independent plant or plants, such that a positive pressure of about 50 Pa, with respect to those surrounding areas, be maintained. In parallel, it is suggested that the ventilation system may be arranged either to produce this pressure in an emergency situation only (*two-stage*) or, preferably, to maintain it at all times (*single-stage*).

The calculations necessary to determine both volume and pressure requirements to meet the duty concerned are tedious and must take into account both those stack and wind effects which may oppose the required air movement. It is, in consequence, appropriate to consider the magnitude of the margins which must be added to any solution produced by the current empiricisms in order to take account of the many variables. There is a clear case here for further investigation by an informed and independent research body such as BSRIA.

Local fume extract systems

Under this heading fall the familiar fume cupboards used in laboratories which, if they are to operate with any success, must be integrated with any inlet or extract ventilation system serving the same space. In general terms, the air volume which is extracted from a cupboard should be such that a face velocity of between 0.25 and 0.75 m/s is produced with the sash fully open (Table 13.9) and a much higher level when the opening is reduced to the working height of between 25 and 50 mm.

An interesting design problem arises in heavily serviced pharmaceutical laboratories where as many as 25 or more fume cupboards may be required in a single room but are not subjected to any predictable pattern of use. As will be appreciated, the requirement for warmed, or conditioned, air make-up may be considerable, with a consequent heavy use of energy, if extraction from the cupboards continues whether they are in use or not.

The arrangement of the extract system might take the form of a fan connected to each cupboard, discharging either individually to atmosphere or into a common discharge duct. The former of these alternatives would produce a forest of unsightly terminals and the

Table 13.9 Air velocities through *open* sashes in fume cupboards

Category of cupboard use	*Velocity* (m/s)
Teaching	0.2–0.3
Research	0.25–0.5
Analytical	0.5–0.6
Highly corrosive or toxic	0.5–0.75
Radioactive*	0.5–2.0

Notes: sash openings are normally 750 mm to 1 m, velocities quoted are through *open* sash.
*Grade of radioactivity determines velocity.

latter, without suitable precautions, might lead to a situation of potential air flow imbalance in the common collecting duct and consequent recirculation to any inactive cupboard. Either solution would be prodigal of energy and of maintenance effort if assured service were to be demanded of all the fans.

A preferred solution would be to provide extraction using a duplicate set of capacity-controlled centrifugal exhaust fans, sited in a convenient plant room, drawing from a common discharge duct provided at the remote end with a dilution entry for outside air. The capacity of each fan would be somewhat in excess of the total extract requirement of all the fume cupboards and the connection to the common duct from each individual cupboard would be fitted with a motorised damper, arranged for on/off switching. A closed loop control system would sense the air flow requirement related to the number of cupboards in use at any one time and adjust the capacity of the duty fan and of the dilution dampers.

As to replacement air, this would be provided from a suitable central plant through one or more variable volume units which, *via* the control system, would modulate the quantity of supply air to match that extracted at any time. With this solution, optimum energy consumption and convenient centralised maintenance would be provided.

Exhaust of industrial fumes

In an industrial context, local extraction of fumes from benches is more effective and more economic than an attempt to deal with them by treatment of the whole volume of a workshop. For a welding bench, a volume of 200–300 litre/s needs to be removed and a convenient way of meeting this requirement is by use of a flexible tube supported from a wall mounted swivel as shown in Figure 13.19. The tube may be connected to an individual fan or, where multiple benches exist, to a header duct and a central fan.

Fabric tube supply systems

The need exists in some industrial buildings for the introduction of large air volumes in circumstances where the process cannot tolerate any significant air movement. A solution which makes use of permeable fabric tubes, in place of the more conventional ductwork plus diffuser assemblies, has been applied with success to tobacco processing and textile spinning workrooms. For a multi-tube arrangement, a sheet metal header duct is connected to the appropriate number of tubes which are inflated and maintained under pressure by a conventional centrifugal fan. Supports take the form of a 'curtain track' above each length which allows a deflated tube to be drawn to the header for disconnection.

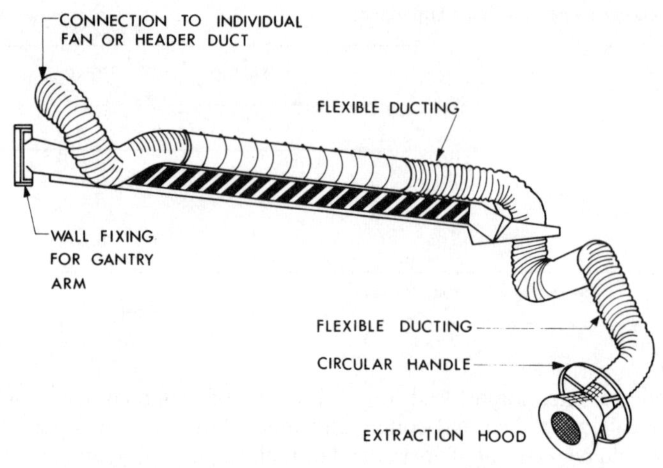

Figure 13.19 Fume extract from a welding bench (Plymovent)

Tube sizes range from 150–750 mm diameter and lengths of 20–50 m are used, dependent upon diameter. The initial velocity of air supply into a tube is between 12 and 15 m/s and a typical pressure drop, for a flow of 500 litre/s through a 300 mm diameter tube, would be about 100 Pa. The efficacy of the system is a function of adequate inflation which, in turn, depends upon a constant air supply and upon a good standard of filtration at the plant. Tubes require laundering at intervals and the support system permits easy removal for this purpose and replacement by a spare.

Plenum systems

In an earlier paragraph of this chapter, it was suggested that a straightforward ventilation system would be designed on the assumption that heating of the building was dealt with by a radiator or other similar plant. The plenum system, which is a method of providing heat using air as the distribution medium, is thus noted here only to show that it has not been forgotten. The reader is referred to the more comprehensive description provided in Chapter 6 (p. 160).

Demand controlled ventilation

It is a generally accepted principle that successful distribution of a supply of ventilating air within a space, from a simple terminal such as a grille, is a function of the design discharge velocity. An increase over this level might lead to draughts, due to impingement of the air stream on walls, etc., and a decrease might lead to 'dumping' of the air near to the terminal. This principle has inhibited any serious application of automatic control, which would vary air volume and hence velocity, during normal operation of a centralised, ducted ventilation system. Boost level pre-occupancy, and passive operation post-occupancy, were about all that was available.

As will be explained in later chapters, the development of more sophisticated terminals in recent years has led to a change of opinion and the concept of *demand controlled ventilation* has been introduced. This method, it is claimed, will allow more effective operation by adjusting the air flow rate to the needs of the occupants at any one time and thus lead to improved air quality, better energy efficiency or both.

The system relies upon control of the supply air volume, or, in some cases, the rate of extraction, using sensors which detect the concentration level of contaminants and decrease, or increase where possible, the ventilation volume to suit. As far as human occupancy is concerned, the level of carbon dioxide in an enclosure may be sensed with comparative ease and is a convenient measure of density of occupation.

Recirculation units

Although the use of recirculation units and equipment should not be considered as providing ventilation in the accepted sense, there are applications where they can be of value. Reference has been made in earlier chapters to temperature gradient within lofty rooms and two diametrically opposite methods have been used to overcome this. For soaring buildings such as churches, a technique used with success in the USA is shown in Figure 13.20. Here, with conventional methods of space heating at low level, heating effect and air movement have been greatly improved by displacement downwards of warm air at high level using a supply drawn from the occupied zone. It has been found that the temperature gradient may be reduced to about 2 K with an air circulation rate equivalent to 3 room volumes per hour.

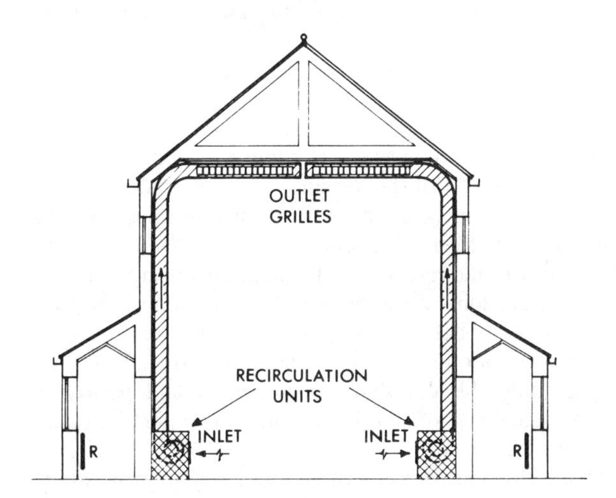

Figure 13.20 Upward air-transfer arrangement in a church

The alternative approach, perhaps more suited to an industrial application, is the suspension of fan units at high level which take in warm air from above them and discharge it to the working level: a unit of the type used is shown in Figure 13.21, the duty recommended being between 1.5 and 2 room volumes per hour.

Lastly under this heading, the large diameter *Punkah* fans which were installed by the thousand in tropical countries before the days of air-conditioning, are being increasingly used in a variety of buildings. They do no more than agitate the air at high level in the room but, in so doing, movement is induced in the lower occupied area which produces evaporation from the skin and consequent cooling. The models used in the past were noisy

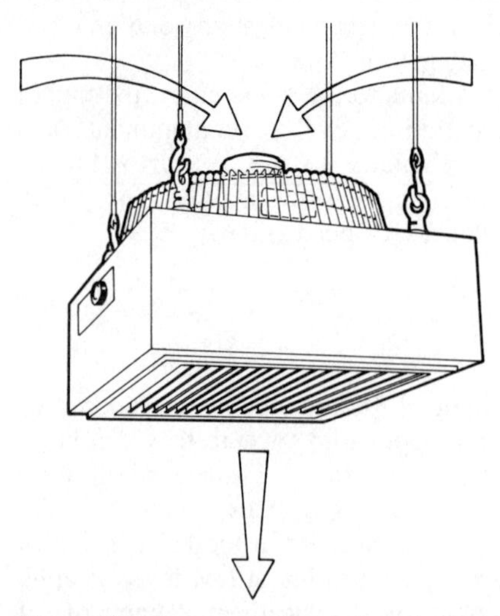

Figure 13.21 Downward air-transfer unit (Powrmatic)

and ugly but modern versions are not only much quieter but greatly improved in appearance.

Ventilation efficiency

To conclude this chapter, it is appropriate to note the term *ventilation efficiency* which has been used increasingly in recent years, as a result of the sensitivity of those non-smokers who may be required to share spaces with those who smoke. This situation has been the subject of considerable research and many published papers.*

The term represents an attempt to quantify the rate at which contaminants are removed from a space and, in particular, from the occupied zone. No simple definition of the term exists but it may be expressed as the ratio between the average concentration of a contaminant within a space and that concentration present at the point of extract. This ratio will depend upon many factors including the distribution, properties and period of discharge of the contaminant, the quality and rate of supply of replacement ventilation air and the pattern of air movement within the space.

For contaminant emissions which are distributed, ventilation efficiency improves with the rate of air supply and with those air movement patterns able to carry the contaminant to the extract positions quickly as may be produced by an effective piston displacement flow. Complete mixing of the air supply with room air produces no more than an average ventilation efficiency and any short circuits in room air flow patterns, supply to exhaust, reduce that efficiency further.

* Shandret, E. and Sandberg, M., 'Air exchange and ventilation efficiency – new aids for the ventilation industry', *Norsk VVS (Norway)*, 1985, **7**, 527–34.
Design and Performance of Mechanical Ventilation Systems – Ventilation Effectiveness. *BSRIA Contract 7190A*, 1987.

Chapter 14

Air-conditioning

The science of air-conditioning may be defined as that of providing and maintaining a desirable internal atmospheric environment irrespective of external conditions. As a rule 'ventilation' involves the delivery of air which may be warmed, while 'air-conditioning' involves delivery of air which may be warmed or cooled and have the moisture content (humidity) raised or lowered.

National and international concern directed at the global environmental effects arising from release of refrigerants into the atmosphere, as described in Chapter 19, and the energy used in mechanical cooling systems have, together, opened the door to a period of fundamental change in attitudes towards the use of air-conditioning in certain buildings.

The 1993 draft proposals for amendment to Part L2 of the Building Regulations included the requirement that mechanical ventilation or air-conditioning 'shall be installed only where it is reasonably necessary'. This matter has been touched upon in the preamble to Chapter 13 but further comment is required here. Whilst statutory regulation may not become effective for some time, there is already a quite discernible trend to question whether air-conditioning is a necessary prerequisite for quality commercial premises.

The recent publication of a *Code of Practice on Environmental Issues* by the Engineering Council has given a lead in general terms and flesh has been put upon the bones of the Code by CIBSE, to suit its particular discipline, by stating:

If the requirement for air conditioning has been fully established, the following principles should be adopted and adhered to:

- *the system should be energy efficient (with due regard being given to the inclusion of cost effective energy saving methods such as free cooling) and also controlled to minimise energy use.*
- *operation and maintenance strategies should be devised and the necessary régimes adopted to deliver economy, efficiency and effectiveness in the working of systems throughout their life cycles.*
- *system design, construction and commissioning should be carried out in accordance with current national and European standards, codes of practice and statutory requirements.*
- *cooling system refrigerants should be used in accordance with the policy laid down in the* CIBSE Guidance Note *on cfcs, hcfcs and halons.*

General principles

The desired atmospheric condition for comfort applications usually involves a temperature of 18–22°C in winter and 21–24° in summer; a relative humidity of about 40–60% and a high degree of air purity. This requires different treatments according to climate, latitude, and season, but in temperate zones such as the British Isles it involves:

In winter. A supply of air which has been cleaned and warmed. As the warming lowers the relative humidity, some form of humidifying plant, such as a spray or a steam injector, with preheater and main heater whereby the humidity is under control, is generally necessary.

In summer. A supply of air which has been cleaned and cooled. As the cooling is normally accomplished by exposing the air to cold surfaces or cold spray, the excess moisture is condensed and the air is left nearly saturated at a lower temperature. Inherent in this process therefore is a measure of dehumidification which counters the increase in relative humidity that results from cooling the air. The temperature of air has then to be increased, to give a more agreeable relative humidity, which can be done by warming or by mixing with air which has not been cooled.

Dehumidifying may also be brought about by passing the air over certain substances which absorb moisture. Thus, in laboratories, a vessel is kept dry by keeping a bowl of strong sulphuric acid in it or a dish of calcium chloride, both of which have a strong affinity for moisture. Silica-gel, a form of silica in a fine state of division exposing a great absorbing surface, is used also for drying air on this principle, but this process is complicated by the need for regeneration of the medium by heat and subsequent cooling, and is not generally used in comfort air-conditioning applications.

Establishment of need

The application of air-conditioning may be considered necessary to meet a variety of circumstances:

- Where the type of building and usage thereof involves *high heat gains* from sources such as solar effects, electronic equipment, computers, lighting and the occupants.
- In buildings which are *effectively sealed*, for example where double glazing is installed to reduce the nuisance caused by external noise.
- The *core areas* of deep-planned buildings where the accommodation in the core is remote from natural ventilation and windows, and is subject to internal heat gains from equipment, lights and occupants.
- Where there is a *high density of occupation*, such as in theatres, cinemas, restaurants, conference rooms, dealing rooms and the like.
- Where the process to be carried out requires *close control of temperature and humidity*, such as in computer suites, or where stored material, or artifacts, require stable and close control of conditions, such as in museums and paper stores.
- Where work has to be carried out in a confined space, *the task being of a high precision and intensive character*, such as in operating theatres and laboratories.
- Where the *exclusion of air-borne dust* and contaminants is essential, such as in micro-chip assembly and animal houses.

In tropical and sub-tropical countries, air-conditioning is required primarily to reduce the high ambient temperature to one in which working and living conditions are more tolerable. In the temperate maritime climate of the British Isles and in similar parts of the world, long spells of warm weather are the exception rather than the rule, but modern forms of building and modern modes of living and working have produced conditions in which, to produce some tolerable state of comfort, air-conditioning is the best answer. Thus we find buildings of the present day incorporating to a greater or lesser extent, almost as a common rule, some form of air-conditioning. This great variety of applications has produced an almost equally great variety of systems, although all are fundamentally the same in basic intention: that is, to achieve a controlled atmospheric condition in both summer and winter, as referred to earlier, using air as the principal medium of circulation and environmental control.

The installation of complete air-conditioning in a building often eliminates the necessity for heating by direct radiation, and it naturally incorporates the function of ventilation, thus eliminating the need for opening windows or reliance on other means for the introduction of outside air. Indeed, opening windows in an air-conditioned building should be discouraged since, otherwise, the effectiveness of the system to maintain conditions will be reduced and the running costs will be increased.

All air-conditioning systems involve the handling of air as a means for cooling or warming, dehumidifying or humidifying. If the space to be air-conditioned has no occupancy, no supply of outside air is necessary, that inside the room being recirculated continually. In most practical cases, however, ventilation air for occupancy has to be included and in the design for maximum economy of heating and cooling, this quantity is usually kept to a minimum depending on the number of people to be served. Thus, in most instances, it will be found that the total air in circulation in an air-conditioning system greatly exceeds the amount of outside air brought in and exhausted. Where, however, it is a matter of contamination of the air, such as in a hospital operating theatre, or where some chemical process or dust-producing plant is involved, 100% outside air may be needed and no recirculation is then possible.

Weather data

With certain designs of plant it may be cost effective to arrange for 100% outside air to be handled, normally during mid-season periods when untreated it can provide useful cooling. Figure 14.1 shows, for various months of the year, the proportion of daytime

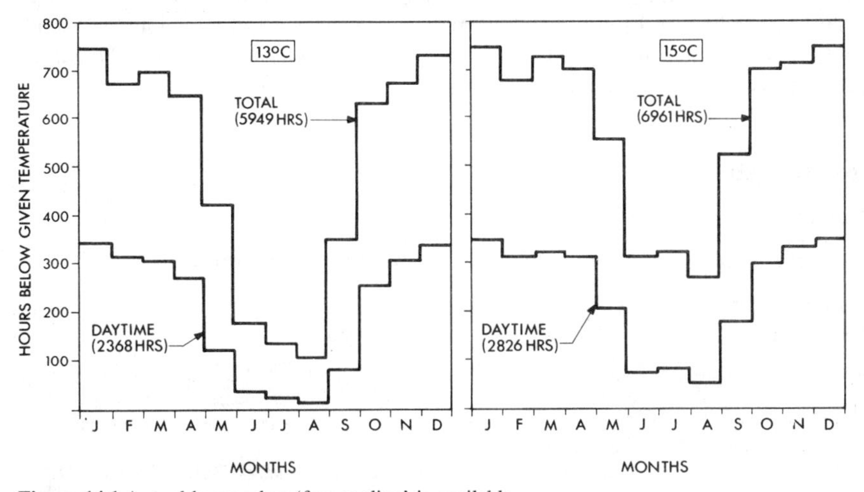

Figure 14.1 Annual hours when 'free cooling' is available

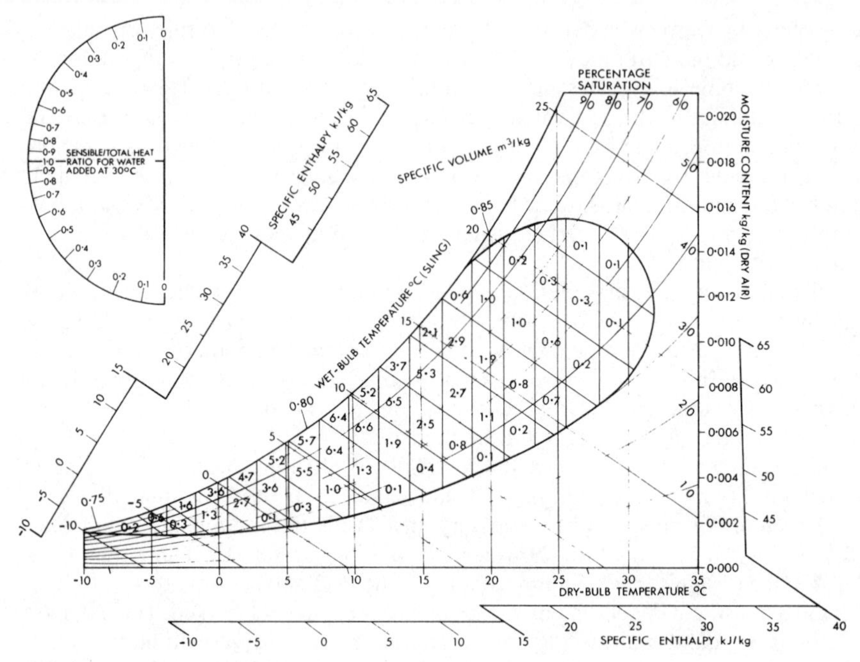

Figure 14.2 Annual percentage frequencies of coincident dry-bulb and wet-bulb temperatures for the London area

hours when the outside temperature at Kew, for the selected CIBSE Example Weather Year,* is below 13°C and is thus available for a cooling duty; a second set of data is also shown in the figure for periods below 15°C. Meteorological data for other stations in Britain show similar availability.

When designing an air-conditioning system it is not enough to consider only the winter and summer peak design temperatures. It is important that the system should operate satisfactorily through the range of annual external conditions. To this end it is necessary to have an understanding of the range and frequency of occurrence of coincident wet-bulb and dry-bulb external conditions. Figure 14.2 shows, on a psychrometric chart, the annual percentage occurrence of external conditions lying within specified limits.†

Traditional systems

Central plant

The basic elements of air-conditioning systems of whatever form are:

Fans for moving air.
Filters for cleaning air, either fresh or recirculated, or both.

* Holmes, M. J. and Hitchen, E. R., 'An example year for the calculation of energy demand in buildings', *Buildings Services Engineer*, 1978, **45,** 186–9.

† Holmes, M. J. and Adams, E. S., *Coincidence of Dry and Wet Bulb Temperatures.* BSRIA Technical Note TN2/77.

Cooling plant connected to heat exchange surface, such as finned coils for cooling and dehumidifying air.
Heater batteries for warming the air, such as hot water or steam heated coils or electrical resistance elements.
Humidifiers, such as by steam injection, water sprays/washers or heated pan type.
A control system to regulate automatically the amount of cooling, warming, humidification or dehumidification.

The type of system shown in Figure 14.3 is suitable for air-conditioning large single spaces, such as theatres, cinemas, restaurants, exhibition halls, or big factory spaces where no sub-division exists. The manner in which the various elements just referred to are incorporated in the plant will be obvious from the caption. It will be noted that in this example the cooling is performed by means of chilled water cooling coils and the humidification is by means of steam injection. A humidifier would be provided only when humidification is required in winter.

In an alternative version of a central plant system, the cooling coil may be connected directly to the refrigerating plant and contain the refrigerating gas. On expansion of the gas in the evaporator, cooling takes place and hence this system is known as a direct expansion (DX) system. It is suitable for small to medium size plants. Humidification could be by means of a capillary washer or water spray into the air stream, but a non-storage type such as steam injection is preferred, due to the risk of *Legionnaires' disease* associated with types which incorporate a water pond to facilitate recirculation within the humidifier. For the same reason, an air cooled condenser could be used as a means of rejecting unwanted heat from the refrigeration machine in place of the cooling tower shown.

In Figure 14.3 it will be noted that there is a separate extract fan shown exhausting from the ceiling of the room. This would apply particularly in cases where smoking takes place, such as in a restaurant, to remove fumes which might otherwise collect in a pocket at high level. Sometimes this exhaust may be designed to remove the quantity equivalent to the outside air intake, in which case the discharge shown to atmosphere from the return air fan would not be necessary.

Motorised dampers are shown in the air intake, discharge and recirculation ducts to allow the proportion of outside and recirculation air flow rates to be varied to effect economy in plant operation. Reduction in energy consumption may also be achieved by transferring heat between the exhaust and intake air streams; this is not illustrated in the diagram but is described in Chapter 17.

Where there is a number of rooms or floors in a building to be served, it is necessary to consider means by which the varying heat gains in the different compartments may be dealt with. Some rooms may have solar gains, and others none; some may be crowded and others empty; and some may contain heat-producing equipment. Variations in requirements of this kind are the most common case with which air-conditioning has to deal and for this a simple central system is unsuitable. The ideal, of course, would be a separate system for each room but this is rarely practicable unless the individual spaces are very large or important.

Zoned systems

A building may be divided into a number of zones for air conditioning purposes. The sub-division could be dictated by spatial constraints, by the requirements for sub-letting or by the hours of use. Each zone may be served by a separate 'central' plant, perhaps located

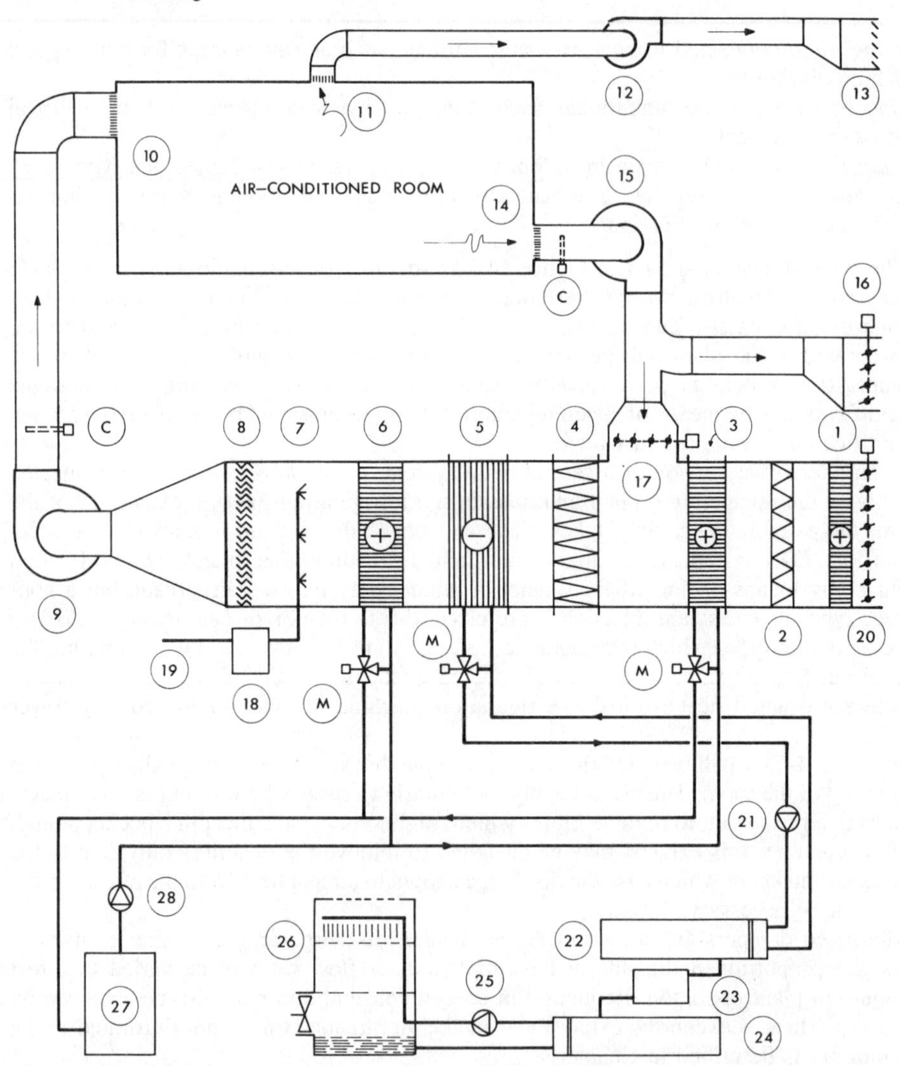

Figure 14.3 Central plant system. 1, frost coil (optional, for fog elimination); 2, pre-filter; 3, preheater; 4, secondary filter; 5, chilled water cooling coil; 6, reheater; 7, direct injection steam humidifier; 8, eliminators (optional); 9, supply fan; 10, conditioned air supply; 11, high level smoke extract; 12, smoke exhaust fan; 13, smoke discharge to outside via louvre; 14, extract and recirculation; 15, main extract fan; 16, discharge to outside via motorised damper and louvre; 17, recirculation; 18, steam generator; 19, water make-up; 20, outside air intake via motorised damper and louvre; 21, chilled water pump; 22, water chiller evaporator; 23, refrigeration compressor; 24, shell and tube condenser; 25, condenser water pump; 26, cooling tower; 27, hot water boiler; 28, hot water pump; 'M' denotes motorised valves; 'C' denotes temperature and humidity control points

in a common plant room, or alternatively one plant located on each floor might be an appropriate arrangement. There is a number of ways in which a central plant, broadly on the lines of that previously described, may be modified in design to serve a number of groups of rooms or zones. To take the simplest case, if such a plant served one large space of major importance and a subsidiary room having a different air supply temperature requirement, it would be possible to fit an additional small heater or cooler, or both, on the branch air duct to the subsidiary room. Separate control of temperature would thus be available.

High-rise buildings, say over 12 storeys, served from a central plant may require plant rooms at intermediate levels to reduce the air quantity conveyed in any one duct and thereby reducing the space taken by vertical service shafts. Similar vertical sub-division would also be appropriate for closed and open piped distributions.

An extension of this principle, to serve two or more zones more or less equal in size from a single plant, may be achieved by deleting the reheater of Figure 14.3; dividing the supply fan outlet into the appropriate number of ducts and fitting a separate reheater to each. Since, however, the output of the central plant cooling coil would have to be arranged to meet the demand of whichever zone requires the maximum cooling, extravagant use of reheat would result and lead to uneconomic running costs.

The plant illustrated in Figure 14.3 is arranged on the 'draw-through' principle, the various components being on the suction side of the supply fan. It is, of course, possible to adapt this sequence to produce a 'blow-through' arrangement with the fan moved to a position immediately following the secondary filter. With such a re-disposition, what is known as a *multi-zone* arrangement may be produced, as illustrated in Figure 14.4. Here, the fan discharge is directed through either a cooling coil or a heating coil into one or other of two plenum boxes; a *cold deck* or a *hot deck*. Each building zone is supplied with conditioned air via a separate duct which, at the plant, is connected to both plenum boxes. A system of interlinked dampers, per zone, is arranged such that a constant air supply is delivered to each zone duct which may be all hot, all cold, or any mixture of the two as required to meet local demand. When providing a mixture, such a plant wastes energy.

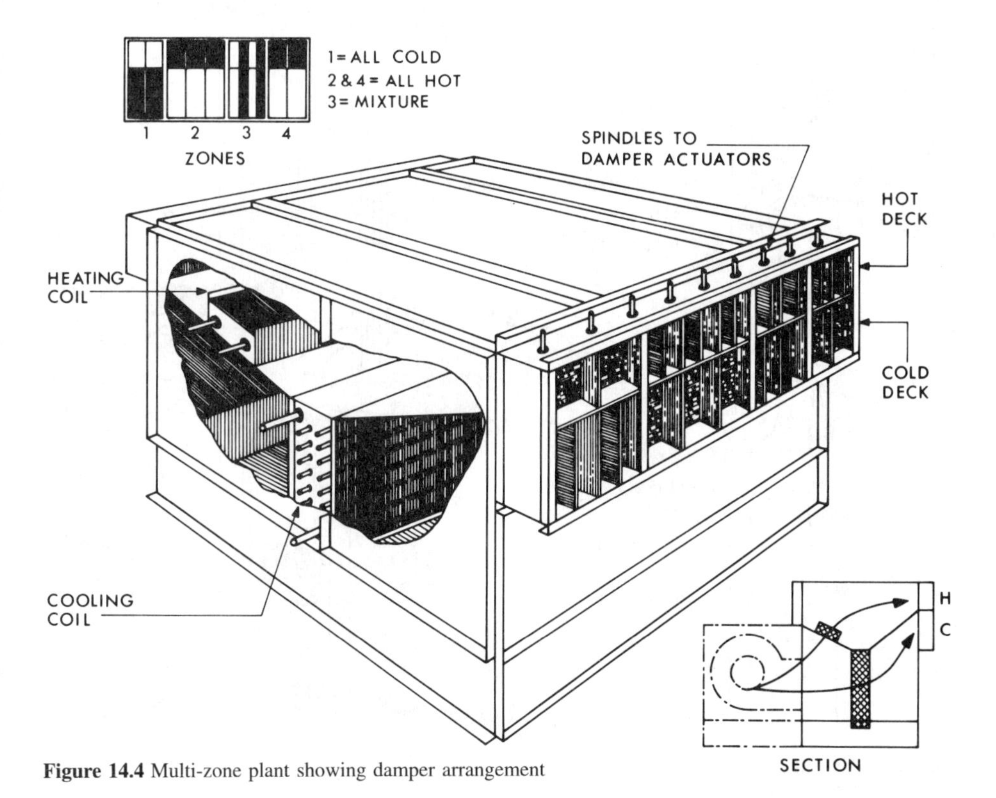

Figure 14.4 Multi-zone plant showing damper arrangement

Zone units (fan coil)

An alternative approach to the problem of providing local control to a variety of building zones having differing demands involves the provision of a recirculating fan-coil unit, having a booster reheater or recooler, within each zone. Outside air, in a quantity suitable to provide for occupancy, is conditioned as to temperature dependent upon outside conditions and corrected for moisture content by means of what is, in effect, a conventional central plant. This supply is delivered to the local zone plant where it is heated or cooled as may be required to suit the zone conditions. Figure 14.5 shows two alternative local plant arrangements, incorporating cooling only, suitable in this case to provide for conditions within the apparatus area of a major telephone switching centre. Note how integration of the air-conditioning system within the structure has been arranged.

For a more conventional application to an office building, Figure 14.6 shows plants arranged in floor service rooms. In case (a), air recirculation from the individual rooms passes through louvres above the doors into the corridor and thence back to the zone plant. While such an arrangement has the great merit of simplicity, coupled with relatively low cost, current thinking and fire regulations disapprove of the use of a corridor *means of escape* as a return air path on the grounds that fire and/or smoke generated in any one

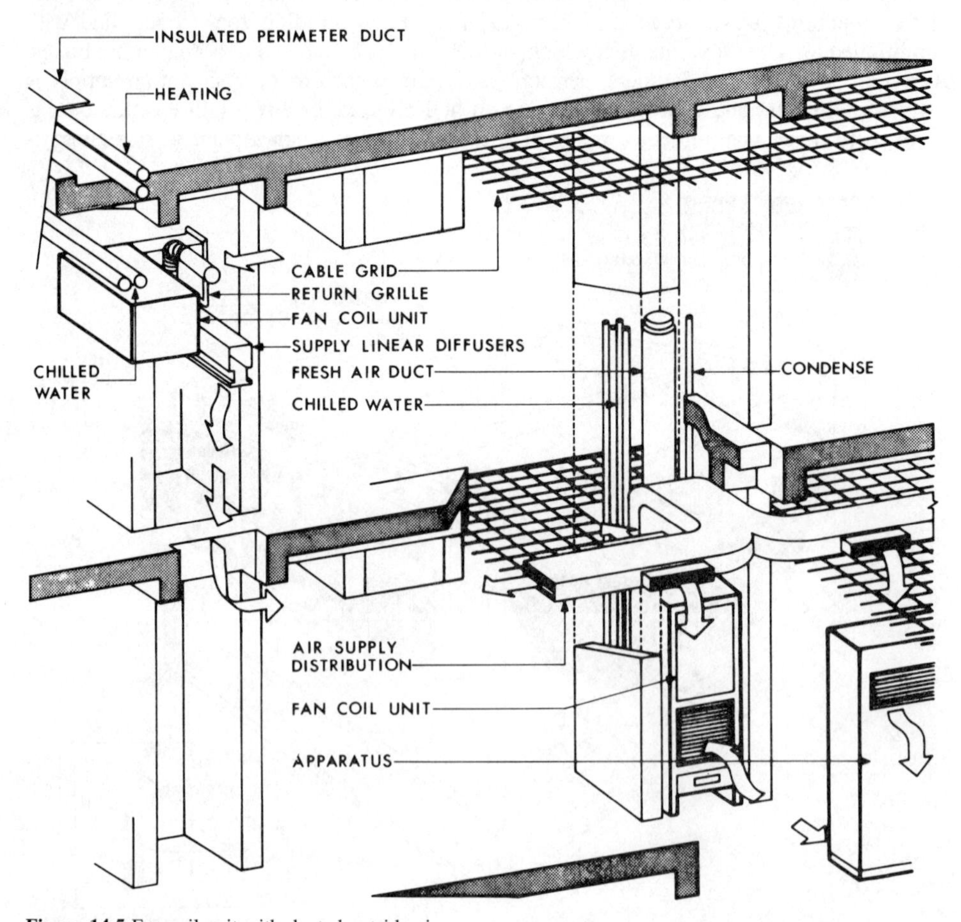

Figure 14.5 Fan-coil unit with ducted outside air

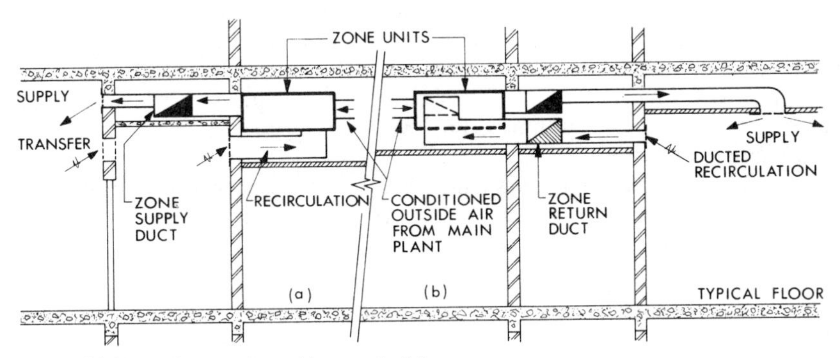

Figure 14.6 Zoned system in multi-storey building

room could be transferred into an escape route and lead to disruption and panic. It is now, in consequence, more usual to provide a quite independent return-air collection system, via duct branches from each room, back to the local zone plant, as shown in case (b).

High velocity systems

Traditionally until the mid 1950s, air-conditioning systems were designed to operate with duct velocities of not much more than about 8–10 m/s and fan pressures of 0.5–1 kPa. With the advent of high-rise buildings and, concurrent with their introduction, demands for improved working environments coupled with *less* space availability for services there was a requirement that tradition be overthrown. This situation led to a radical rethink and to the introduction of a number of new approaches to air-conditioning design using duct velocities and fan pressures twice and more greater than those previously in use.

Whilst the principal characteristic of the new generation of systems relates to the methods adopted for distribution of conditioned air and exploits these to the full, the principles previously described in this chapter remain unchanged. As before, the conditioning medium may be all air or air–water dependent upon a variety of circumstances. Figure 14.7 shows comparative space requirements for the alternative supply media. The low velocity extract ducts associated with the respective supply air

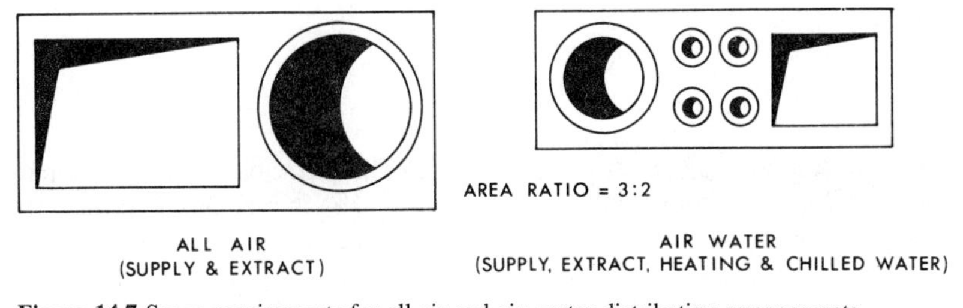

Figure 14.7 Space requirements for all-air and air–water distribution arrangements

quantities will increase the space requirements further in favour of the air–water systems. Added to which, the increase in space cooling loads arising from greater use of electronic equipment increases yet again the spatial benefits of the air–water system since the additional cooling load is handled by the water circuits; the air ducts being unaffected. There is, in consequence, a practical limit to the level of heat gain that can be dealt with satisfactorily by an all-air system.

To complement high velocity supply systems, where space for ductwork distribution is limited, consideration may be given to using a high velocity extract system, but this will impose additional initial cost and subsequent running costs.

All-air systems

Simple systems

These, the most primitive of high velocity systems, differ from their traditional counterparts by the the form of the terminals used to overcome problems arising from noise generated in the air distribution system. To this end, a variety of *single-duct* air volume control devices has been developed to provide for the transition between a high

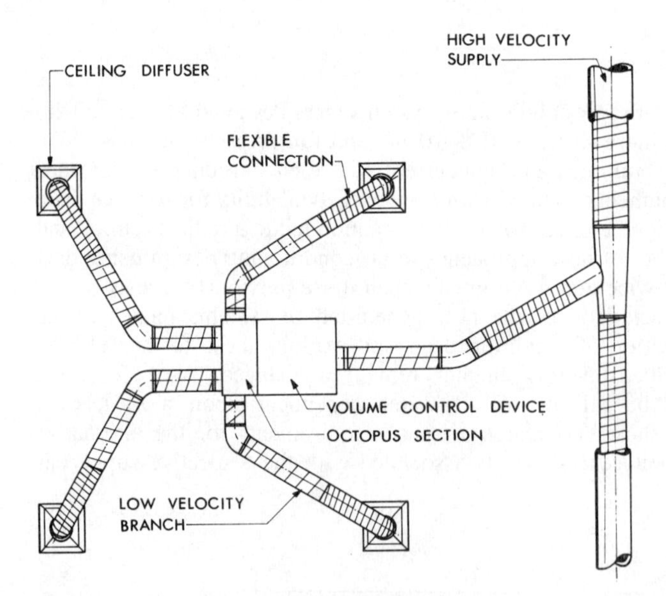

Figure 14.8 Single-duct high velocity terminal box

velocity distribution system and air outlets local to the conditioned spaces. The terminal box, with *octopus* distribution section illustrated in Figure 14.8, is typical of equipment produced for this purpose. It consists, in principle, of an acoustically lined chamber provided with a sophisticated air volume damper or 'pressure reducing valve': such dampers are in some instances fitted with self-actuated devices, others have powered actuators, arranged so that they may be set to provide constant or variable output volume under conditons of varying input pressure. Such devices are a considerable aid in regulating air flow quantities during the commissioning process.

Simple all-air systems will provide adequate service in circumstances where the load imposed is either constant or will vary in a uniform manner for the area served thus allowing temperature control by means of adjustment to temperature of the supply air at the central plant.

All-air induction systems

For the particular case where air is returned to the central plant via a ceiling void and, further, where lighting fittings (luminaires) are arranged such that the bulk of the heat output (which may be as much as 80%) is transferred to this return air, an all-air induction system may be used.

With such an arrangement, conditioned air is ducted to induction boxes mounted in the ceiling void, as shown in Figure 14.9. Each box incorporates damper assemblies or other devices which, under the control of a room thermostat, act to permit the conditioned air flow to induce a variable proportion of warm air from the ceiling void into the discharge stream. Reheat and consequent local control is thus achieved such that, with one type of unit, the cooling capacity may be controlled down to about 45% of maximum. The subsequent introduction of either an automatic switching system, which will minimise the period that heat gain from lighting is available, or the availability of more efficient lamps (both being changes which reduce the potential for reheat) would have a detrimental effect upon the operation of such systems. To counter such eventualities, reheater batteries are available as an optional feature and terminal units which vary the quantity of the primary air supply are also available.

All-air variable volume systems

The traditional approach to air-conditioning design placed, as a first principle, insistence upon the concept of maintaining air discharge to the spaces served at constant volume.

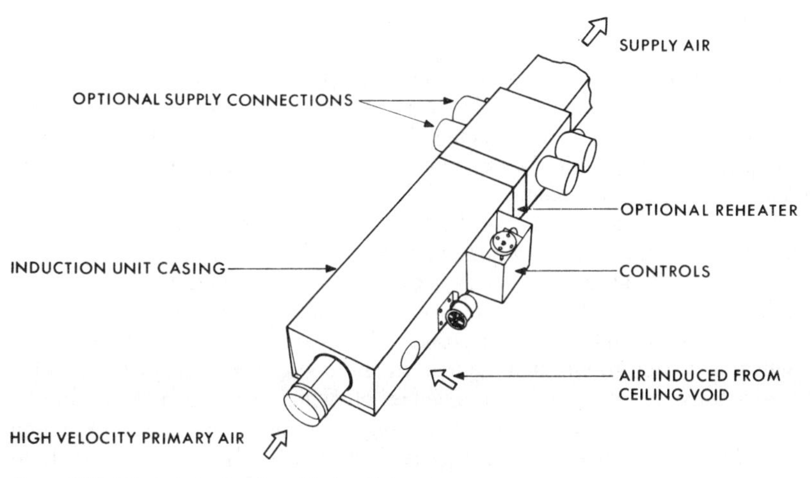

Figure 14.9 All-air induction box (Barber Colman)

Load variations were catered for by adjustment to air temperature. This axiom arose, no doubt, from the known sensitivity of building occupants to air movement and, furthermore, from the relative crudity of the air diffusion equipment then available.

With the advent of terminal equipment not only more sophisticated but also with performance characteristics backed by adequate test data, circumstances have changed. The activities of BSRIA and of a variety of manufacturers in this area must be applauded. Hence the availability of potential for abandonment of the traditional approach.

In principle, the variable volume system may be considered as a refinement to the simple all-air system whereby changes in local load conditions are catered for not by adjustment of the temperature of the conditioned air delivered, at constant volume, but by adjustment of the volume, at constant temperature. This effect may be achieved by means of metering under thermostatic control of the air quantity delivered either to individual positions of actual discharge, as shown in Figure 14.10, or to groups of such positions via a terminal unit of the type illustrated in Figure 14.11.

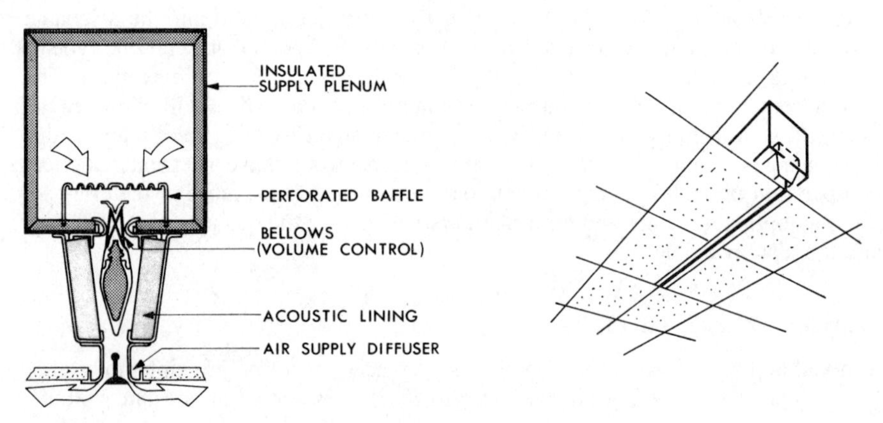

Figure 14.10 Linear diffusers for variable volume (Carrier)

In such cases, a true variable volume arrangement is possible since the effect of reduced output at the terminal units or at the discharge positions may be sensed by a central pressure controller, this being arranged to operate devices which reduce the volume output of the central plant correspondingly. Economies in overall operation in energy consumption and in cost will thus result.

If an adequate supply of outside air is to be maintained and problems of distribution within the conditioned space avoided, volume cannot be reduced beyond a certain level. Good practice suggests that minimum delivery should not be arranged to fall below about 40% of the designed quantity. For different reasons, such a limitation may present problems to both internal and perimeter zones. At internal zones, if lights were individually switched in each room, then the load in an unoccupied or sparsely occupied space could be greatly reduced, in consequence of which such rooms would be overcooled. In the case of perimeter zones, where conduction and solar gain form a high proportion of the design load, it may be necessary to introduce some level of reheat to augment capacity control by volume reduction.

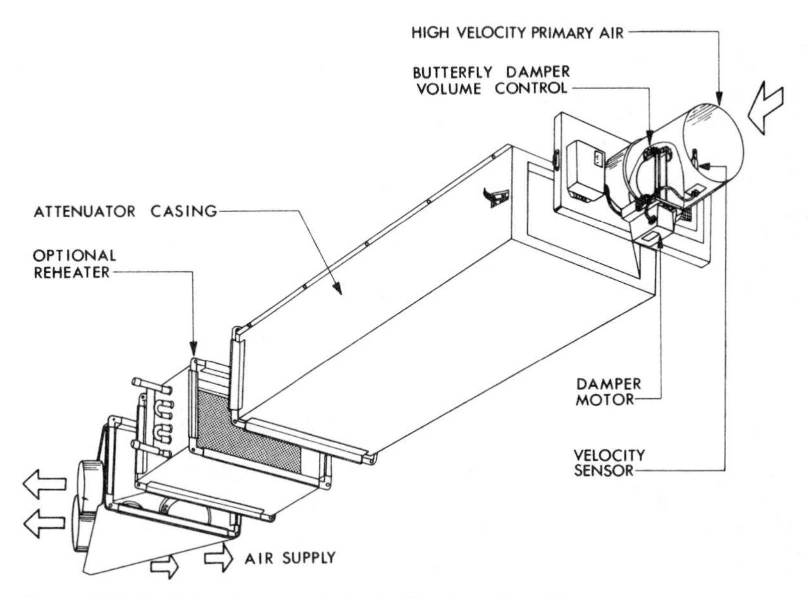

Figure 14.11 Variable volume terminal unit (Waterloo–Ozonair)

A solution for both internal and perimeter zones where the reduced air supply quantity is required to be below that to produce satisfactory air distribution in the space, without causing the cold supply air to *dump* into the occupied area, is to install variable geometry diffusers: a volume flow reduction down to about 25% of the maximum is claimed for this type of outlet. Figure 14.12(a) shows a typical example which operates to maintain the air velocity at discharge to the space by changing the area of opening, thereby ensuring adequate diffusion of the supply air. A similar effect may be obtained using a two-section bypass type air diffuser in which the supply is divided into two passages, one of which is controlled at a constant volume to maintain the velocity of the air stream from the diffuser and hence satisfactory air distribution, as Figure 14.12(b).

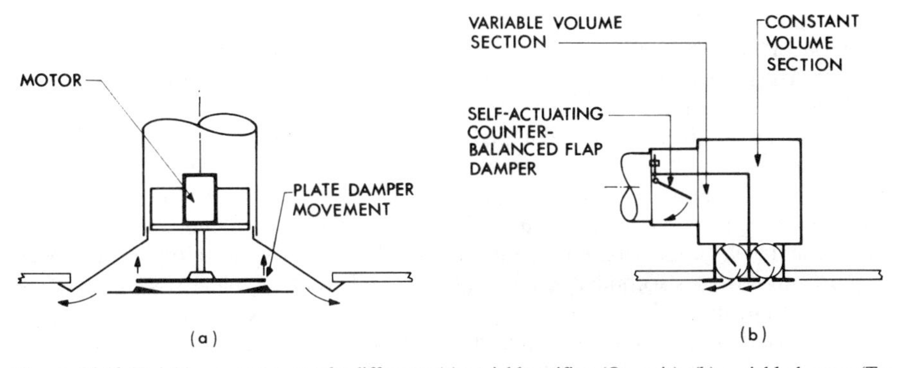

Figure 14.12 Variable geometry supply diffusers: (a) variable orifice (Ozonair), (b) variable bypass (Trox)

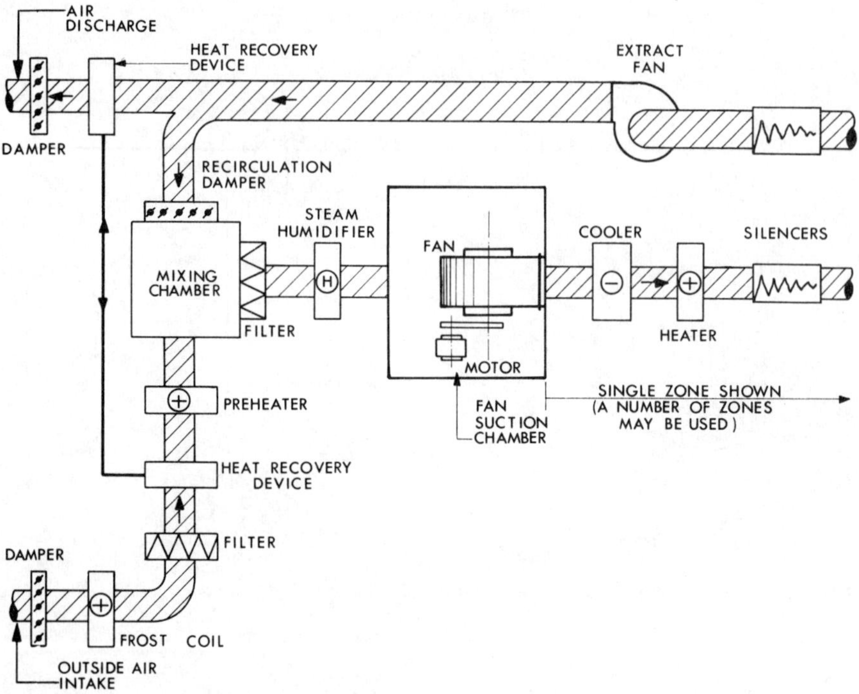

Figure 14.13 Typical plant arrangement for a variable volume system

While most variable volume terminal units can be adapted to incorporate reheaters, the required effect may equally well be achieved at perimeter zones via control of a constant volume perimeter heating system or even space heating units such as hot water radiators either of which may, in any event, be required at windows to deal with down draughts.

The difference in the space heating and cooling loads within internal and perimeter areas may require a two or three zone supply system, with the supply temperature of each zone controlled to suit the particular characteristics of the area served. For single or multi-zone systems, the zone temperatures may be controlled at a constant level, be varied to suit outside conditions (scheduled), or be varied in response to feedback from the controls at each terminal device to provide the optimum supply with a maximum operating economy. Figure 14.13 shows a typical variable air volume system arrangement. To maintain the volume of outside air above the minimum requirement for occupant ventilation, it may be necessary to add air velocity sensors to the plant controls in the outside air intake duct, to actuate the dampers in the outside air, exhaust air and recirculation ducts. This control must be arranged to override any damper controls provided to achieve economy of operation.

The increased requirement for cooling, currently arising from the proliferation of electronic equipment in modern buildings, has led to the development of variable air volume terminals which incorporate a means to provide additional capacity. This has been achieved by introducing a secondary cooling coil in the terminal. Typically, a recirculation fan draws air from the ceiling void through a filter, and discharges this across the coil where it is cooled before being introduced into the space. Two types of fan-assisted variable air volume terminal devices have been developed; one has the secondary fan *in*

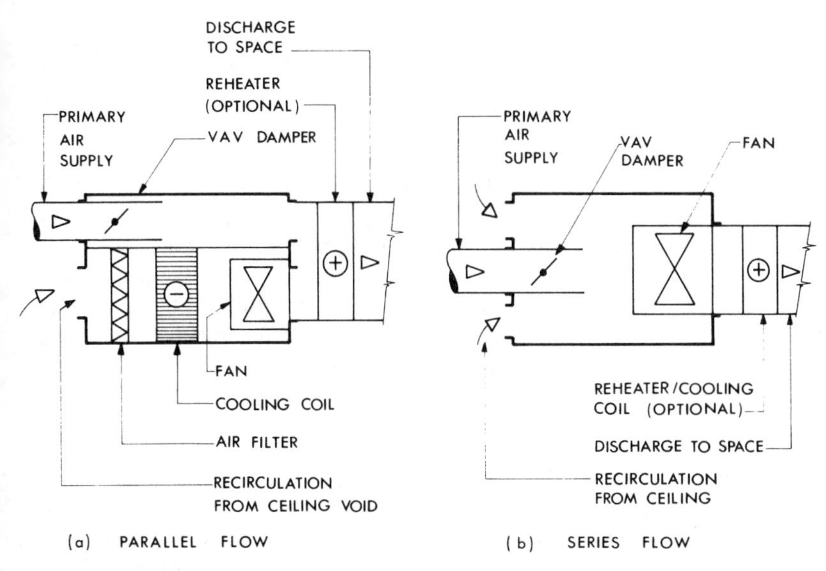

Figure 14.14 Fan-assisted variable volume terminal devices

parallel with the primary air supply and the alternative is an *in series* configuration. Diagrammatic representations of these units are shown in Figure 14.14. The parallel flow type is currently more popular on the grounds that the secondary fan runs only when the cooling load so demands, and thus uses less energy than the series flow terminal where the fan runs continuously. However, the variation in supply air quantity with the parallel arrangement could result in poor air distribution, and the intermittency of the fan operation might cause more disturbance than the continuous running in the series flow unit.

One advantage of the series flow arrangement is that, due to the inherent mixing of primary and recirculated air, the former may be delivered to the unit at a much lower temperature, typically down to 8°C. In consequence, primary air volume and duct sizes may be reduced, which may be an important consideration in refurbishment work. Lower supply temperatures, however, reduce the efficiency of the associated refrigeration machines as explained in Chapter 19.

Since the principle of operation with all VAV systems is to vary the quantity of primary air supplied by the plant, it follows that the associated extract fan must respond to the changes in supply volume to avoid under- or over-pressurisation of the building.

Variable air volume systems are currently a popular choice because the associated energy consumption is lower than that of other equivalent systems. They are particularly suited to buildings subjected to long periods of cooling load, and are easily adapted to changes in office partition layouts. It is quite practical to incorporate constant volume terminal boxes within a variable volume system in circumstances where, for example, it is desirable to maintain a room at a positive pressure. In this application a terminal reheater, either an electric element type or a coil connected to a heating circuit, would normally be installed downstream of the terminal to provide temperature control of the space.

The variable volume principle allows the ductwork to be sized to handle the air quantity required to offset the maximum simultaneous heat gain that could occur and not the sum of the quantities demanded by the peak loads in each of the individual spaces. Extensive use is made of computer analyses to establish the maximum simultaneous design conditions.

All-air dual duct systems

The multi-zone system previously described is arranged to mix, at the central plant, supplies of hot and cold air in such proportions as to meet load variations in building zones. An extension of this concept would be that an individual mixed-air duct was provided for each separate room in the building but this, on grounds of space alone, would not be practicable. The same degree of control may however be achieved by use of the dual-duct system where the mixing is transferred from the central plant to either individual rooms or small groups of rooms having similar characteristics with respect to load variation.

Use is made of two ducts, one conveying warm air and one conveying cool air, and each room contains a blender, or mixing box, so arranged with air valves or dampers that all warm, all cool, or some mixture of both is delivered into the room. Figure 14.15 illustrates such a unit, incorporated therein being a means for regulating the total air delivery automatically such that, regardless of variations in pressure in the system, each unit delivers its correct air quantity. Referred to as 'constant volume control', this facility is an essential part of such a system.

Owing to the fact that air alone is employed, the air quantity necessary to carry the cooling and heating load is greater than that used in an air–water system. Air delivery rates with the dual-duct system are frequently of the order of 5 or 6 changes per hour, compared with the 1½–2 required for introduction of outside air. Owing to the considerable quantity of air in circulation throughout the building, dual-duct systems usually incorporate means for recirculation back to the main plant and this involves return air ducts and shafts in some form.

An advantage of the dual-duct system is that any room may be warmed or cooled according to need without zoning or any problem of change-over thermostats. Furthermore, core areas of a building, or rooms requiring high rates of ventilation, may equally be served from the same system, no separate plants being necessary.

Figure 14.16 shows the plant arrangement of a system in one form, though there are variations of this using two fans, one for the cool duct, one for the warm duct.

To avoid undue pressure differences in the duct system, due for instance to a greater number of the units taking warm air than cool air, static pressure control may be

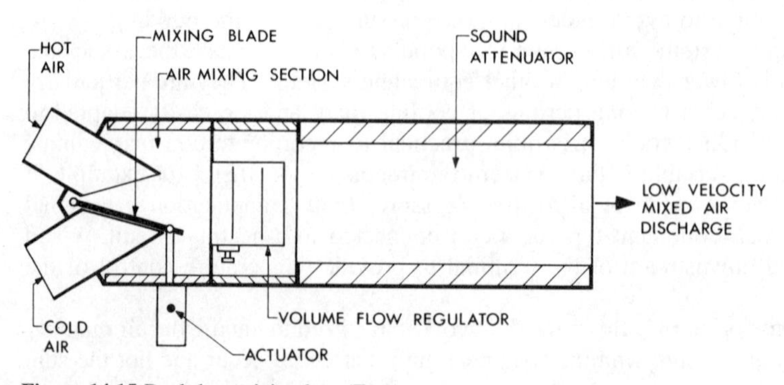

Figure 14.15 Dual-duct mixing box (Trox)

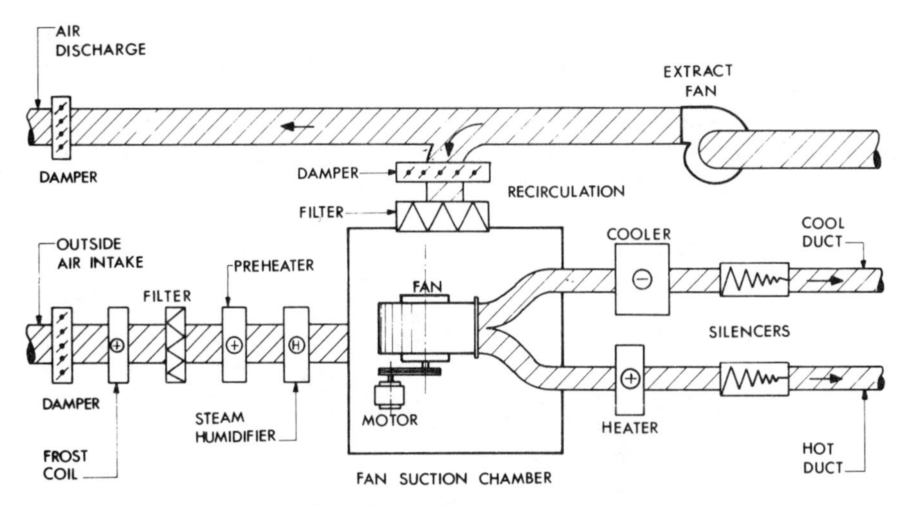

Figure 14.16 Typical plant arrangement for a dual-duct system

incorporated so as to relieve the constant pressure devices in the units of too great a difference of pressure, such as might otherwise occur under conditions where the greater proportion of units are taking air from one duct than from both.

Since dual-duct systems supply a constant air volume to the conditioned areas this overcomes the potential problems arising from maintaining adequate quantities of outside air and from providing satisfactory air distribution experienced with some variable volume systems. The disadvantages of dual-duct systems are high energy use and the need for large shafts and ceiling voids to accommodate the ductwork.

In special circumstances, a dual-duct system may be provided with variable volume terminal equipment to combine the best features of each system. With such an arrangement, when volume has been reduced to the practical minimum, control is achieved by reheat from the hot duct supply. Typically, the dual-duct supply would serve perimeter zones and a single duct supply of cool air would serve internal areas.

Air-water systems

Fan-coil system

From the point of view of economy in building space, the use of water rather than air as a distribution medium for cooling and heating, from plant rooms to occupied spaces, has much to commend it. The *fan-coil* system exploits this saving in space and has the added advantage of offering facilities for relatively simple local temperature control in each individual room. The fan-coil terminals each consist of a chassis which mounts a silent running fan, either centrifugal or cross-flow (*tangential flow*), a simple air filter and either a single water-to-air heat exchange coil or a pair of such coils. They are made in a limited range of sizes and may be fitted, one or more to each occupied room, using either the manufacturer's sheet metal casings or some form of concealment in purpose-designed enclosures. A number of different patterns of fan-coil unit is available to suit various mounting positions and Figure 14.17 shows the form which is probably most familiar, an under-window cabinet type. For mounting horizontally at high level, perhaps above a false ceiling and ducted locally to outlet and recirculation terminals, the type shown in Figure 14.18 is available.

Since each fan-coil unit serves principally to recirculate room air, the necessary volume of outside air to meet the ventilation needs of occupants must be provided quite

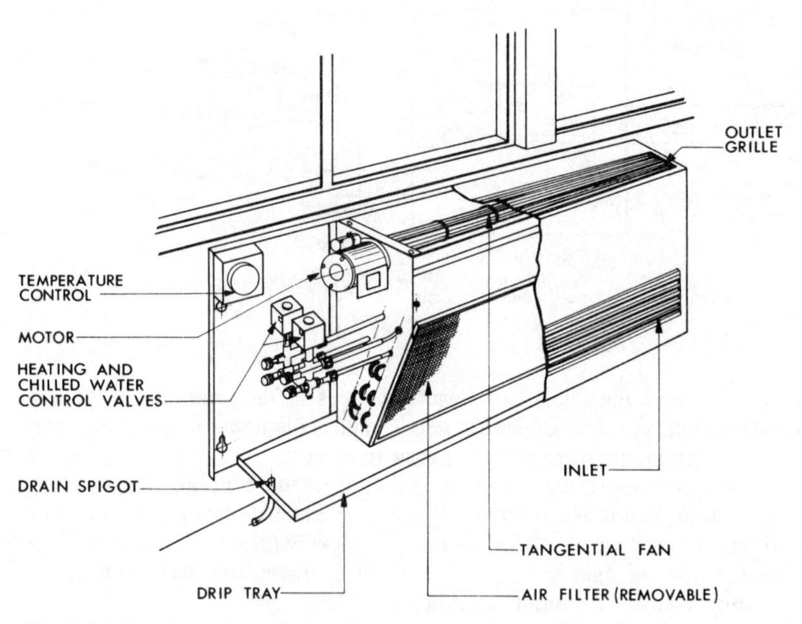

Figure 14.17 Under window fan-coil unit (four-pipe)

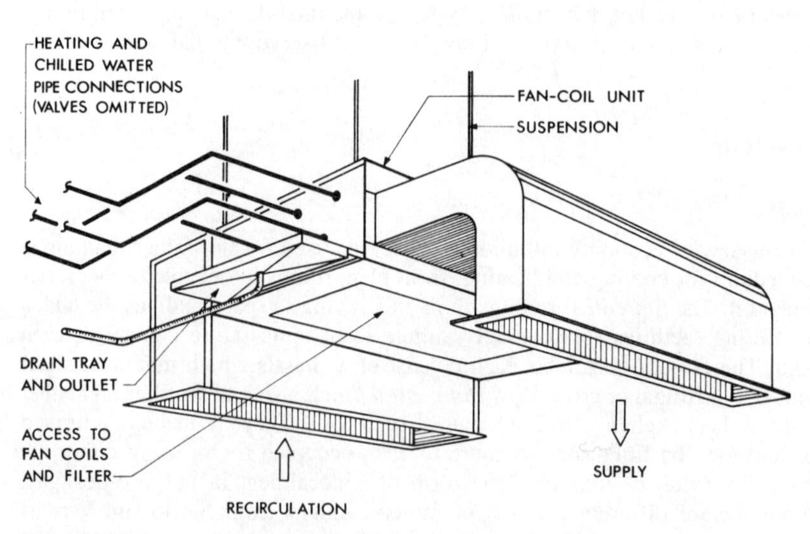

Figure 14.18 Fan-coil unit fitted above a suspended ceiling

independently. This requirement may be met in a number of ways, depending upon the sophistication of the individual system. At one end of the scale this air, pre-filtered and conditioned, is introduced to the space through a ducted system from a central plant, and the exhaust air is ducted to an energy recovery unit of some sort. The diametrically opposite, and obviously much cheaper method (which cannot be recommended), is for each unit to be placed below a window and outside air admitted to it through an aperture in the wall in a manner similar to that shown in Figure 13.7. The hazards of dust and noise pollution to say nothing of unit overload due to wind pressure cannot be exaggerated.

For a high proportion of the year, external conditions in the British Isles are such that rooms on certain elevations of a building may require heating while others require cooling. It is thus desirable to provide units that are able to meet either demand at any time. Such an arrangement is known as a *four-pipe* system, heating flow and return and cooling flow and return pipes being connected to separate coils. A *two-pipe* system provides heated water to a coil in winter and chilled water to the same coil in summer and problems arise in mid-season when some spaces require heating while others are calling for cooling: clearly both requirements cannot be satisfied simultaneously with this two-pipe arrangement. Such a system is suitable only where the period of climate change between summer and winter is short, with little or no mid-season (some parts of the USA have such a climate), or where the internal loads are such as to require only local cooling.

The chilled water flow to the units may be circulated at an elevated temperature, as described later for air–water induction systems, eliminating the formation of condensate on the cooling coils and thus the need to pipe this from each unit to drain. Very basic control of either individual units or groups may be effected by switching the fan motor(s) on and off. The preferred alternative for the climate of the British Isles, however, is to keep

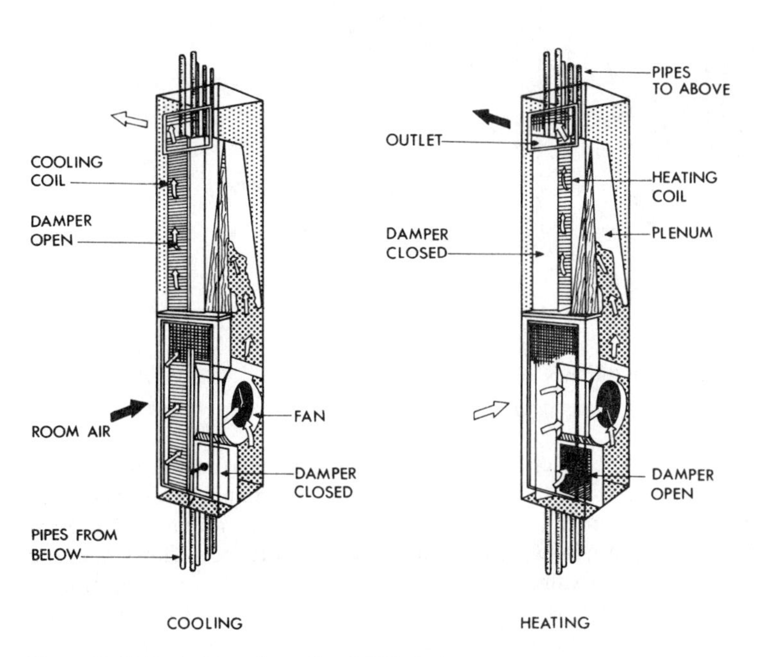

Figure 14.19 Vertical type fan-coil unit (Whalen)

the motor running and to control either the water temperature using automatic valves as Figure 14.17 or the air side using mixing dampers as Figure 14.19.

Fan-coil systems are inherently flexible and are well suited to refurbishment projects since any central air handling plant and duct distribution system will be relatively small in size. The cabinet type of unit, as Figure 14.19, has been designed specifically for application to existing buildings, the component parts being stacked vertically in a small plan area, space being available there also to enclose vertical water piping and a supply air duct.

Air–water induction systems

Although there has been virtually no development in the design of this type of unit for some years, it is appropriate, nevertheless, to outline the principles of their operation and application. As the name implies, the principle of induction is employed in this system as

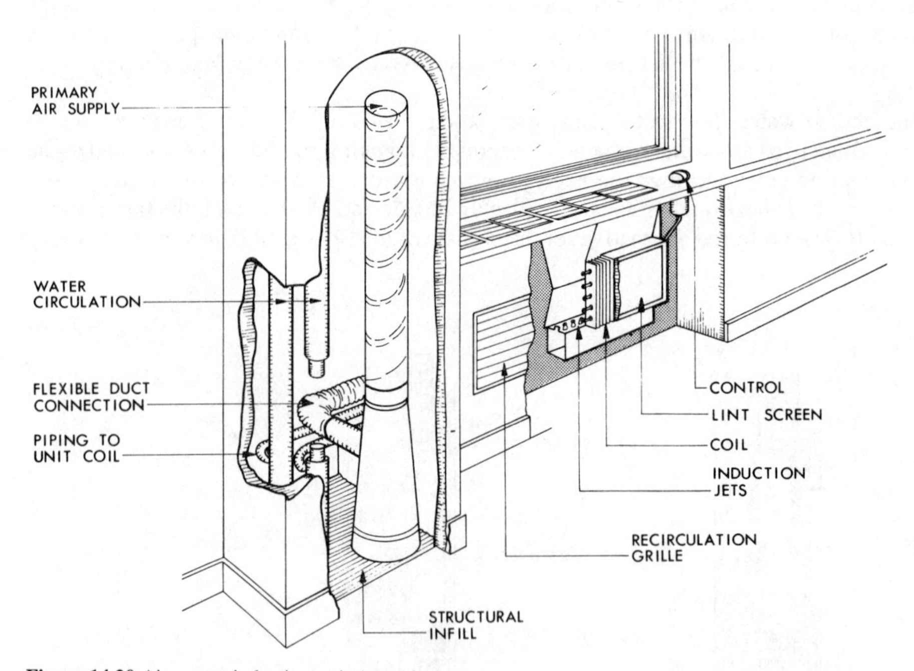

Figure 14.20 Air–water induction unit (two pipe)

a means to provide for an adequate air circulation within a conditioned room. Primary air, conditioned in a central plant, is supplied under pressure to terminal units, generally placed below the window with vertical discharge, each of which incorporates a series of jets or nozzles as shown in Figure 14.20. The air induced from the room flows over the cooling or heating coils and the mixture of primary and induced air is delivered from a grille in the sill. The induction ratio is from three to one to six to one. The primary air supply provides the quantity for ventilation purposes, and the means to humidify in winter and deal with latent loads in summer.

The coils are fed with circulating water which, in the so-called *change-over* system, is cooled in summer and warmed in winter, an arrangement more suited to sharply defined seasons than the unpredictable long springs and autumns of the British climate. In Figure 14.21 (a) and (b), control is achieved by an arrangement of dampers, such that the return air from the room is drawn either through the coils for heating or cooling, or the coils are bypassed to a greater or lesser extent. In the type shown in Figure 14.21 (c), control is by variation of water flow: increase in water flow is required to lower the air temperature during the cooling cycle, and increase in flow is required to raise the air temperature in the heating cycle. There is thus required some means of change-over of thermostat operation according to whether the winter or summer cycle is required.

An alternative method, using the so-called *non-change-over* system, avoids this problem by always circulating cool water through the coils of the induction units and varying the primary air temperature according to weather only. Thus, throughout the year, the heating or cooling potential of the primary air is adjusted to suit that component of demand imposed upon the system, or zone of the system, by orientation, by outside temperature or by wind effect. Any other variant – solar radiation, heat from lighting or occupancy – will necessarily produce a local heat *gain* and the sensible cooling needed to

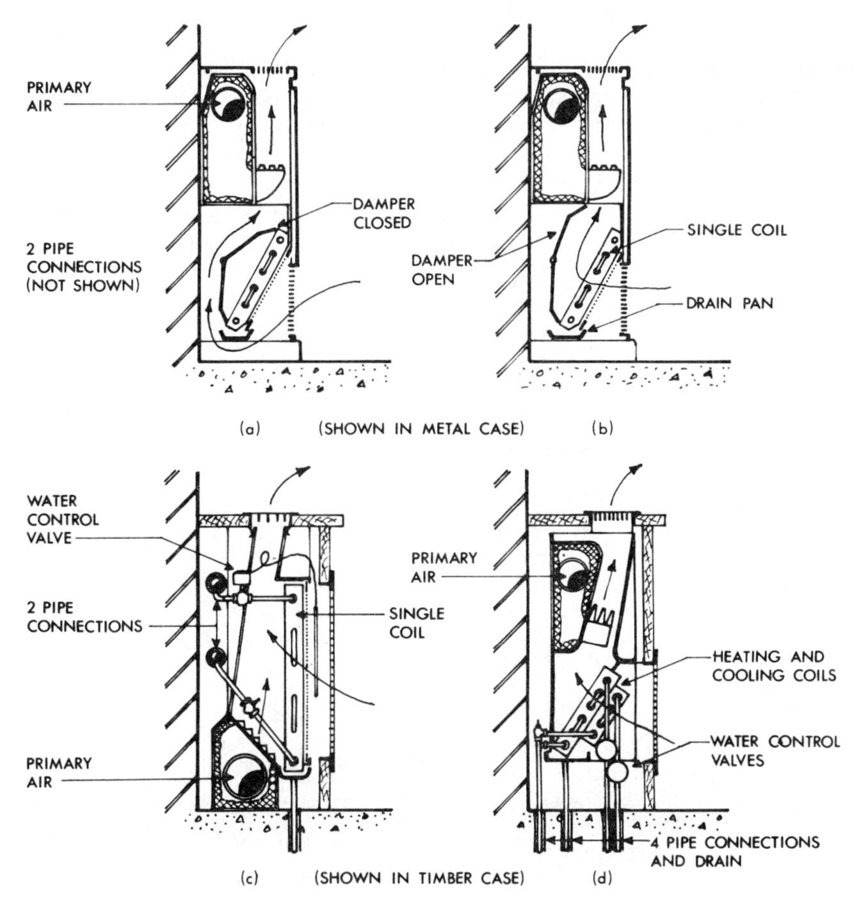

Figure 14.21 Alternative control arrangements for induction units

offset this will be provided by the capacity of the unit coils under local control. In so far as such an arrangement acts as 'terminal recool' in winter, it is uneconomic in terms of energy wastage.

A variation of the two-pipe induction system is the *three-pipe*, in which both warm and cool water are available at each unit, with a common return, and the control arrangement is so devised as to select from one or the other. Likewise, in the main system the return is diverted either to the cooling plant or to the heating plant, according to the mean temperature condition. Such an arrangement suffers from problems related to hydraulic instability due to the changes in water quantities flowing through the alternative paths (see p. 153).

The preferred system for the British climate is the *four-pipe system*, two heating and two cooling, but it is correspondingly expensive. A unit having two coils, one for heating and one for cooling, is shown in Figure 14.21(d).

Induction systems might be expected to be noisy, due to the high velocity air issuing from the jets, but the units have been developed with suitable acoustical treatment such that this disadvantage does not arise in practice.

Morning preheating may be achieved by circulating heating water through the secondary coils allowing the unit to function as a simple natural convector. This avoids the need to run the primary air fan and is therefore energy efficient.

Figure 14.22 is a simplified diagram of an induction system showing the primary conditioning plant, the primary ducting and the water circulation. The heat exchanger shown is for warming the water circulated to the units, and this would be fed from a boiler or other heat source. It will be noted that the chilled water supply to the coils of the induction units is arranged to be in the form of a subsidiary circuit to that serving the main cooling coil of the central plant. Such a system has the advantage of providing a degree of *free cooling* when the outside air supply to the central plant is at low temperature during winter. Furthermore, since the flow to the room units is connected to the return pipe from

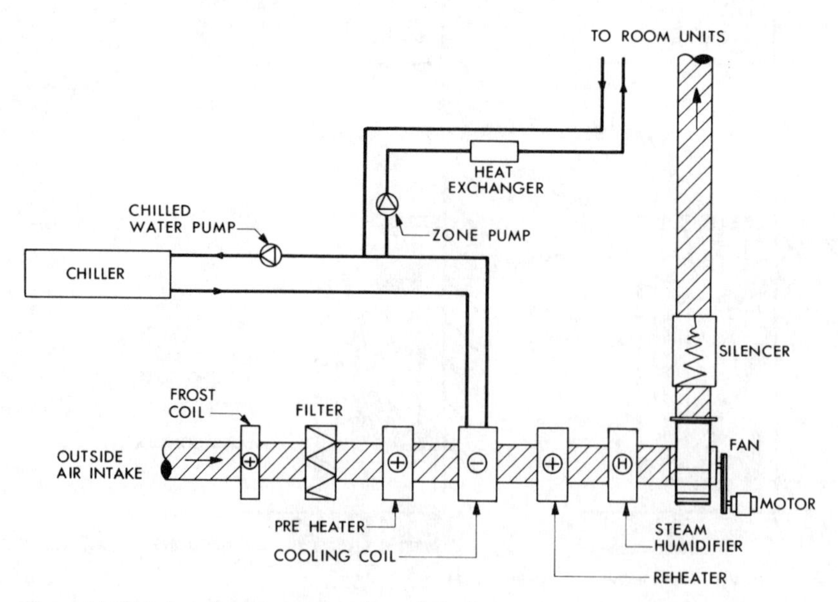

Figure 14.22 Typical plant arrangement for an induction system

the cooling coil it has an elevated temperature and, as a result, this circuit arrangement provides an in-built protection against excessive condensation on the unit coils such that local drain piping may therefore be dispensed with in most cases.

The induction system involves the distribution of minimum primary air, often as little as 1½–2 air changes/hour, and has been widely applied to low-cost multi-storey office blocks or hotels where in either case there is a large number of separate rooms to be served on the perimeter of a building. Current practice suggests that, provided application of the system is confined to perimeter areas not deeper than 4 m, with relatively low occupancy, satisfactory service will result. Interior zones of such buildings that require cooling year round are usually dealt with by an all-air system.

Induction systems inherently cause any dust in the atmosphere of the room to be drawn in and over the finned coil surfaces, and, to prevent a build-up of deposit thereon, some form of coarse lint screen, easily removable for cleaning, is usually incorporated.

Other systems

Upward air-flow systems

The most common arrangement for introducing conditioned air into a space is from supply diffusers or grilles positioned at high level. Becoming more popular however is the use of upward air-flow systems, where the air is distributed within a false floor, often required in any event for routing electrical power and communications networks, and introduced

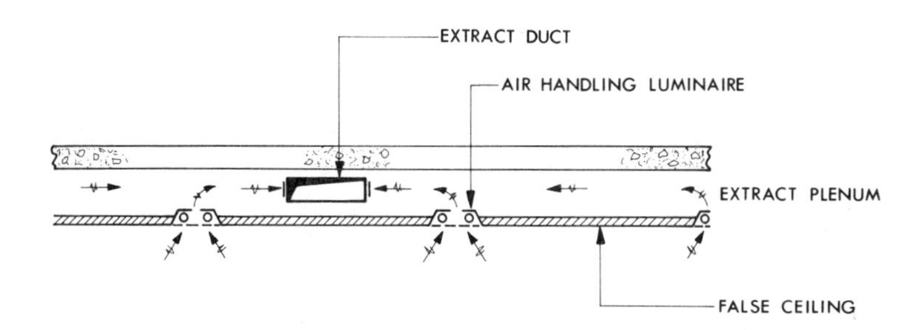

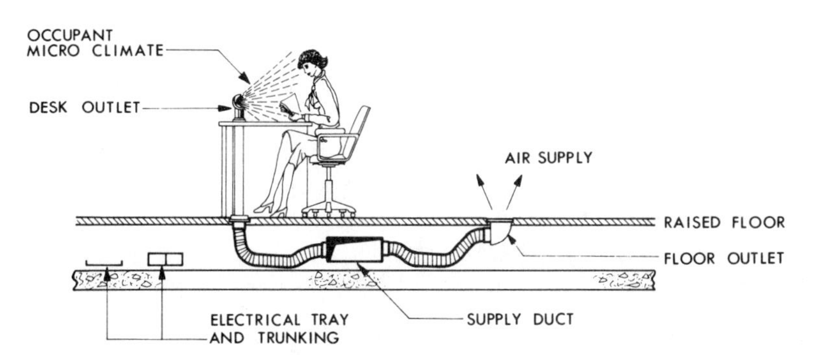

Figure 14.23 Typical upward air-flow arrangements

via floor mounted outlets, perhaps supplemented by desk-top supply terminals. Such systems may function using either the constant or the variable air-flow principle.

Typical arrangements of this type of system are shown in Figure 14.23, from which it may be seen that the desk-top outlet can provide, effectively, a micro-climate for the occupant. It follows therefore that the temperature swing in the general area may be allowed to be slightly greater with this type of system than with one relying on conditioning of the general space. It is claimed that such systems have inherent flexibility to provide for changing the location of terminal devices to suit variations in an office layout. Special air terminal devices have been developed for floor and desk distribution, as described in Chapter 15.

A development of the upward air distribution principle is an arrangement where both the supply and the extract positions are at floor level, with the floor void divided into a supply plenum and a return air space. With this arrangement, no false ceiling is necessary since all servicing may be from low level, including up-lighting. This system uses fan-assisted conditioning modules to filter, cool, heat and humidify recirculated air, the units being either free-standing in the space served or incorporated into service zones. In most cases, they will be supplied with heating, chilled and mains water piping and with a power supply. A typical arrangement of such a system is illustrated in Figure 14.24.

The conditioned air is discharged into the supply plenum to which are connected fan-assisted terminal units, which may be either wall mounted or of an underfloor type, the latter being connected to supply grilles integrated into a standard 600 mm square floor panel. Return air is collected through similar panel mounted grilles into the return air section of the floor void from which it is drawn into the conditioning module. Outside air for ventilation purposes is introduced into the return air section of the floor void and temperature control in the space is achieved by varying the ratio of conditioned and recirculated room air introduced; alternatively, electric trim heaters may be used.

A clear underfloor depth of 200 mm will be required and, typically, up to 300 m^2 may be served from a single module. As will be appreciated, air-tightness around the perimeter

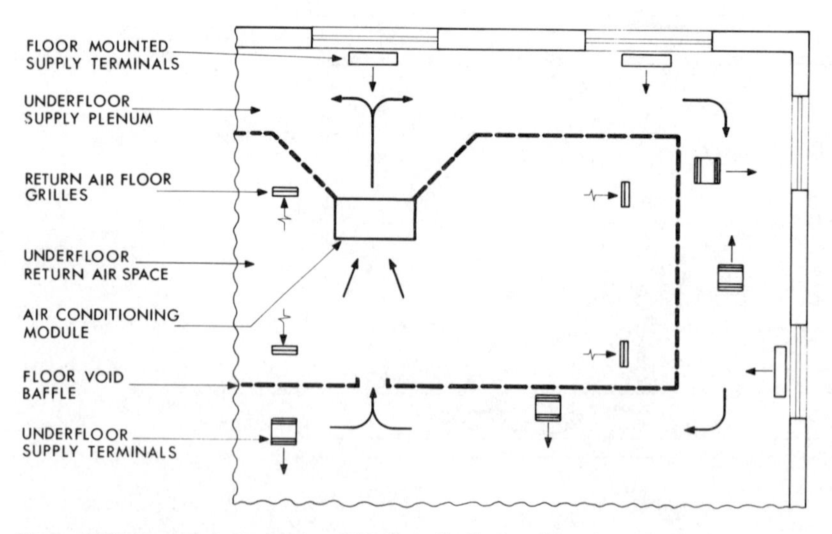

Figure 14.24 Upward air-flow system with floor distribution (Hiross)

and at baffles between the supply plenums and the return air zones is critical to maintain performance. These systems provide good flexibility for changes in furniture layout and all components, including the conditioning modules, terminal units, false floor panels and void baffles are available as a proprietary system. Individual desk-top controls are available, as a proprietary package, to allow the occupant to set the conditions to suit his or her preferred working temperature: some systems also incorporate a presence sensor as an energy saving feature to shut-off the air supply at the workplace when the occupant is away from the desk.

Displacement ventilation

Indoor pollutants are diluted by 'mixing' with outside air when traditional methods of ventilation are employed and the same principle is adopted for dealing with heat gains and losses. An alternative approach is to introduce the supply at one position, and at low velocity, such that it moves in a single direction through the room using a piston effect to take the pollutants, including thermal effects, with it. In such systems, supply air is introduced close to the floor at a few degrees below the room design temperature allowing the input to flow across the floor forming a 'pond'. Heat sources within the room produce upward convective currents resulting in a gentle upward air flow towards high level extract positions. For effective operation, the air in the space should not be subjected to continuous disturbance by rapid movement of occupants neither should there be high rates of infiltration nor down draughts due to poor insulation.

A number of limiting performance and comfort factors impose restrictions upon the use of displacement systems.* The warm and often polluted upward air flow spreads out beneath the ceiling and, since the lower boundary of this layer should be kept above the zone of normal occupancy, application is limited to rooms with high ceilings. Comfort factors which limit the temperature difference between head and feet determine that the air supply temperature should be in the range 18 to 20°C for seated sedentary occupations, and this, in turn, restricts the cooling capacity of the system to 30–40 W/m^2. Air supply terminals should be selected to achieve a uniform air distribution pattern across the floor whilst keeping air velocity low, not exceeding 0.4 m/s close to the point of discharge. Terminals of suitable pattern for this application are illustrated in Chapter 15, p. 414.

Room air-conditioning units

Developed mainly for computer room applications, such units have been produced in recent years to provide close control of both temperature and humidity over wide ranges of load variation. The units may be self-contained, that is with an integral refrigerant compressor, or be served from a central chilled water source. Self-contained units may have remote condensing units or be connected to a water cooling circuit. Heating may be direct electric or by coils served from a central heating circuit.

Humidification is normally by steam injection from electrically heated units integral within the package. Air filtration is a requirement for such applications. Figure 14.25 shows a typical cabinet type unit. Close control of conditions in the space together with the need for quick analysis of component failure has led to the general use of micro-processor controls for these units.

* Jackman, P. J., *Displacement Ventilation.* BSRIA Technical Memorandum 2/90.

For computer room applications, a relatively small quantity of outside air is needed for ventilation purposes and normally no extract system is provided, thereby allowing the supply air to pressurise the space. The air-conditioning units would normally be duplicated, or in larger installations at least one redundant unit would be provided as a standby in case of a unit failure or for use during routine maintenance operations.

Room coolers

This heading covers a separate field in that such units are commonly complete in themselves, containing compressor, air filter, fan and cooling coil. Electric resistance heaters may be incorporated for winter use and, rarely, means for humidification. Fresh air

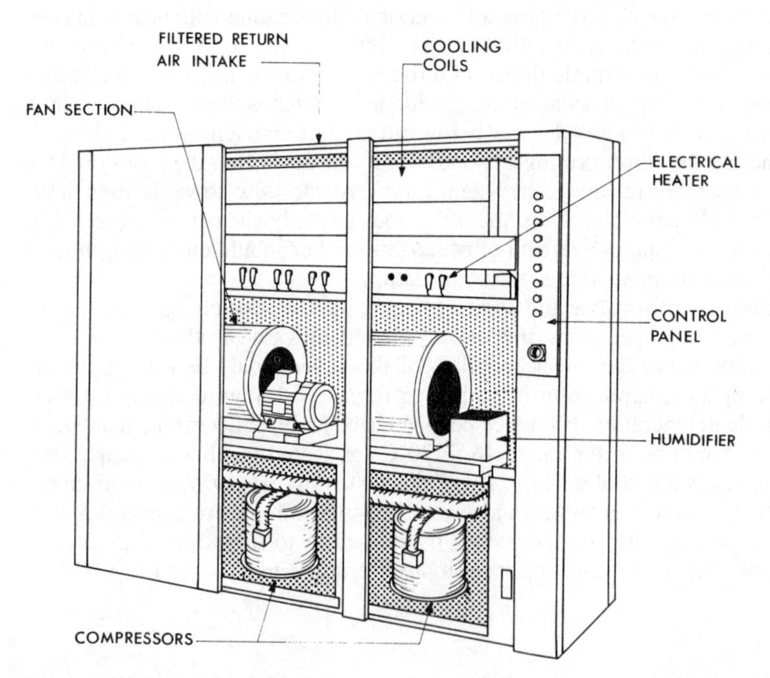

Figure 14.25 Typical down-flow type room air-conditioning unit

may be introduced if required. Being of unit construction, alternatively described as *packaged*, they are not purpose-made to suit any single application and thus may well be economical in first cost. In some cases, a so-called 'split type' of unit may be found where the condenser and compressor are mounted remotely from that part of the equipment which serves the room concerned. Bulk and noise at the point of use are thus much reduced.

Units of small size generally have the condenser of the refrigerator air cooled, but in larger sizes the condenser may be water cooled in which case water piping connections are required. Apart from this, the only services needed are an electric supply and a connection to drain to conduct away any moisture condensed out of the atmosphere during dehumidification. Compressors in most units are now hermetic and are therefore relatively quiet in running.

Sizes vary from small units suitable for a single room, sometimes mounted under a window or in a cabinet, similar to that shown in Figure 14.25 and the range goes up to units of considerable size suitable for industrial application, in which case ducting may be connected for distribution.

Reverse cycle heat pumps

The heat pump principle described in Chapter 19 has been applied to general air conditioning applications. In a similar fashion to the application of induction units and fan-coil systems, heat pumps may be arranged in a modular configuration around the building. Units that operate both as heaters and as coolers are normally installed and these are termed *reverse cycle* heat pumps. Floor, ceiling and under-window types are available; a typical under-window unit is illustrated in Figure 14.26.

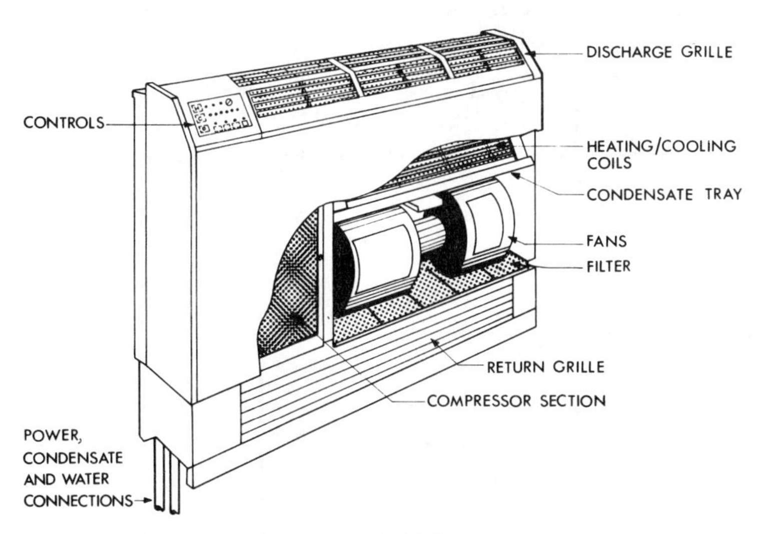

Figure 14.26 Under-window reverse cycle heat pump unit

Each unit incorporates a reversible refrigeration machine comprising an hermetic refrigeration compressor, a refrigerant/room air coil, a refrigerant/water heat exchanger, a cycle reversing valve and a refrigerant expansion device. When the space requires heating, the air coil acts as a condenser, drawing heat from a water circuit through the heat exchanger acting as an evaporator, upgraded by the compressor. When the space demands cooling the air coil becomes the evaporator, the heat being rejected to the water circuit via the water-side heat exchanger acting as a condenser. Figure 14.27 illustrates the components and operating cycles of such a system.

Simultaneous heating and cooling can be provided by individual units to suit the thermal loads around the building. There is a running cost benefit arising from such operation due to condenser heat from those units performing as coolers being rejected into the water circuit, thereby reducing the heat input required from central boilers. A

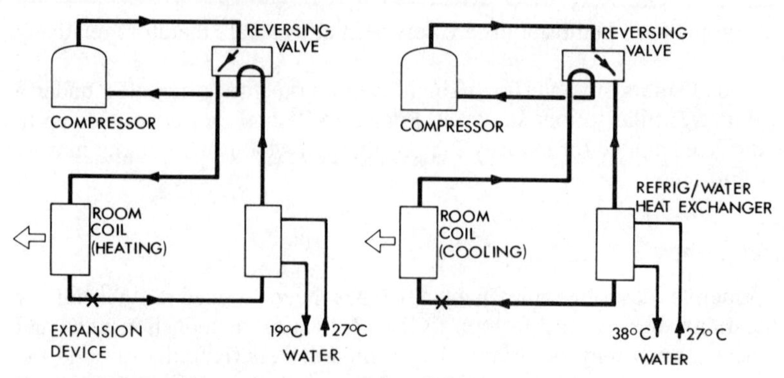

Figure 14.27 Operating cycles for a reverse cycle heat pump system

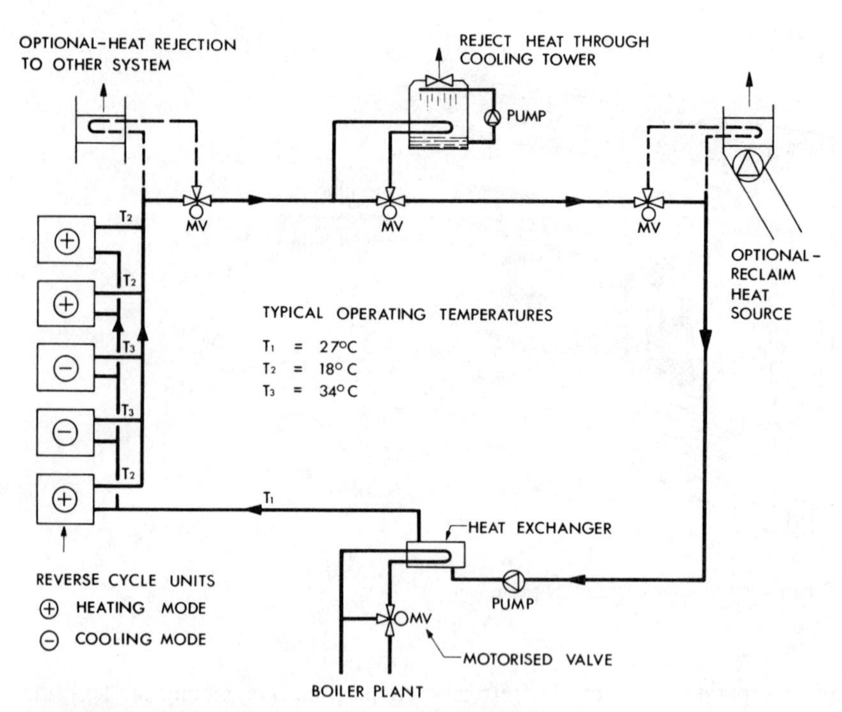

Figure 14.28 System arrangement for a reverse cycle heat pump system

diagrammatic arrangement of the system is given in Figure 14.28, from which it can be seen that a two-pipe closed water circuit is maintained at around constant temperature, typically 27°C, to provide the heat source and heat sink for the heat pumps.

The coefficient of performance of the smaller distributed refrigeration compressors is lower than that obtained from central plant; the ratio being of the order of two to one. However, the distribution losses and the electrical power absorbed by the chilled water pumps associated with a central plant have the effect of reducing the effective difference between the energy requirements of the alternative methods of cooling.

Outside air may be introduced to the space through a central plant either independent from the units, connected to each unit, or alternatively drawn from outside directly into each unit. Air supplied from a central plant is preferred since this provides better control and reduced maintenance.

Temperature control is normally by a thermostat sensing return air to the unit which sequences the compressor and reversing valve. Only coarse temperature control is achieved and control over humidity is poor with such a system. Noise can be a problem with the units located in the space, particularly since the compressors may start and stop fairly frequently.

Larger heat pump units may be used to serve areas such as shops and department stores. Typically, these would operate on the reverse cycle principle, but would use outside air both as the heat source in the heating cycle and as the heat sink for the cooling cycle.

Variable refrigerant volume systems

An understanding of the operating principle of a variable refrigerant volume system will be assisted by reference to the description of the split-system heat pump on p. 546 and Figure 19.24(b). In the case of the variable refrigerant volume system, the external unit, which is usually roof mounted, comprises twin compressors; heat exchangers and air circulation fans. This external unit has refrigerant pipe connections to remote room terminals each of which incorporates a refrigerant-to-air heat transfer coil, a filter and a fan to recirculate room air (Figure 14.29).

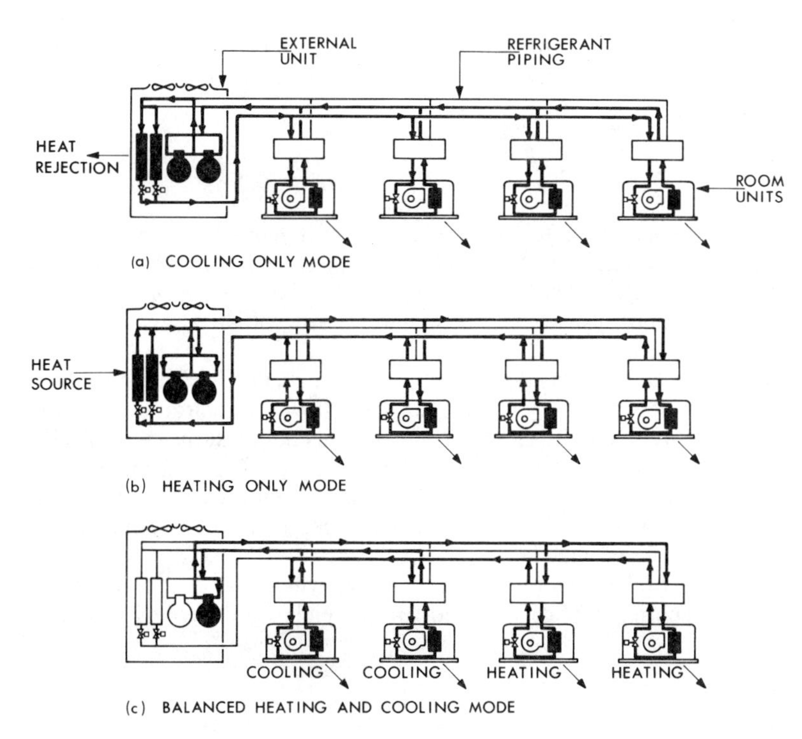

Figure 14.29 Variable refrigerant volume system: modes of operation

In the cooling mode of operation, the heat exchangers of the external unit function as a refrigerant condenser producing liquid which is circulated to the remote room terminals. As the refrigerant liquid passes through the coils where heat is absorbed, evaporation takes place and the gas is returned to the compressors. In the heating mode of operation, the heat exchangers of the external unit will function as a refrigerant evaporator, absorbing heat from the outside air and boiling off the liquid prior to compression. Hot gas from the compressors is then circulated to the room terminals, where useful heat is rejected at the coils and the resultant liquid is returned to the external unit.

The system is also able to operate in a dual mode offering both cooling and heating service to satisfy the requirements of the spaces served, including the thermal balance situation where the room heating and cooling demands are in equilibrium, in which case high thermal efficiencies will result. A three-pipe distribution system is needed for this dual function as shown in part (c) of Figure 14.29.

In all modes, capacity control of the system, as the name implies, is by varying the quantity of refrigerant in circulation by means of speed variation of the compressors, with temperature control in individual rooms achieved by throttling the flow of refrigerant through the coil of the terminal concerned in response to a signal from a thermostat.

The maximum cooling capacity of an external unit is of the order of 30 kW and up to eight room terminals, having typical outputs in the range 2.5–15 kW, may be served from one external unit. There are limitations to the length of pipework between the external unit and the most remote room terminals, normally 100 m, with a maximum height difference of 50 m. The units and terminals, their controls and the distribution fittings are available as a proprietary system.

The extent of the internal piping system gives rise to some concern bearing in mind the potential for refrigerant leakage, the resultant concentration in an occupied space and thence into the atmosphere generally. The refrigerant used in such a system is usually R22, the occupational exposure for which is given in Table 19.1 (p. 521).

Chilled ceilings

A fundamentally different approach to the whole matter of temperature control and ventilation, covered by the general term ‘air-conditioning’, is a system in which surfaces within the ceiling are cooled by chilled-water circulation for the removal of heat gains, leaving to the air-distribution system the sole purpose of ventilation and humidity control.

An essential feature of systems of this type is that the entering chilled water temperature should be above the room dew-point, by at least 1.5 K to allow for control tolerance, in order to avoid any possibility of condensation forming on the cooling surfaces. Typically, chilled ceiling systems have a flow water temperature of 14–15°C and a temperature increase across the exchange device of 2–3 K. The dehumidifying capacity of the air supply is also important for control of the dew-point and, in consequence, a design margin of the order of 20% should be provided.

The cooling surfaces may take any of a number of forms which may be classified into one of the following categories, typical examples from which are illustrated in Figure 14.30:

- Radiant panels.
- Convective panels.
- Chilled beams.

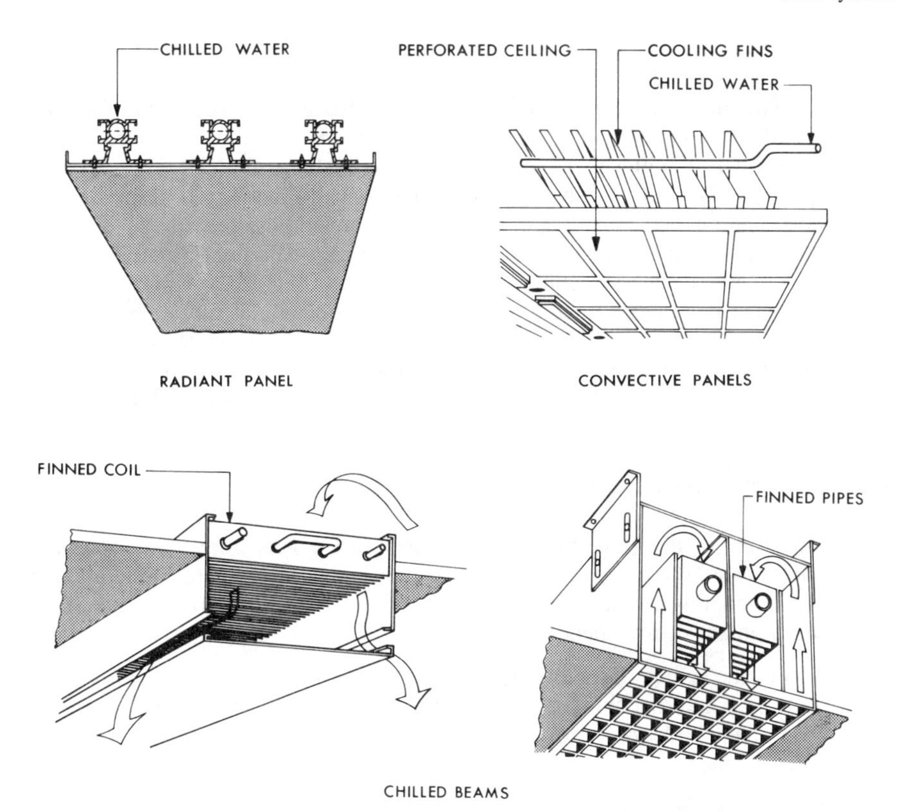

Figure 14.30 Chilled ceiling systems: radiant (Trox); convective (Krantz); chilled beams (Farex, Unilock)

In the case of both the radiant and convective panels, the cooling surface covers large areas of the ceiling. In the former case, where the radiant component provides up to 40% of the cooling effect, pipe coils may be either fixed, via a thermal conducting plate, to the upper surface of the ceiling panel or be embedded within the panel itself in which case the material may be small bore polypropylene in the form of a pipe mat. Radiant panels may be accommodated in shallow ceiling voids, as little as 65 mm for some proprietary designs.

Convective panels take the form of finned pipe coils which are located *within* the ceiling void, typically 250 mm deep, above a perforated or slotted ceiling which has at least 20% free area to the room space. With such a system warm air from the room rises into the ceiling void where it is cooled by the coil and, being now more dense, the air then falls through the ceiling to provide room cooling.

Chilled beams operate in a similar manner to convective panels but in this case the finned coils are concentrated into a smaller unit which may also incorporate a ventilation air supply. These smaller units may be positioned either above the ceiling, with their underside flush therewith or be partly below the ceiling surface to form a dropped beam effect. A minimum depth of ceiling void of about 300 mm is required for installation in this instance. Some later models are suitable for suspension below the ceiling. Others incorporate the facility to introduce ventilation air via nozzles within the unit to increase, by induction, the air flow across the coil and thus extend the sensible cooling capability.

One of the advantages of a chilled ceiling system is that it has capacity to offset high cooling loads without producing an unacceptable air movement within the space, a result which is often difficult to achieve with air-based cooling systems. With all types of chilled ceiling, temperature control is via a simple control valve acting in response to a thermostat, with chosen sections being controlled independently to provide zone control. This same control principle may be applied also to areas of perimeter radiant heating which may be integrated with the chilled ceiling panels to provide for winter weather.

Although convective systems have a higher sensible cooling capacity compared with radiant panels, $160\,W/m^2$ and $120\,W/m^2$ respectively, it is important to recognise that radiant systems reduce the mean radiant temperature in the space and, in consequence, reduce also the resultant temperature sensed by occupants by as much as 2 K.

Alternative methods of cooling

Reference was made in the preamble of this chapter to the need for avoidance, where possible, of methods for dealing with over-heating in occupied spaces which rely upon the use of mechanical cooling plant. There are three obvious routes to follow in pursuit of a solution to the problem of over-heating:

- Examination of the source of the problem with intent to reduce *heat gains* through the structure and from internal sources such as office equipment and lighting.
- Investigation of methods to improve the use and effectiveness of *natural ventilation*.
- Consideration of the use of *unorthodox cooling* techniques.*

Since the first two of these headings have been touched upon in earlier chapters, the following notes relate to remaining items and are to some extent anecdotal.

Unorthodox cooling techniques

Strictly speaking, such methods may be categorised as either *active* or *passive*, the former using some minimum amount of energy and the latter none at all. As might be appreciated, the distinction between the two, in practice, is a matter of degree.

Active (or perhaps passive!) cooling may be achieved, but not exclusively, by one or a combination of the following methods:

- *Night ventilation*, coupled with high thermal mass: low temperature night-time air is passed through the building to cool the structure which, next day, then acts to offset heat gains. (Unfortunately, in modern office buildings with lightweight partitions, false ceilings and carpeted floors, the mass of the actual structure is degraded thermally to 'lightweight' and thus no longer functions well as a store.)
- *Ground cooling:* using stable year-round ground temperature, normally 8–12°C, as a cooling source for circulated water, air or other transfer media.
- *Evaporative cooling* of air streams: passed through a water spray, the temperature of the air may be reduced by 3–4 K. The quite significant associated increase in humidity may not be tolerated in summer.

* International Energy Agency. Energy conservation in buildings and community systems. *Innovative Cooling Systems*. Workshop Report. 1992.

To achieve optimum conditions for effective cooling by such means, detailed examination of the thermal performance of the building fabric and any engineering systems *acting together* must be carried out through the whole range of coincident external and internal conditions within which the building will function. Means for analysis on this scale are not only available but commonplace.

Hollow floor system

A proprietary arrangement, using a combination of night-time ventilation and structural mass, has been developed in Sweden, supply air being passed throug the cores in hollow concrete floor planks before being introduced into the space. To be most effective, both the ceiling soffit and the floor surface should remain 'hard' and uncovered in order not to dilute heat transfer by any covering which might act as an insulating layer. The air, supplied at constant volume to meet the ventilation requirement, is further treated when necessary at a conventional central air handling plant providing filtration and heating or cooling.

The planks, each up to 18 m in length and normally about 1.2 m wide, are pre-cast with five smooth faced cores, 180 mm diameter, per plank. These are modified on site, by core drilling, to provide holes for air inlet and outlet to cores 2 and 4 and cross passages between these and the central core 3. The end holes are plugged by back-filling with concrete to produce the arrangement shown in Figure 14.31, and are then pressure tested to 400 Pa. In use, the air velocity in the cores will be about 1 m/s.

The planks, when laid, provide airways to cover the whole floor area. However, a consequence of the standard plank dimensions is that the system is capable of providing no more than coarse control which imposes a limitation in respect of handling diverse thermal loads across a whole floor. Furthermore, although the high thermal mass provides temperature stability, it is slow in consequence to respond to load changes in the space served. A damper box facility is available which may be fitted such that the air inlet is short circuited to the outlet core, thus increasing the thermal capacity of the air supply.

In summer, the supply air fan will normally run over 24 hours, continuously, to take advantage of diurnal variations in outside temperature. Cool air at night-time will reduce the temperature of the building fabric and, during the day, warmer outside air will be

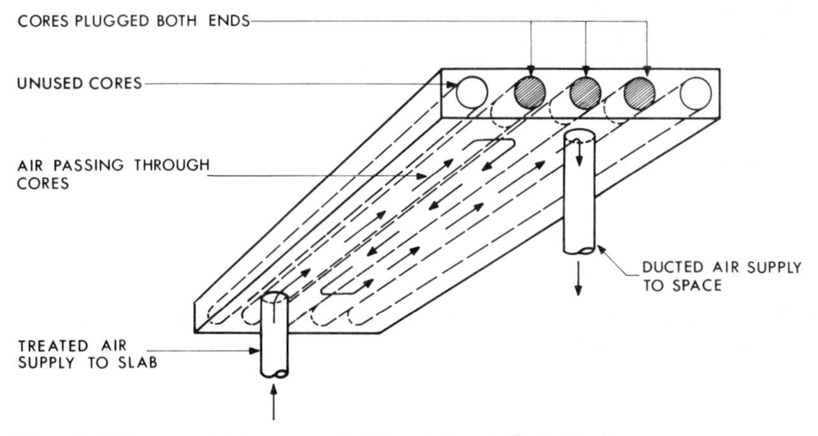

Figure 14.31 Structural slab system with integral airways (TermoDeck)

cooled by the slab which, at the same time, will provide direct radiant cooling. Additional cooling, available at the central plant, may be used to further cool the supply air if required. Typically, air will leave the plant at about 13°C and enter the space at about 3 K below room temperature, a suitable level for use in a displacement ventilation system. In winter, the supply air will be heated, to a maximum of 40°C, and the slab will then perform as a large low temperature radiant surface to supplement the heated air supply.

The concept is a simple one requiring relatively low capital expenditure and little maintenance. Exposed surface duct runs and connections to the core airways may be difficult to integrate aesthetically and, since the performance of the system is inextricably involved with the integrity of the floor structure, great care is required in defining contractual responsibilities. Nevertheless, tests at the Building Research Establishment suggest that, although supplementary cooling from a mechanical source will be required for applications in the climate of the British Isles, use of the system may be economic in energy terms.*

Solid slab system

Pipe coils embedded in a floor slab have been used with some success in a number of installations, using well-established technology and taking advantage of the good energy transport characteristics of water as the heat transfer medium. In the present context, it is the means of cooling the water which is of interest, this being either by a dry air cooler or by a closed circuit water cooler, either of which may be operated at night-time when the outside air temperature is low. The principles of operation of these cooling devices are described under the heading 'Free cooling' on p. 530.

Such a system for an office block in Switzerland serves a building of 8000 m^2 floor area with some 60 km of embedded pipe coils operating in conjunction with roof mounted dry cooler heat exchangers. The water temperature between 8 a.m. and 10 a.m. is reported to be 19°C as is the supply air temperature which is, at most periods of the year, not cooled mechanically. Air delivery to the space is through air handling luminaires which raise the temperature to 1 or 2 K below room level.

Cooling by ground coils

Pipe coils are used in yet another building in Switzerland but in a very different manner since they are buried nearby in earth, some 6 m below the water table, to provide thermal storage, seasonally regenerated, for pre-cooling ventilating air in summer and preheating it in winter. The pipes, which appear to be laid in a parallel grid formation, are of 230 mm diameter plastic and a total length of approaching 1000 m is buried. It would appear that this building is by way of being a research project† and available information regarding performance is meagre as yet.

Summary of systems and application

Figure 14.32 sets out the various air-conditioning system types in common use and identifies their principal characteristics.

To give an indication of the relative merits of the more popular systems, Table 14.1 provides a summary of some of the important design parameters with an indication of how well the various systems are able to satisfy these.

* Willis, S. and Wilkins, J., 'Indoor climate control – mass appeal'. *CIBSE Journal*, 1993, **18**, 25.
† Baumgartner, Th., *Erdwärmenutzung für die Raumklimatisierung*. NEFF-Research Report 390/19922.

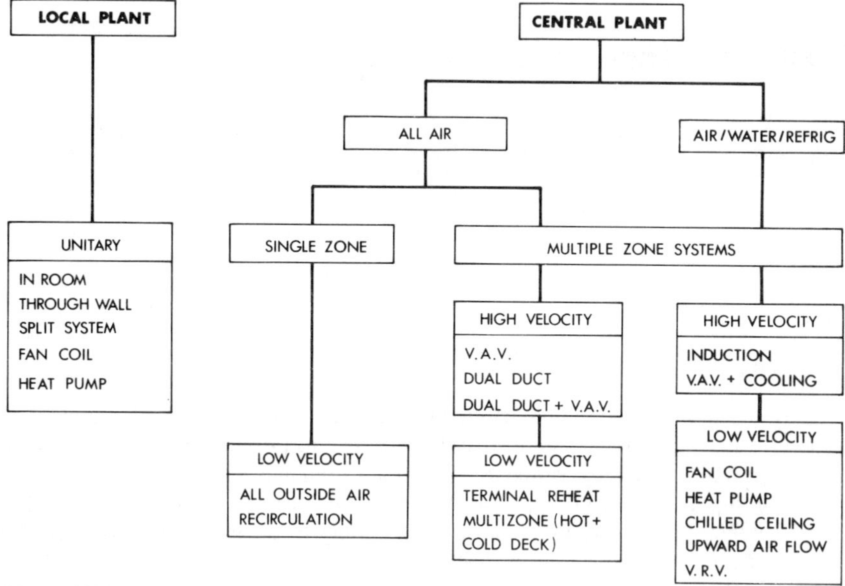

Figure 14.32 Principal characteristics of systems in common use

Table 14.1 Factors, other than thermal, affecting the choice of system

	System type				
	Fan-coil system, 4-pipe with independent air supply	*Single duct VAV with perimeter heating*	*Dual duct VAV*	*Induction system, 4-pipe*	*Reverse cycle heat pump, with independent air supply*
Spatial impact					
plant space	Average	Poor	Poor	Average	Good
riser shafts	Average	Poor	Poor	Average	Average
floor space encroachment	Poor	Average	Good	Poor	Poor
ceiling depth	Average	Poor	Poor	Average	Average
Quality of performance					
temperature control	Good	Good	Excellent	Average	Satisfactory
humidity control	Adequate	Satisfactory	Good	Adequate	Poor
air distribution	Adequate	Satisfactory	Satisfactory	Poor	Poor
noise	Adequate	Good	Good	Adequate	Poor
Costs					
capital	Average	Average/high	High	Average	Low
operating energy	Average/high	Low	Average	Average	Average
maintenance	High	Average	Average	High	High
Flexibility					
to suit partitioning arrangements	Good	Good	Excellent	Poor	Poor
increase cooling load	Good	Average	Good	Poor	Poor
increase ventilation	Average	Good	Excellent	Poor	Average

Chapter 15

Air distribution

Effective distribution of air within an occupied space is the key to successful operation of a ventilation or air-conditioning system. It is of little use to provide plant and ductwork distribution arrangements ideally suited to meet load demands if the methods used to introduce the air supply do not provide for human comfort or process needs.

Successful air distribution requires that an even supply of air over the whole area be provided without direct impingement on the occupants and without stagnant pockets, at the same time creating sufficient air movement to cause a feeling of freshness.

This definition indicates what is probably the key to the problem of successful distribution: that unduly low velocities of inlet are to be avoided just as much as excessively high ones and that distribution above head level not directly discharging towards the occupants will give the necessary air movement to ensure proper distribution over the whole area without draughts. Low level floor supplies, however, introduce air directly into the occupied zone and so need special attention.

To produce satisfactory conditions in the comfort zone of a space to be held at normal temperature, the distribution system should produce an air velocity, at a measurement point 1.8 m above the floor and not less than 0.15 m from a wall, of between 0.1 and 0.25 m/s and never less than 0.05 m/s. Where activity is high and spot cooling needed, as in a factory, a velocity of up to 1 m/s might be acceptable. *Laminar flow*, used in clean rooms, and other special distribution methods are outside the scope of this book.

General principles

There are five general methods of air distribution.

- Upward.
- Downward.
- Mixed upward and downward.
- Mixed upward and lateral.
- Lateral.

The choice of system will depend on:

- Whether simple ventilation or complete air-conditioning is employed.
- The size, height and type of building or room.
- The position of occupants and/or heat sources.
- The location of the central plant, and economy of duct design.
- Constraints imposed by the building structure and internal layout.

Upward system

The air is introduced at low level and exhausted at high level as in Figure 15.1 which shows a section through an auditorium, with mushroom inlets under the seats and riser gratings (see Figures 15.30 and 15.31) in the gallery risers. The air is exhausted around the central laylight in the roof. Such a system could be designed so as to be reversible, i.e. operated as a downward system.

When working upwards, the air generally appears to be somewhat 'dead', due to the very low velocity of inlet (about 0.5 m/s) necessary with floor outlets to prevent draughts in this type of application. When working downwards more turbulence is set up in the air stream, with a greater feeling of freshness.

The upward system is not, however, confined to one with floor inlets. The inlets may equally well be in the side-walls, with extract in the ceiling as before. The limitation of an upward system is that in a large hall it may be difficult to get the air to carry right across without picking up heat *en route* and rising before it reaches the centre.

The upward system is used successfully with simple ventilation systems. When the air is cooled, as in a complete air-conditioning system, it will tend to fall too early, before diffusion, and thus cause cold draughts unless it is introduced through specially designed outlets carefully selected to suit the application. The upward system, however, lends itself to simple extract by propeller fans in the roof in the case of a hall, factory, etc., and is thus generally the cheapest to install.

Another application of the upward system is the swimming pool hall, as shown in Figure 15.2. Here the supply air is introduced through special plastic discharge spouts situated below a large area of glazing and is exhausted by specially treated roof-exhaust units.

On a smaller scale, upward distribution has been successfully applied to computer rooms and offices. In the machine areas of computer suites, where occupancy is transient, air velocities within the occupied zone are less critical than is the ability of the system to maintain close control over temperature and humidity. The large air quantities required in such applications to counter the high heat gain from equipment give rise to air velocities above the recommended comfort criteria since air supplies are distributed to match the positioning of the space loads.

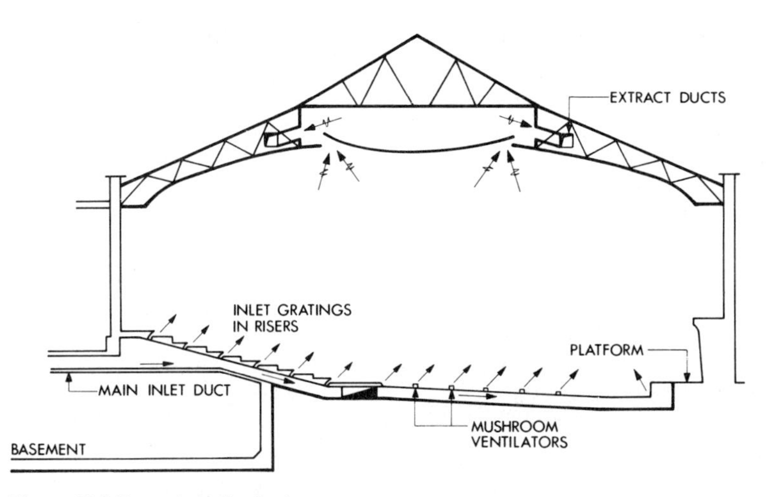

Figure 15.1 Upward air distribution

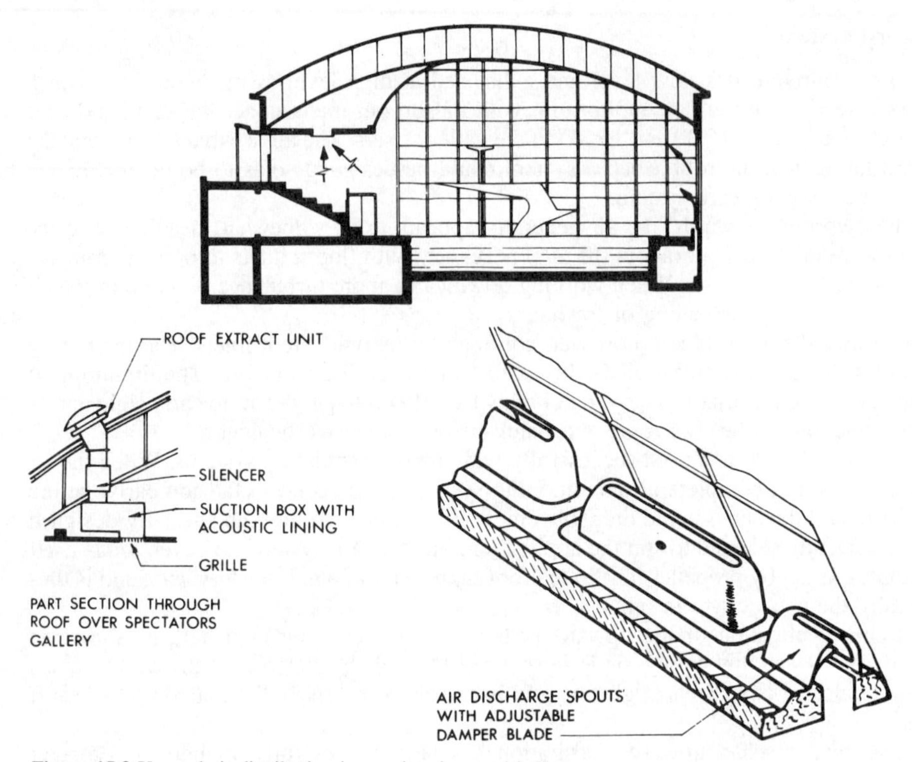

Figure 15.2 Upward air distribution in a swimming pool hall

In office applications air movement is more critical and for floor supply systems it is important that the air is introduced by using a relatively large number of small outlets, the design of these being such as to produce a high induction effect, the supply air mixing quickly with the room air to reduce velocity and temperature differential. A typical arrangement is shown in Figure 14.18. A floor supply system may be supplemented by desk outlets fed from the same plenum thus providing desk bound operatives with a degree of control over their micro-climate. Displacement air-conditioning systems, following the pattern described in Chapter 14, rely for their operation on minimal mixing of supply air with room air and, in consequence, low level side-wall or floor outlets having low discharge velocities are selected for this application. With such systems, perimeter heating loads are dealt with conveniently by an independent system.

On a scale appropriate to single rooms, experimental work has been reported which suggests that for winter use it is possible to introduce warm air via a long slot near floor level, as shown in Figure 15.3, and by this means much reduce the temperature gradient within the room.* Discharge velocities of up to nearly 12 m/s were used without reports of discomfort from occupants.

* Howarth, A. T., Sherratt, A. F. C., Morton, A. S., 'Air movement in an enclosure with a single heated wall.' *JIHVE*, 1972, **40**, 211.

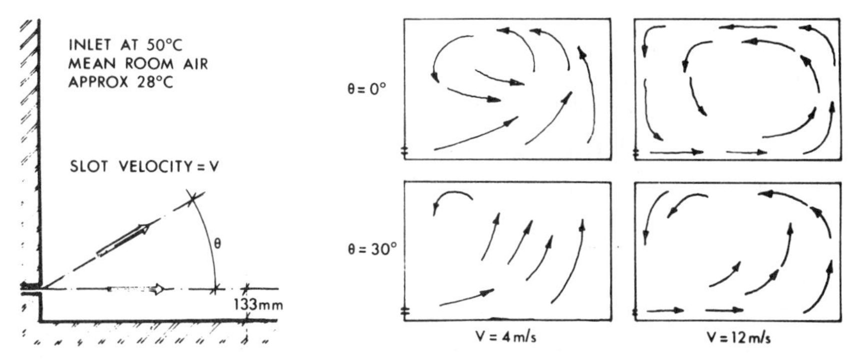

Figure 15.3 Upward air distribution from low level side-wall outlets

Downward system

In this type of system the air is introduced at high level and exhausted at low level, as in Figure 15.4. It is commonly used with full air-conditioning where, due to the air admitted being cooled, it has a tendency to fall. The object of distribution in this case is so to diffuse the inlet that the incoming air mixes with room air before falling. Thus, the inlets shown in the diagram as discharging downwards, in practice deliver in part horizontally at sufficient speed to ensure that the air completely traverses the auditorium. Turbulence is thus caused with the desirable effect already mentioned. On a smaller scale, as applied to an office building, this system appears as in Figure 15.5

Provided the height of room is not abnormal, the extract opening may be at high level as in Figure 15.6. Short circuiting is avoided by the velocity of the inlet air carrying over to the far side of the room. Another possible arrangement is a variation of this, namely, 'downward–upward', as in Figure 15.7(a).

Computer rooms may also be conditioned using a ceiling supply, normally through a perforated ceiling or an equivalent system, with the extract taken out via floor extracts or

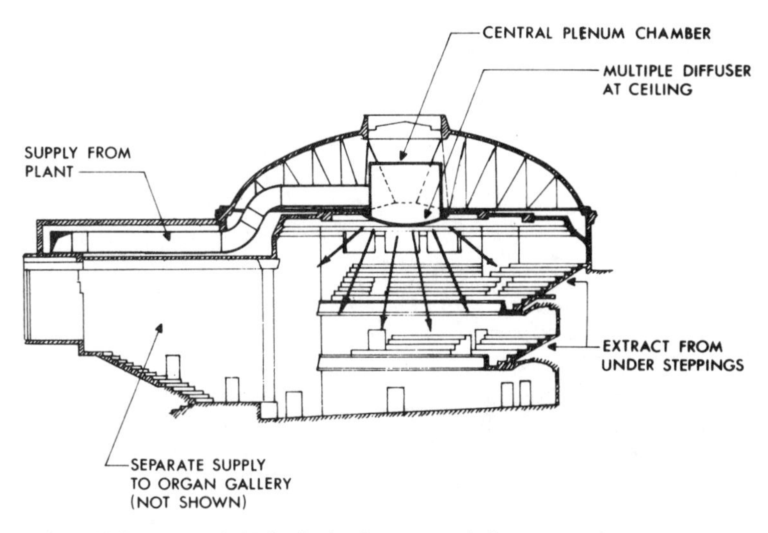

Figure 15.4 Downward air distribution in a concert hall

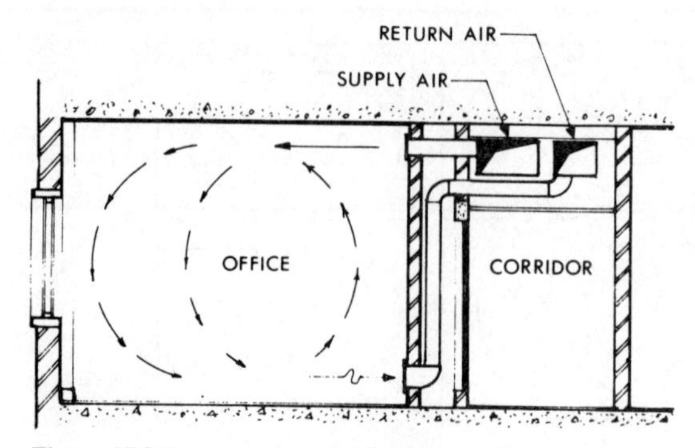

Figure 15.5 Downward air distribution in offices

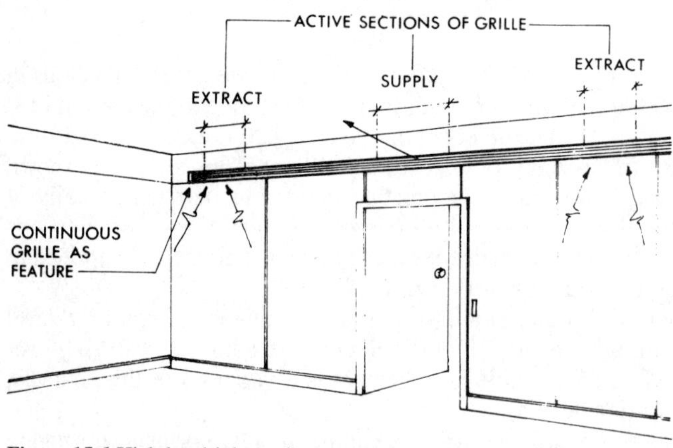

Figure 15.6 High level inlet and extract

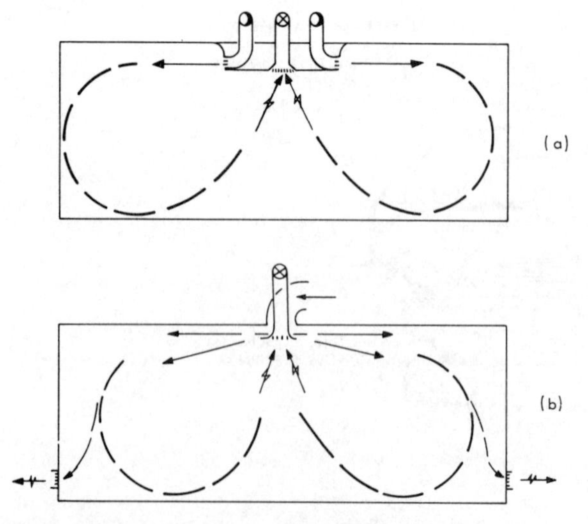

Figure 15.7 Downward–upward and combined arrangements

low level side-wall grilles. The supply arrangement is described then under the heading of *perforated ceilings*.

An application of downward inlet with both downward and upward extract, suitable for rooms of greater height, is shown in Figure 15.7(b). This is usually adopted where smoking occurs and it is necessary to provide some top extract to remove the smoke. In this case the top exhaust is discharged to atmosphere by a separate fan and the low level extract constitutes the recirculated air. The low level extract also serves to ensure that satisfactory air movement is achieved at the occupied level, in cases where the room height is over 4 m. Care must be taken when placing low level extract grilles close to areas where people are seated: such grilles must be selected for a very low velocity through the free area and be well spaced out so that excessive air movement will not occur.

Mixed upward and downward

Such a system is shown in Figure 15.8, illustrating a typical swimming pool hall application. It will be clear from the previous descriptions that the principle of the air distribution is, in effect, an upward system providing good mixing. Normally about 25% of the extract air quantity will be at low level also, the remainder being exhausted at high level.

Mixed upward and lateral

Such a method has been used to describe a system where the air is introduced vertically upwards from beneath a window. The flow pattern is vertical, or virtually so, up to ceiling level and then horizontal across part of the ceiling. Secondary room air is induced into the air stream producing a flow pattern as shown in Figure 15.9. This method of air

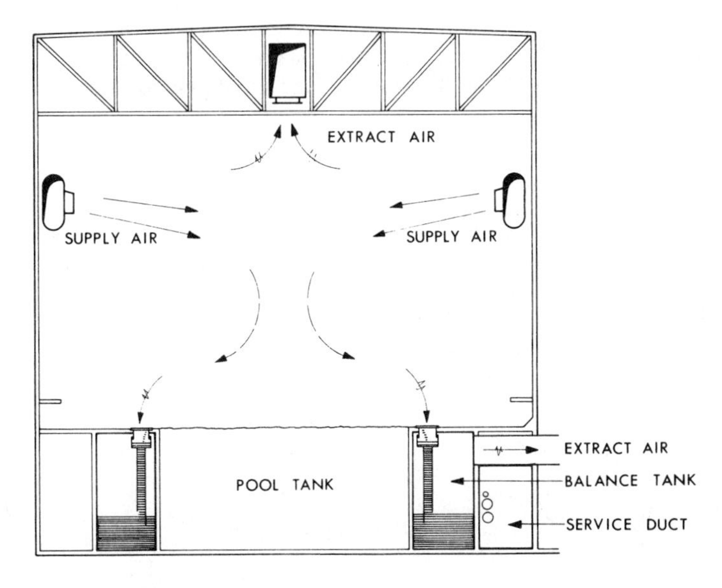

Figure 15.8 Mixed system of air distribution

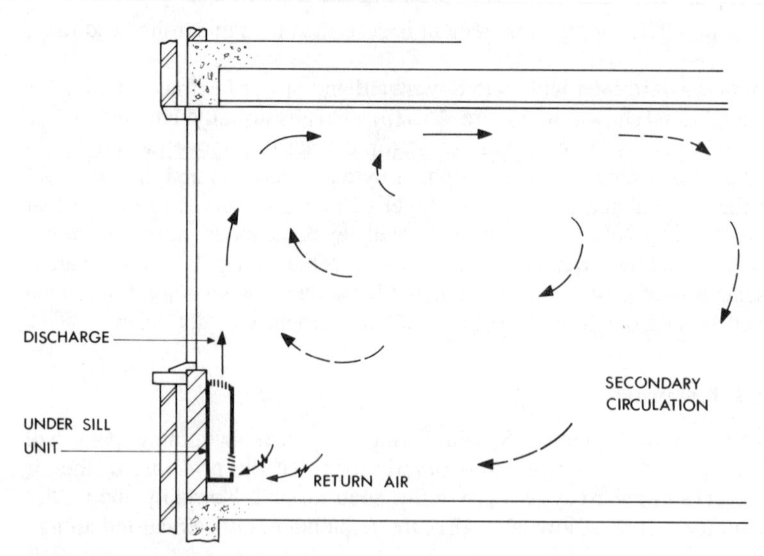

Figure 15.9 Vertical air distribution from sill level

distribution is in common use with induction unit, fan coil and reverse cycle heat pump systems, where the terminals are located at low level under windows. Fixed blade linear diffusers are normally used and the flow may be vertical or angled towards or away from the window. Vertical throw is preferred as this avoids the risk of the air, when cool, 'dumping' into the room. However, where there is a recess formed at the junction of ceiling and window, perhaps where there is a dropped ceiling, the air has to be directed away from the window to avoid interruption of flow at the recess; an angle of 5 or 10° from the vertical is common. With upward flow over glass, care must be taken to ensure that the air stream when cool will not cause the insolated glazing to crack under the conditions of thermal stress then created: the glass manufacturer should be consulted in this respect.

Lateral

This arrangement may sometimes be necessary where dictates of planning preclude more orthodox solutions. Air is introduced near the ceiling on one side of a long low room with a smooth flush ceiling, and is exhausted at the opposite end at the same level taking advantage of the 'Coanda' effect which will be referred to in more detail later (see Figure 15.10). The inlet is at a high velocity and strong secondary currents are set up in the

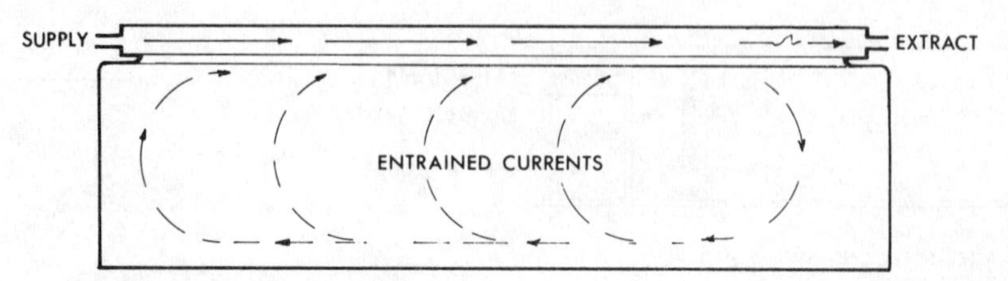

Figure 15.10 Lateral air distribution

reverse direction at the lower levels, as shown. It is these secondary currents which are important in many distribution systems, creating the turbulent mixed flow pattern already mentioned. Care should be taken, however, to avoid high velocity reverse flow at floor level creating draughts.

Distribution for air-conditioning

Air-conditioning usually involves the handling of large air quantities, and one of the chief problems of successful design is how to introduce and extract these quantities without giving rise to complaints of draught or of causing noise. A great variety of methods has been adopted and there are innumerable devices available to suit different conditions or architectural tastes.

In installations involving unusual air distribution patterns, or abnormally high or low air change rates, the system should be tested as a mock-up on site or in a laboratory, such as BSRIA, to ensure that satisfactory air movement will be achieved.

The selection of fixed pattern terminal devices for use with variable flow rate systems must be suitable for the full range of supply volumes. Manufacturers' data are often adequate for the general run of cases, but where a detailed study of the physics of air distribution is required, reference should be made to relevant independent laboratory reports such as those produced by BSRIA.*

Air diffusion terminology

Definitions and recommended terminology used in this subject are given in an ISO Standard,† supplemented in the *Guide Section B3*. The following are relevant to the text:

Isovel. A contour of equal air velocity around a terminal device.
Throw. The distance from the terminal to the position where the velocity has decayed to 0.5 m/s, i.e. the 0.5 m/s isovel.

Normally, velocities for air entering the occupied zone would be limited to 0.25 m/s for cooling and 0.15 m/s for heating.

Spread. The width of the 0.5 m/s isovel (manufacturers may quote for 0.25 m/s isovel).
Drop. The vertical distance from the terminal to the lower part of the 0.25 m/s isovel.

It is also necessary to understand the *Coanda* effect. This occurs when an air stream is discharged along an unobstructed flat surface. The jet entrains room air from one side only and friction loss between the jet and the surface causes the air stream to 'cling' to the surface until the velocity of discharge has decayed sufficiently as a result of entrainment of room air. A projection into, or a large gap in the surface may well destroy the Coanda effect causing the air stream to become detached prematurely from the surface; throw is reduced by about one-third when the Coanda effect does not occur.

Typical air flow patterns from a selection of terminals and applications are given in the *Guide Section B3*.

* BSRIA Application Guide AG 1/74, and Laboratory Report LR 83.
† International Standard, ISO 3258, 'Air distribution and air diffusion-vocabulary'.

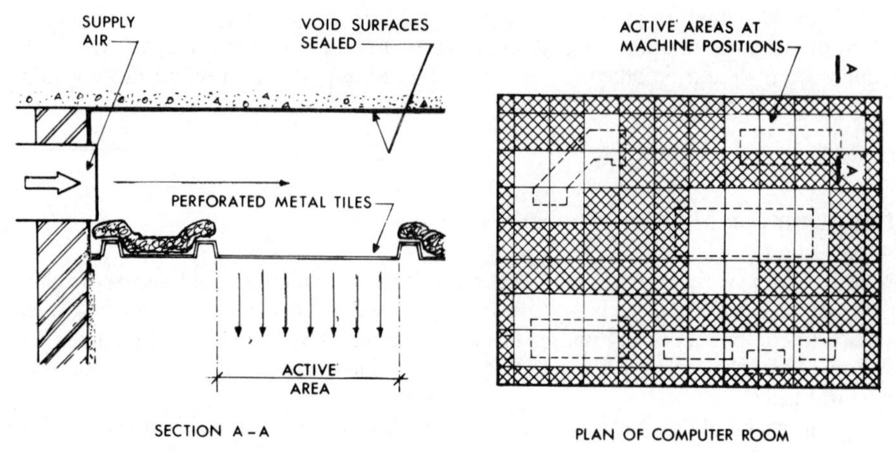

Figure 15.11 Inlet via a perforated ceiling

Perforated ceilings

Probably the most completely diffused form of inlet is the system which employs a perforated ceiling as shown in Figure 15.11. In this case air is discharged into the space above the ceiling, which is usually divided up so as to ensure uniformity of distribution, and the air enters the room through the perforations. Where extremely large air quantities are involved the whole ceiling may be used, but in the normal case only selected areas of panels serve as inlets, acoustic or other baffles being positioned behind the 'non-active' panels. The ceiling inlet system destroys all turbulence, which may be a good thing in some cases but not in others. This system has been applied successfully in law courts, department stores and in confined spaces such as radio commentators' boxes where no other distribution arrangements would be possible. An alternative to a perforated ceiling, but which achieves an equivalent effect, is the use of an array of strip diffusers, installed at say 300 mm intervals. An example is shown in Figure 15.12.

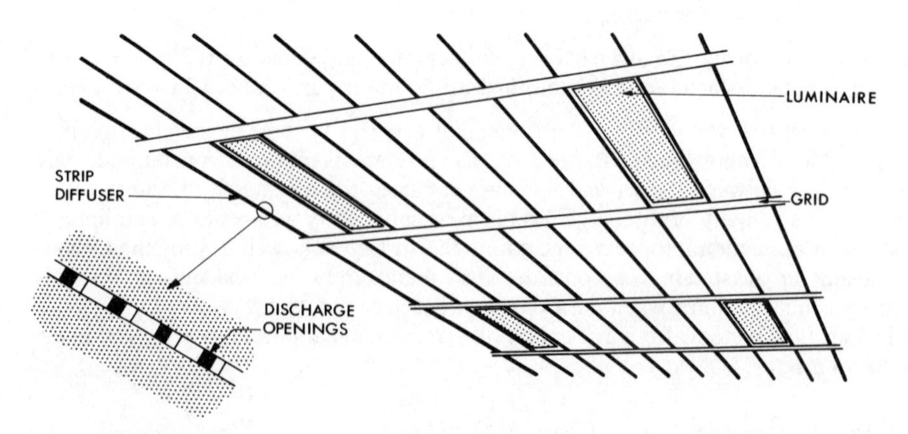

Figure 15.12 Ceiling distribution using multiple strip diffusers

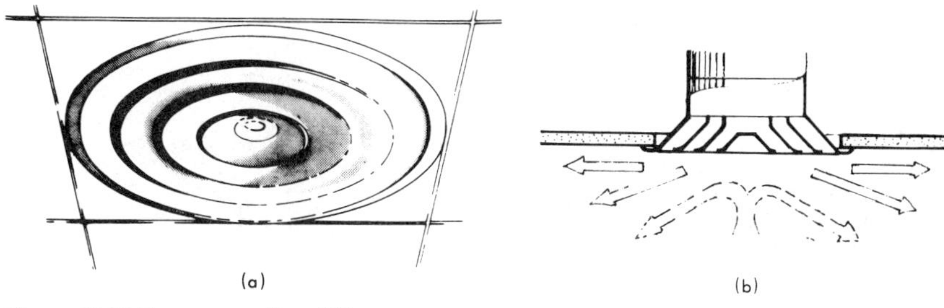

Figure 15.13 Cone-type ceiling diffuser

Cone-type ceiling diffusers

The type of ceiling diffuser shown in Figure 15.13(a) has certain interesting characteristics. Figure 15.13(b) illustrates the kind of flow pattern to be expected from such a unit and it will be noted that entrainment air is drawn up in the centre immediately under the diffuser, being caught up in the nearly horizontal delivery from the unit itself. Due to the fact that air is discharged radially, the air stream velocity falls off rapidly as the distance from the centre increases, and hence this kind of unit may be used with temperature differentials of up to 20 K.

There is a great variety of makes of this type of ceiling diffuser, all under different trade names and some with special features, such as:

- An adjustable arrangement whereby the inner cones may be raised or lowered in relation to the periphery, so causing a variation in the flow pattern to be achieved. Instead of the bulk of the air travelling horizontally, it is possible by this kind of adjustment to make it discharge vertically downwards or at any intermediate flow pattern desired. This might be of advantage in cases where it is desirable to cause the air to descend quickly, such as in a hot kitchen, rather than that it should be dispersed and lost at high level. Terminals of the pattern may be provided with removable cores for cleaning, which would be suitable for applications such as hospital operating theatres.
- In another type, the unit is square instead of circular, this sometimes being necessary to match ceiling tiles, etc.
- In other examples, the diffuser is flush with the ceiling instead of projecting.

Perforated-face ceiling diffusers

This type, shown in Figure 15.14, has been developed from the early *pan* devices and has characteristics similar to the cone pattern: it is often used to blend in with perforated acoustic-tile ceilings. The deflecting pan may be removed if a predominantly vertical air distribution is required. Both the cone and perforated plate types may be fitted with internal blanking pieces to limit the spread of air discharge.

Multi-directional ceiling diffusers

Such units provide facilities for positive directional control of air discharge, the blades being individually adjustable as illustrated in Figure 15.15. They are useful for applications where partition changes may occur or where movement of office machinery may create change in demand for air movement.

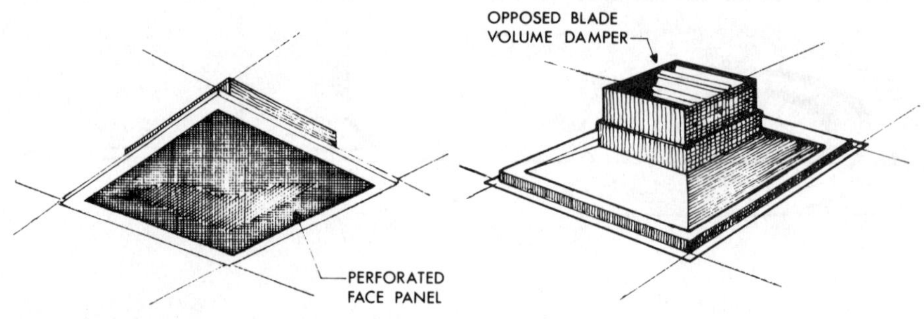

Figure 15.14 Perforated plate-type ceiling diffuser

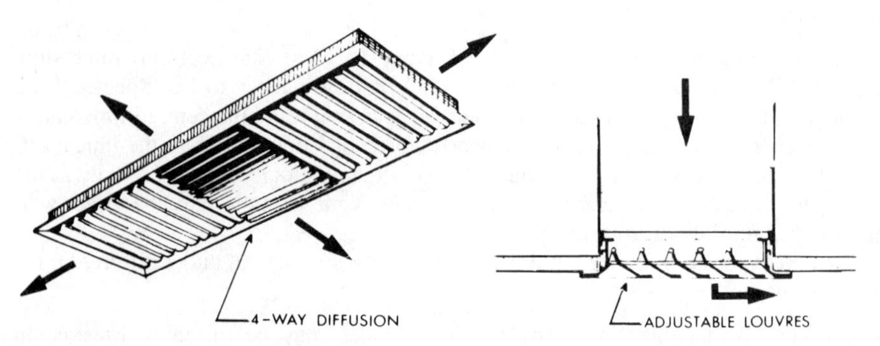

Figure 15.15 Multi-directional type ceiling diffuser

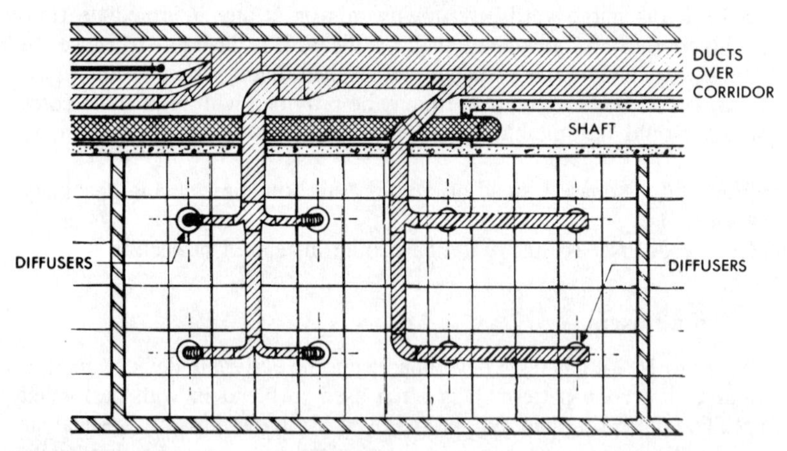

Figure 15.16 Typical layout of ceiling diffusers

A typical layout of ceiling diffusers is shown in Figure 15.16. It will be noted that use is made of the false ceiling space for the concealment of the connecting ducts, the final connections to the diffusers being of flexible ducting, thereby providing dimensional tolerance between the duct and the terminal position. To ensure that air distribution is effective, the room is divided into approximate squares with one diffuser to each. Alternatively, if it be necessary to use a non-symmetrical spacing, segments of diffusers

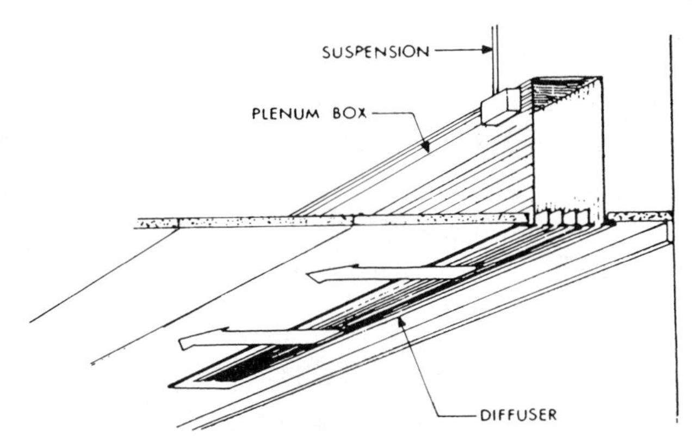

Figure 15.17 Multi-vane type linear diffuser

may be blanked off to avoid, for example, impingement of high velocity air flow at walls which would otherwise create excessive air movement at the occupied level.

Linear diffusers – ceiling

For use in open plan offices and in avoidance of interference with ceiling pattern, the linear diffuser in many different forms is commonly used. In its original form, Figure 15.17, it was an adaptation of the continuous side-wall grille but took advantage of the Coanda effect.

Development of this type of diffuser takes many forms, as shown in Figure 15.18. As may be seen, facilities are available for adjustment of the air flow pattern. Individual sections of the continuous length are supplied by means of air plenum boxes which may, in turn, be integrated with luminaires as Figure 15.19.

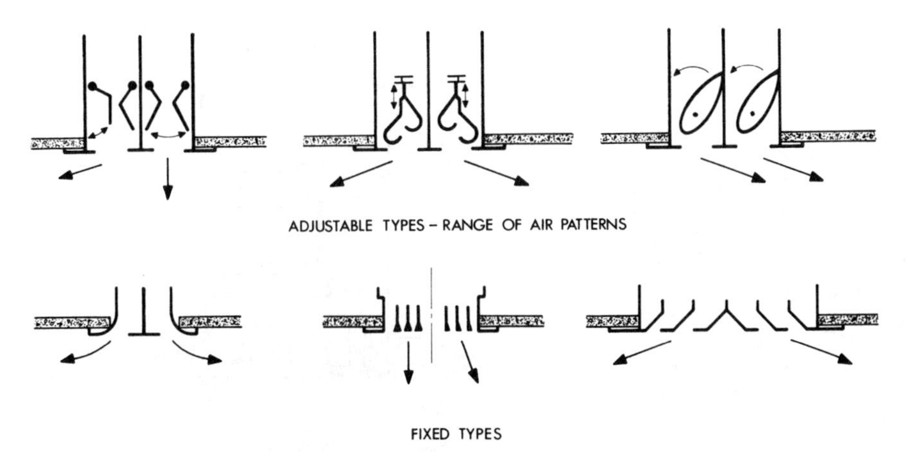

Figure 15.18 Various forms of ceiling-integrated linear diffusers

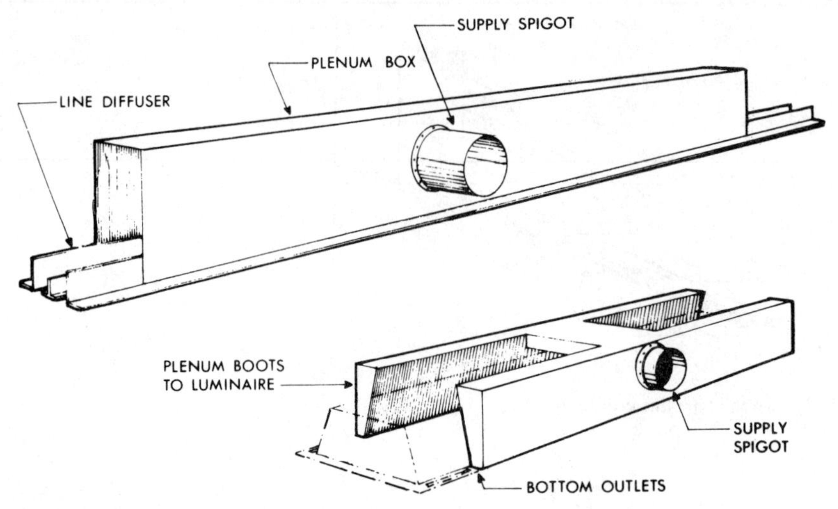

Figure 15.19 Plenum boxes for linear diffusers

Side-wall inlets

Where there is no false ceiling or other means of introducing the air through ceiling diffusers, it is necessary to adopt single wall inlets, these usually taking the form of a series of grilles distributed at intervals along the inner partition wall with ducts in the corridor false ceiling behind. Each inlet is equipped with a grille, so designed as to enable the requisite quantity of air to be introduced without draught.

A common and effective form of grille is that known as the *double-deflection* type, as shown in Figure 15.20. In this form of grille there are two sets of adjustable louvres, one controlling the air delivery in the vertical plane and one in the horizontal plane. The vanes are usually independently adjustable by means of a special tool, and when once set are not altered. A variety of flow patterns can be achieved according to the width of room, aspect ratio and velocity.

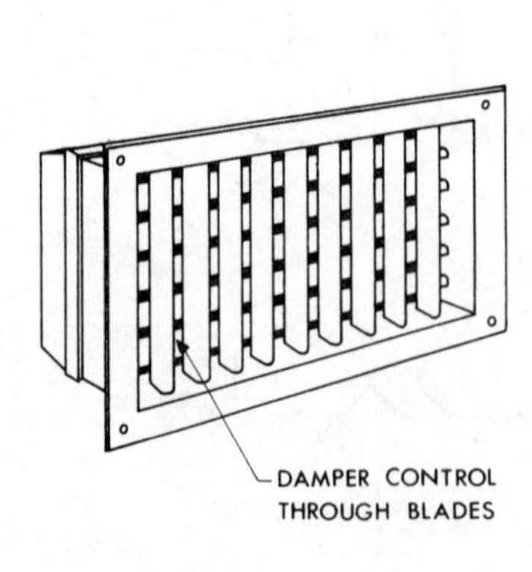

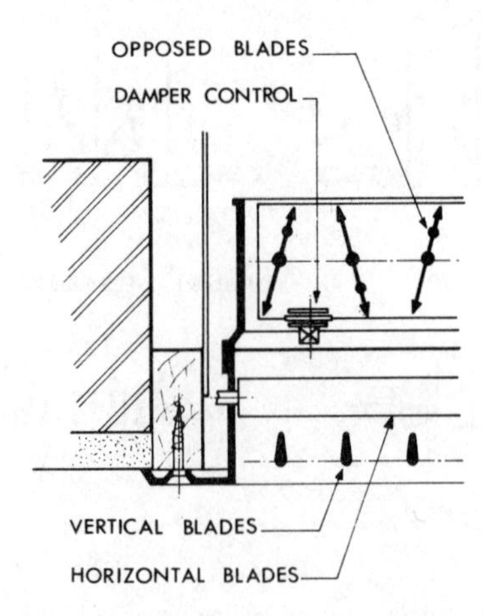

Figure 15.20 Double-deflection type side-wall grilles

For side-wall applications, however, much use is made of the linear diffuser as illustrated, for ceiling application, in Figure 15.17. By careful selection and making use of the Coanda effect by direction of the air pattern towards the ceiling, good overall air distribution may be achieved from such diffusers.

Single deflection type grilles are also available, that is with one set of adjustable blades either horizontal or vertical. Such terminals have limited use for air supply and are more suited to extracts. Fixed single-blade type grilles are also available.

Nozzles

Introduction of air by nozzles was, in the past, usually confined to small compartments such as ship's cabins and to aircraft, where it was desired to obtain the maximum effect under the direct and easy control of the occupant, both in the matter of quantity and direction: terminals of this type are known as *punkah louvres*. Nozzles may however be a useful manner of introducing large quantities of air in some vast arena, where it would be impossible to achieve the throws required by using any other type of terminal. In effect, the more the better. Examples of this kind are the arrangements used to ventilate the Earls Court Exhibition building, the Usher Hall in Edinburgh and, more recently, the Sheffield arena. In this last example, automatic adjustment of the discharge velocity is provided so as to maintain satisfactory air speeds at the occupied levels under both heating and cooling modes of operation.

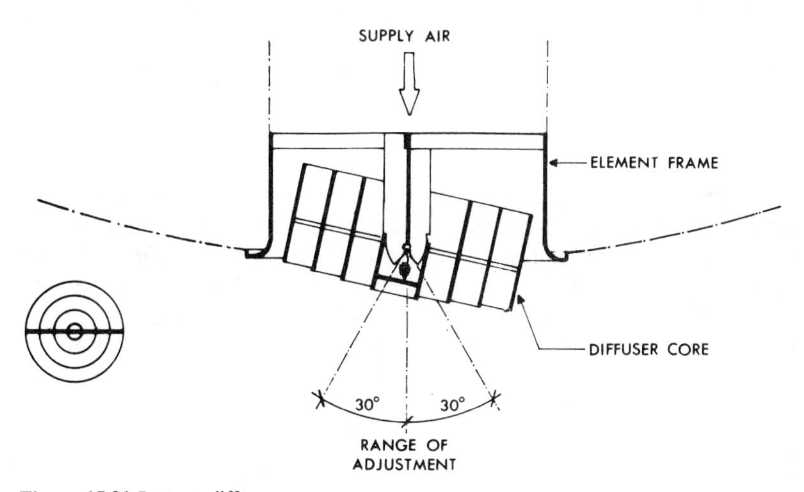

Figure 15.21 Jet-type diffuser

For the auditorium illustrated in Figure 15.4, the Usher Hall in Edinburgh, conditioned air was introduced solely from the central ceiling feature making use of 65 jet diffusers of the type shown in Figure 15.21. Each handled approximately 280 litre/s at an outlet velocity of approximately 3 m/s.* Figure 15.22 shows another but similar approach where *drum* type punkah louvres have been used to introduce large air quantities into a swimming pool hall and similar applications have been used for shopping centres.

* Clark, I. T., 'Air conditioning the Usher Hall, Edinburgh.' *JIHVE*, 1977, **45**, 125.

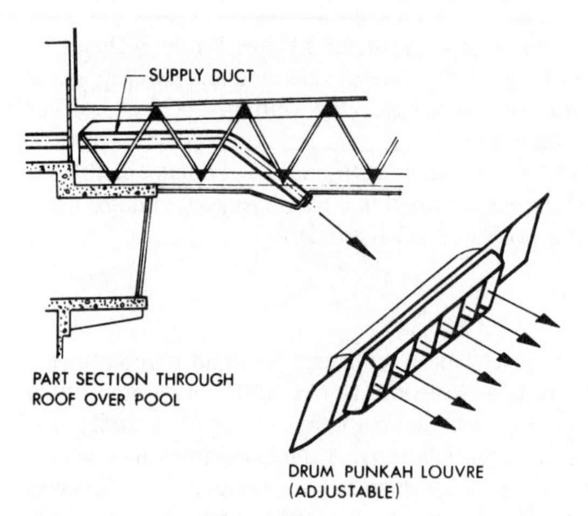

Figure 15.22 Punkah-type louvres in a swimming pool hall

Variable geometry diffusers

Linear and cone-type ceiling diffusers are available with variable outlets which are able to adjust to provide the correct air pattern with varying air flow quantity. Such diffusers are used with variable air volume systems where the variation in air quantity is greater than that for which a fixed device is able to provide a satisfactory distribution within the space. Typical designs for these were illustrated in Figure 14.15.

Floor outlets

The principle to be applied where floor distribution is to be used is 'little and often'. To avoid unacceptable air movement being experienced by the occupants, no more than a relatively small air quantity can be introduced at each outlet. There are effectively two types of floor outlet used in comfort application: the 'twist' pattern and the straight pattern, both being circular and non-adjustable. These are shown in Figure 15.23 together with the resulting air distribution patterns. The limiting air quantity handled by this type

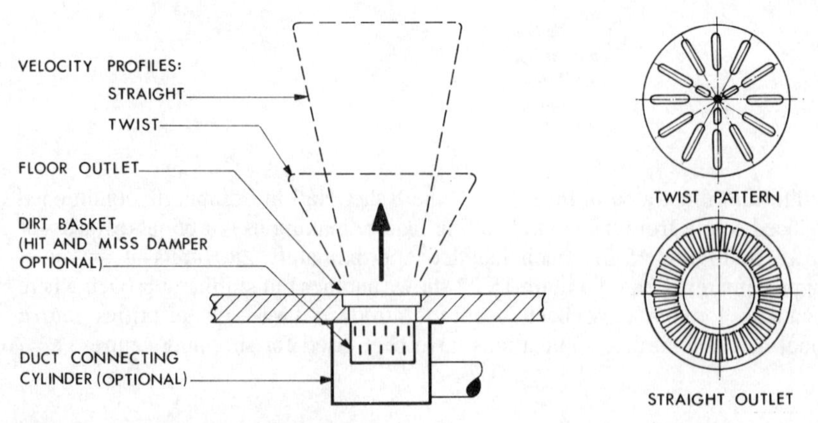

Figure 15.23 Typical floor outlets (Krantz)

of outlet is about 12 litre/s, and the supply temperature should not be lower than 5 K below the occupied zone temperature, and less in the case of displacement ventilation. Outlets should be positioned clear of furniture and not closer than 1 m to the nearest work station.

For computer room applications, where underfloor distribution is employed, the outlets normally take the form of a 600 mm × 600 mm floor plate either perforated or with a fixed linear grille over the face; both types may be fitted with opposed blade regulating dampers. The volume handled by a 600 mm square plate would be of the order of 200–300 litre/s.

Desk-top outlets

There are many designs for these devices, some of which are shown in Figure 15.24, and they are usually used in conjunction with a floor distribution system. The purpose of desk-top terminals is to provide occupants with considerable control over their immediate

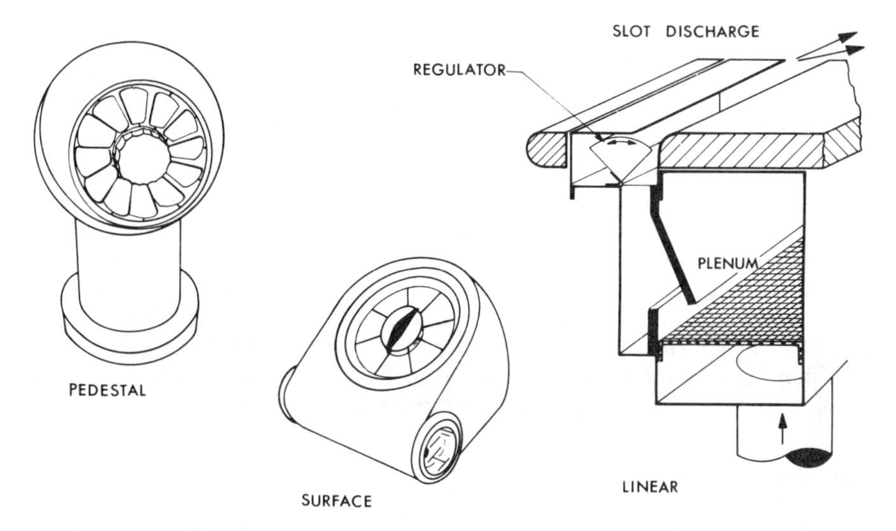

Figure 15.24 Typical desk-top outlets (Krantz)

micro-climate. In practice, the air flow quantity supplied in this fashion would be limited to about 14 litre/s. Similar devices may be used in other applications, such as at the rear of theatre seating.

Low level side-wall outlets

Supply air panels are available for discharging warmed or cooled air, with a low supply-to-room temperature differential, into a space at low level, from a side-wall or column position (Figure 15.25). The principle of design is to introduce a supply at low velocity, evenly distributed over the panel face. Pressure drop across the panel is 50–100 Pa and the maximum face velocity 0.5 m/s.

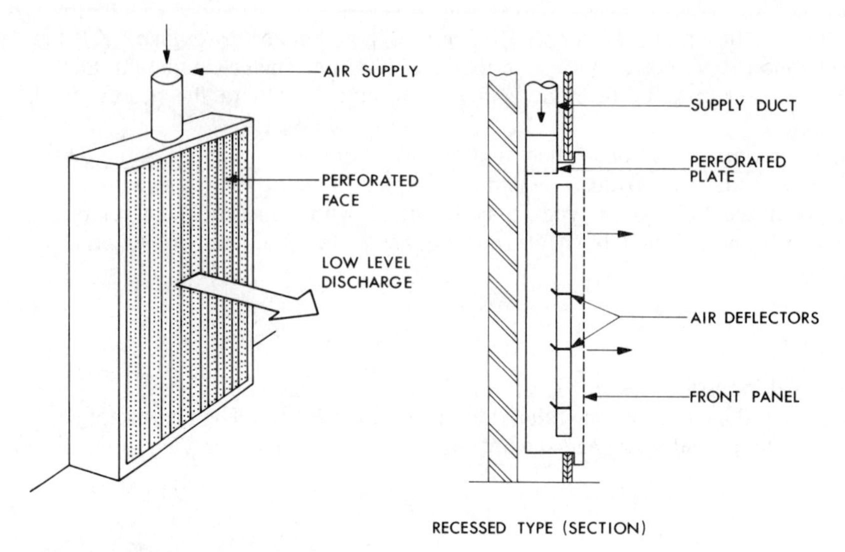

Figure 15.25 Low-level, low velocity side-wall outlets

Volume control

Means for controlling as accurately as possible the volume of air issuing from each grille, diffuser or other inlet, are essential. Probably the commonest form of control is the opposed blade damper. For ceiling diffusers and the like, a variety of multi-louvre dampers is available, or plain butterfly dampers in the ducts may be used. Any damper is a potential source of noise and its location and duty require careful consideration if such is to be avoided. Where there is a risk of a noise problem due to excessive throttling at the terminal, the major part of the regulation must be applied further upstream in the branch ductwork, and where necessary the ductwork will need to be designed to suit.

Selection of supply terminals

The selection of the best number and size of inlet grilles, diffusers and the like, involves a choice to meet a number of variables at the same time:

- Air velocities produced at head or foot level must not exceed those given at the beginning of this chapter.
- If the air entering is cooler than the room, as in an air-conditioning installation, there will be a tendency to fall, best avoided by keeping a reasonable entering velocity with the object of inducing air entrainment from the room; the air mixture then being warmer will have less tendency to fall.
- The velocity of inlet must not be so high that the air will impinge on the wall opposite, thereby causing undue turbulence. The throw should be about three quarters of the distance to the point of impingement with a wall, or opposing air stream.
- The distance apart of inlets, particularly in the case of ceiling diffusers, must be such that the streams from two adjacent units do not collide at such a velocity that a strong downward current results.
- The velocity selected for the grille or diffuser must be such that the sound level produced therefrom is below the design standard for the room.

- Appearance, layout and pattern are all architectural matters, which must also as a rule be taken into account and this sometimes creates a difficulty where the desired spacing or size is not acceptable on these grounds: system performance must take priority.
- The throw must not be directed towards projections that will deflect the air stream from the intended direction.
- The distribution should be arranged to deal with loads at source.

Primary selection should be by use of the design guides published by BSRIA* or the *Guide Section B3* prior to reference to manufacturers' data. These latter are now freely available for all types of equipment and, from a study of them, a variety of alternative solutions to a particular problem may be put down with the object of a final selection being made best suited to meet all the other conditions. The reader is referred to such data for precise information, but a few sample sizes are given in Table 15.1. It will be clear that, with the large number of variables, air distribution design is perhaps more an art than a science but on it to a large measure, as has previously been emphasised, depends the success or otherwise of any air handling installation, particularly in the case of air-conditioning. It is worth noting again here that particular attention must be given to the

Table 15.1 Approximate sizes (mm) of air inlets (representative only)

	Air quantity (litre/s)				
Type	*50*	*100*	*200*	*300*	*500*
Perforated ceiling panels (face size)	300 × 150	300 × 300	500 × 500	600 × 600	1200 × 600
Line diffuser (duct size)	750 × 50	1000 × 90	2000 × 90	3000 × 90	2400 × 165
Ceiling diffuser					
(neck diameter)	100	150	200	250	380
(overall diameter)	330	330	450	600	860
Side-wall grille double deflection type (face size)	200 × 150	300 × 200	450 × 250	450 × 400	660 × 450

Note: above are all selected at approximately the same sound level of 40 dB.

Table 15.2 Typical maximum air change rates based on a cooling differential of 10 K

Device	*Air change per hour*
Side-wall grilles	8
Linear grilles	10
Slot and linear diffusers	15
Rectangular diffusers	15
Perforated diffusers	15
Circular diffusers	20

* Laboratory Reports 65, 71, 79, 81 and 83. Application Guides AG 1/74 and 2/75, and Technical Notes TN 3/76, 4/86 and 3/90. The principal authors are P. J. Jackman and M. J. Holmes.

Table 15.3 Typical maximum cooling temperature differentials

Application	*Temperature difference* (K)
High ceiling large heat gains/under window input	12
Low ceiling air handling luminaires/under window input	10
Low ceiling downward discharge	5
Floor discharge	5

selection of terminals for variable air volume applications to ensure that satisfactory distribution will be achieved through the full range of operating volumes.

The *Guide Section B3* gives useful rule-of-thumb guidance as to both the maximum air change rates that can be achieved using various devices and as to typical maximum cooling temperature differentials (supply air to room) for various applications. This is used here for Tables 15.2 and 15.3, respectively.

Air distribution performance (ADPI)

An air distribution performance index (ADPI), a method of assessing the effectiveness of an installation to meet set comfort criteria, based upon air velocity and air temperature has been established,* but is not in common use in British practice.

High-velocity supply fittings

The use of high velocity air distribution in ducts has been referred to in connection with the induction, dual-duct and variable air volume systems. High velocity air distribution may also be the most practical and economical method both in cost and space for use with any, otherwise normal, ventilation or air-conditioning system where extensive ductwork is involved. This is particularly the case in multi-storey buildings where duct sizes are much reduced by use of higher velocities. Duct velocities up to 30 m/s may be used, although it is more usual to limit these to the range of 15–20 m/s.

At the terminal end, it is necessary to break down the high velocity to low velocity for introduction into the room, and a silencing or attenuating box in some form is required: this subject has been referred to previously and illustrated in Chapter 14.

Combined lighting and air distribution

Mention has previously been made of the use of special fittings which enable some part of the heat generated by lighting to be dealt with at source before it enters the room. Early types of such fittings used 'boots' mounted to conventional lighting enclosures, as Figure 15.19, but air handling functions have been developed in an integrated design, as

* *ASHRAE Handbook*, Fundamentals, 1993.

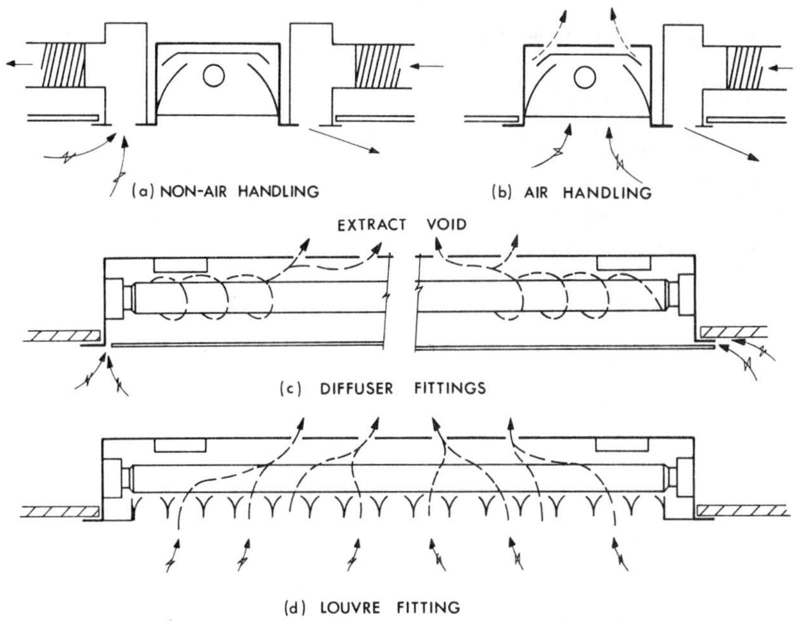

Figure 15.26 Air handling luminaires

previously referred to, Figure 15.26. The performance of the lighting apparatus can be improved by such arrangements, where the air extracted is drawn over the tubes thus producing temperature stability, and the heat entering the room may be reduced very considerably, as described in Chapter 18.

Dust staining

The streaks of dirt that appear sometimes at the point of air discharge from supply terminals are caused either by dust in the room becoming entrained in the air stream, or by dirty supply air. The former is the most usual cause.

Extract or return air grilles

The particular form of grille for extract is unimportant since air approach to a return fitting is no aid to distribution. It may, for instance, be of egg-crate plastic within an aluminium frame as Figure 15.27, or louvred, or any design giving the required free area. Dust collects on extract gratings on the outside, and close mesh or closely placed slats are undesirable as they quickly block up and impede the air flow.

The sizing of extract and return grilles is normally based on limiting noise levels. Recommended maximum grille face velocities are given in Table 15.4, but for critical applications reference must be made to manufacturers' selection data.

Where extraction takes place naturally from one space to another, it may be necessary to use a light-trap grille to avoid direct vision. Transfer grilles may be fitted with fire dampers where necessary. Figure 15.28 shows alternative fittings suitable for this purpose.

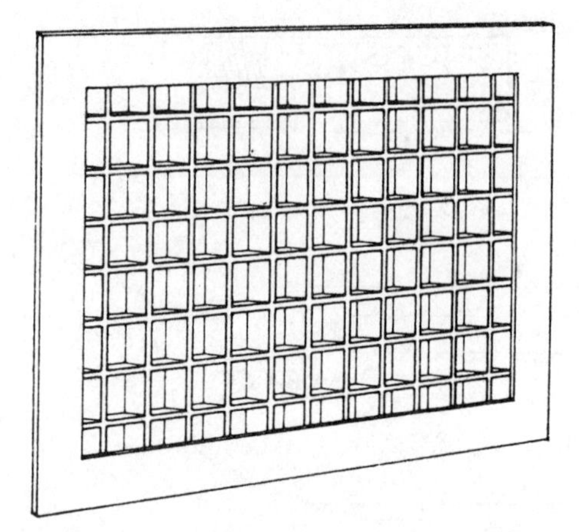

Figure 15.27 Standard 'egg-crate' register

Table 15.4 Typical maximum face velocities for extract grilles

Location	*Face velocity* (m/s)
Above occupied zone	4
In occupied zone	3
Door or wall transfer	1.5
Under-cut doors	1.0

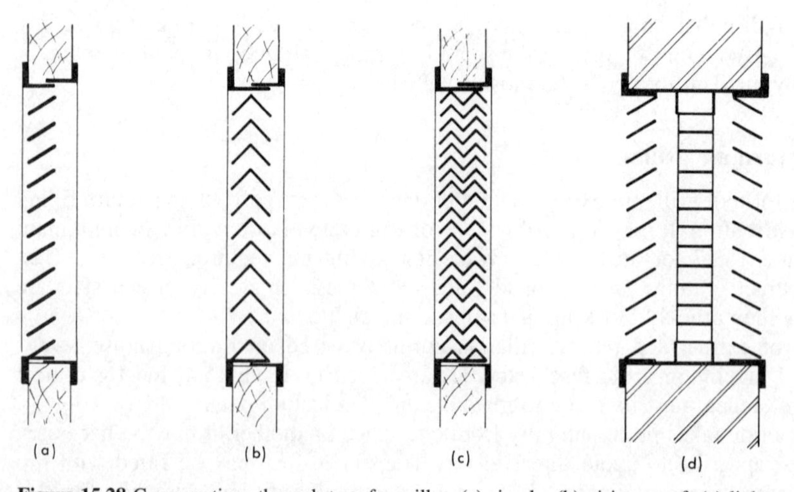

Figure 15.28 Cross-sections through transfer grilles: (a) simple; (b) vision proof; (c) light proof; (d) with fire damper

A method of extracting room air which has been used with some success is illustrated in Figure 15.29. It has been claimed that heat loss and gain to the occupied space is much reduced by passing room return air through the cavity of the double window. A glazing *U* value equivalent to about 0.6 $W/m^2 K$ is produced and, of course, hot and cold radiation, in summer and winter respectively, is reduced proportionately. Although this arrangement, with the blinds outside the glazing, provides the optimum result in terms of energy saving, costs relating to repairs to and cleaning of the blinds may be high. An alternative, which provides an equally effective but more accessible arrangement, is to position the blinds within the ventilated cavity between the inner and outer panes.

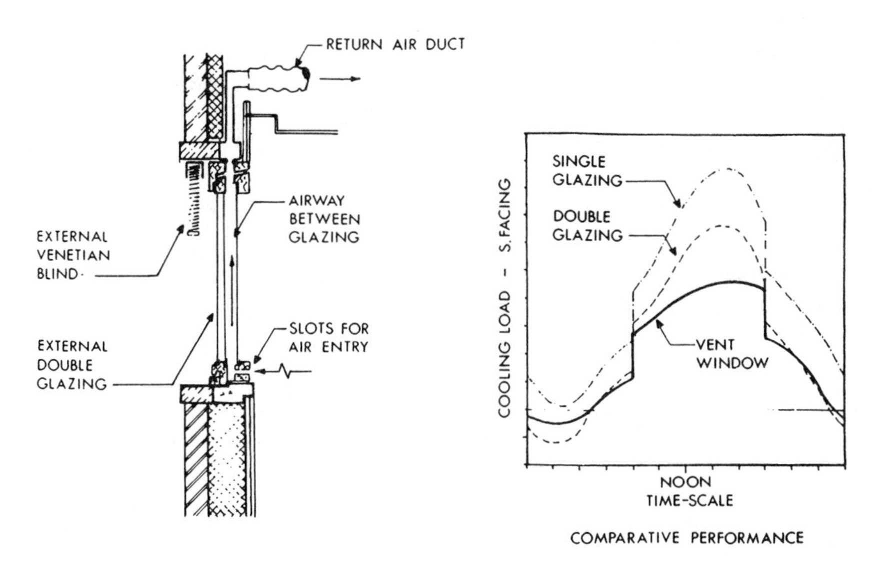

Figure 15.29 Ventilated double window

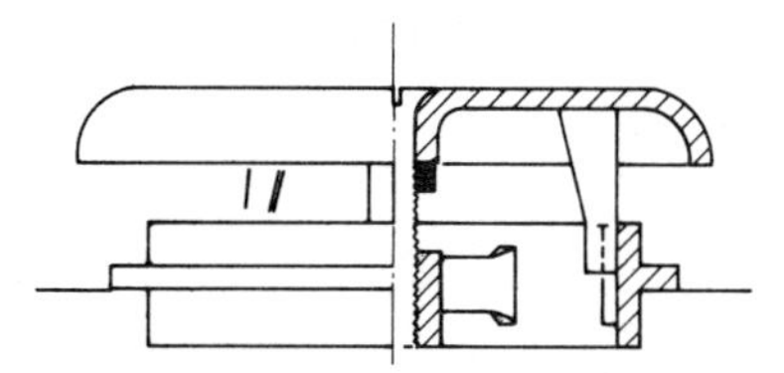

Figure 15.30 Mushroom type floor level ventilator

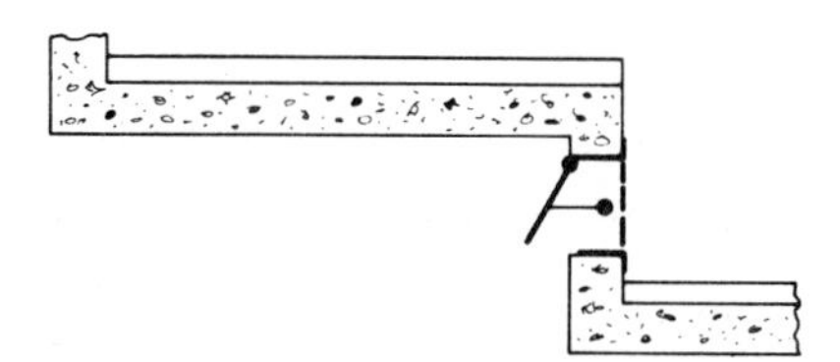

Figure 15.31 Ventilator in gallery steppings

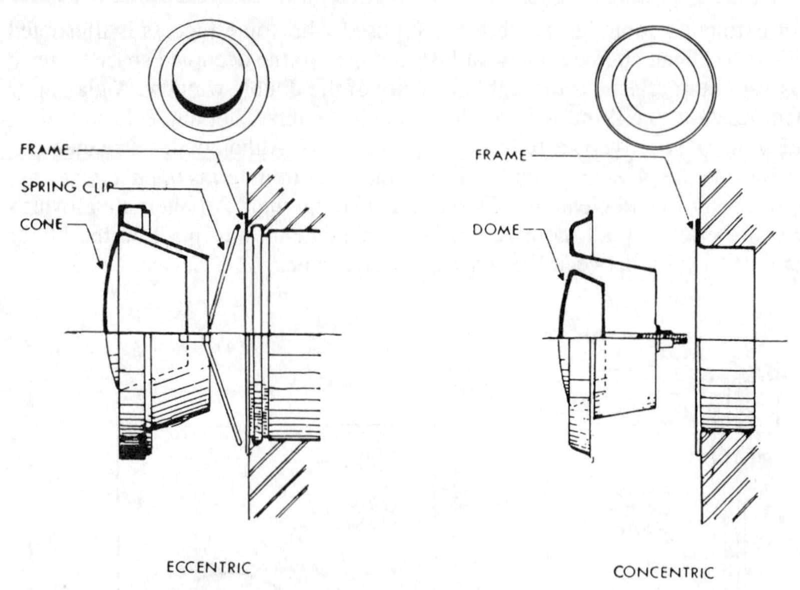

Figure 15.32 Mushroom type wall ventilators

Details of the mushroom ventilator and gallery-riser vent referred to earlier are shown in Figures 15.30 and 15.31. These types are more commonly used as extracts, though they may be used as inlets at low velocities.

Toilet extract

For this purpose, use may be made of the special wall-type mushroom ventilators shown in Figure 15.32. These combine a facility for air volume regulation with a degree of acoustic attenuation.

Chapter 16

Ductwork design

Having decided on the type of air system to be employed, made the calculations of air quantity and temperature, and considered the type and location of the air terminal devices and central plant, it is now necessary to consider in more detail the characteristics of the ductwork system which will convey the conditioned air about the building. Most of the discussion which follows applies equally to ventilating or air-conditioning systems.

One of the fundamental relationships is that between air speed and pressure, given by:

$$p_v = 0.5\,\rho v^2$$

where

p_v = velocity pressure (Pa)
ρ = specific mass of fluid (kg/m^3)
v = velocity (m/s)

for standard air

$$\rho = 1.2\,\text{kg/m}^3$$

thus

$$p_v = 0.6\,v^2$$

Figure 16.1 gives the relationship graphically for air velocities encountered in ventilation work. For other temperatures and pressures, correction is necessary:

$$p_{v2} = p_v(P/101.325)[293/(273 + t)]$$

where

P = alternative pressure (kPa)
t = alternative temperature (°C)

Other properties of air used in deriving the standard data for air flow in ducts are:

temperature = 20°C
atmospheric pressure = 101.325 kPa
relative humidity = 43%

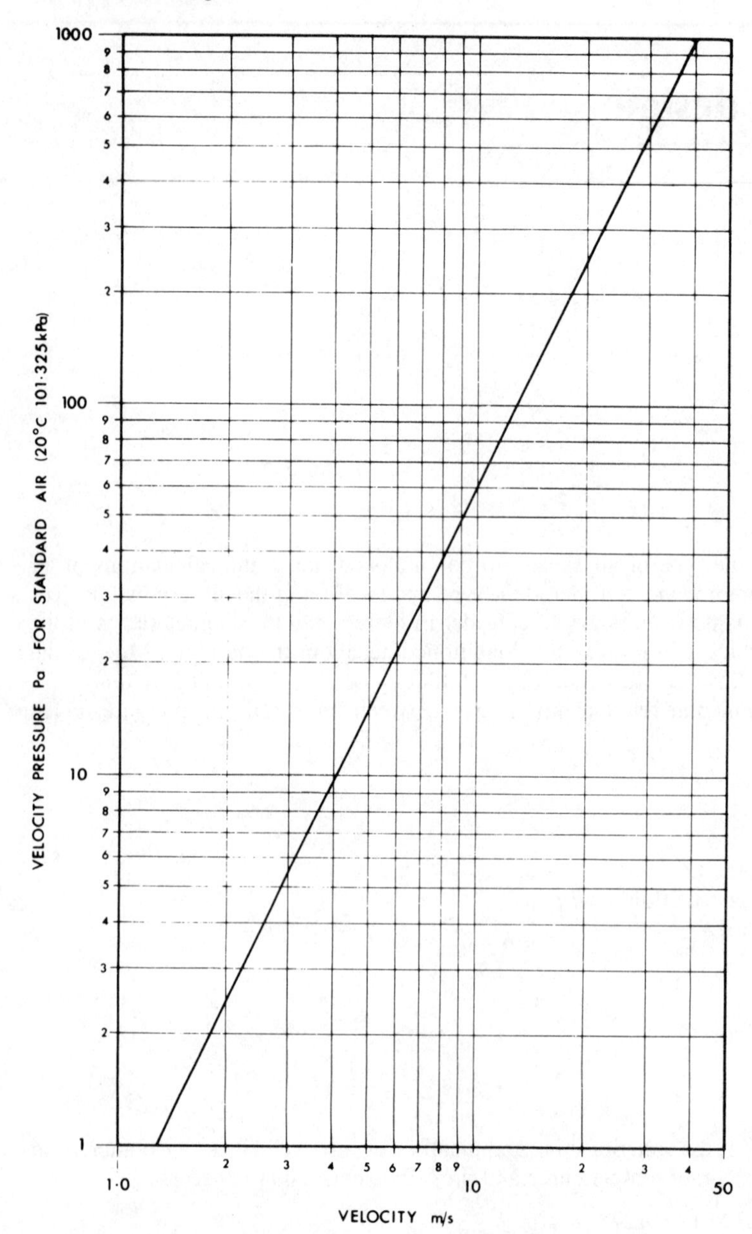

Figure 16.1 Velocity pressure related to air speed for standard air

Ductwork

Ducts for ventilation and air-conditioning are commonly constructed of galvanised sheet steel. Other materials, including *builders' work*, may be used and these are discussed later. The objective in duct design is to provide a system in which the air stream follows the line of the duct with no disturbance to the established velocity profile across the section and that excessive turbulance is avoided. The principles to be followed in the design of such ducts are:

- Avoidance of sudden restrictions or enlargements, or any arrangement producing abrupt changes of velocity.
- Bends to be kept to a minimum, but where required should be radiused, not mitred. Deflectors may be fitted to improve the air flow characteristics.
- Where branches occur, they should be taken off at a gradual angle to avoid abrupt turning.
- Sharp edges should be avoided, as these will be the cause of noise which may travel a considerable distance through the duct system.
- Rectangular ducts should be as nearly as possible square: the more they depart therefrom the more uneconomic they become.
- The ductwork configuration must allow the air flow to be regulated without the need for excessive throttling, which will lead to noise being generated.

These principles apply to all ductwork classifications but are clearly more important the higher the air velocity.

Ductwork classification

The current classification for ductwork is given in an HVCA specification* according to the operating pressure, air velocity and leakage rate. These are summarised in Table 16.1. The air leakage rates quoted are satisfactory for all general ventilation and air-conditioning applications, but where hazardous or obnoxious substances are handled, the ducting must be completely air-tight.

Table 16.1 Ductwork classifications

Duct pressure class	*Static pressure limit*		*Mean air velocity (max.)* (m/s)	*Air leakage limit*[a] (litre/s m^2 *duct surface area*)
	Positive (Pa)	*Negative* (Pa)		
Low	500	500	10	Class A – 0.027 $p^{0.65}$
Medium	1000	750	20	Class B – 0.009 $p^{0.65}$
High	2000	750	40	Class C – 0.003 $p^{0.65}$
	2500	750	40	Class D – 0.001 $p^{0.65}$

[a] p, differential pressure (Pa).

Galvanised steel ducts

The gauges of metal and form of construction commonly used are set out in the HVCA specification* for all of the classifications.

Ducts of galvanised sheet steel are often priced by weight. Sheet metal is made in metric thickness and the mass of metal will thus merely be thickness times area times

* Heating and Ventilating Contractors Association, Specification DW/142.

7800 kg/m^3 (for steel) or whatever other specific mass is appropriate. One of the aims when designing ductwork is to achieve as much standardisation and repetition in the configuration and components as is practical, since this will give a more economic installation.

Rectangular ducts are constructed of flat sheets by bending, folding and riveting and are erected in sections with slip joints or bolted angle ring joints. Larger sizes tend to drum and adequate stiffening is necessary. Sometimes a *diamond break* is used to assist in stiffening, as Figure 16.2, the sheets being pressed to provide, in effect, a very shallow pyramid form; alternative methods include *pleating* or *beading*. Where rectangular ducts are used in high velocity systems, stiffening is particularly important using bracing angles, tie angles, and internal tie rods for larger sizes. Aspect ratios in excess of 4:1 are to be avoided. Sizes should be selected from the range given in the HVCA specification.

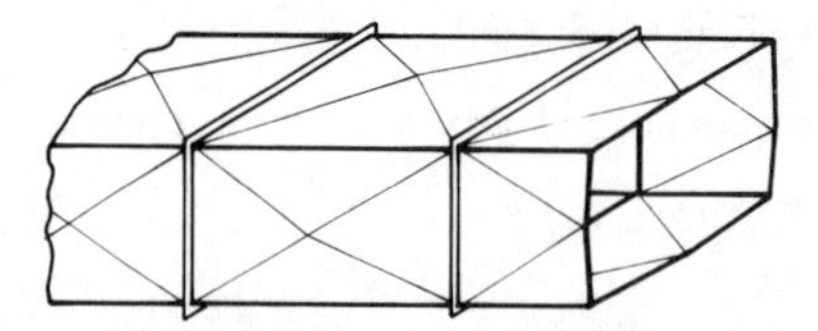

Figure 16.2 Diamond break stiffening for ducting

Circular ducts may similarly be formed from flat sheet with folded or riveted seams and are inherently free from risk of drumming. Traditionally, the use of good quality circular ductwork was largely confined to industrial applications. Mass produced *snap-lock* duct lengths were used for low cost domestic warm air heating, reliance being placed upon jointing tape for air-tightness.

The advent of circular ducts made from galvanised strip steel with a special locking seam, as in Figure 16.3, has led to their adoption generally. Sizes are available from 63 mm up to 1600 mm diameter. Ducts of this type are particularly suitable for high velocity systems due to their rigidity, 'deadness' and air-tightness. Ranges of standard tees, bends, reducers and other fittings are made and these, together with lengths of straight ducting cut off on site as required, facilitate the task of erection compared with the tailor-made methods previously adopted.

Jointing is by means of a special mastic, with some riveting, and in best practice a *heat shrink* plastic sleeve or a *chemical-reaction* tape is used externally to seal the joint.

Alternative methods of jointing ductwork are available, for example where the duct fittings have a groove formed at each end which houses a sealing gasket and locking collar. Connection is made by pushing the duct and fitting together which forces the locking collar into position to provide an 'air-tight' joint. Such forms of construction should be restricted in use strictly to manufacturers' recommendations.

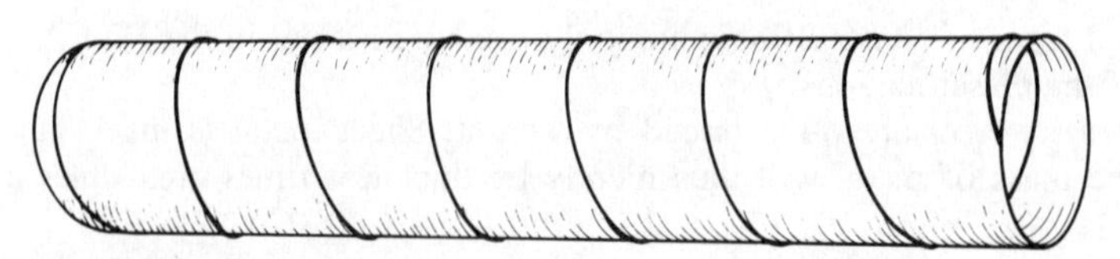

Figure 16.3 Spiral wound circular ducting

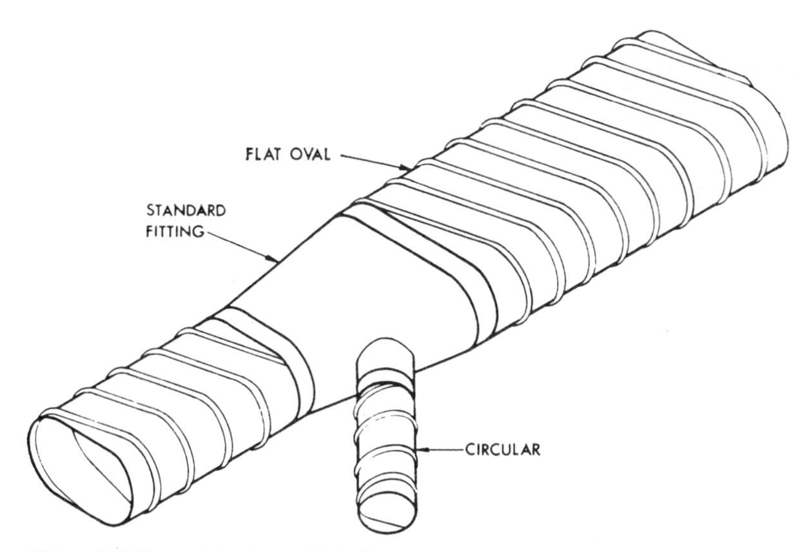

Figure 16.4 Flat oval ducting and branches

Flat oval or spirally wound rectangular ducts are in common use in air-conditioning. These are formed from spirally wound circular ducts to the profile illustrated in Figure 16.4 and are available in sizes from 75 mm × 320 mm to 500 mm × 1640 mm. A limited range of standard tees, bends and reducers is made. Such ducts have the advantage of being mass produced and can be used in conjunction with circular sections where space is limited. Site jointing, as before, is completed by a heat shrink or chemical reaction sleeve.

Builders' work ducts

The suggestion is sometimes put forward, in early stages of design for a building, that main air ducts and rising shafts should be constructed as a part of the fabric. The following advantages are claimed for such an expedient:

- Cheapness in that the space very often exists as a structural wind brace or in association with one or more lift shafts.
- Permanence.
- Reduction in heat gains and losses, due to the heavy construction as compared with metal.
- Rigidity and hence reduction in noise transmission, also as a result of the heavier construction.
- Accessibility for cleaning.

Several of these claimed advantages may have been valid half a century ago when the techniques of sheet metal duct construction were less well developed and when time lag in response to automatic control was of less significance. They must now be questioned and considered in the light of the proven difficulty of producing any form of builders' work construction which is reasonably air-tight.

As a result of structural settlement following completion; later movement as a result of either drying out or temperature changes; and the high permeability of masonry material

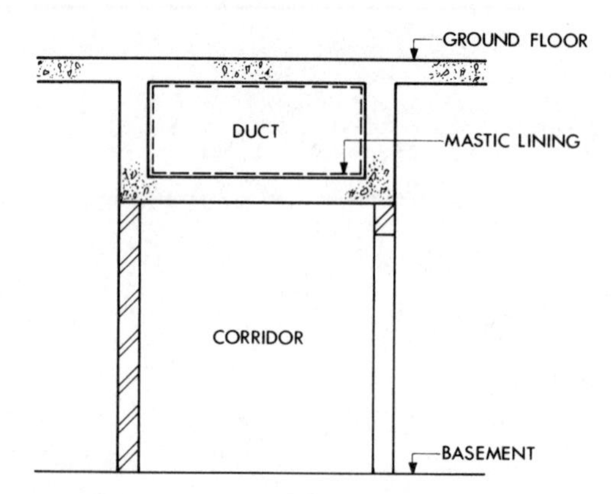

Figure 16.5 Structural air duct over corridor

of all descriptions, builders' work ducts are always suspect. They may be considered reasonably viable only when lined with a compound having some measure of elasticity and which has been applied by a specialist contractor. Following such treatment, air leakage may possibly be reduced but only to a level which is still many times greater than that which would be acceptable for low pressure sheet metal ductwork, as Table 16.1.

Consequent upon these difficulties, the use of builders' work construction should be considered only as a last resort and then for extract ducts alone, at low velocities and pressure differentials. Where the air extracted may be contaminated, such construction should not be used under positive pressure, i.e. on the delivery side of the fan.

Figure 16.5 shows an example of a duct constructed over a basement corridor and serving as a main distribution trunk from the plant to rising shafts about the building: such a construction cannot be recommended since it will, inevitably, not be air-tight.

Ducts of other materials

Other materials used for ducts are:

- *Welded sheet steel* is mainly used in industrial work, or where the air-tightness of the joints is of great importance; such ducts may be 'galvanised after made'.
- *Copper, aluminium, stainless steel.* Ducts constructed of these materials are used in special cases where permanence or a high degree of finish is required, or where special corrosion problems arise.
- *uPVC and Polypropylene* which are chiefly used for chemical fume extraction and tend to be expensive due to the cost of moulds, etc., for fittings and special sections. Reference HVCA specification DW 151. The Laboratory of the Government Chemist has developed a combination of uPVC coated externally with fire-retardant glass fibre reinforced filled polyester resin (GRP) for use with fume extract systems. This is inherently stronger and more durable than uPVC used alone.
- *Glass fibre resin bonded* slabs can be cut and adapted at site, and jointed to form a smooth continuous duct. This technique was favoured in the USA but has not been much used in this country although a HVCA specification exists (DW 191) to codify construction methods. The material is, of course, suitable for the lowest pressures and velocities only and will not, even then, be air-tight.

- *Chlorinated rubber paint or PVC coating* protection to galvanised steel is recommended for use in swimming pool applications.*

A comprehensive list of materials and their application is given in a CIBSE Technical Memorandum.†

Ductwork components and auxiliaries

To achieve a distribution system that will function efficiently, be capable of regulation, and be easily maintained, it is necessary to incorporate a variety of components to assist towards these objectives. The following gives a brief overview of some of these items.

Ductwork fittings

Figure 16.6 serves to illustrate the basic forms that bends can take. *Radius bends* are to be preferred, and where practical a centre-line radius not less than 1.5 times the diameter or duct width should be provided (Figure 16.6(a)). Splitters (as shown in Figure 16.6(b)) should be fitted where it is important to maintain the upstream air velocity profile around the bend, for example at approach to equipment or measuring points. *Mitre bends*, without turning vanes, (Figure 16.6(c)), should not be used since pressure drop and noise generation are high. Mitre bends with turning vanes (Figure 16.6(d)) are quite satisfactory so long as the turning vanes are of the correct design and are constructed to a good standard. Radius bends alone should be used on medium and high pressure/velocity systems.

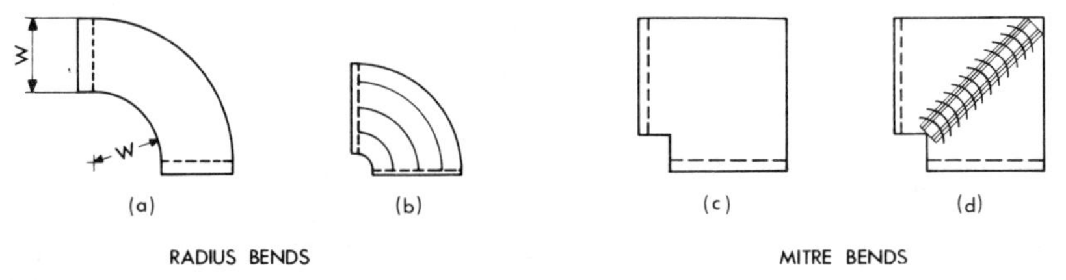

Figure 16.6 Typical ductwork bends

Branch connections should be constructed such that the air flow is divided, or combined in the case of an extract system, with minimum turbulence. Changes of duct size should be made downstream, or upstream, of the fitting, not at the fitting. Typical branch connections are shown in Figure 16.7. Branches on medium and high pressure/velocity systems should be taken off at 45°, or be made to a recommended pattern for such applications.

At a *change of section* in a duct, the aim is to maintain the air stream attached to the duct sides. If the air becomes detached, high turbulence occurs with resulting high pressure loss, noise and a non-uniform air velocity profile. At an expansion, the included angle should be limited to 30°, above this internal splitters should be used, as in Figure 16.8. On duct contractions the angle of taper is less critical, but 40° should be considered a limit for good design.

* *Corrosion of Heat Recovery Exchangers Serving Swimming Pool Halls.* Electricity Council Research Report, August 1961.

† *Design Notes for Ductwork.* CIBSE Technical Memorandum TM8, 1983.

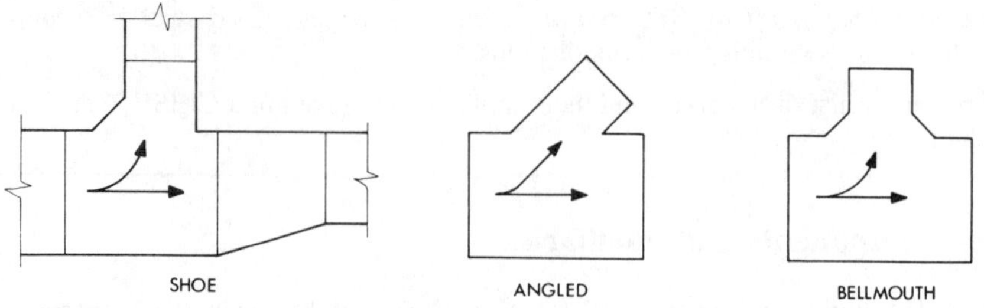

Figure 16.7 Typical ductwork branch connections

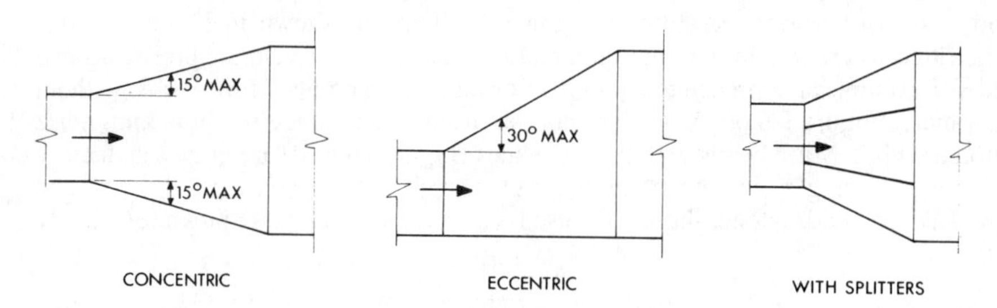

Figure 16.8 Typical ductwork expansion and contraction sections

Fan and plant connections

Catalogue fan performance figures will not be obtained in practice if the inlet air stream is other than uniform and without turbulence. At the fan discharge, all transitions should be gradual, and bends closer than three duct diameters should be arranged to turn in such a manner that the velocity profile from the fan is maintained. Air flow upstream of plant items such as filters, humidifiers, heating and cooling coils must be uniform across the whole cross-section of the equipment, otherwise performance will be reduced to below the rated duty.

There are too many possible combinations of duct configuration to show here, and reference should be made to *CIBSE TM 8*.

Data have been established* to enable the reduction in fan performance arising from poor intake or discharge ductwork configurations to be calculated. This is expressed as an additional system resistance to be added to the ductwork and component pressure loss and is termed *system effect.*

system effect (Pa) = system effect factor × velocity pressure at the fan intake or discharge (Pa)

Some typical system effect factors are given in Figure 16.9.

* *Fans and Fan Installations,* Flakt, 1978.

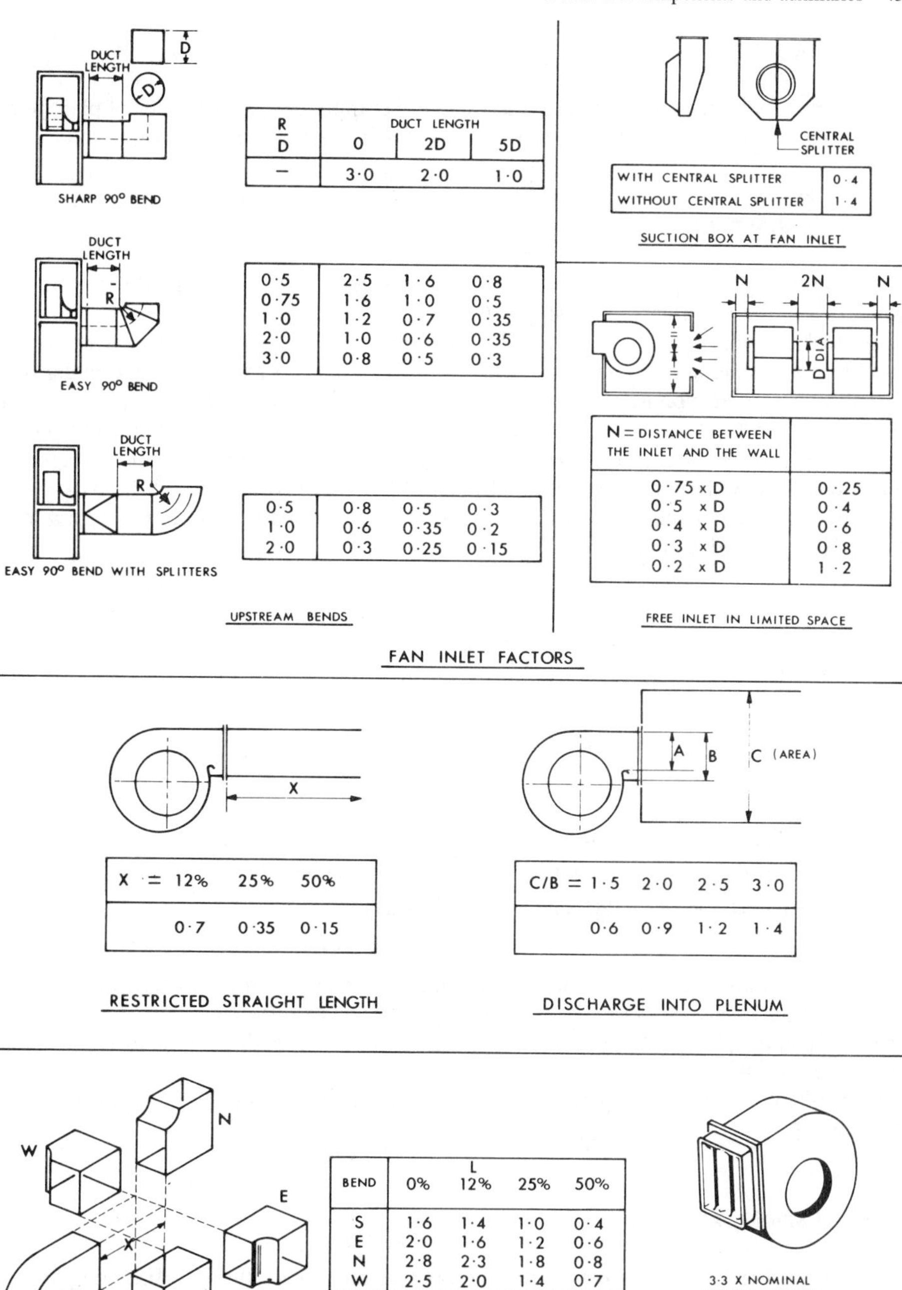

Upstream bends (sharp 90° bend):

R/D	Duct length 0	2D	5D
–	3·0	2·0	1·0

Upstream bends (easy 90° bend):

R/D	0	2D	5D
0·5	2·5	1·6	0·8
0·75	1·6	1·0	0·5
1·0	1·2	0·7	0·35
2·0	1·0	0·6	0·35
3·0	0·8	0·5	0·3

Upstream bends (easy 90° bend with splitters):

R/D	0	2D	5D
0·5	0·8	0·5	0·3
1·0	0·6	0·35	0·2
2·0	0·3	0·25	0·15

Suction box at fan inlet:

WITH CENTRAL SPLITTER	0·4
WITHOUT CENTRAL SPLITTER	1·4

Free inlet in limited space:

N = DISTANCE BETWEEN THE INLET AND THE WALL	
0·75 x D	0·25
0·5 x D	0·4
0·4 x D	0·6
0·3 x D	0·8
0·2 x D	1·2

Restricted straight length:

X =	12%	25%	50%
	0·7	0·35	0·15

Discharge into plenum:

C/B =	1·5	2·0	2·5	3·0
	0·6	0·9	1·2	1·4

Bends:

BEND	L 0%	12%	25%	50%
S	1·6	1·4	1·0	0·4
E	2·0	1·6	1·2	0·6
N	2·8	2·3	1·8	0·8
W	2·5	2·0	1·4	0·7

Figure 16.9 System effect factors

Access openings

Access into ductwork should be provided at every component, at both sides of plant items, and at regular intervals along the run to allow for inspection and cleaning.

Balancing dampers

Permanently set dampers are used in low pressure systems to regulate the air flow quantities at the commissioning stage. These are installed downstream of the fan in the main duct, in every branch duct from the main duct, in every sub-branch serving three or more terminals and at every terminal. The butterfly type is normally used in circular ducts and a multi-bladed type in rectangular sections, with some form of locking devices. The contra-rotating type, commonly called opposed blade dampers, is superior from the regulating point of view, to the type where all vanes rotate in the same direction. A selection of the more popular types of damper is shown in Figure 16.10. Where tight shut-off is essential, it is necessary for the dampers to be felt-tipped and to close on to a felted frame, and with edge seals.

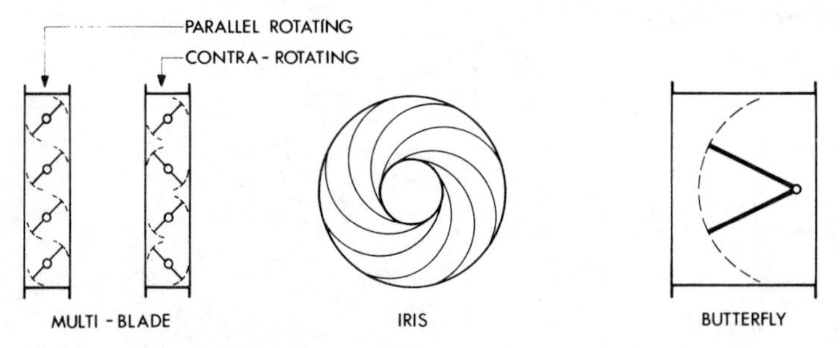

Figure 16.10 Typical balancing dampers for insertion in ducting

Dampers should only be installed in medium or high (pressure) velocity systems where absolutely necessary. Where these must be provided, it will probably be necessary to provide attenuators upstream and downstream together with protection to reduce noise breakout from the duct.

Controllable dampers, adjusted either manually or automatically, are dealt with in Chapter 22.

Fire and smoke dampers

In principle, fire dampers must be installed where ducts pass through fire compartment walls, floors or other elements, with the possible exception of low level penetrations less than 0.013 m^2 in section. Ideally, dampers should be built in to the compartment element. Where this is not practical, the damper should be located close to the element and connected to it by 6 mm thick (minimum) steel plate with flanged joints. Fire/smoke dampers may also be required where ducts penetrate smoke barriers in ceiling voids. The requirements may be relaxed in toilet blocks, where a 'shunt' duct arrangement, as shown in Figure 13.11, may be an acceptable alternative.

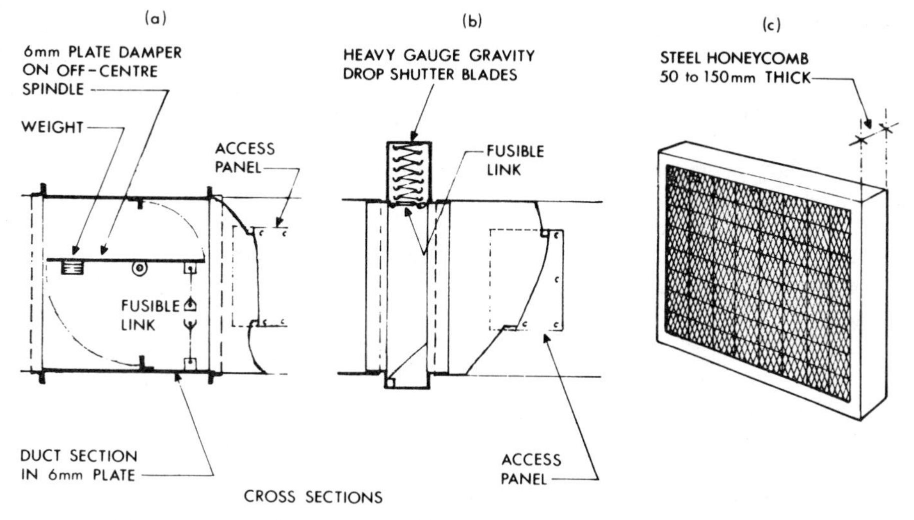

Figure 16.11 Typical construction methods for fire dampers

Figure 16.11 illustrates some typical fire damper constructions. The swing blade type, Figure 16.11(a), consists of a heavy steel casing with the steel damper blade kept open by a fusible link. In the event of a fire the link melts and the damper swings closed. Another pattern, Figure 16.11(b), takes the form of a number of substantial shutter blades retained out of the air stream by a fusible link as before; this is normally referred to as a 'curtain' type. A different form of fire damper is one where a steel honeycomb grid is inserted into the air duct, as Figure 16.11(c), all surfaces of which are coated with an *intumescent paint*. This material has the property of swelling to many times its original volume when heated and thus forming a barrier to air flow.

Combination fire and smoke dampers are designed to function both as a fire damper, operated by a fusible link or thermal actuator, or as a smoke control damper powered by an actuator from a remote signal. Remote indication of the damper position can also be provided.

Smoke release dampers may also be required at air inlets and exhausts. Operation normally would be a fail-safe spring loading, via a remote signal, which would cause the damper to open thereby allowing any smoke in the system to discharge to atmosphere.

Flexible ductwork

This is available in treated fabric on a steel helix of various materials and finishes, and in spirally wound 'bendable' metal of various materials. Selection is made for the particular application. Flexible ductwork is normally used at the final connection to air terminals and at connections to variable air volume and mixing boxes, induction units and the like. The length of these connections should be limited to 600 mm, and never be more than the equivalent of six times the duct diameter.

Whatever the choice of material, it must meet the fire requirements, and the overall air leakage and frictional resistance criteria for the system.

Pressure distribution in ducts

Static pressure

When air is moved in a duct or through a filtering, heating, cooling or humidification plant, a resistance to flow is set up.

The air is slightly compressed by the fan on its outlet side, so setting up a *static pressure* in the duct or plant. This pressure is tending to 'burst' the duct, and may be read by means of a U-tube partly filled with water, connected at right-angles to the air stream at any point in the duct: this is called a *side gauge* (Figure 16.12(a)). On the suction side of the fan the static pressure is negative with respect to the surrounding atmosphere, tending to collapse the duct.

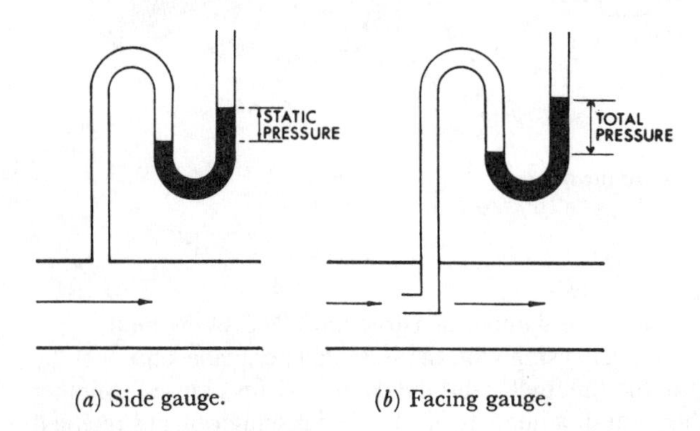

(*a*) Side gauge. (*b*) Facing gauge.

Figure 16.12 Air pressure gauges

As the air proceeds along the duct from the fan, the compression is released gradually until at the end of the duct open to atmosphere, the air is at atmospheric pressure. This falling away of the static pressure proportionately with the length of travel is called the *resistance* of the duct. Similarly all obstructions, such as heaters, filters, dampers, etc., cause a loss of pressure when air is passing through them.

It should be noted that as the static pressure becomes reduced, the air in effect expands such that pressure × volume = a constant (or nearly so, as explained earlier). This expansion therefore signifies an increase in velocity of the air if the size of the duct is unchanged.

Velocity pressure

A fan, in addition to generating static pressure, supplies the force to accelerate the air and give it velocity. This force is termed the *velocity pressure*, and is proportional to the square of the velocity (see Figure 16.1). It is measured by a 'U' tube connected to a pipe facing the direction of air flow in a duct, etc., which is called a *facing gauge* (see Figure 16.12(b)). But, obviously, the pressure so measured will in addition include the static pressure which occurs throughout the duct, as mentioned earlier, and the reading so obtained will thus be the *total* pressure. Thus, the velocity pressure alone may be found by deducting the static pressure from the total pressure reading, or by connecting one side of the U-tube to the facing gauge and the other to the side gauge, provided that the two

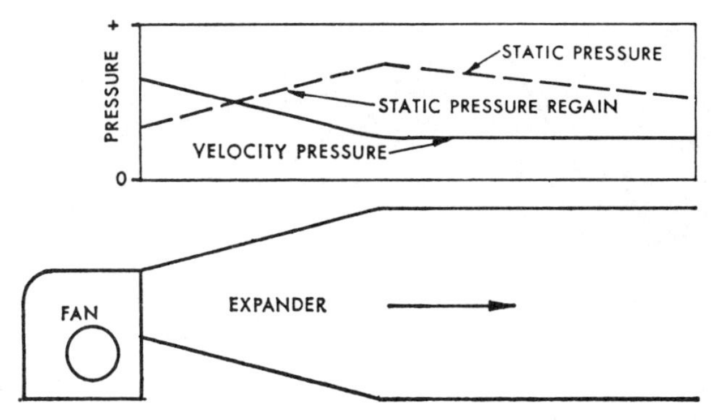

Figure 16.13 Velocity and pressure changes at an expander

gauges are at the same point. A Pitot tube, as illustrated in Figure 16.32, combines a side gauge with a facing gauge in one standard instrument.

If a fan discharges into an expanding duct (Figure 16.13), the velocity will obviously decrease as the distance from the fan increases, and at the same time the velocity pressure will be converted into static pressure (not at 100% efficiency, but about 75% if the expansion is sufficiently gradual). If the fan discharges into a large box (Figure 16.14), from which at some point a duct connects, the fan velocity pressure will be entirely lost in eddies, and at the duct entrance must be recreated by a corresponding reduction in static pressure.

Fan pressure

In all air flow considerations as affecting resistances of ducts, plant, etc., it is the static pressure alone which is of importance, this being the pressure which changes with such restrictions. It is the static pressure set up by a fan which is, therefore, a criterion of its performance. The velocity pressure, if taken as supplementing the fan duty, may be more misleading than useful, owing to the uncertainty of friction losses which occur at points

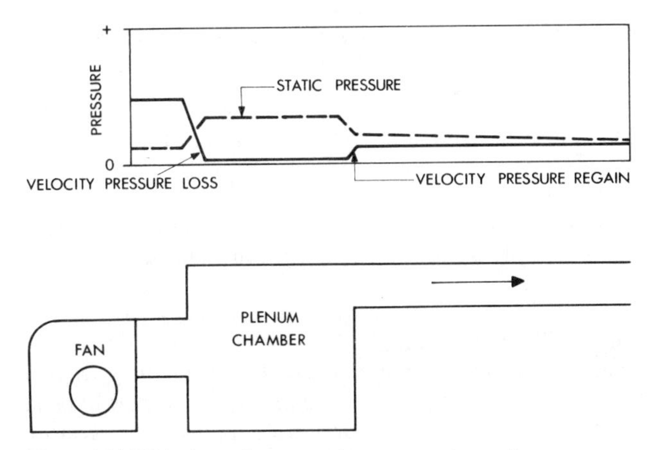

Figure 16.14 Velocity and pressure changes at a plenum box

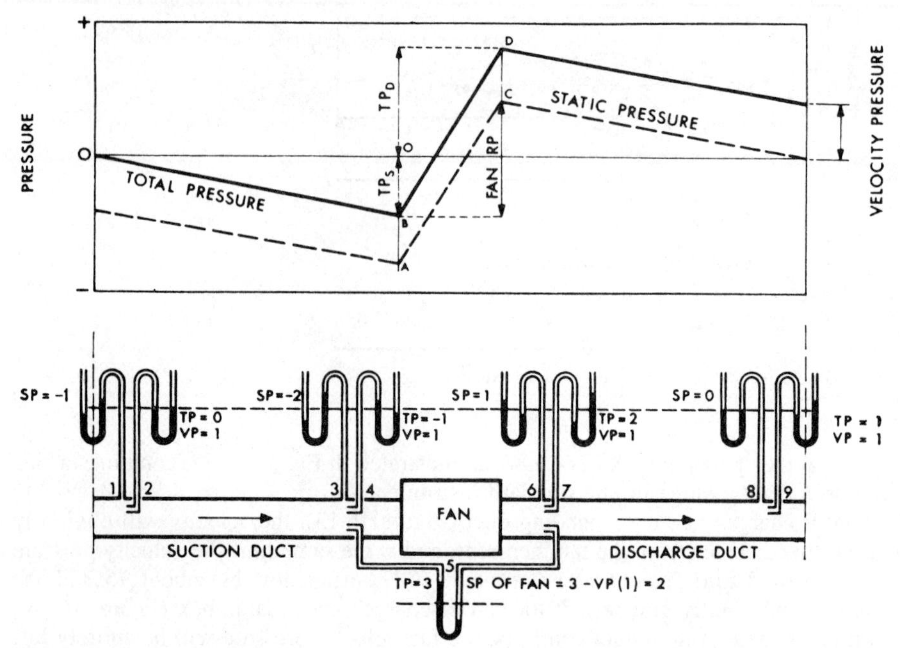

Figure 16.15 Total and static pressure changes at a fan

of varying velocities. The velocity pressure is more generally not recovered, though sufficient must remain at the duct termination to eject the air at the required velocity.

Where, however, by careful design of the fan discharge expander, the velocity pressure is converted to static pressure (probably to the extent of about 75%), this additional pressure may be reckoned as augmenting the static pressure of the fan.

The pressure generated by a fan may be better understood by study of Figure 16.15:

- The total fan pressure is defined as the algebraic difference between the mean total pressure at the fan outlet and the mean total pressure at the fan inlet.
- The total pressure on the suction side, as will be seen from this figure, is TP_S, i.e. the negative pressure *AO* minus the velocity pressure equivalent to *AB*.
- The total pressure TP_D on the discharge side is similarly the static pressure *OC* plus the velocity pressure *CD*.

It has previously been explained that we are concerned only with the resistance pressure set up by the fan and this, as may be seen, is the difference in pressure of points *B* and *C*.

We can arrive at the static pressure by measuring the total pressure of the fan as by the U-tube 5, and deducting therefrom the velocity pressure as given by difference between gauges 6 and 7. Such a method is valid only if the velocity in suction and discharge ducts is the same.

The other U-tubes indicate the pressures at the various points along suction and discharge ducts, and their meaning will be apparent.

If a fan has suction ducting only, the static pressure produced for overcoming friction of ducting will be represented by *OB* (a negative pressure), since the discharge will be at atmospheric pressure. Similarly, if the fan has only discharge ducting, the static pressure will be represented by *OC*, the suction being at atmospheric pressure.

Flow of air in ducts

The static pressure which a fan will produce is utilised as the motive force required to overcome the sum of all resistances to air flow throughout the system of air ducting to which the fan is connected. This book is not the place for a dissertation upon the theory of air flow in ducts save to say that an air stream moving within a long straight duct develops a constant and smooth velocity profile, slower adjacent to the perimeter, due to a drag effect from the walls, and faster at the centre. Any changes in direction; junctions and variations in the cross-section of the duct break up this profile and lead to turbulence and added resistance. Generally speaking, therefore, all resistance to flow results in a drop in static pressure and may be considered as falling into one of three categories:

- That due to the frictional resistance arising from air flow through lengths of straight ductwork made of whatever material.
- That due to individual ductwork fittings such as changes in direction, contractions, enlargements and branches, etc.
- That arising from plant and other built-up items which have characteristics unique to one specialist manufacturer or another.

In many respects, air flow in ducts is analogous to water flow in pipes but an important difference exists in practice in that, whereas pipes are available in a commercial range of diameters, air ducting is more usually purpose made up from flat sheet metal to bespoke sizes and cross-sections. Exceptions to this rule are, of course, spiral-wound circular ducts and the parallel flat-oval sections.

Straight ductwork

To represent resistance, and consequent pressure loss due to air flow in straight circular ducting, the *Guide Section C4* presents a chart, as Figure 16.16. The base scale of the chart is graduated in units of pressure loss (Pa per metre run) and the vertical scale in units of air flow (litre/s). Diameters of circular ducts in mm and consequent velocities in m/s may be read directly. For example, a volume flow of 200 litre/s through a duct of 225 mm diameter, will be at a velocity of 5 m/s and will produce a unit pressure loss of 1.4 Pa/m.

An economic rate of pressure loss takes into account the capital cost of ductwork and balances this against the recurrent cost of energy to develop the fan pressure. Traditionally, a unit pressure loss of between 0.75 and 1.25 Pa/m has been the aim but this may well change as new energy conservation criteria are developed.

The chart, strictly speaking, applies only to ducts manufactured from clean galvanised mild steel sheet: thus, for other materials, pressure loss read from the chart must be corrected, *up or down*, by use of the factors listed in Table 16.2. Since, in the majority of duct installations at low velocity, square or rectangular ducts are used in order to suit building configurations, some method of representing these as *equivalent diameters* to suit the chart is required. Without going into detail, the relationship is complex since both the cross-sectional area and the internal circumference of the alternative shapes must be taken into consideration. A simplified method of calculating a diameter is available, using the expression and values given in Table 16.3. For flat-oval ducts, reference should be made to the 1970 edition of the *Guide Section C4*.

Ductwork fittings

The conventional method of dealing with the pressure loss arising from the presence of all manner of duct fittings, is to evaluate the effect of each separately as a fraction of the

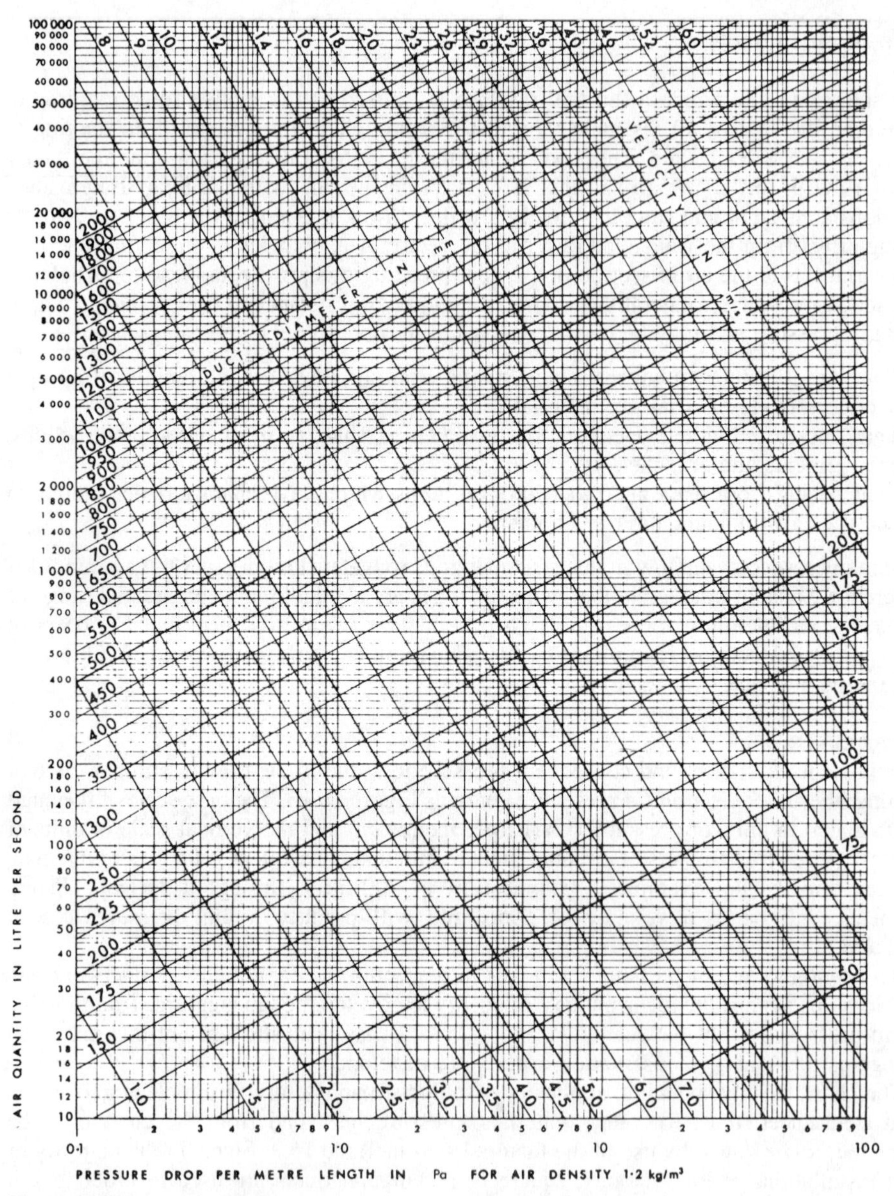

Figure 16.16 Duct sizing chart

Table 16.2 Correction factors for ducts of other materials

uPVC	0.85
Aluminium	0.9
Plastered, smooth cement	1.1
Fair faced brick, concrete	1.5
Rough brickwork	2.5

Table 16.3 Equivalent diameters (*d*) of rectangular ducts (*a* × *b*) for equal volume and pressure drop

$\frac{a}{b}$	x	$\frac{a}{b}$	x	$\frac{a}{b}$	x	$\frac{a}{b}$	x	$\frac{a}{b}$	x	$\frac{a}{b}$	x
1.0	0.908	1.6	0.710	2.2	0.622	2.8	0.557	3.4	0.510	4.0	0.475
1.1	0.865	1.7	0.701	2.3	0.608	2.9	0.548	3.5	0.504	4.2	0.465
1.2	0.829	1.8	0.683	2.4	0.597	3.0	0.540	3.6	0.498	4.4	0.455
1.3	0.798	1.9	0.665	2.5	0.586	3.1	0.532	3.7	0.491	4.6	0.447
1.4	0.770	2.0	0.650	2.6	0.576	3.2	0.524	3.8	0.486	4.8	0.438
1.5	0.745	2.1	0.635	2.7	0.566	3.3	0.517	3.9	0.480	5.0	0.431

$d = 1.265 \left[\frac{(ab)^3}{(a + b)} \right]^{0.2}$ and thence $b = xd$, where b is the shorter side

velocity pressure existing at the position in the system where they occur. A selection of pressure loss factors, ζ, is illustrated in Figure 16.17 for use in conjunction with velocity pressure values read from Figure 16.1, thus:

Factor for a simple 90° bend having an aspect ratio, *H/W*, of 0.5 and a throat radius, *R*, of 0.5	= 0.29
If the air velocity in the duct is 4 m/s then, from Figure 16.1, the velocity pressure	= 9.5 Pa
Thus, pressure loss due to bend = 0.29 × 9.5	= 2.8 Pa

Plant components

For plant components; heater batteries; cooling coils; air filters and the like no generalisation is possible since the resistance to air flow will vary from manufacturer to manufacturer and may well be a function of performance. For a given air flow quantity, pressure loss resulting from such items may well be relatively large. They should be quoted to the designer directly in Pa.

Application

The mean air velocities quoted in Table 16.1 are intended to be broad-band classifications for various qualities of ductwork construction: they are *not* system definitions. For use in design, the data listed in Table 16.4 should be used for conventional low velocity ducting systems. In each case the value quoted is arbitrary but takes general account of the fact that the higher the velocity, the greater the risk of noise disturbance in the space served. For high velocity systems, the situation is rather different since some noise is inherent in the concept in this case: Table 16.5 lists practical values for maximum air velocities in such systems. The routine of carrying out duct sizing calculations depends upon a systematic approach and in all cases it is best to make use of a purpose designed *pro-forma* which, for present and future reference, lists at the very least:

- The type of system.
- The material to be used for the ductwork.
- The method of calculation.

- Any allowances included for commissioning, etc.
- Headings as listed below.

Location section	*Volume*		*Equivalent diameter*	*Rect-angular size*	*Air velocity*	*Pressure drop/m*	*Duct length*	*Fitting*		*Velocity pressure*	*Pressure drop*
	Design	*Revised*						*Type*	ξ		
(a)	(b)	(c)	(d)	(e)	(f)	(g)	(h)	(j)	(k)	(l)	(m)

Table 16.4 Maximum velocities for low pressure ducting systems

Application	*Velocity* (m/s) *Main duct*	*Branch duct*
Domestic	3	2
Theatres, auditoria, studios	4	3
Hotel bedrooms, conference halls, operating theatres	5	3
Private offices, libraries, cinemas, hospital wards	6	4
General offices, restaurants, department stores	7.5	5
Cafeteria, supermarkets, machine rooms	9	6
Factories, workshops	10–12	7.5

Table 16.5 Maximum air velocities for medium and high ducting pressure systems

Air quantity (litre/s)	*Velocity* (m/s) *Medium*	*High*
Less than 100	8	9
100–500	9	11
500–1500	11	15
Over 1500	15	20

Low velocity systems

The most convenient method of sizing ducts in a low velocity system is by using the *equal pressure loss* basis. For this purpose, it is necessary to establish the maximum velocity in the main duct leaving the fan from Table 16.4 and, knowing the air volume to be carried, locate the intersection point of the volume and velocity lines on the chart, Figure 16.16. From this point, a line followed vertically downwards will indicate the unit pressure loss, Pa/m, on the base scale. The main duct diameter may be read, at the same time, from the chart.

If the unit pressure loss so found were to be considered acceptable, the remainder of the duct sections following on (or those in the most disadvantaged duct run, the *index run*, in a system having many such) would be sized using the same unit rate. However, if that rate

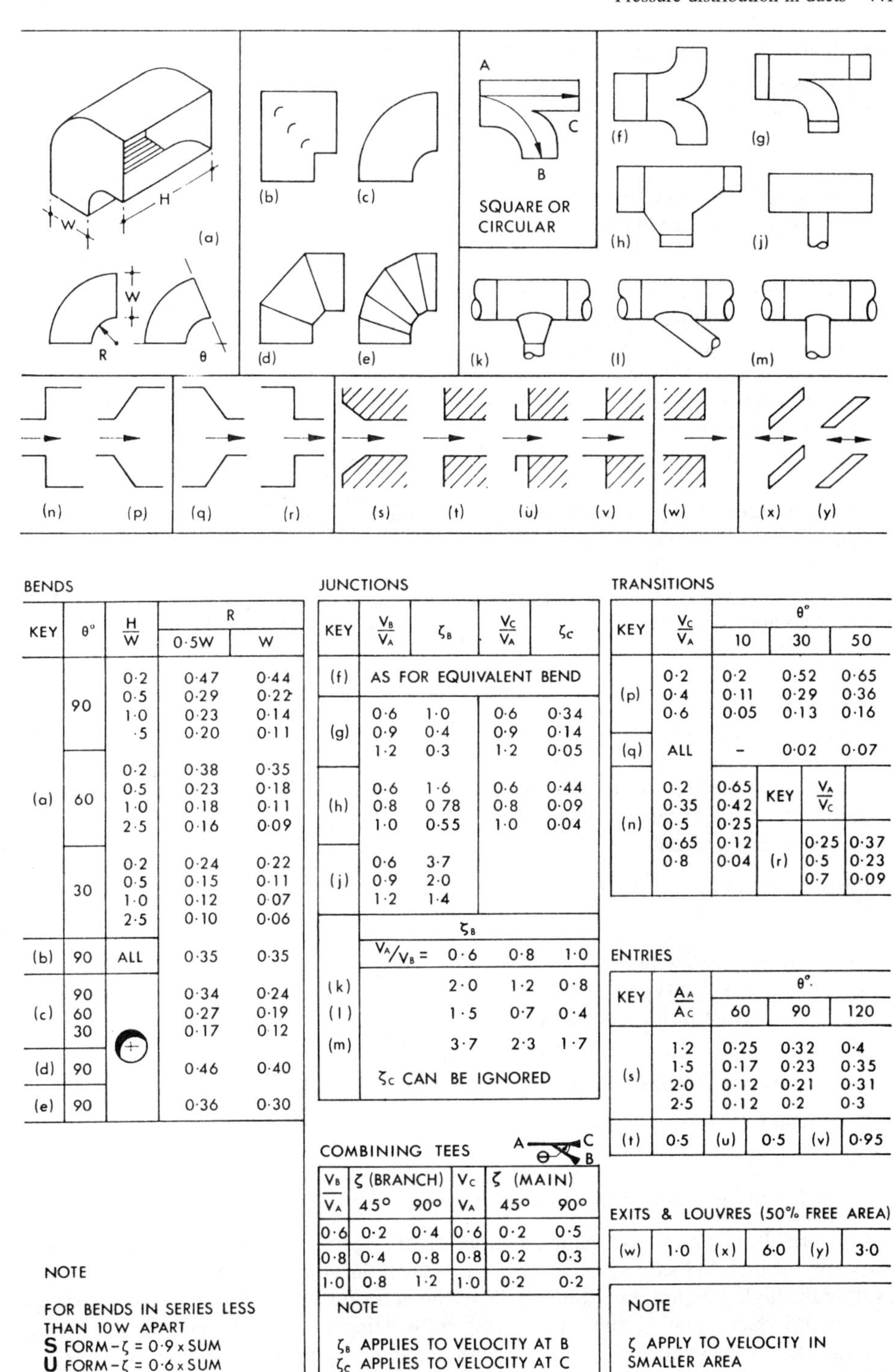

BENDS

KEY	θ°	H/W	R: 0·5W	R: W
(a)	90	0·2	0·47	0·44
		0·5	0·29	0·22
		1·0	0·23	0·14
		·5	0·20	0·11
	60	0·2	0·38	0·35
		0·5	0·23	0·18
		1·0	0·18	0·11
		2·5	0·16	0·09
	30	0·2	0·24	0·22
		0·5	0·15	0·11
		1·0	0·12	0·07
		2·5	0·10	0·06
(b)	90	ALL	0·35	0·35
(c)	90		0·34	0·24
	60		0·27	0·19
	30		0·17	0·12
(d)	90		0·46	0·40
(e)	90		0·36	0·30

NOTE

FOR BENDS IN SERIES LESS THAN 10W APART
S FORM – ζ = 0·9 × SUM
U FORM – ζ = 0·6 × SUM

JUNCTIONS

KEY	V_B/V_A	ζ_B	V_C/V_A	ζ_C
(f)	AS FOR EQUIVALENT BEND			
(g)	0·6	1·0	0·6	0·34
	0·9	0·4	0·9	0·14
	1·2	0·3	1·2	0·05
(h)	0·6	1·6	0·6	0·44
	0·8	0·78	0·8	0·09
	1·0	0·55	1·0	0·04
(j)	0·6	3·7		
	0·9	2·0		
	1·2	1·4		

KEY	ζ_B: V_A/V_B = 0·6	0·8	1·0
(k)	2·0	1·2	0·8
(l)	1·5	0·7	0·4
(m)	3·7	2·3	1·7

ζ_C CAN BE IGNORED

COMBINING TEES

V_B/V_A	ζ (BRANCH) 45°	ζ (BRANCH) 90°	V_C/V_A	ζ (MAIN) 45°	ζ (MAIN) 90°
0·6	0·2	0·4	0·6	0·2	0·5
0·8	0·4	0·8	0·8	0·2	0·3
1·0	0·8	1·2	1·0	0·2	0·2

NOTE

ζ_B APPLIES TO VELOCITY AT B
ζ_C APPLIES TO VELOCITY AT C

TRANSITIONS

KEY	V_C/V_A	θ° 10	θ° 30	θ° 50
(p)	0·2	0·2	0·52	0·65
	0·4	0·11	0·29	0·36
	0·6	0·05	0·13	0·16
(q)	ALL	–	0·02	0·07
(n)	0·2	0·65		
	0·35	0·42		
	0·5	0·25		
	0·65	0·12		
	0·8	0·04		

KEY	V_A/V_C	
(r)	0·25	0·37
	0·5	0·23
	0·7	0·09

ENTRIES

KEY	A_A/A_C	θ° 60	θ° 90	θ° 120
(s)	1·2	0·25	0·32	0·4
	1·5	0·17	0·23	0·35
	2·0	0·12	0·21	0·31
	2·5	0·12	0·2	0·3

(t)	0·5	(u)	0·5	(v)	0·95

EXITS & LOUVRES (50% FREE AREA)

(w)	1·0	(x)	6·0	(y)	3·0

NOTE

ζ APPLY TO VELOCITY IN SMALLER AREA

Figure 16.17 Velocity pressure loss factors for ducting fittings

were thought to be too high or too low then another velocity might be chosen as the datum using the expression:

$$V = Q/A$$

where

V = the chosen velocity (m/s)
Q = air volume leaving fan (m^3/s)
A = duct area (m^2)

In a complex system, some attempt should be made to balance the pressure loss through other duct runs with that of the index run but velocity and possible noise generation would have to borne in mind if any variations from those given in Table 16.4 were extreme. Finally, when all circular duct sizes have been chosen, reference to Table 16.3 should be made and square or rectangular shapes chosen to suit the space available. In carrying out this last routine, it is as well to check that unnecessary minor reductions in shape or size are not being made to suit relatively small changes in velocity, since to do so is uneconomic.

It is sometimes advantageous, when sizing a long run of ducting with air supply grilles along its full length, to take what is called *static pressure regain* into account. This is done by so reducing the velocity in the duct in stages that the regain in static pressure compensates for the frictional loss in the duct up to the point of reduction. This method provides an approximately equal pressure at each grille and this assists in balancing one with another.

High velocity systems
With the use of the higher duct velocities inherent in dual duct, induction unit and variable air volume systems, etc., as in Table 16.5, the use of static pressure regain is much more important, the potential availability being higher. All the static pressure gain calculated, theoretically, at the expanding duct section will not be available since the total pressure there will be reduced due to turbulence and added wall friction. Thus:

$$\text{Static regain} = (p_{v1} - p_{v0}) - \Delta p_t$$

where

p_{v1} = inlet velocity pressure at expander
p_{v0} = outlet velocity pressure at expander
Δp_t = reduction in total pressure

The reduction in total pressure depends upon the design of the expanding duct section and another way of stating the expression above is:

$$\text{Static regain} = f(p_{v1} - p_{v0})$$

where

f = a factor representing the 'efficiency' of the expanding duct section, say, 75%

It is normal practice to use the availability of static pressure regain in sizing the main ducts only of high velocity systems since the branch ducts tend to be self-balancing as a result of the high pressure drop through the terminal fittings, be they double-duct, variable volume, or induction units. The routine adopted in sizing a duct system by this method is first to consider the penultimate section of the main duct and so to select the size that the static pressure regain at the expander is equal to the pressure loss in the final section, repeating this procedure in regression, section by section, back to the plant outlet.

By these means, the entire system will be inherently balanced and the need for dampers in upstream ducts either much reduced or, ideally, avoided altogether. Ductwork for high velocity systems is almost always constructed with spiral-wound circular or flat-oval sections except at plant and final connections to terminals.

Computer design of duct systems

It is now quite common practice for ductwork sizing calculations, other than those for simple distribution networks where an equal pressure drop approach is quite adequate, to be made using computer programs. These are particularly useful when the static regain method is used since this is very tedious to apply manually. An example output is shown in Figure 16.18.

NODES NEAR/ FAR	SIZING LIMITS	DUCT SIZE (mm)	SHAPE CODE	DUCT LENGTH (m)	DIVERSITY	FLOWRATE (l/s)	VELOCITY (m/s)	TOTAL K-FACTOR	PRESSURE GRADIENT (Pa/m)	PRESSURE LOSS IN DUCT (Pa)	/ DAMPER (Pa)	INDEX PATH
10/ 110	FX/	1000/ 800	R	19.20	0.800	10960.	13.67	0.53	1.829	171./	0.	*
110/ 120	/	600/ 600	R	4.30	1.000	3250.	9.02	1.23	1.319	67./	0.	
120/ 121	/	400/ 400	R	0.20	1.000	1050.	6.56	1.08	1.173	29./	7.	
121/ 125	/	740/ 250	F	3.00	1.000	1050.	6.12	0.03	1.060	4./	0.	
125/ 126	MD/ 300	605/ 200	F	2.00	1.000	450.	4.00	1.12	0.621	37./	5. T	
135/ 137	FD/ 300	955/ 300	F	2.00	1.000	750.	2.81	1.38	0.191	37./	0. T	
130/ 140	VL/ 3.0	600/ 600	R	4.25	1.000	1000.	2.78	4.19	0.144	70./	0. T	
210/ 220	/	800/ 800	R	4.30	1.000	4050.	6.32	1.66	0.475	42./	12.	
220/ 221	/	600/ 600	R	0.20	1.000	1500.	4.17	1.39	0.305	15./	14.	
221/ 225	MW/1000	835/ 500	F	3.00	1.000	1500.	4.12	0.00	0.277	1./	0.	
335/ 336	/	735/ 400	F	2.00	1.000	900.	3.47	1.13	0.252	49./	0. T	
335/ 337	/	735/ 400	F	2.00	1.000	900.	3.47	1.13	0.252	49./	0. T	
330/ 340	/	600/ 600	R	4.25	1.000	1200.	3.33	2.61	0.201	69./	0. T	*
110/ 210	SM/	1000/ 800	R	8.90	0.800	8360.	10.44	0.28	1.087	28./	0.	*
210/ 310	/	800/ 800	R	11.10	1.000	6400.	9.99	0.10	1.136	19./	0.	*
430/ 440	/	400	I	2.10	1.000	850.	6.77	0.01	1.283	3./	0.	
440/ 450	/	355	I	2.10	1.000	600.	6.06	0.06	1.207	4./	0.	
450/ 460	/	315	I	3.40	1.000	350.	4.49	1.26	0.796	18./	1. T	
430/ 435	VL/ 3.0	355	I	2.10	1.000	250.	2.53	5.08	0.237	20./	6. T	
440/ 445	VL/ 3.0	355	I	2.10	1.000	250.	2.53	5.08	0.237	20./	3. T	
450/ 455	VL/ 3.0	355	I	2.10	1.000	250.	2.53	4.78	0.237	19./	0. T	

$$*$$ FAN REQUIREMENTS $$*$$

	OPERATING CONDITIONS	STANDARD CONDITIONS
FAN TOTAL PRESSURE	319.9 (Pa)	314.8 (Pa)
FAN STATIC PRESSURE	205.9 (Pa)	202.7 (Pa)
TOTAL FLOWRATE	10960.0 (l/s)	10960.0 (l/s)
AIR POWER	3505.9 (W)	3450.1 (W)

The following TM-8 defined margins are added to give design values

10.0 percent on flow for leakage and balancing
10.0 percent on pressure for flow margin
10.0 percent on pressure for calculation uncertainties

```
*************************************************************
*                                                           *
*          Design duty at standard conditions -             *
*                                                           *
*      12056. l/s at  378. Pa total pressure                *
*                                                           *
*************************************************************
```

Figure 16.18 Sample computer output for duct sizing problem

Fan pressure and fan duty
From the various descriptions of volume flow and pressure loss above it will be obvious that the duct sizing operation is but one part of an exercise to determine the total and the static pressure which must be developed by the connected fan. A summation of the various system losses, those through straight ductwork, through duct fittings and through plant components, with reductions where available to take account of static pressure regain, will go to produce the static pressure against which a fan must operate.

This, of course, is the net or calculated duty and a margin of 10% is usually added to that pressure to allow for minor changes in the duct configuration during the building period, due to revised client requirements and architectural second thoughts. Similarly, margins should be added to the calculated volume of say 5% to cater for ductwork leakage, which is a function of ductwork pressure as noted on p. 425, plus another 5% as an allowance for contingencies in commissioning.

Thermal insulation of ducts

Ducts conveying cooled air in an air-conditioning installation are usually insulated on the outside so as to reduce temperature rise to a minimum, also to prevent surface condensation from the surrounding air. Such insulation must be *vapour sealed*, i.e. sealed against ingress of atmospheric air, otherwise the insulation will quickly become waterlogged and useless. Ducts conveying warm air equally in most cases require insulation to reduce loss of heat. Where ducts carrying heated air are run in the conditioned space only, insulation may be unnecessary.

Transmission of heat through bare metal duct walls varies according to air velocity, the U value rising from 5.6 W/m^2 K at 2 m/s to 7.2 W/m^2 K at 15 m/s. The effect of insulation is to reduce the loss such that air velocity in the duct affects the U value only to a minor extent, and the coefficients given in Table 16.6 may be adopted as covering the loss at any practical air speed. The economic thickness may be estimated for any particular case, and reference should be made to BS 5422.* For practical purposes, the minimum thickness of insulation on ductwork carrying warmed or cooled air is given in Table 16.7.

Materials for duct insulation, internal and external to the duct, must be such as not to support fire. When tested to BS 476: Part 6: 1989 and Part 7: 1992,† the insulating material must have a fire propagation index of performance not exceeding 12, with not more than 6 from the initial period. The surface flame spread, as defined in Part 7, must

Table 16.6 *U* values for insulated ductwork

Thermal conductivity of insulation (W/m K)	*U value* (W/m^2 K), *for given insulation thickness* (mm)		
	38	*50*	*75*
0.04	0.9	0.7	0.5
0.055	1.2	1.0	0.7
0.07	1.5	1.2	0.8

* BS 5422: 1990, *Specification for the Use of Thermal Insulating Materials.*
† BS 476, *Fire tests on Building Materials and Structures,* Parts 6 and 7.

Table 16.7 Minimum thickness for ductwork insulation

Thermal conductivity of insulation (W/m K)	*Insulation thickness* (mm), *for air-to-air temperature difference* (K)		
	10	*25*	*50*
0.04	38	50	63
0.055	50	50	75
0.07	50	75	75

meet the Fire Authorities' requirements. Some suitable materials are glass fibre, mineral wool and polyurethane.

The gain or loss of temperature in air conveyed in a length of duct may for practical purposes be calculated from the following:

$$H = PLU\Delta t$$

where

H = heat loss or gain (W)
P = perimeter of duct (m)
L = length of duct (m)
U = thermal transmittance of bare or insulated duct (W/m^2 K)
Δt = mean temperature differences between inside and outside of duct (K)

The temperature rise or drop along the length of the duct, taking the specific heat capacity of air as unity, is approximately $PL\Delta t/M$ where M is the mass of air in circulation (kg/s). This expression yields results which are sufficiently true for practical purposes, though not strictly so when the difference in temperature along the duct is of such a magnitude as to bring the internal temperature closer and closer to that around it.

Insulated ducts located outside a building must be provided with a commercial weatherproof finish such as bitumen coated roofing felt, with well sealed joints, secured with galvanised wire netting, or some better and more permanent protection. Horizontal ducts fitted externally should preferably be circular or, where they must be rectangular, be provided with a sloping cover to prevent water lying on the upper surface.

Sound control

The increasing amount of mechanical plant in buildings brings with it the problem of noise. We here confine our attention to equipment such as fans, compressors, pumps and boilers, and the channels by which sound may be conveyed therefrom to other parts of the building.

Sound

It is necessary first to consider the mechanism by which sound is transmitted. Vibrations at a point A in Figure 16.19 set in motion the molecules of adjacent air such that a series of waves of compression and rarefaction radiate from the source spherically in all

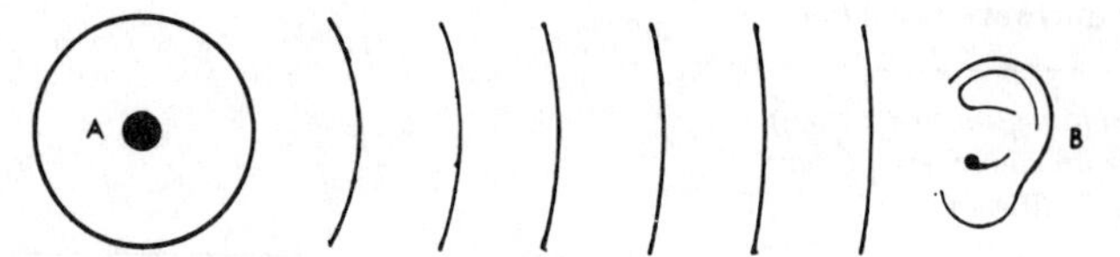

Figure 16.19 Transmission of sound

directions, a part reaching the ear at point B where the eardrum is set in like motion and the brain interprets the sensation as a sound, pleasant or otherwise. A *noise* may be defined as an unpleasant sound, or one that is overloud, or is monotonous or erratic.

The speed of sound depends on the properties of the medium, but may be taken at 331.46 m/s at 0°C and 101.325 kPa (NTP).

Frequency

Sound from point A will most likely be of more than one frequency. The deepest note which the human ear can detect is about 15 Hz, and the highest about 20 000 Hz.

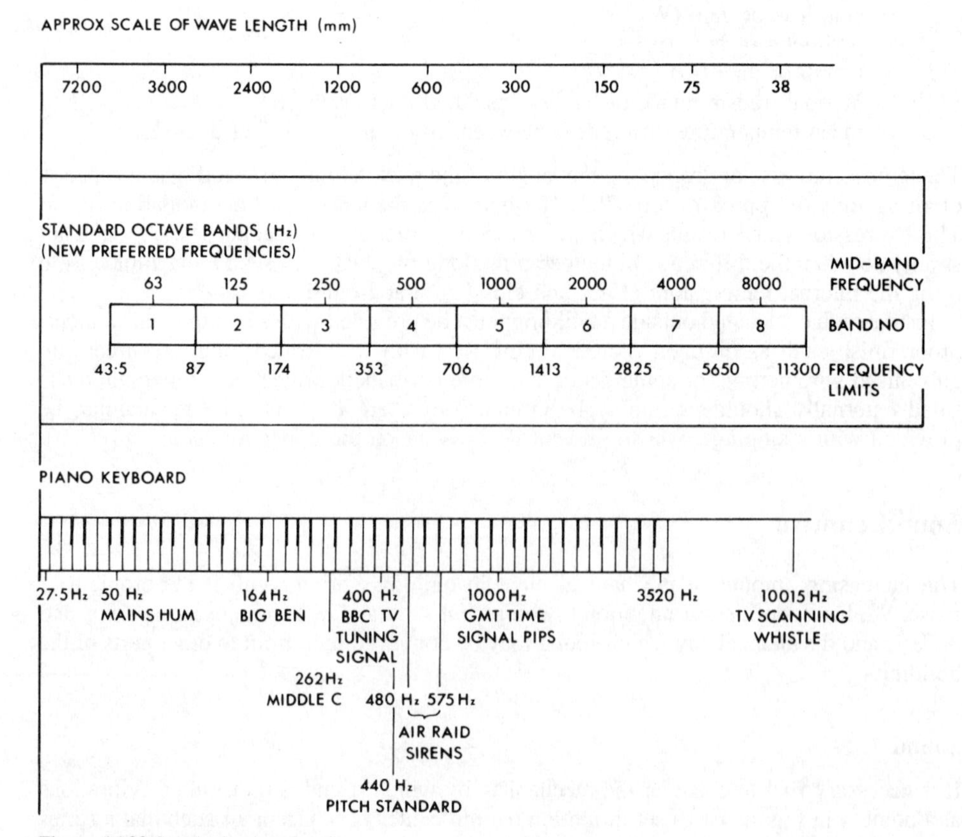

Figure 16.20 Standard octave bands

Frequencies are divided into octaves, each octave being double that of the lower one. The frequency of middle C is 262 Hz; thus C of the octave above is 524 Hz and so on. Figure 16.20 shows standard frequency bands. Most sound sources set up harmonics, i.e. higher frequencies which are multiples of the fundamental.

Amplitude

The amount of energy imparted to the air at point A is termed the *sound power*, and is referred to in terms of *watts*. Sound power may be visualised as the amplitude at the source.

The amplitude, see Figure 16.21, of sound X is greater than Y:X is a loud sound and transmits more energy than soft sound Y. The energy involved is represented by power multiplied by time (W s).

Sound power cannot be measured as such; it can be inferred mathematically from the sound received at a known distance in a sound-proof room.

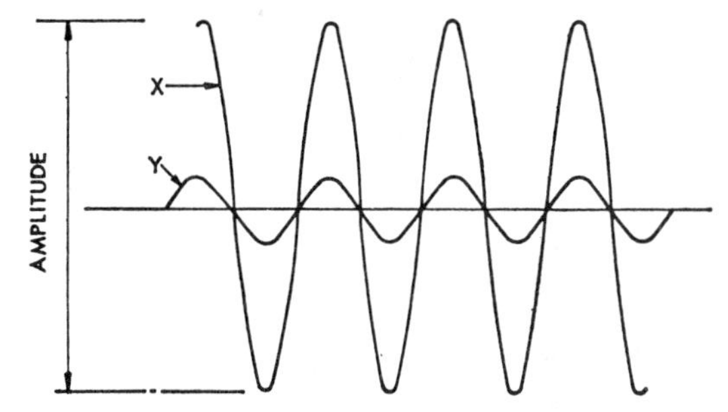

Figure 16.21 Amplitudes of sound power

Sound received

The power of sound received is termed *sound intensity* and is measured in W/m^2. Although the sound produced at point A cannot be measured directly, the sound received by the ear – or by a sound meter – at point B can be expressed in terms of pressure variation.

The amount of variation in pressure of the sound waves reaching the receiver – termed the *sound pressure* – is measured in Pa.

The decibel (dB)

The units referred to above are absolute values and are not convenient for general use. Units of reference have therefore been invented and are termed *bel* and *decibel* (10 dB = 1 bel). They are logarithmic, so that if two sounds differ in magnitude by 1 bel, the magnitude of one is ten times that of the other. Similarly, if the difference is 1 dB, the magnitude of one will be about 26% greater.

The decibel may be defined as $10 \times \log_{10}$ of the ratio between two absolute values such as levels of sound power, sound intensity and pressure.

- Sound power level (with reference to a datum of 10^{-12} watt)

  $= 10\log_{10} W/10^{-12}$ dB

  where W = actual sound power.
- Sound intensity level (with reference to a datum of 10^{-12} watt/m^2, taken to be threshold of hearing)

  $= 10\log_{10} I/10^{-12}$ dB

  where I = actual sound intensity.
- Sound pressure level (with reference to a datum of 2×10^{-5} Pa which corresponds to the reference intensity 10^{-12} watt)

  $= 20\log_{10} P/(2 \times 10^{-5})$ dB

  where P = actual sound pressure level.

Decibels cannot be added arithmetically, but if two sounds of known decibel rating are taken separately, the decibel rating of the two taken together will be:

$$= 10\log_{10} [\text{antilog}(A/10) + \text{antilog}(B/10)]$$

Use of a graph, Figure 16.22, simplifies the task of addition of two dB ratings.

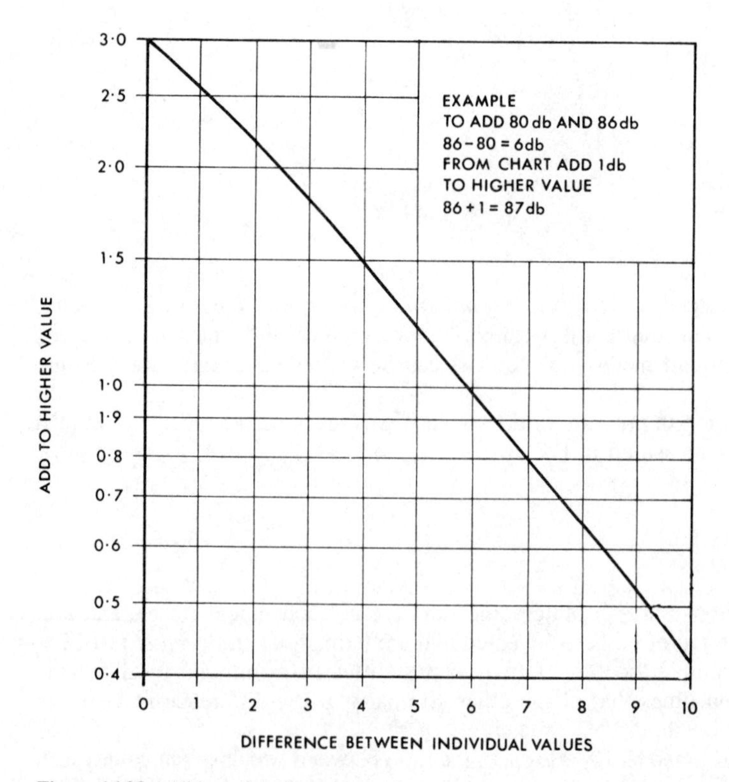

Figure 16.22 Addition of decibel ratings

The phon

This unit is based on the subjective reaction of the ear to loudness and it may be related to sound pressure level in dB by means of data which have been derived by experiments on individuals. Some values on this scale are:

Threshold of hearing	=	0 phon
Whispering	=	20 phon
Average room	=	40 phon
Busy street	=	50–70 phon
Pneumatic drill	=	100 phon

Noise criteria

It is common experience that the human ear can tolerate a noise of low frequency much more than one of high frequency, even though of equal intensity. Thus it is necessary to consider frequency in relation to sound measurement and for this purpose the spectrum of frequencies has been split up into eight octave bands as previously shown in Figure 16.20.

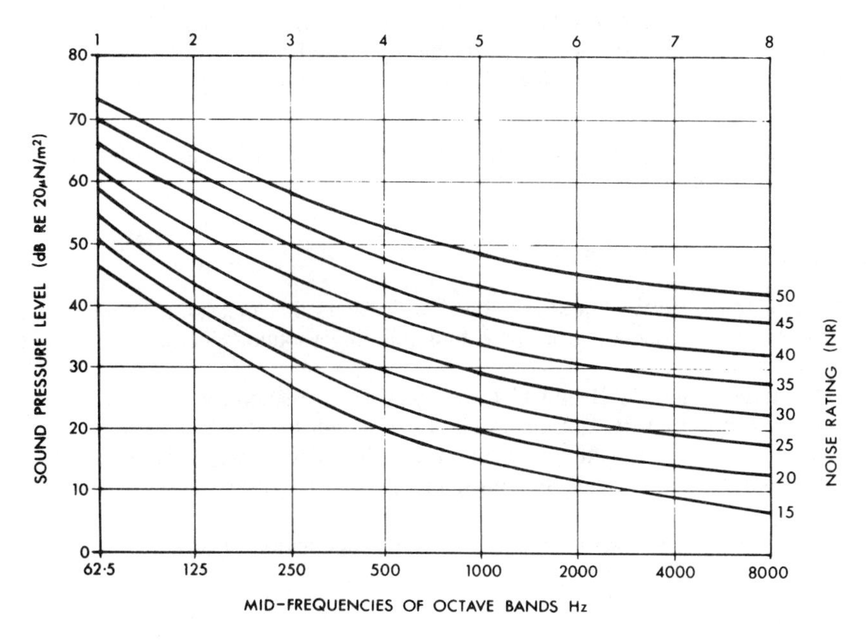

Figure 16.23 Noise rating curves

Noise criteria and noise rating curves (NC and NR) have been established on a subjective basis and each is given in a number corresponding to the sound pressure level in decibels in the sixth octave band (1200–2400 Hz). NC curves were much used in the past but have now been superseded by NR curves for general use. For most practical purposes the two sets of curves may be regarded as interchangeable. Figure 16.23 shows

NR plots of equal tolerance for the eight frequency bands. Acceptable NR values may be classified as follows:

NR 25 very quiet
NR 30 normal living space
NR 35 spaces with some activity
NR 40 busy spaces
NR 45 light industry
NR 50 heavy industry

The *Guide Section A1* gives recommended NR values for various specific applications.

Sound meter

In one form this consists of a microphone coupled to a sensitive milli-ammeter and battery such that variations in sound pressure cause corresponding deflections of the needle of the instrument. By calibrating the dial in decibels a direct reading in dB can be obtained.

As pointed out, however, the ear does not evaluate sounds of different pitch in a linear manner and hence scales have been weighted in an attempt to correct for this by adding filters to the circuit.

- Scale C is as linear, but very high and very low frequencies are suppressed.
- Scale B is as C but with more low frequencies suppressed.
- Scale A is further weighted to exclude all low frequency sound intensities under about 55 dB.

Scale A has now become generally accepted for practical purposes since it appears to give a better measure of human response than scales B or C.

Sound analyser

In effect this is a sound meter, but so equipped with filters that measurement of sound pressure at any range of frequency is possible, thus enabling a sound 'spectrum' to be built up.

In using the three scales A, B and C for subjective assessment, scale A would be appropriate for low noise levels, bearing in mind the greater tolerance of the ear to low pitch sounds, scale B is for medium levels and C for loud noises. The greater the difference between A and C scale readings, the greater the importance of the low frequency component.

The linear scale would be used to obtain a basis for calculation of absolute values.

Noise of fans

The noise produced by fans is often quoted in makers' lists in decibels. It will be apparent from what has been stated that such a rating can only be a measure of the sound intensity level at a given point and the level will vary according to the distance at which the measurement is taken and other conditions. The sound spectrum is also highly important. It is to be assumed that the test takes place in an *anechoic* room, i.e. without reflection or extraneous noise. Given the sound power level of the fan at each frequency range, and ignoring any difference between suction and discharge, it is possible to determine the attenuation resulting from ducts, changes of direction, grilles and the like. If the result

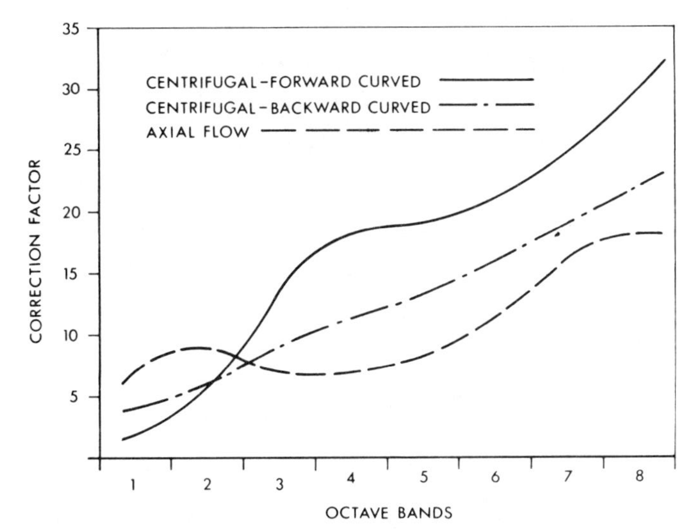

Figure 16.24 Correction factors for sound power of fans

shows that too high a sound level from the fan remains, it is necessary to consider means for greater absorption as referred to later.

Generalised formulae have been published which enable the likely sound power level of a fan to be predicted in the absence of manufacturers' data. One such is as follows:

$$\text{SWL} = 10 + 10\log_{10} Q + 20\log_{10} P$$

where

Q = volume flow (litre/s)
P = static pressure (resistance) (Pa)
SWL = *overall* sound power level (dB)

To establish the individual sound power levels in the various frequency bands, correction factors must be subtracted as read from Figure 16.24.

Noise in rooms

The measurement of noise in rooms should be in conformity with some standard. Readings are commonly taken:

- Not more than 1.2 m from floor.
- Not less than 1 m distance from any grille or other ventilation device.
- Not more than 1 m/s air speed.

There are two components to the sound pressure at a point in a room; direct sound from the source and sound reflected from the surrounding surfaces. The magnitude of the sound pressure level is the decibel sum of the two components, which can be calculated knowing the sound power emitted from the source(s), the location and radiation characteristics of the source(s) and the acoustic properties of the room. When air enters the room from, say, a grille it expands into free space which accounts for some low frequency attenuation; a

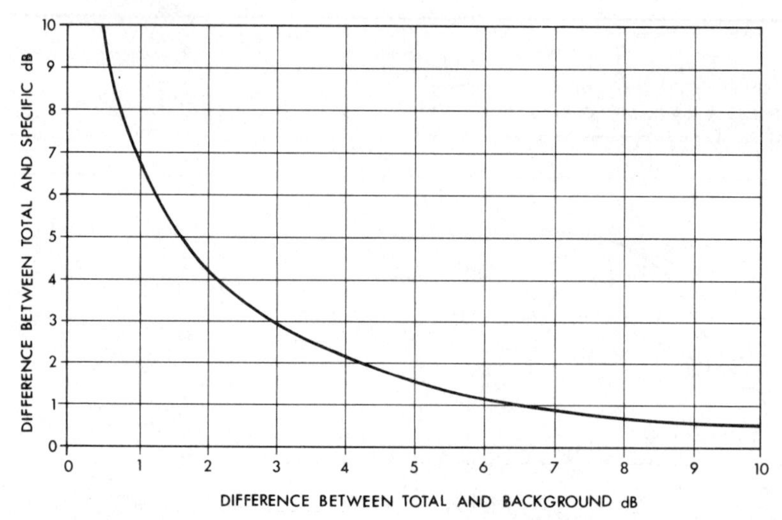

Figure 16.25 Subtraction of decibel values

large number of small outlets will transmit less low frequency noise into a space than a single large grille.

There is the question of background noise which is never absent except in a sound-proof room. It is necessary to measure the dB of the background, and then of the background plus the noise in question. By taking antilogs and the log of the difference, the dB level of the noise can be found (see Figure 16.25). If there is a difference of more than 10 dB the background can be ignored.

The ear must, of course, always remain the final arbiter, as instrumental measurement, however careful, may be misleading. There may for instance be a single monotonous note which has only a slight effect in terms of dB, or there may be an intermittent noise which goes quite unrecorded.

Noise dispersal

The less the noise produced, the easier is the problem of disposing of it. Nevertheless, all plant is liable to cause noise in greater or lesser degree.

Plant noises are of two kinds:

- Mechanical vibration which is transmitted to the building structure and thence to the occupied rooms.
- Airborne noise which likewise can enter the structure but which will be more troublesome if conveyed via ventilation ducts to the rooms.

The first category can be dealt with by mounting such plant on anti-vibration supports comprising springs, rubber blocks and the like, and by installing flexible connections in the ductwork or pipework. However, misaligned flexibles in ductwork may produce turbulence resulting in an increase in sound power level of up to 6 dB or more at low frequencies. Figure 16.26 illustrates the various transmission paths between an air handling plant and an occupied space. Manufacturers' data are available for the various types of systems.

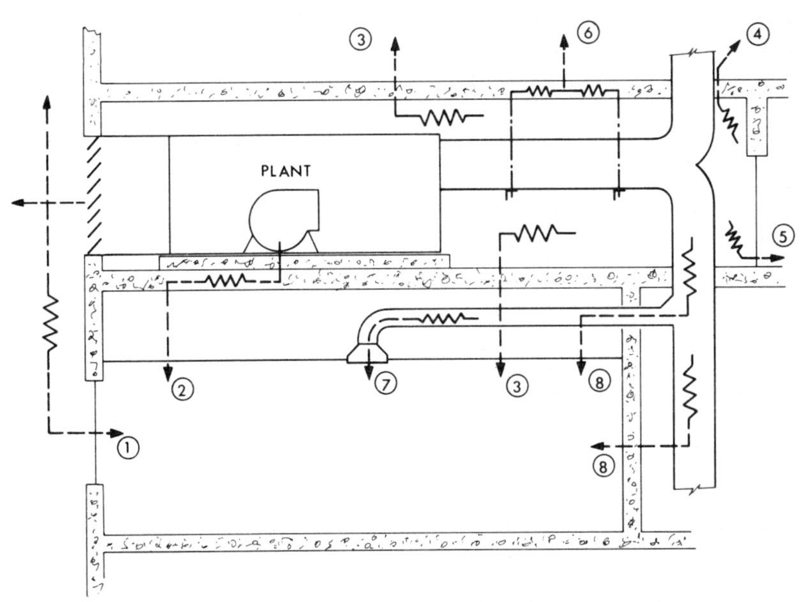

Figure 16.26 Transmission paths of noise and vibration. 1, airborne noise via intake louvres; 2, plant vibration via structure; 3, airborne noise via openings; 4, airborne noise via poor seals; 5, airborne noise via openings; 6, plant vibration via solid duct supports; 7, ductborne noise to terminals; 8, ductborne noise through duct walls

Absorption of sound is achieved by the application to surfaces of so-called acoustic materials such as acoustic tiles, and various soft materials such as glass fibre.

Insulation of sound is achieved by interposing a barrier between the space where the noise is produced and some other space to be kept silent.

In the case of the plant room mentioned in Figure 16.26, the sound may be absorbed by covering the wall, ceiling and other surfaces with suitable material; but this is generally friable and liable to damage, hence the use of insulation might be preferable between say two faces of a hollow partition, or on top of a false ceiling. Usually this problem is not intractable and is covered under the general heading of 'Building Insulation' dealt with in Chapter 2.

Large scale boiler and refrigeration plant rooms give rise to special problems and must be considered in great detail. High mass walls, floors and ceilings are required between major plant rooms and occupied spaces, and special attention is needed to ensure that penetrations do not present a noise leakage path. Where impact is a problem, consideration should be given to the use of a 'floating floor'; that is one in which the slab on which the plant sits is supported above the main structure by a resilient layer or mounting.

Ducts

Considering now the transmission of noise by ducts, we may assume that the noise produced by a fan is conveyed equally via the suction and delivery. The air in passing through the system of ducts will lose some of the noise by attenuation, that is by the pressure waves being damped down due to friction on the duct walls, and at changes of direction.

Table 16.8 Noise attenuation of plain ductwork

Duct item	*Attenuation* (250 Hz)
Straight ducts	(dB/m run)
150 × 150 mm	0.05
600 × 600 mm	0.3
1800 × 1800 mm	0.2
600 mm diameter	0.1
1800 mm diameter	0.03
Square bends	(dB/bend)
150 mm depth	0
300 mm	1
450 mm	3
600 mm	6
Branches	
Acoustic energy is divided in ratio of duct areas	

Secondary noise may be generated in the ductwork by sharp edges, and particularly by dampers.

At the terminal, the grille or diffuser may also generate noise, and this at a point where no treatment is possible; hence correct selection of type and velocity of such items is inherent in good acoustical design.

When the air enters the room from the grille, it expands into free space which again accounts for some attenuation.

The attenuation in ducts depends on length and size: the longer the duct the greater the attenuation; the larger the duct the less the attenuation. The amount of attenuation may be calculated from the data given in Table 16.8 and the accompanying notes, but for fuller treatment the *Guide Section B12* should be consulted.

Attenuation of sound energy at bends is a process of reflection back towards the source, rather than the absorption mechanism of straight duct runs. Radius bends, or mitre bends with turning vanes generally give much less reduction than simple square bends.

Where it is clear that residual sound will be too great, ducts may be lined with absorption material either throughout or for certain lengths, as Figure 16.27. Such material

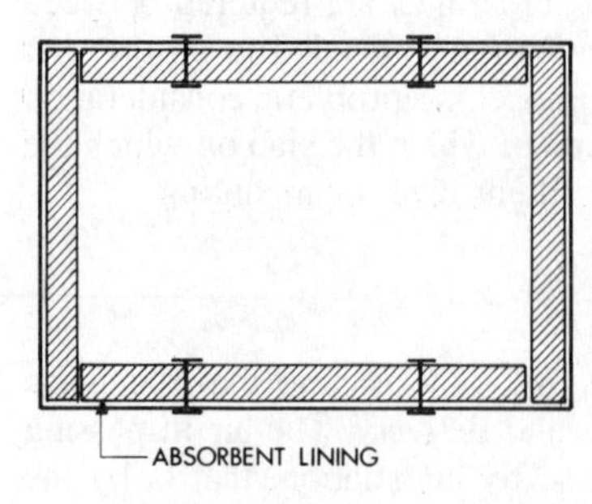

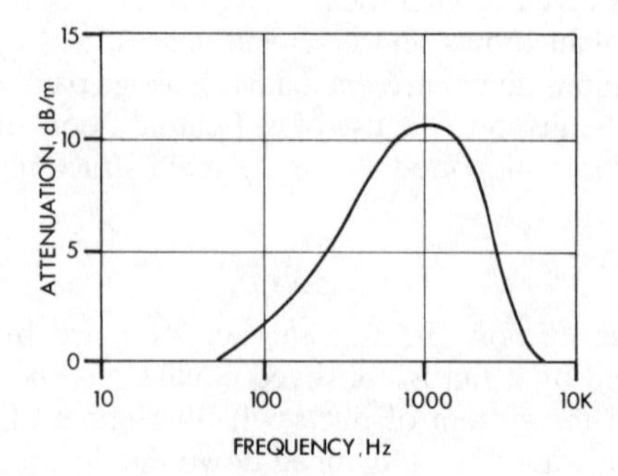

Figure 16.27 Acoustic insulation to ductwork

Table 16.9 Absorption coefficients of some duct lining materials

	Absorption coefficient at			
	125 Hz	*500 Hz*	*2000 Hz*	*4000 Hz*
Fibre glass resin bonded mat				
25 mm thick	0.1	0.55	0.8	0.85
50 mm thick	0.2	0.7	0.75	0.8
Stillite slabs				
25 mm thick	0.05	0.45	0.8	0.75
Polyurethane foam				
50 mm thick	0.2	0.65	0.7	0.7
Fibre board perforated tiles	0.1	0.4	0.45	0.5
Burgess backed fibre glass tiles	0.1	0.6	0.8	0.8

must, of course, satisfy the requirements for fire protection and not erode when subject to air flow.

An approximate prediction for attenuation by use of an absorption lining to ducts is given by the formula:

$$\text{dB/m run} = \left(\frac{D}{A}\right) \alpha^{1.4}$$

where

D = perimeter of lined duct (m)
A = free area of duct (m^2)
α = absorption coefficient of lining material (Table 16.9)

This expression holds for lined lengths up to about 2 m only and has been found to overestimate performance at high frequencies.

Alternatively, or in addition, specially designed attenuators may be used inserted in the run of ducting. These are rated according to dB loss, but the specification of performance should be based on 'insertion loss' not on a test in free space. Attenuators may be built up by splitters or egg crates, as in Figure 16.28, and their design may be considered as if each

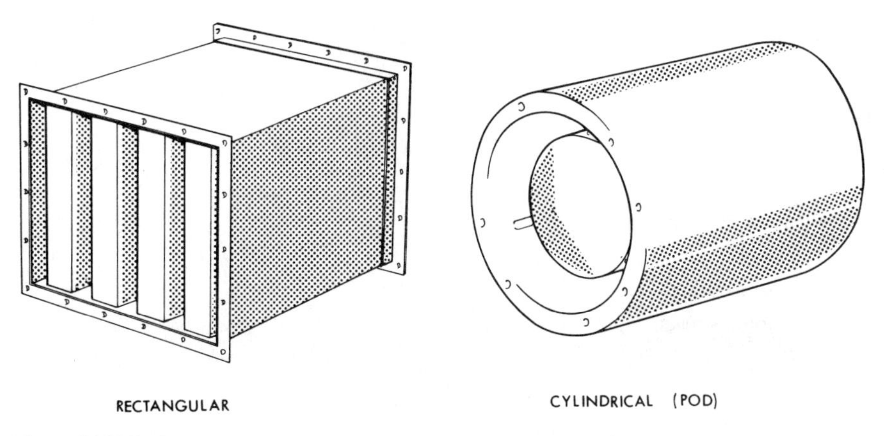

Figure 16.28 In-duct attenuators

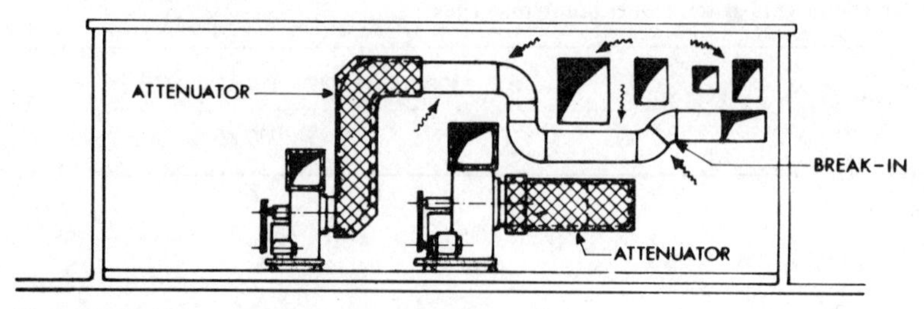

Figure 16.29 Noise 'break-in' within a plant room

section were a small duct lined with absorption material. As attenuation is inversely proportional to the cross-sectional area of the duct, it follows that the smaller the sub-divisions of the absorber the shorter the length required. On the other hand, any absorption device increases resistance to air flow and hence a reasonable balance must be kept.

Where attenuators are fitted in plant rooms, it is particularly important to ensure that they are so sited that *break back* or *break in* of noise is not possible. It is of little use to fit an expensive attenuator to a piece of equipment and then allow a bare sheet metal duct connection to wander across the plant room where noise can re-enter and be conveyed about the building: such equipment should either be followed by an acoustically insulated duct or be fitted immediately adjacent to the plant room wall. Figure 16.29 illustrates this point.

An alternative to the absorption, or dissipative, type of attenuator is being developed. It is commonly known as 'active' attenuation and is based upon the principle that two sound sources of the same frequency and amplitude, but displaced in time, or phase, together produce a reduced noise level. Therefore, introducing a microphone into a duct to pick up the fan noise and feeding it through a phase change device, to be re-introduced into the duct by a loudspeaker, can provide attenuation. This method is said to be most effective at lower frequencies, which is the range where the dissipative type is least effective. A simplified diagram of the system is shown in Figure 16.30.

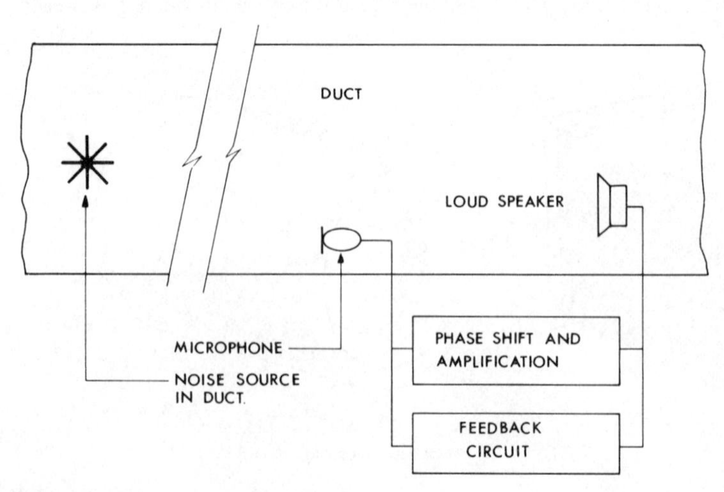

Figure 16.30 Active attenuation diagram

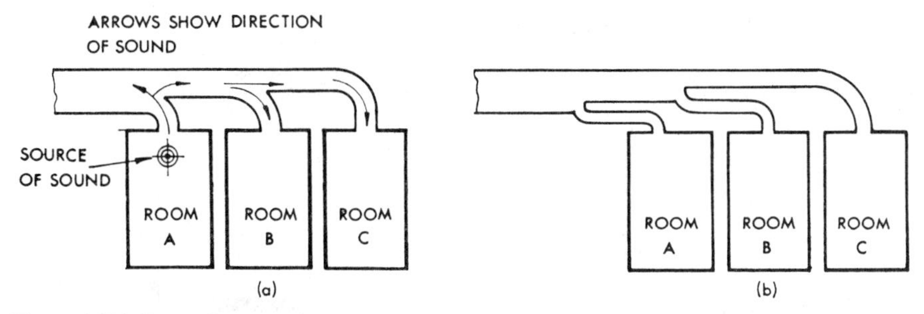

Figure 16.31 Cross-talk attenuation

Another application for attenuators occurs where a number of rooms are connected to a common duct system either inlet or extract, and where it is essential that speech or sounds in one room are not heard in other rooms. In Figure 16.31(a), a sound in room A has but a short path to rooms B and C. In Figure 16.31(b) the branch serving each room is treated acoustically, thereby avoiding any *cross-talk*. It will be noted that the 'shunt-ducts' mentioned in Chapter 13 (Figure 13.11) may act to some extent as cross-talk attenuators. Privacy between spaces is also affected by the background noise level; the higher this level the better privacy attainment. The background may be from external sources, such as traffic and general activity, or internal sources from the ducted systems.

Noise break-out from ducts into acoustically sensitive spaces may be a problem. Rectangular and oval ducts, particularly of large aspect ratio exhibit poor characterisics, whereas spirally wound circular ductwork tends to be more rigid and may be used to reduce break-out. Alternatively the ductwork may need to be insulated acoustically with a dense material.

High-velocity systems

The high pressures much involved in high velocity systems necessitate special attention to acoustical design. The usual procedure is to provide a long silencer in the fan-discharge for absorption of fan noise, especially at the lower frequencies. Noise generated in ducts is dealt with by lining of selected lengths. With the induction system the jets have an important attenuation effect, but some acoustical treatment of the cabinet is often provided. In the case of the dual-duct and variable volume systems the terminal control device (box) is always treated acoustically but in some cases, where the space served has a low NR requirement, further attenuation downstream of the control device may be needed.

Commissioning and measurement

Following completion of the ductwork installation and that of the associated plant, the system has to be commissioned and the performance tested against the design requirements. Commissioning involves setting the plant items to work and regulating the air flow rates within acceptable tolerances. The procedures to be followed in commissioning are given in a CIBSE Code.* A companion to this Code is a BSRIA

* Air Distribution. CIBSE commissioning Code 'A'.

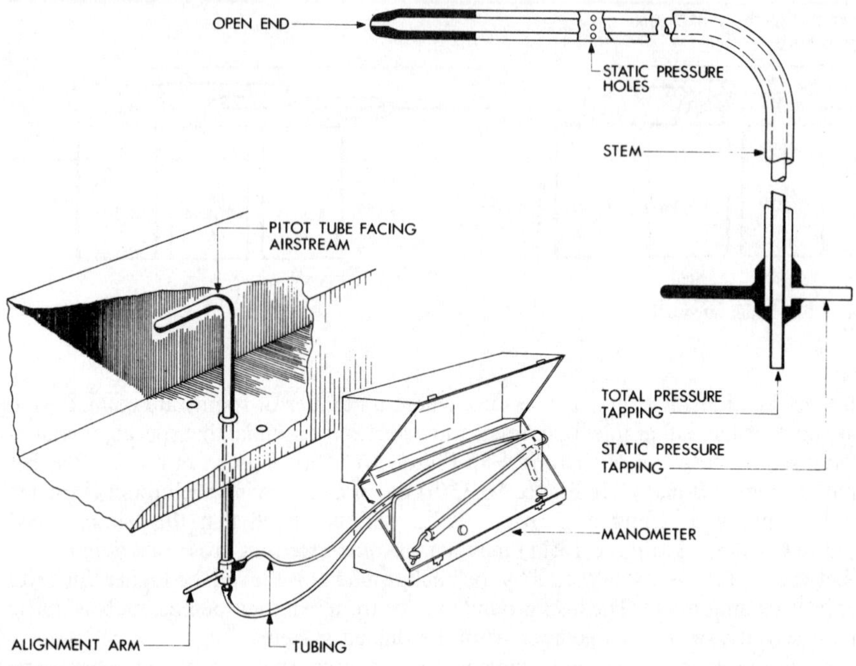

Figure 16.32 Pitot tube in use

Application Guide,* which gives practical advice on measurement techniques and tolerances.

The following describes some of the air flow measuring instruments in use today.

Pitot tube

In its standard form, this comprises two co-axial tubes; the centre tube faces the air stream and receives velocity plus static pressure (total pressure), and the outer tube, which has a series of small holes around the wall, measures static pressure alone. Used with a manometer, as shown in Figure 16.32, velocity pressure may be measured.

Micro-manometer

The pressures at low velocities are slight, as will be noted, and for the purpose of reading them an ordinary 'U' tube or inclined gauge is too insensitive. A micro-manometer, of which one type is shown in Figure 16.33, is therefore necessary. In this an extended 'U' tube is used, being tilted by a micro-adjustment. The level is viewed through the magnifying eyepiece against a cross-wire. The coarse reading is taken on the side scale, and the fine reading on the rotating dial, one revolution of which corresponds to one division of the scale. The liquid may be alcohol and the scale is calibrated in Pa or dPa. Other still more sensitive instruments are made.

* *Manual for Regulating Air Conditioning Installations.* BSRIA Application Guide 1/75.
Regulating Variable Flow Rate Systems. BSRIA Technical Note 1/78.

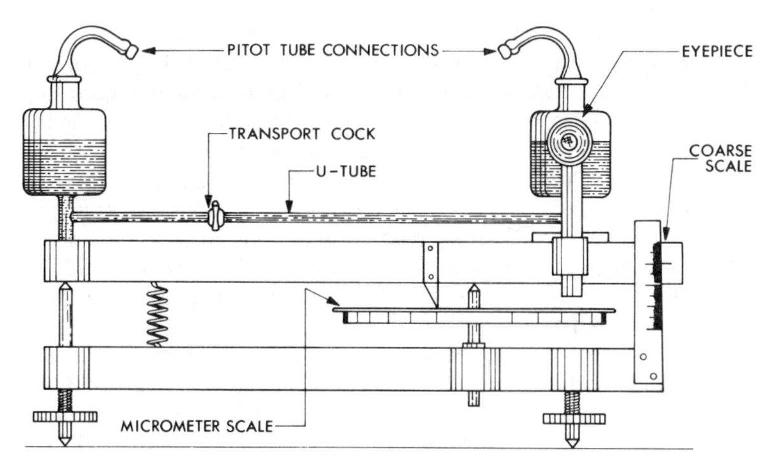

Figure 16.33 Micro-manometer

When measuring air speeds in a duct, however, it must be remembered that the velocity varies over the duct cross-section. In air flow unaffected by fittings and other components the velocity will be greatest at the centre and least at the periphery; such conditions being suitable for flow measurement. Where the flow has a non-uniform velocity profile measurement will be subject to error. In a round duct it is necessary to take readings at a number of points in concentric rings of approximately equal area. The average speed multiplied by the area of the duct will give the volume of air passing. In the case of a rectangular duct the method is similar, except that the duct is divided into equal rectangles.

The rotating vane anemometer

This instrument (Figure 16.34(a)) measures air speed by vanes which revolve as the air impinges on them. The instrument, which is calibrated in metres, serves only to count the

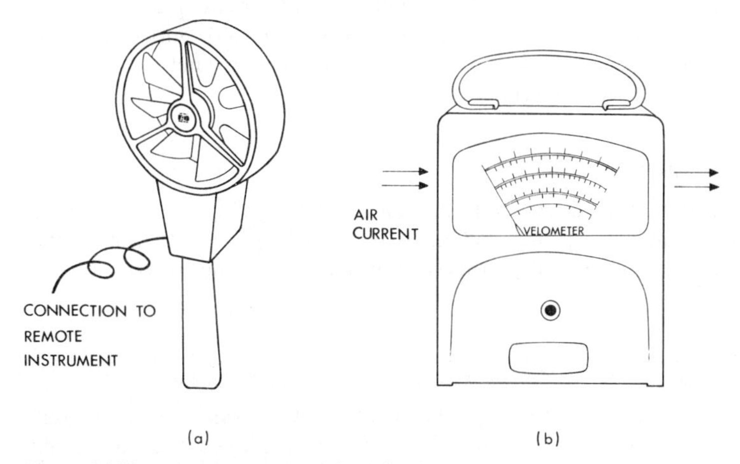

Figure 16.34 (a) Rotating vane and (b) deflecting vane-type anemometers

revolutions over a given time, such as one minute, taken by a stop-watch. The rotating vane is engaged and disengaged from the recording device at the start and finish of the time. After each measurement the reading is returned to zero. The instrument requires to be calibrated periodically. The standard instrument is too insensitive for use below about 0.5 m/s and some instruments are unsuitable for use above about 15 m/s.

For general work it is a useful instrument, chiefly for measuring the air speed from or to openings, etc. In such cases the instrument is placed about 30 mm from the grille and the speed is multiplied by the whole face area (regardless of free area) to obtain the volume in m^3/s. Readings are taken at various points and averaged. It is not so useful for measurements in ducts, as the anemometer has to be introduced through a hole in the duct wall and is then difficult to manipulate.

It must be admitted that the anemometer requires very careful handling, and in the hands of an unskilled operator entirely erroneous results can be obtained. Electronic types are available and may be used for remote reading, where the indicating instrument is connected by cable to the rotating vane. Various measuring head diameters are available, suitable for a range of measurement applications.

The hot wire anemometer

Originally developed for laboratory use, this type of instrument is particularly valuable for measurement of low velocities such as occur in occupied spaces. The measuring head consists of a fine platinum wire which is heated electrically and inserted in the air stream. Air movement cools the wire and, by calibration, current flow may be metered to indicate the velocity sensed. As the instrument is not sensitive to the direction of flow, it is not used for measurement at grilles.

The swinging vane anemometer

An instrument developed for site use, (the 'Velometer'), (Figure 16.34(b)), has been superseded effectively by more modern equivalents. The principle of operation relies on the speed of air to deflect a shutter against a spring. The shutter causes the needle to move over the sector-shaped dial, reading direct in metres per second. It requires no timing and reads from zero to 1.5 m/s, 0–15 and 5–30 m/s, by means of adaptors screwed into the aperture at one end. These may have flexible tube connections, for reading at a distance, which, in turn, may be used with various special mouthpieces fitted to the end of the tube to read velocities in ducts etc. This is a suitable instrument for giving a quick approximate check on adjustments or for exploring velocities over an area.

Diaphragm air pressure gauge

This instrument has a diaphragm and linkage to provide indication of static pressure on a dial. Typically, the scale range is up to 0.25 kPa. It is often helpful to measure static pressure in commissioning induction units, terminal control boxes and across components to check on the pressure loss.

Digital instruments

Modern instruments are battery operated (optional mains power connections for long periods of use are also available) and provide direct digital displays of the measured variable. Some makes are capable of giving dual output for two measured variables,

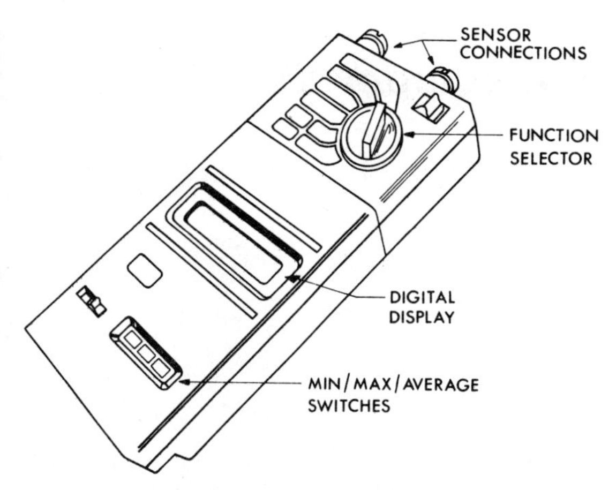

Figure 16.35 Digital multi-purpose instrument

together with minimum, maximum and average of readings taken over a period of time. Typically, humidity, temperature, air speed, pressure and rotational speed measurement are available by fitting different heads to a single instrument, as in Figure 16.35. With memory facilities to provide functions such as datalogging, input codes for location referencing and volume flow rates calculated from velocity measurements, this form of instrument is very convenient for site use.

Chapter 17

Fans and air treatment equipment

The fan is the one item of equipment which every mechanical ventilation and air-conditioning system has in common. A fan is simply a device for impelling air through the ducts or channels and other resistances forming part of the distribution system. It takes the form of a series of blades attached to a shaft rotated by a motor or other source of power. The blades are either in the plane of a disc (propeller fan) or in the form of a drum (centrifugal fan), however a recent development, the mixed flow fan, is a suitable alternative for certain specific applications.

There is as yet no other practicable commercial method of moving air for ventilation purposes, but fans in general suffer from various disadvantages such as low efficiency and noise. It is the latter which is probably the most troublesome to designers, and to which much careful attention must be given if silent running is to be achieved in the system as a whole: factors involved are air speeds, fan speeds, duct design, materials of construction, acoustical treatment of ducts and provision for absorption of vibration.

Fan types and performance

Fan characteristics

A comparison of the operation of fans of various types is best understood by studying their characteristic curves. For this purpose consider a fan connected to a duct with an adjustable orifice at the end, as in Figure 17.1. Pressures are measured by water gauges connected to a standard Pitot tube, see Figure 16.32. The perforated portion gives the static pressure, and the facing tube the total pressure.

If the fan is running with the orifice shut, no air will be delivered. Static pressure will be at a maximum, and velocity pressure nil. As the orifice is opened the static pressure will

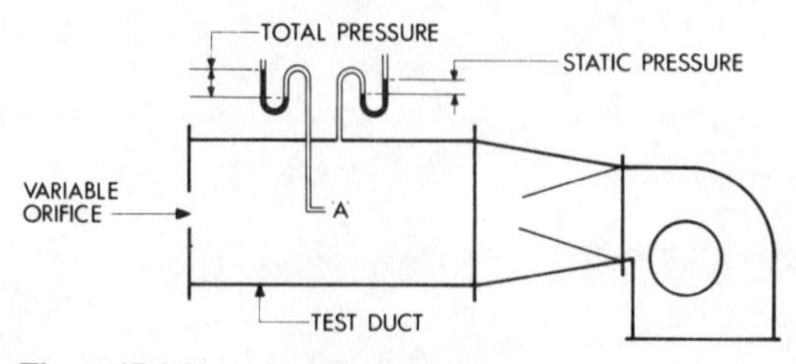

Figure 17.1 Simplified diagram of fan test arrangement

fall and the velocity pressure increase until, with the orifice fully open, the static pressure will be negligible and velocity pressure at a maximum. Over this range the power required to drive the fan will have increased from minimum to maximum, and perhaps will fall away as the total pressure falls off. The power required to drive the fan, if 100% efficient, would be:

$$H_t = V \cdot P$$

where

H_t = total air power (W)
V = air volume handled (m^3/s)
P = total pressure (Pa)

The mechanical efficiency of a given fan will be the ratio between this air power and the actual mechanical power supplied to the fan shaft (fan shaft power): this will depend on design, type of fan, speed, and proportion of full discharge. If the static pressure is used, the efficiency derived will be *static efficiency*: if the total pressure is used the efficiency will be *total efficiency*.

The standard air* for testing fans is taken at a density of 1.2 kg/m^3. Any fan at constant speed will deliver a constant volume at any temperature; as the temperature varies the density will increase or decrease proportionately with the absolute temperature, hence the power input will vary in the same ratio. With increase of temperature the power will be reduced and vice versa. Similarly, decrease of atmospheric pressure (as in the case of a fan working at high altitude) will cause a reduction in power and conversely.

If a fan running at a certain speed be rearranged to run at some higher speed, the system to which it is connected remaining the same, the volume will increase directly as the speed; the total pressure will increase in the ratio of the speeds squared and the power input will increase in the ratio of the speeds cubed.

These relationships are known as the *fan laws* which, for practical purposes, may be expressed for impeller diameter (d), speed of rotation (n) and air density (ρ) as:

Volume flow is proportional to d^3 and n
Fan pressure is proportional to d^2, n^2 and ρ
Fan power is proportional to d^5, n^3 and ρ

Characteristic curves

Fans are of three main types, with sub-divisions as follows:

- Centrifugal type:
  (a) multivane, forward bladed,
  (b) multivane, radial bladed,
  (c) multivane, backward bladed,
  (d) paddle wheel.

  The three types of runner (a), (b) and (c), are shown in Figure 17.2.

- Propeller type:
  (a) ordinary propeller or disc fan.
  (b) axial flow.
- Mixed flow type.

* For details of standard fan testing see BS 848: Part 1: 1992.

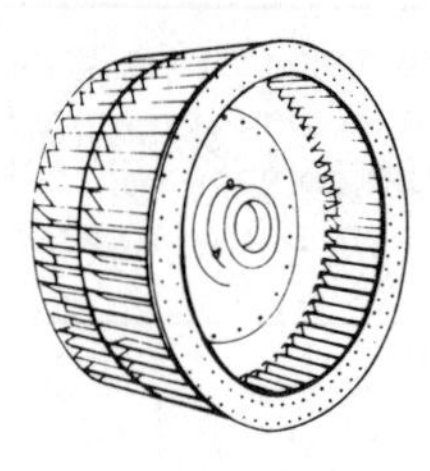

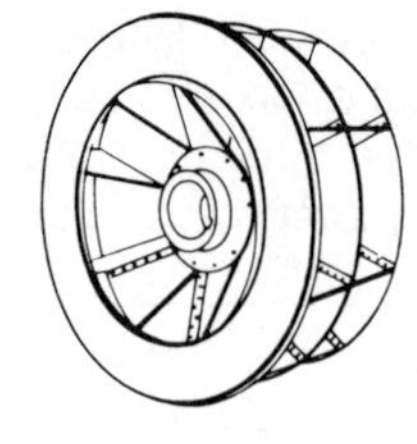

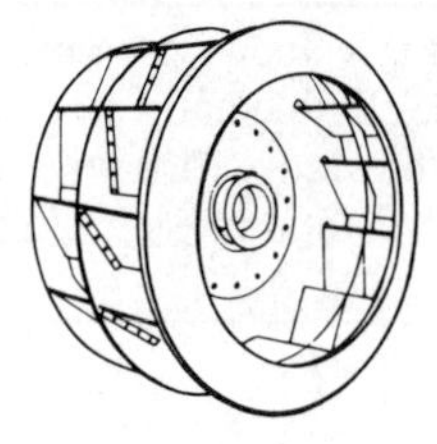

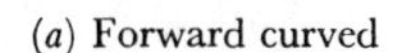

(*a*) Forward curved (*b*) Radial (*c*) Backward curved

Figure 17.2 Types of runner in centrifugal form

Figure 17.3(a)–(f) gives typical curves for the main types. Paddle-wheel fans are not given, as they are now little used in ventilating work on account of their noise, being confined chiefly to dust removal and industrial uses. Radial bladed fans have characteristics which are similar to those of the backward curved type but without the advantage of power limitation. They are not commonly used in ventilation applications.

The curves for static pressure, power input, and static efficiency are drawn from tests at constant speed, as already described. The base of the curve is percentage of full opening of the orifice. The vertical scale is percentage of pressure, efficiency or power input.

Forward curved

It will be noted that the forward-bladed centrifugal fan most commonly used in ventilation reaches a maximum efficiency at about 50% opening, where at the same time the static pressure is fairly high. Fans are generally selected to work near this point. It will also be observed that the power curve rises continuously. Thus, if in a duct system the pressure loss is less than calculated, the air delivered will be more and the power absorbed more, which may lead to overloading of the motor.

Backward curved

This type of fan runs at a higher speed to achieve the same output as a forward curved. The efficiency reaches a maximum at about the same point, and the power, after reaching a peak, begins to fall. This is called a *self-limiting characteristic*, and means that if the motor installed is large enough to cover this peak it cannot be overloaded. This is often useful in cases where the pressure is variable or indeterminate. The pressure curve is smooth without the dip of the forward curved; for this reason this kind of fan is to be preferred where two fans are working in parallel. The forward curved type under such conditions is apt to hunt from one peak to the next, so that one fan may take more than its share of the load and the other much less. The backward-bladed fan is often made with aerofoil blades, so raising efficiency. It is much used in high velocity systems where high pressures are required.

Ordinary propeller

From the curves it will be observed how the pressure falls away continuously, and the static efficiency reaches but a low figure. Thus, this type is unsuitable where any considerable run of ducting is used. To generate any appreciable pressure its speed becomes unduly high, and hence the fan is noisy. Its main purpose is for free air discharge where its velocity curve would rise towards a maximum at full opening. It should be noted that the fan power is a maximum at closed discharge, and, as the motors supplied with

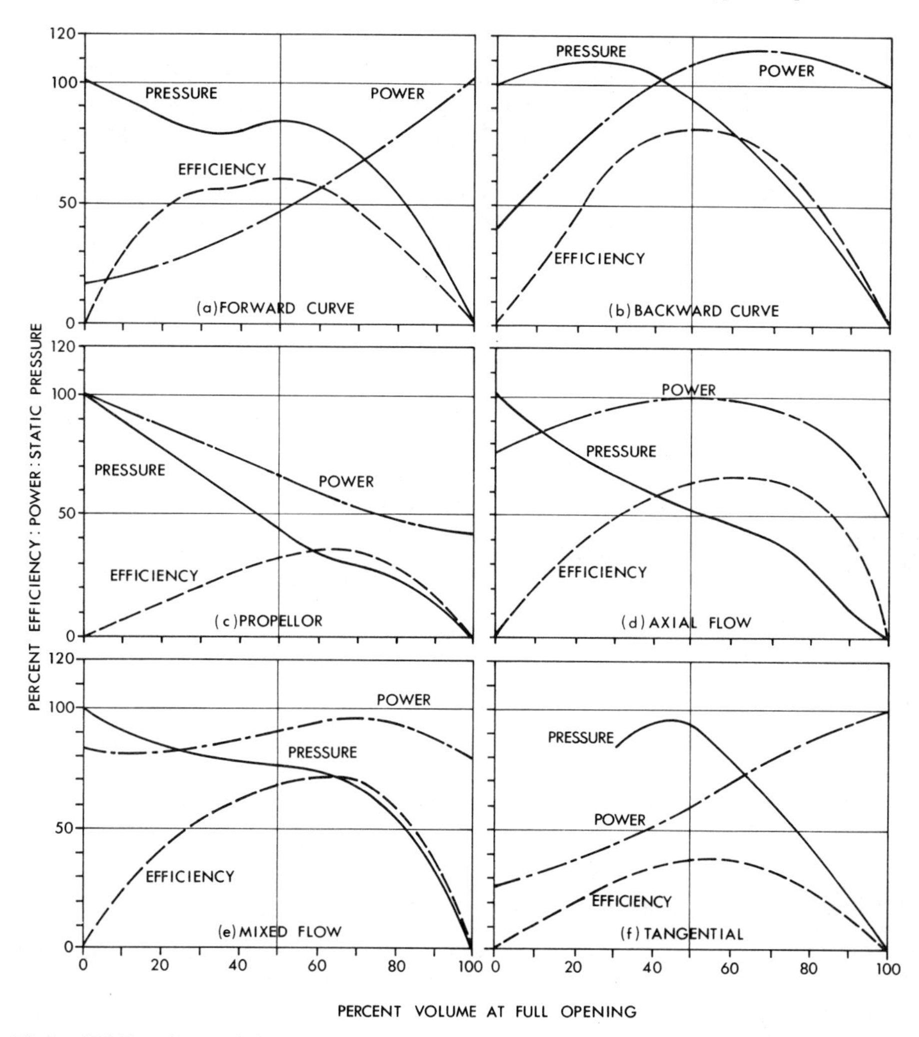

Figure 17.3 Fan characteristics

these fans are not usually rated to work at such a condition, the closing or baffling of the discharge or suction may cause overloading.

Axial flow

This type shows a great improvement over the ordinary propeller fan, both as regards efficiency and pressure. The fan-power curve is self-limiting. Hence these fans may safely be used in conjunction with a system of ductwork, being often more convenient than a centrifugal, particularly for exhausting. They run at higher speed than a centrifugal fan to produce a given pressure, and are liable to be more noisy: this may be overcome, to some extent, by the use of an attenuator. Their limited pressure development may be overcome by installing multiple fans to operate in series.

Mixed flow

Such fans are designed for in-line flow or radial-discharge and comprise an impeller having a number of blades, on a conical shaped hub. The efficiency and pressure developed are generally higher than for the axial flow type, with lower noise levels. Pressures up to 1750 Pa can be achieved with the range of fans currently available. Variable duty can be obtained using either variable speed drives or inlet guide vanes. Either direct drive or belt drive configurations are available. Figure 17.4 shows diagrammatically the operation of a mixed flow fan.

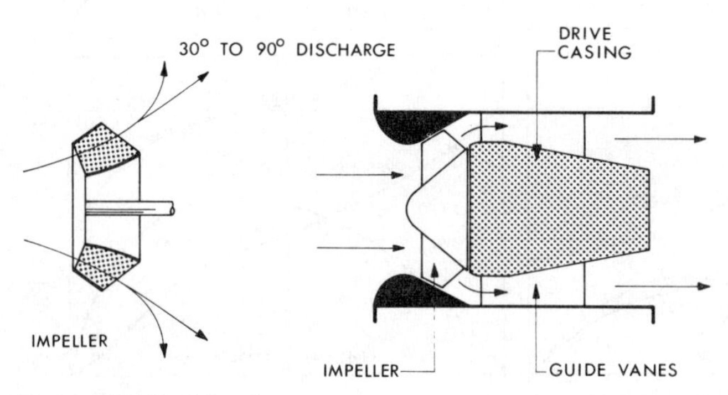

Figure 17.4 Mixed flow fans

The fan curves indicate a non-overloading power characteristic, making it suitable for use with ducted systems. Efficiency is good at about 80% over a limited range of duty, but the pressure characteristic, being similar to that of a forward curved centrifugal, makes it unsuitable for parallel operation.

Centrifugal fan arrangements and drives

Centrifugal fans may be *open* or *cased*. When open they can only be used for exhausting, and the discharge is tangential from the perimeter of the impeller, as might be suitable in a large roof turret.

The usual arrangement is the cased type, and the suction is then either on one side, as in Figure 17.5(a), with a single inlet, or both sides, as Figure 17.5(b), with double inlet. The double inlet double-width fan is useful in packaged air handling plant or where large volumes are concerned as it gives double the capacity of the single inlet with the same height of casing.

Fans are almost invariably driven by electric motor except for cases where there may be a requirement for independent drive from a petrol or diesel engine. Figure 17.5(c) shows a typical arrangement with the fan impeller mounted on a shaft extension of the motor. This is a compact arrangement, but generally used for small or medium-sized fans only.

A motor direct-coupled to a fan with a flexible coupling is illustrated in Figure 17.5(d). The fan shaft runs in its own bearings independently of the motor. This is obviously to be preferred for heavy duty and for large fans. The motor can be removed and replaced without affecting the fan.

The arrangement shown in Figure 17.5(e), where the motor drives the fan via pulleys and vee-belts, has the great advantage that the *motor* speed may be a standard, such as 16

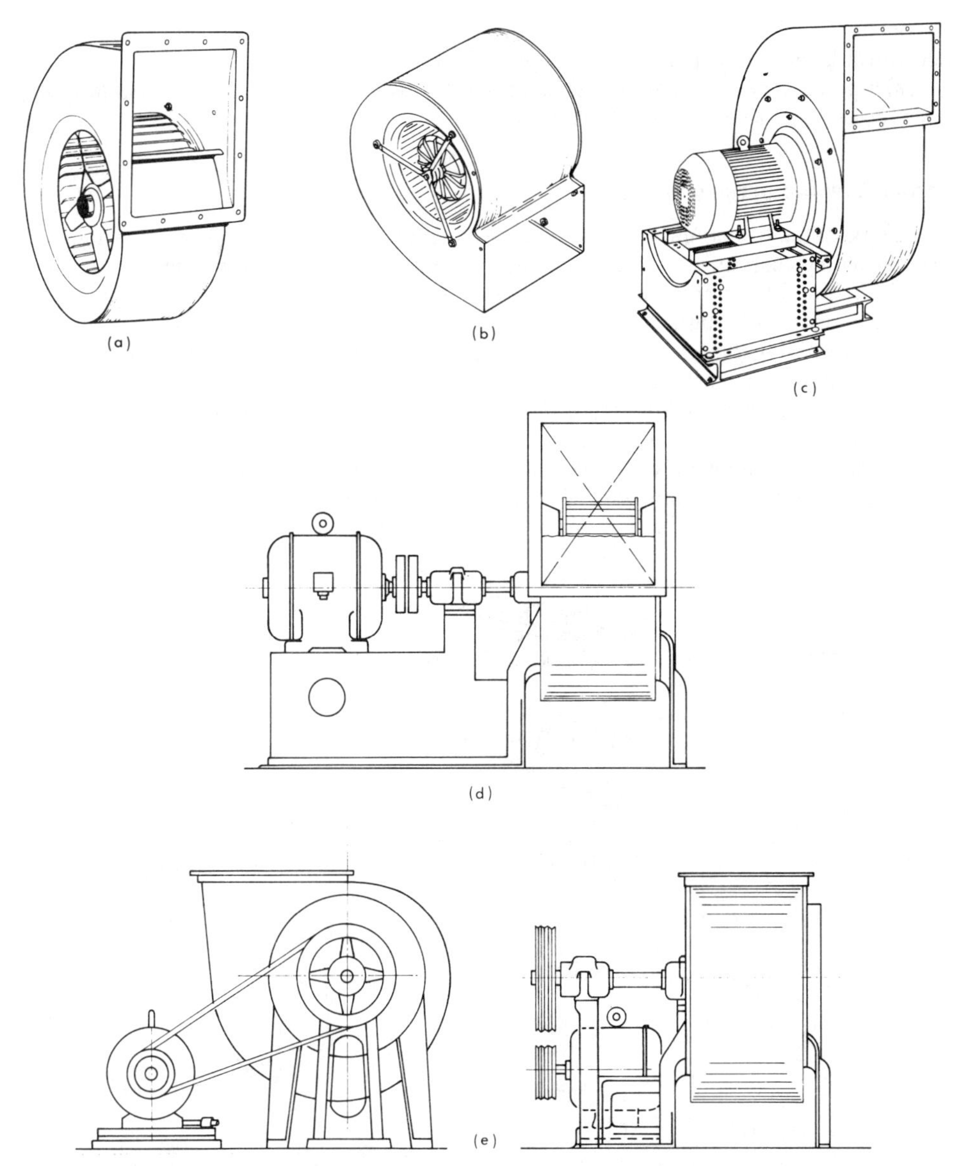

Figure 17.5 Arrangements for centrifugal fans: (a) single inlet; (b) double inlet; (c) close-coupled with motor; (d) flexible coupling in motor drive shaft; (e) belt driven

or 24 rev/s, whilst the *fan* speed is that best suited to the duty. A further advantage is that if on testing it is found that the pressure loss of the system is less or more than allowed for, the fan duty may be corrected by merely changing the pulleys. The belt drive configuration also enables duty and standby motors to be installed to drive a single fan.

It will be noted that in the illustrations a variety of different positions of the discharge opening in relation to the suction eye of the fan is given in each case. It is usually possible to obtain a fan with its discharge at any angle, vertical, horizontal top, horizontal bottom, downwards, and intermediately at an angle of 45°.

Inlet guide vanes, variable speed drive and the recent development* of *disc throttling* may be adopted to give variable duty.

Axial flow fan arrangements

This type of fan can produce pressures up to a maximum of about 1000 Pa, as a single unit within the normal range of noise generation. To achieve such performance, guide vanes would be fitted to improve the operating efficiency by reducing the *swirl* effect.

These fans can be built in one, two or three stages, to obtain increased pressure, the volume remaining the same. Alternatively they may have two sets of blades made counter-rotating. Both direct drive and belt drive types are available.

Such fans are illustrated in Figure 17.6, many are manufactured to permit adjustment of the pitch angle of the blade. Since this angle, for a given fan diameter and speed, determines the volume delivered it follows that adjustment facilities permit output to be matched to the duty required with some precision. Figure 17.6(d) and (e) show how pitch may be adjusted.

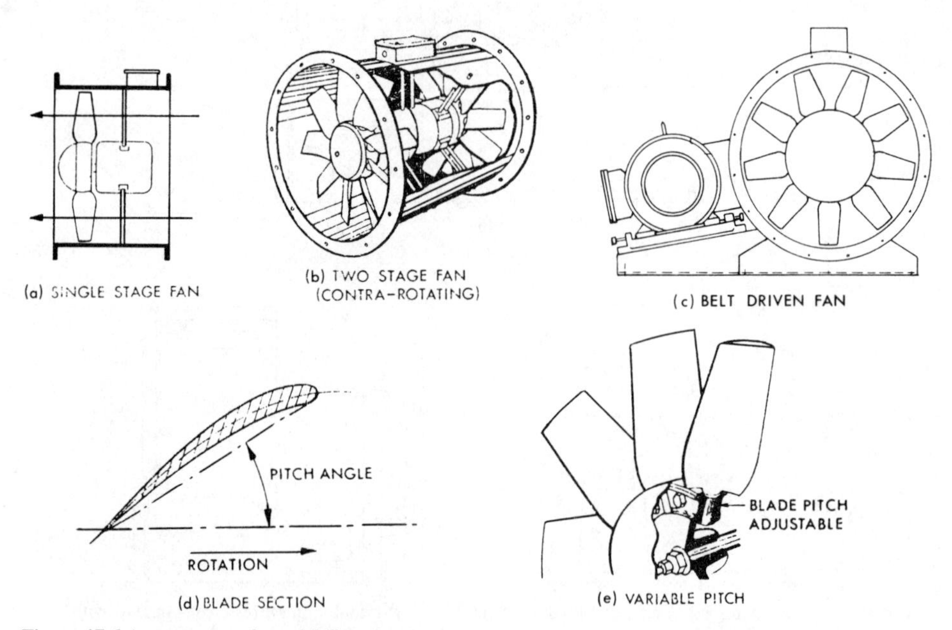

Figure 17.6 Arrangements for axial flow fans

Axial flow fans of the *bifurcated* type, where the air stream is directed around the motor which is enclosed in a protective casing, are suitable for handling corrosive gases, from fume cupboards and the like, and also high temperature gases/air such as experienced in smoke extract during fire conditions.

* *Energy Saving with Centrifugal Forms and the 'Disc Throttle' Variable Volume Controller*, W. T. W. Cory, 1983.

Fan duties

The range of fan types, speeds, pressures, and volumes is too great for any indication to be given here of sizes, duties, power requirements, etc., or of the problem of motor types suitable for fan drives.

Enquiries to fan-makers should give the fullest information possible about any system, as there are many hidden points to be watched in the selection of fans which render mere catalogue reference insufficient.*

Where it is necessary, as in the case of a variable volume air-conditioning system, to exercise control over fan output this may be achieved in a number of ways, which may be summarised conveniently under two headings:

Constant fan speed

- Throttling dampers.
- Inlet guide vanes.
- Variable pitch blade angle (axial flow fans).
- Disc throttling (centrifugal fans).

Variable fan speed

- Eddy current coupling.
- Variable voltage (fixed frequency).
- Variable frequency, variable voltage.
- Thyristor converter (AC to variable voltage DC).†
- Fluid couplings.
- Variable pulley drive.
- Slip-ring motor.
- Switched reluctance drive.

Figure 17.7(a) and (b) shows for constant and variable fan speed drives respectively, the percentage power input to a fan assembly relative to volume flow.

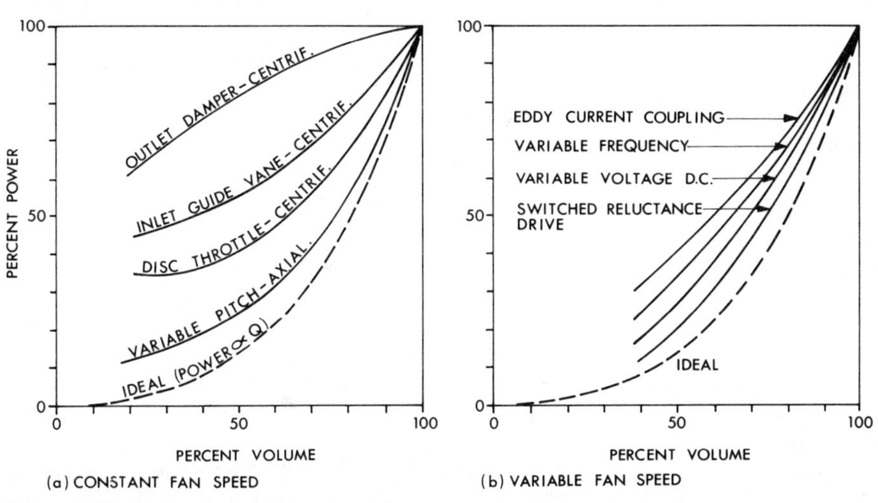

Figure 17.7 Power input to variable volume fans

* Refer to *Fan and Ductwork Application Guide*, published by HEVAC Association.
† AC, alternating current (mains supply); and DC, direct current (supply to motor).

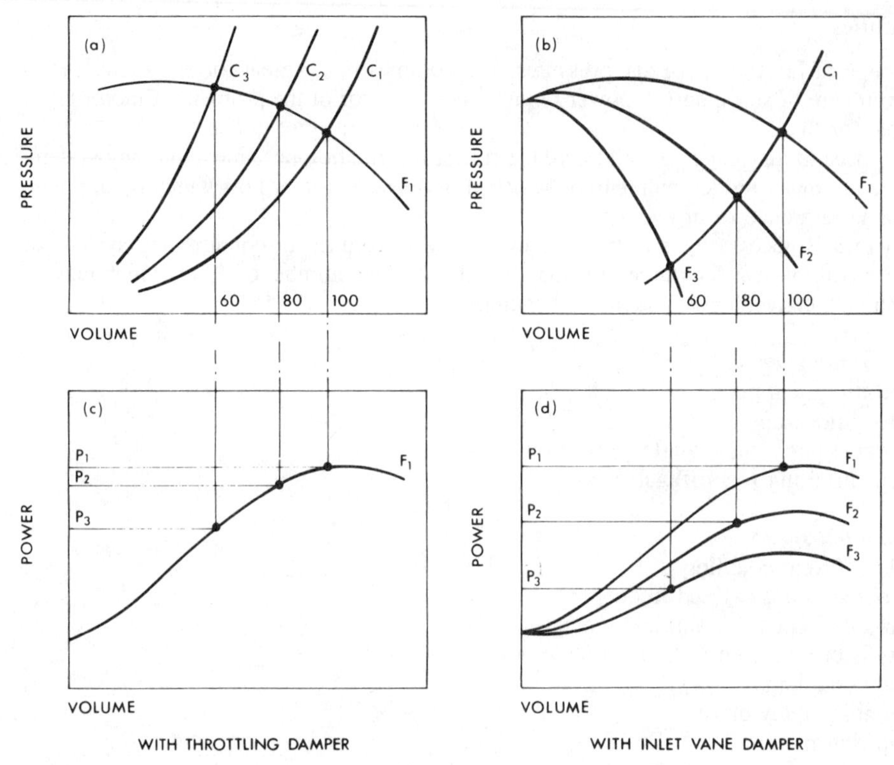

Figure 17.8 Control of fan volume using dampers

Where *throttling dampers* are introduced into the system, volume reduction will be achieved at the expense of efficiency as illustrated in parts (a) and (c) of Figure 17.8. In (a), the full volume operation of a system is represented by the intersection of the fan characteristic F_1 with the system characteristic C_1. The use of throttling dampers changes the *system* characteristic to C_2 and C_3 at 80% and 60% volume respectively. The resultant small savings in power absorbed are shown in (c).

The use of radial *inlet vane dampers* fitted to the eye of a centrifugal fan provides a much more effective means of volume control since such devices act, by changing the air flow pattern at the fan inlet, to modify the performance of the fan impeller in much the same way as would result from change in the speed of drive. This effect is illustrated in parts (b) and (d) of Figure 17.8. As before, F_1 represents full volume operation with the system characteristic C_1: operation of the radial inlet vanes changes the *fan* characteristic to curves F_2 and F_3, which represent 80% and 60% volume, respectively, against a constant system characteristic curve. The resultant dramatic savings in power are shown in (d). Throttling dampers would only be considered for use in very small systems used intermittently.

Where axial flow fans are used with systems demanding volume reduction, the facility of change in *pitch angle* of the blades may be automated. This permits both volume and power to be reduced as required. Figure 17.9 shows the effect of blade pitch angle on power consumption at various duties. As with inlet guide vanes on centrifugal fans, a significant power reduction results.

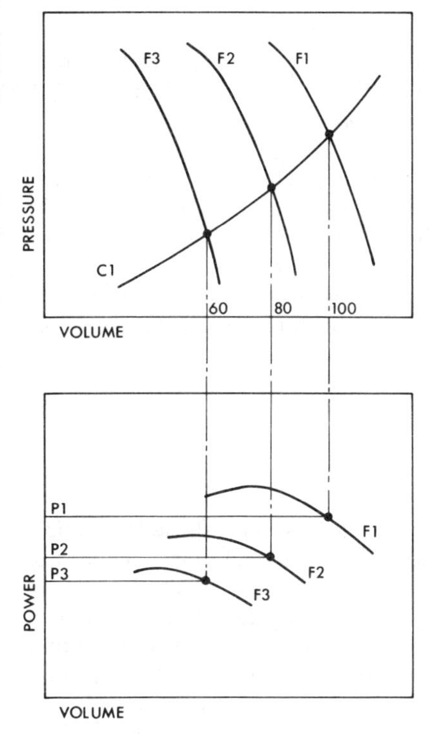

Figure 17.9 Control of fan volume (axial flow) by variable pitch

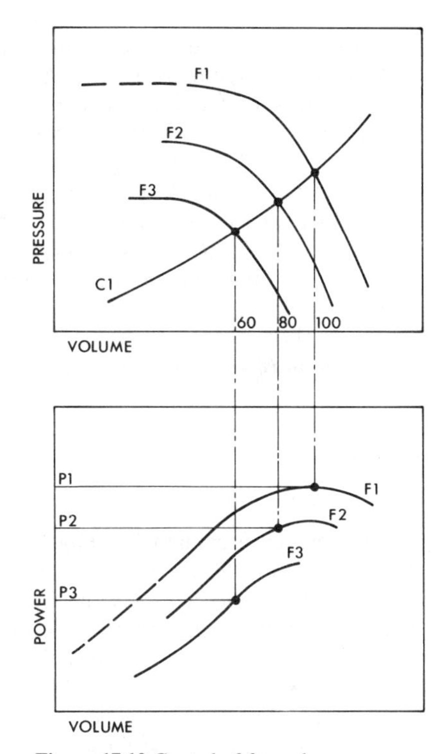

Figure 17.10 Control of fan volume (centrifugal) by variable speed

The *disc throttling* method for control of centrifugal fans alters, in effect, the width of the impeller which varies the volume whilst maintaining the developed pressure. In performance terms this is equivalent to varying the pitch angle of an axial flow fan.

The *variable fan speed* method is well suited to variable air volume systems in comfort air-conditioning applications. In these systems, the torque varies with the square of the speed and the power with the cube of the speed. Usually a drive is supplied capable of working against a torque, with the reduced power and torque requirements at lower speeds taken into account in the design of the motor. Special attention should be given where a constant pressure has to be maintained to all, or to a major part of, the system distribution since, to achieve the required performance over the range of volume variation, the motor characteristics would need to match the particular system requirements. Fan operating characteristics for centrifugal fans with variable speed control are shown in Figure 17.10.

Multi-speed dual wound or pole-change motors may be used where the operating requirements are in clearly defined steps, such as winter/summer or day/night operation.

Position of outside air intake

It might be thought that the purest air supply would be obtainable at the roof of a building, but experience shows that in many instances this is not the case. At roof-level, air is certainly free from street dust, but chimneys or contaminated exhausts from kitchens, toilets or fume cupboards of the same or neighbouring buildings may, with certain states of the wind, deliver fumes into the intake. Particular care must be taken to ensure that drift from cooling towers, or other possible sources of *Legionnaires' disease*, cannot be carried over into air intakes.

An intake at low level, such as near a busy street, would be liable to draw in much road dust and exhaust fumes from motor vehicles. If a point half way up the elevation of a building can be found, this is probably the best. It must, however, be clear of windows where fire or smoke might occur, and particularly of lavatory windows.

There is no general solution to the problem of the outside air intake position, as obviously every case requires examination of orientation, possible sources of contamination, position of air-conditioning plant, and so on.

Air filtration

Air contaminants

Atmospheric air is contaminated by a variety of particles, such as soot, ash, pollens, mould spores, fibrous materials, dust, grit and disintegrated rubber from roads, metallic dust and bacteria. The heavier particles may be such that under calm conditions they will settle out of their own volition. These are termed 'temporary'. The smokes, fumes and lighter particulate matter remain in suspension and are termed 'permanent'. Non-particulate contaminants, such as vapours and gases, also exist in the air. Sulphur dioxide is the most damaging, affecting building fabric, vegetation and artifacts; carbon monoxide and other oxides of nitrogen are also present but normally in small concentrations.

The unit of measurement for dust particles is the *micron* (one millimetre = 1000 microns). The human hair has a diameter of about 100 microns, and the smallest particle visible to the naked eye is about 15 microns. The smallest range of particles we need consider here is of the order of 0.01 to 0.1 μm, which is represented by smokes of various kinds, such as tobacco smoke. The upper range of particle size we need consider is about 15 μm.

Pollution in all its forms, and especially atmospheric pollution, continues to be a subject of increasing public concern, though apart from smoke, fumes and soot it appears doubtful whether much of the other airborne dusts and dirt are susceptible of reduction. Tests are regularly carried out and records kept of suspended pollution material and of sulphur dioxide concentrations.*

Tables 17.1 and 17.2, extracted from the *Guide Section B3*, give typical values for solids in the air for different localities and typical analyses of dust contamination with respect to particle size. These values will vary between locations and with the season; winter conditions normally producing the highest values.

Necessity for air cleaning

If, in a mechanically-ventilated or air-conditioned building, air is blown in without some means for filtration, deposits of dust will be found to occur throughout the rooms and the system of ducts will in itself become coated with solid matter. Heater batteries, cooling coils and fans will also become fouled so that in time the efficiency of the system as a whole will fall off at an increasing rate. Except in certain industrial applications,

* Warren Spring Laboratory, on behalf of the Department of Trade and Industry.

Table 17.1 Typical mass of solids in the atmosphere

Locality	*Total mass* (mg/m^3)
Rural and suburban	0.05–0.5
Metropolitan	0.1–1.0
Industrial	0.2–5.0
Factories or workrooms	0.5–10.0

Table 17.2 Typical analysis of atmospheric dust

	Amount of solid (%)	
Range of particle size, diameter (μm)	*Number of particles*	*Total mass of particles*
30–10	0.005	28
10–5	0.175	52
5–3	0.25	11
3–1	1.1	6
1–0.5	6.97	2
Below 0.5	91.5	1

ventilation and air-conditioning systems therefore invariably include some means for filtration of the air.

The removal of the larger particles is, of course, a simple matter, since any mesh of fine enough aperture will arrest such particles. A plain mesh is however liable to become clogged very quickly, and hence is of little use for the purpose. The finer material and the smokes are, however, much more difficult to arrest and yet it is these which are largely responsible for the staining of decorations, the soiling of shirts and garments and, to some extent, also, the bearing of harmful bacteria. Apart from outside air, recirculated air carries fluff from carpets, blankets and clothes, dust from paper and brought in on shoes and, in an industrial application, any dust resulting from the process.

The greater the degree of filtration, as a rule, the higher the cost of the equipment and the greater the space occupied. The selection of the best filter for a particular application therefore depends on whether great value is placed on a high degree of cleanliness or not. It is perhaps worthy of note that the staining on ceilings close to the point of air supply to a space, seen in many installations after a period of use, is more likely to be due to contamination generated from within the space than to dirt in the outside air.

Tests for filters

The filter efficiency is a measure of its ability to remove dust from the air, expressed in terms of the contaminant concentrations upstream and downstream of the filter, thus:

$$\eta = 100\left(\frac{C_1 - C_2}{C_1}\right)$$

where

η = filter efficiency (%)
C_1 = upstream concentration
C_2 = downstream concentration

Weight or gravimetric test

With this method, a carefully metered quantity of air containing a known quantity of synthetic dust is drawn through a filter paper from the unfiltered intake, and a similar quantity of air is drawn through another filter paper downstream from the filter. These are weighed on an accurate balance and a comparison of the two weights gives the gravimetric efficiency. The heavier particles, as explained, are the most easily collected and these constitute the greater part of the weight, hence even a poor filter will give a high efficiency of perhaps 90% by the gravimetric method. In consequence, the weight test has effectively been superseded.

*Dust spot test**

Using this method, sample quantities of air are drawn, as before, from upstream and downstream of the filter under test through filter papers and the resultant stains are viewed optically. The light penetration is measured by a photo-sensitive cell and a comparison of the relative intensities then provides dust-spot test efficiency. This, known previously as the *blackness* test, is a much more stringent criterion and many filters which may provide a 90% result by the gravimetric method are able to produce no more than 50% following a dust-spot test.

Arrestance tests

These establish the ability of the air cleaning medium to remove injected dusts from an air stream and may be carried out as a part of the dust spot test procedures. Arrestance is expressed as a percentage, calculated as for efficiency, but using mass in place of concentration values. Both the dust spot and arrestance tests may be used for on-site testing (see BS 6540: Part 1: 1985) and may use either atmospheric dust, i.e. that present in the air at the site of the test, or a synthetic dust consisting of 72% by weight fine dust, 23% by weight coal dust and 5% cotton fibres. *Methylene blue* particle sizes closely resemble the distribution found in atmospheric pollution and the test going under this name is similar in principle to the old blackness test and, like it, has been superseded.

Sodium flame test

This is used for testing high efficiency filters, i.e. with a penetration of less than 2%. The method is described in BS 3928: 1969 and involves generation of an aerosol of sodium chloride of particle size between 0.02 and 2 μm. Samples of the air upstream and downstream of the filter are passed through a flame photometer to determine the concentrations captured. Where high filter efficiencies are vital to the installation the quality of the seal between the filter and supporting frame is as important as the efficiency of the filter media. On-site testing of such high efficiency particulate air filter (HEPA) installations is in consequence necessary before the plant is put into service and at frequent intervals during use. Tests used for on-site testing include the Di-octyl-phthalate (DOP) test and the sodium flame test.

* BS 6540: Part 1: 1985.

Practical filters

In selecting a filter it is necessary to know by what method the maker's guarantee of test efficiency has been determined, and what type of dust was used, since it is unlikely that a determination was made under the particular conditions of atmospheric pollution obtaining at the site in question.

The *dust holding capacity* of a filter, that is the mass of dust a filter can retain between its 'clean' and 'dirty' condition, is also an important feature to consider when selecting a filter because this will affect the frequency of maintenance or replacement.

Air filters fall into five main categories as follows:

- *Viscous impingement.* Usually of some form of corrugated metal plates or metal coils or turnings or the like, in each case covered with a viscous oily liquid to arrest the particles on impingement. Other materials used instead of metal are glass fibres, similarly coated with a sticky fluid.
- *Fabric.* The material used in this type of filter is some form of textile, normally glass or synthetic fibres in a random matting bonded together. The dirt particles are in this case arrested partly by being trapped in the interstices of the material, and partly by being caught on the fibres.
- *Electrostatic.* In this system of filtration, dust particles entering the filter are subjected to an electrostatic ionising charge and, on subsequently passing through parallel plates which are alternately charged and earthed, the particles are repelled by the charge plates and adhere to the earthed plates. The electrical charge is supplied from a power-pack containing the necessary transformers and rectifiers to produce the high voltage DC required.
- *Paper or absolute.* This filter uses a special form of paper made usually from woven glass fibre.
- *Adsorption.* Makes use of activated carbon (charcoal), activated alumina or other chemicals to adsorb odours, gases and the like.

Air washers will also act as air cleaning devices, but are unlikely to be suitable for use in modern systems.

Materials used should be inherently non-flammable or so treated that they retain such qualities through their life. Viscous filter liquids should have a flash-point of not less than 177°C. All such material should generate a minimum quantity of smoke and toxic gas*

Relative efficiencies

Figure 17.11 indicates trends of efficiencies which may be expected from the first four types of particulate filter referred to above.

It will be seen that filter 1, depending on impact, has no arrestance on small particle sizes. Filter 2 has a dust spot test efficiency of about 50% for particles of about 1 micron, diminishing rapidly for the smaller particles. The electrostatic filter, curve 3, has a dust spot test efficiency of about 90% while the paper filter 4 achieves virtually 100% for particles of 0.1 μm and above.

All filters vary in efficiency according to the velocity of air through them. Velocities and resistances of various filters are given in Table 17.3.

In the case of fabric and paper filters, owing to the large surface areas involved, designs are usually based on forming the material into a zigzag formation.

* *Test Methods for Ignitability, Smoke and Toxicity of Air Filters.* London Scientific Services, 1990.

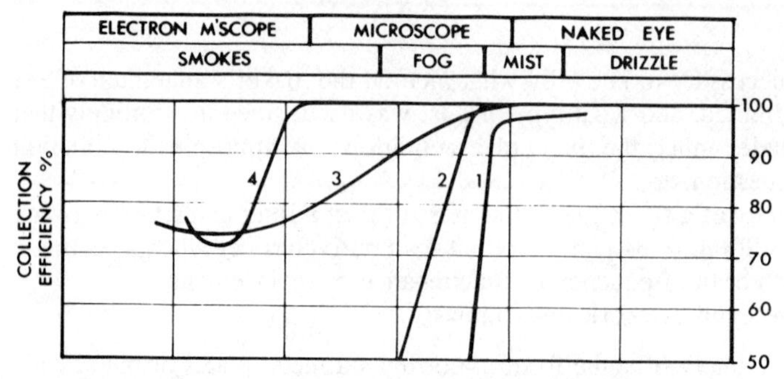

Figure 17.11 Efficiencies of air filters: (1) viscous coated metal; (2) fabric; (3) electrostatic; (4) paper absolute

Table 17.3 Typical characteristics of particulate filter media

Filter type	*Face velocity* (m/s)	*Resistance* (Pa) *Clean*	*Resistance* (Pa) *Dirty*	*Approximate efficiency/arrestance* (%)
Viscous impingement				
cleanable panel	1.5–2.5	20–60	100–160	65–80 arrestance
automatic curtain	2–2.5	30–60	100–190	80 arrestance
Dry fabric or fibrous				
cleanable panel	1.5–2	30–50	75–100	70–80 arrestance
disposable panel	1.5–3.5	45–90	150–250	70–90 arrestance
low efficiency bag	1.5–3.5	25–50	200–250	30–50 dust spot
medium efficiency bag	1.5–3.5	55–140	200–350	50–90 dust spot
high efficiency bag	1.5–2.5	55–140	200–350	Up to 95 dust spot
Automatic roll	2.5–3.5	30–80	160–200	30–45 dust spot
Electrostatic, plus bag filter	1.5–2.5	120–200	250–400	Up to 95 dust spot
Absolute				
low efficiency	up to 1.5	up to 150	up to 400	95 sodium flame
medium efficiency	up to 1.5	up to 280	up to 625	99.7 sodium flame
high efficiency	up to 1.5	up to 280	up to 625	99.997 sodium flame

Filter cleaning

When heavily charged with dirt, the resistance of most filters rises sharply, thus reducing air flow, and hence ventilation rate. The cleaning of filters is achieved in a variety of ways as follows:

Washable filters. These, consisting of a foam plastic element, are contained in metal frames with wire retaining grids as shown in Figure 17.12. The material is flame retardant and, when fouled, is washed in detergent and reused. The media have a long life.

Brush filters. Eliminating the necessity for any special cleaning, these consist of 'flue-brush' elements contained in a segmented frame as shown in Figure 17.13. The element

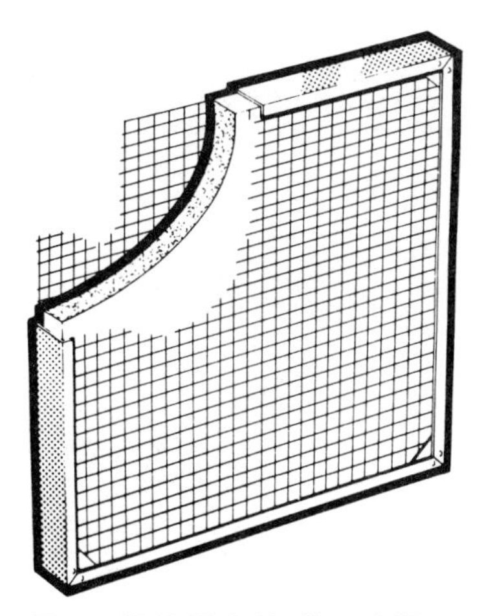

Figure 17.12 Washable filter (ACE)

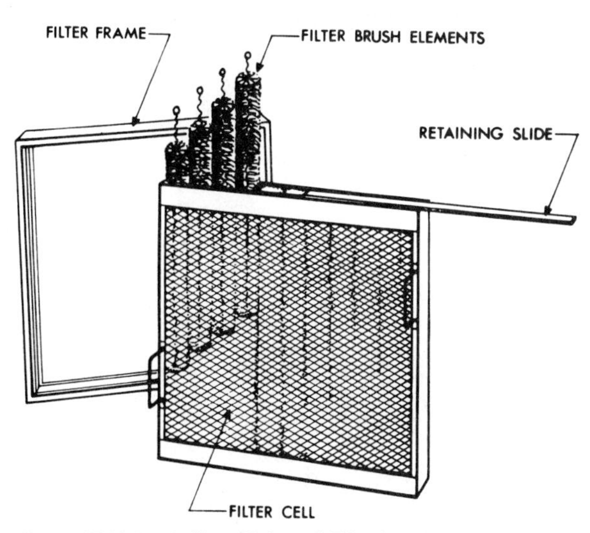

Figure 17.13 Brush filter (Universal Filters)

of hair, nylon, steel or nickel silver wire is removed for either vacuum or other simple cleaning and washed before replacement. The media have an indefinite life.

Fabric filters. In panel or wedge form, as in Figure 17.14, these are mounted in frames and when dirty are thrown away. In another type, the filter material is of a glass fibre or other suitable base, and is supplied in rolls. The roll is horizontal and is gradually wound from one spool onto another on the principle of a camera film, see Figure 17.15; this movement is achieved by motor drive controlled from a pressure differential switch across the filter, or from a timing device. Fresh filtering medium is unrolled only as it is required. The medium may pass over open mesh drums, as shown in the diagram, or be flat and held in place by retaining bars on both upstream and downstream faces. Such equipment may, alternatively, operate with the rolls vertical but the edge sealing arrangements are then less effective and the drive may be less positive due to mechanical problems.

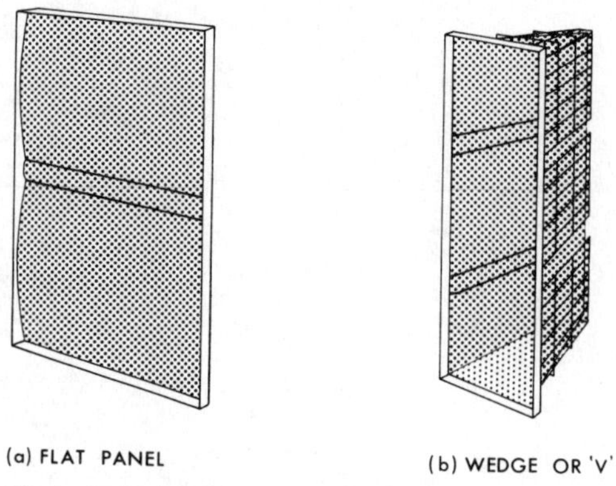

Figure 17.14 Fabric filter, single cell (Trox)

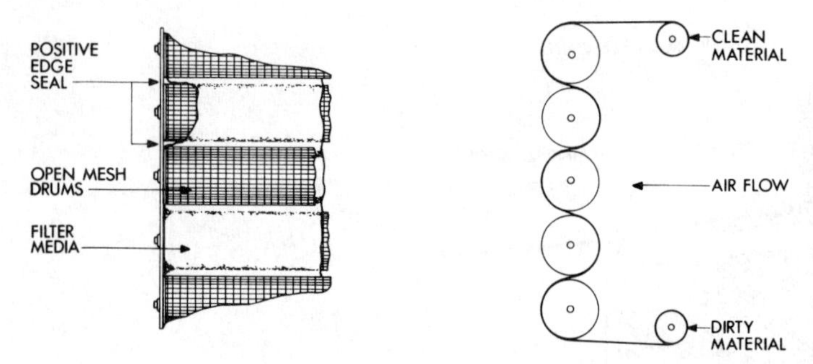

Figure 17.15 Automatic roll type fabric filter (Ozonair)

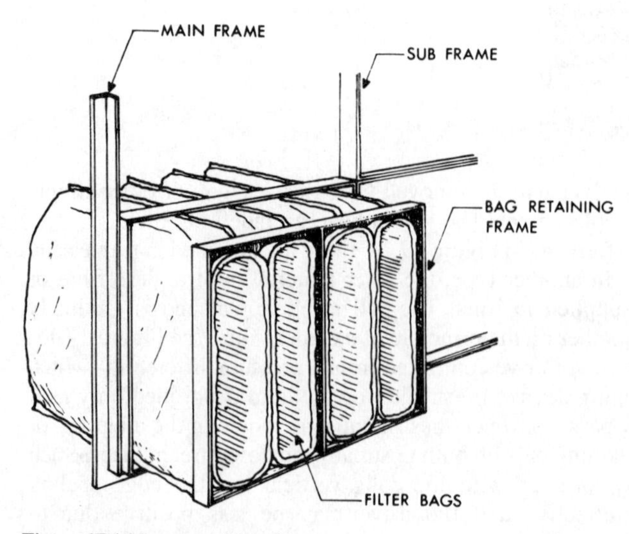

Figure 17.16 Bag type fabric filter

This type of filter will, in theory, require no labour for maintenance, excepting at the long intervals of possibly three to six months for the changing of a complete roll.

Bag filters. The bag filter has a replaceable medium and this has largely taken the place of the 'roll' fabric filter since it has better performance characteristics and requires less skill in maintenance. A common arrangement is shown in Figure 17.16.

Absolute filters. These are supplied in panel form and are discarded when dirty and replaced. To achieve high efficiency the air velocity through the medium is kept relatively low. It is essential that efficient pre-filters are installed upstream of absolute filters to extend the life of the media. A typical cell is shown in Figure 17.17. Special attention must be given to sealing the cells into the frames.

Viscous impingement filters. These may be in the form of cells comprising viscous coated corrugated plates removed for cleaning by hand, which are washed, re-oiled and replaced. They have also been developed on a crude self-cleaning principle and, in one form see Figure 17.18, the zigzag plates are vertical, oil flowing over them from a pump drawing from the base tank at intervals. In another form the cells are on an endless chain dipping into the oil in the base tank and returning for re-use. This type of filter would normally be used where an extremely dirty atmosphere exists, such as found in a heavy industrial area.

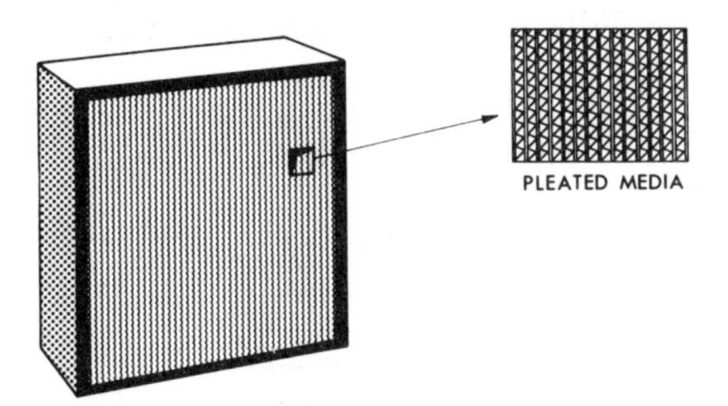

Figure 17.17 Absolute filter, single cell (AAF)

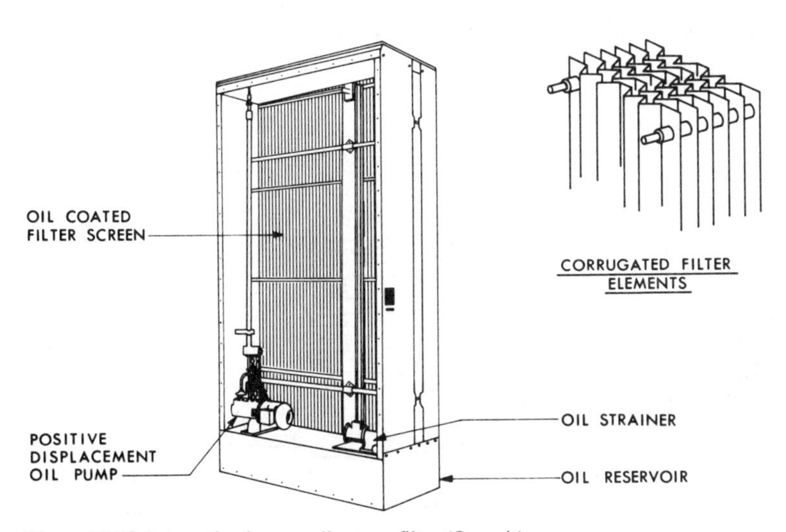

Figure 17.18 Automatic viscous roller type filter (Ozonair)

Electrostatic filter. In this type of filter the dirt collects on the earthed plates and its removal is accomplished by washing with hot-water jets; adequate drainage is therefore necessary. This may be done by hand or automatically, the latter being preferred. Another type uses oiled plates to increase the dust retention properties. After cleaning, a period of drying is necessary before the filter can be put back into use. Pre-filtering is recommended and consideration should be given to an after-filter for use when the electrostatic section is being cleaned or during failure.

Another pattern, which is more compact and requires less maintenance, allows the dust particles to collect on the plates in a thickening layer to be swept off by the air stream as agglomerated flakes on to a roll or bag filter which is an integral part of the filter assembly (see Figure 17.19).

For clearance of fog only, the electrostatic and absolute filters are effective, or a combination of electrostatic plus fabric. Excessive collection of moisture from fog may cause an electrostatic filter to cut-out due to short circuit. A special heater may be used ahead of the filter to ensure that air entering is dry, thus avoiding this trouble.

Adsorption filter. The service life of an activated carbon filter is between 6 months and 2 years, depending on the type of installation and the concentration of contaminants being adsorbed. An efficient particulate filter is recommended for use upstream. Normally, the cells are re-chargeable. Typical cell arrangements are shown in Figure 17.20. The pressure drop across the unit is about 70 Pa and the face velocity of the order 1.5 m/s.

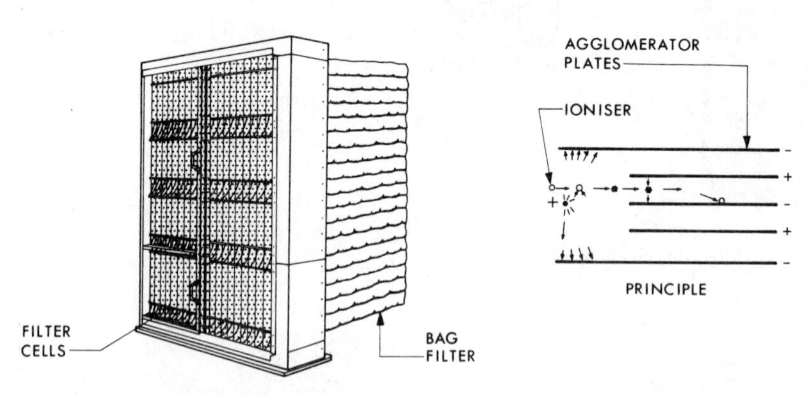

Figure 17.19 Electrostatic filter (AAF)

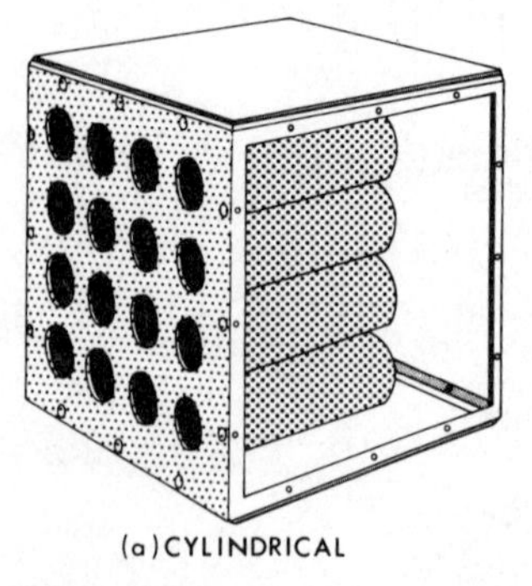

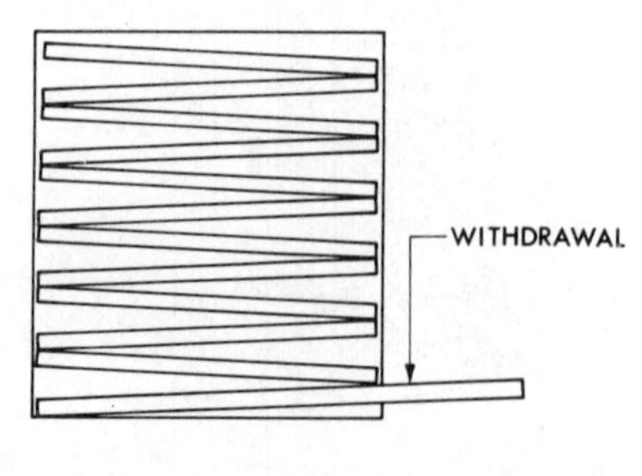

Figure 17.20 Adsorption filter, single cell

Air humidification

During winter months the outside air has a low moisture content and when introduced to a conditioned space tends to reduce the relative humidity. To counter this effect, it is often necessary to introduce moisture into the space to maintain the required conditions for comfort or for a process.

Air humidifiers may be classed as either direct or indirect.

Direct types. Introduce moisture into the space to be treated and their use is normally limited to industrial or horticultural applications: these types are outside the scope of this book.

Indirect types. Introduce moisture into the supply air either at the plant or within the ductwork, dependent upon the type of system.

A further sub-classification of the indirect type is *storage* or *non-storage*. The storage types incorporate a small tank under the humidifying apparatus from which water is introduced into the air stream, in one of a variety of ways. The water level in the tank is maintained by a water supply via a ball valve. With the non-storage type moisture, in the form of steam or atomised water particles, is injected directly into the air stream.

Concern as to humidifier-related illness has changed the approach to selection and use of the associated equipment quite radically. Storage-type units are no longer acceptable for application to systems in hospitals, where steam injection is recommended, a trend seen also in commercial practice. Where storage types exist or are envisaged, particular care in cleaning, maintenance and inspection is required at frequent intervals. Further recommendations relate to the need for biocide treatment of the tank water and drainage thereof to waste, at regular intervals, preferably daily. The tank must be drained and cleaned when the humidifier is to be out of use for any extended period.*

The types of indirect humidifier may be summarised as follows:

- *Non-storage*
  steam injection;
  mechanical separators.
- *Storage*
  spray washers;
  capillary washers;
  pan humidifiers;
  sprayed coils;
  ultrasonic atomisers.

Steam injection

Controlled steam injection may be supplied either from a central boiler plant via a sparge pipe, Figure 17.21, or from a local packaged unit as in Figure 17.22. The former, unless served from an independent steam generator for the purpose, has limited use due to the odour and oil traces characteristic of central boiler installations. The latter type of equipment is mounted on or close to the air-conditioning plant and consists, in essence, of a small electrode steam generator, sized to suit the moisture requirement of the plant, controls and a water supply: steam heated units are also available. The standard equipment

* CIBSE, *Minimising the Risk of Legionnaires' Disease*, CIBSE Technical Memorandum TM13, 1991.
CIBSE, *Legionellosis (Interpretation of Approved Code of Practice: The Prevention or Control of Legionellosis*, CIBSE Guidance Note GN3, 1993.

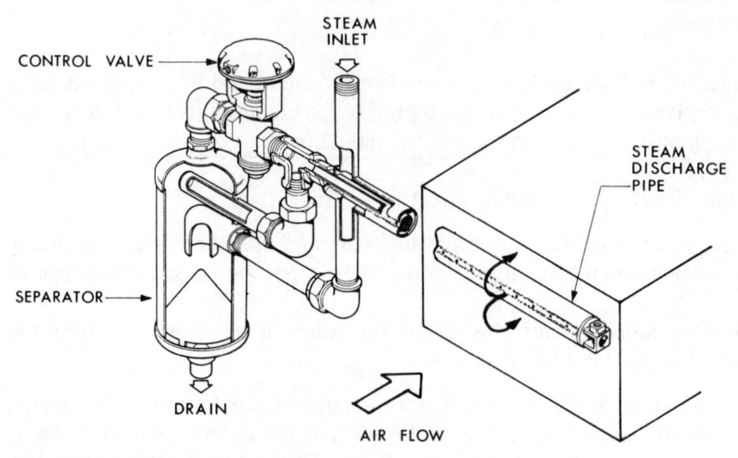

Figure 17.21 Mains steam humidifier, injection type (Spirax Sarco)

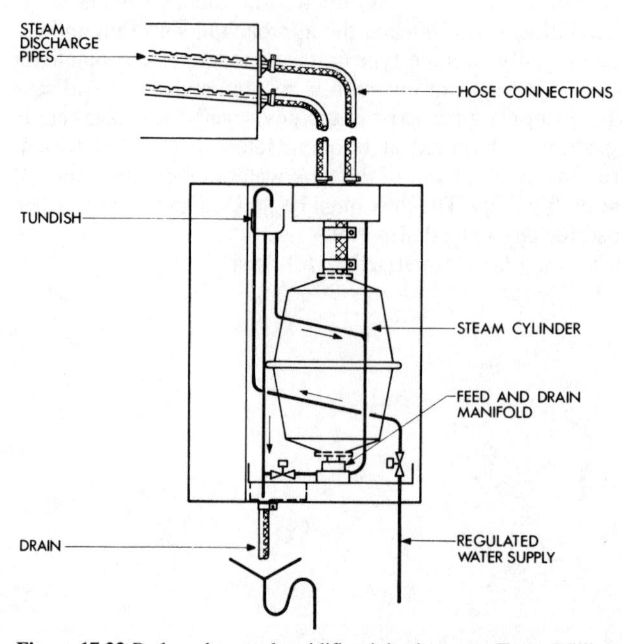

Figure 17.22 Packaged steam humidifier, injection type (Eaton Williams)

available is not suitable where the static pressure in the duct is over 1250 Pa. Good access is required for all humidifiers and sparge distribution pipework should be arranged to be self-draining. Saturation efficiency is about 80%.

Mechanical separators

These commonly operate via use of spinning discs, as illustrated in Figure 17.23. It must be remembered that, in the process of atomisation and evaporation, any mineral salts

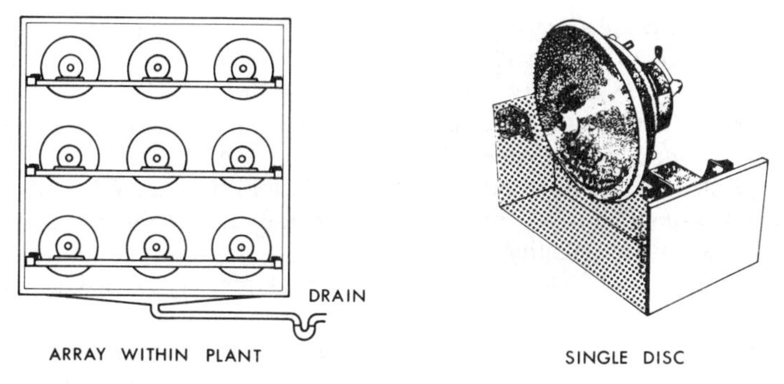

Figure 17.23 Spinning disc type humidifier

dissolved or suspended in the water supply will be released and deposited: some water treatment may be required. Saturation efficiency is claimed to be as high as 90%.

Spray washers

This type of equipment is seldom used in current designs, but since many are still in use a full description is given, together with a figure of appropriate vintage.

An air washer, see Figure 17.24, consists of a casing with a tank formed in the base to contain water. Spray nozzles mounted on vertical pipes connected to a header deliver water in the form of a fine mist. The spray is projected either with or against the air current, or, where two banks of sprays are provided, in both directions, one with and one against the air stream. The water for the sprays is delivered by a pump, under a gauge pressure of between 0.2 and 0.3 kPa, the water being drawn from the tank through a filter. The casing has an access door and internal illumination, and may be of galvanised steel, or constructed of brick or concrete, asphalted inside.

As the mist-laden air is drawn through the chamber it requires to have the free moisture removed, and for this purpose eliminator plates of zigzag formation are provided to arrest the water droplets. Some makes precede these with scrubber plates, down which water is caused to run by a spray pipe at the top in order to flush down dirt which has collected

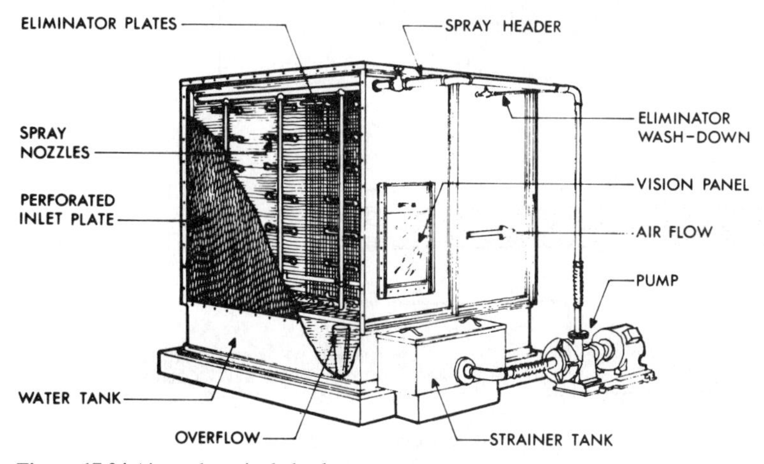

Figure 17.24 Air washer, single bank spray type

from the air. Galvanised eliminator plates are liable to rapid deterioration; better materials are copper and stainless steel. Glass plates with serrated ribs have been used successfully, and they are permanent.

A single bank air washer will not saturate the air more than that corresponding to about 70% of the *wet bulb depression*. A double bank washer may reach 90%. The recirculated water may be cooled or heated for full air-conditioning and dehumidification. Where it is desired to saturate at the dew-point, a two-bank washer is generally necessary with large spray nozzles to pass the necessary quantity of water for the temperature rise allowed. Spray nozzles vary in capacity from 0.05 to 0.25 litre/s, and should be easily cleanable and of non-corrodible metal.

The air velocity through an air washer is usually 2.5–3 m/s. The length is normally about 2.5 m, but is increased with the two or more banks of sprays needed for cooling – sometimes to as much as 4.0 m. On the inlet side straightening vanes, or a perforated grille, are required in order to distribute the air evenly over the whole area. The tank is kept filled by a ball-valve, with a hand valve for quick filling, and there is in addition a drain and overflow pipe. The pump is of normal type, preferably with flexible connections where noise may cause trouble, and arranged to be flooded by the water in the tank, in order to avoid the need for priming.

Apart from removing heavier and gritty material a spray washer has a low air filtration efficiency, but is effective in adsorbing certain gases, such as sulphur dioxide, from the air.

Capillary washers

One arrangement of this type is shown in Figure 17.25. Typically such units comprise one or a series of cells inclined at an angle, or vertical, and containing corrugated aluminium

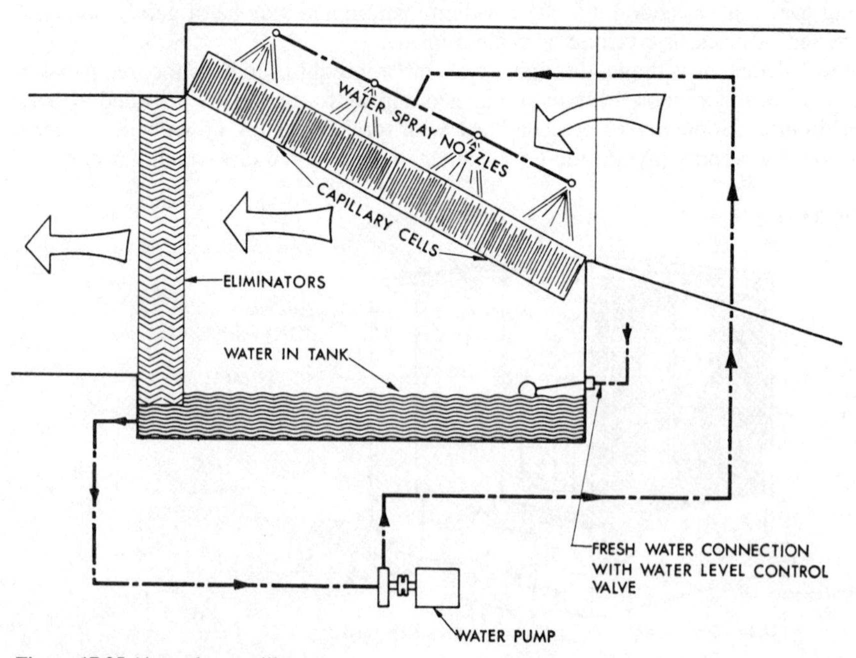

Figure 17.25 Air washer, capillary type

or filaments of glass or similar material which provide an extended surface for water and air contact. Water at low pressure is caused to flow over the cells by flooding nozzles from a pump of low power consumption. The water and air have to negotiate the *striations* of the fill together, and are thus intimately mixed so that saturation up to 90% may be achieved. At the same time it may be taken that the filtering efficiency is of a reasonably high order down to 3 μm. It may equally be used with refrigerated water for a full air-conditioning system. A maximum velocity through the cells of 2 m/s is recommended. Although this form of humidifier is still available from manufacturers, it is not in common use today.

Pan humidifiers

A simple pan humidifier consists of a shallow water tray, replenished by a ball valve and heated to improve its effectiveness by an electric immersion element, or a piped heating supply. Efficiencies are low, and since the heated water would be at a temperature that would present a high risk of bacteriological contamination, this type of humidifier cannot be recommended. This type of equipment is not in common use.

Sprayed coils

Cooling coils fitted in air handling plant, sprayed with water from low pressure nozzles, can provide a convenient and effective means of humidification. This arrangement is described later under the heading 'Sprayed cooling coils' (p. 488).

Ultrasonic atomisers

A recent development generates water vapour from a cold water source using high frequency vibrations to atomise water droplets into the air stream which are then evaporated. Electronic oscillating circuits power transducers, matched at their resonant frequency, to produce high frequency mechanical vibrations just below the water surface which cause water particles to be released from the surface. It is claimed that these humidifiers use less than 10% of the power that would be absorbed by a steam raising system.

General

Indirect humidifiers using a water spray or equivalent should be fitted with eliminator plates downstream to prevent moisture carry over into the ductwork. Treatment of water may be necessary where supplies contain a high degree of temporary hardness or calcium salts. It is worthy of re-emphasis that humidifiers having water storage are subject to risk of bacteria growth and therefore require special attention. Components should be completely drained when not in use for more than a few days. Adequate provision for access, inspection and maintenance is of paramount importance.

Air heating and cooling coils

Air heating coils

When originally introduced these were commonly of plain tubing, as shown in Figure 17.26(a). In order to economise in space and achieve greater output from a given amount of metal, finned heaters are now generally used as in Figure 17.26(b). Construction in the

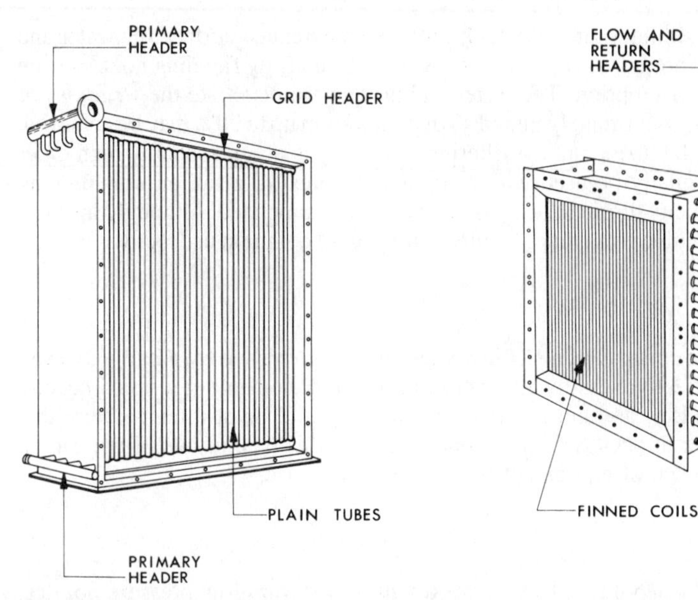

Figure 17.26 Air heater batteries

latter case is commonly with copper tubes and aluminium fins but better practice requires that fins also are of copper in order to avoid corrosion. Plain tubes are less likely to become choked with dirt than finned – a fate all too common. Plain tube heaters are suitable for use in fresh air intakes to prevent wet fog collection on fabric or electrostatic filters. They should in this case be of galvanised steel or other non-corrodible metal, but not copper. Heaters are arranged in stacks or batteries with automatic controls preferably of modulating type to give a steady temperature of output.

Where the heater is warmed by hot water, a constant temperature supply is required from the boiler or calorifier with a pumped circulation. The flow required being relatively large, a control valve of diverter type will preserve the circulation irrespective of the demands of the heater. If the heater is fed by steam, it will require the usual stop valve and steam trap and a means of balancing the pressures so that the condensate may not be held up by a vacuum caused by control valve shut-off. Heaters may be enclosed in sheet-steel casings, 'packaged' with other elements into a prefabricated unit, or built into builders' work enclosures, although this final method cannot be recommended unless the builders' work is lined with sheet metal.

The number of rows of tubes depends on temperature rise, temperature and nature of heating medium, i.e. whether steam or hot water, and air speed. The face area depends on air volume and free area between tubes. This is again determined by the velocity, and it is usual to fix this arbitrarily beforehand, generally between 4 and 6 m/s through free area, or 2.5–3.5 m/s face velocity. For sizing of heaters reference is necessary to makers' data.

Direct electric air heaters are available and suitable for use as *trim heaters* on supplies to individual rooms where a water or steam heating media cannot be provided economically. Electric heaters are also provided on some packaged room air-conditioning units for computer rooms and the like. The case for direct electric heating normally can be made only where the annual demand is relatively small. Heat output is by any number of controlled steps.

Indirect gas fired heaters may also be considered where close control of temperature is not important and where long distribution routes for heating water or steam would otherwise exist, leading to high initial cost and considerable distribution heat loss. A good example of their use is for serving a shopping centre. A flue to atmosphere is, of course, required.

Air cooling coils

Cooling coil surfaces as used in air-conditioning generally perform two functions – to remove sensible heat, and to remove moisture or latent heat: paradoxically, however, they may be sprayed with recirculated water and then used to add moisture or latent heat. The sizing and temperature of operation depend on the sensible:latent ratio. If a low relative humidity is required, the dew-point will be low, and hence the water temperature must be low. If sensible cooling only is required, the coil surface temperature must be kept above dew-point.

In any coil a certain proportion of air fails to come in contact with the cold surfaces, and thus a part may be chilled and dehumidified and a part remain unchanged. By correct selection of coil form the desired ratio may be obtained.

In direct expansion systems, the actual refrigerant gas from the compressor is passed direct into the coils, which form the evaporator of the refrigerating plant.

In a chilled water system, water is circulated through the coil, and this may be used down to about 3.3°C. Below this an anti-freezing mixture such as calcium chloride brine becomes necessary.

In cooling by refrigeration, the higher the evaporator temperature the less power is consumed for a given duty, hence it is desirable to keep the cooling coil temperature as high as possible consistent with the final air temperature required to meet design conditions. This applies whether the coil is used for direct expansion or with chilled water or brine.

A cooling coil surface is therefore usually designed for small temperature differences between water and air; for instance, for air cooled from 24°C to 13°C, water may be at 10°C inlet and 14.5°C outlet (i.e. leaving above the air-outlet temperature). An air washer cannot give a performance comparable to this. Coil face velocities are limited to 2.5 m/s; above 2.25 m/s eliminator plates are fitted to reduce moisture carryover on coils performing a dehumidification function.

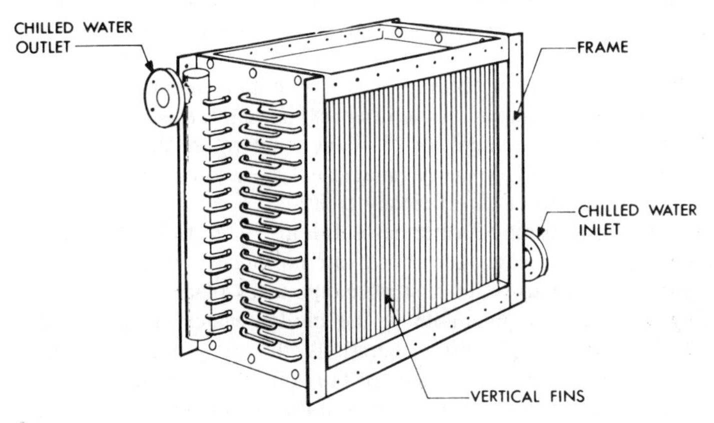

Figure 17.27 Cooling coil (six-row arrangement)

Coolers are generally of finned or block type (as in Figure 17.27), comprising banks of small bore tubes threaded through plates between which the air passes, the whole being of copper which may be tinned after fabrication or, more normally in present day commercial practice, made up with copper tubes and aluminium fins, although the life of this form will be less than the all copper alternative. For applications where coils are to be water sprayed, as discussed later, they must of course be all copper and preferably be tinned. Coils are frequently arranged horizontally, with fins vertical, so as to facilitate drainage of condensation when dehumidifying. A drain pan should be provided and be graded towards a bottom outlet to prevent stagnant water. The connecting drain pipe should have a 'U' trap to maintain a water seal to prevent loss or ingress of air (depending whether the coil is on the suction or discharge side of the fan), see Figure 17.28. The pipe should run to drain via an air gap.

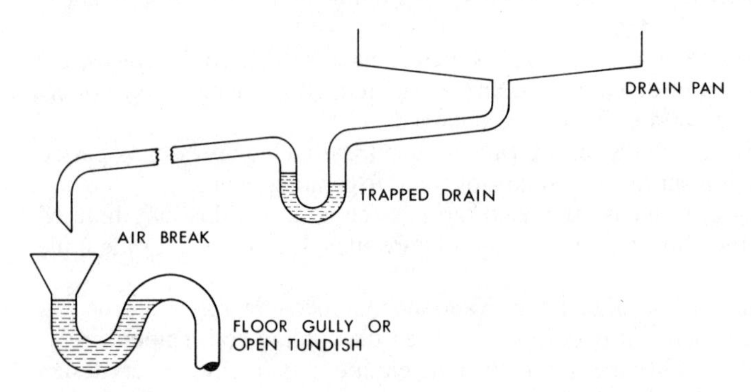

Figure 17.28 Drain arrangements from cooling coil

Sprayed cooling coils

In this application a cooling coil, arranged as normal for a water circulation through the tube internals, is mounted over a water tank and has headers and spray nozzles fitted, facing the upstream face. The nozzles are supplied at low pressure with water drawn by a pump from the tank and returned thereto by gravity after wetting the coil surfaces. A spray pump capacity of 0.75–1.0 litre/s per m^2 of coil face area is usual and, to make good any loss due to evaporation, the tank is provided with make-up from a mains supply via a ball valve.

In winter, during which season the water chilling plant may not be running, use of the spray will enable the coil to act as an adiabatic humidifier and in many respects this is the principal application. As a result of the small air to water temperature differences for which a cooling coil is designed, the surface area available is large and this, when wetted by the sprays, provides for a high saturation efficiency of 80–90%.

With some system configurations, it may further be useful, in winter, to keep the water circulation within the coil active to cater for those circumstances when the air-on conditions are below say 5°C and air to water heat exchange takes place. As a result, a modicum of free cooling will become available in the chilled water circuit for use elsewhere in the system. For certain mid-season operations, the spray may be used to enhance the cooling and dehumidifying performance of the coil when active.

It must be emphasised that care is necessary on the part of the plant owner and operator to ensure that maintenance standards are adhered to and that the necessary precautions against biological and other contamination of the spray water, and thence the conditioned air, are taken. For this reason, sprayed coils are less popular than they were.

Air dehumidifiers

As an alternative to using refrigeration for both cooling and dehumidification, air may be dried by an *absorption* process using a liquid hygroscopic chemical, such as lithium chloride. The process is temperature sensitive and moisture exchange takes place over a coil at a controlled temperature, around 30°C. Specialised packaged plant is manufactured for the process, one such being the Kathabar system, and use is normally limited to industrial process applications, but may be considered for comfort air-conditioning.

Packaged air handling plant

The various components previously described may be built up to provide a complete plant in a number of ways. They may be connected using sheet metal ducts or by incorporation within a masonry chamber: the former method tends to be untidy and space consuming and the latter is somewhat cumbersome and suspect as to permanent air-tightness unless lined with sheet metal.

Current practice favours the use of factory packaged plant where individual components are housed in modular casings for site assembly. The better designs for such modules are so arranged, as shown in Figure 17.29, that they may be assembled in a number of different ways and thus fitted to the building space available.

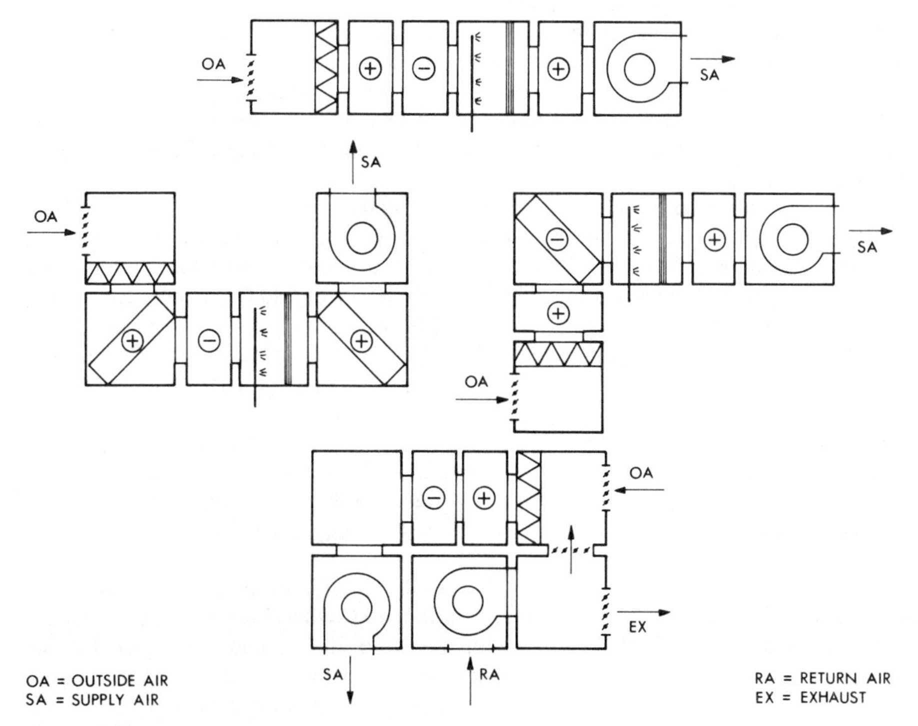

Figure 17.29 Modular packaged air handling plant arrangements

A recent and logical extension of this principle has been the production of these modules in weather protected form such that they, themselves, are plant rooms for, say, roof mounting without the need for architectural enclosure.

In addition to the components shown, packaged equipment is available with air-to-air heat recovery devices built in, and with packaged air cooled refrigeration plant to serve direct expansion cooling coils (these are sometimes operated as heat pumps).

Adequate access for maintenance must be provided between the component elements of the plant; this is particularly important where the process uses water, such as in humidifiers, or where dehumidification may occur on cooling coils. Space must be allowed along the length of the plant to enable control sensing devices to be fitted and sufficient space alongside the enclosure for components to be withdrawn; the width for withdrawal being equivalent to the width of the plant itself unless provision is made for components to be removed in sections.

Air-to-air heat exchangers

Any ventilation or air-conditioning system which takes in outside air, heats and/or cools it and then discharges an equivalent or lesser quantity to waste, offers potential for energy saving. The simplest of plenum ventilation plants, arranged to recirculate 60 or 70% of the air handled, will nevertheless require a heat source to raise the temperature of the remaining outside air supply to whatever level the application may demand. At the other extreme, a sophisticated air-conditioning plant will consume energy in preheaters, re-heaters and zone heaters plus that which will be required as a result of an adiabatic or similar humidification process. In either case, some proportion of the treated air delivered to the building will, by design, quite properly be discarded.

It has been emphasised in earlier chapters that the admission of certain minimum quantities of outside air is necessary for human occupancy and prevention of condensation. For commercial and industrial premises where noxious fumes are generated, in however low a concentration, additional outside air quantities must be supplied beyond the minimum, equivalent to the volume collected for discharge to atmosphere.

To bring the quantities of heat so wasted into perspective, consider a small office block with a floor area of 100 m^2 which has a mechanical ventilation plant to provide a low average quantity of outside air at 2.5 litre/s per m^2, the plant running for 60 hours per week. Over an average winter season, a heat supply of about 280 GJ (equivalent to that provided by burning about 8 tonnes of oil) would be necessary to do no more than raise the ventilation air supply to a degree or so *less* than room temperature. The associated extract ventilation plant would then reject a similar air quantity, at room temperature or warmer, back to outside.

In order to overcome the nuisance of heat gain to rooms from lighting, modern practice allows extracted air to pass over and through luminaires, see Figure 15.26, taking with it up to 70% or so of the electrical input thereto. In our hypothetical example, therefore, the temperature of the air discharged might well be several degrees *above* that held in the office space proper – say 22 or 23°C.

For applications such as hospitals and similar buildings, which ventilate or air-condition without recirculation, in order to avoid contamination, and swimming pools which may be similarly served in order to reduce risk of condensation, the air quantity rejected is far greater per m^2 of floor area. Further, in kitchens and industrial premises, where process heat gain may be high, air will be exhausted at temperatures much greater than those quoted above.

It will be obvious that great scope for energy conservation exists if the heat in exhaust air can be reclaimed and applied, in part at least, as a source of energy to raise the temperature of the outside air used in the parallel supply plants. These same comments apply of course to economies to be achieved in cooling capacity during the summer months since the temperature of air discharged from an air-conditioned building may then be *less* than that of the outside ambient: it may, furthermore, carry less moisture. This aspect assumes more importance in climates which produce extremes of summer temperature.

Available equipment

A variety of types of equipment is available for air-to-air exchange in ventilation or air-conditioning plants. These fall under one of the following headings:

- Plate heat exchangers.
- Glass tube heat exchangers.
- Thermal wheels.
- 'Heat pipe' heat exchangers.
- Run around coils with water circulation.
- Run around coils using refrigeration.

Of these six types, the first three effect heat exchange directly from air-to-air whereas the remainder employ an intermediate circulating medium.

Efficiency of heat reclaim

Before describing the various types of equipment in more detail, the matter of their efficiency in operation requires definition. The data normally quoted derive *temperature* efficiency from the expression:

$$\eta = \left(\frac{t_3 - t_1}{t_2 - t_1}\right) 100$$

where the supply and exhaust flow rates are equal, and t_1 is the temperature of the outside air, t_2 is the temperature of the exhaust air from which heat is to be reclaimed, t_3 is the temperature of the supply air after it has been passed through the heat exchanger and η is the efficiency (%). As will be noted, the expression makes use of dry bulb temperatures and thus, strictly speaking, relates only to sensible heat recovery.

Consider the case of a heat exchanger unit, having a quoted efficiency of 76%, which is applied to a winter reclaim situation handling equal quantities of outside air at –1°C, saturated, and exhaust air at 22°C DB, 15.5°C WB.

Thus,

$$t_3 = [0.76(22 + 1)] - 1 = 16.5°\text{C}$$

For use in summer, when the stated efficiency might have fallen to, say, 66%* taking the same exhaust condition and the outside air supply at 28°C DB, 20°C WB, then:

$$t_3 = 28 - [0.66(28 - 22)] = 24°\text{C}$$

* In winter, as may be seen from Figure 17.30(a), the exhaust side of the heat exchange surface will be wetted for part of the process. In the case of the hygroscopic thermal wheel, however, under the same winter conditions as illustrated, there will be an increase in the moisture content of the incoming air stream, as indicated by S_3 in Figure 17.30(b).

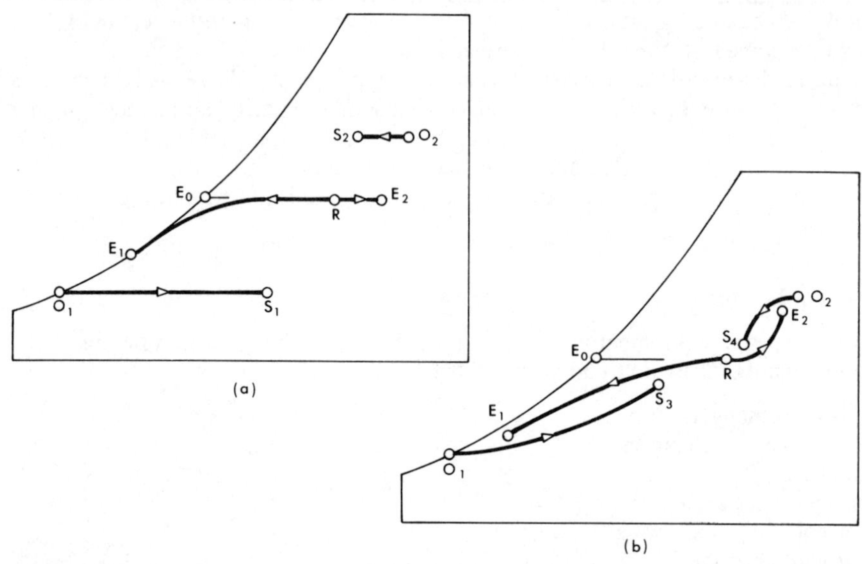

Figure 17.30 Performance of an air-to-air heat exchanger

If these data are plotted on a psychrometric chart as in Figure 17.30(a), the various significant properties may be read as summarised in Table 17.4.

From the expression for efficiency given above and replacing temperatures by the respective enthalpy values the *enthalpy efficiency* may be established. Using the same example the expression may be rewritten in terms of enthalpy thus:

Winter

$$100\left(\frac{S_1 - O_1}{R - O_1}\right) = 100\left(\frac{25.3 - 7.7}{43.4 - 7.7}\right) = 49.2\%$$

Table 17.4 Physical quantities related to state points indicated in Figure 17.30

Position from Figure 17.30		*DB temp.*	*Per kg dry air*	
Point	*Statement*	*(°C)*	*Enthalpy* (kJ/kg)	*Moisture* (kg/kg)
O_1	Winter, outside	−1	7.7	0.0035
O_2	Summer, outside	28	57.0	0.0113
R	Room exhaust	22	43.4	0.0084
S_1	Unit output, winter	16.5	25.3	0.0035
S_2	Unit output, summer	24	52.9	0.0113
S_3	Unit output, winter	16.5	34.8	0.0072
S_4	Unit output, summer	23.5	46.7	0.0090

Summer

$$100\left(\frac{O_2 - S_2}{O_2 - R}\right) = 100\left(\frac{57.0 - 52.9}{57.0 - 43.4}\right) = 30.0\%$$

Since temperature and enthalpy based calculations may produce results which are significantly different in terms of efficiency, it is important to examine critically any data produced by manufacturers in this respect. Where a heat exchanger has equal ability to recover both sensible and latent heat, as illustrated in Figure 17.30(b), the efficiencies in enthalpy terms become:

Winter

$$100\left(\frac{S_3 - O_1}{R - O_1}\right) = 100\left(\frac{34.8 - 7.7}{43.4 - 7.7}\right) = 76\%$$

Summer

$$100\left(\frac{O_2 - S_4}{O_2 - R}\right) = 100\left(\frac{57.0 - 46.7}{57.0 - 43.4}\right) = 76\%$$

Some types of equipment have different characteristics with respect to sensible and latent heat exchange. Taking the previous example and allowing for latent transfer to be at 50% efficiency, the winter and summer results in term of enthalpy would be 67 and 58% respectively.

Where supply and exhaust air quantities are not equal, the notional efficiency will change *pro rata* to the mass ratio. That is, if the supply air quantity were twice that of the exhaust air, the efficiency would be about halved; in the converse case, the efficiency would increase by about 25%; precise figures should be obtained from the manufacturer.

Plate-type heat exchangers

Having no moving parts, this is probably the most simple type of equipment. The two air streams are directed in cross- or counter-flow through a casing which is compartmented to form narrow passages carrying, alternately, exhaust and supply air. Energy is transferred by conduction through the separating plates and contamination of one air stream by the other thus avoided. The form of the casing is arranged to suit the configuration of the transfer surface and to provide for convenience of air duct connections; one example is shown in Figure 17.31. Condensation may occur in the return air passageways and drains are therefore required.

Since the separating plates are normally of metal (aluminium or stainless steel) moisture transfer is not possible and sensible heat only is exchanged. An epoxy or vinyl coating may be applied to aluminium plates for use in mildly corrosive atmospheres such as swimming pools.

Units are available to handle air quantities in the range 60–24 000 litre/s and may be built up in modular fashion to suit individual requirements. Due to the relatively low rate of heat transfer per unit area, the plate surface necessary is large but unit sizes are reasonably compact since air flow passages are kept to minimum width. Temperature efficiencies in the range of 50–80% are claimed, and resistance to air flow is 140–300 Pa at a face velocity of 3 m/s.

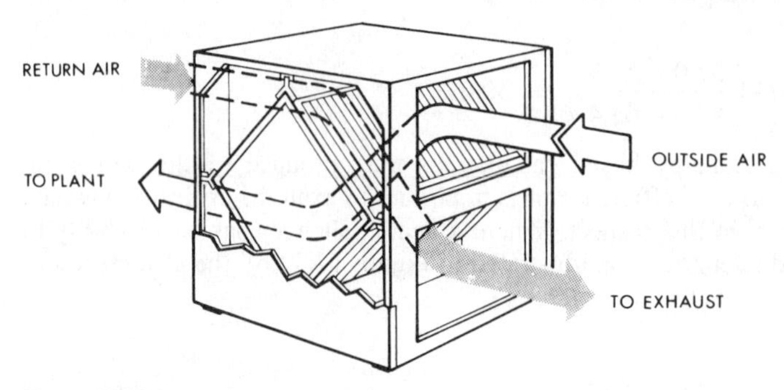

Figure 17.31 Plate type heat exchanger

This type of heat exchanger offers no method of control and therefore a bypass section may be needed to avoid, for example, heating the outside air above the required supply temperature only to have to cool it down again.

Glass tube heat exchangers

The operation is similar to that of the plate heat exchanger; normally the 'clean' supply air would be passed through the tubes and the 'contaminated' exhaust around them to allow for easier cleaning. These units are particularly suited to handling corrosive fumes from laboratories, metal treatment shops, fume cupboards, and for swimming pools.

Units may be obtained to handle up to 16 000 litre/s and with a temperature efficiency of up to 80%. Pressure drop would be of the order of 250 Pa.

Thermal wheels

Constructed on the lines illustrated in Figure 17.32, the *thermal wheel* or *regenerative* heat exchanger consists of a shallow drum containing appropriate packing which is arranged to rotate slowly between two axial air streams, transferring energy between the two. The wheel is mounted in a supporting structure and motor driven at approximately 20 rev/min: the speed may be varied as a means of controlling output.

The media and the form of the heat transfer surface vary as between manufacturers and determine the characteristics of the energy transfer. Sensible heat transfer is obtained from media formed by alternate flat and corrugated metal sheets of aluminium or stainless steel. As for plate exchangers, a protective coating may be applied for use in swimming pool applications and the like. To achieve both latent and sensible heat transfer a corrugated inorganic hygroscopic material may be used; typically these would be produced by an etching process or a lightweight coating of a hygroscopic salt.

Cross-contamination between the two air streams is minimised by so arranging the respective fan positions that the supply air pressure at the recuperator is greater than that of the exhaust stream. By using suitable labyrinth seals and incorporating a *purge sector* which allows for the matrix to be scavanged before supply air passes to the building, it is claimed that contamination is kept to less than 0.1%. *Lithium bromide* as used for treatment of the hygroscopic type matrix is stated to be *bacteriostatic*, i.e. it inhibits the propagation of bacteria. It should be noted, however, that the hygroscopic material may absorb toxic gases, or similar vapours, from the exhaust air which would not be

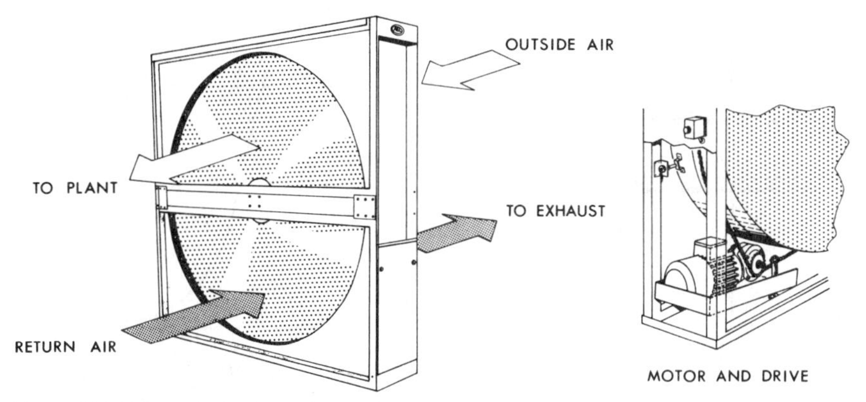

Figure 17.32 Thermal wheel type heat exchanger

completely removed by purging and hence present a risk of contaminating the incoming air.

Wheels are available in sizes up to about 5.5 m diameter to handle air quantities in the range 300–30 000 litre/s but multiple units in the middle of the size range are often more convenient to handle large air quantities. Efficiency may be as high as 85% in sensible heat reclaim and up to 88% is claimed by some manufacturers for transfer of total heat in hygroscopic types; however, efficiencies higher than 85% should be viewed with caution. Resistance to air flow at a face velocity of 3 m/s will be about 150 Pa. The power required to rotate the wheel is quite small being between 60 and 1100 W.

'Heat pipe' heat exchanger

As in the case of the plate type, heat pipe units have no moving parts and are simple in concept. A working fluid is however employed to effect heat transfer. Construction consists of a box enclosure having a dividing partition to separate the supply and exhaust air streams, through which an array of finned heat pipes is assembled.

The 'heat pipe' itself is a by-product of nuclear research developed further in connection with the space programme: in essence it is no more than a super-conductor of sensible heat. Each individual conductor is a sealed tube, pressure and vacuum tight, provided with an internal wick of woven glass fibre normally as a concentric lining to the tube. During manufacture, a working fluid is introduced in sufficient quantity to saturate the wick. The actual fluid used is selected to suit the temperature range required and would typically be one of the common refrigerants.

In operation, heat applied to one end of the pipe will cause the liquid to evaporate and the resultant vapour will travel to the 'cool' end where it will condense, surrendering energy, and the liquid will return through the wick by capillary action to the 'hot' end. Figure 17.33 illustrates this process and shows also how the heat transfer capacity of a pipe may be adjusted by varying the angle to the horizontal at which it lies. This characteristic may be used to match capacity to a given application or, by automation, to provide a means of control. Where the angle of tilt is used in this way, a facility must be provided to reverse the action when the season changes, winter to summer.

An alternative arrangement is where the pipes are installed vertically to transfer heat from a warm lower duct to a cool duct above. With this configuration, movement of the

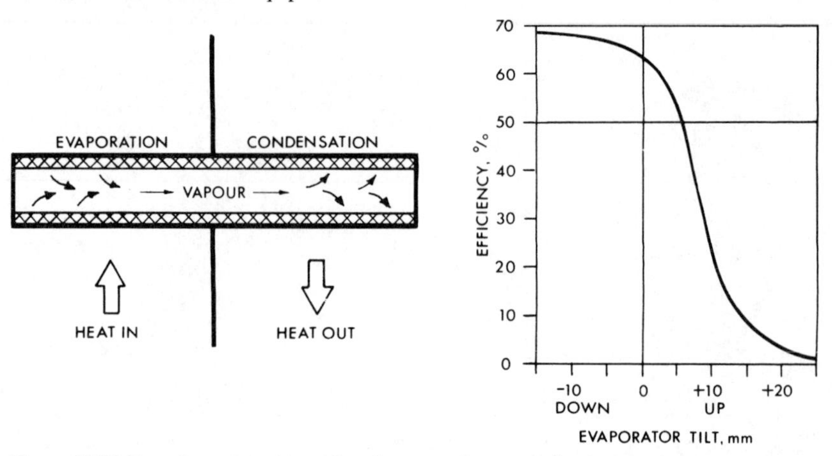

Figure 17.33 Heat pipe orientation and performance (in counterflow)

heat transfer fluid is by phase change; liquid in the lower section absorbs heat and changes to a gas, which condenses, releasing heat in the upper section, causing the liquid to drop to the lower end. Vertical units will not function where the cool duct is below the warmer one.

The capacity of a built-up heat exchanger of given overall dimensions will vary according to the number of rows of heat pipes, the fin spacing and the air velocity. Typically, a six row unit having 55 fins per 100 mm would, for equal supply and exhaust air quantities, have an efficiency for sensible heat exchange of up to 80% at a face velocity of 3 m/s and with a resistance to air flow of 200 Pa. Module sizes range from 150 to 36 000 litre/s. Efficiency of heat exchange is dependent upon the relative direction of air flow in the two ducts. Counterflow gives the higher performance, which is the basis for most published data; parallel air flow will reduce efficiency by about one-fifth of the quoted percentage. Subject to the effectiveness of the division plate and seals between the two air streams there should be no cross-contamination between supply and extract air. This method of heat exchange is seldom used due to the relatively high cost.

Run-around coils (water circulation)

This approach to the problem has the merit of extreme flexibility and is, moreover, founded upon a well understood technology. As shown in Figure 17.34, the basis of the system is a pair of conventional finned tube heating/cooling coils, one fitted in each air stream, connected by a pipework system for pumped circulation of the working fluid, often a 25% solution of *glycol anti-freeze* in water. Table 17.5 gives the freezing point and specific heat capacity of water and ethylene glycol solutions in concentrations 0–40% glycol by mass. The specific heat capacity of the solution affects the efficiency of heat transfer; a reduction in specific heat capacity giving lower efficiency.

The flexibility of the system derives from the obvious ease by which the coils may be connected together; there is no need to disturb the routes of what may be large air ducts to bring them, inlet and outlet for both supply and exhaust, to the heat exchanger. Furthermore, coils may be fitted to any number of exhaust ducts and the heat therefrom collected and distributed to any number of similar coils fitted to supply air ducts. Diversity of energy availability and energy demand between air handling plants may thus be used to best advantage.

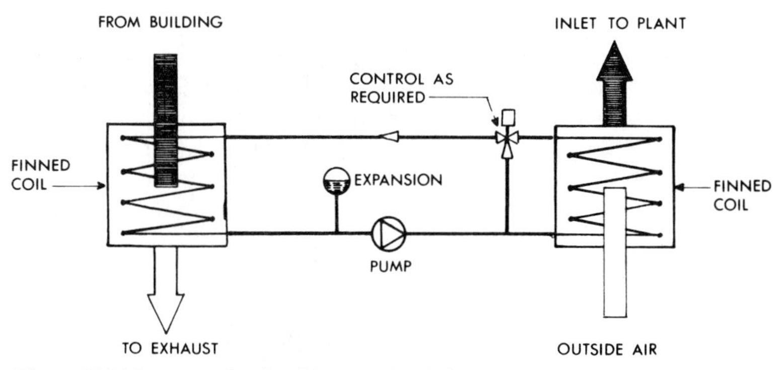

Figure 17.34 Run-around coils with water circulation

Table 17.5 Properties of water/ethylene glycol solutions

Glycol concentration (%)	*Freezing point* (°C)	*Specific heat capacity* (at 10°C) (kJ/kg K)
0	0	4.2
10	−5	4.1
20	−10	3.9
25	−13	3.8
30	−16	3.7
40	−25	3.5

There are, of course, compensating disadvantages, those of most consequence being the need for double heat exchange (exhaust air to fluid and fluid to supply air), the relatively small temperature differentials available for such energy transfer, the need for water pumping power and the matter of heat loss and gain from and to the pipework system.

Direct transfer of latent heat is not possible with this system, but in winter the coil in the exhaust air stream would run wet as would that in the supply air stream during some summer conditions: energy transfer would thus be assisted and efficiency improved as in the case of plate type heat exchangers. In the context of what has been said before, however, heat transfer would be sensible only.

Little more needs to be added with regard to this type other than to emphasise that the small temperature differentials between either air stream and the working fluid will result in deep coils (typically 6–8 rows) and high resistance to air flow with consequent penalties in fan power requirement. The overall efficiency, ignoring fan and pump power, some of which will be recovered in winter but will be a penalty in summer, is not likely to be more than 40–65% at best.

Run-around coils (using refrigeration)

If one considers an exhaust and a supply air duct, separate but not too distant, it is obvious that the evaporation and condensing elements of a refrigeration plant could be fitted within the respective air streams, Figure 17.35. By such means, one of the disadvantages of a water circulating system, i.e. small temperature differentials, could be overcome: in fact, using this 'heat pump' principle (see Chapter 19), the supply air temperature may be raised above that of the exhaust air. The energy required to drive the compressor imposes a penalty but much of this would be recovered as heat to the supply air stream.

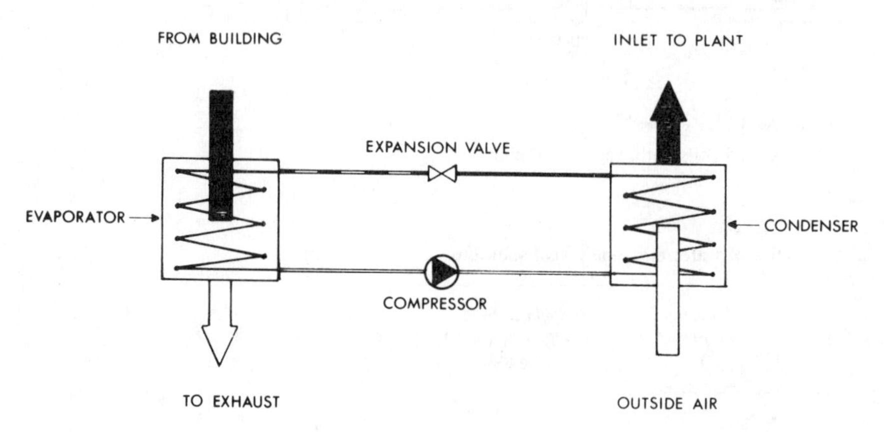

Figure 17.35 Run-around coils using a refrigeration cycle

Many types of packaged air-conditioning plant designed for roof mounting incorporate not only supply and exhaust fans but also air-cooled refrigeration equipment for summer use. In some cases facilities are provided whereby air paths may be redirected during the winter and some part of the refrigeration capacity used as a heat pump drawing energy from exhaust air.

Systems compared

The principal disadvantages of any air-to-air heat recovery system which does not make use of an intermediate fluid for heat transfer is, as previously explained, that two air ducts which may be quite large must be brought together at the heat exchanger. Other problems occur also in analysis but these are amenable to technical rather than spatial solution:

- Any heat exchange element which may at some periods of the year operate with wetted surfaces must be provided with a condensation collection tray and drain facilities. It is important that the configuration of the exchange surfaces permits flow and collection of moisture.
- In extreme weather conditions, a small amount of preheat may be required to prevent freezing of condensed moisture since this could lead to damage to the equipment and an unacceptable resistance to air flow. Preheating may also be desirable for thermal wheel installations to avoid excessive moisture exchange in winter.
- Heat exchangers which have small passages presented to air flow will soon become clogged with dirt unless pre-filters are provided: wash down may be required at intervals. Some manufacturers of thermal wheels claim that the reversal of air flow which occurs in normal operation acts to maintain cleanliness.

In considering the relative economics of alternative methods, account must be taken of annual mean rather than peak load efficiencies and of available means for control of heat exchange, since under certain outside conditions which normally occur during mid-season, the maximum rate of heat transfer available may increase, not decrease, energy consumption. As has been mentioned, the performance of a thermal wheel may be varied by speed change and that of a heat pipe unit by automation of the angle of tilt. The water circuit of run-around coils may be fitted with motorised valves as required, but for plate heat exchangers an arrangement of face and by-pass dampers will be necessary.

Resistance to air flow and consequent increases in fan power must be considered as must the energy consumption by auxiliaries, drive motors, pumps, etc. These, of course, are likely to remain running even when the enthalpy of the supply and exhaust air streams is so close as to lead to minimal interchange. Table 17.6 presents a summary comparing the performance of the various types of equipment based upon the design conditions listed in the first three lines of Table 17.4, with supply and exhaust air quantities equal at a rate of 5000 litre/s.

Indirect evaporative coolers

A development in air-to-air heat exchanger techniques, using established principles, overcomes one of the disadvantages of direct evaporative cooling, namely that of considerable increase in humidity associated with the adiabatic cooling process; some early development work was undertaken at the Department of Engineering Science, Oxford University. In the indirect system the exhaust air is cooled *adiabatically* using a water wash and is used in an air-to-air heat exchanger to cool the incoming air in summer.

Table 17.6 Comparative performance of various types of air-to-air heat exchanger[a]

		Energy reclaim efficiency (%)			
		Temperature		*Enthalpy*	
Type of equipment	*Make*	*Winter*	*Summer*	*Winter*	*Summer*
Parallel plate metal	A	62	62	40	26.5
	B	65	59	41	26
	C	57	53	37	20
Glass tube	C	61	57	40	23
Thermal wheel					
non-hygroscopic	A	74	74	48.5	30
	B	75	75	72	44
	C	74	74	52	37
hygroscopic	A + C	74	74	74	74
	B	70	70	70	70
Heat pipe	C	64	64	43	29
Run-around coil					
water	B	67	61	–	–
water/glycol (25%)	B	64	60	–	–
water/glycol (25%)	C	55	50	35	30

[a] For the winter and summer design conditions listed in Table 17.4. Supply and exhaust air quantity = 5000 litre/s.

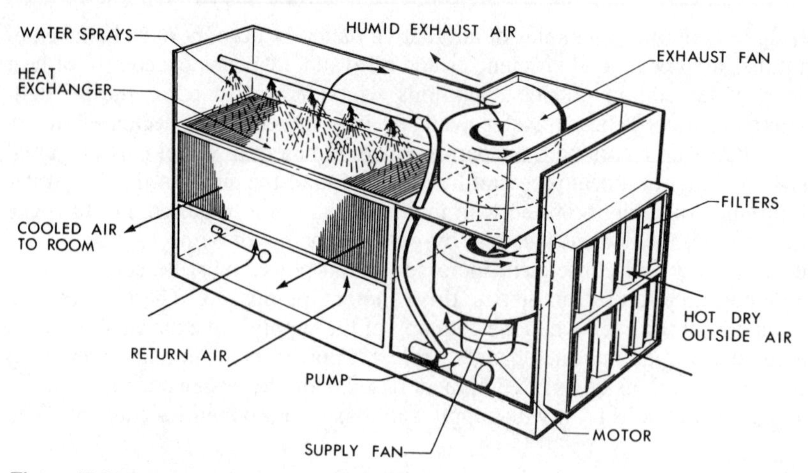

Figure 17.36 Indirect evaporative cooler (PHE Systems)

An example of the heat exchanger configuration is shown in Figure 17.36, from which it will be apparent that the unit may also be used in winter as a 'dry' heat exchanger to heat the incoming air. The application of this type of unit is likely to be limited to overseas climates. The risk of *Legionnaires' disease* must also be considered.

Chapter 18

Calculations for air-conditioning design

It is now proposed to consider the fundamental principles underlying the design of an air-conditioning system. These principles are the same no matter what particular form the system may take, but the degree of accuracy necessary to be achieved will depend upon the application and the sophistication of the controls to be provided. For instance, a microchip manufacturing plant may require minimum tolerance in conditions, whereas less strict limits would be acceptable for comfort conditioning in the case of a department store.

First to be considered here is the general case, as applied to a central air-conditioning system for a single large space, and this is followed by some notes on how these general principles may be applied to certain of the specific types of apparatus already discussed. Design data have been built up around each of the particular forms of equipment mentioned and it would be beyond the scope of this book to explore each one in detail.

Heat gains

The various factors which contribute to the heat gains and losses which occur in a conditioned space have been outlined in Chapters 2 and 3. When designing an air-conditioning system the principal concern is directed towards heat gains, especially during the summer months, although the same system will most probably provide a heating service in mid-seasons and winter also. The reason for this approach is that heat gains present more searching demands than do heat losses.

Sensible heat gains

The *quantity* of conditioned air which must be provided to combat sensible heat gains is directly proportional to the difference between the supply air temperature and that to be maintained in the space. The temperature rise which may be permitted will probably be limited to 6 or 8 K owing to the difficulty of mixing cool entering air with warmer room air without producing draughts. The mass of air flow required to maintain a desired room temperature is thus arrived at very simply by use of an expression similar to that noted in Chapter 13, where:

$$M = H/(c\Delta t)$$

where

M = mass flow of entering air (kg/s)
H = sensible heat gains (kW)
c = specific heat capacity of air (kJ/kg K)
Δt = design temperature rise (K)

Latent heat gains

These do not affect the *quantity* of conditioned air required since they do not cause a rise in temperature. The latent gains are treated quite separately from the sensible gains. The mass flow of air required to deal with the latter will usually be found to produce no more than a small increment in humidity but in an extreme case, limitation of that increment to a acceptable figure may require that the mass be increased and the design temperature rise reduced in consequence.

Psychrometry

Psychrometry is a subject concerned with the behaviour of mixtures of air and water vapour and a knowledge of it is necessary in order to perform any air-conditioning calculations. Some of the general principles were referred to in Chapter 1 but a complete study is outside the scope of this book and is dealt with in many textbooks on thermodynamics and several excellent specialist works.*

Most of the terms which relate to mixtures of air and water vapour were defined in Chapter 1 of this book but a brief list recapitulating those items which are particularly relevant here would include:

Dry bulb temperature, *DB* (°C)
Wet bulb temperature, *WB* (°C)
Dew-point temperature, *DP* (°C)

Vapour pressure (kPa)

Relative humidity, *RH*, and percentage saturation (%)
Absolute humidity or moisture content (kg/kg of dry air)

Total heat, *TH*, or specific enthalpy (kg/kg of dry air)
Specific volume (m^3/kg of dry air)

Barometric pressure

The standard level of atmospheric pressure is 101.325 kPa exactly, corresponding to 760 mm of mercury at 0°C and standard gravity (9.806 65 m/s^2). This level, equating to 1.01325 bar, was the barometric pressure used in calculation of the psychrometric properties of air and water vapour presented in the *Guide Section C1*, and referred to later in this chapter. For practical purposes in air-conditioning design, the data so presented are accurate for situations having barometric pressures between 95 and 105 kPa but, in circumstances which are outside these limits, use must be made of other published data.†

* Goodman, W., *Air Conditioning Analysis.* Macmillan, New York, 1947.
Jones, W. P., *Air Conditioning Engineering.* Edward Arnold, London, 1994.
† *M-C Psychometric Charts for a Range of Barometric Pressures.* Northwood Publications, 1972.

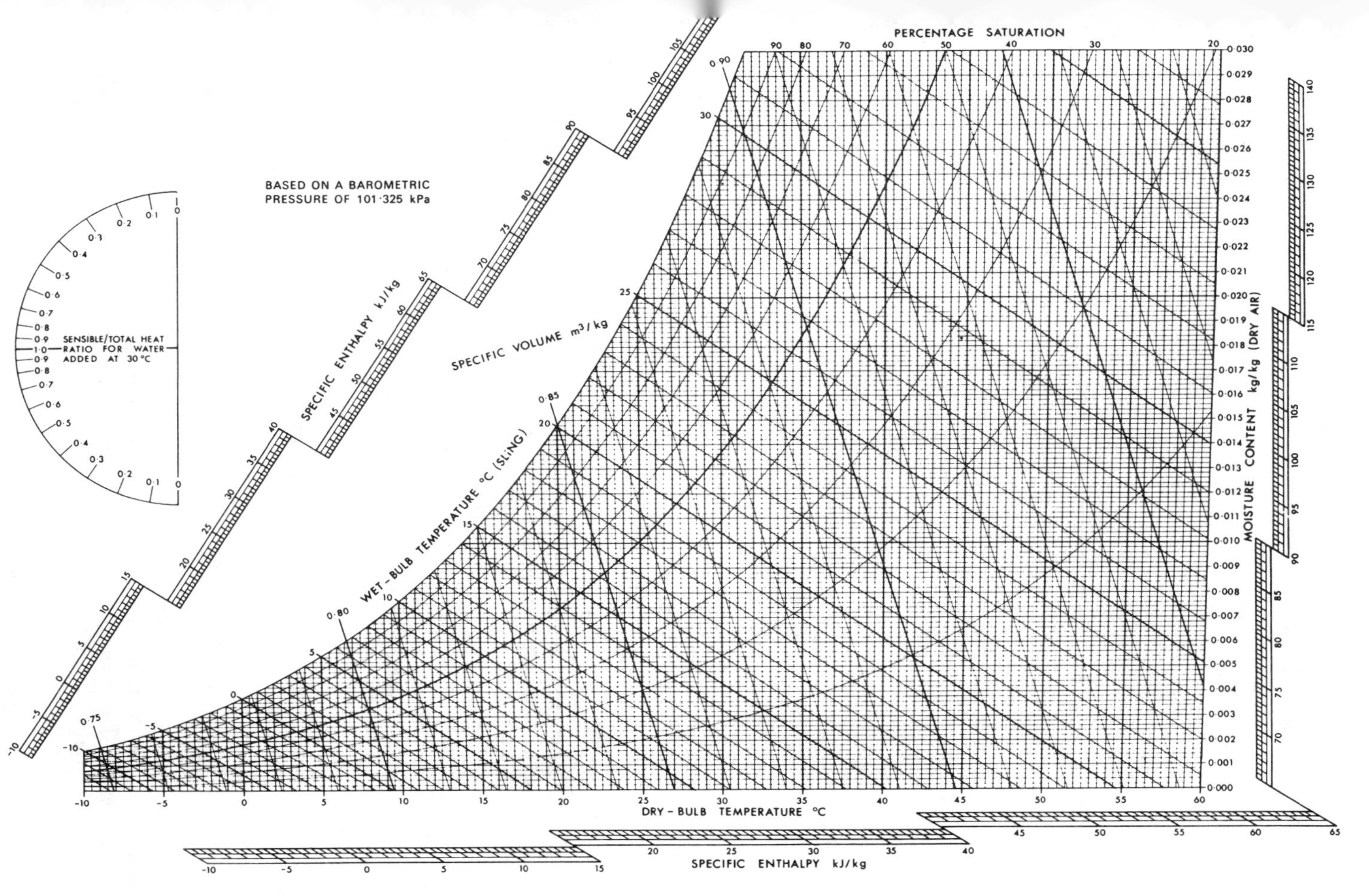

Figure 18.1 Psychrometric chart (as developed for *CIBSE Guide*)

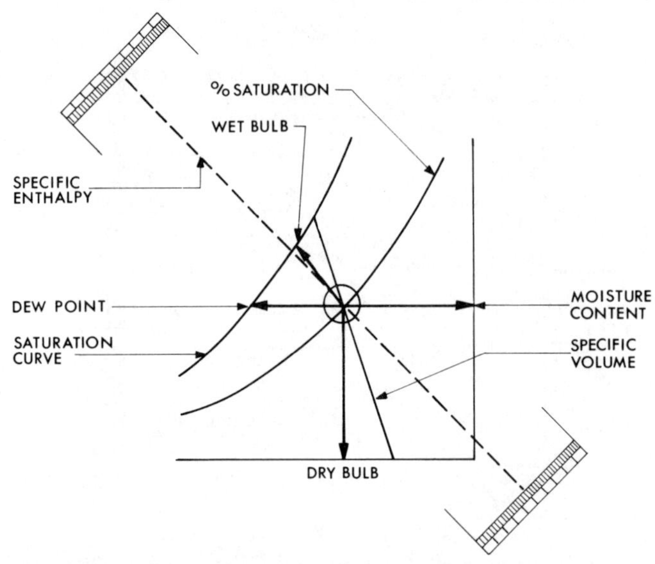

Figure 18.2 Format and method of use for psychrometric chart

Point marked is 20°C DB, 50% Sat. Other data read from the chart are:

W.B.	=	13.9°C
Dew point	=	9.4°C
Moisture content	=	0.0074 kg/kg dry air
Specific enthalpy	=	38.8 kJ/kg
Specific volume	=	0.84 m^3/kg dry air

Psychrometric chart*

The relationship between the various properties of a mixture of air and water vapour may be presented in the form of tables, such as those published in the *Guide Section C1*, or as a chart based thereon, as in Figure 18.1. The principal advantage provided by tables is the level of accuracy offered but this is achieved sometimes only by tedious interpolation. Since it is necessary in design to visualise stages in a process, a chart is often to be preferred as a working tool although a numerical check via tables will serve to 'polish' the conclusions. The arrangement of the co-ordinates on the chart and the method of use is indicated in Figure 18.2, any single *state point* representing values of a number of properties:

Dry bulb temperature. The base of the chart has an evenly spaced scale with divisions at 0.5°C. A vertical line drawn through the state point downwards will meet that scale.

Wet bulb temperature. The saturation line has an evenly spaced scale with divisions at 1°C. A line drawn through the state point, sloping upwards to the left, will meet that scale.

Dew-point temperature. This is read using the wet bulb scale via a horizontal line drawn through the state point to the left, to intersect the saturation line.

Moisture content. The right hand side of the chart has an evenly spaced scale with divisions at 0.001 kg/kg of dry air. A horizontal line drawn through the state point to the right will meet that scale.

* Copies of the chart in pad form are obtainable from CIBSE.

Table 18.1 Properties of dry and saturated air at various temperatures

Temperature (°C)	*Specific volume* (m^3 per kg *dry air*)	*Saturation moisture content* (kg/kg *dry air*)	*Specific enthalpy (total heat)* (kJ/kg dry air)	
			Dry air	*Saturated air*
0	0.77	0.0038	0.00	9.47
2	0.78	0.0046	2.01	12.98
4	0.79	0.0050	4.02	16.70
6	0.79	0.0058	6.04	20.65
8	0.80	0.0067	8.05	24.86
10	0.80	0.0076	10.06	29.35
12	0.80	0.0087	12.07	34.18
14	0.81	0.0100	14.08	39.37
16	0.82	0.0114	16.10	44.96
18	0.82	0.0129	18.11	51.01
20	0.83	0.0147	20.11	57.55
22	0.83	0.0167	22.13	64.65
24	0.84	0.0189	24.14	72.37
26	0.84	0.0214	26.16	80.78
28	0.85	0.0242	28.17	89.96
30	0.86	0.0273	30.18	99.98
32	0.86	0.0307	32.20	111.0
34	0.87	0.0346	34.21	123.0
36	0.87	0.0389	36.22	136.2
38	0.88	0.0437	38.24	150.7
40	0.89	0.0491	40.25	166.6

Source: the *Guide Section C1*: at atmospheric pressure of 101.325 kPa.

Percentage saturation. From a scale at the head of the chart, the curves downwards to the left are at 10% intervals. Interpolation is necessary where a state point falls between the curved lines.

Specific enthalpy (total heat). Detached 'saw-tooth' scales to the left and top and to the right and base are evenly spaced with divisions at 0.5 kJ/kg of dry air. It is necessary to align a rule across both scales and through the state point to obtain a value.

Specific volume. Above the saturation line is a widely spaced scale with divisions at 0.05 m^3/kg of dry air. A line drawn through the state point, parallel to the scale lines will allow interpolation.

Application

Having arrived at the maximum hourly heat gain for the space or spaces to be served, probably the most searching of the seasonal conditions to be met, it is necessary to calculate the conditions to be maintained in the plant and the capacity of the various components (fans, cooling coils, heater batteries, water chillers, and so on).

The routine calculations necessary are best illustrated by application to an example such as the all-purpose hall, in use for concerts and other activities, as shown in Figure 3.13.

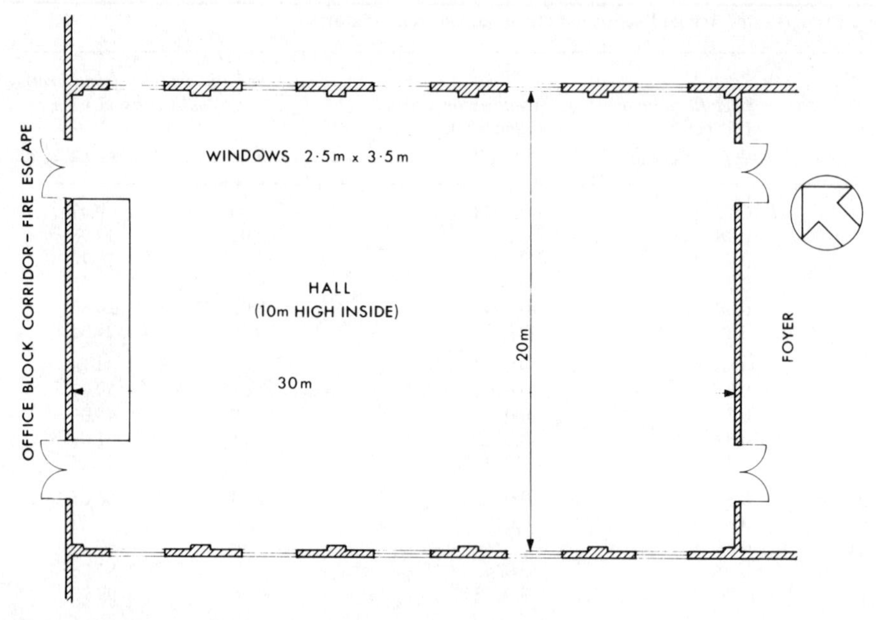

Figure 18.3 Plan of building used in example

This is repeated here for convenience of reference as Figure 18.3. The building characteristics and the bases for the design of the system are:

Building			
Seating capacity	=	500 persons	
Lighting load	=	20 kW	
Volume of hall	=	6000 m^3	
External conditions		*Summer*	*Winter*
Temperature	=	28°C DB	4°C saturated
	=	18.5°C WB	
Enthalpy	=	51.95 kJ/kg	16.70 kJ/kg
*Internal conditions**		*Summer*	*Winter*
Temperature	=	21°C DB	20°C DB
50% saturation	=	14.8°C WB	
60% saturation	=		15.3°C WB
Enthalpy	=	41.08 kJ/kg	42.58 kJ/kg
Outside air per occupant, year round			
Table 1.6	=	8 litre/s	

Summer cooling and dehumidification

The orientation of the example building presents two faces to solar heat gain and it was made clear in Chapter 3 (p. 83) that the south-west glazing will be subject to greatest intensity at 15.00 hours BST in September.

* Dry bulb temperature is used here in preference to resultant temperature. See Chapter 3, p. 83.

Building fabric heat gain

The scope of the calculations set out in the earlier chapter was restricted to an analysis of the heat gains arising from conduction through the structural elements and from solar glazing. The criteria were the outside and inside air temperatures noted above and the results were as follows:

Glazing		
Solar radiation	= 10.7 kW	
Conduction	= 2.8 kW	
Walls		
Solar radiation and conduction	= 1.6 kW	
Roof		
Solar radiation and conduction	= 1.5 kW	= 16.6 kW

Infiltration heat gain

To continue with the example, the matter of infiltration arises which may best be dealt with from experience as an assumed rate of air change.

Allow half an hour change per hour		
Room volume = $(30 \times 20 \times 10)$	$= 6000\,m^3$	
From Table 18.1, specific volume of dry air at room condition	$= 0.83\,m^3/kg$	
Thus mass of infiltrated air:		
$(6000 \times 0.5)/(0.83 \times 3600)$	= 1.0 kg/s	
From Table 18.1, enthalpy of dry air:		
at external condition (28°C)	= 28.17 kJ/kg	
at internal condition (21°C)	= 21.12 kJ/kg	
Thus, by difference, sensible heat gain:		
$1.0\,(28.17 - 21.12)$		= 7.05 kW
From *Basis of Design*, enthalpy:		
at external condition	= 51.95 kJ/kg	
at internal condition	= 41.08 kJ/kg	
Thus, by difference, latent heat gain:		
$[1.0\,(51.95 - 41.08)] - 7.05$		= 3.8 kW

*Internal heat gain**

This arises due to matters which occur within the room. In this case they are confined to heat gains from occupants and to lighting but might, in other circumstances, include motors, or other sources of heat.

From Figure 3.14, for persons at rest:	
sensible gain per occupant	= 88 W
latent gain per occupant	= 30 W

* Sensible and latent heat gains arising from occupants assume here that they are seated at rest. In other circumstances, the gain per occupant might be more but the number, equally, might be less.

Thus, sensible gain = (500×88) = 44.0 kW
and latent gain = (500×30) = 15.0 kW

From *Basis of Design*, sensible gain from lighting = 20.0 kW

Supply air quantity to room

This is calculated from the total of the individual sensible heat gains set out above. It will be noted that no inclusion is made for the gain arising from the supply of outside ventilation air to the room for the benefit of the occupants. This latter is a load on the plant and thus does not affect that air quantity.

Sensible heat gain in room:

$(16.6 + 7.1 + 44.0 + 20.0)$ = 87.7 kW

Design criteria now assumed:

air temperature rise, from supply to room = 6 K
hence, supply air temperature = $(21 - 6)$ = 15°C

From Table 1.1, for dry air at 15°C:

specific heat capacity = 1.02 kJ/kgK

Thus, total air supply mass required:

$87.7/(6 \times 1.02)$ = 14.33 kg/s

Ventilation air

Criteria for the supply of outside air were set out in Table 1.6 (p. 15) and it is assumed that smoking is not allowed in this hall.

From *Basis of Design*:

number of occupants = 500
outside air quantity per person = 8 litre/s

From Table 18.1, specific volume of dry air at room condition = 0.83 m^3/kg

Thus, mass flow of ventilation air:

$(500 \times 8)/(1000 \times 0.83)$ = 4.82 kg/s

Air mixture entering plant

This quantity is the air supply mass required to deal with the sensible heat gain and which includes both the ventilation air and the exhaust air from the hall which is recirculated. The mixture therefore has components having different properties of temperature and enthalpy, etc.

From results above, by difference:

mass of air recirculated = $(14.33 - 4.82)$ = 9.51 kg/s

Thus, temperature of air mixture, dry bulb

$[(9.51 \times 21) + (4.82 \times 28)]/14.33$ = 23.4°C

and enthalpy of air mixture

$[(9.51 \times 41.08) + (4.82 \times 51.95)]/14.33$ = 44.74 kJ/kg

Air mixture leaving plant
It is now necessary to take account of latent heat gain in the room inasmuch as this will affect the ability of the air mixture leaving the plant to absorb unwanted moisture in the room.

Latent heat gain in room as previously:

calculated = (3.8 + 15) = 18.8 kW

From p. 2, latent heat of vaporisation of water at 21°C = 2450 kJ/kg

Thus, moisture increment in room:

18.8/(14.33 × 2450) = 0.00053 kg/kg

From Figure 18.1, moisture content of air at room condition = 0.0079 kg/kg

Thus, moisture of air leaving plant:

(0.0079 – 0.00053) = 0.0074 kg/kg

The energy used to drive the supply fan will be converted into heat and a proportion of this will be transferred to the air stream. Also, the walls of the supply duct between the plant and the room inlet, although thermally insulated, will allow some heat ingress. As a result, the dry bulb temperature of the air leaving the plant must be lower than that required at the room inlet. Such heat gains may be assessed in detail but, for this example, may be taken as 10% of the sensible gain, thus accounting for a 1.0 K temperature rise.

Hence, at the inlet to the supply fan, from Figure 18.5:

dry bulb temperature = (15 – 1) = 14°C

and, from Figure 18.1, air at 14°C dry bulb moisture content = 0.0074 kg/kg, then

wet bulb temperature = 11.5°C
and enthalpy = 32.80 kJ/kg

Cooling capacity
From the various values now available, the amount of cooling capacity required to meet the required conditions for peak gains in summer may be calculated.

Difference in enthalpy, entering to leaving:

(44.74 – 32.80) = 11.94 kJ/kg

Thus, cooling capacity

(11.94 × 14.33) = 171.1 kW

Plant duties
From the results calculated, it is now possible to arrive at the capacities of the various items of equipment which, together, will go to make up the plant for summer use. These are, excluding any margin which might be added in practice:

Supply fan, handling air at 14°C, and specific volume at 0.81 m^3/kg:

(14.33 × 0.81) = 11.6 m^3/s = 11 600 litre/s

Exhaust fan, handling air at 21°C, and specific volume 0.83 m^3/kg (4.82 kg/s exhaust to outside and 9.51 kg/s recirculation):

(14.33 × 0.83) = 11.9 m^3/s = 11 900 litre/s

Cooling coil, 11 600 litre/s with:

air on = 23.4°C DB and 16.1°C WB
air off = 14.0°C DB and 11.5°C WB

Water chiller, excluding any calculated allowance for heat gains in pumps, and water circulating pipes, etc. rounded up = 172 kW

Figure 18.4 shows, graphically on a small section of the chart, the psychrometric changes taking place during the processes outlined in the example. Figure 18.5 illustrates the various conditions on a diagram of the plant and system.

Cooling coils

The example above has assumed that the medium which is circulated through the cooling coil is *chilled water* and that a suitable coil could be found to meet the required duty. Use of chilled water avoids cooling all the air down to a low dew point for dehumidification purposes and then, subsequently, reheating.

Further, advantage is taken of the characteristic of such a coil such that moisture may be deposited on the chilled surface without the bulk of the air being reduced to the same temperature. This is achieved by keeping the surface at the lowest possible temperature consistent with absence of freezing. Had the coil been cooled by direct expansion of a refrigerant within it, a *DX coil*, the calculation would have been similar but the surface temperature of the coil would have been lower.

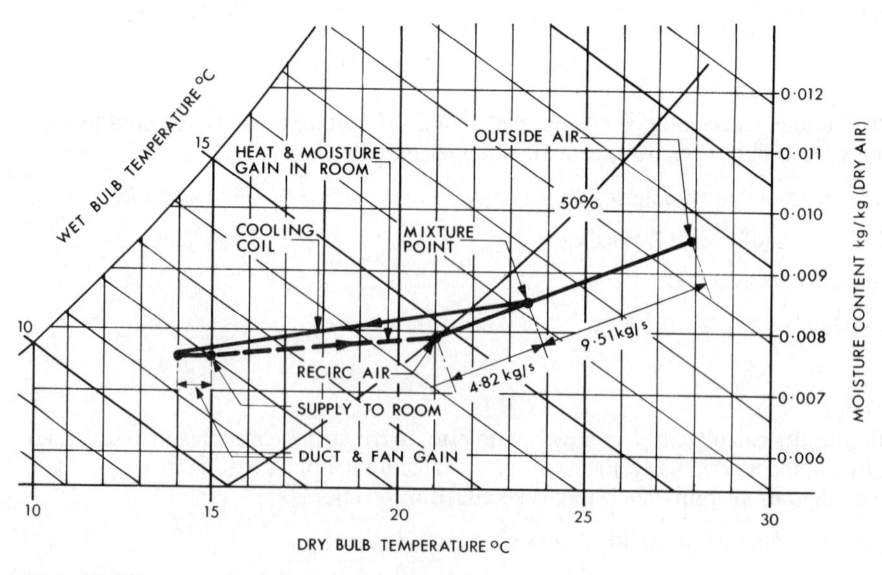

Figure 18.4 Diagram of psychrometric changes (see example calculation)

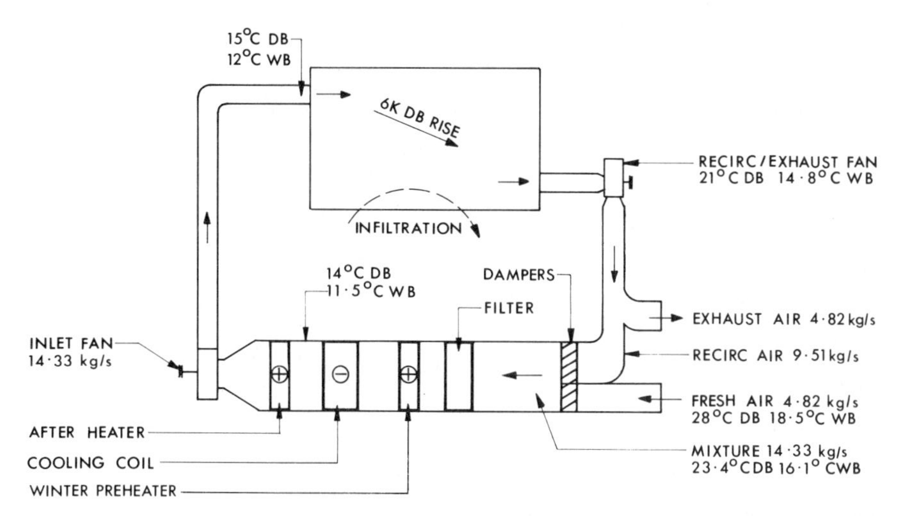

Figure 18.5 Plant conditions when cooling (see example)

In some circumstances, a coil might be arranged to have unchilled and recirculated water sprayed on to the surface in order to improve performance and to provide for adiabatic saturation of the air flow and means for humidification in winter. Concern as to humidifier related illness has led to suspicion of sprayed coils (p. 488), since doubt exists whether plant owners and operators will ensure that the means provided to prevent bacteriological contamination are maintained. Humidification by steam injection is thus often preferred, with saturation efficiencies of up to 80%.

Winter heating and humidifying

The operation of the plant in winter might be with partial recirculation, as in summer, or with 100% outside air. It might be arranged so as to be suitable for the latter method of operation in average spring and autumn weather when neither heating nor mechanical cooling is required. In colder weather, recirculation might be used to promote economy in running cost. The latter arrangement is often made by providing an outside sensor to control motorised dampers and vary the proportion of outside to recirculated air: this is often referred to as an *economy cycle*. In order to avoid the complication of introducing another variable, the two calculations which follow assume that heat losses through the building fabric and those arising from infiltration are offset by a separate system such as hot water radiators. In consequence, the supply air leaving the plant would be at the room temperature, i.e. 20°C.

Hall empty, preconditioning for occupation

The following calculation assumes that the plant will handle 100% outside air when the ambient temperature is 4°C, below which level recirculation would be used. In this example, it is assumed that humidification will be by steam using a packaged injection-type steam humidifier.

Dry bulb temperature and other conditions

in room	=	20°C
mass flow of air	=	14.33 m^3/s
specific heat of air	=	1.02 kJ/kg

Temperature of air entering plant = 4°C

and hence, after heater duty

$(14.33 \times 1.02)(20 - 4)$ = 234 kW

From Figure 18.1, moisture content of:

air at room condition	=	0.00885 kg/kg
outside air	=	0.00505 kg/kg

Thus, moisture added by humidifier:

$(0.00885 - 0.00505) \times 14.33 = 0.054$ kg/s = 196 litre/h

and energy absorbed in humidification:

(0.054×2450) = 132 kW

Hence, total energy absorbed excluding all losses:

$(234 + 132)$ = 366 kW

As an alternative, if permissible, the cooling coil would be pump sprayed and thus would provide adiabatic saturation to the air flow. For the purpose of the example, the figuring assumes 100% coil efficiency, i.e. saturating the air to room dew point but a level of, say, 80–90% would probably arise in practice.

From Figure 18.1, at room condition of 20°C:

60% saturation, dew point	=	12.1°C
enthalpy at dew point	=	34.18 kJ/kg

Thus, duty of preheater

$(34.18 - 16.70) \times 14.33$ = 250 kW

and that of after-heater

$(42.58 - 34.18) \times 14.33$ = 120 kW

Thus, total energy capacity absorbed excluding all losses:

$(250 + 120)$ = 370 kW

Moisture added at sprayed coil (as before)

0.054 kg/s = 196 litre/h

Hall occupied

The following calculation assumes that the air entering the plant is a mixture of outside and recirculated air as for summer operation, outside air conditions being as above. In this example, again it is assumed that humidification will be by steam using a packaged injection-type steam humidifier. The internal heat gains, 44 kW sensible and 15 kW latent, are those due to occupancy and lighting only.

Dry bulb temperature and other conditions:

in room	=	20.0°C
mass flow of air	=	14.33 m^3/s
specific heat of air	=	1.02 kJ/kg

Temperature of air entering plant

[(9.51 × 20) + (4.82 × 4)]/14.33 = 14.6°C

Air temperature rise in room from supply due to 64 kW sensible gain:

64/(14.33 × 1.02) = 4.4°C

Thus, air temperature leaving plant:

(20 – 4.4) = 15.6°C

and main heater duty

(14.33 × 1.02) (15.6 – 14.6) = 14.6 kW

From Figure 18.1, moisture content of:

air at room condition = 0.00885 kg/kg

air entering plant

[(9.51 × 0.00855) + (4.82 × 0.00505)/14.33 = 0.00757 kg/kg

and rise in moisture content in room due to 15 kW latent gain

15/(14.33 × 2450) = 0.00043 kg/kg

Thus, moisture content of supply air:

(0.00885 – 0.00043) = 0.00842 kg/kg

and moisture added by humidifier

(0.00842 – 0.00757) × 14.33 = 0.012 kg/s = 43 litre/h

Energy absorbed in humidification

(0.012 × 2450) = 29.4 kW

Hence, total energy absorbed:

(14.6 + 29.4) = 44 kW

As an alternative, once again if permissible, the cooling coil would be pump sprayed and thus would provide adiabatic saturation to the air flow.

Temperature of air entering plant (as before) = 14.6°C

Moisture content of air mixture (as before) = 0.00757 kg/kg

Enthalpy of air mixture:

[(9.51 × 42.58) + (4.82 × 16.70)]/14.33 = 33.88 kJ/kg

Air temperature rise from supply to room due to 64 kW sensible gain (as before) = 4.4°C

Thus, air temperature leaving plant (as before) = 15.6°C

Rise in moisture content from supply to room due to 15 kW latent gain, as before = 0.00043 kg/kg

Hence, moisture content of supply air (as before) = 0.00842 kg/kg

Since the temperature of the mixture is above the room dew point, no preheat is required for adiabatic saturation.

From Figure 18.1 enthalpy of supply air at 15.6°C and moisture at 0.00842 kg/kg = 37.00 kJ/kg

Thus, difference in enthalpy through plant:

(37.00 – 33.88) = 3.12 kJ/kg

and after-heater capacity

(3.12 × 14.33) = 44.7 kW

Moisture added at spray coil is difference between that of mixture and supply air:

(0.00842 – 0.00757) × 14.33 = 0.012 kg/s = 43 litre/h

It will be noted that, despite the different methods used for humidification and the differences in calculation routine, there are no significant differences between the predictions of energy absorbed nor in the amount of water used. Indeed, what differences there are may have arisen as a result of the difficulty in reading accurate results from Figure 18.1.

Design calculations for other systems

The design calculations for the various other types of systems follow a routine not dissimilar from that which has been outlined earlier but are adapted to suit the particular characteristics in each case. The following paragraphs provide brief summaries of the similarities and differences in approach.

Single-duct terminal reheat systems

The central plant is designed to provide full conditioned air quantities to all areas served, to meet peak demands. No allowance for diversity of load can be made. The terminal reheat equipment is designed to cater for local differences between maximum and minimum load conditions.

Whilst this system has the advantage of offering good temperature control, it is inherently wasteful of energy. Low humidity conditions may arise.

Fan coil systems, ducted outside air

These are designed on similar lines to those applied to induction systems in cases where the fresh air is delivered to the room space via the fan coil unit.

Where fresh air is ducted independently, it may be supplied year-round at constant temperature at, or a degree or two less than, room condition. In this case the fan coil unit, which should be of the four-pipe type (Figure 14.7), will require to be designed to deal with the whole of the heat gains or losses occurring within the space served.

Fan coil systems, local outside air

Such systems which are, in effect, no more than individual room heaters/coolers must be arranged to deal with the whole conditioning load, outside air being provided either directly to the units or via opening windows, etc. No direct control of room humidity can be achieved although dehumidification will occur.

Induction systems

The primary air supply, conditioned at the central plant, is designed to provide an adequate volume for ventilation purposes and humidity control. Temperature is varied to suit external weather conditions and may be further adjusted to take account of solar gain.

The coil or coils in the room units are designed to deal with local sensible heat gains or losses. Chilled water temperature should be selected to reduce the risk of condensation on the coil.

Double duct systems

The central plant is designed to provide a supply of both cool and warm air which are distributed in parallel to individual terminal units. The cool air duct provides an air quantity adequate in volume and temperature to meet the maximum anticipated cooling load of heat gain to the building fabric and from the outside air supply.

The warm air duct provides a supply adequate in temperature to meet the design heat loss, the volume usually being allowed as 75% of that in the cool duct.

Variable volume systems

In the case of the true variable volume system, the air quantity provided by the central plant will be limited to that required to meet the maximum coincident load. It will be reduced from that design volume for all part load conditions. Supply temperature may be constant and determined by the peak cooling load, varied to suit external conditions, or varied to suit the loads actually occurring in the space. This last option requires control feedback from the terminal devices to indicate the actual operating mode.

Normal practice suggests that the air supply to individual rooms should not be reduced by more than 40% and, if minimum load is less than this then some form of temperature adjustment will be necessary, possibly at the central plant but more probably by reheat on a zonal or local basis. The all-air VAV system may be supplemented by fan assisted units, normally installed as an integral part of the control terminal, and served from a chilled water supply. The chilled water temperature must be selected such that any risk of condensation on the coil is reduced.

System diagrams and automatic controls

For diagrams of a variety of air-conditioning systems and notes upon the application of automatic controls to them, the reader is referred to Chapter 22, p. 624.

Chapter 19

Refrigeration: water chillers and heat pumps

For full air-conditioning, some means of cooling and dehumidification is necessary and this, in the majority of cases, is provided by use of a mechanical refrigeration machine or machines. This equipment may be similar to the usual run of plant used for ice-making and cold storage work, etc., except that the temperature to be produced is likely to be higher than that required for such applications.

The energy balance of a refrigeration cycle is such that it may be thought of as a 'thermal transformer', taking in energy at a (relatively) low temperature and discarding it at some (relatively) higher temperature. Where cooling is required, it is the energy at low temperature which is used and that at a higher temperature discarded: in the converse sense, where heating is required, the energy at the higher temperature is used and that at the lower temperature discarded. The latter application is that of a *heat pump* which is often applied in modern practice to make use of a low grade energy source which would otherwise go to waste.

Mechanical refrigeration

This depends upon the principle that a liquid may be made to boil at a chosen low temperature if it is held at a pressure which is reduced to an appropriate level.* To produce boiling, the liquid must be supplied with heat from an external source and this source will thus lose energy and be cooled. Given a suitable liquid, the temperature of boiling may be chosen to suit the required conditions, without resort to unduly low pressures, and although many different substances have been used, complex hydrocarbons have been found to offer the most suitable characteristics as refrigerants. The vapour given off in boiling is compressed, which process adds heat, and the hot vapour is then liquified by removal of that heat, the pressure still being maintained. A sudden release of the pressure is then arranged and, in consequence, the fluid returns to the state in which it began, ready once again to boil at a low temperature. This sequence of events is known as the *vapour compression cycle*.

A refrigeration plant, working on the vapour compression principle, as shown in Figure 19.1, thus comprises these principal components:

* Water at atmospheric pressure boils at 100°C: if that pressure were to be reduced to 1 kPa then the boiling point would be 7°C.

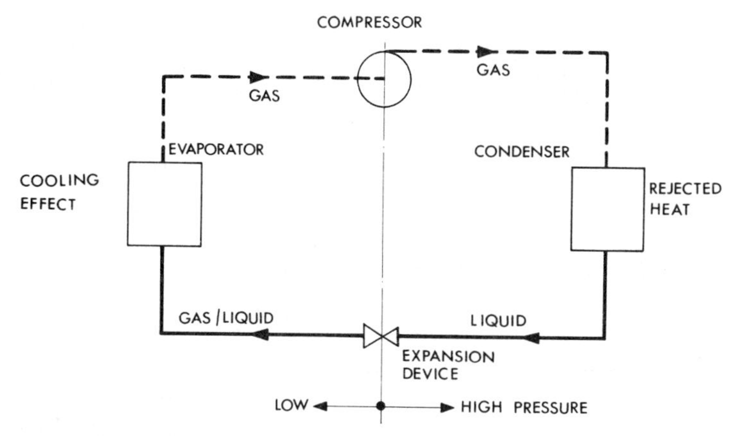

Figure 19.1 Principle of vapour compression cycle

- A *compressor* to apply pressure to the refrigeration medium.
- A *condenser* to receive the compressed gas and liquify it, the latent heat being taken away from the circuit by some external means. One method is to cool the condenser with a water circulation which may then pass to a cooling tower for re-use. Alternatively, the condenser may be cooled by an air current.
- An *expansion device* by which the pressure of the liquid is reduced.
- An *evaporator* in which the medium re-evaporates, extracting heat from whatever surrounds it, e.g. from cooling water or air in an air-conditioning plant or from brine where temperatures below the freezing point of water are needed.

Refrigeration cycle

The vapour compression cycle may best be considered on a pressure–enthalpy (total heat) diagram, as Figure 19.2, which is drawn for the fluid *Refrigerant 134a* (see later text).

Inside the curved envelope, the medium exists as a mixture of vapour and liquid and the increase in enthalpy from left to right on any pressure line within the envelope represents an increase in latent heat. Further, within the envelope, lines of equal temperature (isotherms) are horizontal. Outside the envelope to the left, the medium exists as a liquid below its saturation temperature and there the isotherms are almost vertical. To the right of the saturated vapour curve, the medium exists in the form of a superheated vapour and the isotherms curve downwards. The *critical point* is that at which latent heat ceases to exist: it is not possible to liquify a gas by pressure alone if it is above the critical temperature.

On the diagram, the refrigeration cycle is represented by the outline *A–B–C–D*, the components being:

A–B here the gas is compressed causing a rise in pressure and enthalpy which equals the energy put into the gas by the compressor, all in the superheat region. This takes place at constant entropy.

B–B' represents cooling of the superheated gas in the condenser down to the saturated vapour temperature.

B'–C here latent heat is removed, also in the condenser, and the gas is condensed to liquid.

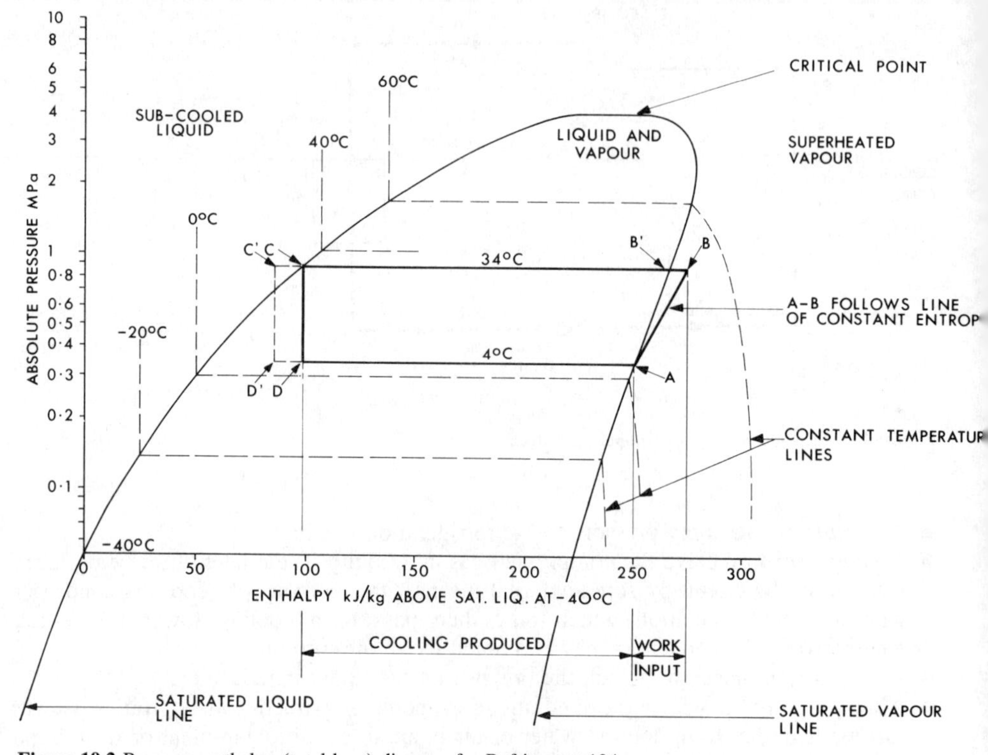

Figure 19.2 Pressure–enthalpy (total heat) diagram for Refrigerant 134a

C–D this is the pressure drop through the expansion device without any change in enthalpy (adiabatic).

D–A represents vaporisation to a dry saturated state in the evaporator, latent heat – represented by increasing enthalpy – being drawn from the water, the air or other medium being cooled. This is the *cooling effect*.

If the condenser were arranged to 'sub-cool' the liquid, say to point *C*', each unit mass of refrigerant in circulation would produce a greater cooling effect (*D*'–*A*) and the cycle would be more efficient in consequence.

The ratio of the cooling effect (as *D–A*) to the energy input (as *A–B*), in terms of enthalpy, is known as the *coefficient of performance*. The smaller the range of pressure (and hence temperature) over which the cycle operates, the less will be the energy expended for a given cooling effect. Hence, for economy in running, it is desirable to design for:

- Evaporator temperatures as *high* as is consistent with other considerations (such as dew-point temperature in an air-conditioning application).
- Condenser temperatures as *low* as possible. When cooling is to be atmospheric, weather records will decide the safe minimum level to be assumed. Maximum cooling is generally required in the hottest summer weather when the condensing arrangements are least efficient and caution is thus necessary in selecting an appropriate temperature.

Should the reader wish to pursue this matter further, a selection of charts depicting the properties of refrigerants in common use is provided in the *Guide Section B14*.

Application of refrigeration

For application to air-conditioning, using a cooling coil, water is pump-circulated through a closed system returning to the evaporator of the refrigeration plant at a temperature which is generally between 7 and 12°C, depending upon the dew point to be maintained: in passing through the evaporator, this water temperature will be lowered by about 4–6 K. In order that the necessary heat transfer may take place, the refrigerant must be at some temperature below that of the leaving water but, at the same time, it must generally be slightly above freezing point. Thus, in a typical case, the following conditions might obtain:

Apparatus dew point	12°C
Cooling coil outlet	10°C
Cooling coil inlet	6°C
Water at evaporator outlet	5.5°C

The refrigerant in the evaporator would in this case be maintained at about 1°C, giving a differential for 4.5 K for heat transfer. As will be appreciated, this small temperature potential means that the cooling surface of a simple tubular type would need to be very extensive: a variety of devices has been developed to augment the transfer rate.

Brine may be used in cooling coils in order to allow lower air temperatures to be obtained (for example to achieve a low dew point condition): the temperatures of the fluid circulating may be −7°C from the evaporator and −3°C returning to it, or lower as required. To achieve such conditions, it is necessary to consider the strength of the brine solution, data for which are available from standard tables (see also Table 17.5 for information on ethylene/glycol solutions)

In instances where cooling for an air-conditioning system is provided from a refrigeration machine by *direct expansion*, the refrigerant is piped directly to cooling coils in the air stream which thus become the evaporator. The surface temperature of the coils is a function of the leaving air temperature required, the form of the coil surface and the velocity of the air flow. Refrigerant temperatures much below freezing point are inadmissible owing to the risk of build-up of ice on the coil surface when dehumidification is taking place. An apparatus dew point of 3°C is normally considered as the practical minimum for such coils if frosting is to be avoided.

Refrigerating media

The factors affecting the choice of a refrigerant will now be clear. A substance is required which can be liquified at a moderate pressure and which has a high latent heat of evaporation. By these means the size of the compressor will be kept to a minimum and the mass of the refrigerant circulated kept relatively small for a given amount of cooling. In addition, compatibility with the type of compressor and the refrigerant system, cost, environmental issues and safety have to be taken into account. The media available include ammonia, carbon dioxide, sulphur dioxide and numerous organic gases.

Ammonia (NH_3), while high in efficiency and low in cost, is not suitable for many air-conditioning applications due to its toxic nature and the serious results which might attend a burst or leak in the system. Carbon dioxide (CO_2) calls for a high power input for a given capacity, and requires the attention of a skilled operator: it is no longer used in the context of air-conditioning. Sulphur dioxide (SO_2) and methyl chloride (CH_3Cl) were used for small plants but although they have good thermodynamic characteristics, one has an unpleasant smell and the other is toxic: neither is now used for application to air-conditioning.

A range of synthetic refrigerants, halogenated hydrocarbons sometimes referred to as *freons*, which are colourless, non-inflammable, non-corrodent to most metals and generally non-toxic, are those now in common use. They may be categorised as falling within one of three chemical forms:

- *CFCs (chlorofluorocarbons).* These have a high ozone-depleting potential (ODP) contributing to the breakdown of the ozone layer, are *banned* by the Montreal Protocol and will cease to be manufactured in the European Community by January 1995. Examples are R11, R12 and R114.
- *HCFCs (hydrochlorofluorocarbons).* These have limited ODP, are classified under the Montreal Protocol as *transitional substances* and are due to be phased out early in the next century. Examples are R22, R123 and R124. R123 is available as a 'retrofit' refrigerant for R11. In this context, 'retrofit' means a fluid which may be substituted into an existing system but will require material changes to equipment.
- *HFCs (hydrofluorocarbons).* These contain no chlorine and therefore have zero ODP and in consequence are not controlled by the Montreal Protocol. Examples are R125, R134a and R152a. R134a is a 'drop-in' refrigerant for R12. Here, 'drop-in' means a fluid which can be substituted directly, requiring replacement of some serviceable components only. R134a is emerging as the preferred refrigerant for most air-conditioning applications.

In addition to an ODP classification, refrigerants are also given ratings for global warming potential (GWP), an index providing a simple comparison with carbon dioxide which has an index rating of unity. The properties and values for the ODP and GWP indices of common refrigerants are given in Table 19.1 together with those for ammonia as a comparison.

In addition, the table gives values for the occupational exposure limit (OEL), in parts per million, which reflect the toxicity level of the various refrigerants. In the knowledge of these allowable concentration levels in occupied areas, it is necessary that adequate rates of ventilation be provided in plant rooms and recommended that refrigerant leak detection be installed.

It is clearly desirable to limit the discharge of any refrigerant to atmosphere and, in consequence, provision should be made for *pump-down* (removal) from machines during maintenance activities. In addition, consideration should be given to limiting the volume of refrigerant gas in a system together with improved standards of design and installation for refrigerant pipework in order to reduce the risk of leakage.*

Air also may be used as a refrigerating medium. One method is to compress it to an absolute pressure of about 1.4 MPa (14 bar) and then, after removing the heat of compression, allow it to expand through a valve. In aircraft, the air cooling cycle is used in quite small turbo equipment running at very high speeds taking advantage of the extremely low temperature of the surrounding ambient for use in the condensing side. For normal land use, bearing in mind the relative cost of plant required, air is not a practical choice as a refrigerant for air-conditioning.

* *CFCs, HCFCs and Halons. Professional and Practical Guidance on Substances which Deplete the Ozone Layer.* CIBSE Guidance Note GN 1, 1993.

Table 19.1 Properties of refrigerants

Properties	Refrigerant: *Ammonia*	*R11*	*R12*	*R22*	*R123*	*R134a*
Gauge pressure (kPa)						
condenser (30°C)	+1060	+24.8	+642	+1100	+8.2	+670
evaporator (–15°C)	+155	–80.6	+81.2	+195	–85.4	+62.6
evaporator (–5°C)	+404	–52.5	+254	+427	–60.5	+249
Boiling point (°C) (standard pressure)	–33.3	+23.8	–29.8	–40.8	+27.8	–26.1
Critical temperature (°C)	1333	198	111	96	184	101
Volume of vapour at –15°C (m^3/kg)	0.509	0.766	0.093	0.078	0.873	0.121
Latent heat of evaporation at 15°C (kJ/kg)	1320	198	162	218	175	187
Theoretical energy input per unit ÷ energy output (kW/kW)	0.211	0.200	0.213	0.216	0.203	0.217
Coefficient of performance (27°C to –15°C)	4.75	5.00	4.69	4.65	4.93	4.61
Ozone depleting potential (ODP)	0	1.0	1.0	0.05	0.014	0
Global warming potential (GWP)	0	1500	4500	510	29	420
Occupational exposure limit (OEL)[b] (ppm)	25	1000	1000	1000	10[a]	1000[a]
Characteristics	Strong irritant Can form an explosive mixture in certain conditions			Odourless Non-irritant Non-inflammable		
Usage	Large plants Piston machines	Reciprocating (piston) rotary, screw and centrifugal plants				

[a] Provisional recommendation by refrigerant manufacturers.
[b] Health and Safety Executive. *Occupational Exposure Limits*. EH40/93.

Water is also used as a refrigerant in a system termed *steam jet* described later.

The refrigerants most suitable for direct expansion into coils in the airway are R22, R134a and certain refrigerant blends, the others all being objectionable owing to their smell, toxicity, inflammability, or inefficiency.

Types of refrigeration plant

Vapour compression

Vapour compression plant is normally classified by the compressor type and thus includes reciprocating, rotary, scroll, screw or centrifugal, either belt driven or directly coupled to a prime mover, normally an electric motor. An *open* compressor has the motor connected through a belt-drive or a coupling requiring an external shaft and seal to contain the refrigerant. *Hermetic* compressors have the motor and compressor as a self-contained assembly with the motor in contact with the refrigerant within the casing, no shaft seal being required. *Semi-hermetic* compressors are similar to the hermetic type, but with access to the compressor for repairs: motor failure would require the assembly to be repaired off-site. *Sealed units* are of the hermetic type with the whole assembly contained within a welded steel shell, used extensively in refrigerators and freezers, and in packaged chiller units for air-conditioning.

Reciprocating

A positive displacement piston machine which can operate over a wide range of conditions, the compressor may have up to 16 cylinders arranged in V or W formation, see Figure 19.3. A typical sealed compressor is illustrated in Figure 19.4.

Capacity control is normally provided by cylinder 'unloading' in steps or by switching multiple compressor and refrigerant circuits; hot gas bypass and speed regulation are also available.

In installations where duty and standby machines are required for security in operation, benefits in efficiency may be obtained by the use of two-speed plant; typically the

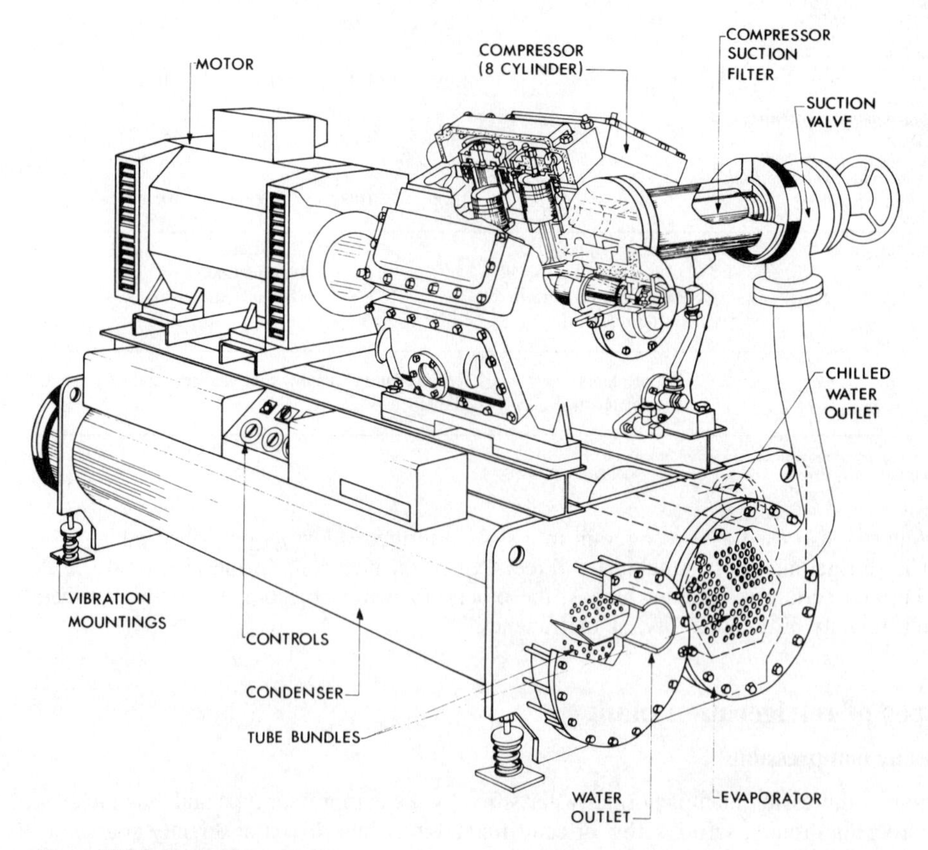

Figure 19.3 Reciprocating compressor

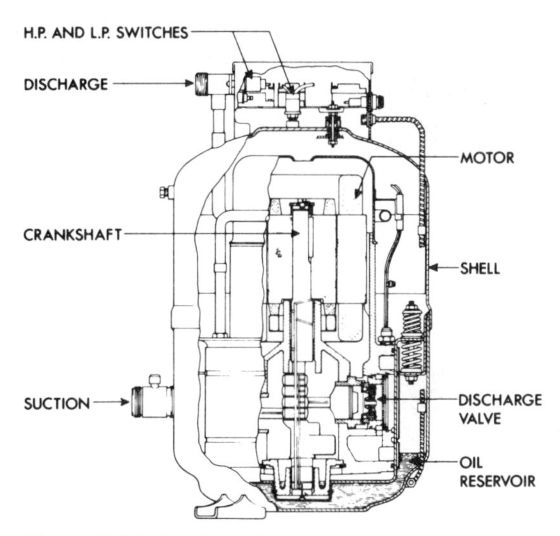

Figure 19.4 Sealed (hermetic) compressor

compressor total efficiency at low speed will be 70% compared with 60% when running at high speed.

Reciprocating machines are available in the range 55–1000 kW cooling duty using R134a and up to 1.2 MW using R22. They may be connected to direct expansion air coils or water chillers on the suction side and to air cooled, water cooled or evaporative condensers on the compressor discharge side.

Hermetic piston-type compressors are widely available up to 120 kW using R22 and are used in combination to make larger capacities. Such machines are obtainable built as a weatherproof unit complete with an air-cooled condenser. A low silhouette arrangement is available for installation on flat roofs. In split systems, where the evaporator and condenser are located apart from the compressor or each other, the length of the connecting refrigerant lines must be kept to within practical limits.

Rotary

In effect another form of piston compressor, but without valves, this type was developed for small scale applications. The assembly is contained in a cylindrical casing and depends for its operation on a shaft eccentric to the cylinder carrying a rotor which when rotated produces the compression effect.

Scroll

A robust and efficient type of machine, based upon the compression effect obtained when an involute spiral is rotated within a second fixed volute and contained within fixed plates at either face, the gas being compressed as the volume is reduced closer to the centre of the scroll, Figure 19.5. Units are available in the range 50–170 kW, operating on R134a and R22, for either air or water cooling.

Screw

Compression is produced by two rotating helical screws, the seal being achieved by oil. A retractable vane enables load variation in a simple manner over a wide range. This form

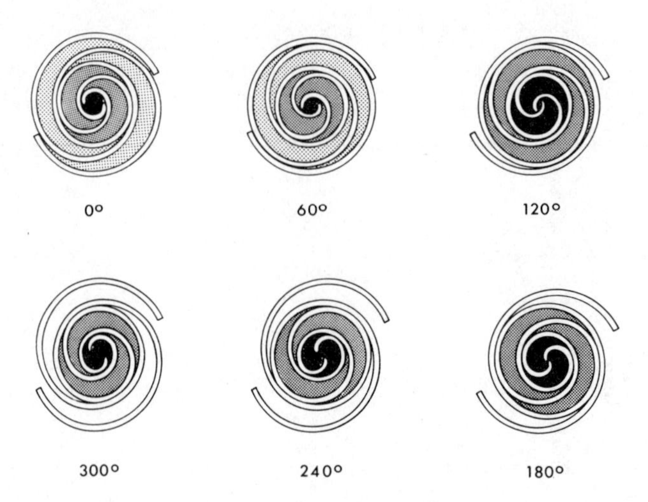

Figure 19.5 Scroll compressor, operating principle

of compressor, Figure 19.6, may be used with a wide range of refrigerants from quite small machines to a cooling capacity in excess of 4.5 MW and offers the advantages of minimum vibration and low noise level. Open, hermetic or semi-hermetic machines are available, depending on the size and the manufacturer. In the smaller sizes, screw compressors are manufactured as sealed units with capacity control by a simple slide valve mechanism down to 25% of full load.

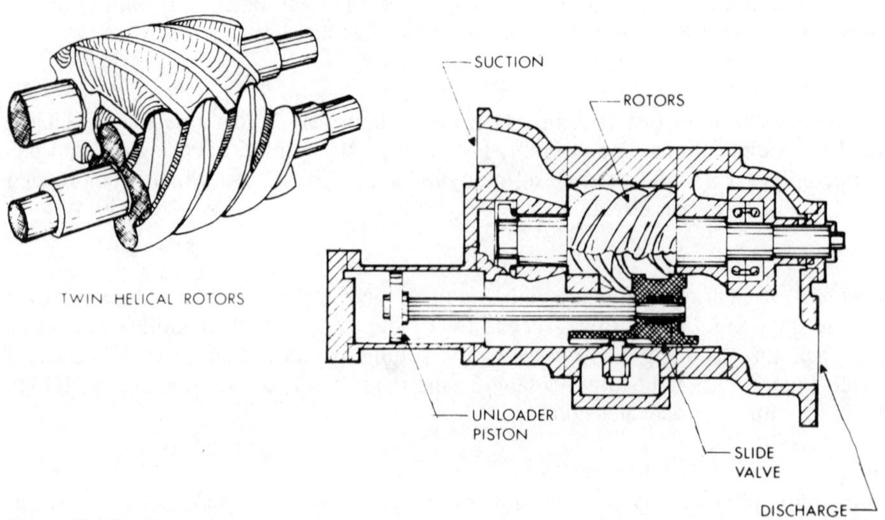

Figure 19.6 Screw compressor

Centrifugal
The centrifugal compressor is used chiefly where large duties are required. The advantages are:

- Saving of space as compared with reciprocating machines.
- Reduced vibration (thus suitable for a roof-top plant chamber).
- Reduced maintenance due to there being no wearing or reciprocating parts.
- Efficient part load operation with a control range down to 10%.

Refrigerants used in centrifugal compressors are R123 or R134a, replacing those using CFCs R11 and R12 in earlier models, which for air-conditioning temperatures operate over a small pressure range, thus reducing slip losses between blades and casing.

An important consideration is the turn-down range, i.e. the ability to follow load variations. The two-stage machine shown in Figure 19.7 has a turn-down to 10% of full load, thus making it highly flexible in operation. Single- or multi-stage machines are available, some driven through gearing to achieve the required speed of rotation. These machines are available in the capacity range of 500 kW to 4.5 MW and are used invariably for water or brine chilling and may be water or air cooled. Capacity control is by throttling at the suction inlet or speed control. These machines are normally of the semi-hermetic or open type with the condenser and evaporator close-coupled to the compressor as one complete unit.

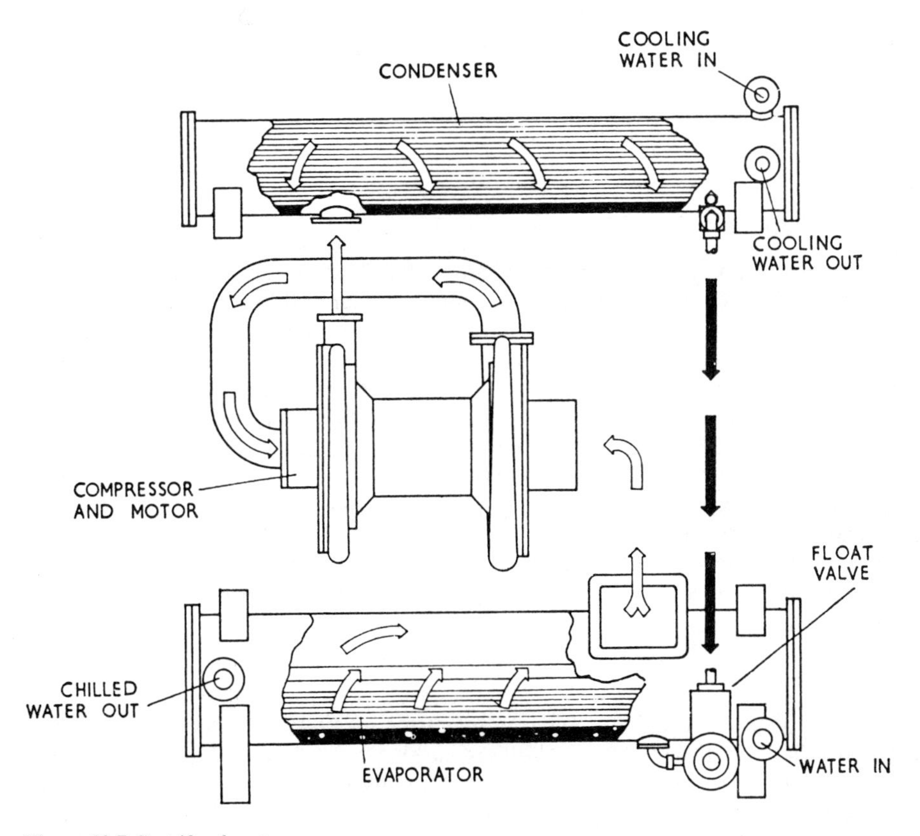

Figure 19.7 Centrifugal system

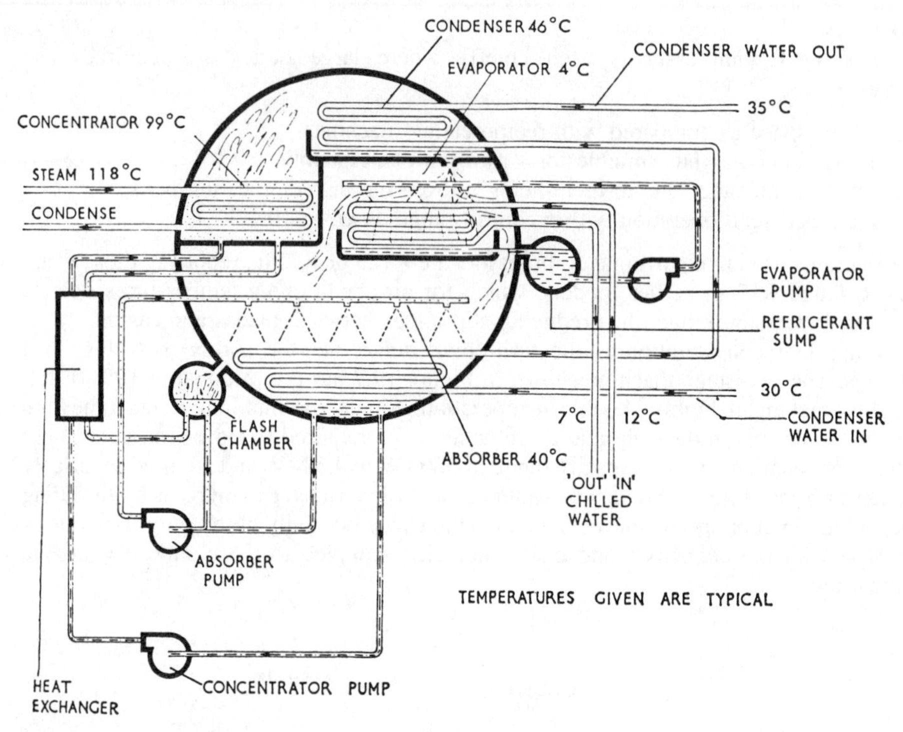

Figure 19.8 Diagram of single stage absorption system

Absorption plant

A single stage absorption plant is shown in Figure 19.8. This type uses either ammonia as the refrigerant and water as the absorption medium, or water as the refrigerant and lithium bromide as the absorption medium. The equipment has no moving parts except pumps. Two-stage plant is also available, incorporating a second concentrator which utilises heat recovery from the first stage of the cycle to provide improved efficiency. The source of energy being either steam, medium temperature hot water or gas, this type of plant is suited to installation where there is a heat source of a suitable grade available, such as with combined heat and power (CHP) systems. Such plant works under a high vacuum and is available in a range from 10 kW to over 5 MW cooling capacity. The heat to be removed by the condenser water with this system is about double that of a vapour compression plant of equivalent capacity. For example, an absorption chiller of 350 kW capacity will use a nominal steam quantity of 220 g/s, with heat rejection of some 900 kW.

Steam-jet plant

An interesting type of refrigerator, using water as the medium, is that shown in Figure 19.9. Its operation depends on the possibility of causing water to boil at low temperatures under high vacua. As noted previously, water at 7°C boils at an absolute pressure of 1 kPa, i.e. about one-hundredth of atmospheric pressure. The absence of any special refrigerant is an advantage and an economy.

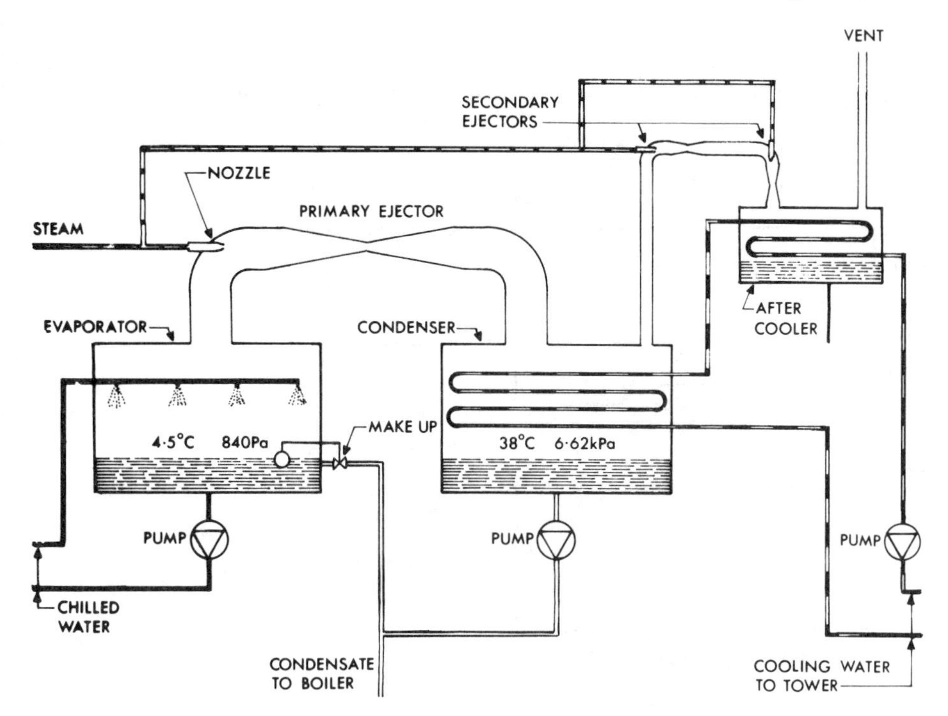

Figure 19.9 Diagram of steam-jet system

The energy input with this equipment is much greater than with the positive compression types owing to the inefficiency of jet compression and the requirement for condensing water flow is about five times that of a vapour compression plant.

A variation of the same system, but using a centrifugal compressor in place of the steam-jet compressor, has also been developed and this avoids the above mentioned disadvantage of high energy input. The great difficulty with both types is the maintenance of the extraordinarily high vacuum for long periods. This method is not widely used, but has been found economical in industrial plants where exhaust steam is available.

Choice of refrigeration plant

The selection of refrigeration plant will depend upon a number of factors, among which are: the location of suitable space; the availability of waste heat; the importance laid upon plant noise and whether condenser heat must be rejected at a distance from the compressor.

Where a refrigeration plant serves a single air-conditioning system of the *all-air* variety, and the two can be located close together, a direct expansion coil or coils may be used subject to the practicality of matching the control characteristics of this type of heat exchange surface to the system load. In other circumstances, for single systems of any *air/water* variety and for all systems which consist of a number of distributed air handling units, a chilled water system is used to transport cooling energy from the evaporator. In this latter case the refrigeration plant as a whole is often referred to as a *water chiller.* A

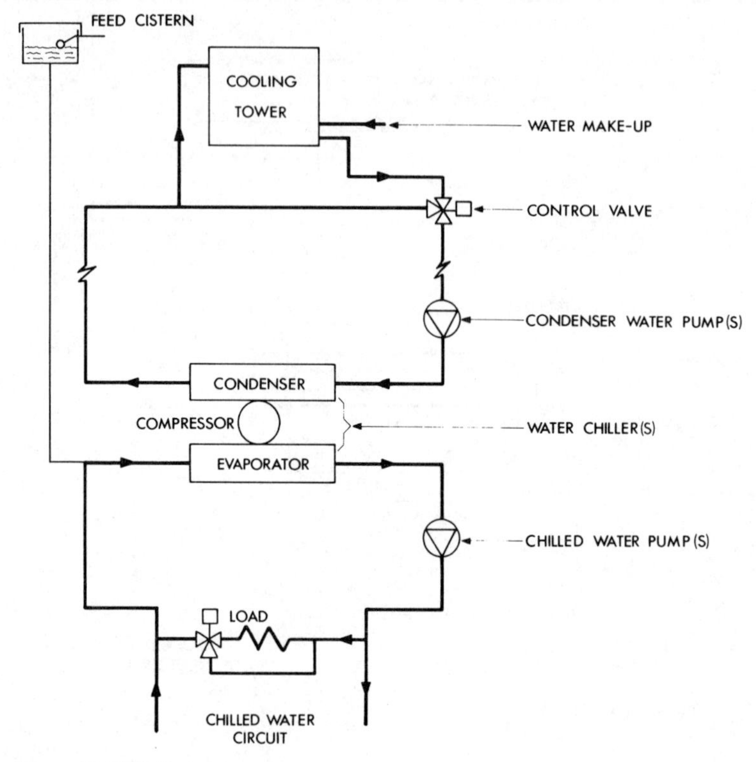

Figure 19.10 Chilled water system

typical arrangement of the principal components of a chilled water system is shown in Figure 19.10.

The recent outbreaks of *Legionnaires' disease* have led designers to favour the use of air cooled condensers, in preference to any water cooling methods, wherever it is possible to so site the refrigeration plant that the necessary large quantities of outside air are freely available for circulation. However, there is an energy penalty for using air cooling, due to the resulting higher condensing temperature, which may be of the order of an additional 20% power absorbed by the compressor motor. Since a cooling tower may be located at a considerable distance from the refrigeration plant, horizontally or vertically, such a combination has obvious attractions as far as flexibility is concerned and it is common practice for the machinery to be sited in a basement and the cooling tower at roof level. It is of course necessary that an open circuit cooling tower should be positioned sufficiently above the refrigeration plant to prevent any drainage or priming problems. Where this cannot be achieved, an intermediate heat exchanger may be installed.

Storage systems

An annual cooling load profile, plotted for almost any air-conditioned building, would show that the maximum load occurs on a few occasions only. This is due to the fact that many of the components of the total are not only seasonal but transitory hour-by-hour. Thus, where the central plant is sized to match the peak load, it will operate at considerably less than its full capacity for the majority of the time. An alternative to sizing a refrigeration plant to meet the peak cooling load would, therefore, be to provide some means to store a cooled medium (water, brine, ice, etc.) and thus allow a somewhat smaller plant to run for longer periods at full output. The stored energy would then be available, as required, to make up the deficit between the plant capacity and the peak requirement. The advantages of such an arrangement are:

- Reduced size of refrigeration plant.
- Reduced maximum demand on power supply.
- Reduced unit energy charge, if run off-peak.
- Inherent standby capacity in the energy store.
- Stability available for control.

The parallel disadvantages are:

- Higher capital cost.
- Increased space requirement.
- Lower coefficient of performance with low temperature or ice storage systems.

The storage system concept may be applied in varying degrees. It is not uncommon in the British Isles for a limited quantity of chilled water to be provided for *peak lopping* and, in this case, capacity would be provided to supplement a water chilling plant for one or two hours in the day, at the time of maximum load. At the other extreme, it would not be impossible to provide sufficient storage to meet a total peak-day cooling load, the plant running during night hours only during the period when an off-peak tariff applies. A further option would be to run the plant continuously over 24 hours and to provide a limited level of storage such that during the whole period of demand for cooling, both the plant output and the stored energy would be used in parallel. Figure 19.11 illustrates these three basic operating modes and indicates the relationship between plant size and stored water quantity.

Whereas the design of a hot water storage system (Chapter 5, p. 135) is able to limit the required volume of the vessels by applying pressure and storing water at an elevated temperature, parallel action (i.e. storage at a temperature much lower than that of usage) is not really practical in the case of chilled water. A temperature of about 4°C is that most usually chosen as the practical minimum but, even at this level, design problems arise since the specific mass of water at that temperature is critical and stratification is unstable.

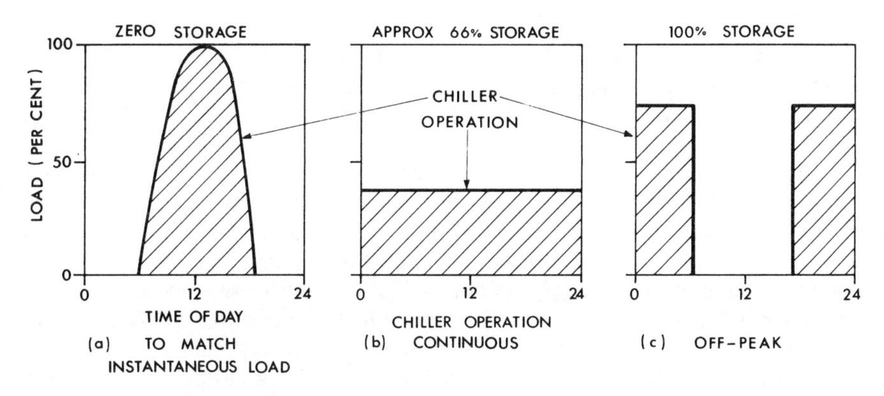

Figure 19.11 Chilled water storage options

As a result, the use of alternative media such as brine, ice and phase-change chemicals must be considered if storage volumes are to be of practical size.

Free cooling

Use may be made of the principle of evaporative cooling in order to produce water at a temperature low enough to be suitable for cooling in an air handling plant. A cooling tower, provided to reject condenser heat in summer, may be usefully employed to provide a source of cooling at times when the outside wet bulb temperature is low enough. A basic arrangement is shown in Figure 19.12 where a closed circuit water cooler is piped into the chilled water circuit: an alternative arrangement would be to use an open circuit tower with a heat exchanger interposed between the open circuit and the closed chilled water circuit, in order to prevent fouling in the closed circuit.

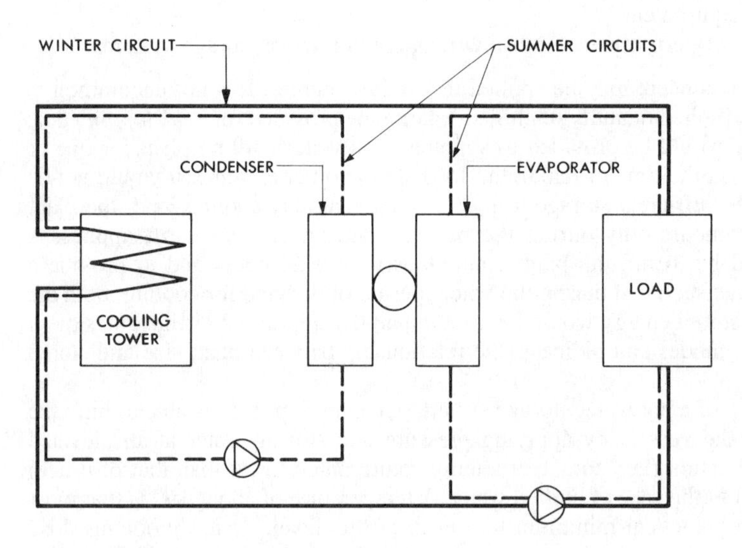

Figure 19.12 Free cooling using an evaporative cooling tower

A system of *dry coil free cooling* may be used to advantage, in particular with systems which operate continuously, such as those which serve computer suites. A water and glycol mixture is circulated by a pump between a fan-assisted dry cooler, similar in form to an air cooled condenser, and a cooling coil in the air stream. Typically, such a system would be designed such that, when the outside air temperature fell to 5°C or below, full cooling would be achieved by this method. As the outside air temperature rose above 5°C, cooling would be introduced progressively from a conventional refrigeration plant to supplement the free cooling effect. A development of this principle is to use the glycol/water circuit and fan-assisted cooler in a dual mode, as a source of free cooling in winter and as a means of heat rejection from the refrigeration machine in summer. The principles of these approaches are shown in Figure 19.13, the equipment being available in packaged form from specialist manufacturers. It is claimed that a pay-back period of two to three years may be achieved on the additional capital cost of the equipment.

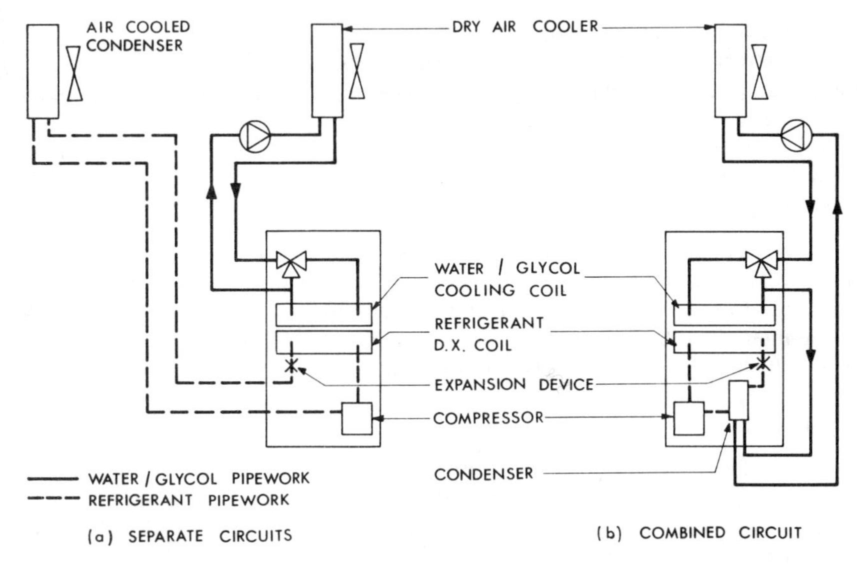

Figure 19.13 Free cooling using a dry coil

The most efficient form of free cooling in this context is by use of the *thermosyphon** principle. This may be applied using either dry coolers or evaporative condensers, the latter being more effective since they cool the refrigerant towards the ambient wet bulb temperature rather than the dry bulb, as is the case with dry coolers. The refrigerant is piped in the normal manner, via the compressor, between a purpose designed evaporator and the condenser, which must be sited at a higher level than the remainder of the system. A by-pass facility is provided around the compressor.

When outside ambient conditions are suitable, the compressor may be switched off and the by-pass put into use, allowing refrigerant to flow between the evaporator and condenser. In this mode, the evaporated refrigerant from the chilled water heat exchanger is drawn up to the condenser, by pressure difference, where it condenses and returns by gravity to the evaporator. In a multi-chiller installation, using this method of operation when outside ambient conditions permit and with a constant cooling demand throughout the year, energy savings of about 50% may be achieved. The additional capital cost of the installation would be of the order of 20%.

Refrigeration plant components

Refrigerant pipework

Some plant configurations will involve the design of pipework carrying the refrigerant between components in the cycle. Special attention must be given to this important aspect of the installation because the fluid in circulation, being volatile, may be in gas or liquid form or a mixture of the two. When dealing with such design, reference should be made to specialist works on the subject, to ensure that the essential basic criteria are met in order to:

* Pearson, S. F., *Thermosyphon Cooling*. Institute of Refrigeration, Session 189–90.

- Minimise oil loss from the compressor.
- Ensure satisfactory refrigerant flow to the evaporator.
- Prevent liquid refrigerant or oil in slugs from entering the compressor (during operation or when inoperative).
- Prevent oil from collecting in any part of the system.
- Avoid excessive pressure loss.
- Maintain the system clean and free from water.

All pipework or plant carrying refrigerant at low temperatures (low pressure/evaporator part of the circuit) or chilled water must be insulated to prevent the formation of condensation and to reduce heat gains. To this end an effective vapour barrier must be applied, or be integral with the insulation, on the outer surface. BS 5422: 1990 gives the recommended thicknesses for given system operating temperatures.

Evaporators

In its most simple form, the evaporator is tubular, the tubes containing the refrigerant and the whole array immersed in the liquid to be cooled. In the *dry system* the evaporator coils are filled with vapour having little liquid present but when operated on the *flooded system*, the liquid refrigerant discharges into a cylinder feeding the coils by gravity. As evaporation takes place, the gas returns to the top of the cylinder and from there returns through the suction pipe to the compressor.

The most common form of evaporator for application to air-conditioning practice is the *shell and tube* type, used with a closed circuit chilled water system serving cooling coils. The water is contained in the tubes and the refrigerant in the shell. A development of the shell and tube type is the direct-expansion shell type evaporator in which the refrigerant is in the tubes and the water in the shell. The tubes may be arranged in two or more circuits, each with its own expansion device and magnetic valve on the liquid inlet to allow step control. Different arrangements of baffles in the shell control the water velocity over the tubes in order to improve heat transfer.

It is generally not necessary to resort to brine for air-conditioning purposes, since chilled water at about 4°C satisfies all normal requirements. Precautions against accidental freezing of the water in the evaporators include low suction pressure cut-outs and low water temperature cut-outs as well as water-flow switches.

In instances where the cooling load may be less than the minimum output of the refrigeration machine and where some severe limitation exists as to the number of starts per hour, it is possible to fit a *hot gas by-pass valve* which will, as required, direct hot gas from the condenser into the evaporator to provide an artificial load. This can be arranged to come into operation automatically at a predetermined point of the capacity controller range and will enable the machine to run continuously even when no true cooling load exists.

Modern water-chilling plant does not generally require the addition of chilled-water storage to act as a ‘flywheel’, because it can be arranged in suitable steps of capacity control to suit the variations in load. Little complication is experienced with piston compressors having four to eight steps of control. Centrifugal units which turn down to 10% can virtually run on pipeline ‘losses’ as can screw-type machines.

Evaporators in the form of *direct expansion (DX)* cooling coils in the air stream, are described in Chapter 17. The refrigerant flow rate is controlled by a thermostatic expansion valve, or a capillary tube on smaller plant; the sensor to control the valve being located in the suction line to the compressor to maintain the correct degree of superheat at the compressor intake and to avoid the damage which would occur if liquid entered.

Coils handling volatile refrigerants present fluid distribution problems and to ensure equal flow of refrigerant through each circuit in the cooling coil assembly, the fluid is passed through a distributor at the coil inlet to divide the flow equally.

Condensers

The evaporative condenser, consisting of coils of piping over a tank from which water is drawn and then circulated to drip over the pipes through which the refrigerant is passed, is the simplest form of condenser. Its use is restricted to cases where the compressor can be sited near to the condenser, to avoid long lines of piping containing refrigerant under pressure. Evaporative condensers are available in sizes up to 1 MW. They give an increase in efficiency because any intermediate exchange of heat is eliminated as compared with a water-cooled condenser connected to a cooling tower. The design must be carefully considered where capacity control is required and where winter operation is envisaged.

Often for air-conditioning applications, the condenser takes the form of a refrigerant-to-water heat exchanger of the shell and tube multi-pass type. Circulating water piping from the condenser to the water cooler with these types traverses the building, and all equipment containing refrigerant is confined to the plant room. Water cooling systems, however, are less popular today because the temperatures at which these operate, around 30°C, are such as to increase the risk of active development of the bacteria of *Legionnaires' disease*. With regular and thorough maintenance, combined with a suitable water treatment, the risk of harmful bacteria development is small.*

This risk has led to an increase in the use of air cooled condensers for all sizes of plant, where the building occupants are particularly susceptible to the disease, as may occur in hospitals. Air cooled condensers are in any case used in most small self-contained or split-package systems and in duties of up to 350 kW in capacity, Figure 19.14. They are often found to be economical in first cost up to about 100 kW cooling capacity. For larger plant,

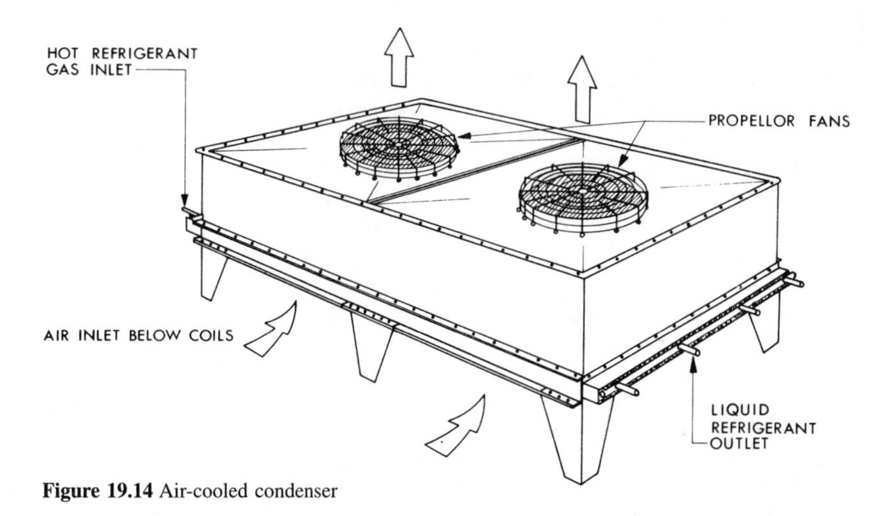

Figure 19.14 Air-cooled condenser

* *Minimising the Risk of Legionnaires' Disease.* CIBSE Technical Memorandum TM13, 1991.
Legionellosis (Interpretation of Approved Code of Practice: The Prevention or Control of Legionellosis. CIBSE Guidance Note GN3, 1993.

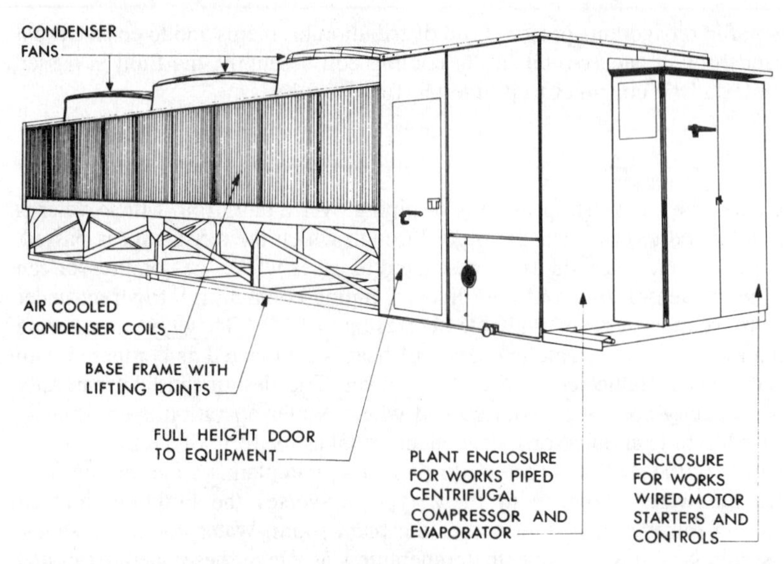

Figure 19.15 Packaged air cooled centrifugal plant (Trane)

the running cost consequence of air cooled compared with that of evaporative water cooled condensers must be considered. The higher condensing temperature that results from air cooling leads to the compressor doing more work to produce the same cooling effect, giving a lower coefficient of performance and in consequence a higher running cost.

Packaged units comprising evaporator, compressor and multiple-fan air cooled condenser are available in large sizes, up to 1 MW cooling, limited only by the largest size that can be transported on a low-load lorry. A typical example is shown in Figure 19.15.

A development of the dry air cooler, which requires a smaller heat exchange surface and a lower air quantity, is realised by pre-cooling of the upstream air using an *adiabatic effect*. The approach air stream is sprayed with a fine mist of mains water, or a pump pressurised supply, the quantity injected being controlled carefully to ensure that it is totally absorbed with no residue. The cooled and humidified air is then drawn over the air cooler by the fan(s) in the usual way. As a further precaution against *L. pneumophilia*, the spray water may be passed through an ultra-violet irradiation unit to provide disinfection. In hard water areas, ionic scale prevention should be used.

Air-cooled condensers are often used in the tropics, despite the fact that they are bulky, in circumstances where the outside air temperature is high. This apparent paradox arises from the simple fact that evaporative condensers and cooling towers require some measure of skilled maintenance, whereas the air cooled condenser requires little attention other than simple basic cleaning and, of course, some lubrication. A further advantage is that it requires no water supply in those places where such may happen to be scarce.

Evaporative coolers

The heat extracted by the refrigerating machine (cooling effect), together with the heat equivalent of the power input to the compressor, raises the temperature of the condenser water by an amount which is dependent upon the quantity of the water which is circulated through the condenser.

The lower the temperature of the condenser water the less power will be required to produce a given cooling effect; and it also follows, conversely, that with a given size of plant the greater will be the amount of cooling possible.

Water from a well or from the main supply will always be the coldest, the former at 12°C and the latter at about 18°C in summer. The quantity to be wasted, however, generally rules out this method. For instance, a 700 kW plant with power input of about 120 kW with a 10 K rise through the condenser would require a flow of $(700 + 120)/(10 \times 4.2) = 20$ litre/s and it is likely that water charges would exceed the cost of current for running the compressor many times over. An alternative is river water, which, subject to Water Authority approval, may be used directly for condenser cooling purposes. Before such a solution is adopted, however, consideration must be given to the need for filtration, the quality of the water and whether the composition is aggressive to the system materials; maintenance will need to be of the highest quality.

Applications exist outside the British Isles, however, where well water is more freely and cheaply available, which produce economical solutions. In one known instance in Europe, such well water is first passed through a pre-cooling coil integral to the air-handling plant prior to use in the condenser. Similarly, in some parts of the Caribbean, clear sea water is available via fissures in the coral which may be pumped in quantities of up to 1000 litre/s through specially designed condensers.

However, cooling the water by evaporation is more generally adopted. Evaporative coolers depend on the ability of water to evaporate freely when in a finely divided state, extracting the latent heat necessary for the process from the main body of water, which is then returned, cooled, to the condenser. In the case stated above, the consumption of water with an evaporative cooler (assuming no loss of spray by windage) would be only $820/2258 = 0.36$ litre/s.

Evaporative coolers divide themselves into two categories, i.e. *natural draught* and *fan draught*. The former is represented by:

The spray pond. In this type the water to be cooled is discharged through sprays over a shallow pond in which the water is collected and returned to the plant. To prevent undue loss by windage the pond is usually surrounded by a louvred screen. Owing to the large area needed for the spray pond system and, moreover, the consequent probability of pollution, this type of equipment is seldom possible for air-conditioning applications.

Natural draught cooling tower. In this, water is pumped to the top of a tower which contains a specially designed 'packing' of plastic or other material which will not support microbiological growth, arranged so as to split up the water stream and present as large an area as possible to the air, which is drawn upwards due to the temperature difference, and by wind. The base of the tower is formed into a shallow tank to collect the water for return to the plant. Again, owing to its size and height, this type of cooler is not frequently used for air-conditioning.

Condenser coil type (evaporative condenser). Where the refrigerating machine is near the point where an outdoor cooler may be used, the condenser heat exchanger may be dispensed with and the refrigerant delivered to coils outside, over which water is dripped by a pump. The water collects in a tank at the base and is recirculated. A louvred screen is usually necessary surrounding the coils.

Fan draught systems are more commonly used owing to the compact space into which they may be fitted. Where possible they are placed on the roof, but if this is impracticable they may be used indoors, or in a basement, with ducted connections for suction and discharge to outside. All fan assisted coolers must be fitted with effective moisture eliminators at the air discharge to minimise the carry-over of moisture which may be contaminated.

The following methods are typical:

Cross-flow type, Figure 19.16. Low silhouette with the water pumped to the top of the tower and discharged over a fill material. Air is drawn in horizontally across the fill to be discharged vertically or horizontally by a fan located at the outlet.

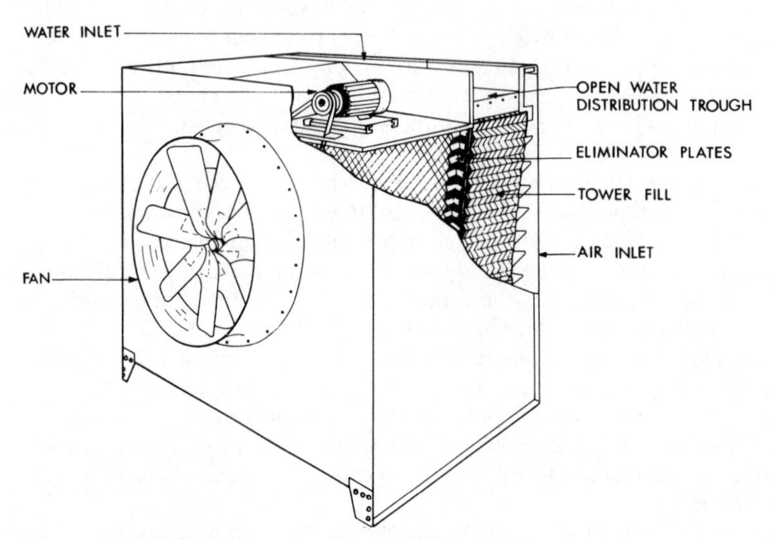

Figure 19.16 Low silhouette cross-flow cooling tower (Heenan Marley)

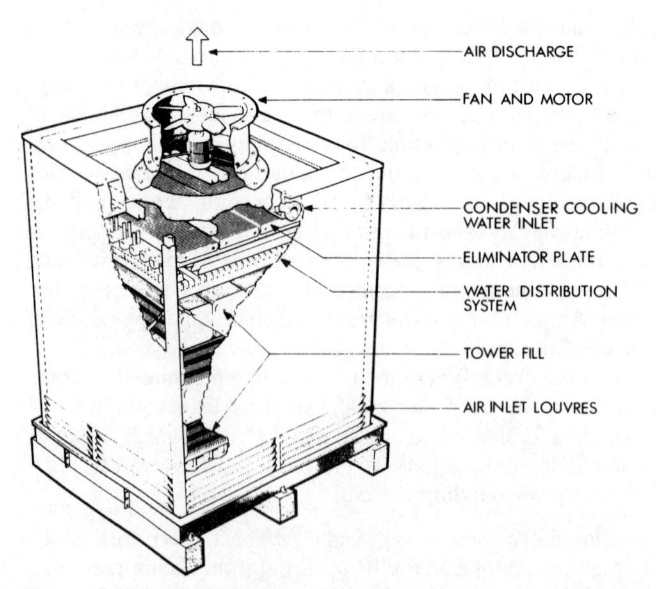

Figure 19.17 Vertical induced draught cooling tower (Heenan Marley)

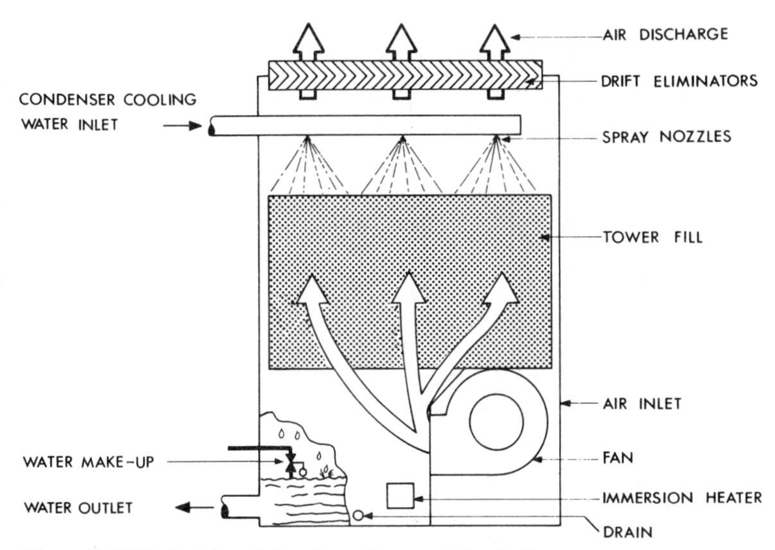

Figure 19.18 Vertical forced draught cooling tower (section)

Induced draught cooling tower, Figure 19.17. In this case the water is delivered to the top by the condenser pump, and cascades over a packing as with the natural draught type. An axial fan is arranged to draw air upwards over the slats at high velocity so that a much reduced area of contact is required.

Forced draught cooling tower, Figure 19.18. This type is similar in function to the induced draught type except that the fan, normally centrifugal, discharges air into the tower at low level pressurising the shell slightly and forcing the air upwards through the packing.

Film cooling tower. Similar to the induced draught type; in this case the water is not sprayed or broken up in any way, but is allowed to fall from top header troughs down slats arranged in egg-crate form. The water remains as a film on the surface of the slats. Higher air velocities than usual are permissible without risk of carrying over of water, and yet air resistance is low due to the open nature of the surfaces.

Closed circuit water coolers, Figure 19.19. In this case the condenser water is not evaporated but circulated through tube bundles which act, as it were, as the packing within the tower. In most other respects such coolers resemble induced draught or forced draught units. An independent water supply is required for the local evaporation circuit which is pump-recirculated over the tube bundles. Such coolers are useful for application to heat recovery systems where fouling of condenser water must be avoided; where the cooler is below the level of the refrigerating apparatus; or where the cooler is used to provide 'free cooling'. Due to the additional heat exchange inherent in the circuit and consequent loss of efficiency, this type of tower tends to be relatively large in size.

Regardless of the type of evaporative cooler, its location must be carefully selected such that the air discharge, which could be contaminated, is not carried into air intakes, or into openable windows, or across public access routes. A high quality of maintenance is essential with such cooling devices and provision must be incorporated for inspection, maintenance, cleaning, water treatment and taking water samples. To reduce the quantity

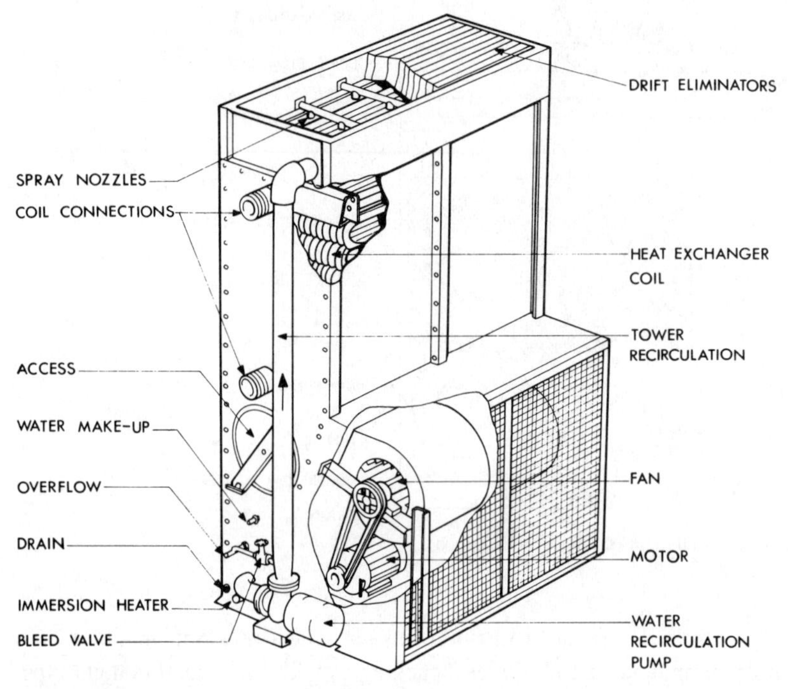

Figure 19.19 Closed circuit water cooler

of water droplets carried from evaporative coolers, eliminator plates should be fitted at the air discharge position.

Rating of cooling towers

The heat to be removed from the condenser cooling water by the cooling tower is equal to the sum of the cooling load plus the heat equivalent of power absorbed by the compressor. An approximate figure of 1.2 kW per kW of refrigeration may be used.

The quantity of water to be passed through the tower is dependent on the temperature drop allowable between inlet and outlet. A usual figure is 5 K and the inlet water flow will then be $1.2/(5 \times 4.2) = 0.057$ litre/s per kW of refrigeration.

The temperature to which the cooling tower may be expected to reduce the condenser water depends on the maximum wet bulb of the atmosphere and the design of the tower: the higher the efficiency the closer will be the water outlet to the wet bulb temperature. A good efficiency will give a difference of about 3 K, so that, if the external design wet bulb is taken at 20°C (refer to Table 3.11), the cooling water will be brought down to 23°C, and, with a 5 K temperature rise through the condenser, the outlet will be at 28°C. This temperature then, in turn, forms the basis for design of the compressor and condenser (e.g. condensing at about 37°C).

Condenser water treatment

As evaporation takes place, there is a continuous concentration of scale-forming solids which may build up to such a degree as to foul the condenser tubes. Similarly, the evaporative surfaces of the cooler suffer a build-up of deposit.

In order to overcome this problem:

- A constant bleed-off of the water is required, the rate of which may be calculated from the known analysis of the water, the evaporation rate and the maximum concentration admissible.
- As is common practice, the water may be treated by a regular chemical dosage which also generally contains biocides to prevent micro-biological growth.

In the case of the closed circuit water cooler, the condenser tubes and the tower tube bundles do not suffer internal fouling but the external surfaces of the latter will, in time, become coated with solids. The need for water treatment of the spray water thus remains.

The total water loss from an evaporative cooler is the sum of evaporation loss, bleed rate and windage loss (water droplets carried from the tower by natural or by induced air movement) which would normally be between 3% and 5% of the condenser water flow rate. It is normal for water storage equivalent to 24 hour usage to be held on site as a reserve against mains failure.

Packaged cooling towers

Whereas past practice was to use structural containment to form the tower proper and to use purpose built internals, water distribution arrangements, etc., cooling towers are now commonly factory fabricated for delivery to site in a minimum number of parts. Whilst this development reduces site works, there is a tendency to design to minimum size also with the result that the depth and capacity of the base tank is inadequate. This condition may lead to problems of overflow when the condenser pumps are stopped or, more seriously, to the formation of a vortex at tank outlet due to an insufficient water depth. To reduce the risk of this problem, only the minimum length of condenser cooling water pipework should be installed above the water level in the cooling tower tank, thereby reducing the quantity of water that will drain into the tank when the pump circulation is stopped; excessive drain-back will cause water loss from the system via the tank overflow.

In all evaporative coolers, water loss by whatever cause is made up by 'fresh' water through a float valve.

Heat recovery

Air-conditioned buildings offer an inherent opportunity for some form of heat recovery system, utilising the heat to be rejected from the condenser of a refrigeration plant. The inner and core areas of such buildings are heat producing on a significant scale, due to lighting and other electrical loads, whilst the perimeter areas, in winter, require the addition of heat. Thus, if the air-conditioning system is such that the refrigeration plant operates in winter to produce chilled water, as in the case of a four-pipe fan-coil system (see Chapter 14), low grade heat is available for immediate use or for storage and use at some later time. The progression of design development is shown in Figure 19.20. Part (a) represents the conventional separatist approach where surplus heat extracted at the condenser of a refrigeration plant is rejected by a cooling tower, a simultaneous demand for heating being provided by a boiler plant. In circumstances where heat rejection plus compressor power match heat demand, the cooling tower and boiler plant could be done

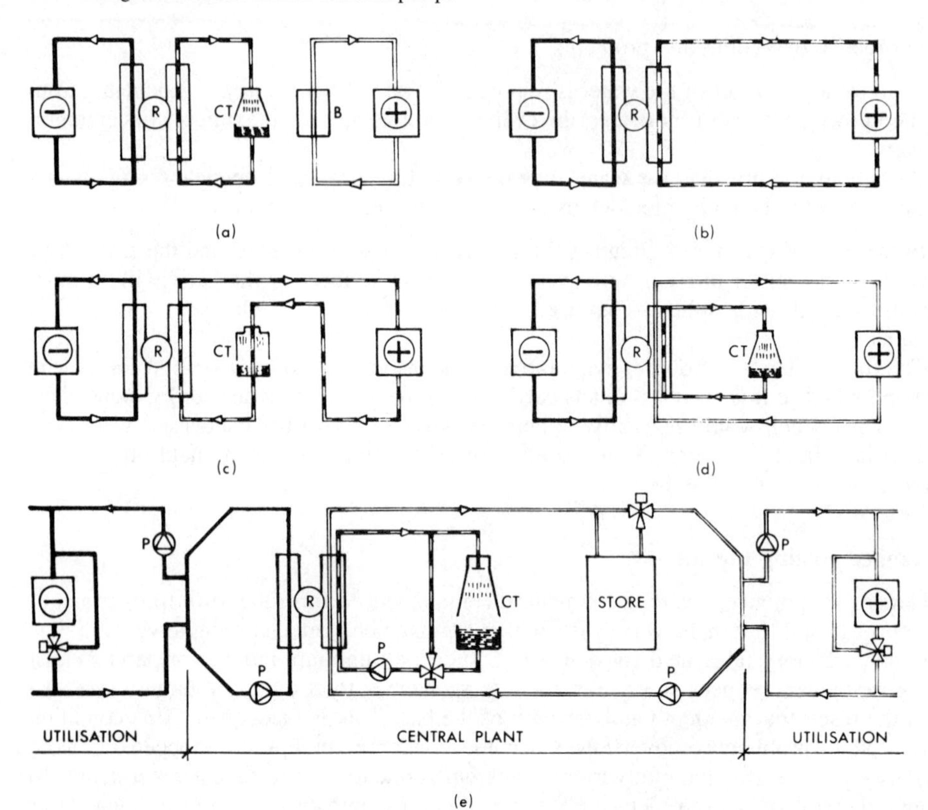

Figure 19.20 Centralised heat recovery systems – development of the concept

away with and the circuit of part (b) substituted: except for the addition here of water circuits, this arrangement is similar in function to that shown in Figure 17.35.

In practice, of course, heat rejection would not match heat demand and, in consequence, the components deleted would have to have alternatives substituted for them to cater for out-of-balance conditions, as shown in part (c) of Figure 19.20. Since water from an open cooling tower carries atmospheric pollutants, it could not be circulated at large throughout the building and thus a closed circuit evaporative cooler is illustrated. The disadvantages of this type of equipment, in terms of efficiency and physical size, are such, however, that what is called a 'double bundle' condenser is introduced as illustrated in part (d). This consists of an over-sized shell which contains two quite separate sets of tubes. Water to an open cooling tower circuit passes through one set in the conventional manner and to the heat recovery circuit through the other.

For the type of commercial building to which central heat recovery systems are applied, there is commonly a heat surplus over the 24 hours of operation and some means of storing the excess, when available, for those hours when a deficit occurs is thus needed. Heavily insulated water vessels may meet this need and could, in turn, be 'topped up' by immersion heaters or pipe coils from a boiler plant if necesary. Figure 19.20(e) shows one form of the finally developed system, including automatic control valves, circulating pumps, etc.

Use of multiple machines

In the case of larger scale projects, not all of the refrigeration plant provided for water chilling purposes would necessarily be equipped with double bundle condensers and used for energy recovery purposes; selection would depend upon circumstances. Taking, for example, a building with which the authors were concerned, the total capacity of the cooling plant was 11.4 MW but only one of the three centrigual machines was used for heat recovery purposes. The equipment chosen produced chilled water at 5.6°C from return at 12.8°C and condenser water was made available at 41.5°C from return at 32.2°C. The power input was approximately 850 kW.

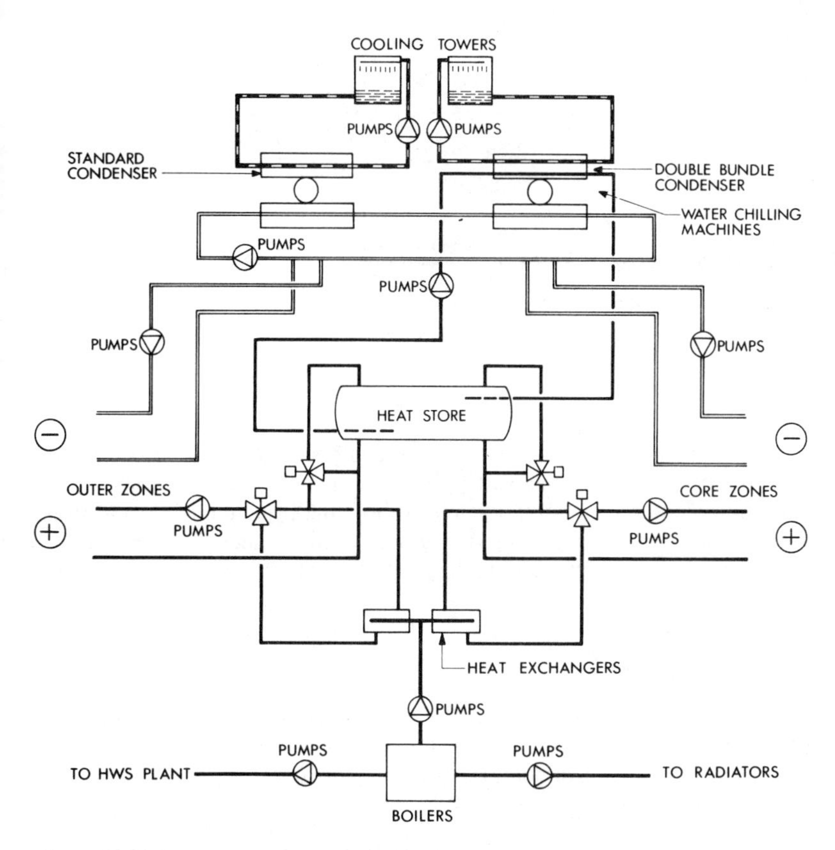

Figure 19.21 Arrangement of central plant heat recovery

Figure 19.21 shows a typical arrangement of plant that may be developed for a large scale installation.

The principle of heat recovery from a refrigerating machine is used in the reversed cycle heat pump system described in Chapter 14, but in that case the refrigeration equipment is distributed through the building interconnected by a water circuit.

Heat pumps

The operating principles of refrigeration equipment have been described where the evaporator is utilised as a source of cooling whilst heat produced at the condenser is either rejected or recovered for use. When similar equipment is utilised in the reverse sense, drawing energy from a low temperature source at the evaporator with the specific purpose of making use of the higher grade output at the condenser, the apparatus is described as a *heat pump*, see Figure 19.22. As may be seen from this diagram the cycle is the same

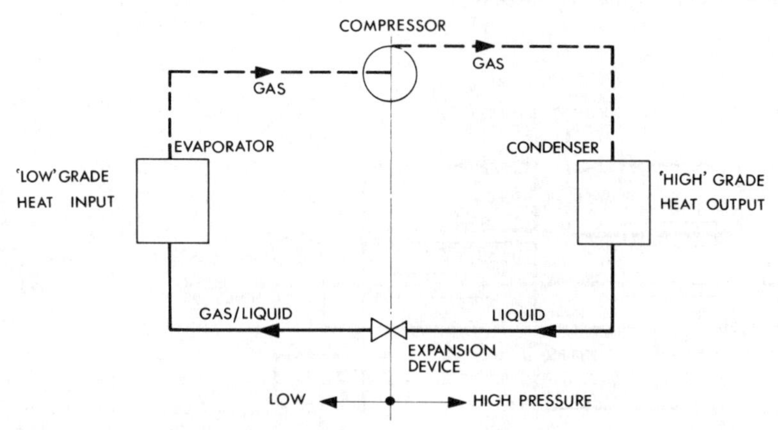

Figure 19.22 The heat pump cycle

as for the refrigeration machine (Figure 19.1) with the component parts described rather differently, but essentially performing in exactly the same manner. Reference should be made to the description of the refrigeration cycle to follow the principle of operation of a heat pump; in which case the compressed gas is passed to the condenser where heat is removed for use, and in the evaporator the refrigerant absorbs heat at a relatively low temperature from the *heat source*.

Coefficient of performance (COP)

This is the term used to describe the *advantage* offered by operation of a heat pump: use of the parallel term *efficiency* is inappropriate when the ideal of 100% is exceeded, such a condition being impossible of achievement. In strict theoretical terms, the coefficient of performance is defined as:

$$COP = T_1/(T_1 - T_2)$$

where T_1 is the condensing, and T_2 the evaporating temperature of the thermodynamic cycle, in degrees kelvin. It is more usual however, in practical terms, to express this coefficient as the simple ratio of energy output to energy input; but it is necessary in this respect to be aware whether the consumption of all appropriate auxiliaries is included in the calculation.

It must be remembered of course that, in comparison with a cooling application, the energy in driving the compressor is an asset as far as a heat pump is concerned; the equation being transposed algebraically:

Cooling

evaporator capacity = condenser capacity – compressor power

Heating

condenser capacity = evaporator capacity + compressor power

It will be appreciated that it is practicable for the same equipment to have a dual function, providing cooling in summer and heating in winter. Such a unit is described in Chapter 14 and the operating principle illustrated in Figure 14.26. One technical reservation which must be appreciated is that, in designing the components for optimum operation for this dual function, these may not be those best suited to either the cooling or heating mode alone and in consequence if the period in one mode far exceeds the other the unit should be selected with a bias towards the more extended period of usage. Typically, a heat pump operating from a source temperature of 5°C and designed for heating only would have a COP of about 3, whereas a reversible machine for the same conditions would have a COP of about 2.6.

As may be seen from Figure 19.23, which represents the output of a reciprocating semi-hermetic machine using refrigerant R22, performance varies with evaporating and condensing temperature, the smaller the temperature difference, the better the performance characteristic. In application to practical problems, it must be remembered that the theoretical performance shown in Figure 19.2 is distorted in practice by superheat and sub-cooling effects: these affect predicted performance which may well be only two-thirds of the theoretical.

In consequence of the requirement that the temperature difference between the condensing and evaporating temperatures (and hence between heat utilisation medium and heat sink) be minimised, applications to space heating are restricted. In winter, where heat

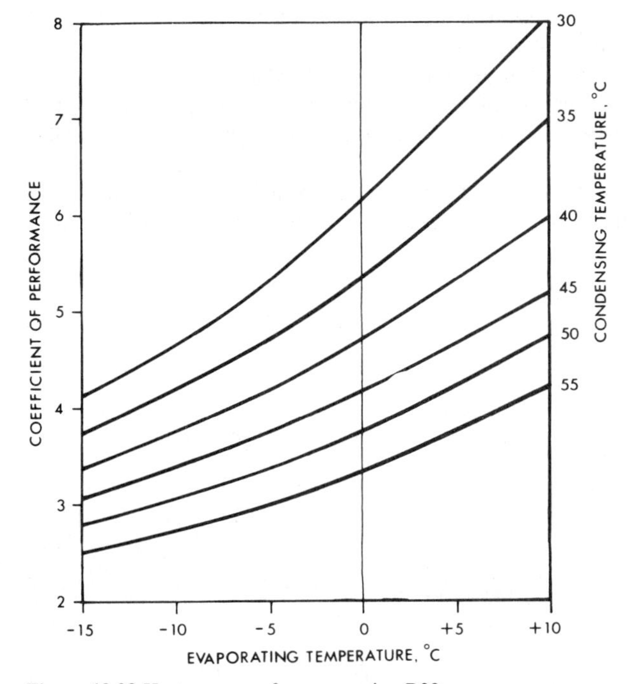

Figure 19.23 Heat pump performance using R22

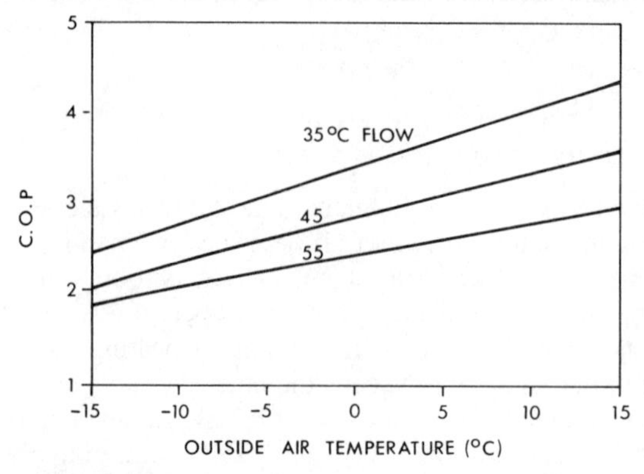

Figure 19.24 Variation of COP with source temperature

output is required to be at maximum, heat sources such as outside air, river water or the earth's crust are likely to be at their annual minimum temperature. This is illustrated in Figure 19.24 which gives the likely order of variation in COP for an air source heat pump. Industrial effluents and power station waste are, however, relatively constant as to temperature throughout the year and it is to these that attention should be directed as well as to any advantage that may arise from collection of solar energy.

The reduction in performance at low source temperature may lead to the need for plant of high capacity, and therefore cost, to meet the heating demand under extreme conditions, in consequence of which, a heat pump may be selected to match only a part of the full design load, say 70%, the remainder being met by a secondary heating source. Such a system is termed *bivalent* and is referred to in Chapter 6. A further consideration with air-source plant is the requirement to defrost the evaporator coil during cold weather, which may account for an overall loss of heat output of up to 10%.

In selecting the alternative energy sources for a bivalent system, it is necessary to consider the flexibility of a combination of a direct supply (electricity or gas) with a storage fuel (oil or coal) with respect to the future reliability of supply of either category of energy source.

There are four basic types of heat pump suitable for use in heating and air-conditioning which may be defined according to the heat sources and method of transporting heat to point of use, namely:

- Air-to-air (air source to air system).
- Air-to-water.
- Water-to-water.
- Water-to-air.

Table 19.2 Typical temperature ranges of ambient energy sources

Source	*Temperature* (°C)
Outside air	−10 to 20
Ground water	8 to 12
Surface water	2 to 6
Ground coils	6 to 12

If the heat source originates from a natural ambient supply, the temperature will vary: typical ranges are shown in Table 19.2.

Compressor drive

Whilst it is common practice to consider the heat pump as being electrically driven, this is not necessarily the case except where small production units are concerned. If a broad approximation of 25% be considered as representative of the conversion of primary energy to electrical power, it is obvious that a heat pump so driven, having a practical coefficient of performance of, say, 2.5, will equate as follows to, say, a boiler fired by natural gas:

Electrical heat pump
COP (primary energy) = 0.25×2.5 = 0.625

Gas fired boiler
COP approx. (primary energy) = 0.62

In consequence, it is proper to consider the case of a heat pump driven by a prime mover such as a gas or a diesel engine in circumstances where the waste heat from the power source may be recovered. Commonly quoted data for prime movers are, at full load:

Shaft power	33%
Waste heat to oil and jacket coolers	30%
Waste heat to exhaust	32%

If a recovery potential of 50% were to be applied to the sources of waste heat and an average 5% energy overhead considered for oil or gas supply, then the following equation results:

Motive power	0.33×2.5	= 0.825
Waste heat	$0.62 \times 0.5 \times 0.95$	= 0.295
Combined COP		= 1.12

In consequence, a clear case exists for further examination of the use of heat pumps driven by other than electricity, the ratio of advantage being of the order of 1.8:1 in favour of this approach. A further advantage of an engine driven heat pump is that both low and high grade heat are available; the former from the heat pump condenser and the latter from cooling of the engine jacket. Notwithstanding these apparent advantages, electrically driven machines have become more popular in commercial applications due mainly to their simplicity, proven record in use and requirement for no more than the very low level of maintenance skills currently available, at little better than 'Leggo' level.

Packaged heat pumps

These are available in a range from small domestic scale units (3 kW) through to equipment having an output in excess of 4 MW heating and 3 MW cooling. At the domestic level, the seasonal COP of an air-source device of around 2.6, coupled with relatively high installation costs and poor reliability, is not an attractive alternative to conventional domestic heating methods for the householder. Equipment is available in a number of configurations, some of the more common types of electrically driven machines being illustrated in Figure 19.25.

On a commercial scale, there is a good case for heat pumps in certain applications, such as in swimming pools, retail stores and industrial drying, where their installation is now

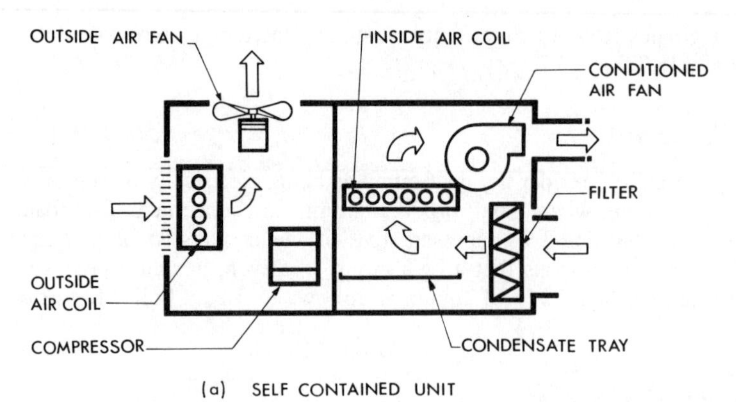

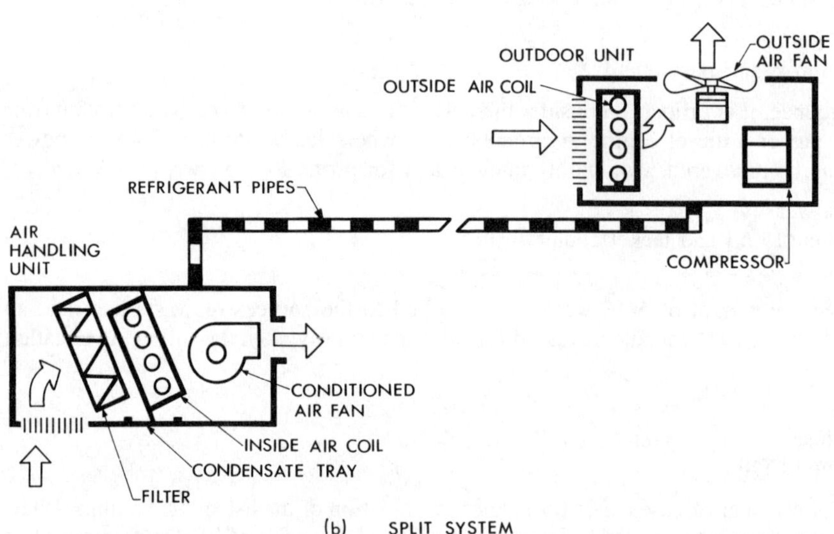

Figure 19.25 Configuration of electrically driven heat pumps

commonplace. In swimming pools the heat pump is used as a heat recovery device, dehumidifying and cooling the warm and moist extract air as the energy source to heat incoming air (during cold weather), pool water and hot water supplies.

The application in retail stores normally takes one of two forms:

- An air-to-air unit, often a reverse cycle device, using outside air as the source to heat or cool supply air as required.
- A water-to-air (or -water) device, extracting heat from condenser cooling water connected to freezer cabinets or cold stores.

In office systems, where air-conditioning is required, the use of distributed reverse cycle equipment, referred to earlier, is more popular than that of the less adaptable equivalent central plant; however, both have been used with satisfactory results. For further reference, the now defunct Electricity Council has published numerous valuable case studies on heat pump applications.

Chapter 20

Hot water supply systems

The provision of hot water for baths, showers, basins, sinks and other points of draw-off may be dealt with either locally or from a central system. In this respect it echoes the 'direct' and 'indirect' methods of providing heating service.

Local systems use either gas or electricity as a heat source, the heater being placed near to the point of consumption. In the case of a central system, the water is heated at some convenient position, which may be relatively remote from the point of consumption, and distributed as may be appropriate by pipework. The heat supply to a central system may be either solid, liquid or gaseous fuel or electrical energy.

Pipework losses

In order to provide a satisfactory level of service to the user, a central system of any size must be arranged to circulate hot water through the distribution pipework. This circulation, by thermo-syphon or by pump, will continue winter and summer although means will normally be provided to stop it when the building is unoccupied. Heat will be lost from this circulating pipework, although insulated, and this loss will persist whether or not any water is drawn off. The ratio between the energy expended in heating up the water supply from cold and that lost through the circulating system represents the efficiency of the system. It is of prime importance, therefore, that the distribution system be kept as compact as is practicable.

Choice of system

For a single isolated draw-off point, some type of local heater would be the obvious choice and, equally, a central system would be preferred for a hotel having bathrooms closely planned both back-to-back and vertically on several floors. As a generality, most institutional buildings, such as hotels, hospitals and educational establishments (whether planned for efficiency or not) are likely to be best served by a central system of some sort because this will, perforce, make available a bulk reserve supply of hot water to deal with peak demands and will at the same time avoid dispersal of those scarce maintenance activities which may be available.

In the case of many other applications, however, the choice is less easy to make and a single building could quite well be served by a mixture of systems. As a simple example, consider a centrally heated eight-storey building consisting of a dozen lock-up shops at ground floor level with six floors of offices above, all occupied by a single tenant, and a large penthouse flat leased separately.

Obviously, all points throughout the building could be supplied from a central plant in the boiler house but this would mean providing a source of heat for 365 days per annum to serve the flat. It would also mean that service to the shops would have to be charged individually to the tenants which might be uneconomic if one or more of them used hot water for cleaning cars (a not unknown habit!).

A sensible solution might be to provide a central system from the boiler plant for the offices, a separate mini-central system for the various hot water needs of the flat and a local heater for each shop. The energy supply to the flat and shops would be via an individual meter per tenant.

Local systems

There are two basic types of local hot water system, fundamentally different in concept. The first makes use of some form of instantaneous heater and has no hot water storage capacity. Except for fuel consumption by a pilot flame in the case of gas firing, this type has no associated heat loss when not in use. The second type incorporates hot water storage, adequate in capacity to meet the local demand, and however well insulated the storage vessel may be, a standing heat loss will occur.

Apart from any question of heat losses, the two types of system will for a given hot water production consume equal amounts of energy. In terms of rate of energy supply, however, an instantaneous heater will impose a greater load since the water outflow must be brought up to the temperature of use, from cold, during the short time of actual demand. With a storage unit, no such limitation exists and the rate of energy supply required will be a function of vessel size and demand pattern.

Instantaneous heaters

The earliest type of heater in this category was the old fashioned bathroom geyser, gas fired. The much refined modern equivalent is made in single and multi-point form and Figure 20.1 shows a well-known example of the latter type. A number of draw-off points

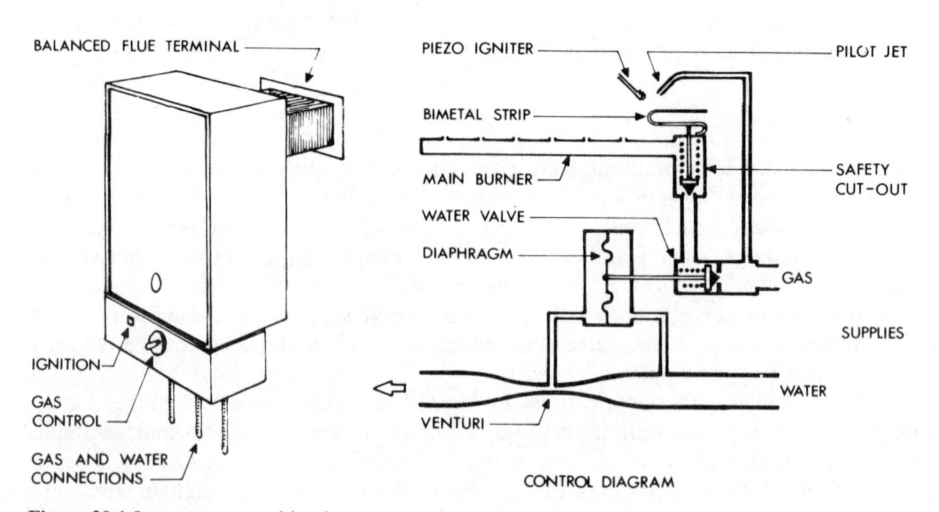

Figure 20.1 Instantaneous multi-point gas water heater (Ascot)

may be served provided that these are within the limit of length permitted for a dead-leg (see p. 582). A gas pilot burns continuously and when a hot water tap is opened water begins to flow and, by the pressure difference across the venturi, the main gas valve is opened, whereby the gas burners are ignited. A safety device cuts off the gas should the pilot light be extinguished. A hot water output of between 0.03 and 0.1 litre/s may be expected.

For all such heaters, whether used continuously or intermittently and wherever fitted, a flue with an outside terminal is essential. Existing heaters which discharge the products of combustion into the room where fitted are lethal and should, when found, be replaced. Models having balanced flues, with inlet and outlet ducts communicating with the outside, are available.

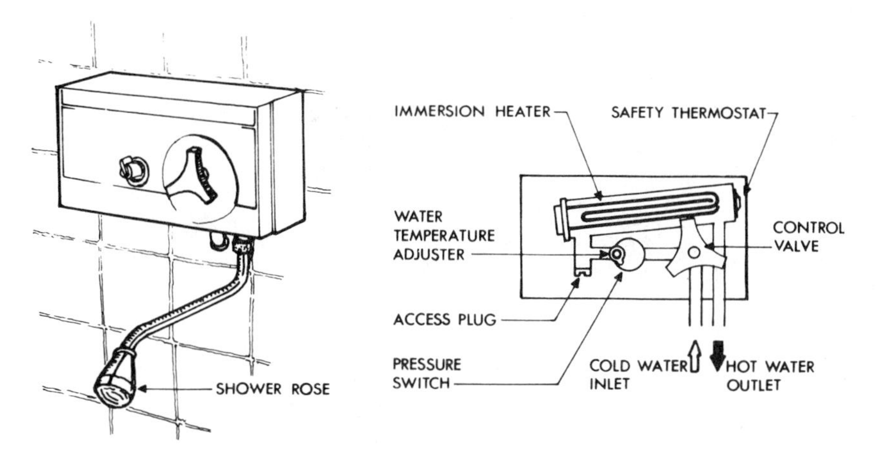

Figure 20.2 Instantaneous single-point electric water heater (Heatrae)

A now familiar innovation is the single-point electric instantaneous heater, designed to serve a wash basin or a shower as shown in Figure 20.2. Various models are available which will provide an outflow of between 0.02 and 0.05 litre/s. Such units are provided with thermostatic controls and a diaphragm pressure switch which permits current to be available only when water is flowing. Preset cut-out and other safety devices are incorporated. Physically, the casing of a unit of this type is small being only 200–300 mm square with a depth of perhaps only 60–80 mm. Electrical loadings are high, at from 3 kW to 10 kW, and since usage of water is immediately adjacent it follows that particular care must be taken in providing electrical protection and good earthing facilities.

In most applications instantaneous heaters, gas or electric, may be connected directly to the cold water main supply but the appropriate water company must be consulted. Where supply via a cold water storage cistern is envisaged, a static head of at least 3 m, equivalent to a pressure of about 30 kPa, is necessary. The rate of water flow from all such heaters is restricted and, in the case of multi-point units, it should not be assumed that a good supply may be obtained from more than one tap at a time. In the case of single point units, water control is normally on the inlet connection and outlet is via an inconvenient swivel spout, permanently open to atmosphere.

Storage heaters (on-peak)

For single point draw-off, local storage heaters are available in the capacity range of about 7–70 litres and are normally provided with 3 kW electrical heating elements. Provided that the storage capacity is not more than 15 litres this type may be connected directly to the cold water main supply. In this application, the simplest form of heater is the free-outlet or non-pressure type illustrated in Figure 20.3. It is the cold water inlet which is the point of control, discharge of hot water being by displacement as the cold supply is admitted. The storage is always open to atmosphere, normally via a swivel spout, and, as the contents expand on heating, a drip may occur at the outlet. To overcome the inconvenience of a swivel arm, the outlet may be piped to a spout fixed to the point of draw-off as shown in Figure 20.4(a), an alternative introduced by one manufacturer combines this spout with the cold inlet water control as Figure 20.4(b).

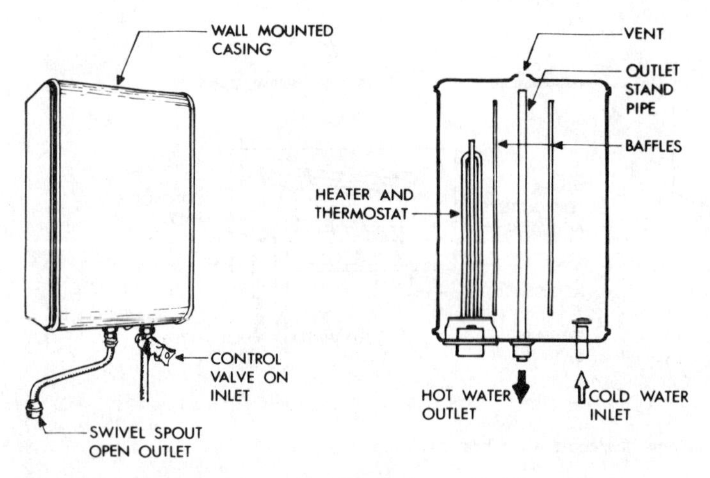

Figure 20.3 Non-pressure-type electric water heater (Sadia)

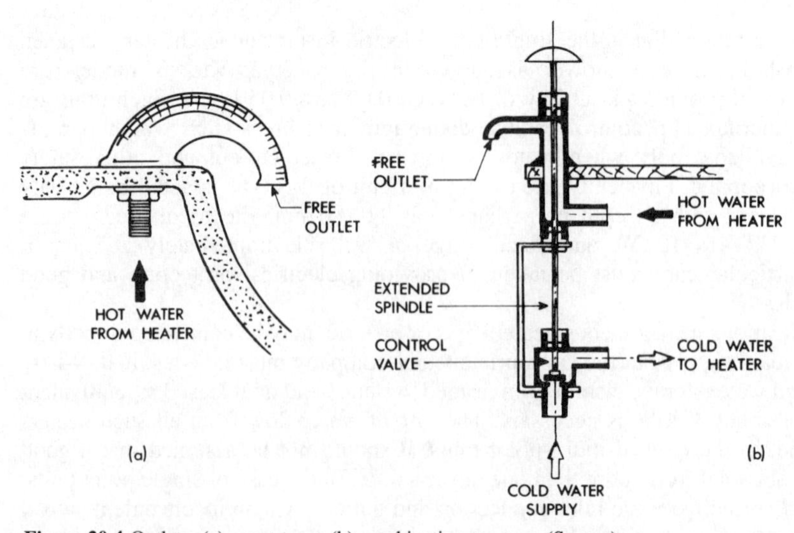

Figure 20.4 Outlets: (a) spout-type; (b) combination tap type (Santon)

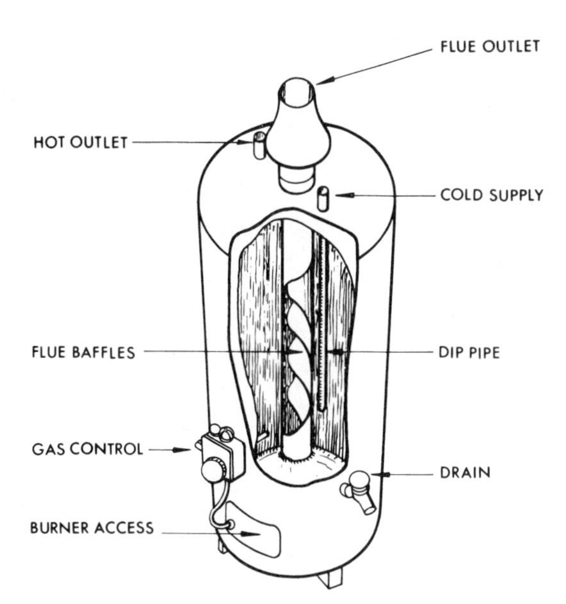

Figure 20.5 Storage-type heater, gas-fired (Lennox)

For multi-point local use, as might arise in each of a number of dispersed toilets in an office building, both gas fired and electrical storage heaters are available, on-peak and off-peak in the latter case. Packaged gas fired units are, in effect, cylinders incorporating a stainless steel heat exchanger, direct fired as shown in Figure 20.5, and are insulated and cased in sheet metal. Such heaters have a water capacity of about 80 litres and a heat input of 8 kW: the products of combustion must be discharged to outside via a conventional or balanced flue arrangement.

For domestic multi-point service from storage, what are known as *circulators* may be used. One such, developed to fit within a domestic warm air heater, is shown in Figure 20.6 and rated at about 4 kW. A cold water supply from a remote storage cistern must be

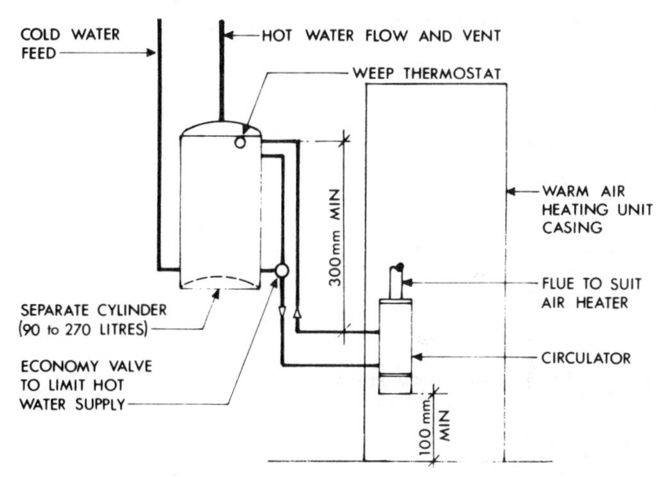

Figure 20.6 Gas circulator with warm-air heating system

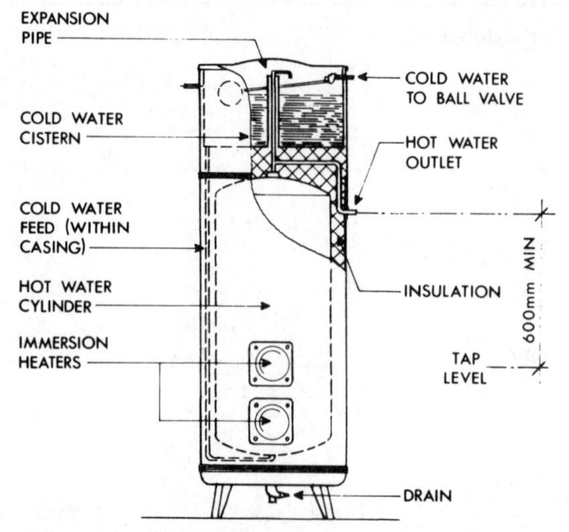

Figure 20.7 Combination electric storage heater with cistern (Sadia)

provided in all such instances as must arrangements for venting. For applications where larger storage or heating capacities are required, gas firing becomes less tidy to apply since the storage cylinder is separated from the fired heat exchanger and the equivalent of a small central system thus arises.

With electrical on-peak heating, local multi-point supply is most conveniently arranged using a so-called combination type unit where heater and cold water cistern are fitted in one casing as shown in Figure 20.7. The heater may be cylindrical or rectangular for easy wall fixing, and capacitites in the range 20–140 litre are available, normally with a 3 kW heating element. The pressure available at the draw-off points will be meagre if the combined unit cannot be fixed well above the level of the taps.

For applications where larger storage or heating capacities are necessary, or where a combination unit cannot be fitted high enough above the draw-off point, pressure type equipment is used, as shown in Figure 20.8. Siting in this case is unimportant and thus the UDB (*under-draining-board*) heater has many applications. All pressure type units require

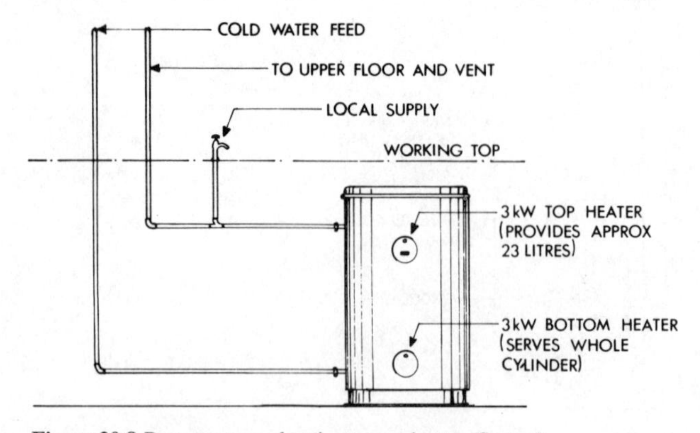

Figure 20.8 Pressure-type electric storage heater (Santon)

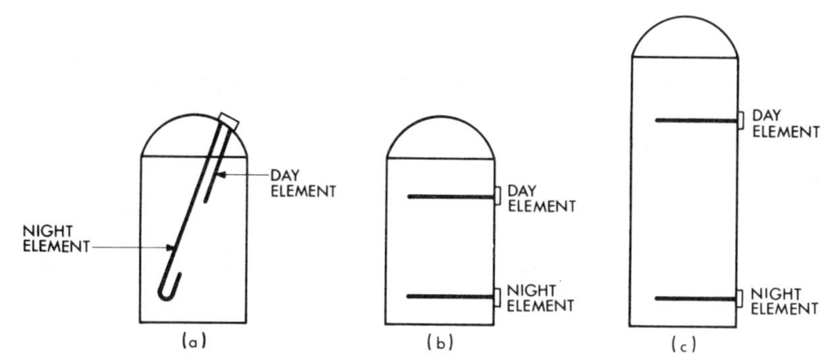

Figure 20.9 Arrangements for electric immersion heaters

a cold water supply from a remote storage cistern and an open vent. Storage volumes cover the range 50–450 litre with heater element capacities to suit.

Local storage heaters are well insulated and are normally fitted with an enamelled jacket ready for mounting in an exposed position. In comparison with an instantaneous heater, at least ten times the space is needed for the same level of service but if sited in a cupboard the difference may not be relevant. A suitably sized storage heater does however offer the advantage of using off-peak current which is not possible with the alternative type.

To make use of the cheap off-peak rate, a storage unit should be fitted with two sets of immersion heater. One set, fitted near the bottom of the vessel, will charge the store overnight and the other, fitted near the top, will be arranged to heat a smaller volume of water at the end of the day when the overnight charge has been exhausted. On a domestic scale, an existing 120 litre cylinder (900 mm × 450 mm diameter) might be converted by using a dual heater as shown in Figure 20.9(a) but the arrangement as Figure 20.9(b) is to be preferred. Better still, for a new installation, would be a 210 litre cylinder (1450 mm × 450 mm diameter) as in Figure 20.9(c). Based on published data, Table 20.1 shows how energy consumption would arise from the alternative arrangements.

Table 20.1 Annual energy consumption (%) for electric hot-water supply

	Single heater			*Dual heater*			
	Standard cylinder size					*Tall cylinder*	
Domestic water use	*Day and night*[a]	*Day*	*Night*	*Day*	*Night*	*Day*	*Night*
Very small	100	50	50	14	86	1	99
Small	100	55	45	21	79	4	96
Average	100	57	43	28	72	7	93
Large	100	60	40	33	67	11	89
Overall average	100	55	45	24	76	6	94
Cost ratio[b]	1.00	0.78		0.62		0.53	

[a] Thermostatically controlled, no time control, normal tariff.
[b] Based upon night rate charged at half day rate.

Central systems

A central system will usually consist of a boiler or water heater in some form, coupled by circulating piping to a storage vessel or vessels. The combination of the two will be so proportioned as to provide adequate service to the draw-off points to match the predetermined pattern of usage. For instance, in a hospital there may be a continuous demand for hot water all day and in this case a small storage capacity with a rapid recovery period (large boiler power) is probably appropriate. Conversely, for a sports pavilion where there may be a single sudden demand following a game, a large storage capacity and a long recovery period (small boiler power) may be adequate.

Direct systems

In the recent past, many systems were *direct* following a traditional pattern in which the water drawn off at the various taps etc. was the same water as that which circulated through the boiler, as shown in Figure 20.10. If the water were hard then a scale deposit would occur on the heating surfaces and thus a special type of boiler was necessary, of a simple type with large waterways and bolt-on access to ample cleaning 'mud-holes'.

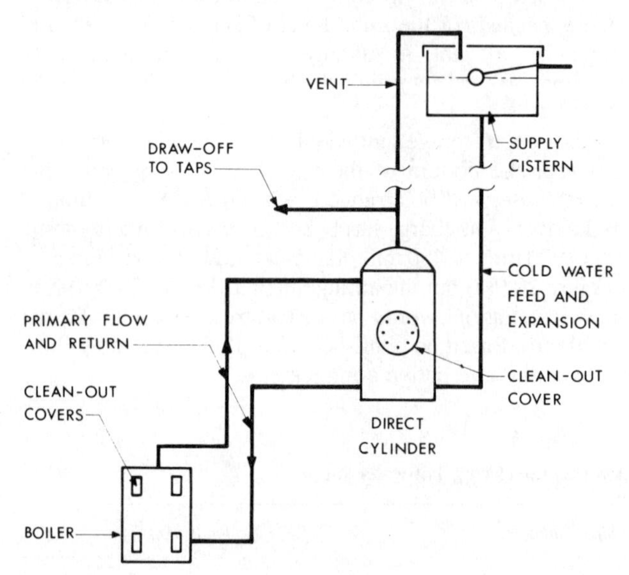

Figure 20.10 Direct hot water supply system

On the other hand, if the water were soft, then an ordinary boiler of cast iron or mild steel would cause discolouration as a result of rusting of internal surfaces. Special boilers were thus made of cast iron subjected to an anti-corrosion treatment such as *bower-barffing* or, where made in mild steel, were treated internally with some form of vitreous enamel coating.

Modern practice, for those *direct* systems now installed, is to use what can best be described as a combined boiler/storage vessel as shown in Figure 20.11, the rate of output of the 'boiler' being enough to heat the storage provided about three times each hour. Such

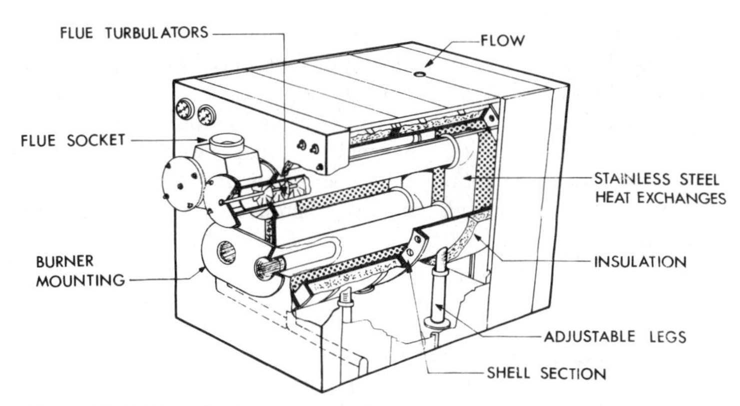

Figure 20.11 Direct fired water storage heater

a unit consists of a cylindrical shell, made in sections and mounted horizontally, which has large bore heat exchange tubes, provided with internal turbulators, arranged in 'U' formation within it. All water-face surfaces are treated with a fluorocarbon coating and the sectional shell may be dismantled for cleaning. It might be that the excellent service provided by this type of heater will lead to a resurgence of interest in direct systems.

Indirect systems

The total separation of boiler water from that drawn off at taps, etc., and a solution to the associated problem of cleaning the internal surfaces of the boiler, are both found when an *indirect* system is adopted. Figure 20.12 illustrates such an arrangement in which any type of heating boiler may be used. The storage cylinder becomes indirect as a result of the provision of heating surface within it to contain the (primary) water circulated from the

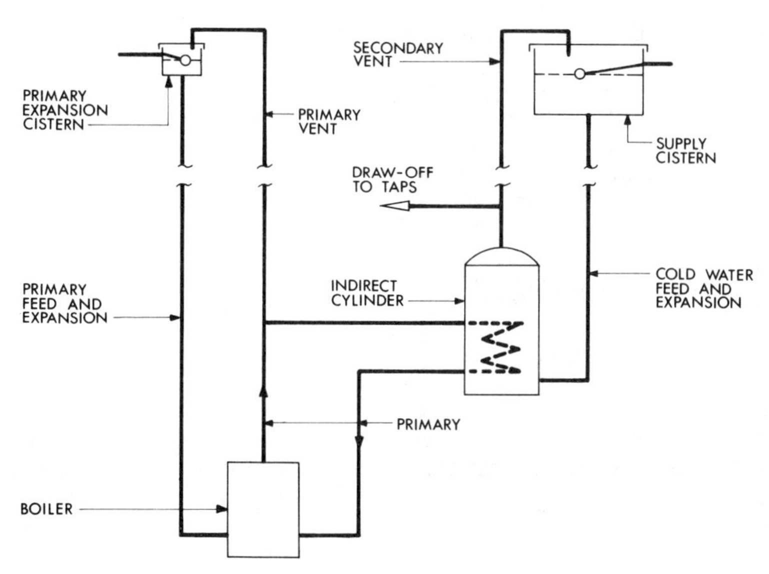

Figure 20.12 Indirect hot water supply system

boiler. The (secondary) water to be heated is outside the heating surface and is thus isolated from the boiler. It will be noted that this separation introduces the need for an additional feed and expansion cistern.

The primary water within the heating surface is recirculated and thus will remain unchanged with the result that scale deposition internally is likely to be negligible. Further, when this primary water is kept at a temperature below about 90°C then, except in areas where the supply has an extremely high *temporary* hardness, there will be little deposit on the external secondary surfaces.

Combined systems

The water content of a boiler serving a *direct* hot water supply arrangement must, for obvious reasons, be quite separate from that within any plant supplying an adjacent heating system. The result is that, unless a duplicate boiler is provided, service will be lost in the event of breakdown and during the necessary annual cleaning. With an *indirect* system however it is common practice to arrange for an indirect cylinder to be served from the same boiler, or group of boilers, which supply the heating system.

Where only one boiler exists, a problem arises in that the primary water supply to the indirect cylinder will be required at a constant temperature year round whereas the parallel supply to the heating system will need to vary in temperature with the seasons. This difficulty is overcome, as shown in Figure 20.13, by running the boiler at a temperature to suit the cylinder and supplying the heating system through a mixing valve as discussed in Chapter 22 (p. 621).

During the summer, when heating is not required, the boiler will be oversized and problems due to cycling and low combustion efficiences may arise. For very small combined systems, it may be sensible to use another source of energy such as electricity

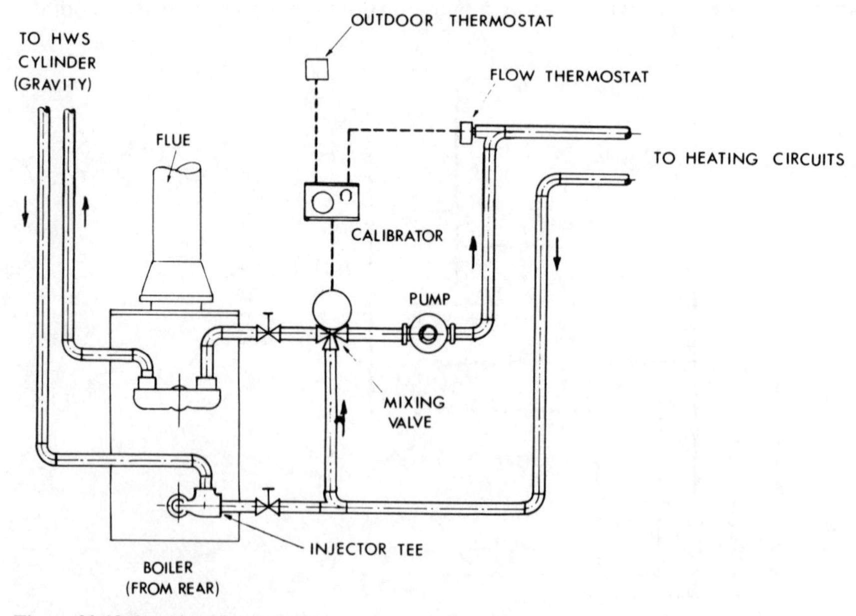

Figure 20.13 Combined heating and hot water supply system

during periods of low overall load: for larger systems, the provision of more than one boiler is to be recommended.

In the case of a group of boilers, one of them might be chosen to be of a size matching the hot water load and be dedicated to serve the indirect cylinder in normal circumstances, although connected in parallel with the other boilers. In the event of breakdown of the dedicated boiler, supply to the cylinder could come from one of the heating units. Conversely, should a heating boiler break down in exceptional winter weather, the hot water boiler could be used to provide additional capacity. As will have been understood from the discussion regarding boiler margins on p. 257, no hard and fast rules can be laid down for plant selection since patterns of use, availability of space and other extraneous circumstances may intervene.

Combination systems

Reference to boilers so described has been made in Chapter 10 but it is the provision of hot water output to a combined *system* which is considered here. Where both heating and hot water are supplied from a single boiler, it is unlikely that both systems will require their full share of output simultaneously, even in mid-winter. For the remainder of the year, as stated above, the boiler will be too large, with attendant penalties.

Despite this situation, however, interaction between the two systems may be such that response by the boiler to demand from either will appear sluggish. The introduction of some form of thermal store between the boiler and the two varying demands should thus improve not only operating efficiency but also response rate. Research has produced results at domestic level which show promise, but comprehensive application data for systems of any significant size are not available.

The design concept used for one make of domestic unit is illustrated in Figure 20.14, the heat source being either a gas fired circulator of the generic type shown in Figure 20.6 or a conventional boiler. The output of the heat source is pump-circulated to the store which has a capacity, varied to suit the size of house served, of between 120 and 200 litre. From the store, a quite separate circulation is directed to the heating system and this feature is turned to advantage when, following a night-time shut down, restarting the heating pump brings the radiators to store temperature in a matter of minutes.

The two coils immersed within the store have external fins to aid heat transfer and are connected in series. Since their water content is small, they are permitted to be supplied

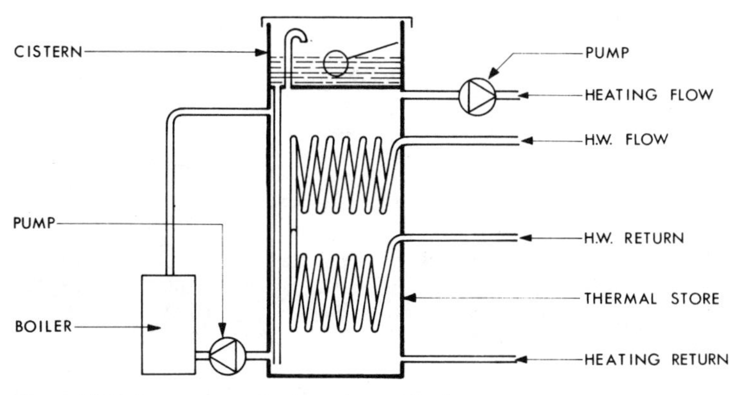

Figure 20.14 Arrangement of a domestic combination system

Table 20.2 Capacities and dimensions of direct and indirect cylinders

Pattern	*BS ref.*	*Dimensions* (mm) *Height*	*Diameter*	*Capacity* (litre)	*Heating surface* (m^2) *Coil*	*Annulus*
Direct	0	300	1600	98	–	–
	1	350	900	74	–	–
	2	400	900	98	–	–
	3	400	1050	116	–	–
	4	450	675	86	–	–
	5	450	750	98	–	–
	6	450	825	109	–	–
	7	450	900	120	–	–
	8	450	1050	144	–	–
	9	450	1200	166	–	–
	9E	450	1500	210	–	–
	10	500	1200	200	–	–
	11	500	1500	255	–	–
	12	600	1200	290	–	–
	13	600	1500	370	–	–
	14	600	1800	450	–	–
Indirect	0	300	1600	96	0.35	–
	1	350	900	72	0.27	–
	2	400	900	96	0.35	–
	3	400	1050	114	0.42	–
	4	450	675	84	0.31	–
	5	450	750	95	0.35	–
	6	450	825	106	0.40	–
	7	450	900	117	0.44	–
	8	450	1050	140	0.52	–
	9	450	1200	162	0.61	–
	9E	450	1500	206	0.79	–
	10	500	1200	190	0.75	–
	11	500	1500	245	0.87	–
	12	600	1200	280	1.10	–
	13	600	1500	360	1.40	–
	14	600	1800	440	1.70	–
Single feed	3	400	1050	104	0.42	0.63
	5	450	750	56	0.35	0.52
	7	450	900	108	0.44	0.66
	8	450	1050	130	0.52	0.78
	9	450	1200	152	0.61	0.91
	9E	450	1500	196	0.79	1.18
	10	500	1200	180	0.75	1.13

Note: based upon a transmission coefficient of 30 $W/m^2 K$, the heating surface listed is stated to be adequate to raise the temperature of the cylinder contents by 55 K in one hour.

with a cold water feed *from the incoming main supply* with the result that outflow pressure is more than adequate to serve a shower head. Provision is made for the hot water discharge to be blended with cold water so that the final supply is at a safe temperature for children. In effect, the store acts as a batch-production instantaneous heater able to provide a hot water outflow, at a rate of about 0.2–0.3 litre/s, for a period of 5 minutes followed by a recovery period of 20–40 minutes in preparation for a further similar outflow.

Cylinders, indirect cylinders and calorifiers

Whilst, in the present context, it is unnecessary to define a *cylinder*, other than to say that standard sizes for such vessels are as set out in Table 20.2, the difference between an *indirect cylinder* and a *calorifier* is less clearly established. Early textbooks and catalogues seem to use the first of the two terms to distinguish heat exchangers for domestic application from those fitted elsewhwere. This usage is confused by the fact that the former had, almost always, an annular heat exchanger whereas the latter were in one form or another of shell and tube construction. Although the annular type unit has now all but disappeared, subsequent references here will draw the traditional distinction.

Indirect cylinders

With an annular heat exchanger, an indirect cylinder could be installed either horizontally or vertically (provided that it had, in the former attitude, been fixed the right way up!). Those cylinders fitted with a helical coil to BS 1566: Parts 1 and 2: 1988, as in Figure 20.15, may be preferred from the point of view of differential pressure, primary to secondary, but there remains some doubt whether or not they suffer from a deficiency in heating surface.

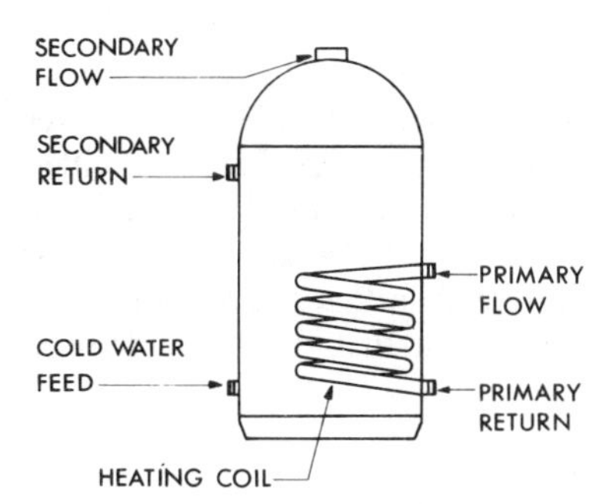

Figure 20.15 Indirect cylinder with helical coil

For use with a low pressure primary medium, and where the storage capacity required is below about 500 litre, indirect cylinders are economical in first cost. Standard sizes are listed in Table 20.2.

Single-feed indirect cylinders

The requirement that a separate feed and expansion cistern be provided to serve the primary circuit of a conventional indirect cylinder has led to the development of special

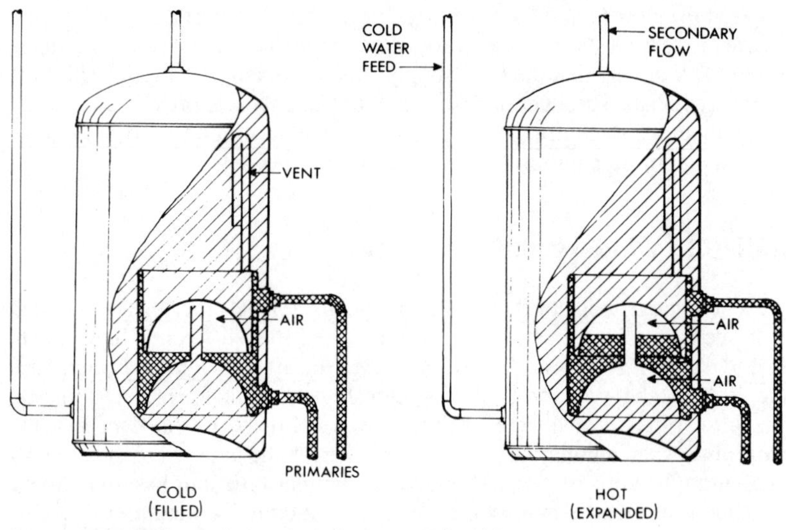

Figure 20.16 Indirect cylinder, single feed type (Primatic)

types, as illustrated in Figure 20.16. Units of this pattern have no moving parts and rely upon air cushions to separate the primary water content from the secondary store, using naturally balancing pressures. Once filled, the air cushions expand or contract in tandem as the primary circuit heats or cools: any excess air is vented via the secondary water content.

Water-to-water calorifiers

For larger units, as covered within this heading, a tubular hairpin pattern heat exchanger or *heater battery* of the type shown in Figure 20.17 is used. The calorifier shell is provided with a flanged *neck* which supports a tube plate and a domed cover, the latter having a partition within it to route the primary water flow in and out of the tubes. Means for withdrawal of the heater battery on a runway are incorporated so that the tubes may be cleaned. The primary pipework connections should be arranged so that withdrawal may be achieved with minimum disturbance to insulation, etc.

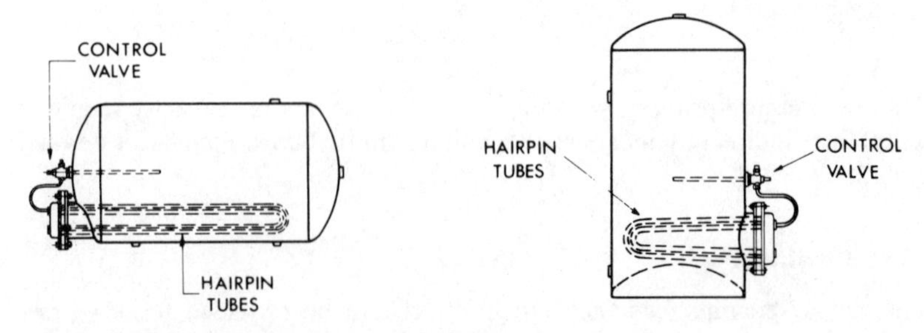

Figure 20.17 Tubular type calorifiers for LTHW and steam primaries

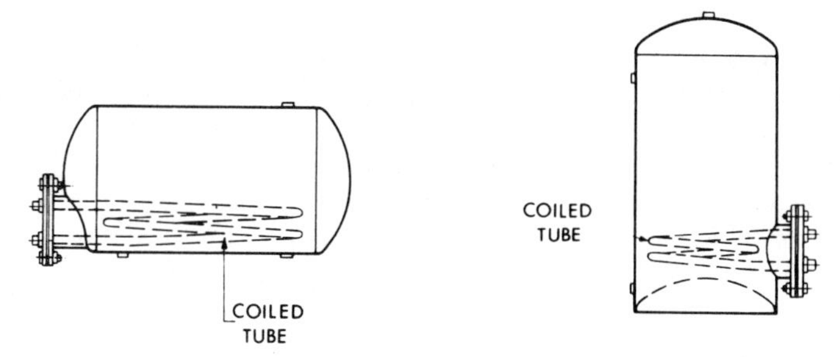

Figure 20.18 Tubular type calorifiers for MTHW and HTHW primaries

For application to medium temperature and high temperature hot water systems, a calorifier will frequently take the form shown in Figure 20.18 in which the primary water follows a single pass in order to maintain a high velocity with the comparatively small volume of water which is in circulation with this type of system. Provision for withdrawal of the heater battery is made as before, the neck being enlarged to suit the alternative tube configuration.

With a micro-bore system of pipework distribution (see p. 150), the type of water-to-water heat exchanger shown in Figure 20.19 may be used, this having a high rate of heat transfer and a head tapping etc. which enables it to be fitted to a standard boss of the type and size provided for electrical immersion heaters.

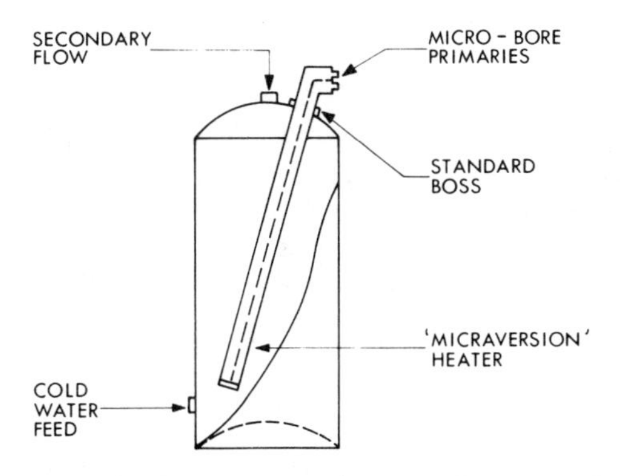

Figure 20.19 Special immersion heater for LTHW micro-bore primaries

Steam-to-water calorifiers

The configuration generally adopted when the heating medium is steam is similar in most respects to that shown in Figure 20.17, with hairpin tubes. The domed cover, here known as a *steam chest*, is provided with an upper connection for the steam supply and, usually, a pressure gauge. Condensate, when formed, leaves by a lower connection and passes to a steam trap set.

Bubble-top calorifiers

Units so described pre-dated by half a century the equipment which has produced such a furore following the introduction of the 1986–9 Model Water Byelaws. A calorifier of this type consists of a vertical cylinder, not unlike the pattern shown in Figure 20.18, but having the shell extended above the hot water outlet to form a sealed air space at the top, equivalent to about 15% of the water volume. The cold water feed derives directly from a pressure main and, when the cylinder is first filled, air is trapped and compressed in the shell extension or 'bubble-top'.

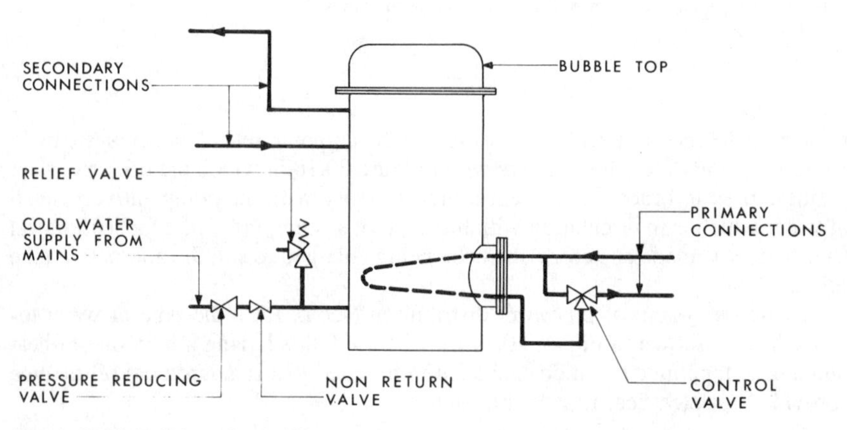

Figure 20.20 'Bubble-top' calorifier

When the water content is heated, it expands, and the volume of expansion is accommodated by further compression of the air within the 'bubble top'. Numerous safety features are incorporated, as shown in Figure 20.20, those on the primary side being varied to suit the heating medium which may be either hot water or steam.

This type of calorifier was developed originally to avoid the need for numerous tank towers on sites near airfields. Application in system design was, needless to say, dealt with by engineers rather than plumbers.

Angelery hot water generators

This equipment is suitable for use with either steam or hot water as the primary medium and deals with the process of supplying hot water in a manner which differs from the conventional. As may be seen from Figure 20.21, the heat exchange surface consists of a series of interlacing coils which provide a high capacity in a small space and an ability to shed scale as they flex in expansion and contraction. While the generator is not strictly an instantaneous heater in the normally accepted sense, it is possible in many applications to dispense with bulk storage completely, the small volume of hot water retained in the shell being adequate to meet an initial demand for a minute or two only. In instances where batch or surge loads occur, storage in an 'accumulator' may be required but, for variations in a normal load, the generator draws on surplus boiler capacity rather than storage.

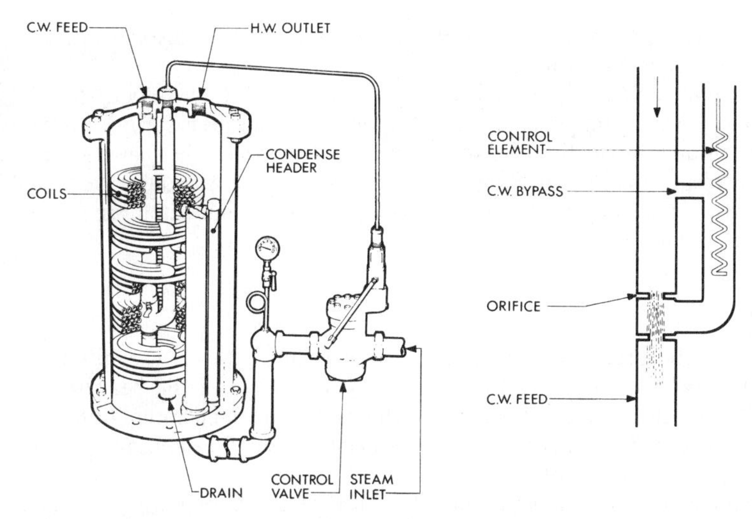

Figure 20.21 Angelery hot water generator (BSS)

Success in operation depends largely upon the performance of the temperature control valve, the sensing element of which is so mounted as to take account of both temperature and rate of demand. A similar principle was advocated following investigations undertaken 30 years ago by the hospital research unit based at Glasgow University.

Plate type heaters

For process applications and for use in hotels and similar institutional buildings such as hospitals, where large quantities of hot water at a constant temperature are required, an application may exist for a true instantaneous heater such as the plate type equipment shown in Figure 20.22. The components of this package include the heat exchanger, a primary circulating pump, a motorised valve and the pipework interconnections. The valve is controlled thermostatically from the secondary side to maintain a constant temperature output at 60°C with a primary water input at 82°C.

The heat exchanger, built up as a 'sandwich' from a number of corrugated plates which provide passages through which the separated primary and secondary water circuits flow, is mounted to a frame and has cover plates which are pulled together by long end-to-end compression bolts: flanges on one cover plate provide for pipework connections. The

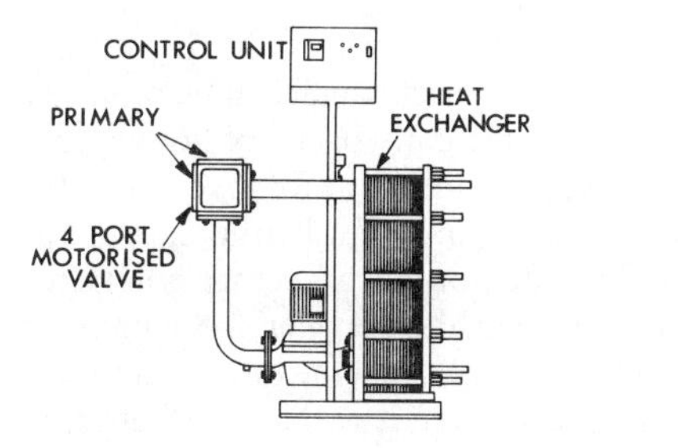

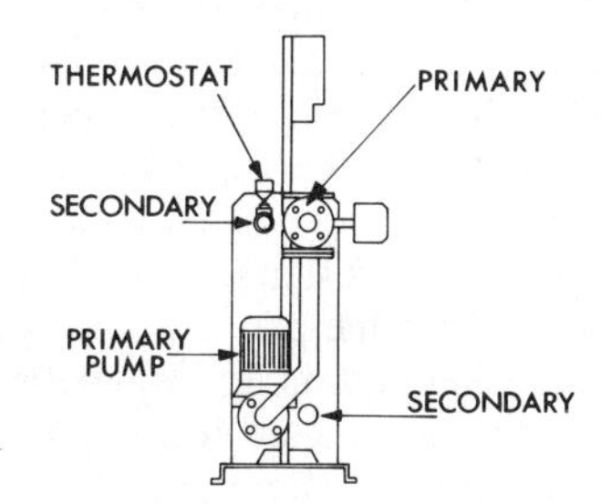

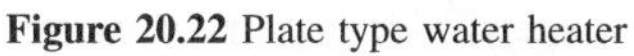
Figure 20.22 Plate type water heater

output of the unit is determined by the number of plates making up the heat exchanger and a range of ratings up to 1115 kW is available.

Since the efficiency of such a unit depends largely upon the high flow velocity of both primary and secondary water, the passages between the plates are necessarily restricted and any deposition on the drowned surfaces will affect performance. The manufacturers claim that the heat exchanger is easy to dismantle for maintenance, as a result of the bolted construction, but use of such equipment in a hard water area with an untreated supply may nevertheless be inadvisable.

Rating of calorifiers

The heating surface (tube area) required is a function of the temperature difference between the primary water or steam and the secondary water and of the heat transfer coefficient, $W/m^2 K$. The question now arises as to what is the true temperature difference bearing in mind that the only known conditions are probably those of the entering primary flow and the entering cold water supply.

The notional design conditions, taking Figure 20.12 for reference, might be, say, 80°C flow and 70°C return for the primary circuit, 70°C for the secondary outflow and 10°C for the entering cold water feed. The mean temperatures would thus seem to be 75°C for the primary and 40°C for the secondary, giving a difference of 35 K. However, this level of difference is very unlikely to remain stable for any length of time in practice.

Following a period of heavy draw-off, the whole of the hot water content of the cylinder may have been used and the difference then might be 60 K or more. At the other extreme, when use of hot water is negligible, and storage temperature has reached 70°C, the difference then might be as little as 5 K. As to the heat transfer coefficient, this will be a function of tube size, spacing and of the level of external scaling: it will vary also according to the vigour of the convection velocities within the shell which will be temperature dependent.

In consequence of all the variables, it is best to consult the calorifier manufacturers and not to rely upon a theoretical approach alone.

Requirements for storage capacity and boiler power

A decision as to the volume of hot water storage and the capacity of the associated boiler plant depends upon considerations which are quite different from those which apply to a heating system. There, as a rule, the quantity of heat which is required under design conditions may be calculated with a fair degree of accuracy and, after adjustments have been made to suit particular circumstances, the use of meteorologial data may be applied to determine the 24 hour pattern of demand.

A hot water supply system, except in the case of an industrial process having a finite time cycle, has to fill a demand which depends upon a user pattern which can be projected only empirically and is, at best, intermittent. For example, the daily usage of hot water in a school, an hotel and an office block with a canteen might be identical for given sizes but the peak demands would differ as to both magnitude and timing. Furthermore, usage in each would not be the same per head, per unit of floor area or per any other criterion.

A natural relationship exists for any particular type of demand between the volume of hot water stored and the boiler power provided to heat it. In general terms, the more generous the storage the smaller the boiler capacity needed since it will have a long time to restore the temperature following draw-off. Nevertheless, such an arrangement carried to excess would lead to an unreasonable delay in recovery of storage temperature if this were to have fallen below normal, due to night-time shut down for example. Conversely, the combination of a small storage volume and a comparatively large boiler would provide quick recovery of temperature after draw-off but probably would be inadequate to meet a sustained heavy demand. A compromise between the extremes must be chosen.

Approximate methods

In many cases a storage capacity equal to the maximum draw-off in any one hour at peak load conditions will be an adequate provision. The associated boiler power may then be sized on the basis that this volume of water will be heated from cold over some longer period such as two or three hours. A rule of thumb basis of this sort is, however, acceptable only in circumstances where no parallel experience with a similar load exists or where no general statistical data are available. A digest of collected data is given in Table 20.3 for the capacities of certain fittings and in Table 20.4 of daily and hourly consumptions of hot water in various types of building. These latter values are necessarily averages but may be used to settle a first approximation for capacity.

Table 20.3 Capacity of various standard fittings

Fitting	*Capacity* (litre)	*Required temperature* (°C)
Bath, average	80–120	40–45
Sink	12–18	50–60
Basin, normal fill	5–10	40–50
Shower rose, 150 mm	0.6 litre/s	40
Shower spray	0.15 litre/s	40

Table 20.4 Hot water consumption

	Consumption per occupant (litre)	
	Per day	*Per peak hour*
Dwellings	30–70	25–45
Factory, toilets only	15–25	5
Hospital		
general	120–175	30
geriatric or mental	80–100	25
infectious	150–250	45
maternity	175–275	50
Hostel		
general	60–100	25
nurses home	100–160	45
Hotel	160–250	30–50
Offices, total	10–20	5
Restaurant	15–40	5
Shops	10–30	5
Schools		
day	15–25	5
boarding	90–140	25

Recent investigations

The approximate method probably overstates the requirement and more recent field work has proposed the alternative approach included in the *Guide Section B4*. There, a series of figures represents the relationship between storage capacity and boiler power for a number of building types. This information may be presented in a rather simpler manner, without any loss of accuracy, by the listings of Table 20.5. The values given do not include any

Table 20.5 Hot water storage at 65°C and boiler power per person and per meal

Building and use	*Validity of data* (persons or meals/day)	*Required storage capacity* (litre per person or litre per meal) *when the recovery period is as follows* (hours)				
		½	*1*	*2*	*3*	*4*
Hostel						
service	80–320	3.0	5.0	7.2	9.7	12.2
Hotel						
service	80–320	7.5	11.0	15.0	16.8	18.5
catering	140–840	2.4	3.4	4.7	5.5	5.9
Office						
service	110–660	0.7	0.9	1.2	1.6	2.0
catering	40–370	2.0	3.5	5.0	6.1	7.1
Restaurant						
service	100–1010	0.3	0.4	0.5	0.6	0.7
catering	100–1010	0.4	0.6	0.9	1.1	1.3
School						
service	360–1600	0.7	0.9	1.1	1.2	1.5
catering	240–1200	1.1	1.6	2.3	2.8	3.2
Shop						
service	50–220	1.5	1.7	1.9	2.4	2.8
catering	60–540	1.0	1.3	1.5	1.8	2.2
Boiler output, kW, required to achieve the recovery rates stated is storage capacity multiplied by:		0.128	0.064	0.032	0.021	0.016

Notes:
1. When storage is to be heated electrically, the rating of the upper immersion heaters (Figure 20.23) should correspond to the boiler output listed: the storage capacities listed are those required above the upper immersion heaters.
2. When occupancy levels or meals served are less than the minimum of the validity band, both storage capacity and boiler output must be increased. The following multiplying factors are proposed: 75% of minimum × 1.2; 50% of minimum × 1.4; 25% of minimum × 1.6.

allowance for the loss of effective storage capacity which results as incoming cold water mixes with the hot water held in the vessel. To provide for this situation, an addition of 25% should be made to the total of the volume calculated but not, of course, to the equivalent total representing the output required of the associated boiler or electrical immersion heater.

Use of electrical energy

Also included in the *Guide Section B4* is a further family of figures, parallel to those noted in the previous paragraph, which purports to show the quantity by which storage capacity

Table 20.6 Additional hot water storage at 65°C required when heating is off-peak

Building and use	*Validity of data* (persons or meals/day)	*Required additional storage capacity* (litre per person or litre per meal) *as between lower and upper immersion heaters (Figure 20.23) when off-peak energy consumption is to be the following proportion of the total*				
		60%	*80%*	*90%*	*95%*	*All*
Hostel						
service	80–320	12.0	18.0	24.0	30.0	36.0
Hotel						
service	80–320	20.0	30.0	40.0	50.0	60.0
catering	140–840	5.0	7.5	10.0	12.5	15.0
Office						
service	110–660	2.0	3.0	4.0	5.0	6.0
catering	40–370	3.0	4.5	6.0	7.5	9.0
Restaurant						
service	100–1010	0.7	1.0	1.4	1.8	2.1
catering	100–1010	1.3	1.9	2.6	3.1	3.8
School						
service	360–1600	1.5	2.3	3.0	3.8	4.5
catering	240–1200	2.5	3.8	5.0	6.3	7.5
Shop						
service	50–220	2.7	4.0	5.3	6.6	8.0
catering	60–540	2.0	3.0	4.0	5.0	6.0

Note: see text for rating of immersion heaters.

must be increased when the greater part of the energy supply is provided by off-peak electricity. It is more convenient and equally valid to use the data listed in Table 20.6 which represent the additional storage capacity, per person or per meal, required *between* the lower and upper banks of immersion heaters, as shown in Figure 20.23. By making the assumption that water is to be stored at 65°C and that a 6½ hour charging period is used (thus allowing a half-hour margin), the required rating in kW of the *lower* immersion heaters may be taken, numerically, as 1% of the total volume stored, i.e. the sum of the components calculated from the listings of Tables 20.5 and 20.6.

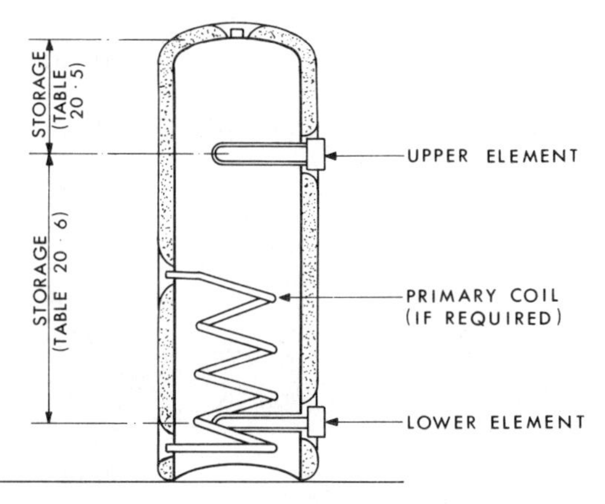

Figure 20.23 Indirect cylinder for off-peak electrical supply

Allowance for heat loss

None of the values listed in Tables 20.5 and 20.6 include any allowance for boiler or heater capacity to make good the heat lost from storage vessels and circulating pipework, primary or secondary. In all cases these losses should be calculated and, where storage vessels are served by a conventional boiler plant, they should be added to the output requirement determined from Table 20.5. For a vessel which is to be heated with an off-peak supply, however, it has been suggested that the calculated total may be dealt with more conveniently if it be converted to an *equivalent stored volume* of hot water. This may be calculated from:

$$V = 3600\,(Lhp)/c\,(t_s - t_c)$$

where

V = notional equivalent stored volume (litre)
L = daytime heat loss from vessels, etc. (kW)
h = daytime running hours (h)
p = proportion of annual load taken off-peak (%/100)
c = specific heat capacity of water = 4.2 kJ/kg K
t_s = temperature of stored hot water (°C)
t_c = temperature of cold water feed (°C) = 10°C

Examples

- A school has 500 pupils and serves 400 meals each day. The water for cloakrooms, etc., is required to be at 55°C and for kitchens, etc., at 65°C. Heat loss is ignored for this example. Recovery periods to be two hours for service and one hour for catering. The demands are to be met by two separate plants.

Cloakrooms
From Table 20.5

Storage required		= 1.1 litre/person
Boiler rating	= 1.1 × 0.032	= 0.035 kW/person

Thus

Storage	= 1.1 × 500	= 550 litre
With margin for mixing	= 1.25 × 550	= 687 litre
Boiler rating	= 0.035 × 500[(55–10)/(65–10)]	= 14.3 kW

Catering
From Table 20.5

Storage required		= 1.6 litre/meal
Boiler rating	= 1.6 × 0.064	= 0.10 kW/meal

Thus

Storage	= 1.6 × 400	= 640 litre
With margin for mixing	= 1.25 × 640	= 800 litre
Boiler rating	= 0.10 × 400	= 40 kW

- An office has a staff of 120 and requires water for cloakrooms at 55°C for a 10 hour working day. Energy supply is to be electrical with 80% of the annual total supply taken off-peak. Recovery period for daytime use to be 2 hours. The heat losses are estimated to be 0.8 kW.

From Table 20.5

Storage required above upper element		=	1.2 litre/person
Upper element rating	= 1.2 × 0.032	=	0.04 kW/person

From Table 20.6

Storage required between elements (i.e. lower)		=	3.0 litre/person

Thus

Upper storage	= 1.2 × 120	=	144 litre
Upper element	= (0.04 × 120) [(55–10)/(65–10)]	=	3.93 kW
Total, including heat loss	= 3.93 + 0.8	=	4.73 kW
Lower storage	= 3.0 × 120	=	360 litre
Stored volume equivalent to heat loss = (3600 × 0.8 × 10 × 0.8)/[4.2 × (55–10)]		=	122 litre
Total storage	= 144 + 360 + 122	=	626 litre
With margin for mixing	= 1.25 × 626	=	782 litre
Lower element	= 0.01 × 626	=	6.26 kW

Storage temperature

It should be noted that the temperatures listed in Table 20.3 are for *hot water use* and do not represent those which are required at the storage vessel. Taking no account of other relevant factors, it is obviously reasonable to store hot water at a temperature higher than that required at the point of use since a smaller volume in storage is then necessary.

It was for many years considered to be good practice to design for, and control, hot water storage temperatures at 65°C (150°F) for all but special applications. Dishwashers, for instance, required an elevated temperature of about 85°C, whereas service to primary schools, old people's homes and prisons was provided at some 15–20 K lower than the normal level. Propositions were made during the early 1970s, with energy conservation in mind, that the traditional temperature levels were not actually needed and figures of the order of 45°C were proposed, with seemingly little thought given to the increased storage volume – and increased heat loss from larger vessels – which would have been required.

These unfortunate suggestions have however been overtaken by the results of investigations into the incidence of *Legionnaires' disease* and the conclusion that it is related to aerosol contamination, some of which has originated in hot water systems. Without going into great detail, it appears that the bacterium is present in most mains water supplies but is dormant at temperatures below about 18°C. It multiplies rapidly between 25 and 45°C and is killed instantly at a temperature of 70°C.

Interim recommendations are, in consequence, that hot water should not be stored at a temperature less than 55°C and, where it has been held at a lower and more critical temperature for any length of time, i.e. overnight, it should be reheated to 55–60°C for an hour prior to exposure and use. Thus, if complex and fallible control cycles are to be avoided, a practical solution is to maintain a storage temperature of 60–65°C whenever the plant is in operation, particularly where night shut-down or any other intermittent heating routine is adopted.

In instances where higher temperatures are necessary, as in the case of dishwashers as mentioned previously, practice in recent years has been to fit such equipment with local electrical or other booster heaters supplied with make-up from the normal hot water system. Any requirements for water at temperatures lower than 55°C may be met by mixing of hot and cold water, which process should be arranged as near to the point of use as is practicable.

Temperature control

Since the volume of hot water stored acts as a cushion, close control of secondary water temperature is neither possible nor necessary. In common with the familiar rod-type thermostat switching arrangements associated with electrical immersion heaters, the simplest forms of control for water temperature are direct-acting, which is to say that expansion of the sensing element provides the motive power for the controller.

In Figure 20.17, a simple and robust type of direct-acting valve is shown, the primary outlet pipework commonly being works mounted: this pattern may be used whether the primary supply is water or steam. The next stage in complexity is a separated direct-acting valve which may be connected by a capillary to the sensing element mounted in the storage vessel. Apart from these simple types, thermostats and control valves may be electrical, electronic or pneumatic and be either two- or three-way pattern.

A disadvantage inherent to both types of direct acting valve, and of other types when misapplied, is that they provide *proportional* control. Thus, with the sensing element immersed in the water store, the valve will be wide open to the primary medium when the store is cold and will close progressively as the temperature rises. Since the temperature difference between the primary medium and the stored water will decrease as the valve closes, sluggish recovery of the last few degrees of storage temperature is inevitable. Any form of *on/off* control is to be preferred for this particular application.

Feed cisterns

For a conventional hot water supply system, a cold water storage cistern is required to supply the hot water storage vessel and replace the water drawn off at taps, etc. A cistern is necessary also, as in the case of a heating system, to accommodate the increase in volume as the water content of the storage vessel and pipework is heated. This expansion takes place via the feed pipe connecting the cistern to the storage vessel and not, as the *Water Supply Byelaw Guide* asserted, through the vent pipe! The feed and expansion pipe from the cistern (or pressurising equipment as noted later) to the storage vessel should, ideally, be so connected to the latter that mixing of the cold supply with the hot water stored is prevented and disturbance to stratification of the latter is minimised. A number of devices has been produced from time to time to achieve this end but with only limited success.

In order that an adequate pressure be provided at the points of use, the cistern should be sited as far above the highest such point as is practicable: in a building of any significant size, a purpose built tank room having easily cleaned surfaces, and not used as an overflow store, is to be preferred. The capacity of the cold water store must depend upon the security of the incoming main supply both as to volume and pressure: if both are

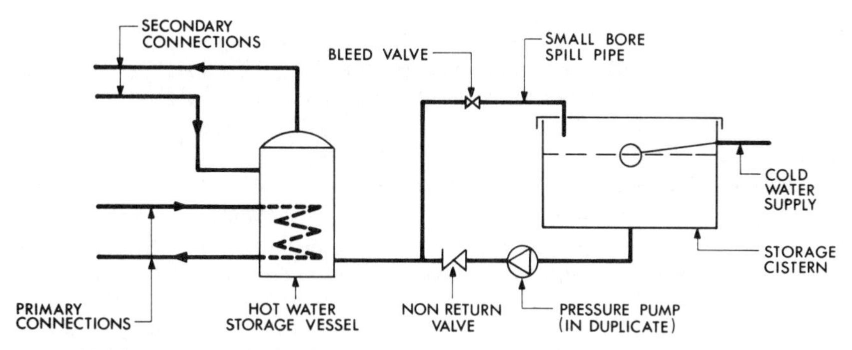

Figure 20.24 Pump pressurisation for hot-water secondary supply

guaranteed then storage for 2 hours of peak usage will probably suffice but if either is suspect, as is often the case, then it would be advisable to plan for three or four times that volume.

Circumstances sometimes arise where either a requirement for the limitation of building height (e.g. a development near an airfield) precludes the provision of an elevated cistern or, more rarely, where the incoming mains pressure is not sufficient to produce an adequate supply to an elevated cistern. It is then necessary to provide suitable cold water storage at a low level and to use this as a supply to some form of mechanical pressurisation equipment. The more suitable types are not unlike those described for heating systems in Chapter 8 (p. 209) and, for simplicity, pump pressurisation as illustrated in Figure 20.24 has much to commend it.

The new Model Water Byelaws require that all cisterns be provided with rigid, close fitting covers which exclude light and that they should have properly made insect screens at all openings: where pipe connections penetrate the cover, the holes should be drilled rather than, as is often the case with present covers or insulation, apparently torn out by wolves.

Unvented hot water systems

Past editions of Model Water Byelaws have included requirements which have inhibited design development of storage type hot water systems in the British Isles. Direct connection to a mains pressure service pipe has not been permitted and provision of an open vent to atmosphere has been mandatory. The use of the bubble-top calorifiers which have been described earlier was generally confined to sites protected by Crown immunity.

The 1986–9 edition of the Model Byelaws no longer proposes that there should be a prohibition of connection to mains pressure service pipes and, as a result, some water companies have relaxed their previous attitudes in order to enable progress to be made. Building Regulations, Section G3, now include requirements covering safety precautions for unvented systems. It is obvious that very careful consideration must be given to these relaxations of previous practice since unvented systems are passing from the sphere of engineering design to one where artisan plumbers will be applying rules of thumb to repetitive domestic installations.

Control packages

It was a requirement under Building Regulations, Section G3 (1985) that *'Any unvented hot-water storage system should be in the form of a proprietary Unit or Package which is the subject of a current British Board of Agrément (BBA) Certificate.'* This meant that the system (in this context the storage vessel) should either be factory fitted with all protective and functional devices or be factory fitted with the former and supplied with the remainder in kit form for installation on site. Further, it was a requirement that all such systems, whether including units or packages, should be installed only by 'competent persons' listed as Approved Installers by the BBA. In the interest of safety, bearing in mind that many such systems may be in domestic premises, it is unfortunate that these very sensible requirements have been diluted in Section G3 of the revised regulations of 1991.

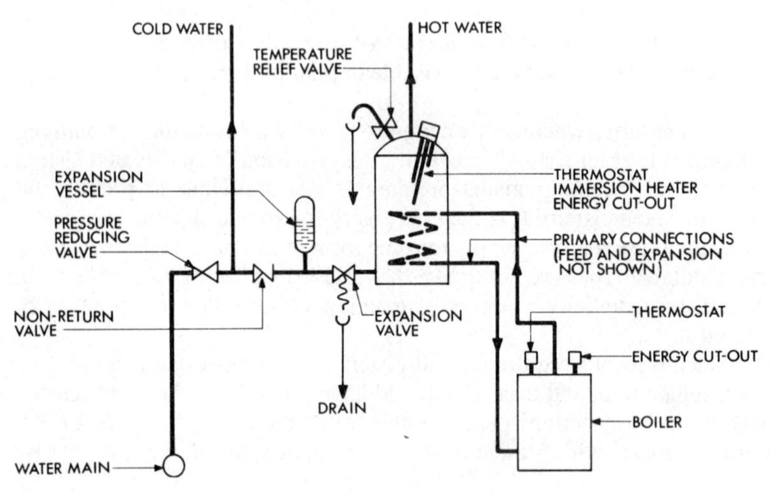

Figure 20.25 Unvented hot water secondary supply system

The various devices shown in Figure 20.25 and listed below are required for all unvented systems and fall within two categories:

Protective

- A thermostat to control the energy source. This should not be set at above 75°C and preferably at 5 K lower.
- A temperature operated cut-out acting on the energy source. This must be factory set at 85°C and be of the manual reset type
- A temperature operated relief valve, factory set to open at 90–95°C. The outlet from this valve must be to a tun-dish and piped thence to a point where discharge of very hot water will not be dangerous.

Functional

- A pressure controller at the connection to the service pipe is required to deal with fluctuations in the supply and to determine the system operating pressure. Good control characteristics at low flow rates and the ability to accept a wide range of inlet pressures are necessary. Either a pressure reducing valve or a pressure limiting valve may be used, factory set.

- A non-return valve in the feed to the storage vessel is necessary in order to avoid back-flow of hot water into the cold water supply.
- A diaphragm type expansion vessel to cater for the expected increase in volume (nominally 4%).
- An expansion relief valve which should not open during normal system operation. The outlet from the valve must be visible and piped to a convenient point. This valve is not a safety device and is not designed to release steam.

It remains to be seen whether the many comfortable advantages of the traditional system, whereby supply is secured to some extent as a result of on-site water storage, building by building, will give way to the supposed economic attractions of the alternative and to a total reliance upon the presumed constant availability of a mains water supply in the future.

Water treatment

A short discussion was included in Chapter 8 (p. 224) to cover the incidence of corrosion in heating systems which have the advantage of reusing and re-circulating the same water content. In the present context of hot water supply, the secondary water content in indirect systems and the whole content of direct systems is subject to continual change. In general terms, the problem in this case relates more usually to the formation of scale rather than to the effects of corrosion. In this respect it is of interest to note that public water supplies to the mainland of the British Isles, south and east of a line drawn between Hull and Bristol (with local exceptions), all fall into either the 'hard' or 'very hard' categories.

A complete discussion of this subject would be beyond the scope of this book and the brief notes which follow are intended only as an introduction to the methods commonly used to treat cold water supplies to hot water systems. Further information may be found in the *Guide Section B7* but specialist advice should be sought where particular problems are known to exist.

Hardness of water

The calcium and magnesium salts present in raw water exist in two forms, as bicarbonates which form what is known as *temporary hardness* and as sulphates, chlorides and nitrates which form what is known as *permanent hardness*. The salts within the temporary category fall out of solution when the water temperature is raised to about 70°C which situation occurs adjacent to heat exchange surfaces even when the storage temperature is below that level. The salts of permanent hardness remain in solution at the temperatures generally encountered in indirect systems but calcium sulphate may cause problems in direct systems at boiler surfaces or where the primary medium of an indirect system is at high temperature.

Commercial installations

Here, dependent upon circumstances, it is probable that a full scale water treatment plant will be installed. This is most likely to be a base- or ion-exchange process which makes use of beds of either natural or synthetic *zeolites* (sodium aluminium silicates). These minerals contain sodium in combination and have the property of exchanging this with the calcium and magnesium salts in a raw water to form sodium bicarbonate which is not hard-scale forming. Natural zeolites (greensands) are impermeable and have a lower capacity for exchange, per unit volume, than the synthetic type. The latter, as a result of their porous structure, must not be used with an unfiltered raw water supply and have a shorter life than the natural material.

The exchange process does not continue indefinitely since the capacity of the zeolite is said to be *exhausted* when all the sodium content has been exchanged: it is then necessary to regenerate the mineral by slow flushing with a strong solution of common salt (sodium chloride) followed by brisk backwash to remove the brine residual and dispose of the calcium and magnesium. It must be emphasised that the process of base exchange, whilst reducing scale deposition, *does not reduce* the total content of dissolved solids in the water.

Dosing

Internal water treatment by dosing boiler make-up water has been a familiar process in the case of industrial plant where the facilities exist to allow the effect to be monitored and the quantities of additive adjusted accordingly. For use at domestic level, it has been necessary to develop simple methods for automatic metering of dosage which require no skill to operate and will not permit overfeed. A number of devices which allow suitable quantities of phosphates and polyphosphates to be injected into an incoming mains water supply is now available for use at domestic level. This injection modifies the hardness salts such that, following treatment, they remain suspended in the water, instead of bonding one to another and forming scale when temperature is increased. The process requires only that the bulk supply of chemicals introduced be renewed from time to time, in the form of powder or spheroid concentrates.

Magnetic treatment, etc.

As an alternative to dosing, use may be made of magnetic treatments which have no chemical effect upon the various hardness salts and do not change the composition of the water. The process involves introduction of a pipeline insert which incorporates a static 'co-axial magnetic circuit' through which the water passes. As a result of this exposure, the particles of hardness salts do not agglomerate but remain freely suspended in the water and thus do not form scale on heated surfaces.

An alternative but not dissimilar process results from exposure of the incoming water to an electrochemical device within which water flow past a zinc–copper combination acts to produce an electrolytic cell. This, in sequence, provides an electrical potential which discourages the particles of scaling salts from attracting one another and thus precludes formation of a layer of scale.

Materials, etc.

Vessels and pipework

The two materials most commonly used in construction of hot water systems have been galvanised steel and copper. Selection of one or the other has often been made following study of data such as that set out in Table 20.7, the results representing the limits within which steel could be used. As a result of the ease with which it is manipulated and the simpler fabrication techniques which are associated with it, copper piping is now used

Table 20.7 Limits for use of galvanised steel

	Temporary hardness greater than:			
pH value	mg/litre	Parts per 100 000	Grains per imp. gallon[a]	Grains per US gallon
7.3	210	21	15.0	12.3
7.4	150	15	10.5	8.8
7.5	140	14	9.8	8.2
7.6	110	10	7.7	6.4
7.7	90	9	6.3	5.3
7.8	80	8	5.6	4.7
7.9 and over	70	7	4.9	4.1

[a] 1 grain per Imperial Gallon = 1 degree Clark.

almost universally. Hence, since a mixture of metals is particularly undesirable where the associated water is subject to continuous change, with a consequent continuous release of dissolved oxygen, the use of galvanised steel for storage vessels is no longer usual. Copper vessels having thin shells are now very often provided with sacrificial aluminium anodes to protect them against corrosion.

It is common practice in Europe to use steel vessels, copper lined, for hot water storage and, of recent years, *thermo-glazing* has been similarly employed, this taking the form of a ceramic enamel coat applied internally to the steel shell. The use of such vessels may become more familiar in the British Isles where non-vented systems are used since all-copper construction, to meet the higher pressures which will then obtain, may be highly priced. A range of thermo-glazed units in factory made insulating casings, as shown in Figure 20.26, has been on the market for some time.

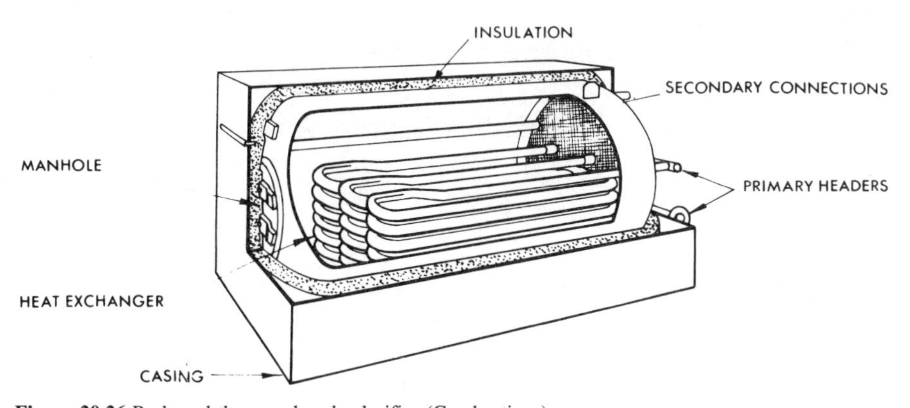

Figure 20.26 Packaged thermo-glazed calorifier (Combustions)

Insulation

Heat losses from all the components of a hot water system continue throughout the year, often day and night. To suggest that, for many systems, as much energy is dissipated in this way as is used to heat the water drawn off at taps is no exaggeration. In consequence, a very good standard of thermal insulation to pipework is recommended. Storage vessels for domestic use may be factory insulated with 50 mm sprayed urethane foam or fitted with a standard 80 mm thick quilted jacket. Larger size vessels for the commercial sector should be insulated to at least as good a standard, and be finished with a protective sheet metal casing.

In addition to the energy wastage taking place, the heat given off by pipes is objectionable in summer. Where such pipes are routed in ducts alongside a cold water distribution system, the heat added to the latter will not only make it unpalatable for drinking purposes but might create the very temperature in which *L. pneumophilia* will thrive.

Solar collectors

The most common application of solar heating in the British Isles has been for preheating the cold water supply to domestic hot water systems. Various forms of solar collector have been developed, as shown in Figure 20.27 which shows, firstly, a typical general arrangement and, secondly, various types of collector ((a) pipe and fin; (b) water sandwich; (c) semi-water sandwich; (d) parabolic focusing channel): there are other types of course but all are derivatives from those illustrated. The materials from which collectors have been constructed are diverse: all copper; all aluminium; all plastic and any number of mixtures. Since, to avoid damage by frost, the working fluids circulated have contained corrosive additives, metal mixtures should have been avoided: a combination of copper piping in the system proper and an all aluminium collector is a unique specification for disaster.

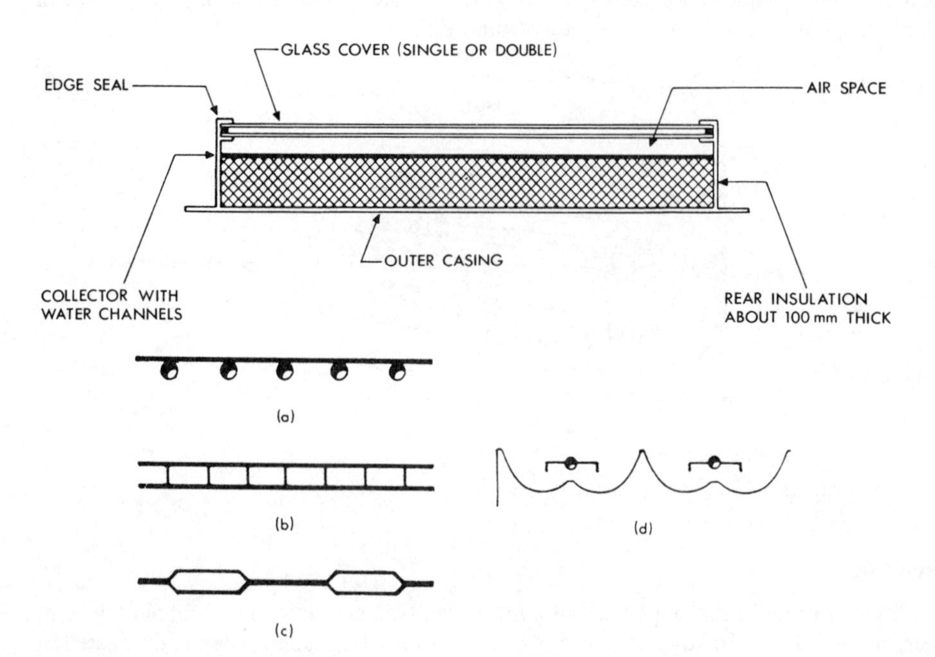

Figure 20.27 Basic types of solar collector

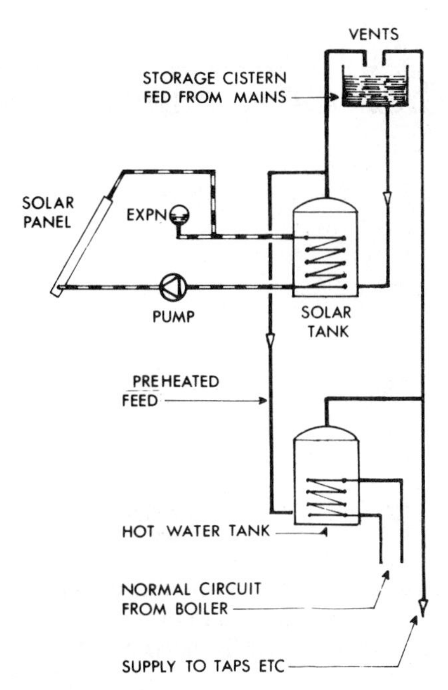

Figure 20.28 Simple solar hot water preheat system

In general, the efficiency of solar collectors is low with respect to the energy available, a common level being between 20 and 40%. Thus, although a collector may *intercept* up to about 3.5 GJ/m^2 of energy per annum, the most that can be expected to be of use is 1.4 GJ/m^2.

More important than low efficiency however is the question of risks to health arising from concentrations of *L. pneumophilia* within the solar store. As has been emphasised previously, 25–45°C is the temperature range for rapid multiplication of the bacteria and a residence time of one or two days is adequate for a high concentration to develop. This temperature and this time are typical of the conditions within the many domestic solar stores arranged as shown in Figure 20.28. It is open to question therefore whether it is sensible either to construct more such systems of this nature or to continue to use those which exist.

Chapter 21

Piping design for central hot water supply systems

Piping arrangements for hot water supply systems fall naturally into four categories and are best considered within these, as follows:

- The primary pipework which provides for circulation of water between the energy source and the storage vessel or vessels. This circulation is related to the maintenance of the required temperature in the store and, except in that sense, is independent of the rate of outflow at draw-off points.
- The secondary outflow pipework which is required to pass hot water in the quantity demanded by the draw-off points when taps, etc., are opened.
- The cold water feed pipe to the hot water store which is required to pass make-up water equivalent to the hot water quantity drawn off. This pipe, originating from either an elevated cistern or a mains pressure service pipe, conveys the motive force which causes hot water to flow at draw-off points.
- The secondary circulating pipework which, in conjunction with the outflow pipework, provides means whereby hot water is constantly available near to the draw-off points.

Primary circulations

Primary pipework (direct system)

To ensure that circulation between energy source and storage is brisk, taking account of the probability that the available pressure difference will be small, the primary flow and return pipework should be sized generously. In addition, it is sensible in hard water areas to make allowance for the inevitable reduction in diameter due to scale deposition and thus, even for the smallest system, piping should not be less than 25 mm galvanised steel or 28 mm copper.

The pipe size required may be determined using exactly the same methods of calculation as those described in Chapter 7 for a heating system operating with *gravity* circulation, taking the circulating head as the difference in pressure between the hot rising and the cold falling columns. The effective height is from the centre of the boiler to the centre of the cylinder, as shown in Figure 21.1, and a temperature difference, flow to return, of 20 K may be assumed for purposes of calculation. As a first approximation, the rates of energy flow through given sizes of pipe, as listed in Table 21.1, may be used. These are based upon a boiler flow temperature of 60°C and a height of 3 m, the travel

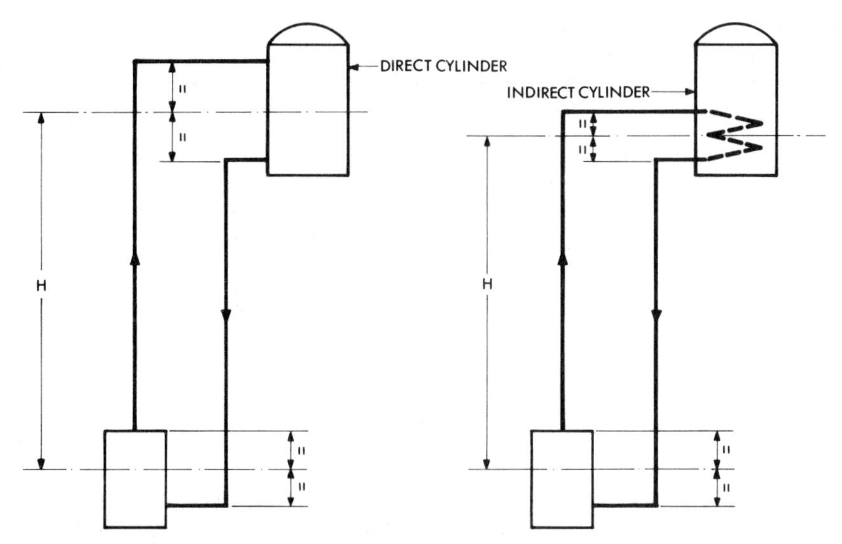

Figure 21.1 Effective heights for gravity circulation in primaries

being 15 m and incorporating eight bends. The tabulated figures for energy flow are 20% less than the theoretical in order to allow for scaling in the pipes.

It will be appreciated that, although the flow temperature from the boiler may be controlled at a relatively constant level, the temperature of water returning to it from the store will vary according to the rate of draw-off and, in consequence, the circulating head will change also. The most active circulation will occur during and after a period of heavy draw off, i.e. when the lower part of the store is certain to be cold, and the least active after a period of no draw-off when the contents of the store are nearing boiler flow temperature.

This situation suggests that supply and demand are in equilibrium at all times, which is generally the case, and that the variation in circulating head will offer a form of temperature control. The exception arises when, following a period of 'stagnation' at near to the required storage temperature, a quick response to sudden heavy draw-off is required and the necessary rapid circulation is delayed by system inertia.

Where a direct system incorporates the type of storage boiler shown in Figure 20.11, no primary pipework is needed and the response rate to demand is always extremely rapid.

Primary pipework (indirect system)

While the advisability of sizing pipework generously, where circulation is to be by gravity, applies to indirect systems also, in this case no allowance need be made for scale deposition when selecting pipe sizes since the water content of the primary circuit is constantly recirculated and does not change. The need for a brisk circulation remains, nevertheless, since the temperature of the stored secondary water is raised through some form of heat exchanger whereas, in a direct system, it is heated within a boiler. It is current practice to use a pumped circulation through a primary circuit except for the smallest domestic installations.

Table 21.1 Pipe sizes and energy flow (kW) for primary circuits in copper (BS 2871: Table X) and galvanised steel (BS 1387: heavy) pipes

Copper pipes			*Steel pipes*		
Pipe size	*System type*		*Pipe size*	*System type*	
	Direct	*Indirect*		*Direct*	*Indirect*
28	5	7	25	4	7
35	8	10	32	8	13
42	12	18	40	11	19
54	22	31	50	19	31
67	35	51	65	35	59
76	47	66	80	53	82

Since the primary circuit will, in all probability, originate from a boiler plant which also serves a heating system, it is also probable that the flow temperature therefrom is fixed at 80°C. In any event, however, it will be advantageous to provide a primary flow temperature at a reasonably high level in order to economise in heat exchanger surface. The temperature difference, flow to return, may again be about 20 K or, when a pumped circulation is used, 10–15 K.

For a gravity circulation, the pipe sizes required may again be determined using the same methods of calculation as those described in Chapter 7. As a first approximation, the data in Table 21.1 may be used, these being based upon a flow temperature of 80°C and other particulars as before, plus an allowance equivalent to 3 m of pipe for the coil within the storage cylinder. The pipework arrangement and effective height are shown in Figure 21.1.

In a domestic size system, where a single boiler serves both heating and hot water supply, the adoption of a mixed pumped/gravity circulation may offer advantages. Figure 20.13 illustrates an arrangement where use is made of an injector to accelerate the circulation in the hot water primary circuit at times when the heating pump is running.

Where the primary circuit is to be pumped, it is necessary that it be arranged to be quite separate from any space heating circuit served from the same energy source. For a low-rise

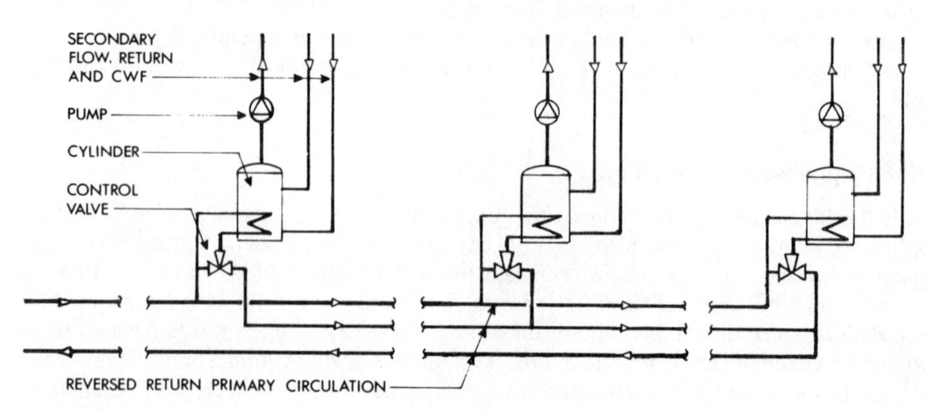

Figure 21.2 Dispersed indirect cylinders with central-source primaries

institutional building or group of buildings, it may be more economic to provide a number of dispersed hot water stores fed from a constant temperature primary distribution system, as shown in Figure 21.2, rather than a widespread secondary hot water circuit. Such an arrangement may offer the facility to serve other equipment requiring a constant temperature supply such as ventilation heater batteries, etc., and would be quite separate from any parallel circuit which, at a temperature varied to suit weather conditions, might serve heating apparatus. A separate cold water cistern will, of course, be required for each of the dispersed stores.

Secondary outflow and return pipework

The simplest form of outflow pipework from a secondary system is a series of dead-legs as shown in Figure 21.3. The length of these must however be strictly limited for three very good reasons:

- If the length were too great, an undue time would elapse before hot water reached the tap and, furthermore, cold water would have to be run to waste.
- Following draw-off, a long dead-leg would be full of hot water which, being unused, would be an energy wastage.
- If draw-off were intermittent, an excessive quantity of water would stagnate in the pipework and might be held on occasions within the critical temperature band for rapid development of *L. pneumophila*.

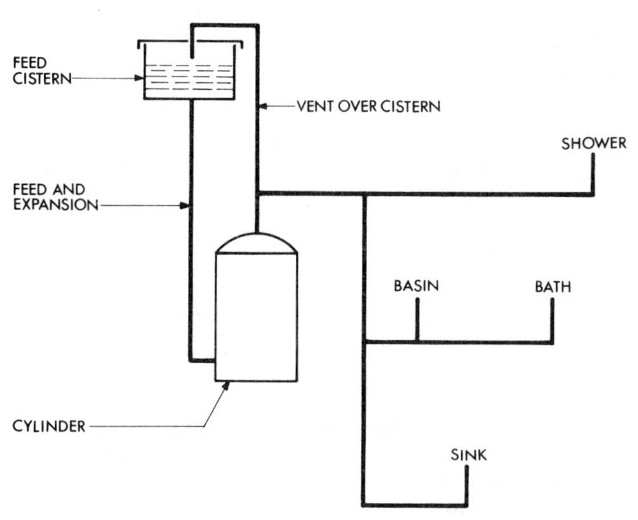

Figure 21.3 A dead-leg system

Water companies are concerned principally with limitation of wastage and Model Byelaws have for many years required that dead-legs shall not exceed the lengths listed in Table 21.2. In the joint interests of health and energy conservation, it would seem appropriate for these permitted lengths to be greatly reduced.

Table 21.2 Maximum length of dead-leg permitted by water companies

Largest internal diameter (mm)	*Nominal pipe sizes*		*Maximum length* (m)
	Steel	*Copper*	
Not exceeding 19	15	15	12
19 to 25	20	22	8
Exceeding 25	25	28	3

Outflow from draw-off points

Fundamental to the design of secondary outflow pipework is the quantity of hot water required at the various points of draw-off. A table in the preceding chapter gave details of the capacity of a variety of standard fittings and the rates of water flow which are necessary to produce this capacity in a reasonable time are listed in Table 21.3. These are the data used to produce the values quoted in the *Guide Section B4*.

Table 21.3 Discharge rates and pipe sizes for draw-off points

Fitting		*Rate of flow* (litre/s)	*Size of connection* (mm)
Bath	(private)	0.3	20
	(institutional)	0.6	20–25
Basin	(bib tap)	0.15	15
	(spray tap)	0.05	10
Shower	(spray)	0.15	15
	(100 mm rose)	0.4	15
	(150 mm rose)	0.6	15
Sink		0.3	20

Where there are only a few draw-off points to be served, as in a private house, it is reasonable to assume that at some time all the taps, etc., will be in use simultaneously, although this is unlikely. With any sizeable system it would be extravagent to make such an assumption since, as the number of taps served increases, it becomes less probable that they will all be open at once.

For instance, a hot water tap at a wash basin may be open for 30 seconds but at least a minute will elapse before the tap is opened again. In a row of ten basins, this would mean that a maximum of about three taps is likely to be open simultaneously even if people are queuing to use them. The type of building, the type of fitting and the pattern of use are all important matters for consideration. Thus, a group of showers in a sports club house, at the end of a match, is likely to be in full use at the same time but, in a hotel, it is probable

Table 21.4 Comparative demand units

	Category of application		
Fitting	*Congested*	*Public*	*Private*
Basin	10	5	3
Bath	47	25	12
Sink	43	22	11

that only a small proportion of the baths provided will be filling at the same time. While special cases must always be considered on their merits, a generalised approach for application to a wide spectrum of buildings is necessary.

Various attempts* have been made to establish a basis which may be used with confidence and a method based upon the theory of probability was developed for the *Guide Section B4*, adopting the concept of a scale of *demand units*, a wash basin tap being taken to represent unity. No values can, unfortunately, be allocated to showers and basin spray taps since they require water flow for the whole period of fitting use. Otherwise, the type of application is taken into account in this scale by weighting the units according to assumed intervals of use varying from 5 to 20 minutes for a basin and from 20 to 80 minutes for a bath. From this approach, a much simplified list of 'demand units' may be produced as Table 21.4, with three categories of use:

'Congested' (where times of draw-off are regulated).
'Public' (normal random usage).
'Private' (infrequent or spasmodic).

Using a simple diagram of the outflow system, the individual and the cumulative 'demand unit' totals may be marked up against each section of the pipework and then, using Figure 21.4 which is based upon probability of simultaneous use, these may be converted to flow rates in litre/s. For showers and basin spray taps, the whole flow rate, as given in Table 21.3, must be added to the individual and cumulative totals.

Similarly, flow to fittings such as wash fountains and washing machines which may require a specific pattern of supply must be dealt with separately according to type and the makers' flow requirements.

It cannot be over-emphasised that a method such as that noted above is a statistical approach to a human problem and is therefore no more than a useful indication of a route to follow in design. There is no substitute for either a broad experience or a specific knowledge of the pattern of use in other similar applications.

Cold water feed pipework

The cold water feed to a hot water store is an integral part of the secondary outflow pipework. However, it is convenient to deal with it separately here, since it may originate either from an elevated storage cistern or, as now permitted by the 1986/9 Model Water

* Bull, L. C., 'Simultaneous demands from a number of draw-off points', *JIVHE*, 1956, **23**, 445.

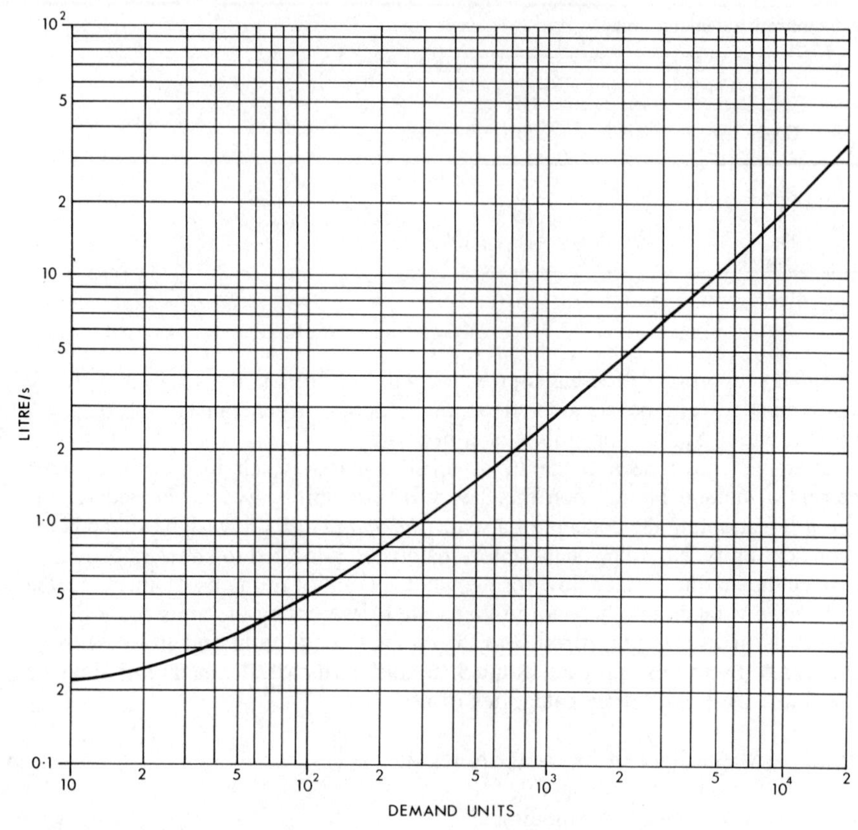

Figure 21.4 Demand units and flow probability

Byelaws, from a mains pressure service pipe. In either case, the head of water (or the mains pressure) provides the motive force which produces outflow at the points of draw-off. The relationship between these alternatives is given in Table 21.5.*

As a result of export effort, various manufacturers have gained experience in the field of equipment suited to unvented hot water systems and designers who have practised abroad are familiar with most of the potential advantages and problems. Nevertheless, no fully accepted practice exists which has refined experience in Europe and elsewhere to meet British needs.

The working pressures for such systems fall into two groups, one relating to a level of between 300 and 400 kPa and the other to a reduced level of up to 250 kPa. The determining factor is, to some extent, the type of storage vessel used since the most substantial grade of the familiar thin-wall copper cylinders (Table 20.2) is rated for a working pressure of only 245 kPa or 25 m head of water. That is of course not to say that non-standard cylinders cannot be made to order nor that other types such as steel vessels with vitreous enamel or similar linings are not available.

* Tables 21.5, 21.7 and 21.8 are grouped together at the end of this chapter on pp. 599 and 600.

As an aside, it is worth remembering that the design flow rate for a 15 mm basin tap may be obtained with an applied pressure of less than 10 kPa. At the opposite extreme of experience, a 13 mm nozzle supplied at the pressure noted above, 250 kPa, will throw a jet of 20 times that quantity for 20 m!

Piping design for outflow

Taking the system illustrated in Figure 21.5 as an example of feed from an elevated cistern, the hot water flow rate through the cold water feed and the secondary outflow pipework would be determined as shown in Table 21.6. Many designers mark up the 'demand units' and flow rates on a diagram of the system but the sequence of working is more easily illustrated by the tabular approach used here.

If the cold water feed were from a mains pressure service pipe, the procedure followed would be just the same. In all probability, the cold water supply arrangements would be served from the same intake into the building and the duty of the pressure-reducing valve would thus be based upon the total combined demand from all draw-off points, hot and cold.

Once the water quantities to be carried by each section of outflow pipework have been established, the next step is to consider either the head of water available from the cistern or the residual of the working pressure selected where feed is from a mains service pipe.

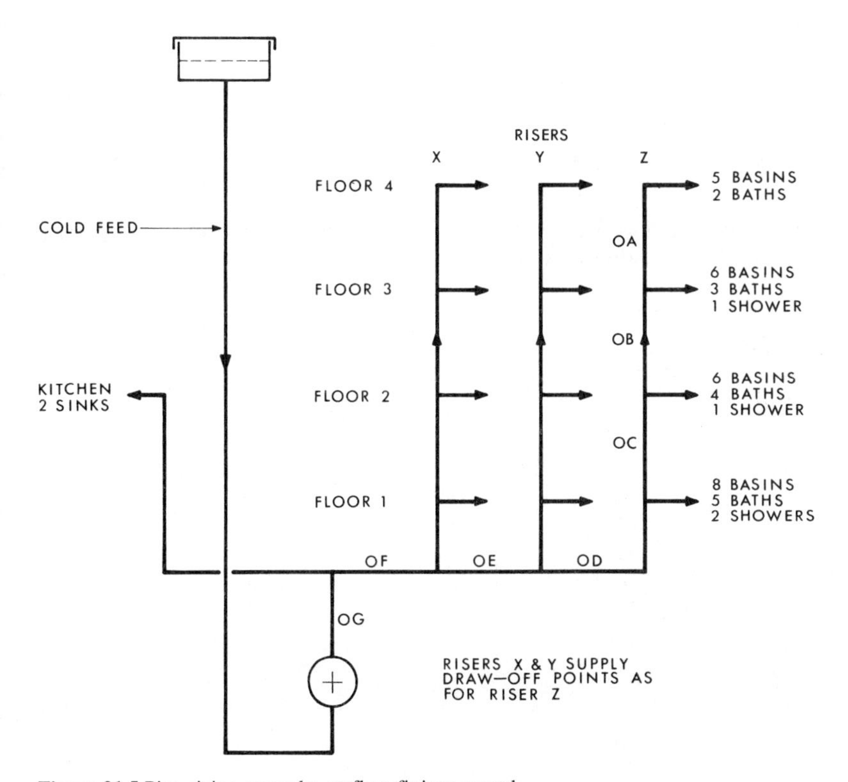

Figure 21.5 Pipe sizing example: outflow fittings served

In either event, the worst case is that presented by the highest draw-off point in the building. Progressing down a multi-storey building, a greater head or pressure will be available on lower floors which may lead to splashing problems there unless steps are taken to throttle the supply.

Tables 21.7 and 21.8 have been compiled from data included in the *Guide Section C4* for water flowing at 75°C in copper pipes to BS 2871: Table X and in galvanised steel pipes to BS 1387: heavy respectively. Bearing in mind the other potential variables in this application, errors arising from applying these tables to designs where water at other temperatures or other grades of copper or galvanised pipe are to be used are negligible. It will be noted that flow rates may be read from these tables against either head loss (m water per m run) or pressure loss (kPa per m run), whichever is more convenient for the application.

Example

Taking the system shown in Figure 21.5 as an example, physical data and the flow rates previously calculated may be added as shown in Figure 21.6.

Clearly, the worst case is that of the fourth-floor taps on riser *Z* which are both the most distant from the cistern and only 3 m below minimum water level, i.e. they have only 3 m head of water available to cover frictional resistance and provide outflow. The travel to this point is:

Cold water feed	=	24 m	
allow for fittings	=	3 m	27 m
Outflow main	=	12 m	
allow for fittings	=	2 m	14 m
Riser Z	=	11 m	
allow for fittings	=	2 m	13 m
Total	=		54 m

Hence, the available unit head = 3/54 = 0.055 m/m run and, interpolating from Tables 21.7 and 21.8, the following listing may be made:

	Pipe size (mm)	
Riser Z	*Copper*	*Steel*
Branch	28	25
OA	28	25
OB	35	32
OC	35	40
OD	42	40
OE	42	50
OF	54	65
OG	54	65
CW feed	67	65

Table 21.6 Application of demand unit calculation: worked example

Floor	Fittings		Demand units (public)				Flow (litre/s) From Figure 21.4	Totals
4	5 basins	×	5	=	25			
	2 baths	×	25	=	50	75	0.4	0.4
3	6 basins	×	5	=	30			
	3 baths	×	25	=	75	105	0.51	
	1 shower	×	0.15 litre/s				0.15	0.66
2	6 basins	×	5	=	30			
	4 baths	×	25	=	100	130	0.56	
	1 shower	×	0.15 litre/s				0.15	0.71
1	8 basins	×	5	=	40			
	5 baths	×	25	=	125	165	0.66	
	2 showers	×	0.15 litre/s				0.30	0.96
Kitchen	2 sinks	×	0.3 litre/s				0.6	0.6

Item	Section	Demand units (cumulative)	Flow (litre/s) From Figure 21.4		Showers and sinks		Totals
Riser	OA	75	0.4		–		0.4
	OB	180	0.74	+	1 × 0.15	=	0.89
	OC	310	1.0	+	2 × 0.15	=	1.3
Mains	OD	475	1.45	+	4 × 0.15	=	2.05
	OE	950	2.5	+	8 × 0.15	=	3.7
	OF	1425	3.45	+	12 × 0.15	=	5.25
Total	OG		5.25	+	2 × 0.3	=	5.85

The example could be continued by considering the lower branches of riser Z since a greater head of water will be available progressively. The head lost due to friction on this longest run, as far as the junction to each branch, would thus be deducted from the total available and the surplus used to size the pipes of the branch. The same process would then be applied to the other risers X and Y and to the branches from them, the run to each being shorter than that to the furthest branch on riser Z.

Re-examination is particularly appropriate for the long branch to sinks in the kitchen since the head available there is 9 m and the loss through the cold water feed pipe and main section OG is only about 2 m (33 × 0.055 = 1.82 m). Thus a head of 7 m is available at the branch which, allowing for 5 m equivalent length there for fittings, represents a unit availability of 7/(29 + 5) = 0.2 m/m run. The pipe size required, in consequence, for a flow of 0.6 litre/s, would be 22 mm in copper or 25 mm in galvanised steel, one size less in each case than if the original unit figure of 0.055 m/m run had been used.

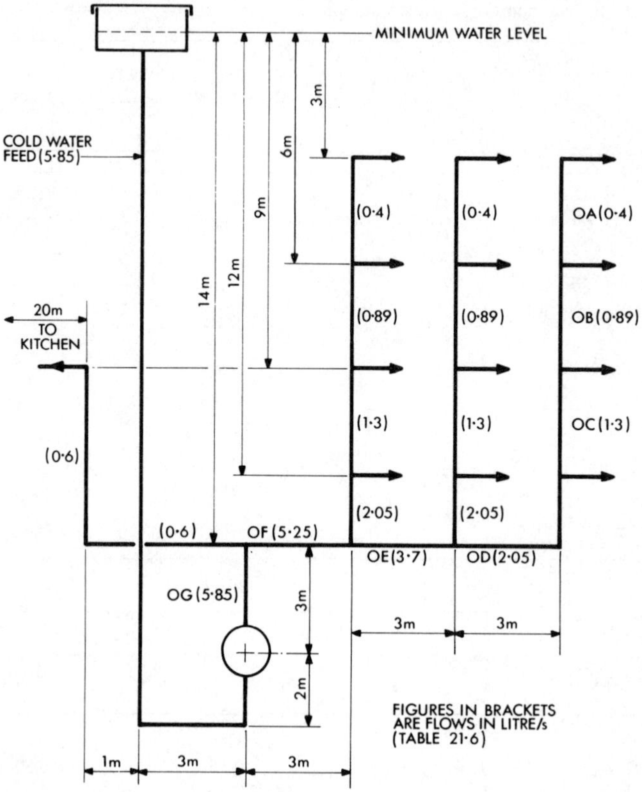

Figure 21.6 Pipe sizing example: dimensions

For a system of modest size, such refinement in calculation may not be worthwhile bearing in mind the many variables. An experienced designer may be content to use a single value for unit head loss except where this would lead to obviously uneconomic design. However, in the case of large systems or those in buildings having either inadequate cistern height or a variety of floor levels, exhaustive examination is necessary.

Piping design for secondary returns

Note has already been made of the limitations which should be placed upon the length of dead-legs and thus, where hot water draw-off points are dispersed widely, it is necessary to consider whether to provide:

- A local instantaneous heater to some or all draw-off points.
- A number of dispersed storage vessels with a dead-leg distribution from each. (These might be served either by local energy sources or from a primary distribution system as shown in Figure 21.2.)
- A secondary circulation system.

It is not possible to generalise as to either the most practical or the most economic solution since each application must be considered on its merits. A secondary circulation, with which we are concerned here, offers facilities for heating towel rails and linen cupboard coils, etc., at times when a heating system may be shut down but, since there are continuous heat losses from the circulating pipes even when well insulated, is wasteful of energy.

The water quantity to be circulated through secondary pipework is a function of the heat emission from the system and of a chosen temperature drop, normally 10–12 K for a gravity circulation and about 5 K where a pump is to be used. The heat emission derives from both outflow and return pipework plus the useful output of any towel rails or linen cupboard coils connected to them. Piping emission may be calculated using data from Table 21.9 and the sectional emissions are collected and totalled to produce sectional loadings in exactly the same way as for a heating system.

Table 21.9 Theoretical heat emission from horizontal steel (BS 1387) and copper (BS 2871) pipes

Pipe size (mm)		*Emission* (W/m run)			*Thickness of insulation* (mm)
		Bare		*Insulated*	
Steel	*Copper*	*Steel*	*Copper*		
15	15	42	31	8	25
20	22	51	43	9	25
25	28	62	53	9	32
32	35	75	64	10	32
40	42	84	75	12	32
50	54	102	93	13	32
65	67	125	112	16	32
80	76	143	125	17	32
100	108	179	171	18	38

Notes:
Mean temperature difference = 40 K.
Insulation conductivity = 0.04 W/m K.

The sizes of the outflow piping will be known as a result of an earlier calculation but the water quantities representing heat emission will be small in comparison with those required for draw-off. Thus, the pressure loss arising from circulation in these pipes is likely to be low and, in conventional design, the return pipes will be a size or two smaller than the equivalent flow pipes. Thus, emission from the returns may be taken either from assumed smaller sizes or, by way of an approximation, as being two-thirds of that of the associated flows.

If a gravity circulation appears practicable, reference would be made to data such as that listed in Table 9.3 (p. 244) and, for flow at storage temperature (65°C) and return at say 55°C, the available circulating head would be 50.25 Pa per m height. Alternatively, if a pumped circulation were necessary as a result of either system size or a preponderance of mains pipework below cylinder level, then a unit pressure drop of about 60 Pa per m of total travel may be taken as a first approximation. This figure includes a 25% allowance for bends and other resistances and takes account of the 'over-size' flow pipework.

Example

Using Figure 21.7 as an example, this being the same system as was shown in Figure 21.6 but with return pipework added. Each branch circulation serves a towel rail and a linen cupboard coil per floor, rated together at 0.5 kW, and a drying coil at 200 W is fitted in the kitchen (6.2 kW in all). The sizes of the outflow pipework have already been calculated and heat emission may thus be set out as shown in Table 21.10.

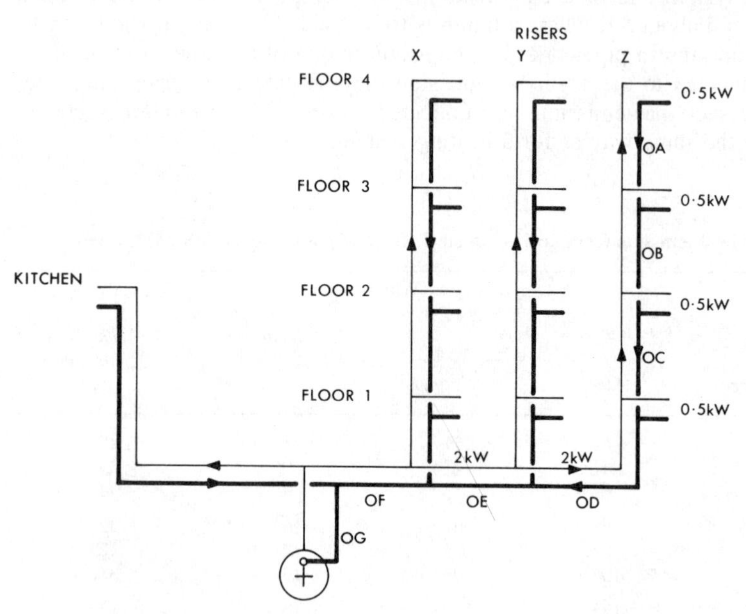

Figure 21.7 Pipe sizing example: secondary returns

Table 21.10 Heat emissions from piping: worked example

Section	Pipe size	Emission (W/m)	Length (m)	Emission (W) Sectional	Emission (W) Total
OA	28	9	3	27	
OB	35	10	3	30	
OC	42	12	3	36	
OD	42	12	2	24	117
Riser X					117
Riser Y					117
OD	42	12	3	36	
OE	42	12	3	36	
OF	54	13	3	39	
OG	67	16	3	48	159
					510
				Returns as ⅔ of flows	340
				Total (W)	850

For an extensive circulation system, it would be necessary to make a full calculation in order to apportion the mains emission, using the method described in Chapter 9. In the present case, however, the towel rails are uniformly disposed and the emission represents only 0.85/7.05 = 12% of the useful heat output. It would, in consequence, be reasonable to apportion the pipe losses pro rata.

By reference to Figure 21.8, an energy based chart for water flow in copper pipes, prepared on the same basis as Figure 9.2 (p. 229), the listings as columns 5, 6 and 7 of Table 21.11 are produced, those in column 7 being the proportion of the available pressure which is absorbed in the flow mains to each floor.

Assuming circulation to be by gravity, the available pressure for floor 1, which is the worst case in this context, would be 5 × 50.25 = 251 Pa and, by way of comparison, that for floors 2, 3 and 4 would be 402, 553 and 704 Pa respectively. From these pressures, the small proportion (Table 21.11) absorbed in the flow pipework must be deducted, e.g. the balance available at floor 1 is 251 – 45 = 206 Pa. Hence, the unit pressure available for the return pipework from floor 1 is 206/27 = 7.6 Pa/m run.

This unit availability may be used as for the flow piping and the pipe sizes noted in column 8 of Table 21.11 selected. As a result, again by reference to Figure 21.8, the listings in columns 9–11 in Table 21.11 are produced, those in column 11 being the proportion of the available pressure which is absorbed in the return mains to each floor.

Table 21.11 Pressure loss in secondary circulation: worked example

1	*2*	*3*	*4*	*5*	*6*	*7*	*8*	*9*	*10*	*11*
					Flow piping				*Return piping*	
				Pressure loss				*Pressure loss*		
Section	*Loading* (kW/K)	*Length* (m)	*Pipe size* (mm)	*Unit* (Pa/m)	*Section* (Pa)	*Total* (Pa)	*Pipe size* (mm)	*Unit* (Pa/m)	*Section* (Pa)	*Total* (Pa)
OG	0.69	5	54	1.9	9.5	9.5	42	6.5	32.5	32.5
OF	0.67	5	54	1.7	8.5	18.0	42	6.2	31.0	63.5
OE	0.45	5	42	3.2	16.0	34.0	35	7.5	37.5	101.0
OD	0.22	7	42	1.0	7.0	41.0	28	6.5	45.5	146.5
OC	0.17	4	35	1.5	6.0	47.0	22	13.7	54.8	201.3
OB	0.11	4	35	0.3	1.2	48.2	15	42.5	170.0	371.3
OA	0.06	4	28	0.7	2.8	51.0	15	15.0	60.0	431.3
Floor 1	0.06	5	28	0.7	3.5		15	15.0	75.0	
OG–OD	–	22	–	–	41.0	44.5	–	–	146.5	221.5
Floor 2	0.06	5	28	0.7	3.5		15	15.0	75.0	
OG–OC	–	26	–	–	47.0	50.5	–	–	201.3	276.3
Floor 3	0.06	5	28	0.7	3.5		15	15.0	75.0	
OG–OB	–	30	–	–	48.2	51.7	–	–	371.3	446.3
Floor 4	0.06	5	28	0.7	3.5		15	15.0	75.0	
OG–OA	–	34	–	–	51.0	54.5	–	–	431.3	506.3

Notes:
1. Pressure loss as read from Figure 21.8 (p. 592).
2. Equivalent lengths for pipe fittings are assessed on a per cent basis.

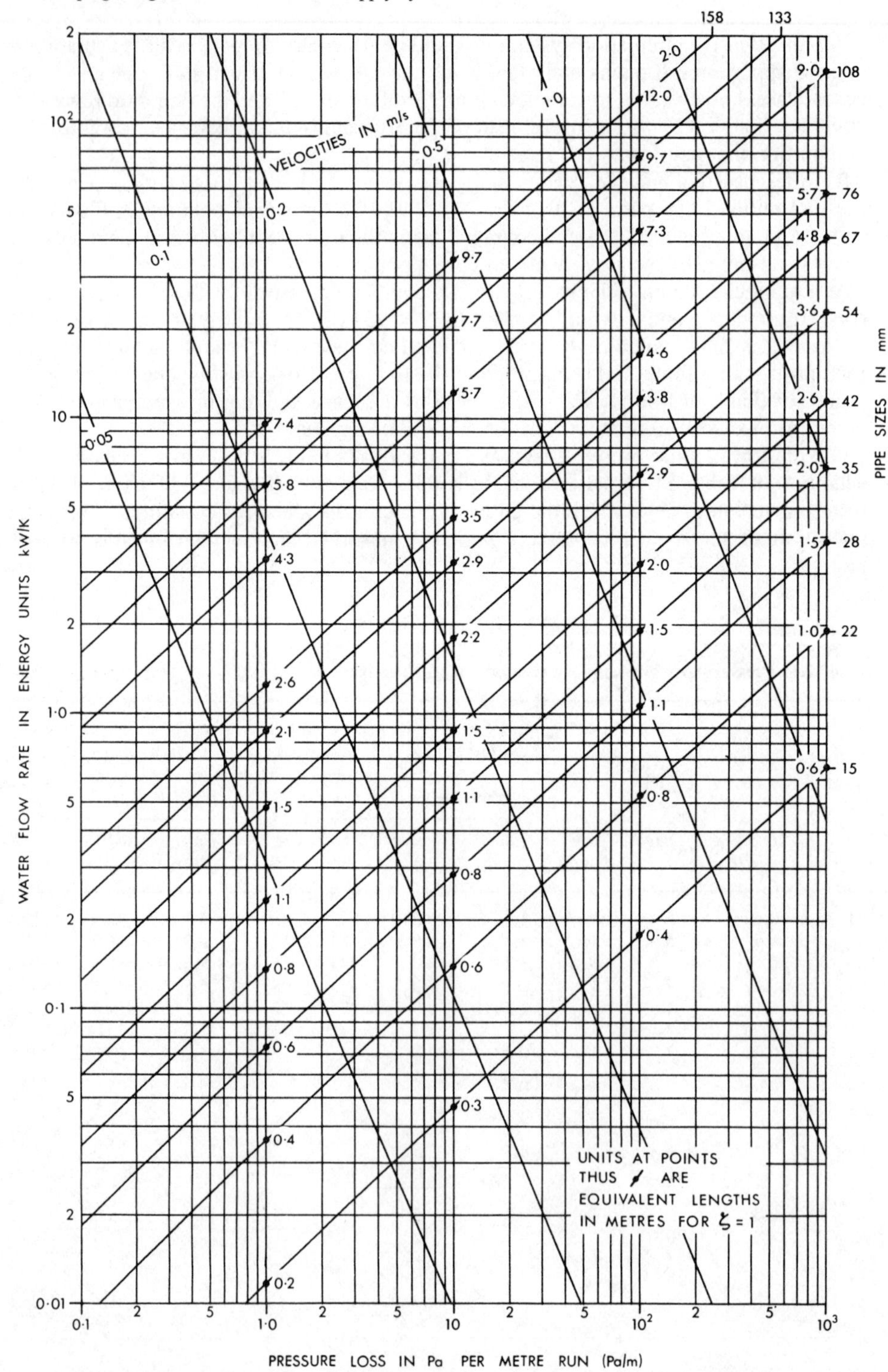

Figure 21.8 Sizing chart in energy units (kW/K) for water flow at 75°C in copper pipes (BS 2871: Table X)

It will be noted that an exact balance is not possible between available pressure (206 Pa) and pressure loss (222 Pa) for either floor 1 or any other floor. It is rarely practicable to achieve a precise match using commercial sizes of pipe but the consequences, in this application, are not very significant when of the magnitude illustrated. The result would be that the system came into hydraulic balance at a temperature difference varying by a degree or less from the arbitrary 10 K chosen.

If it had been decided for some reason that a pumped circulation could be provided to a system similar to that considered here, then a temperature drop of 5 K might have been chosen, thus doubling the water quantities previously calculated. The index circuit in this case would then be the most distant, that at floor 4 of riser Z, and the pressure loss through the flow pipes to that extremity would then be about 180 Pa. Assuming that the return pipes were one size smaller than those previously selected, with a minimum of 15 mm, the pressure loss through them would be 3650 Pa giving a total of 3.7 kPa.

It might be thought that the disparity between the pressure loss totals for flow and return is too great and that larger pipes should be chosen for the latter. However, a circulating pump rated to produce the required delivery of 0.33 litre/s against a pressure head of (say) 4 kPa represents the lowest end of the scale of performance offered by many manufacturers of small domestic heating pumps. This situation serves only to emphasise that the temperature differential across a pumped secondary circulation is in many cases an unsatisfactory design criterion.

Outflow via return pipework

It is, of course, fallacious to suppose that all the water flowing from a draw-off point has arrived there via the pipework designated as 'outflow'. In fact, as illustrated in Figure 21.9, when a tap is opened at X, water will flow to it from the cylinder through both the flow and the return piping. The water in the return pipe travels in the opposite direction to that which is usual, the pressure causing circulation being very much less than that provided for outflow by the height of the cistern. The proportion arriving at the tap from each direction will be dictated by the relative resistance to flow offered by the alternative routes.

Consideration of this situation has led to a belief, not yet proven for a sufficient number of designs, that, bearing in mind the small temperature difference across the secondary

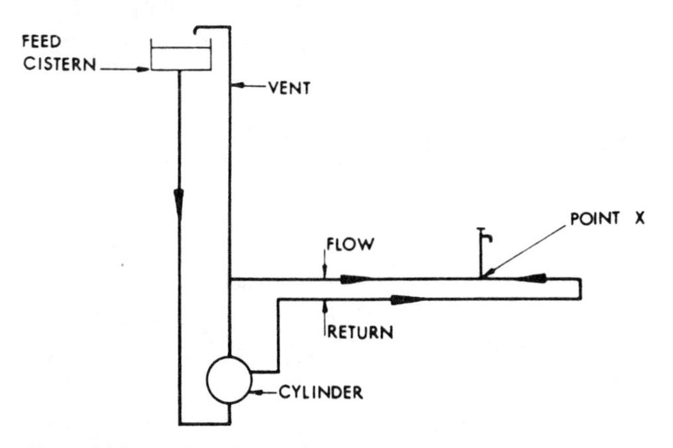

Figure 21.9 Direction of water flow towards an open tap

circuit when pumped, there may often be a good economic case for ceasing to think about 'flows and returns' as such. This approach requires that, during the design process, draw-off points are allocated alternately to one of two parallel circulating pipes and that 'demand units' are totalled accordingly. When this process is followed to a logical conclusion, the two pipes will carry loads which are approximately equal and will then be sized, section by section, to suit. The storage cylinder is fitted with two outflow connections, side by side at the top, and one pipe is connected to each. For circulation purposes, all pipework having been sized, it remains only to calculate pressure losses for the smaller water quantity required and to select a suitable pump.

System arrangements

In very many instances, the arrangement of a piping system is dictated by the configuration of the building rather than technical considerations. Furthermore, it is very often the case that hybrid arrangements suit the needs of particular areas in a single building. There is, in principle, nothing wrong with a mixture of arrangements provided that the design follows an easily discerned logic, that no hydraulic interactions take place, that problems related to excessive pressure at draw-off points are avoided and that facilities for air venting and drainage are considered as the design is developed. It is worth emphasis that the presence of air, and the necessity to dispose of it, are perennial problems with any hot water system. More practical difficulties arise in use from air venting, or the lack of it, than from almost any other cause.

Various specific system arrangements can be identified and these are illustrated in Figure 21.10, the two fundamental sub-categories being up-feed and down-feed, with and without high level storage. It should be noted here, however, that although the use of high level hot water stores may provide a solution to a variety of problems, access to them and maintenance of them may create other difficulties which must not be overlooked.

Up-feed systems

For the sake of easy comparison, item (a) in the figure shows a simple up-feed system as instanced also by the example considered earlier in this chapter. This is probably the most common arrangement but has a disadvantage in that the longest runs of pipe very often serve the draw-off points subject to the least pressure head. Air venting at the top of each riser is sometimes possible but otherwise reliance must be placed upon release of air through the highest tap of each riser which, in that case, must be above the end of the circulation, as shown.

In order to cater for periods when draw-off is heavy and supplies to the upper floors might tend to be reduced, an up-feed system with subsidiary high level storage at the top of each riser may be used, as shown in item (b) of the diagram. The assumption is made that a peak demand will not last for more than a minute or so and that the high level stores will not be emptied before the demand ends. The sizing of outflow pipework for such a system may be simplified by ignoring the top one or two floors and selecting sizes for the piping to serve the lower floors only.

Down-feed systems

The least satisfactory of such arrangements is as item (c) of the diagram, and arises from building arrangements which require that both the outflow and the return main are run not

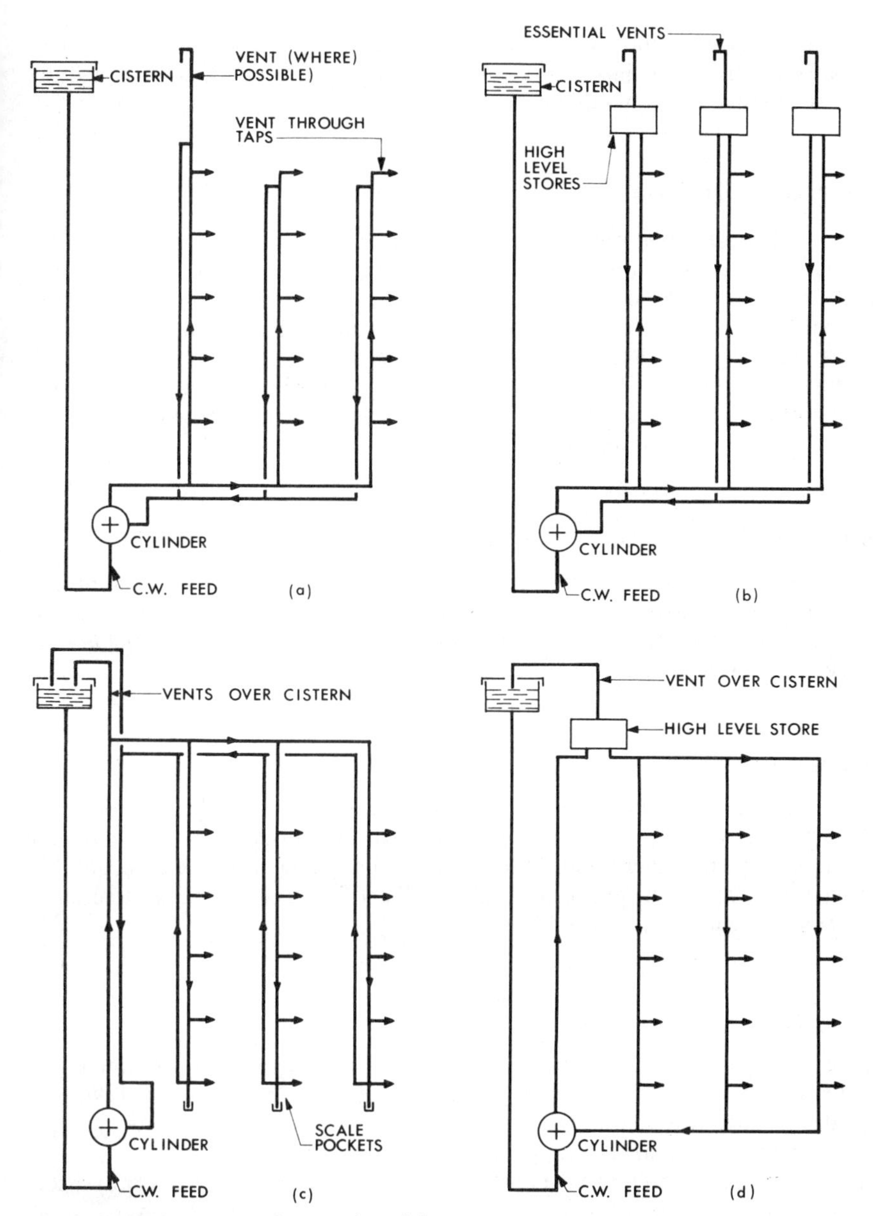

Figure 21.10 Arrangements for secondary piping systems

far below the cold water cistern. At times of heavy demand, the outflow main will be emptied of water, air will be admitted through the vent and a spluttering spasmodic supply will result. This configuration should be avoided except in circumstances where it is possible to fit the cistern very much higher than the circulating mains.

To overcome this problem in other circumstances, a method has been evolved whereby a single hot water store is introduced at the top of the rising outflow main, as shown in item (d) of the diagram. Here again, the assumption is made that supply to the upper floors derives from the high level store, where the effective travel is no more than the pipe length between that store and the points of draw-off. For the lower floors, it is asssumed that supply will be assisted by flow through the return piping.

Capacity of high level stores

This is necessarily based upon yet a further assumption, in this case by allowing for a 2–3 minute supply to those draw-off points which have been taken to be fed from the high level store. Thus, if a supply of 2 litre/s has been assumed to be fed downwards, the store capacity would be 360 litre. Since the high level stores are auxiliary to the main storage cylinder, the capacity of the latter should not be reduced.

It is the use of a series of such empirical assumptions, in parallel with more reasoned consideration of diversity of use, that has led to the reappraisal, previously mentioned, of the roles of the traditional 'flows and returns'.

Air venting of high level stores

It is essential that each vessel be provided with a separate open vent pipe taken up to a suitable height and it should be noted that an automatic air release valve is not an acceptable substitute. This provision is necessary to provide against collapse (*implosion*) of the vessel when water is drawn from it.

Tall buildings

Bearing in mind the comments made earlier regarding problems of excess pressure at draw-off points, the case of very tall buildings is worth consideration. The preferred design solution is, as for other building services, the introduction of intermediate plant rooms at intervals over the height of the building. Plant for the hot water system might thus be disposed as shown in Figure 21.11.

Unvented systems

Some of the comments made under previous headings will not apply to systems fed from a mains pressure service pipe and, similarly, some of the problems associated with the traditional design approach will no longer arise. These, however, are succeeded by other difficulties particular to the unvented system.

The most important side effect, in the context of piping design, is the necessity to ensure that the various safety devices fitted to the storage vessel will remain 'drowned' at all times, that is to say that they must always remain responsive to water temperature. It is, in consequence, necessary to take particular precautions to prevent a hot water cylinder from being drained either by leakage or by draw-off from a circulation below the level of the vessel.

The flow connection from the cylinder, being at the top, does not commonly present any difficulties and drainage via the cold water feed may be prevented by arranging for it to approach the cylinder at a level above the various sensors and fitting a vacuum relief valve. Thus, the residual problem rests with the secondary return connection and this may be solved by either of the two methods shown in Figure 21.12.

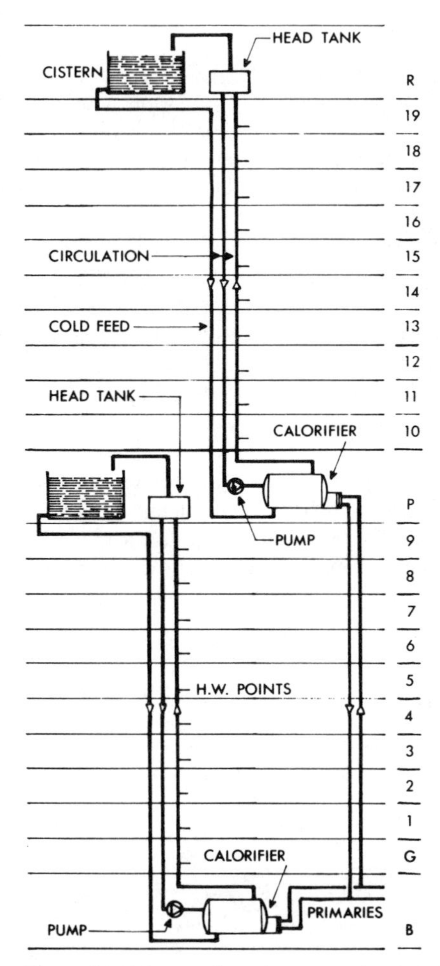

Figure 21.11 Intermediate plant rooms in tall buildings

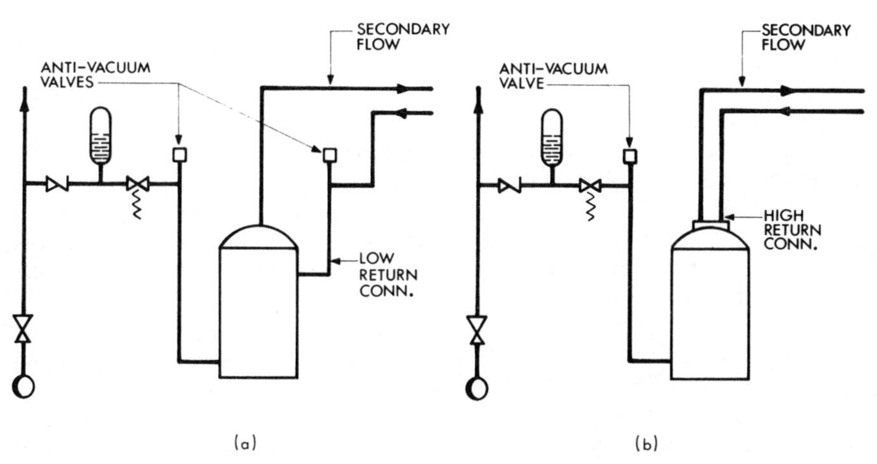

Figure 21.12 Unvented systems: piping arrangements to prevent drainage

Circulating pumps

In instances where a pump is required to assist in the secondary circulation it may be fitted, as was the case for a heating circuit pump, in either the flow or return pipework, Figure 21.13(a) and (b). There are a number of compelling reasons however why the former should be chosen since, with the pump so sited:

- Any pressure residual, small though it may be, is an addition to the cistern head at draw-off points and not a deduction from it.
- A non-return valve fitted in the circulating pipework with bypass connections to the pump, as noted later, does not inhibit a parallel outflow through the return pipe.
- The feed and expansion pipe, fitted as it must be to the storage vessel, is at pump suction and thus any prospect of air being drawn into vents at system high points is avoided.

Since the water quantity flowing to serve draw-off points is many times greater than that in circulation to counter heat loss, it is necessary to provide some means whereby a circulating pump will not inhibit outflow. This is achieved by fitting a non-return valve in the outflow pipe, with the pump connections taken to either side as shown in Figure 21.13.

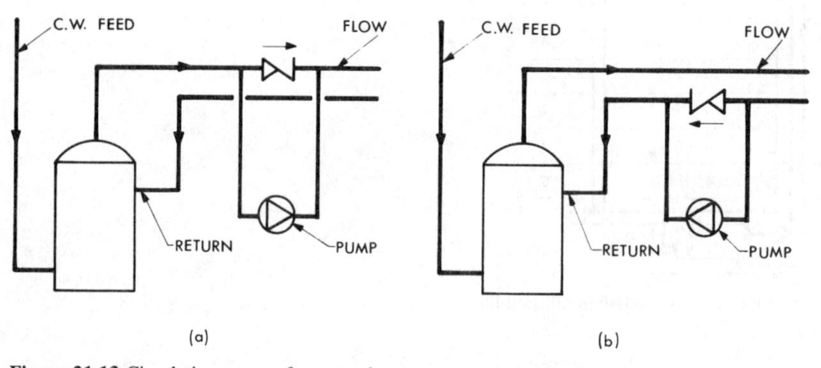

Figure 21.13 Circulating pumps for secondary systems

As in the case of heating systems, centrifugal-type pumps are best suited to the circulation of the comparatively small quantities of water involved. The pump body should be of a pattern which may be opened for cleaning and removal of any scale and should preferably be made of a copper alloy which is compatible with the pipe materials used in the system. Submerged rotor pumps are generally unsuitable for service in the secondary pipework of hot water systems.

Water hammer

This phenomenon results from the shock waves which occur when water flow is checked suddenly and has often, in the past, been associated with the use of percussion type basin

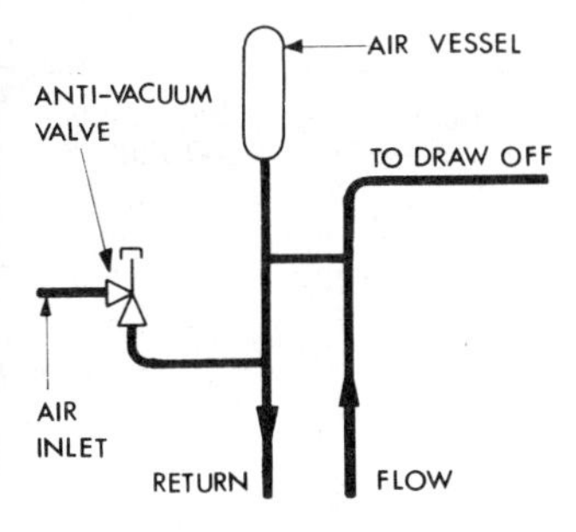

Figure 21.14 Unvented systems: prevention of water hammer

taps on pressure main cold water supplies. Unvented hot water systems are, in continental practice, seemingly subject to this problem and the introduction of a small air vessel at the top of each riser, as in Figure 21.14, an adaptation of the traditional cure, is likely to be effective.

Table 21.5 Pressure in kPa and head of water in m at 10°C

Head (m)	*0*	*0.1*	*0.2*	*0.3*	*0.4*	*0.5*	*0.6*	*0.7*	*0.8*	*0.9*
0	0	0.98	1.96	2.94	3.92	4.90	5.88	6.86	7.84	8.82
1	9.8	10.8	11.8	12.7	13.7	14.7	15.7	16.7	17.6	18.6
2	19.6	20.6	21.6	22.5	23.5	24.5	25.5	26.5	27.4	28.4
3	29.4	30.4	31.4	32.4	33.3	34.3	35.3	36.3	37.3	38.2
4	39.2	40.2	41.2	42.2	43.1	44.1	45.1	46.1	47.1	48.0
5	49.0	50.0	51.0	52.0	52.9	53.9	54.9	55.9	56.9	57.8
6	58.8	59.8	60.8	61.8	62.7	63.7	64.7	65.7	66.7	67.7
7	68.6	69.6	70.6	71.6	72.5	73.5	74.5	75.5	76.5	77.4
8	78.4	79.4	80.4	81.4	82.4	83.3	84.3	85.3	86.3	87.3
9	88.2	89.2	90.2	91.2	92.2	93.1	94.1	95.1	96.1	97.1

Table 21.7 Flow of water at 75°C in copper pipes (BS 2871: Table X)

Head loss (m of water per m run)	*Water flow* (litre/s) *in pipes of stated outside diameter* (mm)									*Pressure loss* (kPa per m run)
	15	*22*	*28*	*35*	*42*	*54*	*67*	*76*	*108*	
0.01	–	0.13	0.25	0.46	0.77	1.56	2.80	3.97	10.4	0.1
0.02	–	0.19	0.37	0.67	1.13	2.29	4.11	5.81	15.2	0.2
0.03	–	0.23	0.47	0.84	1.42	2.87	5.14	7.26	18.9	0.3
0.04	–	0.27	0.55	0.99	1.66	3.36	6.02	8.50	22.1	0.4
0.05	0.11	0.31	0.62	1.12	1.88	3.80	6.80	9.60	25.0	0.5
0.06	0.12	0.34	0.69	1.24	2.08	4.20	7.51	10.6	27.6	0.6
0.07	0.13	0.38	0.75	1.35	2.26	4.57	8.17	11.6	30.0	0.7
0.08	0.14	0.40	0.81	1.45	2.44	4.92	8.78	12.4	32.2	0.8
0.09	0.15	0.43	0.86	1.55	2.60	5.24	9.37	13.2	34.3	0.9
0.10	0.16	0.46	0.92	1.64	2.75	5.56	9.92	14.0	36.3	1.0
0.12	0.18	0.51	1.01	1.82	3.04	6.14	11.0	15.5	40.1	1.2
0.14	0.19	0.55	1.10	1.98	3.31	6.67	11.9	16.8	43.6	1.4
0.16	0.21	0.59	1.19	2.13	3.56	7.18	12.8	18.1	46.8	1.6
0.18	0.22	0.63	1.27	2.27	3.80	7.65	13.6	19.2	49.8	1.8
0.20	0.23	0.67	1.34	2.40	4.02	8.10	14.4	20.4	–	2.0
Equivalent lengths for $\zeta = 1.0$	0.6	1.0	1.5	2.0	2.5	3.6	4.7	5.6	8.9	Equivalent lengths for $\zeta = 1.0$

Notes:
Values of ζ for fittings: bends, tees, reducers, enlargements = 1.0.
Screw down valve or tap = 10.0; connections to cistern and cylinder = 1.5.

Table 21.8 Flow of water at 75°C in galvanised steel pipes (BS 1387: heavy)

Head loss (m of water per m run)	*Water flow* (litre/s) *in pipes of stated outside diameter* (mm)									*Pressure loss* (kPa per m run)
	15	*22*	*28*	*35*	*42*	*54*	*67*	*76*	*108*	
0.01	–	0.10	0.19	0.41	0.63	1.21	2.47	3.85	7.84	0.1
0.02	–	0.14	0.27	0.59	0.91	1.72	3.52	5.48	11.1	0.2
0.02	–	0.18	0.33	0.72	1.11	2.12	4.33	6.74	13.7	0.3
0.04	–	0.20	0.38	0.84	1.28	2.46	5.01	7.79	15.8	0.4
0.05	0.10	0.23	0.43	0.94	1.44	2.75	5.61	8.72	17.7	0.5
0.06	0.11	0.25	0.47	1.03	1.58	3.02	6.15	9.56	19.4	0.6
0.07	0.11	0.27	0.51	1.11	1.70	3.26	6.65	10.3	21.0	0.7
0.08	0.12	0.29	0.54	1.19	1.82	3.49	7.11	11.1	22.4	0.8
0.09	0.13	0.31	0.57	1.26	1.94	3.70	7.55	11.7	23.8	0.9
0.10	0.14	0.32	0.61	1.33	2.04	3.90	7.96	12.4	25.1	1.0
0.12	0.15	0.36	0.66	1.46	2.24	4.28	8.71	13.5	27.5	1.2
0.14	0.16	0.39	0.73	1.61	2.47	4.71	9.59	14.9	30.3	1.4
0.16	0.17	0.40	0.76	1.67	2.55	4.87	9.93	15.4	31.3	1.6
0.18	0.18	0.44	0.81	1.80	2.75	5.25	10.7	16.5	33.8	1.8
0.20	0.19	0.46	0.86	1.89	2.90	5.54	11.3	17.5	35.6	2.0
Equivalent lengths for $\zeta = 1.0$	0.4	0.6	0.8	1.1	1.4	1.9	2.7	3.3	4.7	Equivalent lengths for $\zeta = 1.0$

Notes:
Values of ζ for fittings: bends, tees, reducers, enlargements = 1.0.
Screw down valve or tap = 10.0; connections to cistern and cylinder = 1.5.

Chapter 22

Automatic controls and building management systems

Controls theory and practice is an extremely broad and complex subject. It is possible here to give only a general overview of the more important principles and to outline some of the more common practices. For further details, the reader is recommended to consult either a specialist text or a more comprehensive digest.*

The controls philosophy must be developed together with that of the systems design since, in the final installation, the two have to operate in harmony. However accurate the selection of equipment for a given application, if the controls are sub-standard as to design, installation, commissioning or maintenance then the performance of the overall system is unlikely to meet the design requirements.

The popular adage *keep it simple* applies to controls perhaps more particularly than to any other aspect of heating and air-conditioning. There is no benefit gained if the designer produces a system of controls which is beyond the comprehension of those personnel responsible for the operation and maintenance of the completed installations. By way of example, a full building management system might be operated as no more than an expensive time switch if that is the level of understanding of the operators: a control system must be designed for the user and not to satisfy the ambitions of the designer.

There has been a very considerable development in the application of controls since the introduction of computer technology. This has been particularly marked in the case of building and energy management systems, where significant benefits in the conduct of plant operation, maintenance and economy in running are available. The topic is dealt with towards the end of this present chapter.

The main objectives of a controls system may be summarised as:

- Safe plant operation.
- Protection to the building and system components.
- Maintenance of desired conditions.
- Economy in operation.

It is essentially the desire to achieve energy savings that may lead sometimes to a proliferation of controls; nevertheless, such an objective is fundamental to good practice and may include:

- Limiting plant operating periods.
- Economical control of space conditions.
- Efficient plant operation to match the load.
- Monitoring system performance.

The importance of economical use of energy is emphasised by some of the basic provisions incoporporated into recent legislation.

* 'Automatic controls and their implications for systems design'. *CIBSE Applications Manual*, 1985.

Building regulations

The current Building Regulations, 1985, (Part L), make it mandatory for the first time that space heating or hot water systems in buildings, other than dwellings and very small installations, are provided with automatic controls, such that:

- Space temperatures are controlled by thermostats.
- The temperature of hot water heating systems is varied according to the outside temperature (weather compensated).
- Systems are provided with a timeswitch (or optimum start control) to ensure that they operate only when the building is occupied.
- Multiple boiler plant is controlled in the most efficient manner.
- Hot water supply storage is thermostatically controlled, and on larger systems (over 150 litre and not heated by off-peak electricity) the heat supply is shut off when there is no demand.

Elementary components

The nature of heating and air-conditioning systems is such that, for the majority of the period of operation, plant and system capacity will exceed demand and the order of this excess varies with time: steady state conditions may be assumed never to occur. It follows therefore that, if the plant were to be uncontrolled, the conditions in the occupied space would be outside the desired range in consequence of which some means of control is a fundamental requirement.

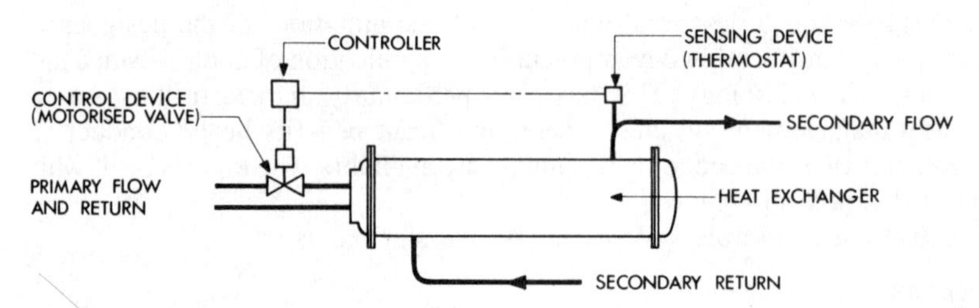

Figure 22.1 Simple heat exchanger controls

A simple control system, Figure 22.1, comprises a *sensing device*, to measure the variable, a *controller*, to compare the measured variable with the desired set-point and to send a signal to the *control device*, which in turn regulates the input. In such a system, a thermostat (sensing device) in the flow pipe from a heat exchanger measures the temperature of the water (controlled variable) and signals the information to the controller. The controller compares the flow temperature with the desired temperature (set point) and passes a signal to the control valve (control device) to open or close, thereby regulating the amount of heat introduced to the heat exchanger. This is an example of *closed loop* control, where feedback from the controlled variable is used to provide a control action to limit deviation from the set-point. An *open loop* system has no feedback from the controlled variable; an example of this is given later in connection with heating system controls.

System types

There are many classes of control system, but they may be grouped conveniently under headings:

- Direct-acting.
- Electric/electronic.
- Pneumatic.

The simplest form of controller is *direct-acting*, comprising a sensing element, say in a room or in the water flow and which, by liquid expansion or vapour pressure through a capillary, transmits power to a bellows or diaphragm operating a valve spindle. The most common example is the thermostatic radiator valve which, when installed in the supply pipe to a radiator, convector or other heat emitter (as in Figure 22.2), serves the purpose of an individual room controller.

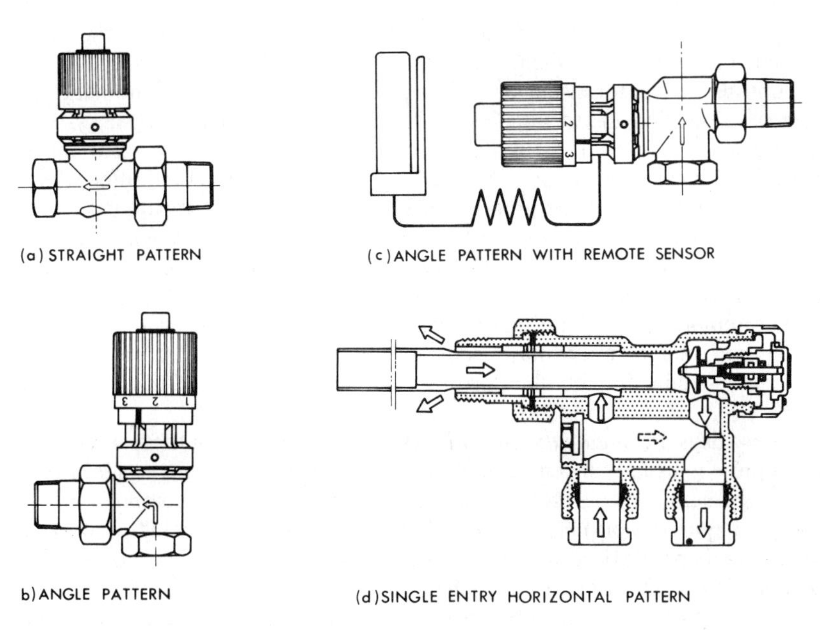

Figure 22.2 Thermostatic control valves for radiators, etc.

Direct-acting thermostats have little power and their control band can be somewhat wide, although they have been considerably improved in recent years. Direct-acting thermostatic equipment gives gradual movement of the controlling device and thus may be said to *modulate*.

The most common control system is the *electric/electronic*, which may be found in domestic applications (thermostat and motorised valve) through to large commercial and industrial installations. The basic functions in an electric system are switching and resistance variation. Switching may be achieved by closing metallic contacts, using a tilting mercury switch, or a relay (electro-magnetic switch) which is an electrical holding device that uses the magnetic effect produced by an energised/de-energised coil. To produce a variable signal, a moving arm, in a sensing device such as a thermostat, travels across a potentiometer in a 'bridge' circuit, which would be used, for example, to provide movement of a modulating control valve or damper. The simplest control device is the solenoid valve consisting of a coil within which an iron 'plunger' slides to give linear movement.

Electronic controls operate at 24 volt or less and use smaller strength signals from the sensing elements; typically thermocouples or resistance thermometers which have no moving parts. The circuitry incorporates amplifiers to magnify the signal but this requirement is reduced with thermistors because these give a greater range of resistance for a given temperature change. Extensive use is made of solid state components and printed circuit boards. Electronic systems provide accurate control and, being free from mechanical parts, are now very reliable. Where computer software logic is built into the controller the system may be described as *direct digital control (DDC)*.

For large installations, *electric/electronic* systems of control are by far the most common, and may be applied to perform any number of desired functions in a great variety of ways. The simplest electrical system comprises an on–off thermostat connected to an electrically operated valve or damper such that, when the thermostat calls for heat, the valve or damper is opened and, likewise when satisfied, the valve or damper is closed.

Modulating controls enable, for instance, a mixing valve to be so adjusted automatically that it finds some mid-position and so supplies the desired flow-temperature to the system. For example, for a quick heat-up in the morning, the control system may be so arranged as to call for full heat for the first hour or so according to external weather, after which the water temperature will be reduced proportionately to the external temperature.

The wiring to electrical controls and to the circuit boards for electronic controls, together with the associated switches, indicator lights and displays can become complex and it is usual to centralise these components in a pre-wired control panel. This panel may also accommodate remote temperature and humidity indicators and recorders and other instruments if required; motor starters may also be installed into such panels. It is essential that all components within and on the face of control panels are permanently and accurately labelled.

Less popular these days are *pneumatic controls*. The compressed air for these, if derived from a central supply, is usually taken through a reducing valve at a gauge pressure of about 100 kPa. If the supply were provided independently, a small air compressor and storage cylinder is required, normally duplicated, automatically maintaining a constant gauge pressure of perhaps 400 kPa in order to give a good storage of air for sudden demands and supplying again through a reducing valve.

Compressed air for pneumatic controls must be provided from oil-free compressors, preferably duplicated, and air coolers and driers, also in duplicate, must be fitted at the discharge to a storage vessel and pressure reducing valve. Piping must be run with great care in order to provide adequate drainage of the water which will condense out of the air: the operation of the control system would be upset by any water or oil carry-over and traps must be provided at all low points. In consequence of these necessary precautions, it will be understood that it is very unlikely that a supply from an industrial compressed air plant will be suitable for service to a control system. Where plant and control devices are distributed around a site, a central compressor plant may be used, providing a high pressure air supply to an air storage vessel and pressure reducing valve at each plant room location.

Air is supplied to each sensing device and by the action of some form of mechanical movement from it, variable amounts of controlled air bleed away thereby changing the air pressure in the supply line which is the signal transmitted to the control device. Pneumatic systems would in modern practice be considered only for industrial plant or larger air-conditioning installations incorporating multiple air handling plants serving high velocity systems, such as induction units, variable air volume or dual duct, and then only in the rare circumstances when a high quality of maintenance can be guaranteed.

A combination of electrical and pneumatic controls may be useful where room control is required. Pneumatic room thermostats are sometimes cumbersome but may be replaced by electrical instruments which in turn control the air supply to pneumatic valves, etc., as before. This special application will be apparent, and need not be discussed further.

Other forms of control system operation such as hydraulic, using water or oil, are not used today in building services work.

Sensing devices

Siting of sensing elements is critical to the achievement of good control. In pipework or ductwork, sensors must be so arranged that the active part of the device is immersed fully in the fluid and that the position senses the average conditions. Where necessary, averaging devices serpentined across the full cross-section of a duct should be used. Sensors should either be protected from the radiant effects of local heat exchangers, such as heater batteries in an air handling unit, by polished shields left open or perforated away from the radiation, or they may be positioned in a bleed-off duct.

In sensing space conditions, the device must not be in direct solar radiation or be located on a surface not representative of the space conditions such as on a poorly insulated outside wall. Local effects from heat sources, radiators or office equipment for example, will also give unsatisfactory results. Where necessary aspirated sensors should be used to improve sensitivity to *air* temperature.

Temperature

Thermal expansion of metal or gas or a change in electrical characteristics due to temperature variation are the common methods of detection. Figure 22.3(a) shows a simple bi-metallic type thermostat having closing point contacts; a two-wire type would, for example, open or close a circuit to stop or start a motor; a three-wire type would, typically, open one circuit and close another to start a motor or operate a valve in reverse direction. The sealed bellows type, Figure 22.3(b), is filled with a gas, vapour or liquid,

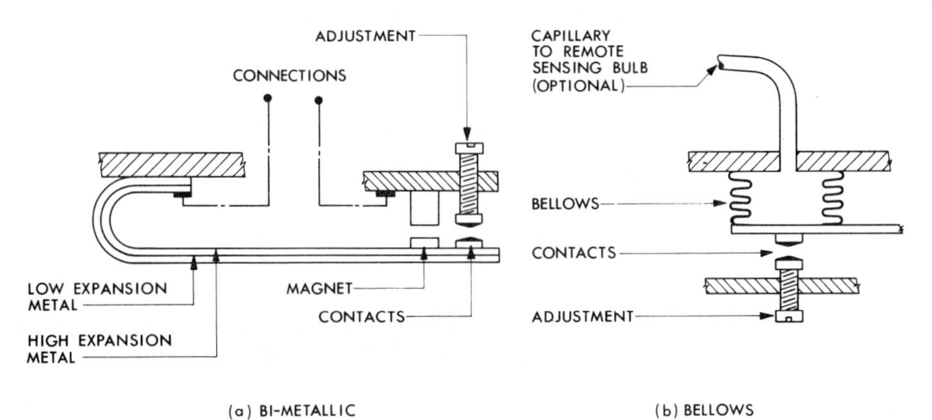

Figure 22.3 Electrical bi-metallic and sealed bellows thermostats

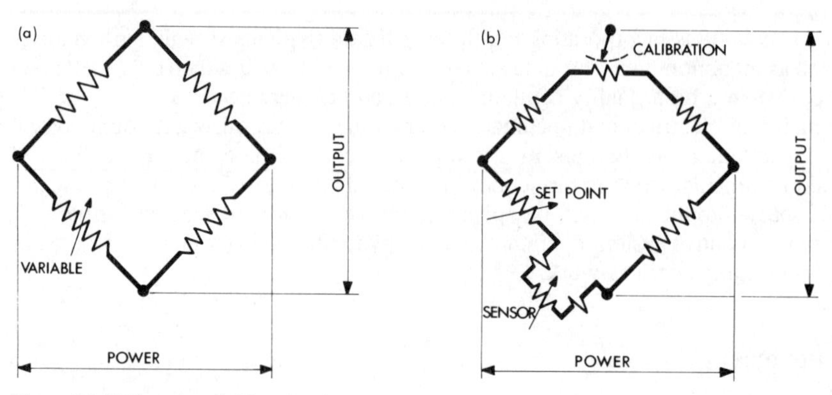

Figure 22.4 Wheatstone bridge circuits

which responds to change in temperature by variation in volume and pressure causing expansion or contraction. Remote sensing elements operate on this principle, the remote sensing bulb being connected to the bellows by capillary tube.

Electronic sensing elements have no moving parts. The resistance bulb type, normally a coil of nickel, copper or platinum wire around a core, produces a variation in electrical resistance with change in temperature. *Thermistors*, which are semi-conductor devices, also produce a change in resistance, but inversely with respect to temperature change such that resistance decreases with increase in temperature; the non-linear output may be corrected using linearising resistors in the circuit. *Thermo-couples* comprise two dissimilar metal wires joined at one end; a voltage proportional to the temperature difference between the junction and the free ends results.

Development of the simple *Wheatstone bridge* circuit, Figure 22.4(a), is used in electronic controls. When the resistances are in balance there is zero output but as one or more of the resistances changes, the circuit becomes unbalanced which results in an output signal proportional to the change. A bridge circuit showing calibration and set-point adjustment is shown in Figure 22.4(b).

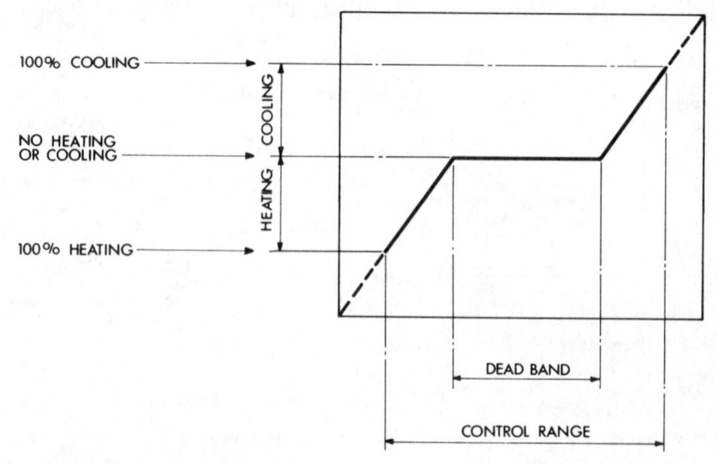

Figure 22.5 Dead-band thermostat operation

Dead-band thermostats have been developed to reduce energy use, on the principle that for comfort applications a variation in temperature, of perhaps 3–4°C, is acceptable. This type of thermostat has a wide dead-band through which a change in temperature produces no change in the output. Figure 22.5 shows an example of such an application.

Humidity

Fabrics which change dimension with humidity variation, such as hair, nylon or wood, are still in use as measuring elements but are unreliable and require considerable maintenance: hygroscopic plastic tape is now more common. These media may be used to open or close contacts or to operate a potentiometer, as in the case of a thermostat. An illustration of the principle is given in Figure 22.6.

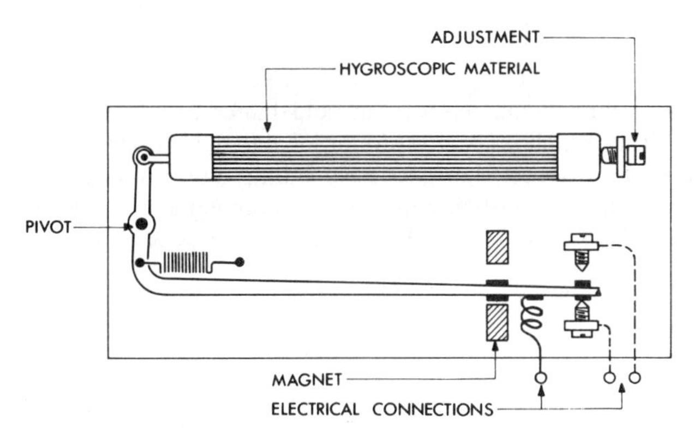

Figure 22.6 Electrical humidistat

For electronic applications, use is made of a hygroscopic salt such as *lithium chloride*, which will provide a change in resistance depending upon the amount of moisture absorbed. These are relatively cheap but are slow to respond to change and are easily damaged. More robust but considerably more expensive are the solid state sensors which use polymer film elements to produce variations in resistance or capacitance.

Wet and dry bulb thermostats working on a differential principle were once popular but are no longer used.

Pressure

Bellows, diaphragms and Bourdon tubes are typical of the sensing devices used and, of these, bellows and diaphragms acting against a spring are the most common. Such equipment can be sensitive to small changes in pressure, typically 10 Pa. The pressure sensing motion may then be transmitted directly to an electric or pneumatic control device.

In electronic systems, the diaphragm or the sensing element is connected to a solid state device having the characteristic that, when distorted, its resistivity changes; this is known as the *piezo-electric* effect.

Flow

There are many methods used to detect fluid rate. In water systems the most common is to detect pressure difference across a restriction to flow, such as an orifice plate or a venturi and extensive use is made of calibrated valves.

Various devices are available to sense air velocity in ductwork. In larger cross-sections, where the velocity may vary across the duct area, an array of sensing devices is required to establish an average value.

Paddle blade switches are used in water circuits, normally as a safety feature. For example, a flow switch of this type would normally be incorporated in the chilled water circuit connected to the evaporator of a refrigeration plant and be arranged in such a manner that the plant would not start until the switch sensed that water flow was established. By this means, damage to the equipment would be avoided consequent upon freezing of the static contents of the evaporator.

Enthalpy

Normally only used in air-conditioning plant, temperature and humidity sensors feed signals to a controller, the output from which is a signal proportional to the enthalpy of the air. Control devices are available which accept signals corresponding to the enthalpies of two air streams, typically outside air and exhaust air and, depending upon the relationship between these two values, a control action on dampers or heat exchangers is initiated.

Control devices

The most common components used in the field are control valves, for steam and water, and control dampers for air systems. The selection and sizing requires an understanding of both the devices and of the system characteristics. The system to be controlled and the associated flow rates would be sized at the peak design load, but would operate for most of the time at some partial load. The control device, therefore, has to provide stable control over the full range of operating conditions.

The movement of a valve or damper is determined by an actuator which is the component that responds to the signal from the controller. The actuator characteristics which are of importance are *torque* (the ability to cause movement of the control device) and *stroke period* which is the period of movement between the limiting positions (open to closed and vice versa). Selection of the actuator type will depend upon the choice of control system.

Electric motors

For control mechanisms, these range between 5–50 VA and operate via reduction gearing to give a high torque/low speed characteristic: they normally operate using a single phase supply at 240 V or lower to suit the system. Two-position motors are of the unidirectional spring-return, or unidirectional three-wire signal type, and are used when the speed of movement produced by a solenoid is too fast.

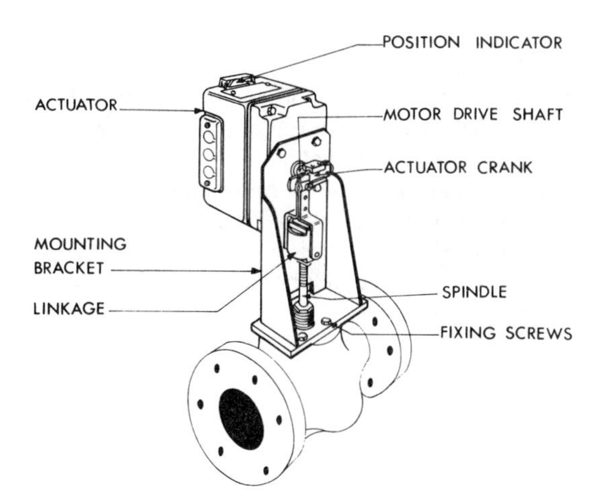

Figure 22.7 Electrical control valve

Modulating control requires a reversible motor that can be held at any position through the movement: either reversible induction or shaded-pole motors may be used. A typical arrangement driving a control valve is illustrated in Figure 22.7. The motor may be provided with cam operated auxiliary switches to open or close at any set position of the movement. An auxiliary potentiometer may also be driven to provide a signal to another control device.

For the operation of louvres or dampers, the power required may be considerable and the motor must be chosen to suit. Sometimes the louvres will require to be sectionalised, each section being worked by one motor with linkage. Again, such damper motors may be on/off, i.e. open/closed, or they may modulate to give settings in any intermediate position.

Solenoids

These devices operate on the electromagnetic principle with the armature directly connected to the control device, normally two-position and suitable for small sizes (Figure 22.8). Modulating control may be achieved by applying a variable voltage to the coil, the spindle movement being proportional to the supply voltage.

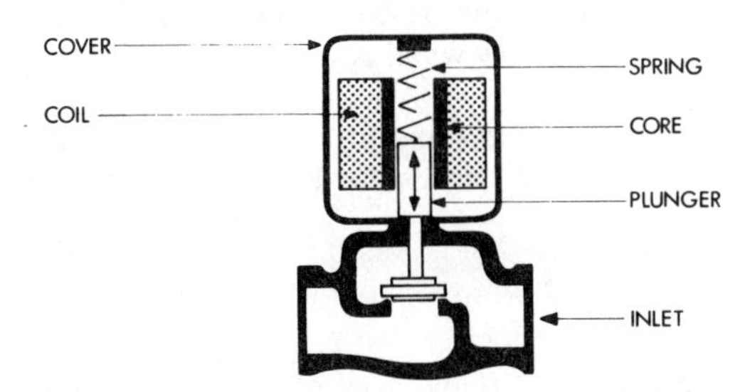

Figure 22.8 Solenoid valve

Pneumatic actuators

Pneumatic valves (Figure 22.9) operate by means of a diaphragm or copper bellows connected to the valve spindle. The air supply connects to the top of the chamber (in the direct acting type) and has a small orifice plate, or needle valve, allowing only a small quantity of air to pass. The diaphragm top also connects to the thermostat or other sensing device and when the pilot valve in the latter shuts, the air pressure builds up on top of the diaphragm and depresses the valve spindle to close or (in the reverse-acting type) to open the valve. When the pilot valve opens, the pressure on the diaphragm is released and the spring around the valve spindle forces the valve up again.

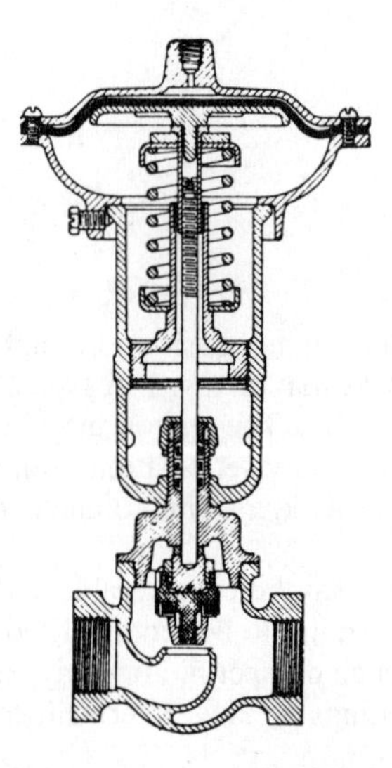

Figure 22.9 Pneumatic diaphragm valve

The air from the pilot valve discharges to atmosphere. Any movement of the pilot valve causes the main valve to find a similar intermediate position, so giving a 'floating' control over the complete range. A pneumatic three-port valve operates in the same way.

Each valve should have a control cock and pressure gauge. These are usually centralised on a board along with the thermometers and other instruments, so that the operator may see the condition at a glance. The pressure gauges are connected to the diaphragm top in each case, so giving an indication as to whether the valve is open, shut, or in a mid-position. A main pressure gauge on the air supply will show whether or not the compressor is functioning.

Pneumatic damper motors, Figure 22.10, work on the same principle as for valves, except that the diaphragm is larger in order to give the necessary force to operate the damper.

Valves

Selection of the correct type of valve to provide for stable control requires detailed knowledge of the application and is specified in terms of the *valve characteristic* and

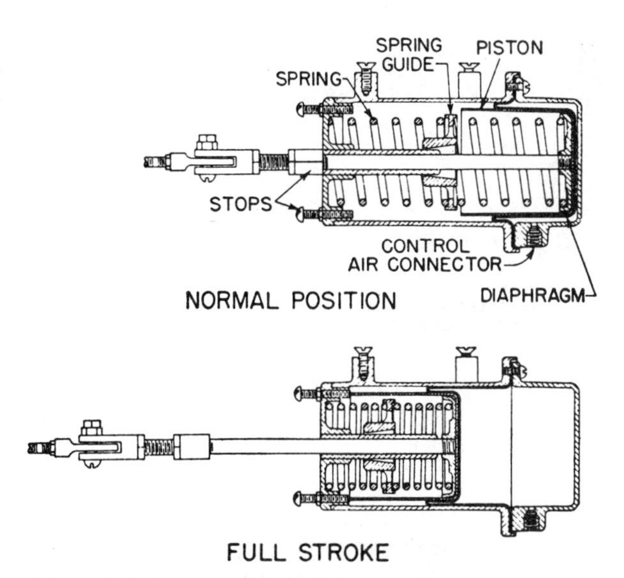

Figure 22.10 Pneumatic damper operator

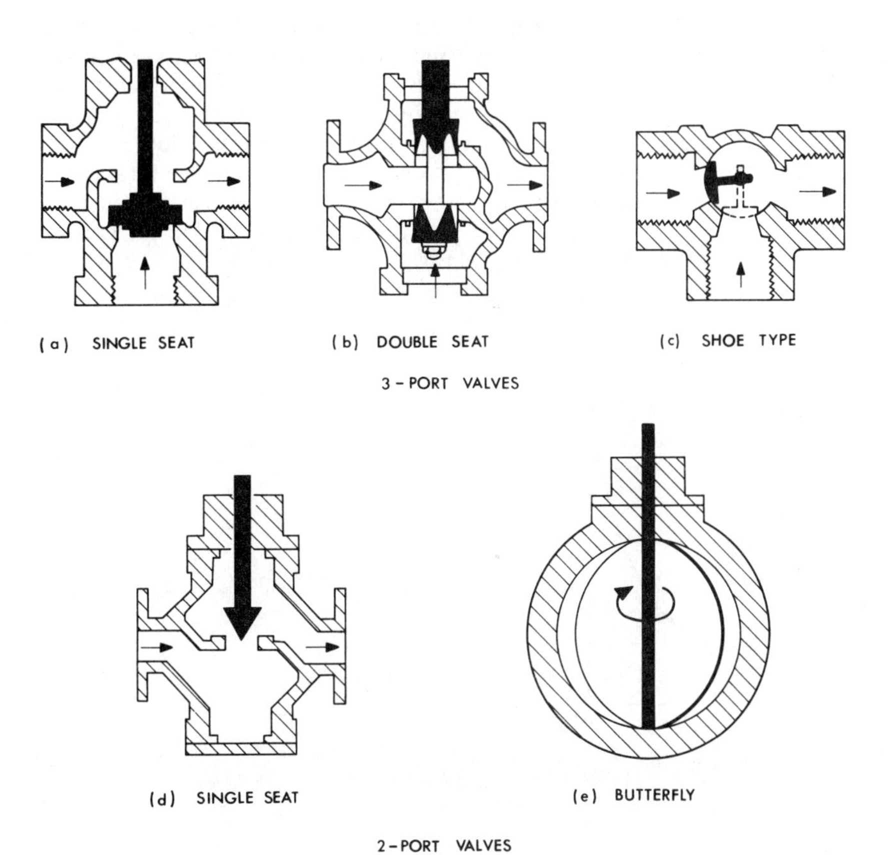

Figure 22.11 Typical valve types

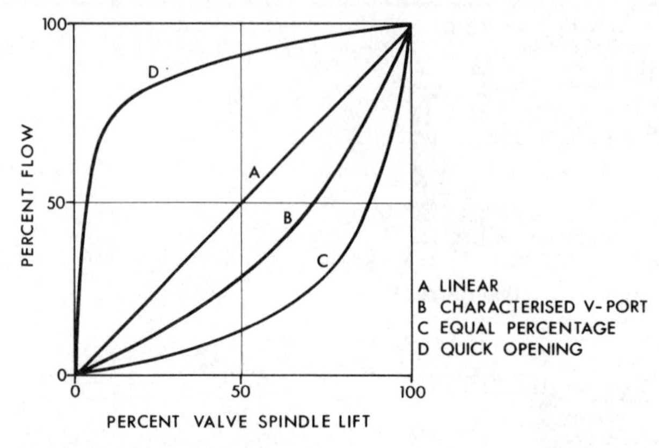

Figure 22.12 Control valve characteristics

authority. Valve characteristic, determined by the design of the plug, (Figure 22.11*), is a function of valve lift and flow rate; the four main characteristics are shown in Figure 22.12.

The choice of valve characteristic depends upon the output characteristic of the controlled equipment. For steam applications, since heat output is effectively directly proportional to flow rate, a *linear* characteristic valve would normally be suitable. Heat exchangers having water as the primary medium, however, have very different characteristics. Figure 22.13 shows heat output to primary heating water flow rate for a typical air heater battery. An *equal percentage* valve would be most suitable for this type of application, producing a closer relationship between valve lift and heat output.

Valve authority is a function of the pressure drop across the valve and of the pressure drop across the remainder of the circuit as given by

$$N = P_1/(P_1 + P_2)$$

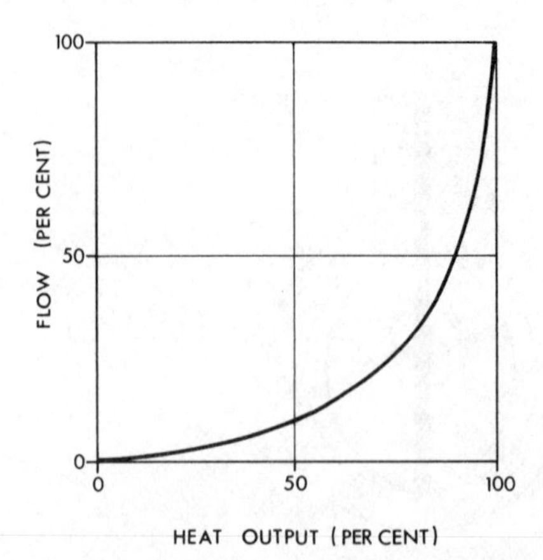

Figure 22.13 Typical hot water air heater battery characteristic

* Reproduced from CIBSE *Applications Manual for Automatic Controls*.

where

N = valve authority
P_1 = pressure drop across the open valve
P_2 = pressure drop across the remainder of the circuit.

The most common applications are represented in Figure 22.14. For stable control, the valve authority should be not less than the following values.

Mixing = 0.3
Diverting = 0.5
Throttling = 0.5

Application of the above criteria would normally lead to control valves being selected at one, or maybe two, pipe sizes smaller than the pipeline diameter. Three-port valves should always be fitted in the *mixing* mode, that is with two inlets and one outlet. As illustrated in Figure 22.14, such valves may be applied to either a mixing or a diverting duty but many of the patterns available are not suitable for use in the *diverting* mode, that is with one inlet and two outlets.

Selection of two-port valves follows similar criteria as for the three-port variety. However, more detailed consideration is necessary in connection with the remainder of the circuit, because as the valves close the flow rate and pressure distribution around the

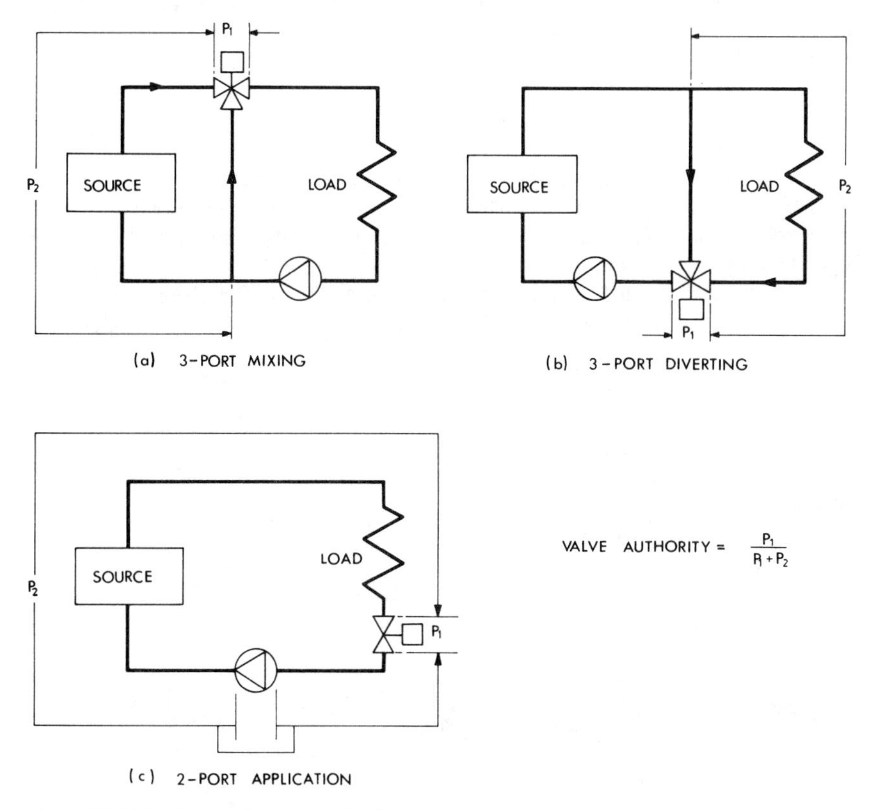

Figure 22.14 Basic control valve applications

circuit will vary: this is particularly important since the installation is likely to have many two-port valves in the same pumped circuit. As the control valves throttle the flow through the controlled circuits the pump, if operated at constant speed, would tend to develop an increased pressure which in turn would affect the authority of the valves. Added to which, this increase in pressure might affect the ability of the valves to open and it may be necessary, therefore, to use variable flow pumping with a system controlled by two-port valves.

Steam control is a special application of the two-port valve. As the valve closes, the area under the seat decreases causing an increase in steam velocity, which tends to counter the decrease in volume flow. The limiting condition, for practical purposes, is when the downstream absolute pressure is 60% of the upstream pressure. When sizing a steam valve, therefore, the pressure drop at full load is taken as 40% of the upstream absolute pressure, except when this pressure is less than 100 kPa, when impractical conditions arise from this rule of thumb and a smaller pressure drop is taken, with resulting loss in effective control; reference to manufacturers' data is necessary in these circumstances. The heat exchanger surface area must, in consequence, be suitable for a steam supply lower than the mains pressure.

Butterfly valves are, in effect, two-port control devices, but have poor control characteristics equivalent to that illustrated as *quick opening* in Figure 22.12. Application is limited to two-position (open/closed) but standard equipment cannot be guaranteed to give total shut-off: special valves with liners are available for this purpose.

Noise emanating from control valves should not be overlooked in the selection of components and in the design of the configuration of connecting pipework. The cause may be either unique or a combination of mechanical vibration, turbulence or cavitation.

Dampers

Control of air flow follows the same principles as for water systems, and the function may be either modulating or two (or more) – position. There are two forms of control damper, parallel and opposed blade type, and their inherent characteristics are shown in Figure 22.15. Dependent upon the damper authority (defined above for valves) the damper characteristic will vary when installed in a system (Figure 22.16). The closest to linearity is obtained where, for opposed blade dampers, the authority is 0.05 and, for parallel blade

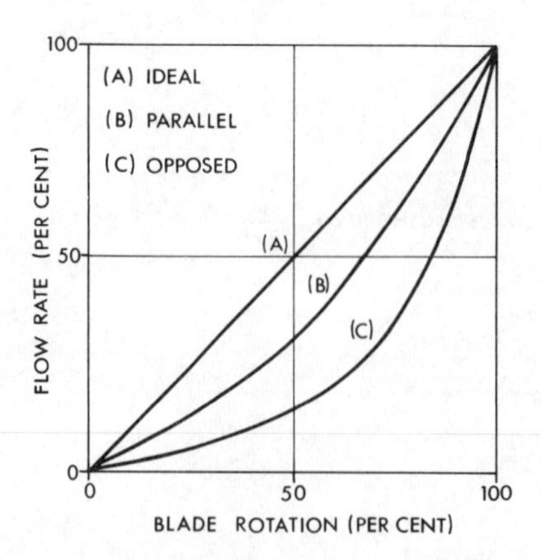

Figure 22.15 Control damper characteristics

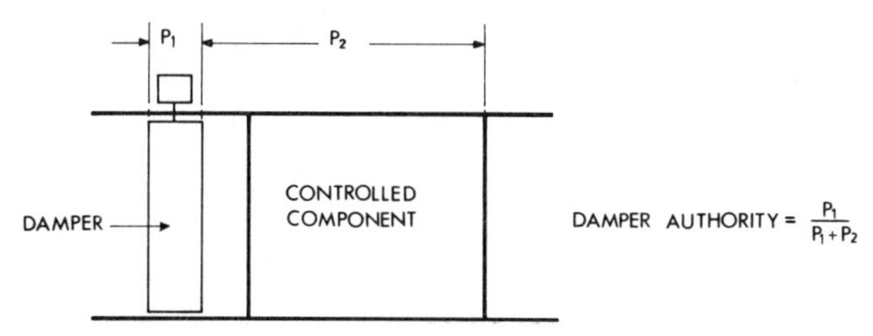

Figure 22.16 Basic control damper application

type, 0.2. To limit the pressure drop in a system, therefore, opposed blade dampers, sized for an authority of about 0.05, would normally be selected.

When used for controlling outside/exhaust/recirculating air in an air handling plant (Figure 22.17) the outside air damper (A), and the exhaust damper (B), would normally be opposed blade type, dimensioned at the duct size. However, for abnormal applications where, for example, high resistance plant components are to be installed, a check should be made from first principles. The function of the recirculation damper (C) is to impose a resistance in the bypass duct equivalent to the static pressure difference between the supply and extract ducts and to have a characteristic to complement that of the outside air damper in maintaining constant the mixed flow to the fan. This damper will of necessity have a high authority, up to 100%, and since the characteristic of the parallel blade type is better at higher authority (pressure drop in this case having no influence on running costs) the parallel blade type would be preferred. Where the pressure to be absorbed by the damper is high, due to the operating conditions of the plant, an additional resistance, such as a perforated plate, may be incorporated in the bypass to reduce the authority of the damper.

Another common application of control dampers is to modulate the rate of flow through a cooling coil to provide what is termed *face and bypass* control. The principle is

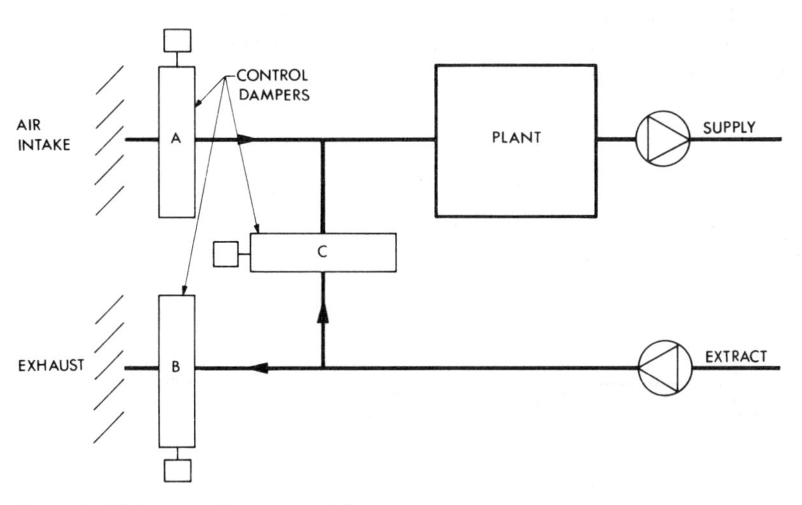

Figure 22.17 Outside/exhaust/recirculating air control

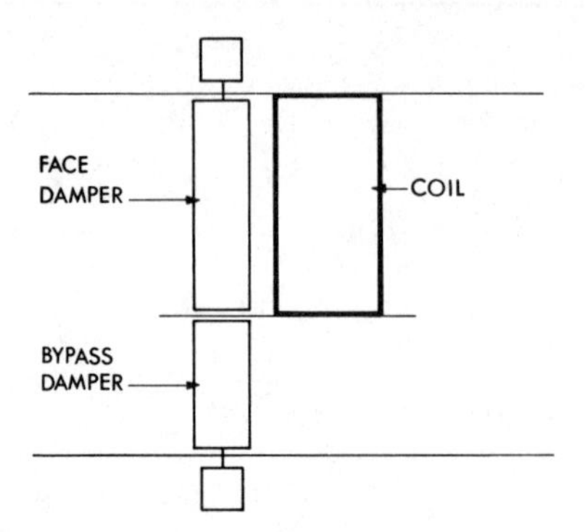

Figure 22.18 Face and bypass control

illustrated in Figure 22.18. In sizing the dampers, the same logic as above may be applied, to provide constant pressure drop across the combination throughout the range of operation and hence not affect the flow rate in the system. As for outside/recirculating air dampers, the characteristics of the face and bypass dampers must complement one another. It would normally be the case that the face damper would be the same size as the coil, but the bypass damper may have to be smaller than the associated duct.

Controller modes of operation

There are various ways in which a controller can cause a control device to operate in response to a signal from a sensing device. The most common modes that may apply are described here.

Two-position control

A typical application is on/off switching, in which the sensing device, perhaps a simple switching thermostat, provides two signals, for example opening contacts below a set-point and closing them above a set-point. The device may be arranged to operate in the opposite way and is then called reverse acting.

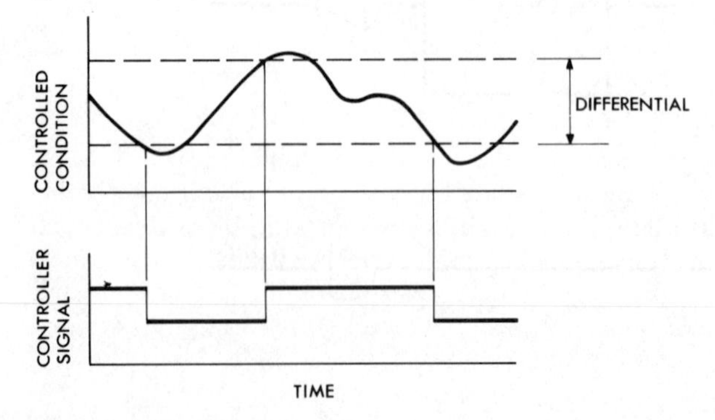

Figure 22.19 Two-position control

The interval between the switching actions, an inherent characteristic of the device, is normally referred to as the differential gap. Where a differential gap is too wide, an accelerator (in the case of a thermostat, a heating element energised when the thermostat calls for heat thereby anticipating the response) may be used to reduce the range, or swing, of the controlled condition. The swing between the controlled space temperature may be greater than the differential gap in the sensing device due to slow response of the controls and the thermal inertia inherent in the building. Figure 22.19 illustrates the performance of such a method of control.

On/off control would give quite acceptable results where the controlled variable has large thermal inertia, such as a hot water service storage calorifier or a space heated by a mainly radiant source, where close control is not critical.

Step-control

It is sometimes necessary to operate a series of switching operations in sequence from one sensing device. For example, when multiple refrigeration compressors have to be started in turn, with increasing cooling load as sensed by a change in chilled water return temperature, or when air heating is accomplished by an electrical heater with multiple elements switched in steps. See Figure 22.20.

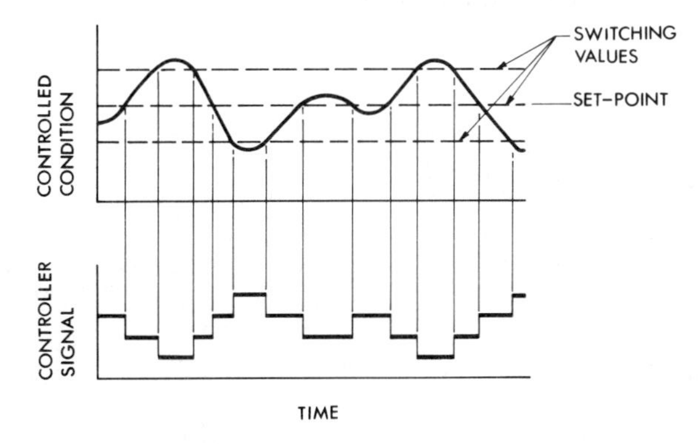

Figure 22.20 Step control

Proportional control

With this form of control the output signal from the controller is proportional to the input signal from the sensor. Figure 22.21 illustrates proportional control action from a start condition and shows that, initially, the controlled variable will be driven towards the set-point at such a rate as to cause overshoot. On the return cycle, the overshoot will be less and this oscillation will continue until stable conditions exist; but if the system is unstable the hunting will continue indefinitely. With proportional control there will nearly always be an off-set from the desired set point. Proportional band, the deviation in the controlled variable necessary to produce the full range of control action, may normally be varied on the controller.

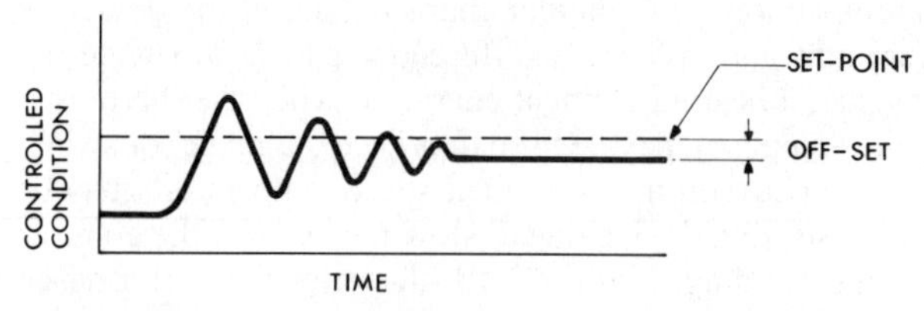

Figure 22.21 Proportional control

Proportional control is to be preferred to an on/off approach where the thermal capacity of the controlled variable is low and where the primary rate of heat exchange is fast, as in the case of air heating and cooling batteries and non-storage water heat exchangers.

Floating control

With floating control, there is normally a neutral zone around the set-point, within which no control action occurs: the control device remains in the last controlled position. When the variable moves outside the neutral zone, a signal causes the actuator to move at a constant rate in a direction corresponding to whether the variable is above or below the set-point, see Figure 22.22. An example would be a control device motor running at a

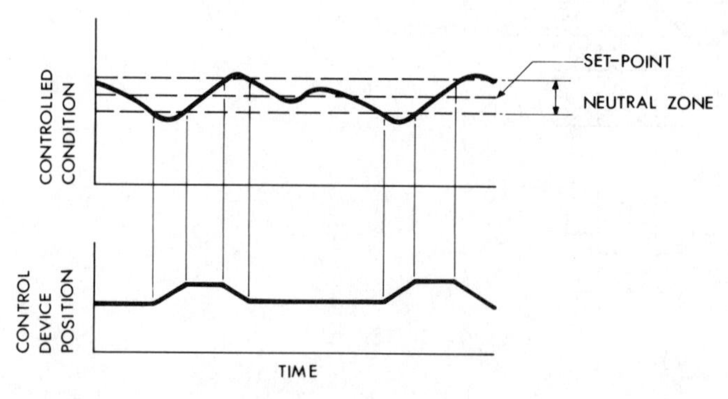

Figure 22.22 Floating control

constant but slow speed; the full stroke being at least two minutes. A multi-step controller might also be used to initiate a multi-speed actuator action.

Other forms of modulating control avoid off-set and may be preferred to proportional or floating control in some applications.

Integral control

Seldom used alone, this is an important addition to other forms of control, particularly to the proportional mode. With integral action there is continuous movement whilst deviation from the set-point persists such that the *rate* of movement is a function of the amount of deviation from the set-point.

Derivative control

This mode involves a further development of integral action such that the controller output is a function of the rate of change of the controlled variable. This form of control, like the integral mode, would not normally be used alone, but in combination with others.

Proportional plus integral

Sometimes referred to as proportional with reset (or abbreviated to PI) this combination gives stable control with zero off-set, as shown in Figure 22.23. So long as there is deviation from the set-point, the controller will continue to signal a change until zero error exists. This approach would be applied, for example, to space temperature control in circumstances where the load fluctuated widely over relatively short periods of time and where, in consequence, a simpler type of controller could not produce a stable condition without the proportional band being wide beyond acceptable limits. In addition, this mode is used more generally for applications where close control is required.

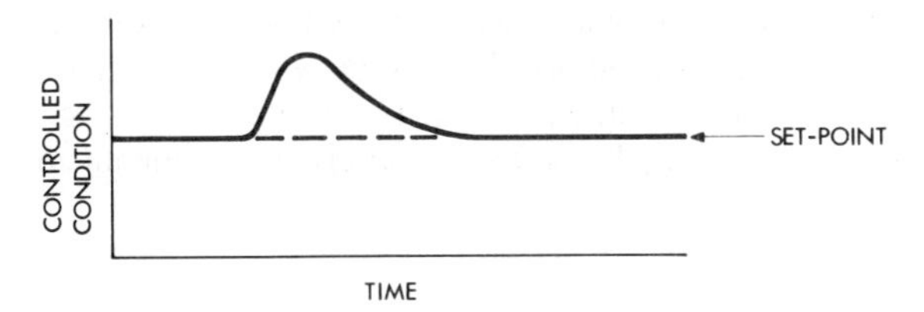

Figure 22.23 Proportional plus integral control

Proportional plus integral plus derivative

Abbreviated to PID, this mode of control would be used where there are sudden and significant load changes and where zero off-set from the desired set-point is required. There is seldom a case for such control in heating and air-conditioning work.

Systems controls

Common to all systems is the requirement, by legislation, that they be started and stopped by some form of time switch. There are two basic types; the simple on/off switch and optimum start (and stop) control.

On/off time switch

Simple devices of this type are suitable for small systems, those where the number of days of use is limited, such as church heating, or those where the thermal inertia of the building is so high that little practical purpose would be served by varying the system start-up time with respect to external temperature. Time switches for other than domestic scale buildings would normally be of the 7-day type, enabling each day of the week to be programmed separately. Microprocessor based systems can provide for a complete year, including Bank Holidays, etc., to be entered to memory at one time. To reduce the heat-up period, an additional feature to time switch control, known as *fixed time boosted start*, is the ability to run the system at maximum output for a period until either the desired internal space conditions are achieved or the end of a fixed preheat period is reached, at which time the controls change to normal mode under thermostatic control.

Optimum-start control

This system, developed jointly by the Property Services Agency and controls manufacturers to reduce energy consumption, serves to delay the start time of a heating or air-conditioning system until the latest possible time to give the shortest preheat period, normally at full output, which will achieve the desired conditions at the start of the occupancy period. The system of control monitors the inside and outside temperatures, see Figure 22.24, and takes account of the thermal response of the building and the system. Modern equipment is self-adapting, in that it can monitor its own performance (in achieving correct conditions at the start of the working period) and take corrective action to improve these from one day to the next. Optimum-stop control is also available to switch plant off as soon as possible, consistent with maintaining acceptable temperatures at the end of the occupied period. This is quite satisfactory for heating systems, but there is concern over its application to plenum or air-conditioning systems where stopping the plant in advance of occupants leaving the building also stops the supply of outside air to the space.

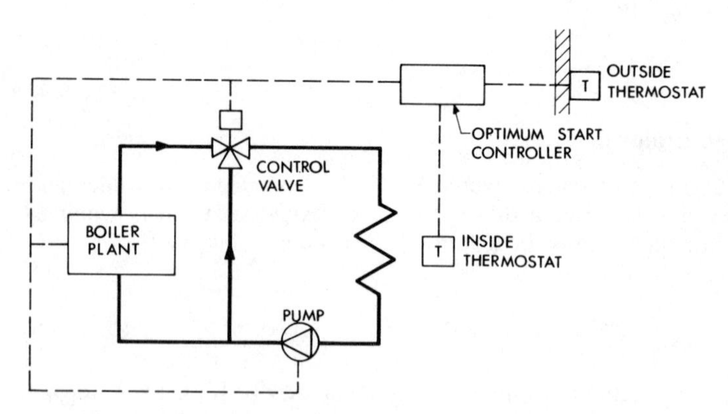

Figure 22.24 Optimum-start control system

The characteristics of the optimum-start principle are shown in Figure 22.25 from which it can be seen that the switch 'on' time is related to the fall in internal temperature. It also shows that the outside temperature affects the rate of fall of inside temperature, and also the rate at which the heating system can heat up the space. Generally, wet heating systems have a slow response compared with air based systems, and in consequence need to be switched 'on' earlier. In hotter climates optimum start may be applied to space pre-cooling, but this facility has no practical benefit in the climate of the British Isles.

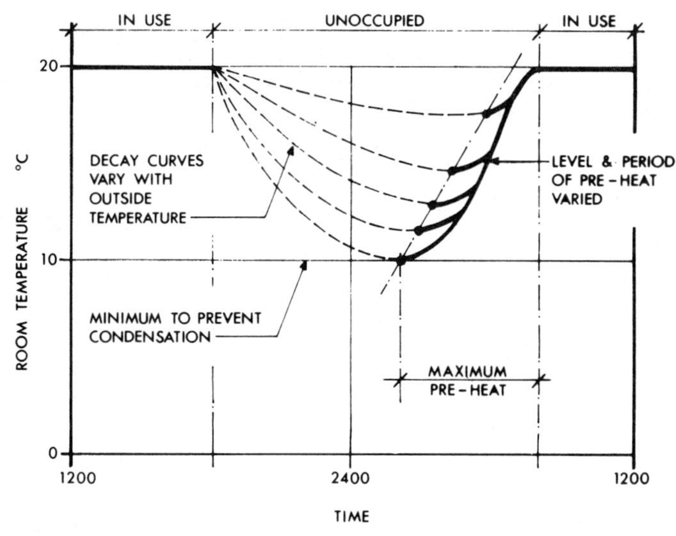

Figure 22.25 Characteristics of optimum-start control

Regardless of the method employed for system start-up, separate frost protection is an essential feature for all systems in order to protect the conditioned spaces from condensation risk (normally 10°C is the lowest acceptable temperature) and to protect plant components from freezing.

Heating system controls

Figure 22.26 illustrates a simple heating system with optimum start, plus conventional heating *compensator* controls. In addition to optimum-start provisions, as shown in Figure 22.24, this includes thermostats to sense external temperature and that of water

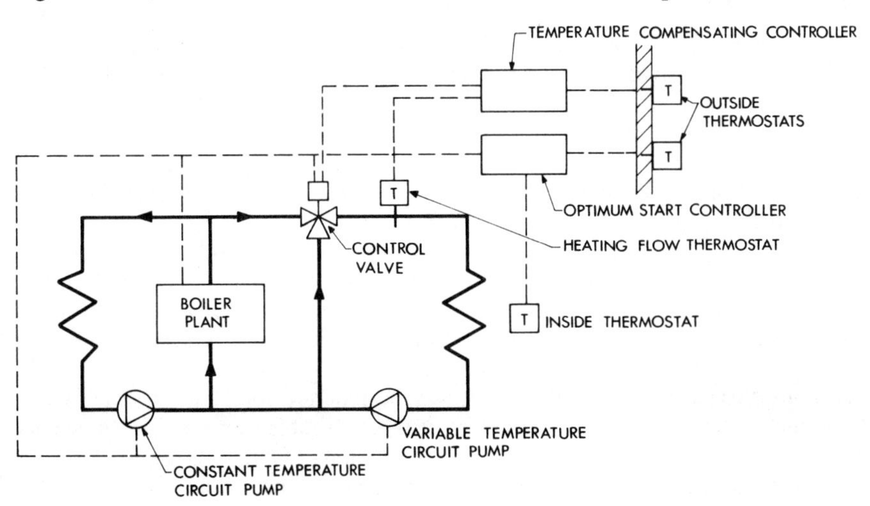

Figure 22.26 Basic heating system controls

in the flow main to the system. By means of calibrating mechanisms, the temperature of water leaving the control valve is adjusted to suit the outside air temperature. In cold weather, balance between the two thermostats will require the water to be at a high temperature and, conversely, in warm weather balance will be achieved with water at a lower temperature. At any point between the two extremes the water temperature is adjusted proportionately to the outdoor temperature. Refinements have been introduced into this system by way of a heating element in the external unit, so that the effect of wind is taken into account. The compensator controller effectively resets the desired temperature set-point for the system flow water appropriate to the outside conditions obtaining. If there are a number of main circuits to various parts of the building, it is usual for each to have its own compensator, circulating pump and mixing valve. This is an example of *open-loop control*, where there is no feedback from the controlled variable, namely the space temperature.

Zone control

Control of the heating system for a building or series of buildings by *zone control* is based, in effect, on the assumption that certain areas have common characteristics and hence similar heat requirements. For instance, in a rectangular building with north and south aspects on the long face, it may be that all rooms on the south aspect will have similar heating requirements and may be served from one zone, likewise north facing rooms may be served from a second zone. The southerly aspects will require less heat than the northerly at times and therefore will need independent mixing valve control to provide different flow water temperatures. Similarly, in a tall building, due to the greater exposure of the upper floors, it might well be desirable to zone the upper floors separately from the lower ones. Further zoning may be required depending on the elevational treatment, leading to four or more zones for the building.

Individual room control

The ultimate refinement is to be achieved by control of the heat emitter in each room. In buildings where there is significant variation in room loads, due perhaps to machinery or occupancy levels, individual room control will be required to enable uniform temperatures to be maintained. The alternative is, of course, to assume that the occupants can control the output from heat emitters manually, dependent upon conditions at any one time: in practice, however, this is seldom satisfactory because openable windows are used as the control device with the consequent waste of energy.

A combination of compensator control plus individual room control has considerable merit, since the zone compensation provides a limit to energy use and the room controls give further savings together with a capability to off-set local heat gains. Clearly, with such a combination system zoning is less critical and larger areas may be served from each zone, resulting in initial cost savings.

Whereas compensated temperature circuits may be required for systems which serve radiators, radiant panels and natural convectors, it is undesirable to serve fan convectors and heater batteries from these circuits. A constant temperature circuit is required for fan assisted systems and for serving the primary heat source to hot water service storage calorifiers and other types of heat exchanger.

Boilers

Whereas in the past, with a hand-fired coke boiler, a crude form of *damper–regulator* sufficed to control the air admitted below the grate for combustion, this meant that the boiler often operated for long periods at much below its rated output. Such a routine is bad for the boiler, particularly with oil firing, due to the likelihood of condensation being caused under such conditions with corrosive effect on the boiler itself. It is better practice to control the boiler in such a manner that it runs at a constant high temperature, to reduce the tendency for condensation to occur, and then to mix the return water with the flow, by means of a mixing valve, for the purpose of serving the heating system. Where heat at a constant temperature is required this would be served direct from the boiler with a separate pump, as indicated in Figure 22.26.

The control of the boiler may also be under the influence of a time switch, as described previously. Alternatively, it might be considered preferable simply to reduce the temperature at night (*night set-back*), which involves a second thermostat, the time switch changing over from one to the other. In other than very small systems, multiple boilers would be installed, which provide standby in case of failure of one unit and allow each unit, when in use, to operate closer to full output and peak efficiency, rather than on low load with consequent inefficiency.

Boilers are commonly provided by the manufacturer complete with integral safety and thermostatic controls. Multiple boiler installations are piped in parallel and controlled in sequence; more boilers operating as the load increases. Provision should be made for the sequence to be varied to give equal use of all units through the life of the plant. Although straightforward in concept, this form of control can present problems during partial load and it is important to consider the operating temperatures through the full range of output. Some of the points to consider with multiple boilers are those related to ensuring that:

- Return water temperature is high enough to prevent condensation occurring in the flueways.
- Resulting flow temperatures from sequence operation are within acceptable limits (excessively high temperatures can occur).
- Flow rates through individual units are above the manufacturers' recommended minimum.
- Flow rates are balanced through each unit.

It may be necessary to introduce individual pumps to each boiler; or to use a primary circulation through the boilers, from which secondary circuits are connected to the various heat emitters or heat exchangers. The plant controls must always be considered in conjunction with the remainder of the heating system.

Domestic heating

Such evidence as is available suggests that the most effective method for control of domestic heating is the thermostatic radiator valve coupled, of course, with full use of a time programmer to provide for intermittent boiler shut-down at night or during other periods when the dwelling is unoccupied.

Heat exchangers

Storage calorifiers and non-storage heat exchangers would normally be provided with a modulating valve, fitted to the primary supply, and controlled from a thermostat located in the shell in the case of a storage calorifier, or in the secondary flow pipe from a non-storage heat exchanger. In the case of a heat exchanger serving a heating system, it is possible to link these controls with a compensated controller to vary the flow temperature according to outside conditions. A direct acting control valve may suffice where simple constant temperature is required.

Air-conditioning system controls

Automatic controls are an essential part of any air-conditioning system. The wide variety of purposes for which such plants are required, the number of different systems which it is possible to select, and the multitude of types of control equipment available require a complete book in themselves to cover adequately.

It is possible here to indicate only the main principles involved, some of the chief component items of apparatus commonly used being described elsewhere (Chapter 17). From this it may be possible to understand some of the problems concerned and how they may be tackled. Other permutations and combinations of system and equipment may be built up to suit any variety of circumstances.

The main types of plant will be considered and reference to the diagrams will be necessary along with the following descriptions. For further information the reader is directed to the *Guide Section B3* and to the CIBSE Application Manual AM1/1985 *Automatic Controls and their Implications for Systems Design.*

Central system with cooling coil

The control arrangement in this case is illustrated in Figure 22.27 which shows a typical array of sensing devices, control valves and motorised dampers, including some the desired safety features, to provide full control over temperature and humidity in the conditioned space under all outside air conditions. This figure relates in principle to the example air-conditioning calculation in Chapter 18 (p. 511) with humidification by steam injection.

In an all-air system, the quantities of outside and recirculated air, respectively, may be varied in any combination between full outside air, with no recirculation, and a mixture in any ratio subject to the provision of the minimum quantity of outside air which will satisfy the ventilation requirements of the occupancy. The mixture ratio is controlled to provide the most economic level of plant operation as determined by the condition of the outside air relative to the required condition of the supply air.

A dry bulb thermostat or an enthalpy sensor is used to sense the outside air condition, the latter being used more generally nowadays in preference to a wet bulb thermostat. Modulating dampers in the recirculated, exhaust and outside air ducts are controlled to provide the desired mixing ratio and, for reasons of energy economy, the plant may operate on full recirculation during building heat-up in winter.

In summer, the dampers will normally be set to introduce a minimum outside air quantity and the cooling coil will be controlled to maintain the required dew-point for the supply air and, during this season, moisture will be removed from the air for the majority of outside conditions. The final dry bulb temperature will be controlled by the after-heater. During mid-seasons, full outside air may be used to take advantage of the available 'free cooling' potential which is available using air at outside temperature. The supply air dew-point and the final supply temperature to the space will be controlled as for the summer situation.

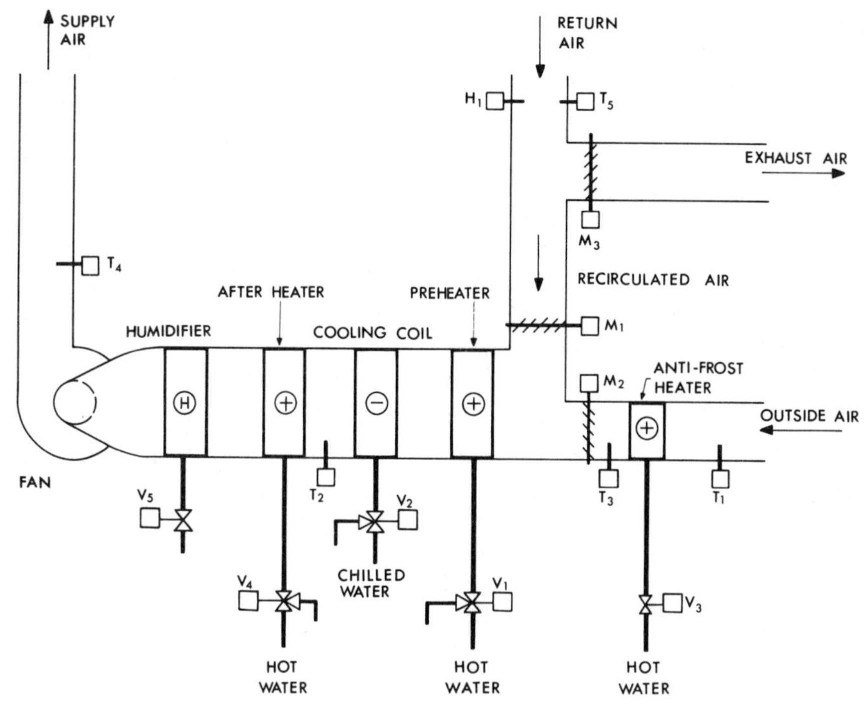

Figure 22.27 Control principles of a central plant system with cooling coil

In winter, the dampers may modulate to mix outside and recirculated air to provide the most economical ratio such that, whenever outside conditions permit, the mixture will provide air at the desired supply temperature and thus eliminate the need for preheating. The humidifier will be operated as required to provide the correct moisture content in the supply to the space.

Central system with cooling coil bypass

This is a development of the previous system where, in order to avoid wasteful use of cooling and reheating in the event of prolonged periods of less than peak load together with closer control of room humidity, a bypass may be introduced around the cooling coil as shown in Figure 22.28. In this case, dampers in the recirculated air duct and in the bypass duct will operate together to allow the correct quantities onto and around the cooling coil.

Central system with sprayed coil

This type of plant, arranged as illustrated in Figure 22.29, is controlled in a manner similar to that with an independent humidifier, as discussed previously. But in this case the coil is sprayed with recirculated water to provide adiabatic humidification. Since the sprayed coil has no facility to provide capacity control, the plant dew-point is controlled either by

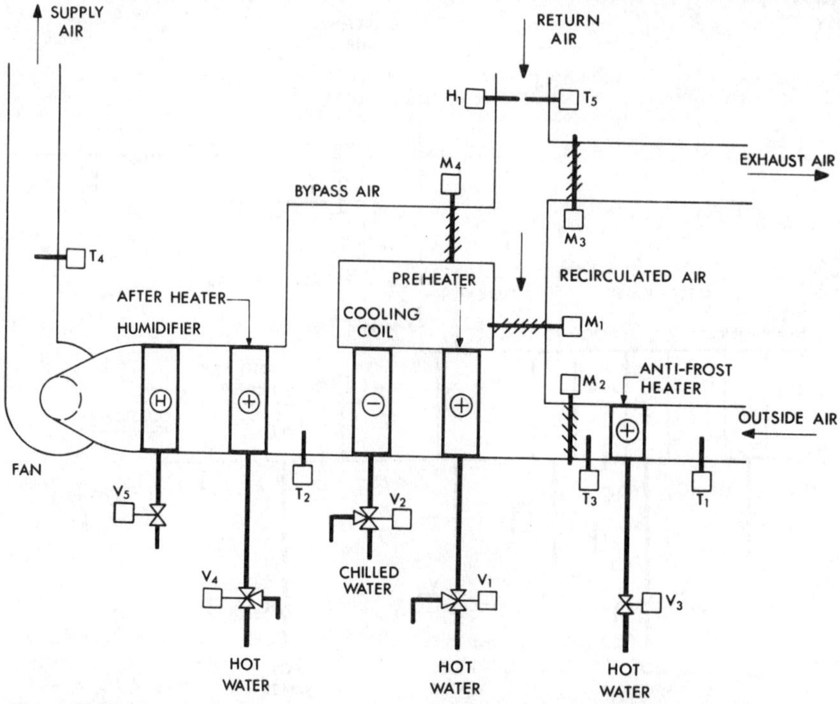

Figure 22.28 Control principles of a central plant system with cooling coil by-pass

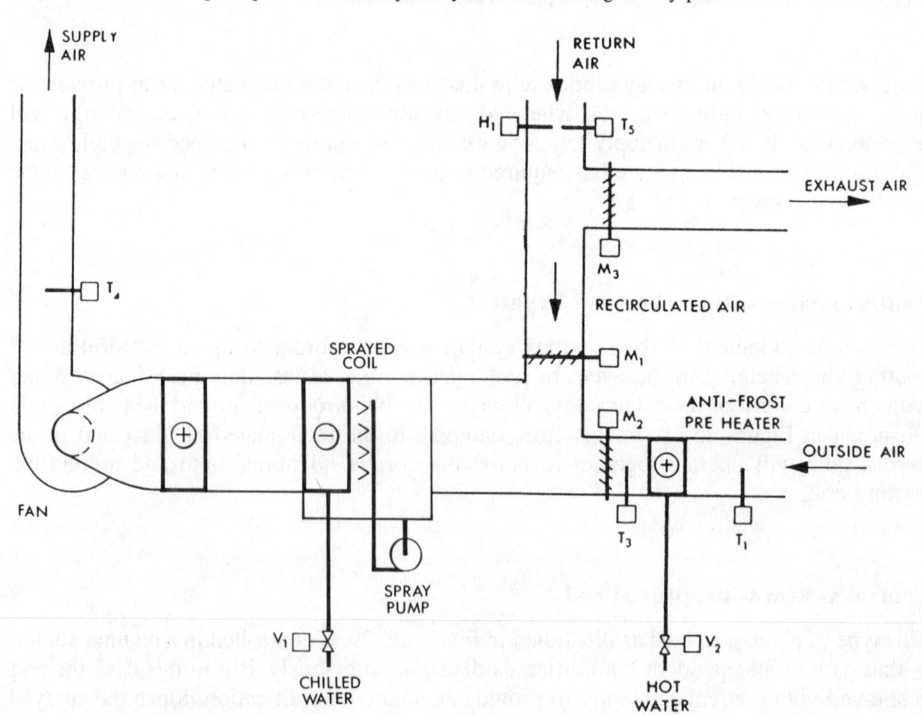

Figure 22.29 Control principles for a central plant system with a sprayed coil

adjustment to the mixing ratio of outside to recirculated air or by the introduction and control of a preheating coil upstream of the sprayed coil. The second arrangement produces a dry bulb temperature onto the sprayed coil which will allow the adiabatic cooling effect to produce the desired moisture content off the coil.

Zoned system

In this case the central plant delivers air at a constant outlet condition and the zones are controlled individually.

Terminal reheat system

In this case, all or the bulk of the required reheating to achieve control of dry bulb temperature is transferred from the central plant to individual rooms within the conditioned space. Thus, the output of the central plant is controlled to meet the peak cooling load likely to occur at any one time in any individual room.

Within each room, a thermostat will be arranged to control admission of heat to the reheater incorporated in the terminal unit. This reheater would usually take the form of a hot water coil provided with a control valve, but electric resistance heaters are sometimes used.

If room load variations are small, a conventional central plant reheater may be used to deal with the common reheat load. In such cases, the temperature of the conditioned air delivered may be such that it can be discharged into all rooms without creating problems and reheat may thus be provided by a perimeter heating system fitted with suitably responsive room-by-room controls.

Fan-coil systems

The principal difference between those systems previously described and a fan-coil arrangement is that where a ducted air supply is provided, this handles outside air only and not the full conditioned air quantity.

Where fan-coil units are large and are, in effect, themselves recirculating air-conditioning plants serving significant building zones, as shown in Figure 22.30, the primary air supply from the central plant will be controlled at constant condition year-round. Controls will thus be as described for a central plant system, the reheater being omitted. Each zone plant will be provided with cooler and reheater coils fitted with control valves and these will be operated separately or, more usually, in sequence by a thermostat in the fan-coil outlet. This may be reset as before by a further thermostat fitted in either the conditioned space or the return air duct.

Fan-coil units of smaller size, where fitted in individual rooms, will be served by a central system controlled much as described for induction units in a later paragraph. Individual units may be controlled by room thermostats operating a valve or valves in the water connections which may be two-, three- or four-pipe. Manual on/off control to the fan is often provided: automatically controlled fan switching, however, may cause annoyance due to the change in noise level.

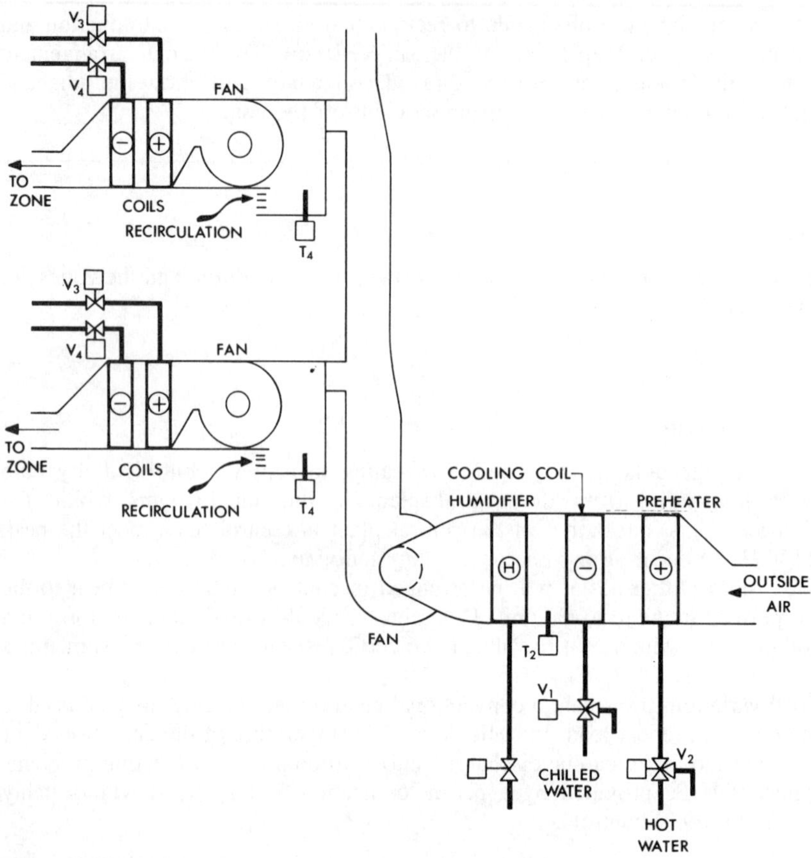

Figure 22.30 Principles of a zoned system with large fan-coil units

Where individual units are provided it is preferable and often more practical to introduce ventilation air from the central plant into the space via terminals independent of the units. This arrangement also enables the air flow quantities to be balanced more accurately.

Induction system

There are two principal methods of control applied to induction systems, the 'changeover' and the 'non-changeover'. The former is more appropriate to climates having distinct seasons and need not concern us here. Two-, three- and four-pipe units are available to provide progressively better availability for control of space conditions: the two-pipe system was the more common in Great Britain. However, induction unit systems are seldom used these days; the preference being for either fan-coil or variable air volume systems.

The arrangement shown in Figure 22.31 illustrates the controls of the non-changeover type having two-pipe water connections, typical of the systems installed in many office developments during the 1970s. Water temperature is kept constant at say 10°C both

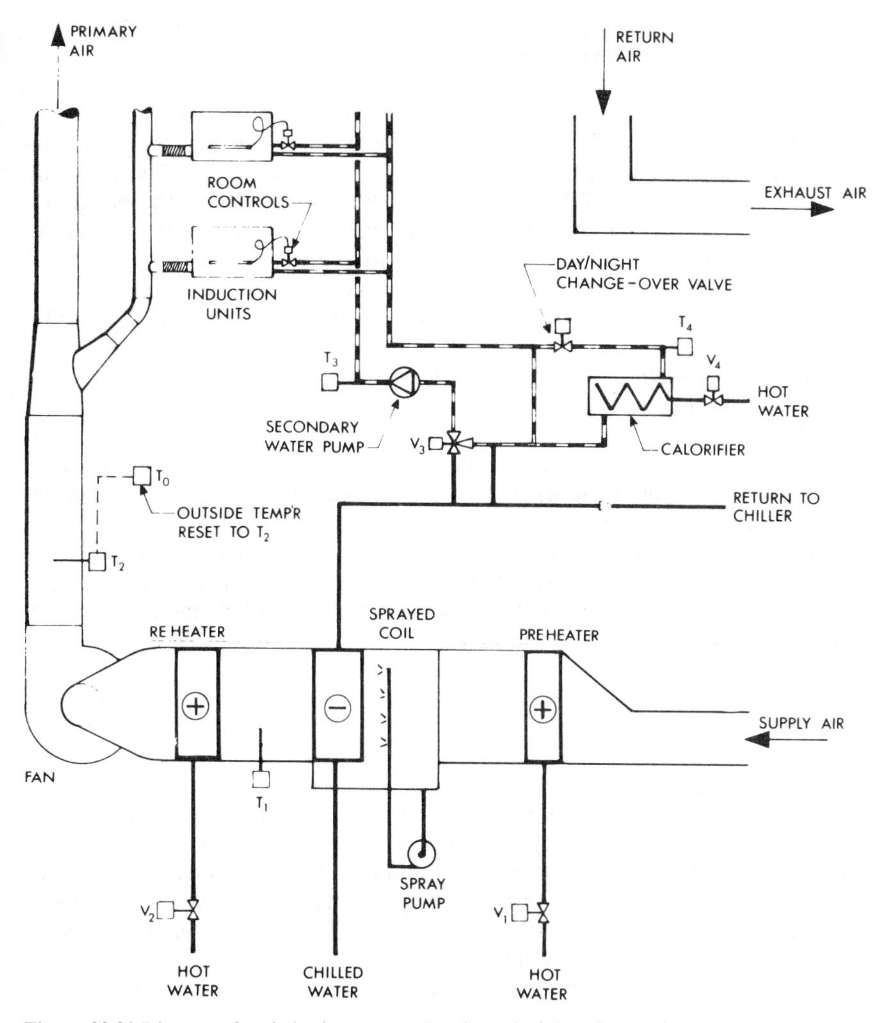

Figure 22.31 Diagram of an induction system showing principles of control

in summer and winter by means of the mixing valve in the 'secondary' water circuit as shown. The water is cooled by a water chiller in summer, but in winter by the entering cold outside air passing over the cooling coil thereby cooling water in circulation to the induction units, a process known as 'free cooling'. The whole of the chilled water is circulated through the cooling coil without control, thus ensuring that in summer the dew-point (about 7°C) of the primary air to the units is below the temperature of the coils in the room units, in order to prevent the latter from gathering condensation.

The primary air temperature is maintained at about 16°C in summer but in winter is varied according to the weather, by the outside thermostat shown, which may be adjusted to take some account of solar radiation, up to about 60°C.

To maintain warming during winter nights without the fan running, the calorifier is brought into service allowing the units to act as natural convectors.

Each induction unit in the system shown has a direct-acting control valve in the water circuit with the actuating element in the induction air stream from the room. Thus, in summer, assuming no heat gains in the room, the valve will be closed, the air alone maintaining room temperature. When sun or internal heat gains cause the room temperature to rise, the valve will open and the coil will lower the circulating air temperature. In winter the primary air will warm the room but, should the temperature mount too high, the control valve will open to admit cool water, so correcting the temperature. To avoid variation in pressure, should a large number of control valves close, a constant pressure differential control valve is included as a bypass.

An alternative arrangement provides local control by means of a damper arrangement within the unit which adjusts the direction of the induced air such that part or all is diverted past the cooled coil. This damper may be controlled manually or automatically by thermostat. Some such units make use of the pressure of the primary air to operate the damper but others rely upon an independent supply of compressed air to actuate the controller.

Moisture content is controlled in summer by the chilled coils and in winter by the humidifier, to a constant dew-point. Alternatively, a sprayed coil system may be used, controlled in the same manner, but only where a high quality of maintenance can be guaranteed.

Dual-duct system

Figure 22.32 illustrates the principle of the control system applying to a dual-duct system where two fans are included, one for the warm duct and one for the cool.

The temperature of the cool duct is varied according to the external temperature from say 7°C in summer to say 16°C in winter, or to a more limited schedule where the internal gains dominate the room loads. In summer it is cooled by the chilled water coil: in winter the heater comes into use as necessary. The temperature of the warm duct is also varied according to the outside temperature from, say, 21°C in summer to, say, 38°C in winter.

It will be noted that, in the arrangement shown, the return air is brought into the fan suction chamber on the side nearest the warm duct, the return air being normally nearer to the warm duct condition, thereby assisting winter heating particularly where the return air is drawn through luminaires.

Other controls indicated are concerned with maintaining air pressures within certain limits by damper operation. Dehumidification in summer is performed by the cooling coil, without specific control and, in winter, humidity is increased as necessary by sprays or steam injection controlled from return air conditions. Vane control dampers are illustrated in the figure, but there are many forms of fan control available, as described in Chapter 17.

An alternative arrangement for a dual duct plant is given in Figure 14.21 which illustrates a single fan configuration. Equipment controls will be generally as previously described.

The blender units are thermostatically controlled from the rooms individually, or in groups, by adjustment of an internal mixing damper or dampers.

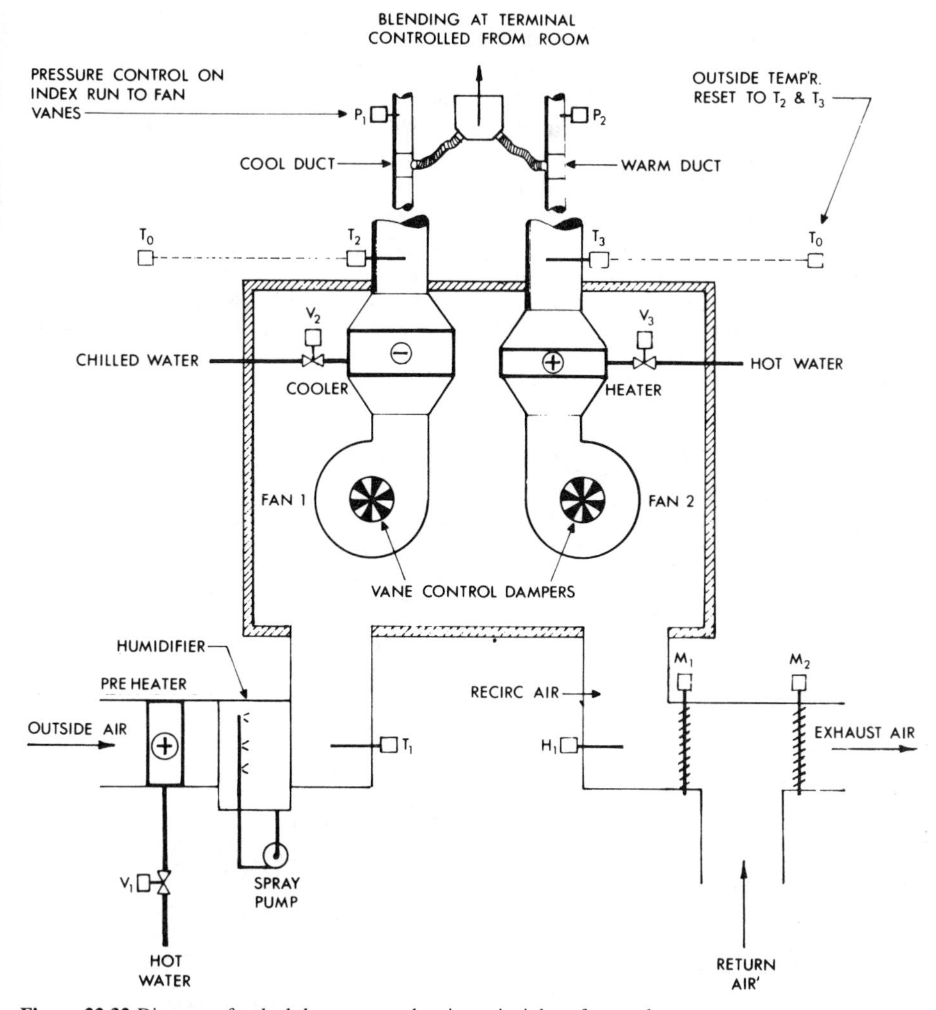

Figure 22.32 Diagram of a dual duct system showing principles of control

Variable volume system

A true variable volume system, in its simplest form, will have a central plant controlled in most respects in a manner similar to that first described here. Particular attention will be paid however, in the interests of energy saving, to control the fan operation such that volume variation at terminal units is reflected in volume reduction at the central plant. Fan volume control is achieved by sensing the static pressure in the supply duct and controlling the fan to maintain that pressure constant. The best position for locating the sensor can only be found by testing during the commissioning process by simulating varying load conditions, but a practical rule-of-thumb for most systems is two-thirds along the index run from the supply fan. The methods of fan control are dealt with in Chapter 17.

In the simple form of the true system, control of space conditions and in particular those in internal zones of deep plan buildings is achieved by variation in the quantity of air supplied rather than by any changes in temperature. This control, by room thermostat operation upon electrical or pneumatic devices, acts to reduce the volume of conditioned air delivered either to a zone or an individual room where the load is at less than the design peak. Such devices may be integral to the actual terminal diffusers or incorporated in regulators serving a number of such diffusers.

Good practice suggests that the volume of air supplied to meet the designed full load cannot be reduced under control by more than about 40% if room distribution difficulties are to be avoided, unless variable geometry supply diffusers are used to maintain air velocities. In consequence, control by terminal reheat may be necessary should local demand for cooling fall below that level: hence, room or zone terminals may be supplied

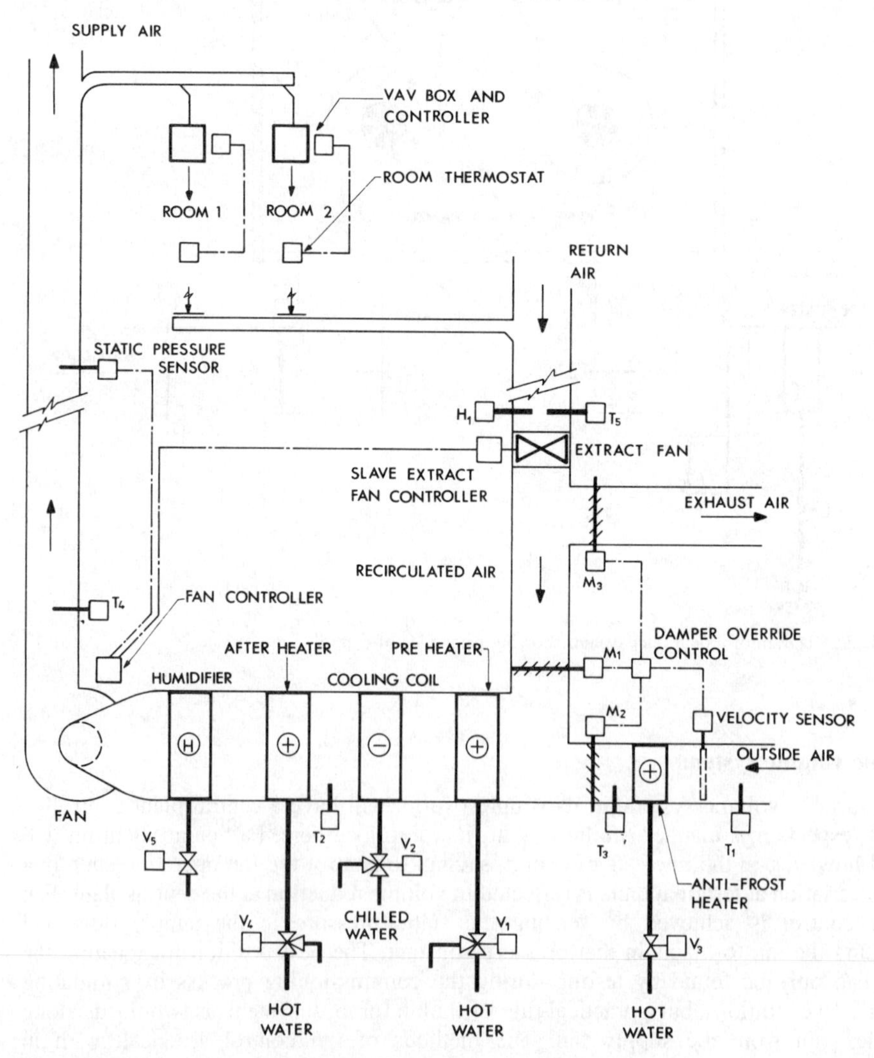

Figure 22.33 Variable air volume system with fan assisted terminals

from a dual-duct system or be themselves provided with reheat batteries supplied by hot water or, rarely, with electrical resistance heaters. For perimeter zones, the necessary reheat may be controlled via any form of convective heating system there installed if this can be arranged to provide adequate response. Modern control systems allow the perimeter heating and variable volume box controls to be sequenced, room-by-room if so desired.

Of particular importance is the supply of adequate quantities of outside air to each room. Unless additional controls are incorporated, the outside air quantity will vary proportionally with the supply quantity, which may lead to inadequate outside air provision and high running costs. To overcome this, an air velocity sensor is installed in the intake duct to control the motorised dampers and ensure that the ventilation air quantity is maintained at or above the minimum requirement.

In Chapter 14 the concept of fan assisted control terminals was introduced as a means to provide additional cooling to supplement that inherent in the air supply. A valve in the chilled water supply to the cooling coil, and in some parallel arrangements the fan, is controlled by a thermostat sensing room temperature. A typical system is shown in Figure 22.33.

It is also necessary to control the extract fan as a 'slave' to the supply fan in order to maintain the correct balance between the supply and extract air streams over the full operating range and to achieve maximum energy savings.

Systems which are controlled to vary the volume of air delivered to rooms or zones while the central plant volume remains constant do not offer the same facilities for energy conservation.

Reverse cycle heat pump system

The control of individual heat pump units is described in Chapter 14. To supplement these local units, it is normal to install a central air handling plant to provide ventilation air to the conditioned spaces; this may be taken to be of the same configuration as for fan-coil systems.

Smoke control

Although plant start and stop functions have been dealt with previously in this chapter, an additional feature to be incorporated, common to ventilation and air-conditioning systems where air is distributed to and from occupied spaces, is provision for smoke control during a fire situation. The requirements vary dependent upon the Fire Authority concerned and the particular characteristics of the building, but common to the majority of installations is the need to provide the facility for the Fire Brigade to be able to switch the plant 'on' and 'off' from a convenient position close to a main entrance to the building; this is called the 'Fireman's panel'. Separate control over the switching of supply and extract systems is required. In a fire situation, the air handling plant would normally be arranged to stop automatically, except where individual systems were designed specifically for smoke-movement control in which case these would be switched automatically into a predetermined fire control mode.

Chilled water system control

Contrary to the general practice with heating circuits, it is not usual to vary the temperature of the chilled water flow from the evaporator to cooling coils. However, there

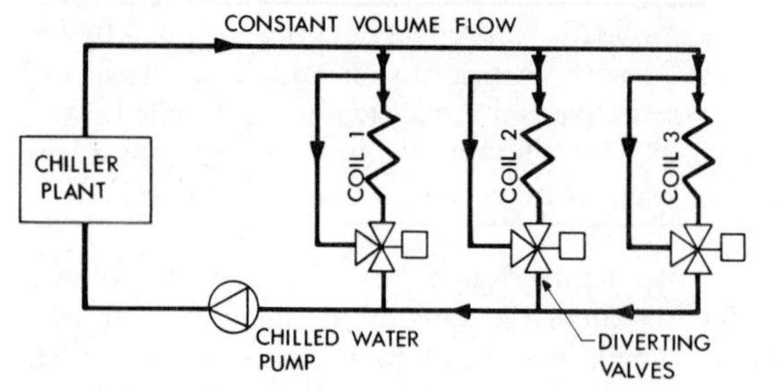

FLOW QUANTITIES

COIL		
COIL	1	100 UNITS
	2	200
	3	100
TOTAL		400 CONSTANT

Figure 22.34 Three-way valve control to a cooling coil

can be very large variation in cooling loads at zone or central plants dependent upon the time of day as the sun moves around the building. As a result, depending on the magnitude of this variation, it may be more economical to use a variable water flow system in preference to constant flow. Where constant flow is used, control of chilled water to the coil will be by three-way modulating control valves as Figure 22.34, but where variable flow is preferred, two-way modulating valves would be used.

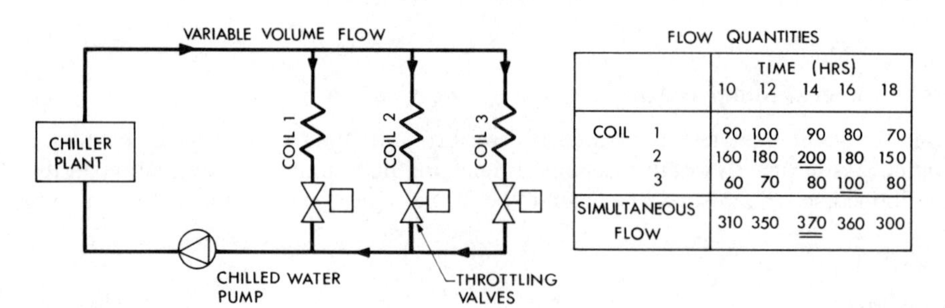

FLOW QUANTITIES

		TIME (HRS)				
		10	12	14	16	18
COIL	1	90	100	90	80	70
	2	160	180	200	180	150
	3	60	70	80	100	80
SIMULTANEOUS FLOW		310	350	370	360	300

Figure 22.35 System with two-way control valves

A typical system incorporating two-way control valves is shown in Figure 22.35, from which it can be seen that the maximum pumped water quantity is proportional to the maximum simultaneous cooling load, and not to the sum of the maximum coil loads: in consequence, smaller pumps may be used thus reducing both installation and running costs. With a variable flow system, provision must be incorporated to ensure that the flow rate through the chiller evaporator can never fall below the manufacturer's recommended minimum and thus incur a risk of freezing.

Chiller control

Single chillers will normally be supplied by the manufacturer with integral safety and capacity controls. Multiple chillers may be arranged in either series or parallel with respect to the chilled water circuit, see Figure 22.36. The points to be aware of when selecting the more appropriate arrangement are that:

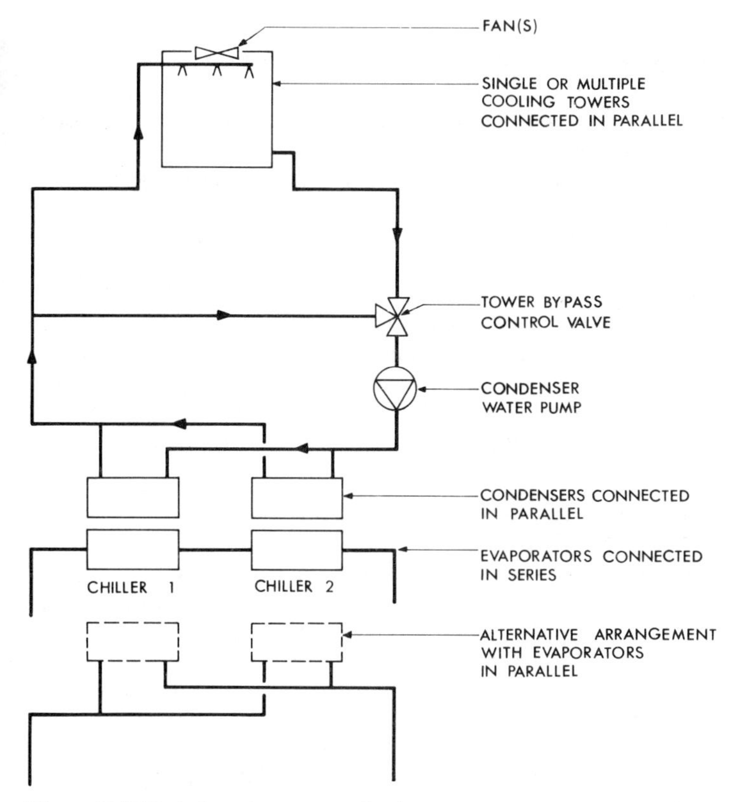

Figure 22.36 Typical condenser water circuit

- Since the frequency of starting will be limited, the minimum output (turn down) of one machine must be less than the lowest operating load of the system.
- In a parallel arrangement, it must be established that the sequence control will not lead to sub-cooling and that the water flow rates through each evaporator are regulated to the correct quantity.
- In a series arrangement, it must be established that the capacity and characteristics of the machines are compatible with the differing flow and return water temperatures which will result from the operating mode.
- Although a series arrangement may permit control of water flow temperature to be more stable, the hydraulic resistance of the system will increase with consequent higher pumping costs. (It is not usual to provide arrangements where more than two machines operate in series.)

Normally, with reciprocating machines, capacity would be adjusted by proportional control, sensing the temperature of the return water. With screw or centrifugal machines, PI control would normally be used.

Cooling towers

A typical arrangement for a condenser water circuit, connecting chiller condensers to either cooling towers or an equivalent, is shown in Figure 22.36. The cooling effect of the towers is controlled either by sequence switching of the fans on a multiple unit or by speed control of a single fan. A bypass valve is normally provided in the heat-rejection circuit, which operates in conjunction with the fan controls on start-up and sometimes during colder weather, in order to maintain a constant return water temperature to the condensers. To ensure economy in energy, the condenser water temperature may be varied with respect to the imposed load and thus retain a practical minimum between the condensing and evaporating temperatures of the chillers. Attention must be given to the hydraulics of the condenser water circuit to avoid unintended cavitation and all water based cooling plant must be provided with frost protection.

Where water is used for condenser cooling, a number of chilling machines may be connected together, in parallel, into a single circuit to be pumped over multiple cooling towers or their equivalent. Where air cooled condensers are used, however, each chilling machine will have a dedicated cooler. Fan speed control, fan switching or modulating dampers may be used to control either the condensing pressure or temperature. Normally, PI action would be installed for control purposes.

Building management systems

The introduction of computer and data communication technologies has given rise to dramatic developments in management systems, probably the most significant single change in the heating and air-conditioning industry in recent years. Advances in micro-electronics, and the reduction in cost of components coupled with improved reliability, has led to widespread use of these systems, from quite small installations to multiple site applications and with varying levels of sophistication. Flexibility in available systems has led to different approaches to application and this, in turn, has given rise to descriptions such as *energy management system, building energy management system, building automation system, supervisory and control system*; the authors prefer building management system (BMS) as the generic title.

The basic functions of building management systems may include:

- Initiation of systems control functions.
- Continuous monitoring of systems.
- Warning of *out of limit* conditions (alarm).
- Initiation of emergency sequences.
- Logging of significant parameters.
- Monitoring and recording energy use.
- Condition monitoring and fault analysis.
- Planned maintenance and other housekeeping functions.
- Tenant billing.

Systems may be further enhanced by the use of modern database software, graphics and word processing techniques to provide opportunities for applying BMS to total building management functions. The BMS, therefore, can provide many benefits but these are dependent upon the capability of the systems operators and their motivation to make use of the data available. The BMS should of course be designed in the knowledge of the operator's objectives and of the level of skill available: unfortunately this is not always possible. Some of the benefits which can result, however, are:

- Lower energy consumption.
- Improved system reliability.
- Savings from programmed maintenance.

- Reduced number of watch keeping operatives.
- Improved building management.

To balance these advantages, the initial cost of a BMS may be high but this is very much linked to the extent of the systems and the degree of sophistication. Added to this is the additional cost of BMS maintenance and, in the first few years after installation, those costs associated with analysing system performance in order to optimise control settings for minimum energy use and for setting up auditing programmes and the like. Comprehensive training of the operators is vital if the potential of the BMS is to be realised and this may require basic software skills where a facility for creating new routines for fine tuning of the control functions is incorporated.

Hardware

The principal components of a BMS may be described briefly as follows.

Central processor
A computer containing the software and memory for operating functions and providing the interface with the system peripheral equipment. Batteries are normally provided for back-up in case of mains power failure.

Transmission network
The wiring or data 'highway' links between the system components, sometimes known as local area networks (LAN). Critical links are normally duplicated along alternative routes for security reasons. Telephone lines may be leased for the purpose of interfacing between buildings remote from one another, with auto-dialling facilities introduced for automatic remote monitoring whereby the status of plant may be accessed at regular intervals and recorded at a central location. *Modems* (modulator/demodulators) are the interface between the BMS and the telephone system, required at each end of the telephone link.

All BMS cabling should be protected from extraneous electrical interference and fibre optic technology may be used for data transmission.

Data gathering panels
Panels located close to plant, through which control signals and data are communicated.

Outstations
Processors located close to plant, capable of handling a high volume of signals from different components.

Sensing devices
For every point to be monitored, a sensing device; switch; auxiliary contact or the like will be required.

Data inputs

Digital – two position switching signals for status and alarm.
Analogue – values measured by varying voltage or current in the form of a pulsed signal.
Control – signals to component driving mechanisms.

Keyboard
A means to input instructions.

Printer
An output device for permanent record of alarms; status; trend logging, etc.

Visual display unit (VDU)
A screen display of system schematics or a summary of status; can give instantaneous values (dynamic display).

Keyboards, printers and VDUs may be duplicated at different locations around a building; typically in the control centre, in the engineer's office, in the maintenance department and at a central security station (for out of normal hours monitoring). Different levels of access into the system would be available to limit the facilities appropriate to the user code. An example of an installation network is shown in Figure 22.37 to show the disposition of terminal devices.

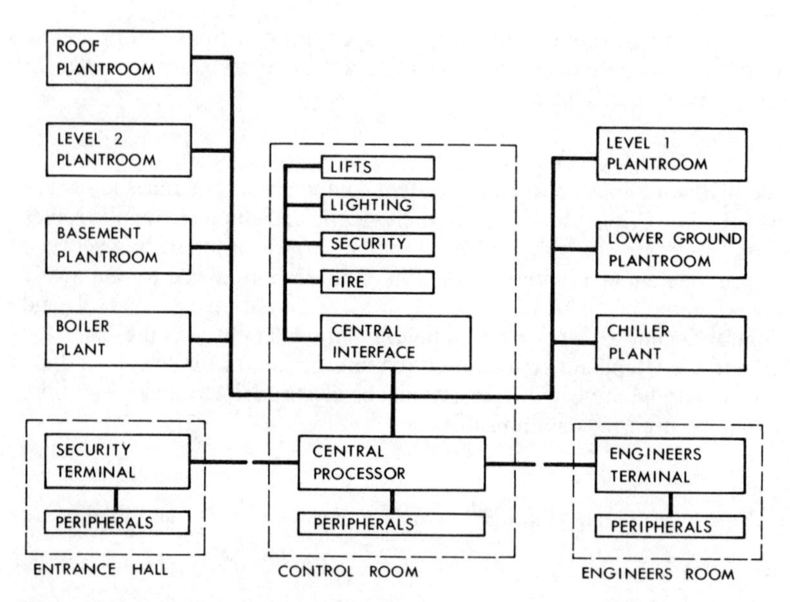

Figure 22.37 Typical building management system network

Software
Computer programs which set the logic, procedures and rules concerning the operation of any data processing system. Becoming increasingly important is the need to connect together components and systems by different manufacturers. Although the merit of using the same transmission system for both control functions and life safety systems is in some doubt, there is considerable benefit to be gained from controls manufacturers using a common protocol standard. As it is, integration between different systems, for example BMS and lift, security and fire alarm systems, is restricted to *gateway* access which in effect is a 'black box translator' which enables transfer of signals between the different systems.

Central intelligence

The early systems had central intelligence, in which all signals were transmitted and received through a central processor. Communication between the central station and data gathering panels was *hard-wired*, that is with individual connections between each unit. This type of system may be described as two-level (Figure 22.38): data are collected from plant controllers and sensors at data gathering panels, and transmitted to the central processor; any required return signal is transmitted back to the controls via the reverse

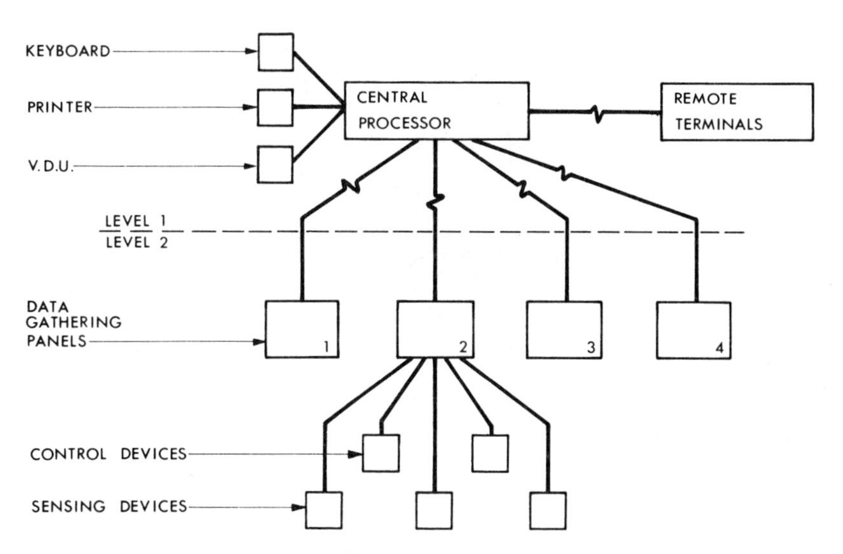

Figure 22.38 Central intelligence BMS

route. This system configuration is still in use today, the difference being that communication is no longer necessarily hard-wired. *Multiplex* systems are used, in which a single communication network connects all panels with signals pulsed between the components at intervals of milliseconds. A dedicated wiring system would normally be used for multiplexing circuits in new installations; however, for retrofit projects, consideration may be given to using existing power cabling for transmitting the signals. *Power line carrier* systems operate by superimposing high frequency signals on to the power waveform.

Distributed intelligence

As systems increased in size, the concept of having all communication with a central processor led to congestion of the data flowing to and from a single location, and to the need for excessive computing capacity. A natural development was to have as much as possible of the communication and data processing dealt with close to the plant. *Intelligent outstations* strategically located around the building, with only limited communication with the central computer for monitoring and data gathering purposes, came into use. This led to 'three-level' systems as shown in Figure 22.39 and the principle may, of course, be extended further to four and more levels, communicating between buildings sometimes many miles apart.

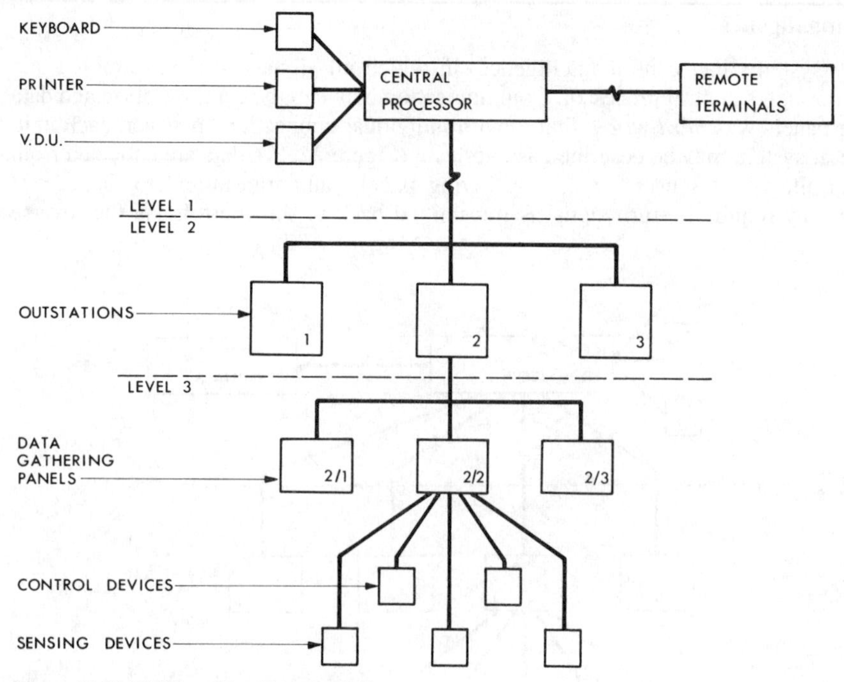

Figure 22.39 Distributed intelligence BMS

In the three-level system, intelligent outstations are introduced between data gathering panels and the central processor. In most systems the outstations are able to operate independently of the central processor and provide all the necessary control and switching functions. The system is therefore suited to a staged expansion from, say, a two-level system initially, later extended to a three-level, incorporating the central processor.

Commissioning of distributed intelligence systems is more flexible since plant can be set up to run from the outstations, the final commissioning of the central processor being

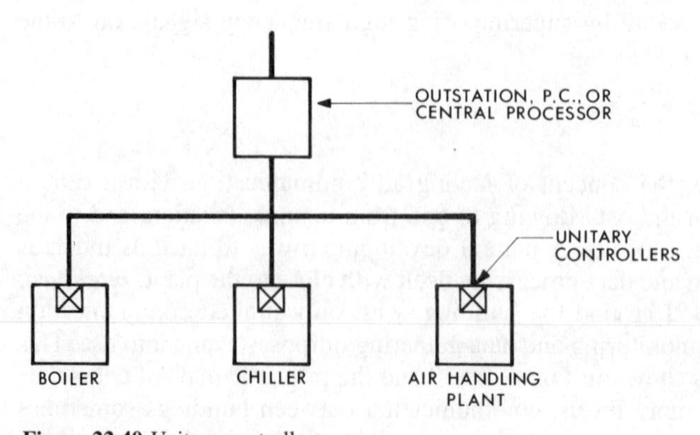

Figure 22.40 Unitary controllers

arranged to follow once outstations are operating satisfactorily. The central unit in a distributed system provides the same facilities as with the central intelligence alternative, including re-programming of the outstations.

A further development, illustrated in Figure 22.40, is for individual items of plant such as chillers, air handling units, boilers and terminal control units to be delivered to site complete with pre-wired *unitary controllers* suitable for either stand-alone use or for full integration with a BMS. This provides intelligent plant control which has been tested and proven before leaving the manufacturers' works and, in consequence, reduces site wiring and control panel work.

PC (personal computer) based systems

Many manufacturers are now marketing the distributed intelligence type of system using standard PCs. This enables a system to be built up floor-by-floor or plant-by-plant, as required and at low initial cost, with the flexibility to link many PCs together to produce a comprehensive system for the building. In addition to the task of monitoring and controlling plant, the PCs may be used for other business functions.

Long-distance communication

It is not uncommon for communication over long distances to be achieved satisfactorily using telephone system networks. Signals are converted from, and to, the digital pulses of the controls system into the analogue form required by the telephone system by *modems* (an acronym for modulator/demodulator).

Software

These are computer programs which provide the system with intelligence for operational and management functions: new programs are being developed all the time by the systems manufacturers. Only tried and tested software should be specified since development invariably takes a considerably longer time than the suppliers anticipate and, in consequence, many systems never achieve their full potential.

Current tried and tested software includes multi-level alarm initiation, time switching, optimum start/stop, optimum reheat, damper control, plant cycling, system controls, restart after power failure, electrical demand monitoring, load shedding, maximum demand limitation, data logging and summation and alarm priority. Other systems such as lighting control, lift control, security and fire detection may be *stand alone* and linked to BMS for alarm monitoring purposes, or be totally integrated with BMS. This latter arrangement is not normally the case for reasons previously outlined here.

Benefits in use

One of the major difficulties experienced on many projects, particularly the larger and more complex, is the achievement of satisfactory completion including correct commissioning and performance testing. At best, it is likely that the systems will be commissioned to satisfy the bases of design: controls set points, for example, would be based upon theoretical steady state calculations, modified perhaps by the judgement of the commissioning engineer, to reflect the dynamic response of the plant under the particular operating conditions obtaining at the time of the commissioning.

Using a BMS, it is possible for the systems and controls to be finely tuned with the usage patterns of the building and to the actual thermal response of the building elements. This may be undertaken on site or at a remote location, utilising telephone links, which may provide an opportunity for the designer to monitor performance. This work may be assisted by the use of computer simulation techniques,* to check the performance against design parameters and to test alternative controls strategies. The system performance over a range of different control philosophies may be simulated through the weather conditions and operating routines of a typical year in order to establish the optimum control set-points and plant operating periods for minimum energy use. It is only after such work is complete that the client will gain maximum benefit from his significant investment in the building management system.

The use of computer design methods for system sizing is now fairly common and many of these design programs have been extended into the more complex area of performance simulation. One aim of these new generation models is to simulate the performance of the building envelope and also the way in which the building, with its engineering systems, interacts with climate and occupancy patterns in a real situation. The increasing availability of relatively low-cost but very powerful microprocessors will enable BMS eventually to have on-board simulation facilities running in real time. This will have two major benefits: firstly, it will enhance significantly the monitoring capability of the BMS in being able to raise alarms immediately if there should be significant departure from the design performance concept and, secondly, in self-learning adaptive control.

Table 22.1 Energy saving using a building management system

	Energy saving (%) relating to:	
Item	*Inefficient system*	*Efficient system*
Part load efficiency	9.5	2.0
Optimum start	7.5	2.0
Temperature control	7.0	2.0
Optimum stop	5.0	1.0
Holiday scheduling	4.0	–
Pumping/distribution	2.0	0.5
Miscellaneous	2.0	0.5
Staff awareness	5.0	1.0

Source: DOE/PSA M and E *Engineering Guide*, Volume 2, May 1987.

The magnitude of potential energy savings arising from the use of a BMS is dependent, of course, upon the type and condition of the installations before the addition. Energy savings of up to 40% may be achieved where the BMS is introduced into an existing installation which was poorly controlled and maintained. Even so, compared with an efficiently operated system without BMS, the addition may offer potential savings of up

* BRE, IEA (21C), *Empirical Validation*, Ref. IEA 21RN327/93, January 1993.

to 10%. Table 22.1 provides an indication of the order of savings related to a range of operating functions.

Future developments may be expected in all areas of BMS systems. In the development of software, for example, standard protocols will in time be established which will allow full exchange of data between different manufacturers' systems. In respect of hardware, improvements are anticipated to operator terminals in graphic displays, touch responsive screens and the use of active video images which will allow users the facility to see on screen the results of keyboard commands.

Chapter 23

Running costs

The cost of operation of any system providing space heating, ventilation, air-conditioning or hot water supply will depend upon a number of variables including:

- Fuel consumption.
- Power consumption.
- Water consumption.
- Maintenance and consumables.
- Labour.
- Insurance and similar on-costs.
- Interest on capital and depreciation.

The amenity value of various fuels may enter into these considerations. In some cases, such as an industrial application, convenience and cleanliness might be unimportant whereas in a bank or office block they would merit first priority. Similarly, the character and anticipated life-span of the building may bear upon choice and a fuel which would be suitable for a well-equipped solid structure would not match the expendibility of, say, a system-built school.

When selecting systems for a building it is necessary that both the initial cost of the installation and the operating costs be calculated for all the options to establish the most appropriate balance to suit the client's particular circumstances. In some cases initial costs may be the principal factor, whereas for others low operating costs may be the priority; in most cases, however, there will be an optimum choice somewhere between these two extremes. In addition, there are other factors to be taken into account such as reliability, safety and environmental issues.

Energy costs over the last few years have been relatively stable compared with the situation which existed following the 'energy crisis' of 1973. Since fossil fuel is a finite resource, it is likely that energy prices will begin to rise again in real terms in the foreseeable future, but there is no way of predicting when and by how much prices will rise. Figure 23.1 shows how energy prices have increased since 1972. It has been estimated that approximately 55% of national annual consumption of primary energy in the British Isles is used in building services and that about 50% relates to the domestic sector. Economy measures over the past 20 years or so have succeeded in reducing energy use but there is a lot more yet to be achieved.

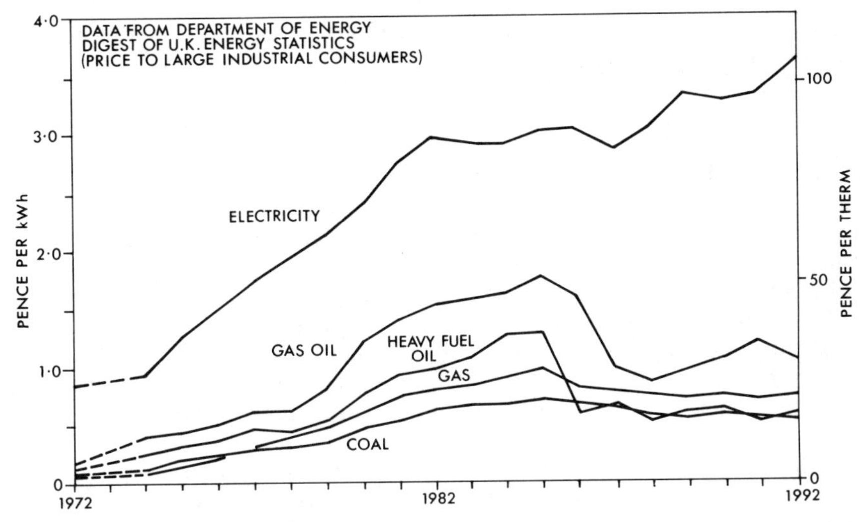

Figure 23.1 Energy prices since 1972

Dependent upon the viewpoint taken in respect of energy use, assessment of the effectiveness of a design may be expressed in terms of:

- Energy cost.
- Energy demand, at the site.
- Primary energy consumption.

Decisions based on fuel and electricity prices are, of course, subject to the commercial and political climate at the time and the results are in consequence variable. Calculations may also be expressed in terms of energy use at the site, or energy demand, in watt or joule energy units. Energy demand expressed in W/m^2 has been adopted by CIBSE to establish whether a building and its systems satisfy set energy demand targets. The CIBSE *Energy Code* sets out a method for calculating demand and target values, which distinguishes between thermal and electrical demands in order to enable other factors such as national reserves to be considered.

Primary energy

This is the quantity contained in fossil fuel, coal, oil and natural gas, etc., from which an overhead must be deducted to cover consumption in production, transmission and distribution in order to arrive at the net supply received by the actual consumer. Typical net to gross values may be taken as:

Electricity	27%
Coal	98%
Gas	94%
Oil	93%

After allowance has been made for combustion inefficiency the proportion of primary energy actually remaining for use in building systems is approximately as follows:

Electricity	27%
Coal	59%
Natural gas and oil	61%

The order of these figures must be borne in mind when considering the merits of conservation measures and those of alternative sources of energy.

Various methods are available for making energy consumption calculations, ranging from very approximate manual estimates through to detailed thermal modelling using computer simulation techniques. It is generally accepted that heating, hot water service and ventilation systems may be assessed with reasonable accuracy using manual methods, but that air-conditioning systems require analysis by some more refined form of computer analysis.

The various methods available may be used with reasonable confidence to compare design solutions. Use of the results to forecast actual energy use must be approached, however, with caution. There are numerous factors which may affect such a forecast adversely, such as the relevance of the weather data assumed, the method of operation or control of the systems and the quality of the maintenance (or lack of it) to which they have been subjected. All these aspects will have a marked effect on energy use.

Space heating

For residential buildings, the *BRE Domestic Energy Model* (BREDEM) provides the standard method of estimating energy use in dwellings and, to assess the effectiveness of design solutions, increasing use is made of *Energy Ratings*. Two national rating schemes, supported by the Energy Efficiency Office of the Department of the Environment, are the *National Home Energy Rating* (NHER) scheme and the MVM *Starpoint* scheme. Both are computer models which provide a simple numerical energy rating, from 1 to 10 for NHER and from 1 to 5 for Starpoint, based upon the energy cost. The higher the rating, the more energy efficient the design.

For all other types of building the degree-day method which has been in use for many years is still the standard manual calculation for heated and naturally ventilated buildings and provides results of acceptable accuracy. Particular care must be taken, however, where the systems vary significantly from the accepted traditional designs, in which case recourse to computer modelling methods would be the preferred alternative. The degree-day method is a function of heat loss, the ratio of normal to peak load which will apply over the period considered, hours of use, internal heat gains, thermal characteristics of the building fabric and system efficiencies.

Heat losses

The totals calculated for design purposes will be in excess of those used as a basis for an estimate of energy consumption. This is due to the fact that if one considers any building, heating design must be such that on each external aspect, sufficient warmth may be provided to maintain a satisfactory internal temperature. In practice, air infiltration resulting from wind will occur only on the windward side; other aspects, i.e. the leeward side, will exfiltrate.

The improvement in the thermal transmission properties of the building fabric in recent years has reduced heat losses by conduction and, in consequence, infiltration losses are a higher proportion of the total; in excess of one-third of the total in the example in Chapter 2. The heat loss used for energy calculations may thus be taken at 15–20% lower than the total calculated for design purposes.

Any losses from piping in a central system which do not contribute to the building heat requirements must be calculated as a separate exercise and added to the net heat loss figure. It should be remembered that, with certain types of system, mains losses may be constant throughout the heating season and thus disproportionately high. An example would be a fan-convector system, controlled by room thermostat switching of the fan motor, fed from constant temperature heating mains.

Proportion of full-load operation

This will be a variable factor depending on the weather. Obviously no system will be called upon to operate at 100% output (based on winter external design temperature) during the whole season. The routine of establishing the proportion of this full load which may be assumed for the purpose of calculation, and how it varies for different parts of the country, is the basis of the method to be described.

Degree-day method

The British degree-day* assumes that in a building maintained at 18.3°C, no heating is required when the external temperature is 15.5°C or over. The difference between the external daily mean temperature and 15.5°C is then taken as the number of degree-days for the day in question. Thus, for an external daily mean temperature of 5.5°C, there are 10 degree-days and if this mean were constant for 7 days then 70 degree-days would be accumulated. Monthly totals are published for various locations which may be used to compare trends from month-to-month or be summed to give the total for the season or for the year. Degree-days for use with hospital installations are used to a base temperature of 18.5°C.

The degree day should not be regarded as an absolute unit, nor should it be pressed beyond its usefulness for comparative purposes. For instance, the effect of low night temperatures may give exaggerated degree-day readings even though the buildings may not be heated at night.

If it were assumed that full load on a heating system takes place when the outside temperature is at –2°C, the maximum degree-day may be taken as 17.5 in any one day. For a heating season of 30 weeks (October/April), there is a possible total of 3710 degree-days and, for 39 weeks (September/May) a possible of 5778 degree-days. Traditionally the 30 week season has been used but, with rising standards, the 39 (or nominally 40) week season is often considered to be more appropriate, particularly for application to residential heating.

* *Fuel Efficiency Booklet No. 7*, 1985. Energy Efficiency Office.

Table 23.1 Degree-day totals for the British Isles, 20-year averages[a] (1972–91), for a base temperature of 15.5°C

Area	*Annual total*	*Seasonal totals*	
		Sept./May	*Oct./April*
Thames Valley	2072	1977	1807
South Eastern	2341	2195	1973
Southern	2261	2111	1890
South Western	1897	1794	1601
Severn Valley	1984	1882	1722
Midlands	2482	2310	2064
West Pennines	2354	2201	1977
North Western	2544	2346	2077
Borders	2628	2386	2083
North Eastern	2504	2329	2074
East Pennines	2437	2270	2032
East Anglia	2384	2234	2013
West Scotland	2579	2367	2087
East Scotland	2650	2420	2119
NE Scotland	2773	2516	2190
NW Scotland[b]	2674	2369	2031
Wales	2243	2066	1826
N. Ireland	2457	2262	1994
Dublin	2322	2135	1897
Cork	2027	1864	1650
W. Coast Ireland	2188	1979	1738

[a] Totals for Republic of Ireland are a 9-year average (1961–70).
[b] Readings available since October 1991.

Degree-day data are compiled monthly by the Meteorological Office and are published by various organisations.* The values listed in Table 23.1 are for a 20 year average to 1991 for stations in the United Kingdom and for a 9 year average to 1970 for those in the Republic of Ireland.

Period of use

Here it is necessary to make assessments of the period of occupancy of a building and the length of time during which the heating system will be at work. It is clear that a finite distinction must be made between the two and although the former may be defined with ease, the latter will depend upon the particular characteristics of each individual building structure and the system associated with it. It is the response of the structure to heat input and the varying abilities of any one of a variety of systems to match that response which is important, as has been discussed in earlier chapters.

For the heating system, the hours of use may be considered in terms of periods per year, per week or per day. For the yearly use, as has been explained previously, a figure of 30

* 'Degree-Day tables for 15.5 and 18.5°C', *BSRIA Engineering Services Management.*

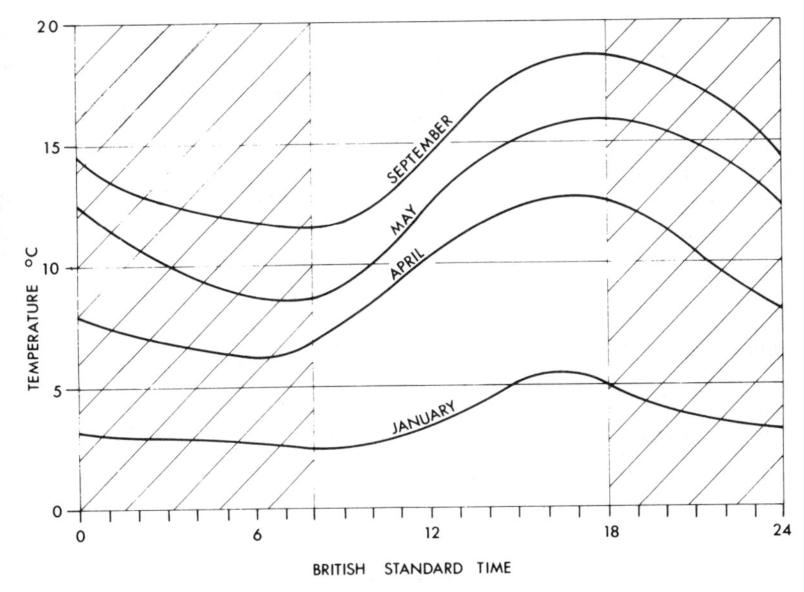

Figure 23.2 Diurnal variations in outside temperature for Manchester: September and May curves displaced to suit time base

weeks is that most commonly assumed for commercial and industrial premises. Since, as illustrated in Figure 23.2, the diurnal external temperature range in May and September is such that early morning and evening heating may well be required in houses, flats, hotels and hostels, etc., a 40 week season is likely to be more appropriate in these instances.

The weekly use again depends upon the kind of building and type of occupancy. Domestic premises, hotels and the like require service over seven days; commercial and industrial buildings need not be heated for more than five days and no more than one or two day usage is normally considered for churches, village halls and the like. For daily use, similar conditions apply and it will be appreciated that continuous heating over 24 hours each day is very rarely required. The shorter the period when service is provided, however, the greater will be the no-load losses or, where plant operation is to be intermittent, the greater the requirement for preheating.

Hence, if energy supply is to be by oil or gas firing or by some form of direct electrical heating, on-peak, no night-time operation of the system will be necessary except perhaps in the most severe weather when heat might be needed to prevent condensation. The plant will be required nevertheless to provide preheating prior to occupation. On the other hand, with solid fuel firing by an automatic means, some level of intermittent operation will usually be necessary in order to keep the fire alive overnight and a no-load loss will thus arise with little heat input to the building. This loss will not be absolute of course and some minor reduction in preheat time may be possible simply because the water content of the boiler will not be completely cold.

Equivalent full load hours

The period during which the heating system is in use, as distinct from the period of occupancy of the building, may be expressed in terms of *equivalent hours of full load* *operation.* A number of different methods leading to evaluation of the quantity by manual calculation has been devised and an outline of one of these is presented in the *Guide Section B18.* The underlying theory* takes account of variations in the level of adventitious heat gains within a building, the period of occupation, the thermal capacity of the structure and the ability of the heating system to respond to demand. Allowance is made also for the fact that the degree-day data compiled by the Meteorological Office relate to the maintenance of an internal temperature of 18.3°C, as has already been explained: for other internal temperatures, an adjustment is necessary.

As far as the internal heat gains are concerned, it is suggested that buildings may be classified as shown in Table 23.2 and that the thermal capacity of the structure be categorised as light, medium or heavy, although it is appreciated that judgement will be required to interpret these distinctions:

Light. Single-storey factory type buildings with little partitioning.
Medium Single-storey buildings of masonry or concrete with solid partitions.
Heavy. Multi-storey buildings with a lightweight facade and solid partitions and floors.

Table 23.2 Approximate temperature rise due to internal heat gains

Building characteristics	*Temperature increment* (K)
Large areas of external glazing; much heat-producing equipment; dense occupancy	5–6
One or two of the above characteristics	4–5
Traditional with normal levels of glazing, equipment and occupancy	3–4
Small glazed areas; little or no heat producing equipment; sparse occupancy	2–3
Residential and dwellings	5–8

Note: these data should not be used for purposes outside this present context.

With regard to hours of system use, the method allows for continuous running over 24 hours or, alternatively, for operation as is appropriate to periods of occupation from 4 to 16 hours per day.

The routine set out in the *Guide Section B18* may be simplified, with no loss of accuracy, by selection of the appropriate degree-day figure from Table 23.1 and an estimated value of internal heat gain from Table 23.2. The former is then corrected by multiplication using values chosen in turn from each of Tables 23.3–23.5 in order to represent the various aspects of the problem.

For example, consider a building in East Anglia having a traditional heavyweight structure and provided with a slow response heating system such as hot water radiators.

* Billington, N. S., 'Estimation of annual fuel consumption', *JIHVE*, 1966, **34**, 253.

The system is operated intermittently over 30 weeks per annum for a 5 day week and occupation during a 12 hour day. The inside and outside design temperatures are 20°C and –3°C, respectively, and thus:

Degree-days (Table 23.1)	=	2013
Temperature increment (Table 23.2)	=	3°C
Factor for building type, use and temperatures (Table 23.3)	=	1.22
Factor for intermittent use over 5 days of a slow response system in a heavy building (Table 23.4)	=	0.81
Factor for occupation over 12 hours in a heavy building (Table 23.5)	=	1.03

Hence,

Equivalent full load operation
= 2013 × 1.22 × 0.81 × 1.03 = 2049 hours/annum

Table 23.3 Factors relating building characteristics to inside and outside design temperatures

Temperature increment from Table 23.2 (K)	*Inside design temperature* (°C)											
	19				*20*				*21*			
	Outside design temperature (°C)											
	–1	–2	–3	–4	–1	–2	–3	–4	–1	–2	–3	–4
2	1.40	1.34	1.28	1.22	1.46	1.40	1.34	1.28	1.52	1.45	1.39	1.33
3	1.27	1.21	1.16	1.11	1.34	1.28	1.22	1.17	1.40	1.34	1.28	1.23
4	1.14	1.09	1.04	0.99	1.21	1.16	1.11	1.06	1.28	1.22	1.17	1.12
5	1.01	0.96	0.92	0.88	1.09	1.04	0.99	0.95	1.16	1.11	1.06	1.02
6	0.88	0.83	0.80	0.76	0.96	0.92	0.87	0.84	1.04	0.99	0.95	0.91
7	0.76	0.72	0.69	0.66	0.83	0.80	0.76	0.73	0.92	0.88	0.84	0.81

Table 23.4 Factors relating structural mass to mode of plant operation and system type

Building structure	*Continuous plant operation (24 hours)*		*Intermittent plant operation*[a]			
			Slow response system		*Fast response system*	
	7-day week	*5-day*[a] *week*	*7-day week*	*5-day*[b] *week*	*7-day week*	*5-day*[b] *week*
Heavy	1.0	0.85	0.95	0.81	0.85	0.71
Medium	1.0	0.80	0.85	0.68	0.70	0.56
Light	1.0	0.75	0.70	0.53	0.55	0.41

[a] With night-time shutdown. [b] With weekend shutdown.

Table 23.5 Factors relating structural mass to period of occupation

Building structure	*Period of actual occupation*			
	4 hours	*8 hours*	*12 hours*	*16 hours*
Heavy	0.96	1.0	1.03	1.05
Medium	0.82	1.0	1.13	1.23
Light	0.68	1.0	1.23	1.40

It is a simple step from this point to calculate the net annual heat requirement, i.e. for a heat loss of 500 kW:

$$\text{Annual load} = (2049 \times 500)/1000 = 1025\,\text{MWh}$$

or

$$= 1025 \times 3.6 = 3690\,\text{GJ}$$

Annual fuel or energy consumption

In order to determine annual fuel consumption, and hence cost, it is necessary in each case to take account of both the properties of the fuel and the *seasonal heat conversion efficiency*. The latter, which includes allowance for plant operation at less than full load, may be read from that part of Table 23.6 appropriate to the plant under consideration. In addition, consideration must be given to the *utilisation efficiency* of the heating or other system (as distinct from that of the plant) which will depend upon the facilities offered for control, the disposition of equipment and other kindred aspects. Suggested values, based upon data included in the *Guide Section B18*, are given also in the two parts of Table 23.6. It will be appreciated that these efficiency values are necessarily figures for guidance: they should be varied in circumstances where it is known that either plant or system performance is significantly better or worse than the average.

Taking, as a single example, a boiler plant fired by light grade fuel oil (class E), which operates to serve the system considered above, the relevant facts may be marshalled as follows:

Calorific value of fuel (Table 11.4) = 41 MJ/kg
Specific gravity of fuel (Table 11.4) = 0.93
Annual load (previously calculated) = 3690 GJ

Routine of system operation = intermittent
System type = automatic central radiator

Heat conversion efficiency (Table 23.6) = 0.65
Utilisation efficiency (Table 23.6) = 0.97

Thus

$$\text{Consumption} = (3690 \times 1000)/(41 \times 0.93 \times 0.65 \times 0.97)$$
$$= 153\,500\ \text{litre/annum}$$

Table 23.6 (a) Heating systems (%)

	Intermittent			*Continuous*		
System description	*Heat conversion efficiency*	*Utilisation efficiency*	*Seasonal efficiency*	*Heat conversion efficiency*	*Utilisation efficiency*	*Seasonal efficiency*
Automatic central radiator or convector systems	65	97	63	70	100	70
Automatic central warm air systems	65	93	60	70	100	70
Fan assisted electric off-peak heaters	100	90	90			
Direct electric floor and ceiling systems, non-storage	100	95	95	100	95	95
District heating						
warm air	75	90	67.5			
radiators/ convectors				75	100	75
Electric storage radiators				100	75	75
Electric floor storage systems				100	70	70

Notes:
Allow for rekindling on intermittent solid fuel plant.
Allow for fuel oil preheating where required.

(b) Hot water systems (%)

System description	*Heat conversion efficiency*	*Utilisation efficiency*	*Seasonal efficiency*
Gas circulator/storage cylinder[a]	65	80	52
Gas and oil fired boiler/storage cylinder[a]	70	80	56
Off-peak electric storage with cylinder and immersion heater	100	80	80
Instantaneous gas multi-point heater	65	95	62
District heating with local calorifier[ab]	75	80	60
District heating with central calorifiers and distribution[ab]	75	75	56

[a] Make separate allowance for mains losses.
[b] Heat conversion efficiency in summer may reduce depending on sizing of heat generators.

Running of auxiliaries

In addition to the fuel consumption, allowance must be made for usage of electrical power by a variety of auxiliary equipment such as circulating pumps and pressurising equipment, etc. In particular, boiler ancillaries must be considered; oil preheaters, transfer pumps, burners and fans; gas boosters and burners; solid fuel stokers, transporters and ash handling equipment. In making estimates of the running times of such equipment in instances where the heat generating plant is operated intermittently, an inclusion must be made for the preheating period which will average about two hours where optimum start controls are used and three hours or more in the case of simple time switching.

As a rule, for the lower range of boiler sizes up to say 1 MW, the cost of the electric current for burner operation is not very significant and is often ignored. For a boiler of this size, when burning oil requiring preheating:

Hourly fuel consumption	= about	0.03 kg/s
Temperature rise (say 10–80°C)	=	70 K
Specific heat capacity	= about	2 kJ/kg

Thus, approximately

Heat required = $0.03 \times 70 \times 2$	=	4.2 kW
Burner fan and pump, etc.	= say	2.0 kW
Induced draught fan (if any)	= say	3.0 kW
Transfer pumps and sundries	= say	1.0 kW
Total		10.2 kW

Running hours for the auxiliaries will be greater than those of boiler output but even if the total were to be half as much again, the consumption noted above would represent only about 1% of the overall energy input. In the case of solid fuel firing, the same order of power consumption for auxiliaries would apply but for gas firing, using a packaged burner with an associated pressure booster, the power consumed would be only about a third of that amount.

Similarly, the current used for driving circulating pumps, etc., is of relatively small magnitude and indeed most of the energy paid for emerges as heat somewhere in the system and so is not altogether lost. A boiler plant of 1 MW capacity would probably be associated with a heating system requiring pumping power of the order of 2 or 3 kW, running continuously throughout the season and thus perhaps equivalent to another 1% of the overall energy input.

As a very broad approximation, pending a proper calculation from manufacturers' ratings of the various drives, etc., an allowance for a power consumption equivalent to 5% of the overall energy input as fuel should cover all auxiliary equipment.

Direct heating systems

For direct systems of convective type, methods similar to those used for calculation of energy consumption in central plants may be used, using appropriate values from Table 23.6, provided that equivalent arrangements for thermostatic and time switch control are provided. If the system were to be of radiant type, a true comparison must be based upon equation of demand to maintenance of dry resultant temperature as outlined in Chapter 2. An estimate may then be made of running cost taking the time when the heaters are likely to be in use and multiplying this by their rated capacity. There is often no true temperature control, in the accepted sense, provided for such systems, and data from comparable installations are probably the best guide.

Mechanically ventilated buildings

Essentially, the thermal energy requirements for mechanically ventilated buildings may be taken to be the same as for the equivalent heated and naturally ventilated spaces, with limited correction and with some additions. It is assumed that no cooling or humidification is to be provided.

Heat requirements

The following items should be considered in turn:

- Reduction in air infiltration rate, possibly due to sealed windows.
- Additional load due to ventilation air, i.e. the quantity of outside air introduced by the ventilation system.
- Reduction in load due to incidental gain to the air from the fan (fan gain).
- Allowance for the benefit arising from the use of any heat recovery device.
- Correction to the effective length of the heating season.

The first two items above will be self-explanatory.

A mechanical ventilation system operating in a heating mode will benefit from heat gain to the air as the electrical power driving the fan is down-graded to heat. The quantity of heat dissipated to the air stream will depend upon whether the fan motor is within or outside of the air stream. Table 23.7 gives approximations adequate for the purpose.

Table 23.7 Approximation of fan heat gain to an air stream

Location of motor	*Motor size (kW)*	*Heat gain to air stream (%)*
In air stream	All	100
Out of air stream	Up to 4	75
Out of air stream	Above 4	85

The application of heat recovery devices to air systems, together with the thermal efficiencies of available equipment, is described in Chapter 17. The benefit in thermal energy terms is in effect to reduce the additional load imposed upon the supply plant due to the introduction of outside air. Any increase in system pressures arising from the imposition of heat exchange equipment in the supply and extract systems must be included in the calculations together with any power requirements for additional pumps or drive motors associated with the heat recovery devices.

Since the ventilation plant may run year-round, the effective period over which heating is introduced may be extended to offset the effect of cold days outside of the normal heating season. In consequence, the degree-day totals may have to be increased, perhaps to the annual totals (see Table 23.1).

Power requirements

The system components driven by electric motors would normally include the supply and extract fans and any drives associated with the heat recovery equipment. Taking the supply fan as an example the power requirement is given by:

$$W = VP_t/(1000\,\eta)$$

where

W = absorbed power (kW)
V = volume flow rate (litre/s)
P_t = fan total pressure (kPa)
η = fan, motor and drive efficiency

The annual power consumption for the supply fan is given by $W \times$ number of hours in operation which will, of course, include for the preheat periods where appropriate.

Conclusion

It must be appreciated that manual calculations of the type outlined earlier in this chapter for heating and ventilating plants, relying upon application of degree-day data, etc., cannot be expected to provide an accuracy in absolute terms of better than *plus or minus 20%.* The variables encountered and the assumptions made preclude a greater precision. The results however are valuable in a comparative sense and may be used with a much improved level of confidence when considering the relative merits of a variety of energy sources: this is the application for which they are intended.

Air-conditioned buildings

Energy predictions for air-conditioned buildings take on a different dimension from those for heated only buildings. As well as heat energy, the designer is concerned with cooling, dehumidification and humidification and in consequence, simple averaged temperature relationships between inside and outside are no longer an adequate basis for calculation. It is necessary to take account of the coincident values for dry-bulb and wet-bulb temperatures and of solar radiation; wind speed may be important but its variation with time is not normally taken into account since it is less significant than the other parameters.

It would be possible to make manual calculations but, for practical purposes, these would be limited to 12 average monthly conditions. However, averages over such long periods may give misleading results since, for example, a month's average conditions may indicate no heating or cooling, whereas the actual span of conditions may give rise to a requirement for both heating and cooling. An improvement on this approach is the concept of *banded weather data* details of which are given in the *Guide, Section A2,** where the weather in each month is divided into groups of days corresponding to ten equal intervals (bands) for a given parameter. The calculation process, nevertheless, would be tedious and liable to error and would be further complicated by the need to calculate not only the room heating and cooling loads for each set of conditions but also the corresponding plant psychrometric processes.

* Petherbridge, P., and Oughton, D. R., 'Weather and solar data', *BSERT,* **4** (4), 1983.

Such energy calculations can only be carried out with any accuracy and consistency by computer simulation methods, a science of modelling the behaviour of real systems. These techniques enable the calculations to model the interactions between building fabric, particularly the influence of thermal mass on dynamic response, room heating and cooling loads, external climate, simultaneous psychrometric processes at the plant, control methods, plant efficiencies and other factors. In the program a logical mathematical description may be prepared to account for all the significant variables, including user operating patterns for intermittent or continuous occupancy, use of equipment and lighting. It should be recognised that such sophisticated programs require substantial amounts of data and that care is needed to ensure that the data are appropriate to the design issue which is being examined.

These programs require considerable computer power;* but the ever decreasing price-to-performance ratio of modern computers over recent years has made such detailed analyses not only possible but cost effective also. For further details reference should be made to one of the software houses specialising in thermal modelling work. Simulation programs would normally use weather tapes giving hour-by-hour values including dry bulb and wet bulb temperatures, solar radiation, wind speed and direction and illuminance. These are available for a number of locations.

CIBSE has taken a part in the development in the British Isles of the concept known as an *Example Weather Year*,† which is claimed to be the most suitable consecutive 12 months weather record, from the Meteorological Office archives, for the purpose of predicting future energy requirements. The year is taken to start at the beginning of the

Table 23.8 Computer simulation study: monthly energy consumption (MJ) of a central plant

	Central plant		*General auxiliaries*		
Month	*Boilers*[a]	*Chillers*[b]	*Fans*	*Pumps*	*Other*
January	117 617.9	187.9	7810.4	3464.9	22 270.9
February	117 319.0	238.1	7995.8	3529.4	27 747.1
March	102 906.7	184.5	9209.5	4333.0	29 293.4
April	62 596.7	474.4	7955.9	4831.9	20 924.3
May	781.1	4 670.4	7683.2	5287.5	9 513.4
June	526.4	11 916.2	9492.3	6699.6	0.0
July	413.0	10 999.1	9335.9	6691.7	0.0
August	489.3	13 457.6	9055.7	6394.0	0.0
September	1 159.3	5 825.4	8744.4	5856.1	0.0
October	69 668.1	724.5	8795.9	6049.4	15 948.1
November	75 706.1	346.3	8155.0	4109.2	12 804.1
December	100 533.8	196.1	7103.3	3101.6	19 976.3
Total	649 717.0	49 220.4	101 375.2	60 378.0	158 477.6

[a] Gas fired. [b] Electrically driven.

* Quick, J. P. and Irving, S. J., *Computer Simulations for Predicting Building Energy Use.* CIBSE Conference, Dublin, 1982.

† Irving, S. J. *The CIBSE Weather Year: Weather Data and its Application*, CIBSE Symposium Proceedings, 1988.

Table 23.9 Computer simulation study: monthly heating/cooling loads and utilisation of heat recovery (kWh)

(a) Rooms

	System energy input to rooms			
	Sensible		*Latent*	
Month	*Heating*	*Cooling*	*Heating*	*Cooling*
January	19 334	8 513	966	993
February	19 117	9 227	1134	924
March	16 586	11 123	1045	1302
April	10 052	9 568	692	1322
May	1	9 241	217	1414
June	3	13 905	49	1909
July	1	12 912	45	1984
August	1	13 379	43	1902
September	3	10 949	61	1801
October	11 133	11 196	484	1551
November	12 383	9 685	437	1404
December	16 198	8 006	821	924
Total	104 802	127 704	5994	17 430

(b) Equipment

	Room heaters and coils			*Central plant*	
Month	*Room heating units*	*Air heating coils*	*Air cooling coils*	*Boiler load*	*Chiller load*
January	19 014	2352	100	23 503	105
February	18 889	2444	127	23 466	133
March	16 416	1931	98	20 181	103
April	9 877	818	255	11 764	268
May	0	91	2869	101	3012
June	0	60	8023	66	8425
July	0	47	7301	52	7666
August	0	56	9362	61	9830
September	0	136	3640	150	3822
October	11 112	817	394	13 122	413
November	12 334	942	186	14 604	195
December	16 040	2181	104	20 043	109
Total	103 682	11 875	32 459	127 113	34 081

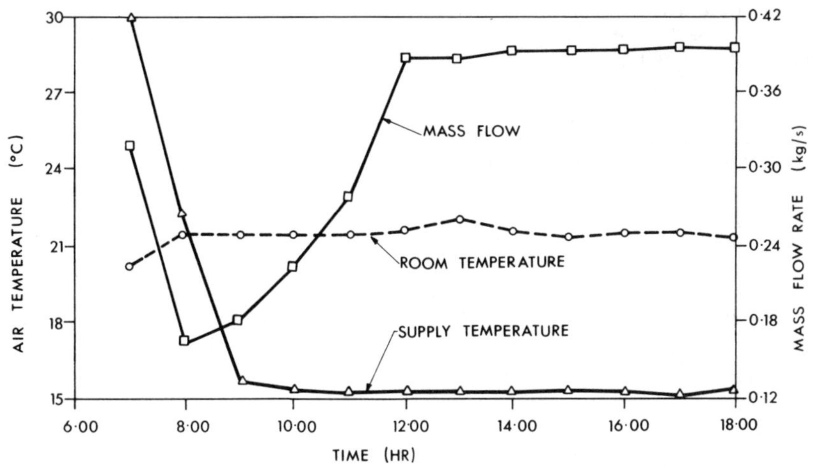

Figure 23.3 Simulated flow rate, supply air and room temperatures for a VAV system

heating season, i.e. 1 October. Currently data, in the form of weather tapes, are available for fifteen locations in the British Isles:*

Aberdeen	Camborne	Kew
Aberporth	Dundee	Manchester
Aldergrove	Eskdalemuir	Newcastle
Birmingham	Glasgow	Norwich
Bristol	Heathrow	Sheffield

Further description of computer simulation methods would have little added benefit but selected output from one such program is given in Tables 23.8 and 23.9 to provide examples of the level of information which may be made available by the use of such techniques. The monthly and annual energy consumptions given in the tables may be extended to fuel tariff and maximum demand analysis if required, to relate energy use to running costs.

Other variables may also be analysed by these methods, such as variation in flow rate to a room served from a VAV system, supply air and room air temperatures (Figure 23.3).

Hot water supply

The consumption of energy for a central system may be considered in two parts:

- Heat losses from storage vessel(s) (Table 23.10), circulating pipework (Tables 6.10 and 6.11), towel airers, linen cupboard coils, dry coils, etc.
- Actual hot water drawn off by the users.

It is worthy of note that these two components of the total are often of the same order of magnitude.

* Facet Ltd, Marlborough House, Upper Marlborough Road, St. Albans, Herts AL1 3UT, UK.

Table 23.10 Heat loss from storage cylinders (insulated with 75 mm glass fibre sited in an ambient temperature of 25°C)

Capacity of cylinder (litre)	*Heat loss from cylinder*	
	Watt	MJ/annum
150	40	1 300
250	55	1 800
300	60	2 000
450	80	2 500
650	100	3 300
1000	135	4 300
1500	180	5 600
2500	250	7 600
3000	280	8 800
4500	360	11 300

A knowledge of the type of building will determine whether heat losses are continuous for 24 hours per day, as in a hospital, or for some lesser period such as 8 hours per day in a school. Similarly the days per annum will vary. For intermittent operation, the heat loss from storage vessels and pipework will continue after the use of the system has ceased, perhaps until the water temperature is in equilibrium with the surrounding air depending upon the period of close-down.

The heat consumption of the water actually drawn off, taken from cold at say 10°C to hot at say 60°C, will be derived from data of water consumptions of comparable buildings. Table 20.4 (p. 565) provides data in this respect. The total of the two components will then form the basis of the sum in which the calorific value of fuel and the seasonal efficiency are taken into account as for heating (see Table 23.6(b)).

Energy targets

Part 2 of the *CIBSE Energy Code* sets down procedures to enable designers to compare calculated energy demands for thermal and electrical consumption with energy targets. Part 2(a) for heated and naturally ventilated buildings and Part 2(b) for mechanically ventilated, whilst Part 2(c) for air-conditioned buildings is in the drafting process.

The procedures given in the *Code* require calculations to be based on set formulae and on data given in the publications, thereby eliminating possible errors arising from engineering judgements. It is emphasised in the *Code* that the calculation procedures are not intended to forecast energy consumption, but are a means of comparing a calculated demand with an equivalent energy target.

The method for calculating demand is similar in principle, but not the same, as that described previously for space heating systems. Energy target values are based upon the physical characteristics of the building, external wall and floor areas, number of storeys and storey height and established constraints related to the type of building and hours of use. Clearly, for a design to be deemed satisfactory, the predicted energy demand must be less than the target value.

Energy audits

An energy audit is fundamentally a management tool to control energy use and costs. It is essential, in the first place, to analyse energy use; where, how much and in what form it is being expended. Then, consumption must be monitored at regular intervals, monthly being often the most convenient period, through the heating season and, where appropriate, through the cooling season. The results should then be compared with a suitable indicator for the same periods, such as degree days. Once the operator has a full understanding of how energy is being used, targets may be set with the aim of reducing usage and followed by detailed studies with intent to identify further savings. At this point it may be beneficial to consult an expert in the field.

The Energy Efficiency Office of the Department of the Environment and CIBSE have both published guidance on the subject,* and the *DHSS Encode, Volumes 1 and 2* are useful references. For general information upon the efficient use of energy in buildings, the reader is referred to the excellent publications produced by the Building Research Energy Conservation Support Unit at the BRE.

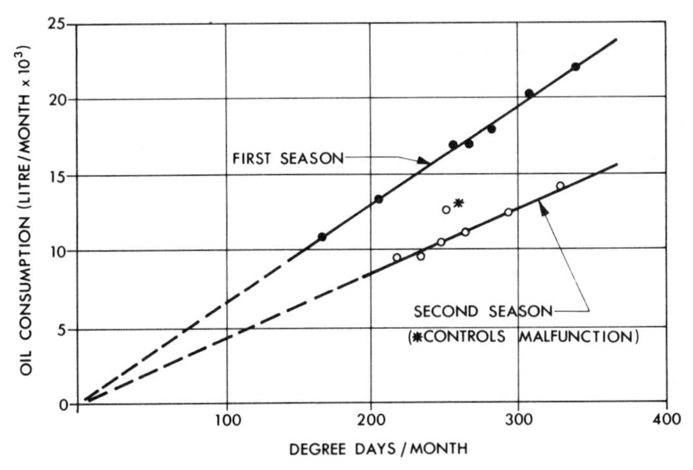

Figure 23.4 Fuel oil consumption relative to degree-days

In carrying out such an audit, it is necessary to establish that, commensurate with the cost of meeting the objective, the following criteria are met in the most energy efficient manner:

- That the most economical source of energy is used and at the best commercial tariff.
- That all energy is converted efficiently.
- That distribution losses are minimised.
- That patterns of demand are optimised and plant controls are compatible.
- That energy recovery equipment is provided.

A graphical representation of readings taken will identify significant variations from an established trend. Figure 23.4 shows a typical example of fuel oil usage against

* Fuel Efficiency Booklet No. 1. *Energy Audits*, Energy Efficiency Office, 1989.
CIBSE Application Manual No.5, *Energy Audits and Surveys*, 1991.

degree-days for two heating seasons, and as can be seen a straight line graph should be obtained. An alternative approach is to calculate for each month a *litre/degree-day* which should give a constant figure, subject to allowable practical tolerances. An annual litre/degree-day ratio, or an equivalent for different fuels, will provide also a means to compare the performance of one building with another. This will quickly identify poor performance, for whatever reason, and establish where improvements should be investigated as a first priority.

Energy content of fuels

It is useful to bring together a summary of energy availability from the various sources and Table 23.11 lists gross heat quantities provided by various fuels.

Fuel prices are not given here because these will vary with the area of the country and with market forces of the day. Prices can be readily obtained from the appropriate supply authorities.

Table 23.11 Energy available from alternative sources

Fuel	*Unit of delivery*	*Heat content per unit of delivery* (MJ)
Solid		
coal	kg	28–30
Oil		
kerosene (C2)	litre	36.6
gas oil (D)	litre	37.9
light (E)	litre	40.3
medium (F)	litre	40.7
heavy (G)	litre	41.2
Gas		
natural	therm	105.5
propane	litre	0.095
butane	litre	0.122
Electricity	kWh	3.6

Environmental audits

Environmental issues are of growing public concern. The construction and use of buildings have a major impact upon both local and global policies, not the least of which is concerned with management of the energy consumed by engineering services. The burning of fossil fuels not only depletes a finite resource but also produces carbon dioxide, thereby contributing to *global warming*, sometimes referred to as the greenhouse effect. Table 23.12 lists information on the rate of carbon dioxide emission by combustion of various common fuels.

Many of the other environmental issues arising from the subject of heating and air-conditioning of buildings have been dealt with elsewhere in this book. To enable an assessment to be made of the impact that a building and its engineering services will have

Table 23.12 Carbon dioxide emission from fuels*

Fuel	*Co$_2$ emission* (kg/kWh delivered)
Electricity	0.72
Coal	0.34
Oil	0.29
Natural gas	0.21

* Values taken from BREEAM data.

on the environment, a rating system has been developed based upon a point scoring system known as BREEAM, the *BRE Environmental Assessment Method.* A number of publications on the BREEAM method are available from BRE and others are in the course of preparation.

Maintenance

There are various levels of maintenance which may be applied to building services, the two most common being:

- *Corrective.* The majority of operations are carried out on breakdown or fall-off in performance, backed up sometimes with specific tasks undertaken on a regular basis.
- *Preventative.* Planned procedures are undertaken at regular intervals related to statistical failure rates of equipment with intent to extend the life of the plant overall to a maximum and to minimise the risk of breakdown. Work is carried out to a predetermined schedule enabling resources and material purchase to be planned in advance.

Labour

All new plant may be assumed to be fully automatic. Thus estimates of labour based on hand-stoking, or even the filling of hoppers and the clearing of ash from boiler plant are very rarely relevant.

Below a certain size of plant, the occasional attention required for fully automatic systems is so small as to be no more than a very part-time occupation, and it cannot be given a value. What that size is it is difficult to say, as there comes a point, such as in a hospital or a factory, where operators must be employed anyway, even though their duties are more in the nature of shift technicians or mechanics than involving actual labour. The cleaning of boilers, etc., keeping the log, turning pumps and fans, etc., off and on and minor maintenance jobs, such as oiling and greasing, are still required. This may be worked on a three-shift basis which with weekends and holidays involves four men with a possible fifth as stand-in.

However, if costing must be applied, an assessment can be made at current rates of remuneration for the kind of attention mentioned above.

Contract maintenance

Average maintenance of small installations is often covered by a maintenance contract with the installer or fuel supplier, involving two or three visits a year and costing a nominal sum per visit.

In the medium range of installations it is now common practice to employ one of the maintenance contractor firms to include for example checking, cleaning of boilers, lubrication, the cleaning of calorifiers, changing filters on air handling systems, cleaning and checking operation of air treatment equipment, adjustment and attention to controls and building management systems, inspection and routine activities associated with refrigeration plant and general overhaul. It is often the case that a specialist contractor will be employed to look after items such as controls and refrigeration equipment. Annual charges will vary according to the size of the installation. In the larger scale plants such as are dealt with by hospital authorities, estate companies and industrial concerns, the maintenance staff may be part of the organisation, probably operating on a planned maintenance basis. Building management systems are now used for plant asset records and in planning, scheduling and recording maintenance activities.

Taken by and large, maintenance of heating plant may be expected to be covered by an allowance of the order of 10% of the fuel cost per annum, and, in the case of air-conditioning systems, an allowance of the order of 20% of the annual energy cost. Clearly these figures depend to a large extent on the level of sophistication of the systems and the relative ease of access to plant. Reference may be made to the Building Maintenance Information Service for further details.

Insurance

Currently, insurance and annual inspection is required for plant including steam boilers and receivers, pressure vessels and certain types of ventilation equipment. The same requirements do not apply to low pressure heating systems nor to general ventilation and air-conditioning plant, but this matter is under review by the Health and Safety Commission. Nevertheless, it is customary for the owner of any sizeable system to take out insurance as a matter of self-protection, and to cover against fracture, burn-out and accidents of all kinds.

Investment appraisals

All major decisions concerning proposals for schemes relating capital expenditure to revenue return, ranging from new installations, to modernisation and to energy conservation measures, should be subject to some form of economic appraisal. The capital cost may relate to all scales and conditions of building, from the dwelling to the largest industrial complex and from present stock, having significant useful life remaining, to concepts as yet undeveloped from a basic client brief.

Similarly, revenue return may arise from static measures such as orientation, building form and materials of construction or from the dynamic characteristics, life cycle and maintenance needs of equipment. For a full understanding of the impact of engineering systems upon the building, space for plant, fuel storage, service shafts and the like should be a part of the analysis. Obviously, the effect of taxation cannot be ignored, nor indeed can the impact of investment grants, capital allowances, accelerated depreciation allowances and so on. Each of these facets of the problem acts and interacts with the remainder.

Life cycle costs

Life cycle costing methods may be used to compare alternative solutions and would include the total initial cost of the installation, operating costs through the life of the plant, including energy and maintenance, and any disposal cost at the end of its useful life. Analysis may involve component replacement at intervals, if the total period used for comparison exceeds the expected life of any component part. Many economic appraisal methods may be used to produce long term cost forecasts to assist in the decision making process. Recourse to this method of analysis would be made where there is a high initial cost difference, or high energy cost difference, or where there is significant disparity between the expected life of the options.

The financial return from *static* energy conservation inclusions, orientation, building form and shape, solar exclusion, thermal insulation and the like may be assumed to relate directly to the remaining whole-life of the building. In the case of *dynamic* action, however, different criteria obtain, but there is limited information available of an historical nature with respect to this subject area; in particular on operating costs. One important aspect is the expected useful life of plant and equipment. Table 23.13 brings together information from various sources for a range of systems. The life of any component will depend to a large extent upon the working environment, the quality of manufacture, level of maintenance, hours of use and its suitability to the mode of operation. As a consequence, a spread of years is offered for guidance in the table.

Payback period

Probably the most common term used when assessing the viability of a proposal to replace an existing scheme with one to be justified on grounds of running cost savings, such as energy saving methods, is *simple payback period*, i.e. the number of years required for a capital expenditure to be recovered through annual income or, in the context of energy conservation, annual savings. Nevertheless, using raw cost data, payback is a rather crude concept for use in decision making since it takes no account of the fact that capital, if invested elsewhere, would earn interest. For example, if an energy conservation measure to reduce annual running cost by £2000 required capital expenditure totalling £8000 this would give a simple payback period of £8000/2000, i.e. 4 years.

Present value

The concept of *present value* (PV) has much to commend it since it is easy to understand and can handle staged expenditure or known changes in the pattern of annual saving. In brief, present value analysis converts all outgoings – including capital expenditure, running costs, repairs and where appropriate all income, to their equivalent values as measured at a single point in time, usually the present. The analysis relies upon the fact that a pound today is worth more than a pound tomorrow – inflation aside – since if today's pound were invested it would have earned interest by tomorrow. Thus, stating the converse, a pound at some future date is worth less than a pound today: the value will have been reduced – or discounted – in proportion to the rate of interest assumed.

Most relevant works of reference contain tables listing present values over a wide range of time periods for selection of interest or discount rates. The public sector, for example NHS Hospital Trusts, uses a discount rate (sometimes referred to as the *Test Discount Rate*, TDR) of 6% in the fourth quarter of 1993, but a rate of 8% is in common use in the nationalised industries. Public sector rates are reviewed quarterly but are revised much less frequently. Organisations in the private sector select a level which is most appropriate

Table 23.13 Economic life of equipment

Type	*Item*	*Life* (years)
Boiler plant	Steam and HTHW	
	shell and tube	15–25
	water tube	25–30
	cast iron	25–30
	Medium and LPHW	
	steel	15–20
	cast iron sectional	15–25
	electrode	30–40
	Incinerators	15–20
Boiler auxiliaries	Combination controls	15–20
	Boiler electrodes	5–10
	Feed pumps	15–20
	Feed water treatment	15–20
	Oil burner	15–20
	Atmospheric gas burner	20–25
	Non-atmospheric gas burner	15–20
	Solid fuel handling	10–15
	Fans	15–20
	Instrumentation	10–20
Chimney	Steel	8–15
Heating equipment	Heat exchangers	20–25
	Radiators, cast iron	20–25
	Radiators, steel	15–20
	Suspended ceiling heating	15–20
	Unit heaters, gas or electric	10–15
	Unit heaters, hot water or steam	15–20
	Fan convectors	15–20
Liquid distribution	Pipework	25–30
	Pumps, base mounted	20–25
	Pumps, pipeline	10–15
	Pumps, sump	5–15
	Pumps, condensate	15–20
	Tanks	15–30
	Valves	20–25
Refrigeration plant	Reciprocating, large	15–20
	Reciprocating, medium or small	10–15
	Centrifugal	20–25
	Absorption	20–25
	Cooling towers, depending on material	10–25
	Air-cooled condensers	15–25
	Evaporative condenser	15–25
Air-conditioning units		10–15
Heat pumps		15–20
Air-handling plant	Package units, small	10–15
	Cooling coils	15–20
	Heating coils	15–20
	Air washers	15–20
	Fans, centrifugal	20–25
	Fans, axial	15–20
Air distribution	Ductwork	25–30
	Dampers	15–20
	Fans, propellor	10–15
	Fans, roof mounted	15–20
	Grilles, diffusers, etc.	25–30
	VAV or DD control terminals	15–20
	Induction unit or fan coils	15–20
	Insulation	20–25
Controls	Pneumatic	20–25
	Electric/electronic	15–20
	Self-contained	10–15
	Motors and starters	15–20

to their business objectives: a discount rate of around 10% is common. Whilst the required rate of return on an investment may be set at a level higher than the discount rate, it is common practice for the same figure to be adopted for an assessment of viability and for a comparison of options. There are two important aspects which require emphasis, the first being that the method assumes constant money value in real terms over the period considered, i.e. that there is either zero inflation or that inflation is at a common level across the board. The second is that the analysis is sensitive to both the discount rate and the life cycle of the component parts. Examples of the application of present value techniques are given in the *Guide Section B18* and in the *CIBSE Energy Code Part 2.*

Energy savings viability charts

It is possible to use the tabulated figures for PV to produce what, for want of a better description, may be called viability charts, plotting the ratio between capital cost and annual savings due to conservation measures against a time base. Figure 23.5(a) shows how such a chart would appear for a number of different discount rates from zero to 20%.

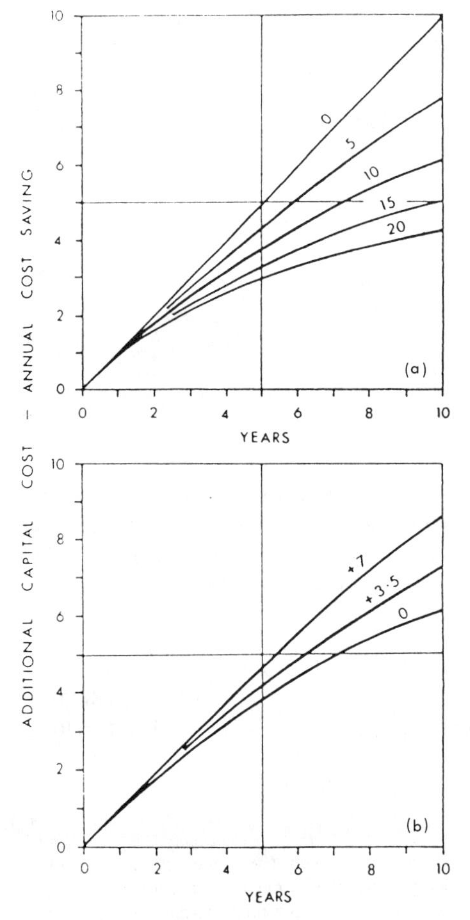

Figure 23.5 Viability charts showing effect of changes in: (a) discount rates; and (b) energy cost premium

Figure 23.5(b) takes a discount rate of 10% as a base and shows how the relationship will vary if energy costs inflate disproportionately to the general pattern.*

Three curves are shown, representing zero, plus 3.5 and plus 7% per annum. An excess rise of 3.5% per annum represents a doubling in 20 years and an excess rise of 7% per annum represents a doubling in 10 years.

To cater for the situation where the energy saving material or equipment has a life shorter than that of the building to which it is provided, the capital outlay figure may be adjusted using data extracted from a further set of calculated factors as in Table 23.14. The first and second columns here show how the present value of a single sum varies with the passage of time: the fourth column lists multipliers for capital sums according to the life cycle of the material or equipment. These multipliers are no more than successive additions from the second column, e.g. the 20 year cycle factor is the sum of the initial outlay, unit plus the 20 and 40 year values: 1.0 + 0.149 + 0.022 = 1.171.

Table 23.14 Present values of a single sum, discounted at 10% and life cycle multipliers for capital sums

n (years)	*Present value of a single sum*	*Life cycle* (years)	*Multiplier for capital cost*
Present	1.0		
5	0.621	5	2.630
10	0.386	10	1.623
15	0.239	15	1.310
20	0.149	20	1.171
25	0.092	25	1.101
30	0.057	30	1.057
35	0.036		
40	0.022		
45	0.014		
50	0.009		
55	0.005		

A developed chart for a discount rate of 10%, plotted to a logarithmic base scale, is shown in Figure 23.6 with added curves for excess premium in annual energy costs. The scale on the right hand side is provided as an aid to interpolation between the curves. The broken lines illustrate the following applications to practical problems:

Example 1

Capital cost of energy saving measure having a 60-year life as for the building — £8000

Estimated annual saving at present day prices — £1540

Ratio = 8000/1540 = 5.2

* At the time of compiling this revision, energy prices have remained effectively in line with the level of general inflation for a number of years. Relative high energy cost inflation rates could return in the future, and in consequence it is considered valid to include the methodology for carrying out cost appraisals.

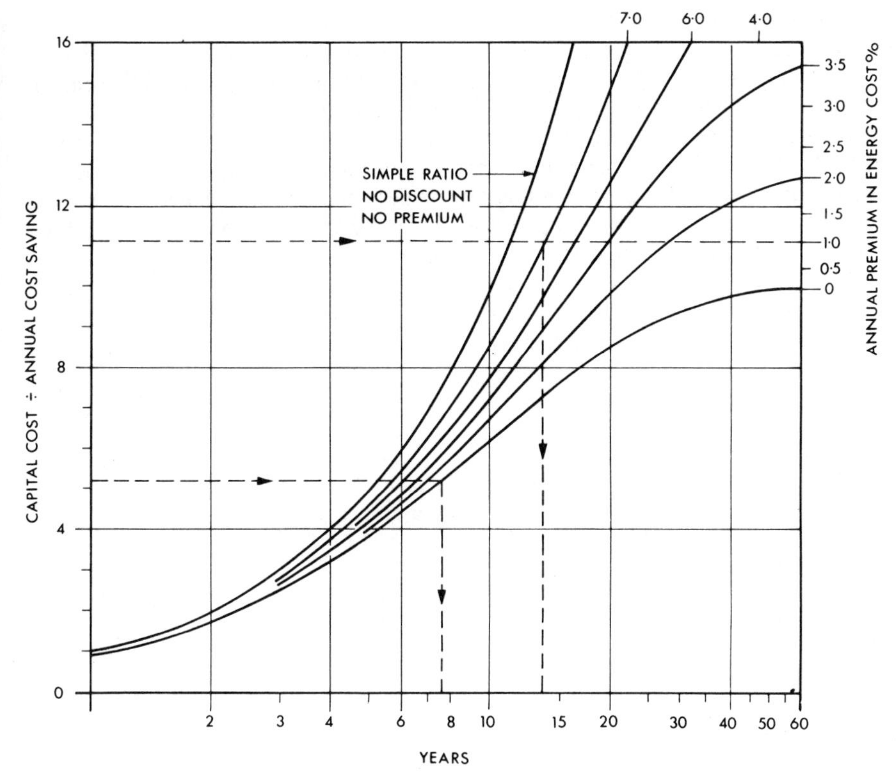

Figure 23.6 Viability chart for 10% discount rate

From the chart, even if energy costs do not rise disproportionately, the measure will be viable at 7.6 years and profitable thereafter.

Example 2

Capital cost of energy saving equipment requiring renewal every 15 years	£2000
Multiplier from table	1.31
Estimated annual saving at present day prices	£236

Ratio = (2000 × 1.31)/236 = 11.1

From the chart it may be seen that:

- The equipment will never be viable unless energy costs rise disproportionately to the general level of inflation.
- If energy costs rise disproportionately at 7% per annum, the equipment will be viable at 13.5 years.
- The equipment will be viable at the end of the building life (60 years) if energy costs have risen disproportionately by about 1% per annum over the whole period.

To conclude this brief attempt to explain the impact of economic factors upon energy conservation measures, the authors feel bound to add that they cannot accept that cost equations are the end of the story. These take no account of amenity values, thermal and visual comfort, aesthetics, contentment, productivity and quality of life. Some at least of these aspects should be quantified and introduced as weighting to the results produced by soulless mathematics. There remain, furthermore, the fundamental issues of national and international importance that the remaining reserves of energy be husbanded and that CO_2 emissions be limited to minimum practical levels. Economic considerations should perhaps be regarded as of secondary importance.

Chapter 24

Combined heat and power (CHP)

The concept of providing heat and electrical power from a single station is by no means new. Many private hospitals and other public buildings in the British Isles were equipped, 80 or more years ago, with plant able to fulfil the dual function. At a time when interest was growing in the provision of district heating, immediately after the Second World War, a matter which aroused considerable debate was, if heat were to be provided from an electrical generating station, which was the primary output and which the recovered waste product?

The two earliest editions of this book, of 1936 and 1945, included a chapter entitled *Combined Electrical Generating Stations* which described plants installed to the designs of the original authors prior to 1939. It is of passing interest to note that, for a hospital site, the electrical requirement was provided by four alternators, having a total output of 500 kW, driven by reciprocating steam engines. Space heating and domestic hot water systems were supplied from calorifiers fed with exhaust steam from the engines, topped up as necessary with live steam from the boiler plant. In contrast, the second project was equipped with four compression ignition engines driving DC generators with a total capacity of 1.8 MW. Space heating and domestic hot water systems were supplied with waste heat from the engine jackets and exhaust systems, topped up as necessary from shell type hot water boilers.

In later editions of the book, since interest in large scale on-site generation had lessened, this last chapter in the text changed both in emphasis and title and then concentrated, with some reservations, upon the thermal and piped distribution aspects of group and district heating systems. In the last few years, however, it would seem that the wheel of fashion has turned full circle, possibly as a result of the 1983 Energy Act which encouraged owners of small generating plants to run these in parallel with the public mains supply system and export surplus capacity thereto, thus enabling them to recoup a marginal proportion of the owning and operating cost.

Basic considerations

Heat generated by a conventional boiler plant is normally obtained at a reasonably high thermal efficiency whilst delivery of electrical power to the end user is very often at an average efficiency, with respect to raw energy, of not much more than 26%. The intentional wastage of heat at cooling towers is a feature of most generating stations: some small attempts have been made to use the cooling water for heating but nothing in the British Isles compares with the co-ordinated use of energy resources elsewhere in Europe or Scandinavia.

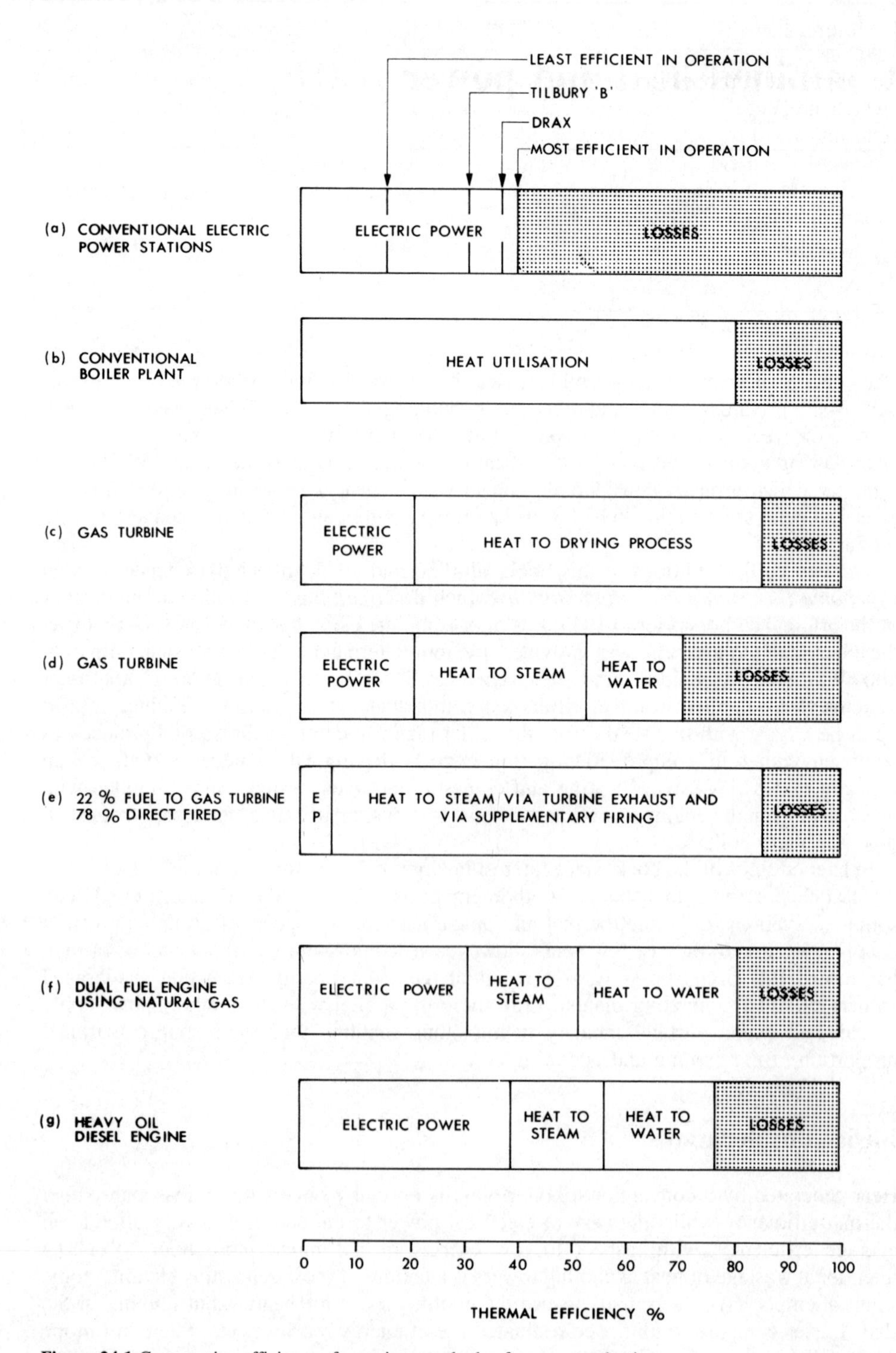

Figure 24.1 Comparative efficiences for various methods of energy production

It is sufficient to say that the least thermally efficient of power stations could be made the most efficient if use were to be made of the waste heat. Further, it is obvious that, when an appreciable proportion of the nation's supply of electricity is generated in stations which are out-dated and have efficiencies as low as 20%, the cost of power to the consumer will remain high. Figure 24.1 illustrates the comparative efficiency of various forms of electrical, thermal and thermal-electric energy production.

A combined heat and power installation will necessarily comprise three fundamental items of plant:

- A prime mover.
- An electrical generator/alternator.
- Equipment for heat recovery.

To these essential items must be added:

- Automatic running controls.
- Safety and monitoring controls.

and, perhaps,

- Supplementary heat and power sources.

Figure 24.2 illustrates, in schematic form, how the various components of a simple system might be arranged. Similar groupings of components have been built for many years and the concept of independence from public supplies has always appealed to planners. When local generation has been employed, it has usually been for one of two reasons, either to ensure a continuity of supply where essential to the functional use of the building and process or to provide complete security of an electrical supply in all circumstances, including national disaster or insurrection.

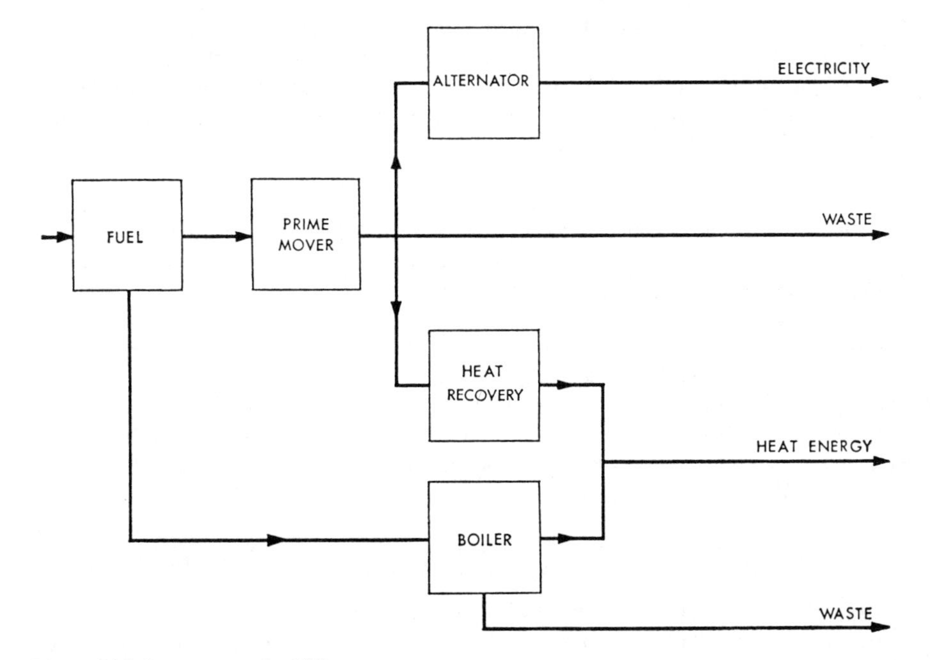

Figure 24.2 Components of a CHP system

Prime movers

At this stage, it would be as well to consider the form and availability of the heat rejected by some of the normally encountered prime movers since it is an essential to the design of CHP systems that the marked differences in these characteristics are not only appreciated but put to use. The more important are:

- The heat to power ratio at both full and part load.
- The level or levels of heat rejected.
- The energy input rate throughout the working range.

The first of these characteristics has the greatest influence on selection of equipment since the exercise will be centred on matching the heat-to-power ratio of the plant to the building energy-demand profiles. However, it is not enough to ensure that a match exists at full or any other loading but rather to maintain a balance for as much of the year as possible.

Suitable prime movers fall into three general categories: steam turbines, gas turbines and reciprocating engines. Of these, the steam turbine has the highest heat-to-power ratio at some ten or more to one. It follows that such equipment is most suited to applications which have large thermal demands *vis-à-vis* electric power requirements: these will probably lie in an industrial field, where a large process load may be accompanied by a lesser electrical power requirement. Gas turbines have heat-to-power ratios of the order of three to one but the manner of application can increase this ratio materially by adding supplementary direct firing to the waste heat boiler. The third group includes industrial gas engines, diesels and spark-ignition engines which are in many cases derived from the automotive field: items in this group have heat-to-power ratios of between one and a half and two to one. In exceptional circumstances, of course, the load profiles of a single building or a building complex may be matched more accurately by the heat-to-power characteristics of a CHP group made up from a combination of prime movers.

The following two paragraphs provide no more than brief notes on the characteristics of steam and gas turbines since they are more appropriate to either large site areas or to industrial applications and are thus not in context with current financial limitations or the size of the single site building projects which are the subject of this chapter.

Steam turbines

Electrical generating stations on a national scale employ high pressure steam boilers and sophisticated techniques to obtain the maximum possible electrical supply from the fuel input, small though it is in relative terms. However, thermal-electric stations of comparable generating capacity, designed to produce heat as well as power, would be designed for quite different pressure and operating conditions and would, in general, tend to employ less complex and cheaper steam raising plant. In such cases, back pressure turbines are most likely to be used or, where some proportion of higher pressure steam is required for process use, this may be obtained by pass-out from the turbine at an intermediate pressure. Typically, 4 kg/s of steam per MW would be available at a pressure of about 200 kPa which represents the heat-to-power ratio of 10:1 mentioned previously.

Gas turbines

The full-load thermal efficiencies of typical open-cycle gas turbine sets are such that the generation of electricity without subsequent heat recovery, leads to an efficiency of less than 20%. In fact, industrial gas turbine sets have full load heat ratios varying from about

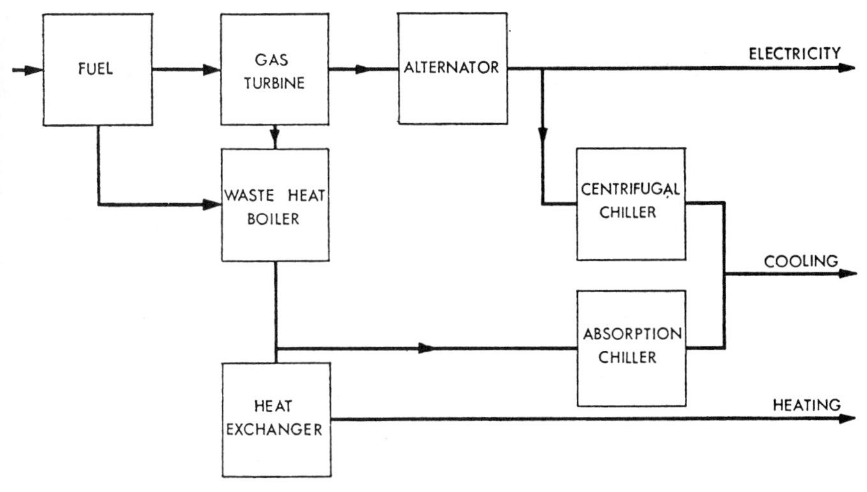

Figure 24.3 Line diagram of a gas turbine total-energy system

five or seven to one, which represent an efficiency range of 14–20%. However, the exhaust temperature of a gas turbine at full load will be between 450 and 550°C and thus, using two or more stages of heat transfer to utilisation circuits, an overall thermal efficiency of approaching 85% may be achieved. By their very nature, gas turbines tend to be noisy and plant rooms thus require a good deal of acoustic treatment. A line diagram of a gas turbine heat recovery source is given in Figure 24.3.

Reciprocating engines

Whilst the open-cycle gas turbine has the great merit of rejecting waste heat in a single fluid stream and at a relatively high temperature, the shaft power produced for generation of electricity is comparatively low. The various forms of reciprocating engine, on the other hand, are able to convert an equivalent amount of fuel into approximately double the shaft power obtained from a gas turbine. Among the types of reciprocating engine generally suitable for combined heat and power (CHP) systems are the following:

- Industrial spark-ignition gas engines manufactured for stationary use and of rugged construction, with first class in-built facilities for such maintenance as is necessary.
- Industrial diesel engines having similar characteristics to the above, many initially designed for use as prime movers for standby generators.
- Automotive derived engines, i.e. lorry (diesel) and motor car (petrol) engines which have been modified and de-rated for this application.

The first two of these categories are both familiar and proven: they need little further comment except to add that their operational life is long, extending in known instances of as much as 50 000 hours. Figure 24.4 illustrates how the output of a large industrial engine might be employed. The production-line engines of the third category will normally have been modified as to pistons, valve gear, ignition equipment and sometimes cylinder heads

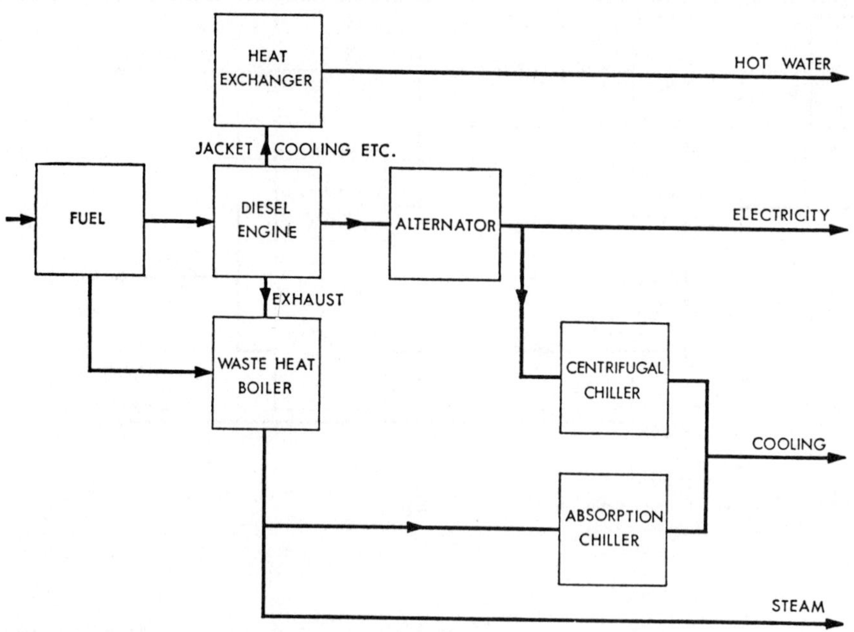

Figure 24.4 Line diagram of a diesel engine CHP system

to suit operation with natural gas as a fuel. Since they will run as stationary units in relatively clean plant rooms and at constant speed, their life may be expected to exceed that of their road-service equivalents by a considerable margin.

Generators and controls

Equipment for electrical generation, normally taking the form of three-phase 415 V alternators, will not be discussed here at any length since electrical equipment is not within the scope of this book. The leading characteristics of the two generic types available, *synchronous and asynchronous*, must be noted, however, since they are fundamental to system economics and operation. The former are significantly more expensive in first cost and require a complex control system to maintain a standard frequency: for parallel operation with the public mains supply, further synchronisation equipment is essential. Synchronous machines are, however, battery started and thus self-sufficient: they do not normally require power factor correction and they may be used in the role of standby generators without modification except to switch gear.

An asynchronous machine is similar to an induction motor with the *excitation current*, at a low leading power factor, taken from the public mains supply. In fact, they are often used in the role of a motor for starting the prime mover, taking over as a generator when the running speed is reached and the prime mover is able to provide the required power. For this starting duty, care must be taken that the 'motor' is not overloaded. In small sizes, induction generators are some 3–4% more efficient than the alternative but capacitors for power factor correction are required. A major advantage therefore is simplicity, since no complex synchronisation gear is necessary. A disadvantage however is that, unless a suitable auxiliary supply is available to provide the excitation current, an asynchronous machine cannot act as a standby generator in the event of failure of the public mains supply.

Generation equipment of any type must be provided with all those controls appropriate to the starting, running and shut-down sequences and also with the necessary sensors and actuators to suit the monitoring circuits associated with mains failure, etc. The neutral earth connection, required when generation is wholly independent of the public supply, must be isolated by a safety interlock when exporting power and running in parallel with the public mains supply.

Heat recovery equipment

Heat recovery from reciprocating machines may be better appreciated by reference to Figure 24.5 which is a Sankey diagram representing energy input and the various components of energy output. As will be seen, over 50% of input is recoverable at a temperature adequate for use in a low pressure hot water heating system. If a suitable use for low grade heat is available, such as a swimming pool or a preheater for domestic hot water, a further 10% of input may be recovered by use of a condensing heat exchanger although this may not be economic in many cases and in any event should never be incorporated when the fuel has any sulphur content.

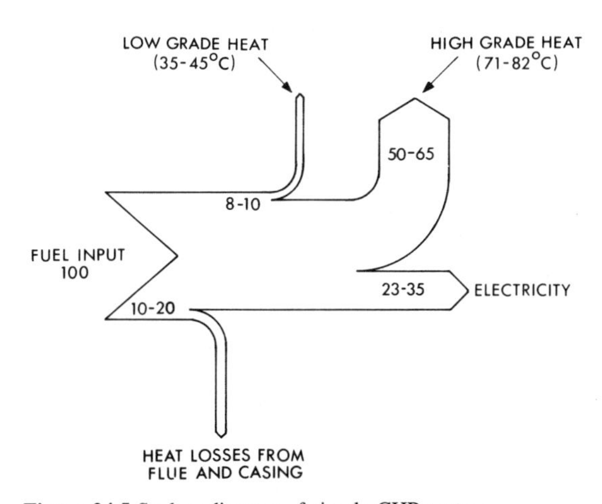

Figure 24.5 Sankey diagram of simple CHP system

It will be appreciated that the water jackets of most engine types are not designed to withstand the pressure which would be imposed upon them if they were connected directly to a building heating system. It will be necessary therefore to interpose a *water-to-water* heat exchanger in the circuit and this will, usually, be of the *shell-and-tube* type mentioned previously in Chapter 8. The exhaust gases, similarly, must pass through a heat exchanger, *gas-to-water* in this case, which may, in some instances, serve a dual role and be arranged to act as a silencer to attenuate exhaust noise from the engine, as Figure 24.6. A secondary terminal silencer on the exhaust line may also be required.

The primary (heat recovery) circuit of the water-to-water heat exchanger will in most instances be pumped through the engine jacket and then be boosted in temperature when passed, in series, through the gas-to-water unit. The secondary circuit will be that of the

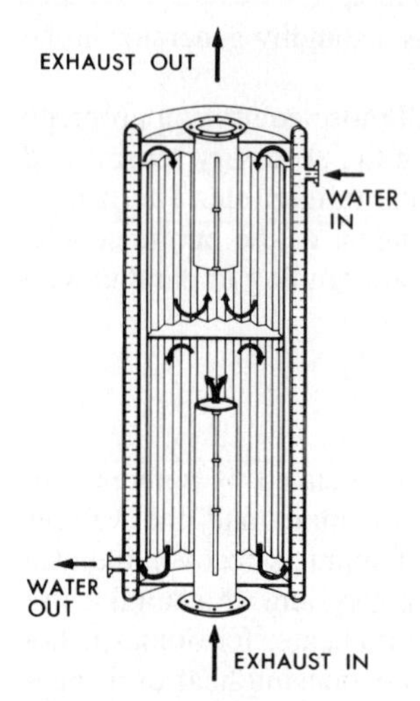

Figure 24.6 Combined exhaust heat exchanger/silencer

building heating system as illustrated to the left hand side of Figure 24.7. For use in summer or at other times when the heat recovered would be in excess of requirements, the recovery circuit must be arranged to permit dumping of the surplus at a dry cooler or similar blown unit. It is as well, also, that the gas-to-water heat exchanger be provided with a bypass, not shown in the figure, on the exhaust side which should be well protected from accidental human contact.

The piping configuration used for connection of the heat recovery system to the utilisation circuits must be selected to suit the pumping and other hydraulic arrangements.

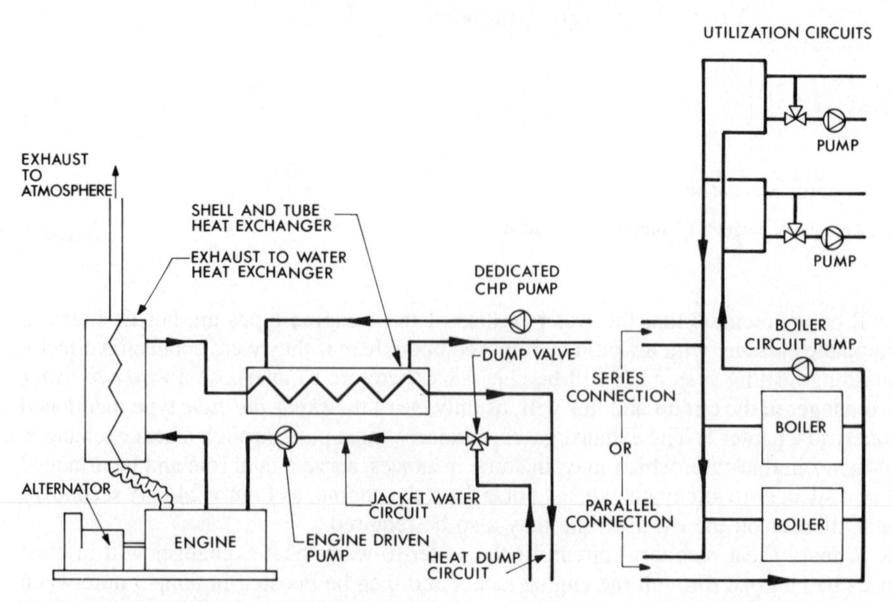

Figure 24.7 Circuit diagram of pipe connections

In general, since the CHP system is unlikely to have sufficient capacity to meet the peak thermal load in mid-winter, boiler plant of some sort, for emergency standby or parallel operation, will need to be incorporated in the central plant and means for dealing with this will need to be considered also. When adding heat recovery equipment to an existing system it is sensible to involve the original designer in order to avoid complications. The right hand side of Figure 24.7 shows the alternatives of series and parallel pipe connections between the two systems. To avoid complication, details irrelevant to the operation of the circuits such as isolating valves etc. are not shown in this figure.

In most instances, the output of the CHP alternator will be connected, for normal running, to a discrete section of the electrical distribution system and, if the central switch gear is properly designed, sub-divided and identified, connection should not present any problems. If, however, a significant volume of power export to the public supply is envisaged or if the alternator is to be used for emergency standby purposes, connection will be less simple and, once again, if an existing system is to be extended, it would be sensible to involve the original designer.

Packaged plant

It is fortunate that, in this particular field, a number of suppliers are able to offer standard design packages made up from engines, alternators and heat recovery components which are complementary one to another and pre-wired complete at works to a central control

Table 24.1 Duties, etc., of available packaged plant (1993)

Energy (kW)			*Ratio*	
Input	*Output*			
Gas	*Electrical*	*Thermal*	*Output/ input*	*Thermal/ electrical*
135[a]	38	70	0.80	1.84
190[c]	54	97	0.79	1.80
260[c]	75	130	0.79	1.73
300[d]	88	140	0.76	1.59
310[a]	95	150	0.79	1.58
400[c]	110	191	0.75	1.74
490[c]	145	265	0.84	1.83
503[d]	147	240	0.77	1.63
555[b]	175	270	0.80	1.54
610[a]	185	320	0.83	1.73
740[c]	220	385	0.82	1.75
840[b]	285	400	0.82	1.40
850[a]	290	420	0.84	1.45
990[c]	300	445	0.75	1.48
1190[b]	385	580	0.81	1.51
1398[c]	404	529	0.67	1.31
1598[c]	469	728	0.75	1.55
1775[b]	575	870	0.81	1.51
1980[c]	560	900	0.74	1.61
2037[c]	605	755	0.67	1.25
2328[c]	710	1039	0.75	1.46
2380[b]	770	1165	0.81	1.51
2717[c]	805	1007	0.67	1.25
3105[c]	945	1384	0.75	1.46

[a–b] Data are from different manufacturers and may not have been presented on exactly the same basis.

panel. Table 24.1 lists the ratings of a selection of the packages available. In the abstract, some such packages may not be particularly well laid out either for maintenance purposes or appearance but this fault is probably a result of either an insufficient turnover to pay for refinement in design or of the inevitable cost cutting in a competitive market.

Plant room accommodation

To avoid noise nuisance to building occupants, steps should be taken to isolate any plant room accommodating CHP equipment, with particular emphasis upon doors which should be heavy and well fitting. As is noted at the end of Chapter 12, openings to admit both combustion air for the prime mover and ventilation air for removal of heat generated by equipment generally should be provided with sound-attenuating louvres. If site circumstances so dictate, it may be desirable to site the equipment at roof level in which case anti-vibration mountings and a sectional acoustic enclosure will certainly be required.

CHP systems – criteria for selection

The principal matter which should be taken into account when either considering a variety of components or selecting a packaged CHP system is the probable load factor, *vis-à-vis* the building loads, electrical and thermal, to which it is to be connected. Generally speaking, the output of the equipment or system should be such that use can be made of the heat recovered throughout the day during the summer months and for a large proportion of all hours during the winter months.

It is generally agreed that if the plant cannot run for a minimum of 3000 hours per annum (an overall load factor of 33%) it cannot possibly be viable: an annual load factor of between 0.5% and 0.7% (4500–6000 hours per annum) would provide a simple pay-back period of three to four years. In consequence of these limitations, a plant running time of about 10 hours per day in summer and 18–24 hours per day in winter would be necessary.

Taking account of these minimum recorded running hours, it must be remembered that repeated recycling of the prime mover will shorten its working life very considerably due to additional wear during the first few minutes after start-up, particularly with cold lubricating oil. The control arrangements should limit multiple starts to less than six per hour. Similarly, in the interests of maintaining an optimum efficiency, modulation of demand on the prime mover should be avoided.

In consequence of these problems, it would seem that, for the generalised case, it is the thermal rather than the electrical load which is the determinant of plant size and that this may well be of the order of a third of the rating of any associated boiler plant. If it be proposed that the size of the CHP plant be increased so that the alternator is large enough to be used as a standby generator, it is almost certain that the machine would be too large to meet the normal electrical load of the building. In these circumstances, although undesirable from the economic point of view, the use of two smaller duplicate units would be desirable so that one could run continuously for the CHP duty and the second be used only when there is a peak demand or need for stand-by equipment.

Maintenance and life cycles

The economics of operating an in-house CHP plant will depend not only upon the cost differentials of energy supplies but also upon establishing realistic life cycles for the various components of the plant and taking note of the costs of day-to-day maintenance. The most vulnerable item of equipment is the prime mover and it is generally accepted that industrial machines will require a major overhaul about once every 3 years whereas automotive derived machines will need similar attention about every two years. Dependent upon circumstances, it has been estimated that day-to-day maintenance will cost, in 1993, between about 0.5 and 2.0 pence per kWh generated.

Bearing in mind that many building owners and occupiers do not employ, or wish to employ, the staff having the necessary skills, it is fortunate that many of the suppliers of standard design packages are able to offer a range of maintenance agreements ranging from a full service including emergency call-out to a minimum of major overhauls only. The former of these may have attractions but the higher charge might be such that the marginal cost advantage of adopting a CHP system would disappear.

Alternative systems

The plant described under previous headings has been devoted to heat recovery for piped space heating and domestic hot water systems. An alternative does exist however in the form of the packaged plant shown in Figure 24.8. As may be seen, this consists of a package having two compartments, one to accommodate the prime mover and heat recovery equipment proper and a second to house components of a warm air plenum system, including recirculation and supply fans, a filter and a mixing chamber for either admitting outside air or dumping return air.

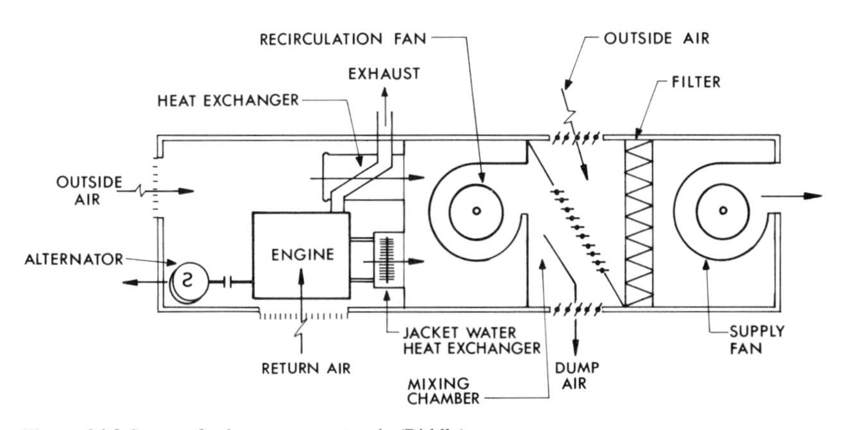

Figure 24.8 System for heat recovery to air (Biddle)

As shown, the package is designed for roof mounting and includes an acoustic enclosure but, presumably, other configurations are possible. The manufacturers claim that very high energy efficiencies are obtained at full load, the exhaust heat exchanger being designed as a condensing unit, and plant radiation losses reclaimed. Modulation of thermal output is by dumping of unwanted return air at the mixing chamber thus reducing annual efficiency considerably.

The package may be modified by the addition of a 'water heating pack' which will convert 50% of the available recovered heat for export from the package to water-using systems. The manufacturers suggest that modulation of output may then be achieved by proportioning the heat recovered between the air and hot water systems.

The application of this type of unit would seem to be confined to sites having a high electrical load which is more or less constant over the whole year. The prime mover and alternator, often in duplicate, would be sized to meet that load, thus showing an economy over purchase from the public mains supply. Heat recovery would be an adventitious bonus.

Conclusion

The introduction of small scale, single site, CHP systems as described in this chapter* is a considerable advance upon the grandiose proposals advanced in the recent past. The previous, seventh, edition of this book referred to projects of the magnitude then envisaged and offered the opinion that they were doomed to failure. Hindsight confirms that there was truth in the paragraph which ended that edition:

> The more sensible alternative lies with the large number of relatively small industrial estates and science parks throughout the British Isles, each having access to real 'hands-on' experience of heat and power generation. Encouragement of properly engineered exercises on this scale, where high quality construction of the heat distribution system will not absorb a disproportionate cost, would be more productive than a greater expenditure upon grandiose cost-limited white elephants. Real energy savings would result, industrial competitiveness would be improved and return upon capital expenditure would be available rather than a long term bad debt. Small, in this context, is financially sound as well as beautiful.

* For further information on this subject, the reader is referred to the many excellent publications produced by the Energy Technology Support Unit (ETSU) at Harwell and, in particular, the following: Heap, C. C. and Grey, R., *Guidance Notes for the Implementation of Small Scale Packaged Combined Heat and Power.* Good Practice Guide, GPG 1, 1990.

Appendix I

Temperature levels

Degrees Celsius to degrees Fahrenheit

°C	*0°*	*1°*	*2°*	*3°*	*4°*	*5°*	*6°*	*7°*	*8°*	*9°*
	°F	°F	°F	°F	°F	°F	°F	°F	°F	°F
0	32.0	33.8	35.6	37.4	39.2	41.0	42.8	44.6	46.4	48.2
10	50.0	51.8	53.6	55.4	57.2	59.0	60.8	62.6	64.4	66.2
20	68.0	69.8	71.6	73.4	75.2	77.0	78.8	80.6	82.4	84.2
30	86.0	87.8	89.6	91.4	93.2	95.0	96.8	98.6	100.4	102.2
40	104.0	105.8	107.6	109.4	111.2	113.0	114.8	116.6	118.4	120.2
50	122.0	123.8	125.6	127.4	129.2	131.0	132.8	134.6	136.4	138.2
60	140.0	141.8	143.6	145.4	147.2	149.0	150.8	152.6	154.4	156.2
70	158.0	159.8	161.6	163.4	165.2	167.0	168.8	170.6	172.4	174.2
80	176.0	177.8	179.6	181.4	183.2	185.0	186.8	188.6	190.4	192.2
90	194.0	195.8	197.6	199.4	201.2	203.0	204.8	206.6	208.4	210.2
100	212.0	213.8	215.6	217.4	219.2	221.0	222.8	224.6	226.4	228.2
110	230.0	231.8	233.6	235.4	237.2	239.0	240.8	242.6	244.4	246.2
120	248.0	249.8	251.6	253.4	255.2	257.0	258.8	260.6	262.4	264.2
130	266.0	267.8	269.6	271.4	273.2	275.0	276.8	278.6	280.4	282.2
140	284.0	285.8	287.6	289.4	291.2	293.0	294.8	296.6	298.4	300.2
150	302.0	303.8	305.6	307.4	309.2	311.0	312.8	314.6	316.4	318.2
160	320.0	321.8	323.6	325.4	327.2	329.0	330.8	332.6	334.4	336.2
170	338.0	339.8	341.6	343.4	345.2	347.0	348.8	350.6	352.4	354.2
180	356.0	357.8	359.6	361.4	363.2	365.0	366.8	368.6	370.4	372.2
190	374.0	375.8	377.6	379.4	381.2	383.0	384.8	386.6	388.4	390.2
200	392.0	393.8	395.6	397.4	399.2	401.0	402.8	404.6	406.4	408.2
210	410.0	411.8	413.6	415.4	417.2	419.0	420.8	422.6	424.4	426.2
220	428.0	429.8	431.6	433.4	435.2	437.0	438.8	440.6	442.4	444.2
230	446.0	447.8	449.6	451.4	453.2	455.0	456.8	458.6	460.4	462.2
240	464.0	465.8	467.6	469.4	471.2	473.0	474.8	476.6	478.4	480.2
250	482.0	483.8	485.6	487.4	489.2	491.0	492.8	494.6	496.4	498.2
260	500.0	501.8	503.6	505.4	507.2	509.0	510.8	512.6	514.4	516.2
270	518.0	519.8	521.6	523.4	525.2	527.0	528.8	530.6	532.4	534.2
280	536.0	537.8	539.6	541.4	543.2	545.0	546.8	548.6	550.4	552.2
290	554.0	555.8	557.6	559.4	561.2	563.0	564.8	566.6	568.4	570.2
300	572.0	573.8	575.6	577.4	579.2	581.0	582.8	584.6	586.4	588.2

Appendix II

SI unit symbols

Quantity	*Unit*	*Symbol*	*Common multiples or sub-multiples*	*Multiplier*	*Symbol*
Length	metre	m	kilometre	$m \times 10^3$	km
			millimetre	$m \times 10^{-3}$	mm
Area	square metre	m^2	hectare	$m^2 \times 10^5$	ha
			sq. millimetre	$m^2 \times 10^{-9}$	mm^2
Volume	cubic metre	m^3	litre	$m^3 \times 10^{-3}$	(litre)*
			cu. millimetre	$m^3 \times 10^{-9}$	mm^3
Time	second	s	hour	s × 3600	h
Velocity	metre per second	m/s			
Acceleration	metre/sec^2	m/s^2			
Frequency	hertz (cycle per sec)	Hz			
Rotational frequency	revolutions per sec	s^{-1}			
Mass	kilogram	kg	tonne	$kg \times 10^3$	t
			gram	$kg \times 10^{-3}$	g
			milligram	$kg \times 10^{-6}$	mg
Density (specific mass)		kg/m^3			
Specific volume		m^3/kg			
Mass flow rate		kg/s			
Volume flow rate		m^3/s	litre per second	$(m^3/s) \times 10^{-3}$	litre/s
Momentum		kg m/s			
Force	newton	N	meganewton	$N \times 10^6$	MN
			kilonewton	$N \times 10^3$	kN
Torque		Nm			
Pressure (and stress)	pascal	Pa	megapascal	$Pa \times 10^6$	MPa
			kilopascal	$Pa \times 10^3$	kPa
			bar	$Pa \times 10^5$	b
			millibar	$Pa \times 10^2$	mb
Viscosity					
Dynamic	pascal second	Pa s	centipoise	$Pa\ s \times 10^{-3}$	cP
Kinematic	centimetre^2/sec	cm^2/s	centistoke	$(cm^2/s) \times 10^{-2}$	cSt
Temperature	Kelvin	K			
	degree Celsius	°C			
Heat					
Energy, Work, Quantity of heat	joule	J	gigajoule	$J \times 10^9$	GJ
			megajoule	$J \times 10^6$	MJ
			kilojoule	$J \times 10^3$	kJ
Heat flow rate (power)	watt	W	gigawatt	$W \times 10^9$	GW
			megawatt	$W \times 10^6$	MW
			kilowatt	$W \times 10^3$	kW
Thermal conductivity		W/m K			
Thermal resistivity		mK/W			
Specific heat capacity		kJ/kg K			
Latent heat		kJ/kg			

*This book does not use *l* as a symbol for litre.

Appendix III

Conversion factors

Imperial units to SI

		SI	
Unit	*Imperial*	*Exact*	*Approximate*
Length	1 inch	25.4 mm	25 mm
	1 foot	0.3048 m	0.3 m
	3.28 feet	1 m	
	1 yard	0.9144 m	0.9 m
	1 mile	1.609 km	1.6 km
Area	1 sq. in	645.2 mm^2	
	1 sq. ft	0.092 m^2	
	10.77 sq. ft	1 m^2	
	1 sq. yd	0.836 m^2	
	1 acre	4046.9 m^2	
Volume	1 cu. in	16.39 mm^3	16 mm^3
	1 cu. ft	28.32 litre	28 litre
	35.32 cu.ft	1 m^3	
	1 pint	0.568 l;itre	0.6 litre
	1 gallon	4.546 litre	4.5 litre
Mass	1 pound	0.4536 kg	
	2.205 pounds	1 kg	
	1 ton	1.016 tonne	1 tonne
Density	1 lb/cu.ft	16.02 kg/m^3	
Volume flow rate	1 gall/minute (g.p.m.)	0.076 litre/s	0.075 litre/s
	1 cu. ft/minute (c.f.m.)	0.472 litre/s	0.5 litre/s
Velocity	1 foot/minute	0.0051 m/s	
	197 ft/minute	1.0 m/s	
	1 mile/hour	0.447 m/s	0.5 m/s
Temperature	1 degree Fahrenheit	0.556°C	
	t = 32°F	t = 0°C	
Heat	1 British thermal unit (Btu)	1.055 kJ	1 kJ
	1 'Old' therm (100 000 Btu)	105.5 MJ	100 MJ
	1 Unit of electricity (kWh)	3600 kJ	
Heat flow rate	1 Btu/hour	0.2931 W	0.3 W
	1 horsepower	745.7 W	750 W
	1 ton refrigeration (12 000 Btu/hour)	3.516 kW	3.5 kW

Imperial units to SI (*continued*)

Unit	*Imperial*	*SI*	
		Exact	*Approximate*
Intensity of heat flow rate	1 Btu/sq. ft hour	3.155 W/m^2	3 W/m^2
Transmittance (*U* value)	$\frac{1 \text{ Btu}}{\text{sq. ft hour °F}}$	5.678 W/m^2 K	6 W/m^2 K
Conductivity (*k* value)	$\frac{1 \text{ Btu inch}}{\text{sq. ft hour °F}}$	0.1442 W/m K	
Resistivity (1/*k*)	$\frac{1 \text{ sq. ft hour °F}}{\text{Btu inch}}$	6.934 m K/W	
Calorific value	1 Btu/lb 1 Btu/cu. ft	2.326 kJ/kg 37.26 J/litre or 37.26 kJ/m^3	2.5 kJ/kg
Pressure	1 pound force per sq. in (lb f/sq. in) 1 inch w.g. (at 4°C) 1 inch mercury (at 0°C) 1 mm mercury 1 atmosphere (standard)	6895 Pa or 68.95 mbar 249.1 Pa or 2.491 mbar 33.86 mbar 1.333 mbar 101 325 Pa	7000 Pa 70 mbar 250 Pa 2.5 mbar 34 mbar 1 bar
Pressure drop	1 inch w.g./100 ft	8.176 Pa/m	
Latent heat of steam (atmospheric pressure)	970 Btu/lb	2258 kJ/kg	2300 kJ/kg
Latent heat of fusion of ice	144 Btu/lb	330 kJ/kg	
Steam flow rate	1 lb/hr 8 lb/hr	0.126 g/s –	 1 g/s
Heat content	1 Btu/lb 1 Btu/gall 1 Btu/cu. ft	2.326 kJ/kg 0.2326 kJ/litre 0.0372 kJ/litre	
Thermal diffusivity	1 ft^2/hr	2.581 × 10^{-5}m^2/s	
Moisture content	1 lb/lb 100 grains/lb	1 kg/kg 0.014 kg/kg	

Index

Absolute filters, 475, 476, 479
Absorbing glasses, 76
Absorption coefficients, duct lining materials, table, 455
Absorption plant, refrigeration, 526
Access openings, ductwork, 432
Accumulators, steam, 116
Acoustic insulation, ductwork, 454
Acoustics, *see* Sound
Activated carbon filters, 343
Activity, human, 7
Adiabatic effect, dry air coolers, 534
Admittance (Y value), steady state, 60
Adsorption filters, 475, 480
Adsorption process, dehumidification, 489
Air:
  conditioned, 85
  distribution and movement, 341
  dry and saturated, properties, tables, 159, 505
  and fuel combustion, 333
  humidity, 342
  movement of, 7, 341
  properties, tables, 159, 505
  purity, 7–8, 342–3
  as refrigerating medium, 520
  removal from steam heating systems, 155, 157–8
  standard, for fan testing, 463
  temperature, ventilation air, 85
  volume, 7
Air bottles, high temperature systems, 224
Air change, and ventilation allowance, 42–3
Air change rates, 20, 42, 339; table, 417
  and ventilation, table, 339
Air cleaning, 472–3
Air-conditioned buildings, running costs, 656–9
Air-conditioning, 365–99
  definition, 365
  distribution of air, 407–22
  factors, 399
  general principles, 366–8
  high velocity systems, 373
  need, 366–7
  other methods, 390
  summary, 398
  traditional systems, 368–75
  unorthodox techniques, 396
  and weather data, 367–8
Air-conditioning design:
  application, 505–14
  calculations, 501–15
  heat gains, 501–2
  other systems, 514–15
  psychrometry, 502–5
  system diagrams and automatic controls, 515
Air-conditioning systems:
  central system with cooling coil, 624–5
  central system with cooling coil by-pass, 625
  central system with sprayed coil, 625–7
  chilled water system control, 633–4
  chiller control, 634–5
  controls, 624–36
  cooling towers, 635–6
  dual-duct system, 630–1
  fan-coil system, 627–8
  induction system, 628–30
  reverse cycle heat pump system, 633
  smoke control, 633
  terminal reheat system, 627
  variable volume system, 631–3
  zoned system, 627
Air-conditioning units, economic life, table, 666
Air contaminants, *see* Contaminants
Air cooling coils, 487–9
Air dehumidifiers, 489
Air diffusion, terminology, 407
Air distribution, 400–22
  for air-conditioning, 407–22
  combined lighting/air distribution, 418–19
  economic life, table, 666
  general principles, 400–7
  outlets, 407–22
  performance, 418
  terminology, 407
Air filters:
  cleaning, 476–80
  efficiency, 475–6
  practical, 475
  tests, 473–4
Air filtration, 472–80
  air cleaning, 472–3
  air contaminants, 472
  filter tests, 473–4
  filters, 475–80
Air flow, in ducts, 437–9
Air grilles, extract and return, 419–22
Air handling plant, packaged, 489–90
  economic life, table, 666
Air heater batteries, 486
Air heating coils, 485–7
Air heating systems, 111, 159–61
  high velocity system, 161
  plenum system, 160–1
Air humidification, 481–5
  capillary washers, 484–5
  mechanical separators, 482–3
  pan humidifiers, 485
  spray washers, 483–4
  sprayed coils, 485
  steam injection, 481–2

Air humidification – *continued*
ultrasonic atomisers, 485
Air infiltration:
summer, 85
winter, 41–7
application, 46–7
natural, table, 42
through window cracks, 43–6; table, 45
whole building, 54
Air intakes, fans, 471–2
Air pressure gauges, diaphragm, 460
Air supply:
to boiler houses, 332–3
for human emissions, 335–7
to occupied rooms, criteria, 340–3
for other reasons, 337–9
rates, table, 15
Air-to-air heat exchangers, 490–500
comparisons, table, 499
Air valves, 223
Air velocities:
fume cupboards, table, 361
measurement, 14
Air venting, 223–4
high level stores, central hot water systems, 596
Air volumes:
natural ventilation, stack effect, table, 347
wind, table, 348
Air washers, 483–5
Air-water systems, induction, air-conditioning, 381
All-air systems, air-conditioning:
dual duct, 370
induction, 375
simple, 374
variable volume, 376
Aluminium ducts, 428
Aluminium radiators, 187
Ammonia, in refrigeration, 519
Amplitude, sound, 447
Anemometers:
hot wire, 14, 460
rotating vane, 459–60
swinging vane, 460
vane, 14
Angelery hot water generators, 562–3
Apartment heaters, 193
Arrestance tests, air-filters, 474
Ash, 312
handling, 293
in liquid fuels, 294
Assmann thermometers, 14
Atmospheric burners, gas firing of boilers, 282–3
Atomisers, ultrasonic, 485
Attenuators, 455–7
cross-talk, diagrams, 457, 461
in-duct, diagram, 455
Auditoria, ventilation, 340
Automatic controls:
air-conditioning system controls, 624–36
and building management systems, 636–43
Building Regulations, 602
control devices, 608–16
controller modes of operation, 616–19
elementary components, 602
heating system controls, 621–4
sensing devices, 605–8
systems controls, 619–21
system types, 603–5
Automatic roll fabric filters, diagram, 478
Automatic stokers, 272–5
Auxiliaries:
energy consumption, table, 657
running of, and costs, 654
Axial flow fans, 465, 468

Bacharach scale, 289
Backward curved fans, 464
Bacteria, 226, 473
see also Legionnaires' disease
Bag filters, 476, 478, 479
Balanced flues, 261, 331
Balancing dampers, 432
Balancing valves, 242
Barometric pressure, calculation, 502
Baseboard convectors, 191
Basements, external distribution of heating, 164
Bellows, expansion joints, 251, 252
Bitumen residue, 293
Bituminous coal, combustion, 333
Bivalence, 140–1
Blinds, and summer heat, table, 93
Block storage radiators, 124–5
BMS, *see* Building management systems
Bodily comfort, *see* Comfort
Body odour, 1, 336
Boiler auxiliaries, economic life, table, 666
Boiler firing:
gaseous fuel, 281–7
miscellaneous burner equipment, 287–8
oil fuel, 276–81
oil fuel regulations, 280
solid fuel, 272–6
Boiler houses, 262
air supply to, 332–3
roof-top, 332
Boiler plant, economic life, table, 333
Boilers:
balanced flue, 261
cast iron, 263
condensing, 259–61
controls, 629
efficiency, 256–7
electrode, 268–9
firing, 272–86
and firing equipment, 255–89
fittings and mountings, 270–1
hot water, 564–70
and hot water supply, 261–2
input and output, 256–7
instrumentation, 289
margins, 257–8
miscellaneous burner equipment, 287–8
packaged models, 258–9, 267–8
power, 255–6
selection criteria, 257–62
shell type, 266–7
steel, reverse flow, 265
steel-sectioned, 264–5
thermal storage, 259
types, 263–9
water tube, 268
see also Dry core boilers; Wet core boilers
Booster heaters, 128
Bower-barffing, 554
Branch connections, ducts, 429–30
BRE Domestic Energy Model (BREDEM), 646
BREEAM (BRE Environmental Assessment Method), 663
Brush filters, 476–7
Bubble-top calorifiers, 562
Builders' work ducts, 427–8
Building insulation, 25–7
Building management systems (BMS), 636–43
benefits, 641–3
central intelligence,
distributed intelligence, 639–41
hardware, 637–8
long-distance communication, 641
PC-based systems, 641
software, 638, 641
Building materials, table of properties, 3
Building Regulations, 27, 29–30
Part L1, 28

Part L2 under revision, xi
and automatic controls, 602
and energy conservation, 66
Buildings:
extraneous influences (winter), 18–19
orientation, 81–2
shape and solar effects, 80–1
in summer, 65–93
in winter, 18–64
see also Tall buildings
Bunker underfeed stokers, 275–6
Burner controls:
gas-fired, 284–5
oil-fired, 280–1
Burner equipment, for boilers, 287–8
Burners, gas-fired, for boilers:
atmospheric, 282–3
controls, 284–5
dual-fuel, 285
gas boosters, 285–6
packaged, 283–4

Calculations, for air-conditioning design, 501–15
Calorifères, 189
Calorific value, and combustion, 314–15
Calorifier-condensate cooler units, 213–14
Calorifiers:
bubble-top, 562
non-storage, 212–14
rating, 564
ratings and mountings, 214
steam-to-water, 561
water-to-water, 560–1
Canopies, kitchen ventilation, 356–7
Capillary washers, humidification, 484–5
Car parks:
contaminants in, 8
ventilation, 340
Carbon, 290, 312
Carbon dioxide (CO_2), xi, 662; table, 663
and refrigeration, 519–20
ventilation rates needed, table, 335
Cast iron boilers, 263–4
Cast iron radiators, 183–4
Ceiling diffusers, 409–11
Ceiling heating, electrical, 106
Ceiling panels, embedded, 179–80
Ceilings:
chilled, 394
and kitchen ventilation, 357–9
perforated, 408
Central air-conditioning system, controls, 624–7
Central intelligence, building management systems, 639
Central plant:
air-conditioning, 368–9; diagram, 370
heating/cooling loads, table, 658
Central processors, building management systems, 637
Central stores:
capacity, 122–4
control, 138
electrical storage heating, 117
equipment for, 130–8
Central systems, hot water supply, 554–9
combination systems, 557–9
combined systems, 556–7
direct systems, 554–5
indirect systems, 555–6
piping design for, 578–600
Centrifugal compressors, 525
Centrifugal pumps, 200–1
application, 205–6
characteristics, 204–5
for condensate, 204
construction, 206–7
mountings, 207
for water systems, 201–3
Change-over system, induction air-water systems, 384
Charge controllers:
central stores, 138
room stores, 129
Chartered Institution of Building Services Engineers, *see* CIBSE
Chilled beams, 394
Chilled ceilings, 394
Chilled water storage, 528–30
Chilled water system control, 633–4
Chiller control, 634–5
Chimney loss, 315
Chimneys, 318–29
construction, 326–9
economic life, table, 666
gaseous fuels, 325
heights, 322–4
materials for, 329
mechanical draught, 324–5
solid/liquid fuels, 319–25
see also Combustion and chimneys
Chlorinated rubber paint, for ducts, 429
Chlorofluorocarbons (CFCs), in refrigeration, 520
CHP, *see* Combined heat and power
CIBSE:
Air distribution, 457n
CIBSE Energy Code, 29
CIBSE Guide, xi
Circulating pressures, gravity hot water systems, table, 244
Circulating pumps, central hot water systems, 598
Circulators, gas, 551–2
Cisterns, hot water supply systems, 570–1
Clean Air Act, 322–4, 342
Cleaning agents as contaminants, 8
Climate, maritime, 65
Clinker, handling, 293
Closed stoves, 96–7
Clothing, and human comfort, 7
Coal:
energy from, table, 662
net to gross values, 645–6
properties, tables, 307, 308
Coal tar series, 293
Coanda effect, air diffusion, 407
Coefficient of linear expansion, property of materials, table, 3
Coefficient of performance (COP):
cooling effect, 518
heat pumps, 542–5
Coils:
air cooling, 487–9
air heating, 485–7
Coke, unavailability, 291
Coking stokers, 274–5
Cold draw, 253
Cold water feed pipework, central hot water supply, 583–5
Combination boilers, 259
Combined heat and power (CHP), 671–82
basic considerations, 671–3
CHP systems - criteria, 680–2
components, figure, 673
generators and controls, 676–7
heat recovery equipment, 677–80
plant room accommodation, 680
prime movers, 674–5
reciprocating engines, 675–6
Combustion air, oil fuel boiler firing, 278
Combustion and chimneys, 312–33
air supply to boiler houses, 332–3
calculations, 316–18
chimneys, 318–29
flue-gas disposal, 329–32

Combustion and chimneys – *continued*
process, 312–15
reactions, table, 313
Comfort, bodily, determination, figure, 16
Commercial premises:
air distribution, figures, 404
heating, factors in choice, 113
natural ventilation, 344–5
Compressor drives, heat pumps, 545
Compressors, refrigeration, 521–6
centrifugal, 525
reciprocating, diagram, 522
sealed (hermetic), diagram, 523
Computer rooms, air distribution, 403, 405; plan, 408
Computers, heat from, 87–8
Concert hall, downward air distribution, diagrams, 403
Concrete boilers, 269
Condensate:
centrifugal pumps for, 204
disposal, 157, 158–9
flow, data, table, 250
pipe sizes, 250
Condensate handling equipment, 214–18
Condensate return pumping units, 217–18
Condensation, 37–41
application, 39–41
avoidance in heating systems, 96
interstitial, 38–9
vapour pressure gradients, diagram, 39
Condenser coil type, evaporative condensers, 535
Condenser water treatment, refrigeration, 538–9
Condensers:
air-cooled, 533–4
refrigeration components, 533–4
water circuit, figure, 635
Condensing boilers, 259–61
Conditioned air, 85
Conduction, 6
losses in summer, 82
losses in winter, 20–2
Conductivities, 22
in materials, table, 3
Conductivity, thermal, 6
of pipe insulating materials, figure, 162
Cone-type ceiling diffusers, 409
Construction features, atypical (winter), 31–7
Contaminants:
and air supply, 337, 338–9, 472
atmospheric solids and dust, tables, 473
indoor, man-made, table, 8
limiting values, table, 338
released by fuel combustion, table, 336
Contract maintenance, 664
Control:
direct heating systems, 109
economic life, table, 666
Control devices, 602, 608–16
dampers, 614–16
electric motors, 608–9
pneumatic actuators, 610
solenoids, 609
valves, 610–14
Control packages, unvented hot water systems, 572–3
Control systems:
direct-acting, 603
electric/electronic, 603–4
pneumatic, 604–5
Controller modes of operation, 616–19
derivative control, 619
floating control, 618
integral control, 618
proportional control, 617–18
proportional plus integral, 619
proportional plus integral plus derivative, 619
step-control, 617
two-position control, 616–17
Controls, *see also* Automatic controls; Systems controls
Convection, 6
formula, 172–3, 174
Convective heating, combined with radiant, 183–9
Convective panels, chilled ceilings, 394–5
Convector heaters, 189–96
continuous natural convectors, 190–1
electrical, 107–9
emissions, table, 198
forced fan convectors, 192–3
gaseous, 101–4
general, 195–6
natural convectors, 189–90
skirting convectors, 191
trench convectors, 191
unit heaters, 193–5
Conversion factors, table, 685–6
Cooling:
alternative methods, 396–9
factors affecting choice of system, 396
Cooling coils, air-conditioning system, 487–9
by-pass, controls, 625, 626
controls, 624–5
Cooling effect, 518
Cooling loads, rooms and equipment, table, 658
Cooling techniques, unorthodox, 397–9
Cooling temperature differentials, table, 418
Cooling towers:
closed circuit, 537, 538
contaminants, 8
controls, 635–6
crossflow type, 536
film, 537
forced draught, 537
induced draught, 537
packaged, 539
rating, in refrigerating plant, 538
Copper ducts, 428
Copper pipes:
condensate flow, table, 250
flow of liquefied petroleum gases, table, 305
sizing chart for water flow, 229
water content, table, 142
water flow, table, 600
Cornish boilers, 266
Corrected effective temperature, 10
Corrosion, 224–6
Costs, *see* Running costs
Cross-talk attenuation, diagrams, 457, 461
Curtain walling, diagram, 46
Curves of fans, 463–6
Cylinders:
direct, capacities and dimensions, table, 558
indirect, 559–60
indirect, capacities and dimensions, table, 558
liquid fuels, 295

D factors, floor heating, table, 122
Dampers:
balancing, 432
control device, 614–16
for fans, 469–70
fire and smoke, 432–3
motorised, 369
Data gathering panels, BMS, 637
Data inputs, BMS, 637
Dead-band thermostats, 606–7
Dead-leg systems, pipework, 581–2
Decibel ratings, addition, figure, 448

Decibel values, subtraction, figure, 452
Decibels, 447–8
Decrement factor (*f*), steady state, 61
Degree-day method, running costs, 647–8
Dehumidifiers, air, 489
Dehumidifying, 366
  and summer cooling, 506–11
Delivery:
  liquefied petroleum gas, 303–4
  liquid fuel, 294
  North Sea natural gas, 302
  solid fuels, 291
Demand controlled ventilation, 362–3
Dense phase handing, of solid fuel, 292
Density, *see* Specific mass
Design temperatures:
  normal, in heat emitting equipment, 172
  various emitters, 173
  water heating systems, table, 139
Desk-top outlets, air-conditioning, 415
Diaphragm air pressure gauges, 460
Diesel engines, combined heating/power, diagram, 676
Diffusers, air-conditioning, 409–11, 412, 413, 414
Digital instruments, 460–1
Dip-pipes, high-temperature hot water systems, 145
Direct heating, 94
  categories of system, 95
  control of systems, 109
  electrical, 104–9
  gaseous fuels, 98–104
  liquid fuel, 97–8
  running costs, 654–5
  solid fuel, 96–7
Disc throttling, of fans, 471
Discharge selection, air distribution, 416–18
Discount rates:
  effects, figure, 667
  viability, chart, 669
Displacement ventilation, 389
Distributed intelligence, BMS, 639–41
Diversity of use, effect on heat, 88–9
Domestic chimneys, 326
Domestic heating controls 623
Domestic warm air systems, 102–3
Dosing, of boiler make-up water, 574
Double duct systems, design calculations, 515
Double glazing and double windows, 32
Down-feed systems, piping design, 594–5
Downward system, air distribution, 403–5
Drains, from air coiling coils, diagram, 488
Draught:
  chimneys, 320–2
  mechanical, 324–5
Drop, air diffusion, 407
Dry air, properties, table, 159
Dry coil free cooling, 530–1
Dry-core boilers, 117, 132–3
  capacity, 123
Dry resultant temperatures, 48
  index of comfort, 11
  table of values, 15
Dual-duct system:
  air-conditioning, 380
  controls, 630–1
  factors, table, 399
Dual-fuel burners, gas-fired, 285
Ducts:
  builders' work ducts, 427–8
  circular, 426
  flat oval, 427
  galvanised steel, 425–7
  high-velocity systems, 442–3
  low-velocity systems, 441–2
  and noise, 453–7
  other materials, 428–9
  rectangular, 426
  thermal insulation, 444–5
Ductwork, 424
  acoustic insulation, diagram, 454
  classification, 425
  design, *see Ductwork design*
  economic life, table, 666
  flexible, 433
  galvanised steel, 425–7
  kitchens, 359
  noise attenuation, table, 454
Ductwork design, 423–61
  air flow in ducts, 437–9
  commissioning and measurement, 457
  computer design of duct systems, 443
  ducts, 424–9
  ductwork components and auxiliaries, 429–33
  noise dispersal, 452–7
  pressure distribution in ducts, 434
  sound control, 445–52
Ductwork fittings, 429
  flow of air, 437, 439
Dufton, A.F., 9
Dust spot tests, air filters, 474
Dust staining, air discharge points, 419
Dwellings, ventilation, 340

'Economy 7 boilers', *see* Wet-core boilers
Effective temperature, 9–10
Electric motors, control device, 608–9
Electric storage heaters, 552
Electric water heaters, diagrams, 549, 550
Electricaire, *see* Warm air units
Electrical energy, use in hot water supply systems, 566–7
Electrical heating:
  convective, 107–9
  radiant, 104–7
Electrical off-peak storage systems, 96
Electrical storage heating, 116–38
  capacity of central stores, 122–4
  capacity of heat stores, 117–19
  capacity of room stores, 119–22
  methods, 116–17
Electricity:
  energy available, table, 662
  as fuel, 304–5
  net to gross values, 645–6
  space heating, 306
  tariff structures, 116, 305–6
Electricity supply, to storage equipment and installations, 138
Electrode boilers, 268–9
Electronic controls, 603–4
Electrostatic filters, 475, 476, 480
Embedded ceiling panels, 179–80
Embedded floor panels, 180–3
Embedded panels, water systems, 241
Emissions, *see* Heat emission
Emissivity of radiation, 7
Emulsification, boiler firing, 278
Energy:
  and alternative sources, table, 662
  comparative efficiencies, diagram, 672
  conservation, *see* Energy conservation
  content, in fuels, 662
  costs, 644

Energy – *continued*
prices, table, 645
savings by BMS, table, 642
Energy Act 1983, 671
Energy audits, 661–2
Energy conservation:
energy-saving canopies, 357
energy-saving ceilings, 359
energy savings viability charts, 667–70
fuel savings, 19, 57
primary influences, 67–8
summer, 66–85
winter, 19
Energy consumption:
central plant, table, 659
for electric hot water, table, 553
primary, diagram, 67
and running costs, 652–3
Energy cost premium, figure, 667
Energy production, comparative efficiencies, table, 672
Energy ratings, 646
Energy recovery, kitchen canopies, 356–7
Energy-saving, *see* Energy conservation
Energy targets, 660–3
Engineering Council, *Code of Practice on Environmental Issues*, 365
Enthalpy:
psychromatic charts, 505
sensing devices, 608
of steam, 154
Environmental audits, 662–3
Environmental temperature, 11, 47
Equilibrium conditions, gravity circulation, 245–6
Equipment:
economic life, table, 666
heating/cooling loads, table, 658
for room stores, 124–30
Equitorial comfort index, 11
Equivalent temperature, 9
Escape routes, smoke control, 360
Ethylene glycol/water solutions, table, 497
Eupatheoscope, 9
Evaporative coolers, 534–8
indirect, 499
Evaporative cooling, 396
Evaporators, refrigeration components, 532–3
Excess air, and combustion, 313
Exhalation, human, and air supply, 335
Exhaust heat exchanger/silencer, diagram, 678
Expanders, velocity and pressure changes, 435
Expansion joints and loops in piping, 251–2; table, 254
Expansion, pressurisation by, 207–8
External design temperature, summer, table, 90
External distribution systems:
basements and subways, 164
indirect heating, 163–6
overground, 163
underground ducts, 164–6
External walls, *U* values, 61
Extract grilles, table of velocities, 420
Extract or return air grilles, 419–22

F factors, floor heating, table, 122
Fabric filters, 475, 476, 477, 478
Fabric tube supply systems, 361–2
Face and by-pass controls, of dampers, 615–16
Facing gauges, 434
Factories, *see* Industrial buildings
Fan coil systems, air-conditioning, 373–5
controls, 627–8
design calculations, 514–15
factors, table, 399
Fan coil units, 372–3
Fan convectors, 192–3
Fan-diluted draught, flue gas disposal, 329
Fan heat gain, and air stream, table, 655
Fan heaters, *see* Storage fan heaters
Fan laws, 463
Fan-powered extract units, 350
Fan pressure, ducts, 435–6, 444
Fan speeds, constant and variable, 469
Fanger's comfort criteria, 10
Fans:
air filtration, 472–80
air heating and cooling coils, 485–9
air humidification, 481–5
air-to-air heat exchangers, 490–500
and air treatment equipment, 462–500
axial flow, 465, 468
centrifugal, 463, 466–8
characteristics, diagram, 465
duties, 469–71
mixed flow, 463
noise, 450–1
outside air intake, 471–2
packaged air handling plant, 489–90
propeller, 463, 464–5
total and static pressure changes, diagram, 436
types and performance, 462–71
ventilation, 349–52
Fans and plant connections, ductwork, 430–1
Feed cisterns, hot water supply, 570–1
Feolite, 124, 125, 132
Fibrous plaster radiant ceilings, 179
Films, vapour resisting, table, 39
Filters:
absolute, 475, 476, 479
adsorption, 475, 480
air, *see* Air filters
bag, 476, 478, 479
brush, 476–7
paper, 475
viscous impingement, 475, 476, 479
washable, 476, 477
Fire and smoke dampers, 432–3
Fires:
luminous, 98
open, 96
Firing of boilers, 272–86
Fixed carbon, in solid fuel, 290
Flash point, liquid fuels, 294
Flash steam, calorifier-condensate cooler units, 213–14
Flash steam recovery, 156
Flat roofs, *U* values, 62
Flats, heating in multi-storey blocks, 112–13
Flexible ductwork, 433
Floor heating, 128–9
capacity, 120–2
electrical, 106–7
Floor outlets, air-conditioning, 414–15
Floor panels, embedded, 180–3
Floors:
insulation, 36
U values, table, 31
Flow, sensing devices, 608
Flow heater boilers, 269
Flow rate, supply air and room temperature, diagram, 659
Flue gas:
analysis, 289, 314
disposal, 329–32
temperature, 289, 313–14
Flues:
balanced, 261, 331
materials for, 329

multi-appliance, domestic, 330–1
multiple, 327
single-flue construction, diagram, 327
Fluidised bed boilers, 288
Force, units, 2
Forced convectors, electrical, 108–9
Forced (fan) convectors, 192–3
Forward curved fans, 464
Four-pipe induction system, air-water, 386
factors, table, 399
Free cooling, 629
air-water induction system, 384
refrigeration plant, 530–1
table, 367
Freons, in refrigeration, 520
Frequency (sound), 446–7
Fuel analysis, in construction, 313
Fuel consumption
annual, for various regimes, table, 56
running costs, 652–3
savings, diagram, 57
Fuel handling, *see* Fuels: handling
Fuel oil:
air required for combustion, 333
consumption, and degree days, figure, 661
properties, table, 309
Fuel savings, 19; diagram, 57
see also Energy conservation
Fuels:
air needed for combustion, table, 333
capacity, 290
choice, 114–15
electricity storage and handling, 304–6
energy content, 662
handling
liquid, 299–301
solid, 292–3
heat content, table, 662
liquefied petroleum gas, 303–4
liquid fuel, 293–302
miscellaneous, 306–7
North Sea natural gas, 302–3
relative costs, figure, 119
solid fuel, storage and handling, 290–1
storage and handling, 290–311
Full-load operation, proportion, and running costs, 647
Fume extract systems, local, 360–1
Fumes, industrial, exhaust of, 361
Fundamentals, 1–17

Furnace linings, oil firing, 280
Furnishings, contaminants in, 8

Galvanised steel:
ducts, 425–7
limits for use, table, 575
Gas:
energy available, table, 662
natural and manufactured, properties, table, 309
pressurisation, 209–11
see also Natural gas
Gas boosters, gas-fired boilers, 285–6
Gas-fired convectors, 101, 102
Gas-fired radiant tubes, diagrams, 100
Gas fires, 98
Gas pressurisation sets, 142, 209–11
Gas turbines, and CHP, 674–5
Gas water heaters, diagram, 548
Gaseous firing of boilers, 281–7
Gaseous fuels:
chimneys for, 325
combination calculations, 317–18
heating, 98–104
Gaseous media, direct heating, 111
Gauges, facing and side, 434
Generators and controls, CHP, 676–7
Glass:
absorbing, 76
heat transmisson, table, 21
low emissivity coatings, 78
reflective, 78
'smart' windows, 78
solar control and heat transfer, 76–8
U values, 32–3
Glass fibre ducts, 428
Glass tube heat exchangers, 494
Glazing:
solar and conduction gain, 83–4
solar gain through, 75–8
thermal transmittance, table, 33
U values, 32–3
Global warming, 662
Globe thermometers, 12–13
Globe warning potential (GWP), 520
Gravimetric tests, air filters, 474
Gravity circulation, water systems, 243–6
Greenhouse effect, 662
Grilles, and air distribution, 419–22

Ground coils, system of cooling, 398
Ground cooling, 397
Ground floors, 31–2
Ground water, and underground pipe ducts, 164
Gun-type pressure jet burners, oil fuelled boiler firing, 278–9

Halls of residence, *see* Public buildings
Handling of fuels:
liquid, 299–301
solid, 292–3
Hardness of water, 573
Hardware, building management systems, 637–8
Health and Safety at Work Act 1974, 27
Heat:
emission, *see Heat emission*
latent, 66
machine, 88
from occupants, 86
office machinery and computers, 97–8
sensible, 66
units, 2
unwanted, 337
unwanted by humans, 336
Heat emission:
convectors, table, 198
floor panels, table, 182
horizontal pipes, tables, 169, 170, 171
from pipework, 162–3, 235; tables, 589, 590
radiators, tables, 187, 197
Heat emitting equipment, 172–98
combined radiant and convective, 183–9
convective, 189–96
criteria, 172–4
radiant heating, 174–83
Heat exchangers:
air-to-air, 490–500
controls, 623–4
glass tube, 494
'heat pipe', 495–6
plate-type, 493–4
run-around coils, 496–8
systems compared, 498–9
thermal wheels, 494–5
Heat flow, units, 2
Heat gains:
application, 82–5
calculations, 501–2
internal, 54–5
latent, 502
miscellaneous, 85–9

Heat gains – *continued*
sensible, 501
solar, 68–85; diagram, 70
and temperature rise in buildings, table, 650
Heat generators, hot water systems, 142
Heat input, from incidental sources, table, 55
Heat loss:
from body, 1
cost, 646–7
hot water supply systems, 568–9
winter, 19–27; example, 26
Heat meters, 220–3
evaporative, 221–2
hot water systems, 220–1
hours-run meters, 222–3
steam systems, 220
water flow meters, 222
'Heat pipe' heat exchangers, 495–6
Heat pumps:
cycle, diagram, 542
economic life, table, 666
packaged, 545–6
and refrigeration plant, 542–6
Heat reclaim, efficiency, 491–3
Heat recovery:
equipment, and CHP, 677–80
and refrigeration, 539–41
use of multiple machines, 541
Heat recovery to air system, diagram, 681
Heat requirements:
mechanically ventilated buildings, 655
miscellaneous allowances, 52–5
Heat stores, capacity, 117–19
Heat stress, 1
Heat transfer, fundamentals, 172–4
Heaters, oil storage, 299
Heating, direct and storage, preheat for, 60
Heating elements, in floors, diagram, 120
Heating equipment, economic life, table, 666
Heating loads, rooms and equipment, table, 658
Heating methods, 94–115
categories, figure, 95
direct and indirect systems, 94
factors affecting choice, 111–15
options, indirect, table, 95
Heating system controls, 621–4
boilers, 623
domestic heating, 623
heat exchangers, 623–4
individual room control, 622
zone control, 622
Heating systems, period of use, and efficiency, table, 653
Height factors, allowance for in heating, 52–3; table, 54
High level stores, capacity and air venting, 596
High rise buildings, *see* Tall buildings
High temperature hot water, 143–6
data for pipework, 170
High temperature panels, electrical, 105
High velocity systems, 442–3
air-conditioning, 375
air heating, 161
fittings, 418
and sound, 457
Holden heat-house, 276
Hollow floor systems of colling, 398
Home heating, choice factors, 112–13
Hopper stokers, 275–6
Horizontal projections, *see* Projections
Hospitals, ventilation, 340
see also Public buildings
Hot draught tubes, dry-core boilers, 132
Hot water:
consumption, table, 565
storage capacity, and boiler power, 564–70
storage, tables, 566, 567
Hot water boilers, *see* Boilers
Hot water supply:
boiler capacity, 261–2
running costs, 659–60
Hot water supply systems, 547–77
central systems, 554–9
choice, 547–8
cylinders, indirect cylinders and calorifiers, 559–64
design temperatures, table, 173
feed cisterns, 570–1
heat meters, 220–1
local systems, 548–53
materials etc, 574–6
period of use, efficiency table, 653
requirements for storage capacity and boiler power, 564–70
solar collectors, 576–7
unvented hot water systems, 571–3
water treatment, 573–4
Hot wells, steam piping, 158, 159
Hot wire anemometers, 14, 460
Hotels, *see* Public buildings
Hours-run meters, 222–3
Human activity, 7
Human biology, and environment, 1
Human comfort, 4–9, 16
criteria, 4–9
determination, 16
Human emissions, air supply for, 335–7
Humidification, and winter heating, calculations, 511–14
see also Air humidification
Humidifers, 342, 369, 482, 483, 485
Humidistats, diagram, 607
Humidity (moisture content), 8–9; table, 8
air ventilation, 342
induction systems, 630
measurement, 13–14
psychrometric charts, 504
sensing devices, 607
Hybrid circuits, piping arrangements, 152–3
'Hydraulic' design, 241
Hydraulic gradients, water systems, 239
Hydrofluorocarbons (HFCs), refrigeration, 520
Hydrogen, 312
Hypocausts, 111
Hypothermia, 1

Immersion heaters, 553, 561
Indirect cylinders, *see* Cylinders, indirect
Indirect evaporative coolers, 499–500
Indirect heating, 94
air, 159–61
external distribution system, 163–6
gaseous media, 111, 159–61
liquid media, 109–10, 139–53
pipework, heat emission, 162–3
steam, 153–9
vapour media, 110–11, 153–9
water, 139–53
Indirect heating systems, 139–72
options, 95
piping design, 227–54
Induction system, air conditioning
controls, 628–30
design calculations, 515
factors, table,
Industrial buildings:
heating choice factors, 114

ventilation, 340, 345–7
Industrial fumes, exhaust of, 361
Industrial warm air systems, 97–8, 103–4
Infiltration, *see* Air infiltration
Infra-red heaters, 99, 104
Inlet vane dampers, fans, 470
Inlets, air distribution, 412
sizes, table, 417
Instantaneous heaters, hot water supply, 548–9
Instrumentation, boilers, 289
Insulating materials:
resistances, table, 64
thermal, table, 27
thermal conductivity, figure, 162
Insulation:
of buildings, 25–7
hot water systems, 575–6
methods of applications
to floors, diagram, 36
to roofs, diagram, 35
to walls, diagram, 34
under-hearth, in boilers, 280
wall cavities, 33–5
Insulators, 6
Insurance, and running costs, 664
Intermittent operation of heating, summer, 91–3
Interstitial condensation, 38–9
Investment appraisals, 664–70
energy savings viability charts, 667–70
life cycle costs, 665
payback period, 665
present value, 665–7
Ionisation, 9
Isotherms, summer, British Isles, figure, 90
Isovels, 407

Kata thermometers, 14
Kathabar system, 489
Kerosene, 293
Kitchens:
canopies, 356–7
ventilation, 340, 354–9

Labour costs, maintenance, 663
Ladder circuits, 152
Laminar flow, 400
Lancashire boilers, 266
Latent heat, 2
Lateral movement, air distribution, 406–7
Lean phase handling, solid fuel, 292
Legionnaires' disease (*L. pneumophila*), 342, 369, 500, 528, 533, 569, 576, 577
Legislation:
on ventilation, 339–40
winter buildings, 27–30
Level indicators, in tanks, 299
Life cycle costs, investment appraisals, 665
Light fuel oil, 293
Lighting and air distribution, combined, 418–19
Lighting, heat from, 87
Linear diffusers, ceiling, 411
Liquid distribution, economic life, 666
Liquid fuel heating, 94, 97–8
Liquid fuels, 293–302
chimneys for, 319–25
combustion calculations, 316–17
delivery, 294
handling, 299–301
storage, 294–8
Liquid media, indirect heating systems, 109–10
Liquefied petroleum gas
delivery and storage, 303
flow in pipes, table, 305
pipework, 304
properties, table, 310
Lithium chloride, 607
Local systems, hot water supply, 548–53
instantaneous heaters, 548–9
storage heaters, 550–3
Long-distance communication, BMS, 641
Loops and offsets, thermal expansion, 253–4
Louvres, natural air inlets, 346–7
Low emissivity coatings, on glass, 78
Low level side-wall outlets, air conditioning, 415
Low temperature hot water, 141–2
data for pipework, table, 169
Low temperature panels, electrical, 105–6
Low temperature warm water, 140–1
Low velocity systems, ducts, 441–2
Luminaires, air handling, 419
Luminous fires, 98, 104

Machines, heat from, 88
Magazine boilers, 272
Magnetic treatment, hot water supply, 574
Maintenance:
and CHP systems, 680–1
and running costs, 663–4
Manchester, outside temperature, table, 649
Manholes, in tanks, 298
Materials:
hot water systems, 574–6
insulation, resistances, table, 64
vapour resistivities, table, 38
see also Properties of materials
Measurement, methods, 12–14
Mechanical atomisation, boiler firing, 277
Mechanical draught, chimneys, 324–5
Mechanical refrigeration, 516–19
Mechanical separators, humidification, 482–3
Mechanically ventilated buildings:
heat requirements, 655
power requirements, 656
running costs, 655–6
Mechanised design, water heating, 241
Medium temperature hot water, 142–3
data on pipework, table, 170
Metal radiant ceilings, 177, 178
Metal radiant panels, 174–6
Metals, properties, table, 3
Meters, heat, *see* Heat meters
Methyl chloride (CH_3Cl), 520
Micro-bore system, water heating, 150–1
Micro-manometer, 458–9
Micron, measurement of dust particles, 472
Milne, A.A., quoted, xiv
Miscellaneous allowances, heat requirements, 52–5
Mixed flow fans, 466
Modular boilers, 258
Modular ceiling shells, ventilation, 357–8
Moisture:
direct heating appliances, table, 96
in masonry materials, 22–3
unwanted, 337–8
unwanted by humans, 336
Moisture content, *see* Humidity
Montreal Protocol, 520
Movement, air, *see* Air: movement
Moving bed stokers, 275
Multi-directional ceiling diffusers, 409–11
Multi-storey flats, *see* Flats; Tall buildings
Multiple zone systems, water heating, 240

Municipal waste, *see* Waste products
Mushroom ventilators, 421–2
MVM Starpoint scheme, 646

National Home Energy Rating (NHER), 646
Natural draught cooling towers, 535
Natural gas, 302–3
 air required for combustion, 333
 delivery and pipework, 302–3
 flow in steel pipes, 303
 flue gas disposal, 329–32
 net to gross values, 645–6
 properties, table, 309
Natural ventilation, 343–7
Neat flame burners, 282
Neutral point piping systems, 242
'Nightstor', *see* Dry-core boilers
Night ventilation, 397
Nitrogen, 312
Noise dispersal, 452–7
 see also Sound
Noise rating curves, figure, 449
Noise, transmission paths, figure, 453
Non-change-over system, induction air-water systems, 385
Non-homogeneous constructions, 31
Non-storage calorifiers, 212–14
North Sea natural gas, *see* Natural gas
Nozzles, air-conditioning, 413

Occupancy, and energy consumption, 67
 see also Period of use
Occupants:
 contaminants from, 8
 heat and moisture, table, 336
 heat emission, 86
Occupational exposure limit (OEL), 520
Octave bands, standard, diagram, 446
Off-peak storage systems:
 electrical, 96
 hot-water storage, table, 567
Office machinery:
 contaminants from, 8
 heat from, 87–8
Offices, *see* Commercial buildings
Offsets and loops, thermal expansion, 253–4
Oil:
 flow in steel pipes, table, 300
 net to gross values, 645–6
Oil filled radiators, electrical, 107
Oil-fired industrial warm-air unit, diagram, 97
Oil fuel firing of boilers, 276–81
 safety precautions, 286
Oil storage requirements, table, 295
Oil tanks, 294–5
 capacities, table, 296
 fittings, 298–9
Oils, energy available, table, 662
On-off time switches, control systems, 619
Opaque surfaces, solar gain through, 72–5
Open fires, 96
 and ventilation, 343–4
Open vents, hot water boilers, 270–1
Operation, continuous versus intermittent, 56–60
Optimum-start controls, 620–1
Outflow via return pipework, central hot water systems, 593–4
Outlet valves, oil tanks, 299
Outlets, air distribution, 414–16
Outstations, BMS, 637
Overground distribution of heat, 163
Oxygen, 312
Ozone-depleting potential (ODP), 520

Packaged air cooled centrifugal plant, diagram, 534
Packaged air handling plant, 489–90
Packaged boilers, 258–9, 267–8
Packaged burners, gas-fired boilers, 282, 283–4
Packaged cooling towers, refrigerating plants, 539
Packaged heat pumps, 545–7
Packaged plant, CHP, table, 679
Packaged room coolers, 393
Paddle blade switches, 608
Painting of radiators, 188–9
Pan humidifiers, 485
'Panelite' floor panels, 181, 182
Paper filters, 475
Parallel single-pipe circuits, water heating, 234
Partitions and party walls, *U* values, 62
Payback period, investment appraisals, 665
PC-based systems, building management systems, 641
Percentage saturation, 9, 505
Perforated ceilings, 408
Perforated-face ceiling diffusers, 409
Period of use, of buildings:
 annual fuel or energy consumption, 652–3
 and running costs, 648–55
Perkins, high temperature systems, 143, 144
Petroleum gas:
 liquefied, 303–5
 properties, table, 310
Petroleum oil, 293
Phon, The, 449
Piezo-electric effect, 607
Pipe-in-pipe systems, heat distribution, 165–6
Pipes, water content, table, 142
Pipework:
 for combined radiant and convective heating, 183
 design, *see Piping design*
 economic life, table, 666
 expansion, 251–4; table, 251
 heat emission, 162–3; tables, 169–70, 171
  for heat storage, 259
 hybrid circuits, water heating, 152–3
 liquefied petroleum gas, 304
 losses, hot water supply system, 547
 materials, hot water supply systems, 574–5
 for modular boilers, diagram, 258
 North Sea natural gas, 302
 primary calculations, 578–81
 reverse-return circuit, water heating, 151–2
 secondary outflow and return pipework, 581–94
 single-pipe circuit, water heating, 147–9
 steam, data, table, 171
 steam heating systems, 157–9
 system arrangements, 594–600
 tables, 169–71
 thermal expansion, 251–4
 thermal insulation, 162–3
 two-pipe circuit, water heating, 149–51
 water heating systems, 146–53
Piping design:
 calculations, 234–8
 for central hot water supply systems 578–600

indirect heating systems, 227–54
parallel single-pipe circuits, 234
single-pipe circuits, 232–3
steam systems, 246–50
two-pipe circuits, 233–4
water systems, 227–46
Pitch angle, fan blades, 470–1
Pitched roofs, *U* values, 62
Pitot tubes, 458
Plant:
commissioning and testing, 68
size and pre-heat, 59–60
Plant components, ductwork, flow of air, 439
Plant connections, ductwork, 430
Plant room accommodation, and CHP, 680
Plaster, fibrous, radiant ceilings, 179
Plate-type heat exchangers, 493–4
Plate-type water heaters, 563
Plenum boxes, velocity and pressure changes, 435
Plenum system:
air heating, 159, 160–1
and ventilation, 362
Pneumatic activators, control systems, 610
Pneumatic controls, 604–5
Pollution, *see* Contaminants
Polyethelyne tubes, 181
Polypropylene ducts, 428
Polythene tanks, liquid fuels, 295
Pour point, liquid fuels, 244
Power requirements, mechanically ventilated buildings, 656
Preheat:
for direct and storage heating, 60
and plant size, 59–60
variations in times, table, 59
Preheat capacity, 57–9
profiles, diagram, 58
Present value, and investment appraisals, 665
Pressed steel radiators, 184–6
Pressure:
loss, secondary circulation, table, 591
sensing devices, 607
units, 2
Pressure, steam, *see* Steam pressure
Pressure atomisation, boiler firing, 277
Pressure distribution in ducts:
application, 439–44
fan pressure, 435–6
static, 434
velocity pressure, 434–5
Pressure jet, boiler firing, 277
Pressure loss, in chimneys, table, 320
Pressure loss factors, single resistances, water flow, table, 230
Pressure reducing sets, steam pressure, 219
Pressurisation of system, 207–11
application of, 211
by expansion, 207–8
by gas, 209–11
by pump, 209
Primary energy, running costs, 645–6
Primary pipework, central hot water supply
direct, 578–9
indirect, 579–81
Prime movers, and CHP, 674–5
Process equipment, heat from, 88
Projections from buldings, effect on cooling load, diagram, 79
Propane:
air needed for combustion, 333
storage, 304
Properties of materials, 204;
table, 3
Psychrometers, 13–14
Psychrometry
calculations, 502–5
charts, 503–5
Public buildings, heating, 113
Pumps, and auxiliary equipment, 199–226
air venting etc, 223–4
centrifugal, 200–7
condensate handling equipment, 214–18
corrosion, 224–6
heat meters, 220–3
non-storage calorifiers, 212–14
pumps, 199–207
steam pressure reduction, 218–19
system pressurisation, 207–11
see also Heat pumps
Punkah fans, 363–4
Punkah louvres, 413, 414
PVC coatings for ducts, 429

Quantities, used here, 2
Quartz lamp heaters, 105

Radiant ceilings, 174, 177–80
Radiant heaters:
electrical, 104–7
gaseous fuels, 98–100
Radiant heating:
and convective heating, 183–9
equipment, 174–83
Radiant panels, 174–6
chilled ceilings, 393
Radiant strips, 176
Radiant tubes, 99–100
Radiation, 6–7
formula, 173
Radiator connections, gravity circulation, 245
Radiators
aluminium, 187
cast iron, 183–4
emissions, table, 197
painting, 188–9
pressed steel, 184–6
siting etc, 187–9
storage, capacity, 119
see also Storage radiators
Reciprocating compressors, 522–5
Reciprocating engines, and CHP, 675–6
Reciprocating pumps, 199
Recirculation units, ventilation, 363–4
Reflective glasses, 78
Reflectors, behind radiators, 189
Refrigerant pipework, 531–2
Refrigerants 134a, 517, 518, 519–21
properties, table, 521
Refrigerating media, 519–21
Refrigeration, 516–46
application, 519; and run-around coils, 498
choice of plant, 527–31
cycle, 517–19
heat pumps, 542–6
heat recovery, 539–41
mechanical refrigeration, 516–19
refrigerating media, 519–21
Refrigeration plant:
absorption plant, 526
choice, 527–31
components, 531–9
economic life, 666
reciprocating compressors, 522–5
steam-jet plant, 526–7
types, 521–7
use of multiple machines, 541
vapour compression, 521–2
Regulating valves, water heating, 242
Reheaters, 371
Relative humidity, 9, 342
Relay points, steam pipework, 158
Resistance factors, wind effects, 348
Resistances, insulation materials, table, 64

Respiration, human, and air supply, 335
Reverse cycle heat pump system, air-conditioning, 391
controls, 633
factors, table, 399
Reversed-return systems, water heating, 239–40
Ringlemann charts, 289
Rooftop boiler houses, 332
Roofs:
insulation, 36; diagram, 35
solar and conduction gain, 84–5
solar gain, 74
Rooflights, single glazed maximum areas, 28
Room air-conditioning units, 389; figure, 390
Room control, heating system, 622
Room coolers, 390
Room heaters and coils, heating/cooling loads, table, 658
Room stores:
capacity, 119–22
control, 129–30
equipment, 124–30
electrical storage heating, 116
Room temperature profiles, 77
Rooms, heating/cooling loads, table, 658
Rotary compressors, 523
Rotary cup burners, oil fuelled boiler firing, 279–80
Rotary gear pumps, 199
Rotating vane anemometer, 459–60
Run-around coils:
using refrigeration, 498
water circulation, 496–8
Running costs, 644–70
air-conditioned buildings, 656–9
energy targets, 660–3
hot-water supply, 659–60
investment appraisals, 664–70
maintenance, 663–4
mechanically ventilated buildings, 655–6
period of use of building, 648–55
primary energy, 645–6
space heating, 646–8

Safety precautions, gas- and oil-fired burners, 286
Sankey diagram, CHP system, 677
Schools, ventilation, 340
see also Public buildings
Screw compressors, 523–4
Scroll compressors, 523, 524
Secondary outflow and return pipework, central hot water supply, 581–94
Secondary return pipework, central hot water supply, 588–93
Semi-exposed surfaces, diagram, 29
Sensing devices, 602, 605–8
building management systems, 637
enthalpy, 608
flow, 608
humidity, 607
pressure, 607
temperature, 605–7
Series circuits, 152
Shading:
correction factors, table, 92
devices, table, 77
structural, 78–82; table, 79
Shell type boilers, 266–7
SI unit symbols, table, 684
Side gauges, 434
Side-wall inlets, 412–13
Single-duct terminal reheat systems, design calculations, 514
Single-duct VAV, factors, table, 397
Single-pipe circuits, water heating, 232–3
Sizing charts:
saturated steam flow, 248
for water flow, 228, 229
Skirting heaters, 107, 191
Sling psychrometers, 13–14
Slip tiles, embedded ceiling panels, 180
Sludge valves, 298–9
Small-bore hot water systems, 148–9
'Smart' windows, 78
Smoke:
control, 360, 633
dampers, 432–3
indication, in boiler plant, 289
problems, 327–8
Sodium flame tests, air filters, 474
Software, Building maintenance systems, 638, 641
Sol-Air temperature, 11, 48, 72
Solar collectors, hot water supply, 576–7
Solar control films, 78
Solar gain, cooling load (summer), tables, 92, 93
Solar heat:
gains, summer, 68–85
through glazing, 75–8
through opaque surfaces, 72–5; table, 74
on vertical surface, diagram, 71
Solar radiation, incidence, 70–2
Solar thermometers, 12
Solenoids, control device, 609
Solid fuel boiler firing, 272–6
Solid fuel direct heating, 96–7
Solid fuels, 290–3
chimneys for, 319–25
combustion calculations, 316–17
delivery, 291
handling, 292–3
storage, 291
Solid slab system, cooling, 399
Sound, 445–6
amplitude, 447
analyser, 450
control, ductwork, 445–52
fan noise, 450–1
frequency, 446–7
meters, 450
noise criteria, 449–50
noise dispersal, 452–3
noise in rooms, 451–2
Space heating:
electrical, 306
running costs, 646–8
Sparge pipes, 136
Specific heat capacity, gases, 4
units, 2
Specific heat, property of materials, table, 3
Specific mass (density), property of materials, table, 3
units, 2
Specific volume, psychrometric charts, 505
Split-casing circulating pumps, 142
Spill-valves in pumps, 209
Spoiler by-passes, centrifugal pumps, 203
Spray ponds, 535
Spray washers, humidification, 483–4
Sprayed coils, air-conditioning, 485, 488–9
controls, 625–7
Spread, air diffusion, 407
Sprinklers and storers, 274
Static pressure in ducts, 434
Steady state, for winter heating, 19, 60–1
Steam:
flash steam recovery, 156
generation, 154

pipework data, tables, 171
pressure, *see* Steam pressure
properties, 154
removal of air, 155
saturated, properties, tables, 167–8
Steam boilers, fittings and mountings, 271
Steam drums, high temperature hot water systems, 145
Steam heating systems, 110–11, 114, 153–9
characteristics, 153–7
metering, 220
piping, 157–9
piping design, 246–50
principles, 246–50
velocities, table, 247
Steam humidifers, 481–2
Steam injection, humidification, 481–2
Steam-jet plant, refrigeration, 526–7
Steam pipework, data, table, 171
Steam pressure, 154, 155
reduction, 156–7, 216–17, 218–19
Steam-to-water calorifiers, 561
Steam traps, 214–17
super-lifting, 218
Steam turbines, and CHP, 674
Steel, *see also* Galvanised steel
Steel boilers, 264–5
Steel ducts, 428
Steel pipes:
condensate flow, table, 250
cross-sectional areas, steam, table, 247
flow of liquefied petroleum gases, table, 305
natural gas flow, table, 303
oil flow, table, 300
saturated steam, sizing chart, 248
water content, table, 142
water flow, sizing chart, 228
water flow, table, 600
Steel radiators, 184–6
Stevenson screens, 13
Stoichometric condition, 313
Stokers (for boilers):
automatic, 272–5, 291
hopper and bunker underfeed, 275–6
Storage capacity, hot water, 564–70
Storage cylinders, heat loss, table, 660
Storage fan heaters, 116, 120, 127–8
Storage heaters, hot water supply systems, 550–3
Storage heating:
central, 117
electrical, 116–38
heat store capacity, 117–19
room store,
Storage materials, heat capacities, table, 125
Storage of fuels, *see* Fuels: storage
Storage radiators, 116, 119, 235–7
Storage systems, refrigeration, 528–30
Storage temperature, hot water, 569
Stoves, closed, 96–7
Straw boilers, 287
Structural shading, 78–82
effect on energy consumption, table, 79
Sub-atmospheric systems, steam systems, 158
Subjective temperature, 10
Subways, external distribution of heating, 164
Sulphur, 312
content of liquid fuels, 294
Sulphur dioxide (SO_2)
pollutant, 323
in refrigerators, 519, 520
Summer, and buildings, 65–93
air conditions, 366
Summer cooling and dehumidifcation, calculations for air-conditioning, 506–11
Sun path diagrams, 68–70; figure, 69
Sun shading, computer plot, 80
Super-lifting steam traps, 218
Surface factor (*F*), steady state, 61
Surface resistance, 19, 20–1; table, 21
Surface temperatures, 24–5
Sweat, *see* Body odour
Swimming pools, upward air distribution, diagrams, 402, 405
see also Public buildings
Swinging vane anemometers, 460
Switches, *see* Control systems
System arrangements:
air venting, high level stores, 596
capacity high level stores, 596
central hot water supply piping, 594–600
circulating pumps, 598
down-feed systems, 594–6
tall buildings, 596
unvented systems, 596–7
up-feed systems, 594
water hammers, 598–9
System pressure condition, 242–3
System pressurisation, 207–11
Système International, temperature, 5
Systems controls, 515, 619–21
on-off time switch, 619
optimum-start control, 620–1

Tall buildings:
central hot water supply systems, 596
plans, and solar heat, 81
ventilation, 371
wind currents, diagram, 44
Tank rooms, liquid fuel, 295–8
Tanks, liquid fuel, 294–5
fittings, 298–9
Tariff, *see* Electricity
Telescopic joints, expansion offsets, 252
Temperature:
air, 341
computer plot, without air-conditioning, 76
control, *see* Temperature control
corrected effective, 10
dew point, 504
dry bulb, 90, 368, 504
dry resultant, 11
effective, 9–10
effective temperature chart, 10
environmental, 11, 47
equitorial comfort index, 11
equivalent, 9
Fanger's comfort criteria, 10
and humans, 4–5
measurement, methods of, 12–13
sensing devices, 605–7
sol-air, 11, 48, 72
of steam, 154
subjective, 10
summer, 66
thermal indices, 9–11
wet bulb globe, 11, 90, 368, 504
wet resultant, 11
Temperature control, 55, 89
hot water supply, 570
Temperature difference, and natural ventilation, 347
Temperature differences, summer, 89–90
Temperature differences, winter, 47–52
applications, 46–7, 51–2
inside, 47–8
miscellaneous allowances, 52–5

Temperature differences, winter – *continued*
outside, 48–50; map, 49
ratios, 50
values, table, 51
Temperature levels, table, 683
Temperature limiting controllers, radiator systems, 222
Temperatures, day, July, table, 73
Terminal diffusers, air systems, 161
Terminal equipment, indirect systems:
liquid media, 110
vapour media, 111
Terminal reheat systems, air-conditioning, controls, 627
Terminals, flues and chimneys, 331–2
Test discount rate, (TDR), 665
Tests for air filters, 473–4
arrestance, 474
dust spot, 474
sodium flame, 474
weight (gravimetric), 474
Theatres, ventilation, 340
Thermal balance point, 68
Thermal centrelines, gravity circulation, 245
Thermal conductivity, property of materials, table, 3
Thermal expansion, 4
provision for, in piping, 251–4
Thermal indices, 9–11
Thermal inertia, of structures, 20
Thermal insulation
ducts, 444–5
materials, table, 27
pipework, 162–3
Thermal Insulation (Industrial Buildings) Act 1957, 27
Thermal response (time lag), 56
Thermal storage, and boiler selection, 259
Thermal storage cylinders, 117, 135–8
capacity, 163–4
Thermal transmittance coefficient, *see U* values
Thermal wheels, 494–5
Thermistors, 606
Thermo-glazing, 575
Thermometers, 12–13
Kata, 14
Thermostatic control valves, for radiators, 603
Thermostats, 605
Thermosyphons, 531
Three-pipe induction system, air-water systems, 386
Throttling dampers, for fans, 470
Throw, air diffusion, 407
Timber:
as fuel, 306–7
properties, tables, 3, 310
Time switches, on-off, 619
Toilets, ventilation, 340, 422
Transmission network, BMS, 637
Transmittance coefficients, *see U* values
Trench convectors, 191
Trim heaters, 486
Tubular heaters, electric, 108
Turbines, steam, and CHP, 674
Two-pipe circuit, water heating, 149–51, 233–4
Two-pipe gravity circulation, 245

U values, 23–4
floors, table, 31
glass etc, 32–3
insulated ductwork, table, 444
maximum permissible with single glazing, table, 28
for normal exposure, table, 61–2
for typical uninsulated constructions, table, 63
Ultrasonic atomisers, 485
Underfeed stokers, 274
Underfloor warming, *see* Floor warming
Underground ducts, external distribution of heating, 164–6
Underhearth insulation, boilers, 280
Unit heaters, 193–5
Unitary controllers, BMS, 640–1
Units, used here, 2–4
Universities, *see* Public buildings
Unorthodox cooling techniques, 397–9
Unvented hot water systems, 571–3
piping arrangements, 596–7
Up-feed systems, piping design, 594
UPVC ducts, 428
Upward air-flow systems, air-conditioning, 387, 401–3

Vacuum pumps, removal of air in steam systems, 158
Valves:
control devices, 610–14
hot water boilers, 270
regulating, water systems, 242
Vane anemometers, 14, 459–60
Vaporisation, boiler firing, 277
Vapour, resistivities of common materials, 38
Vapour compression, 521–2
Vapour compression cycle, 516–17
Vapour media, indirect heating, 94, 110–11
Vapour pressures, 4; table, 40
Vapour seals, 26
Variable fan speed, fans, 471
Variable geometry diffusers, 414
Variable refrigerant-volume systems, air-conditioning, 393
Variable volume systems:
controls, 631–3
design calculations, 515
Velocities in ducting systems, tables, 441
Velocity of water in pipes, 231
Velocity pressure:
in ducts, 434–5
relation to air speed, figure, 424
Vent pipes, 298
Ventilated double window, diagram, 421
Ventilation, 334–64
air change rates, table, 339
air supply, 85, 335–9
criteria of air supply, 340–3
demand controlled, 362–3
efficiency, 364
of kitchens, 354–9
legislation etc., 339–40
mechanical inlet and extract, 353–4
mechanical inlet, natural extract, 352–3
methods, 343
natural inlet and extract, 343–9
natural inlet, mechanical extract, 349–52
recirculation units, 363–4
special applications, 360–3
summer, 85
Ventilation allowance, 20
and air change, 42–3
Ventilators:
floor, diagrams, 421
mushroom, 421, 422
Vertical temperature gradients, diagrams, 53
Vessels, materials, hot water supply, 574–5
Vibration, transmission paths, diagram, 453
Viscosity, quality of liquid fuels, 294
Viscous impingement filters, 475, 476, 479
Volatiles, in solid fuel, 290

Volume:
air, *see* Air: volume
steam, 155
units, 2
Volume control, air distribution, 416

Walkways, overhead, protecting external pipework, diagram, 164
Wall cavity fill:
insulation, 34
thermal resistance, 33–5
Walls:
solar and conduction gain, 84
solar gain, table, 74
Warm air systems, industrial, 97–8
Warm air units, 117, 123, 130–1
Warmed floors and walls, 116
Washable filters, 476, 477
Waste products, as fuel, 307
properties, table, 311
Water:
flow, *see* Water flow
hardness, 573
properties, table, 167
for thermal storage, properties, table, 136
treatment of condenser water, 538–9
velocity in pipes, 231
see also Humidity; Moisture
Water chillers, 527–8
Water circulation, run-around coils, 496–8
Water content, steel and copper pipes, table, 142
Water/ethylene glycol solutions, table, 497
Water flow, 229–30
direction, diagram, 593
meters, 222
in pipes, 229–30, 600
through single resistances, 230–1
sizing charts, 228, 229
Water hammer, 598–9
Water heaters, plate type, 563–4
Water heating and humidification, calculations, 511–14
Water heating systems, 109–10, 139–53
characteristics, 139–46
high temperature hot water, 143–6
low temperature hot water, 141–2
low temperature warm water, 140–1
medium temperature, 142–3
piping arrangements, 146–53
Water heating systems and piping design:
applications, 231–41
footnotes, 241–3
gravity circulation, 243–6
piping arrangements, 146–53
principles, 227–31
Water-to-water calorifiers, 560–1
Water treatment:
additives, table, 225
hot water supply systems, 573–4
Water tube boilers, 268
Weather data:
and air-conditioning, 367–8
maritime climate, 65
Weight tests, air filters, 474
Wet bulb globe temperatures, 11, 13
Wet core boilers, 117, 133–4
capacity, 123
Wet resultant temperatures, 11
Wheatstone bridge circuits, 606
Whole building air infiltration, 54
Wind and air volumes, for natural ventilation, table, 348
Wind-chill indices, 11
Wind currents, diagram, 44
Wind effects, and natural ventilation, 348–9
Windows:
air infiltration through cracks, 43–6; table, 45
framed, thermal transmittance, 33
single-glazed maximum areas, 28
'smart', 78
U values, 32–3
ventilated double, diagram, 421
Winter, buildings in, 18–64
air-conditioning, 366
external design temperature, table, 50
Wiredrawing, steam systems, 156–7
Wood, *see* Timber

Y factors, in off peak consumption, table, 119
Y value (admittance), 60

Z factors, in design heat loss, table, 118
Zeolites, and water treatment, 573–4
Zone control, heating systems, 622
Zoned systems
air-conditioning, 369–71
controls, 627